Chemie der Pflanzenschutz- und Schädlingsbekämpfungsmittel

Herausgegeben von R. Wegler · Band 6

Chemie der Pflanzenschutz- und Schädlingsbekämpfungsmittel

Band 6

Insektizide · Bakterizide · Oomyceten-Fungizide
Biochemische und biologische Methoden · Naturstoffe
Insecticides · Bactericides · Oomycete Fungicides
Biochemical and Biological Methods · Natural Products

Herausgegeben von R. Wegler

Mit 105 Abbildungen und 92 Schemata

Springer-Verlag Berlin Heidelberg New York 1981

Professor Dr. Richard Wegler
Auf dem Forst 2, D-5090 Leverkusen 1

ISBN 3-540-10307-4 Springer-Verlag Berlin Heidelberg New York
ISBN 0-387-10307-4 Springer-Verlag New York Heidelberg Berlin

CIP-Kurztitelaufnahme der Deutschen Bibliothek
Chemie der Pflanzenschutz- und Schädlingsbekämpfungsmittel/hrsg. von R. Wegler. –
Berlin, Heidelberg, New York: Springer.
NE: Wegler, Richard [Hrsg.]
Bd. 6 Insektizide, Bakterizide, Oomyceten/Fungizide, biochemische und biologische Methoden, Naturstoffe. – 1981
ISBN 3-540-10307-4 (Berlin, Heidelberg, New York)
ISBN 0-387-10307-4 (New York, Heidelberg, Berlin)

Herstellung: Brühlsche Universitätsdruckerei, Gießen
2152/3140-543210

Vorwort

Der vorliegende 6. Band könnte die Überschrift „Pflanzenschutzforschung auf neuen Wegen“ tragen. Trotz der verbreiteten, meist auf mangelndem Informationsstand beruhenden Voreingenommenheit gegen den chemischen Pflanzenschutz, nimmt dessen Bedeutung ständig zu. Das beruht auf der Notwendigkeit einer sicheren und rationellen landwirtschaftlichen Produktionssteigerung, die durch das rapide Wachstum der Weltbevölkerung erzwungen wird. Zwar ist die landwirtschaftliche Produktivität auch in einigen Entwicklungsländern während der letzten 15 Jahre merklich gestiegen, sie hält aber vielfach mit der Vermehrung der Bevölkerung nicht Schritt.

Der Pflanzenschutz ist nicht einseitig auf bestimmte Methoden festgelegt, sondern bedient sich unvoreingenommen aller wirtschaftlich und ökologisch vertretbaren Wege und Verfahren. Hierbei wird auch die Forschung nach neuen Wirkstoffen und chemischen Verfahren auf sehr breiter Grundlage weiterbetrieben, wenn auch unter merklicher Steigerung der Forschungskosten. Jeder neue Wirkstoff muß einen Fortschritt gegenüber einem eingeführten Handelsprodukt darstellen. Wichtige Kriterien sind hierbei neben der Wirksamkeit die Umweltverträglichkeit, unter anderem günstigere toxikologische Eigenschaften, Sicherheit in der Anwendung und Wirtschaftlichkeit. Wie schon in Band 3 (Beitrag Haug) dargestellt, sind gerade auch diejenigen Kosten für Forschung und Entwicklung, die der langfristigen Unbedenklichkeit und Sicherheit der Produkte dienen, überproportional gestiegen.

Immer stärker fordern Öffentlichkeit und Politiker den Ausbau rein biologischer Verfahren des Pflanzenschutzes, von denen angenommen wird, daß sie gänzlich risikofrei und ohne Nachteile seien. Aber auch hier gibt es unübersehbare ökonomische und ökologische Grenzen und Risiken. Alternative Methoden rechtfertigen sich nicht schon dadurch, daß sie etwas anderes anbieten. Sie müssen beweisbare Vorteile, ohne wesentliche ökonomische und eventuell auch ökologische, hygienische oder arbeitstechnische Nachteile erbringen.

Zwischen den Verfahren der rein chemischen Schädlingsbekämpfung und der rein biologischen Schädlingsbekämpfung stehen die biochemischen und biotechnischen Methoden. Weltweit dürfte der Forschungsaufwand für diese Arbeitsrichtungen dem, den die chemische Industrie für die Insektizidforschung betreibt, nicht nachstehen. Der hohe Forschungsaufwand für die biotechnischen und biochemischen Verfahren wird vielfach damit begründet, daß die Risiken bei der Anwendung und für die Umwelt geringer, und daß sie billiger seien. Hierbei wird oft übersehen, daß jedes neue Verfahren, gleich welcher Herkunft, in einer echten Kalkulation alle Kosten bis zur Einführung und den gesamten Aufwand für die spätere Praxisanwendung umfassen muß. Hierzu gehören die oft jahrzehntelang erfolglosen Versuche, einschließlich aller Instituts- und Personalkosten, unabhängig vom Kostenträger. Ausbringung, Überwachung und Beratung durch höher qualifizierte Fachkräfte müssen ebenfalls in die Kalkulation eingehen, auch wenn es sich um staatlich bezahltes Personal handelt. Bis heute bilden allerdings solche praktischen Einsätze noch die Ausnahme. Sie werden in verschiedenen Ländern auf 0,1 bis 1,3 % der mit chemischen Mitteln behandelten Fläche geschätzt. Die wissenschaftlichen Erkenntnisse, die die Forschung nach biochemischen und biotechnischen Verfahren erbracht hat, sind bewundernswert. Sie sollten daher in einer Auswahl nach den bis heute erfolgreichsten und vielversprechendsten Methoden in diesem Band vorgestellt werden.

Die in anfänglicher Begeisterung gehegte Hoffnung, daß die Pflanzenschutzmittel der sogenannten „Dritten Generation", z. B. Pheromone oder Wachstums- und Entwicklungshemmer von Insekten, leicht überschaubar und relativ einfach zu handhaben seien, hat sich leider mit fortschreitender wissenschaftlicher Erkenntnis ganz und gar nicht erfüllt. Auch die ursprüngliche Hoffnung, daß keine Resistenzprobleme entstehen könnten, war nicht gerechtfertigt.

Da die Chemie der Pheromone verschiedener Art und Wirkungsweise in der Aufklärung und Anwendung am weitesten fortgeschritten ist, wird dieser Teil der biochemischen Methoden hier besonders eingehend behandelt. Eine erste Einführung in die Chemie der Sexuallockstoffe der Insekten hat schon Eiter (†) in Band 1 gegeben.

Die biologische Kenntnis über Art und Wirkungsweise der Pheromone ist Voraussetzung zum Verständnis der chemischen Forschungsergebnisse auf diesem Gebiet. Die Kapitel über diesen Fragenkomplex wurden deshalb hintereinandergeschaltet und von zwei international anerkannten Forschern ausführlich behandelt. Für die Beurteilung der biotechnischen Methoden spielen natürlich auch die in der bisherigen praktischen Anwendung gewonnenen Erfahrungen eine bedeutende Rolle. Sie haben unerwartet große Schwierigkeiten ergeben. Darüber hinaus zeigt sich aber auch – wie schon im biologischen Kapitel deutlich wird –, daß nicht die Lockwirkung eines einzigen spezifischen Insektenhormons die Populationsdynamik steuert, sondern daß hierfür das Zusammenwirken eines ganzen Komplexes von chemischen und anderen Signalen mit sehr verschiedenartiger biologischer Funktion erforderlich ist. Damit stellt die gleichzeitige Bekämpfung mehrerer Insektenarten in der Praxis bei gemeinsamem Auftreten an Nutzpflanzen ein oft sehr schwieriges und teilweise fast unlösbares Problem dar. Für Prognoseverfahren, die für den integrierten Pflanzenschutz erforderlich sind, gewinnt die Anwendung von Pheromonen zunehmend an Bedeutung. In solchen „monitoring systems" können Pheromone wichtige Dienste leisten, wie das gerade auch am Beispiel des Forstschutzes deutlich wird.

Die Anwendung von Juvenoiden, die in die Entwicklung der Insekten eingreifen, ist ebenfalls zu den biochemischen Methoden zu rechnen. Der vorliegende Beitrag international bekannter Fachleute zeigt, daß die sichere Steuerung bzw. Verhinderung der Insektenentwicklung mit Hormonen und Hormonmimics dieses Typs in der Praxis noch schwieriger sein dürfte, als die Anwendung von Pheromonen.

Durch eine zufällige Beobachtung hat die rein chemische Forschung zu einem neuen Wirkungsprinzip für die Insektenbekämpfung geführt, nämlich zur Verhinderung der Chitin-Synthese. Forscher der Firma Duphar, bei der diese Entdeckung gemacht wurde, geben in einer zusammenfassenden Übersicht über dieses Arbeitsgebiet eine kompetente Darstellung. Der Beitrag zeigt, wie umfangreich die Vorarbeiten und Prüfungen aller Art für die Einführung eines Insektizids mit zuvor nicht bekanntem Wirkungsspektrum sein müssen.

Phosphorsäureester haben, insbesondere nach der weitgehenden Anwendungsbeschränkung für einige chlorierte Kohlenwasserstoffe, ihre bedeutende Stellung auf dem Insektizidmarkt behauptet. Eine Ergänzung der im Springer-Verlag erschienenen Neufassung des Kapitels über Phosphorester-Insektizide aus Band 1 erschien daher gerechtfertigt. Eine Autorin, die die vorausgegangenen Fassungen bereits mitgestaltet hat und in der Forschung der organischen Phosphorchemie tätig ist, hat den hier vorliegenden Beitrag übernommen.

Erstmalig wurde ein Beitrag über bakteriell bedingte Pflanzenkrankheiten und deren Bekämpfung aufgenommen, der von Mitarbeitern der Firma Ciba-Geigy zur Verfügung gestellt wurde, und der zeigt, wie schwierig heute noch die Bekämpfung von bakteriellen Infektionen der Naturpflanzen ist.

Ebenfalls von Mitarbeitern der Firma Ciba-Geigy wird über Fortschritte in der Bekämpfung von Oomyceten berichtet. Dem Wunsch der Autoren, nochmals einen kurzen historischen Überblick über die entsprechenden Fungizide zu geben, wurde entsprochen. In Band 2 und 4 finden sich bereits umfangreiche Darstellungen über die Geschichte aller Fungizide, dort aber in deutscher Sprache.

Eine moderne Forschungsrichtung befaßt sich mit der Suche nach natürlich vorkommenden insektiziden, fungiziden und herbiziden Wirkstoffen, ihrer Isolierung, Konstitutionsaufklärung und, wenn möglich, Synthese und systematischen Abwandlung. Diese Arbeiten werden von Bund und Ländern gefördert und fügen sich dem Trend der öffentlichen Meinung, in der Schädlingsbekämpfung „natürliche" Wirkstoffe zu verwenden. Über solche Wirkstoffe mikrobieller Herkunft berichten Autoren der Bayer AG, die auf diesem Forschungsgebiet arbeiten. Bislang ist die praktisch verwertbare Ausbeute dieser Arbeitsrichtung noch bescheiden, doch mag der Beitrag Denkanstöße für weitere Arbeiten in diesem Bereich geben. Er vermittelt zugleich eine kritische Übersicht. Die weitaus meisten natürlich vorkommenden, biologisch aktiven Verbindungen sind höchst komplizierter Natur und damit wenig geeignet, synthetisiert zu werden. Darüber hinaus sind sie Licht und atmosphärischen Einflüssen gegenüber wenig beständig, was für ihr Umweltverhalten, nicht aber für ihre praktische Anwendung als positiv gelten kann. Besonders zu beachten ist aber, daß derartige natürliche Wirkstoffe häufig extrem toxisch sind.

Der vorliegende Band wird durch einen umfassenden Beitrag fortgesetzt und abgerundet, der als Band 7 Anfang 1981 erscheint und die aus Pyrethruminhaltsstoffen abgeleiteten und entwickelten Verbindungen zum Inhalt hat.

Nicht für alle vom Herausgeber angestrebten Themen konnten geeignete Autoren gewonnen werden. Manche Forschungsinstitutionen zögern, aus der Besorgnis heraus, der Forschungskonkurrenz auf einfache Weise einen neuen Forschungsstand mitzuteilen, ihren Mitarbeitern eine Publikationsgenehmigung zu erteilen. Desto mehr danke ich allen Autoren aus Universitäten, Forschungsinstituten und besonders der Chemischen Industrie für ihre Mitarbeit. Ich bin davon überzeugt, daß eine solche Mitarbeit insgesamt zu einer Wissenserweiterung führt, aus der die Firmen selbst letztlich den meisten Nutzen ziehen, besonders aber Firmen, deren Mitarbeiter durch ihre Beiträge sich einen hohen Stand an Spezialwissen angeeignet haben.

Die Zweisprachigkeit des Bandes beruht nicht zuletzt auf dem verständlichen Wunsch einiger Autoren, die durch die Mitarbeit bedingte zusätzliche berufliche Belastung nicht auch noch durch eine korrekte Übersetzung in die englische Sprache zu erhöhen.

August 1980 R. Wegler

Inhaltsverzeichnis

Biology of Pheromones

J. A. A. Renwick and J. P. Vité

Chemistry of Insect Pheromones

H. J. Bestmann and O. Vostrowsky

Die praktische Verwendung von Insektenpheromonen

M. Boneß

The Use of Juvenoids in Insect Pest Management

J. P. Edwards and J. J. Menn

Pflanzenschutzpräparate mikrobieller Herkunft

K. Bauer, D. Berg, E. Bischoff, H. v. Hugo und P. Kraus

Insektizide Phosphorsäureester

Ch. Fest

Bacterial Plant Diseases and Their Control

T. Egli and E. Sturm

Recent Advances in the Chemical Control of Oomycetes

T. H. Staub and A. Hubele

Benzoylphenylurea Insecticides

W. Maas, R. van Hes, A. C. Grosscurt and D. H. Deul

Mitarbeiterverzeichnis

Dr. K. Bauer, Sparte Pharma – Institut für Biochemie, Bayer AG, D-5600 Wuppertal 1

Dr. D. Berg, Sparte Pflanzenschutz – Chemische Forschung, Bayer AG, D-5600 Wuppertal 1

Prof. Dr. H. J. Bestmann, Institut für Organische Chemie, Universität Erlangen-Nürnberg, Henkestr. 42, D-8520 Erlangen

Dr. E. Bischoff, Sparte Pharma – Institut für Biochemie, Bayer AG, D-5600 Wuppertal 1

Dr. M. Boneß, Sparte Pflanzenschutz, Bayer AG, D-5090 Leverkusen-Bayerwerk

Dr. habil. H.-H. Cramer, Sparte Pflanzenschutz, Bayer AG, D-5090 Leverkusen-Bayerwerk

Dr. D. H. Deul, Agrobiological Laboratory “Boekesteyn”, Duphar B.V., Noordereinde 63, 1243 ZG ’s-Graveland, The Netherlands

Dr. J. P. Edwards, Slough Laboratory, Ministry of Agriculture, Fisheries and Food, London Road, Slough SL3 7HJ, England

Dr. T. Egli, Ciba-Geigy AG, CH-4002 Basel

Dr. Christa Fest, Pflanzenschutz – Chemische Forschung, Bayer AG, D-5600 Wuppertal-Elberfeld

Dr. A. C. Grosscurt, Agrobiological Laboratory “Boekesteyn”, Duphar B.V., Noordereinde 63, 1243 ZG ’s-Graveland, The Netherlands

Dr. R. van Hes, Agrobiological Laboratory “Boekesteyn”, Duphar B.V., Noordereinde 63, 1243 ZG ’s-Graveland, The Netherlands

Dr. A. Hubele, Ciba-Geigy AG, CH-4002 Basel

Dr. H. von Hugo, Sparte Pharma – Institut für Biochemie, Bayer AG, D-5600 Wuppertal 1

Dr. P. Kraus, Sparte Pflanzenschutz – Anwendungstechnik – Biologische Forschung, Bayer AG, D-5090 Leverkusen

Dr. W. Maas, Agrobiological Laboratory “Boekesteyn”, Duphar B.V., Noordereinde 63, 1243 ZG ’s-Graveland, The Netherlands

Dr. J. J. Menn, Zoecon Corporation, 975 California Ave., Palo Alto, CA 94304, USA

Dr. J. A. A. Renwick, Boyce Thompson Institute, Ithaca, NY 14853, USA

Dr. T. H. Staub, Ciba-Geigy AG, CH-4002 Basel

Dr. E. Sturm, Ciba-Geigy AG, CH-4002 Basel

Prof. Dr. J. P. Vité, Forstzoologisches Institut, Universität Freiburg/Breisgau, Bertholdstr. 17, D-7800 Freiburg

Dr. O. Vostrowsky, Institut für Organische Chemie, Universität Erlangen-Nürnberg, Henkestr. 42, D-8520 Erlangen

Dr. Brünhilde Zeller, Sparte Pflanzenschutz, Bayer AG, D-5090 Leverkusen-Bayerwerk

Biology of Pheromones

J. A. A. Renwick

Boyce Thompson Institute, Ithaca, NY 14853, USA

J. P. Vité

Forstzoologisches Institut der Universität, D-7800 Freiburg i. Br.

Contents

1. Introduction

Early research on pheromones of insects and subsequent identifications of active compounds were often accompanied by promises of new and exciting methods for control of specific insect pests. However, in spite of some limited success, recent investigations on the use of pheromones in pest management systems have done much to dispel the early optimism. Some skepticism about the potential of pheromones may well be justified, but many conclusions have been based on misconceptions or incomplete information. A full understanding of an insect's biology, behavior and ecology appears to be necessary to evaluate each situation.

The outstanding feature of pheromones for pest management is their function as highly specific messages to signal orientation of individuals towards others of their species, towards their host plants, or away from unfavorable situations. Efficiency in mating, locating prey or colonizing hosts can be shown to depend on pheromones, but the behavior of a pest population is more complicated than might be indicated by observations on individual insects. Orientation actually depends on intrinsic conditions of each insect, and occurs in a sequence of behavioral phases ("Reizkette") [1] that are influenced by other stimuli in a variable manner. The complexity of these communication systems demands a better understanding of how pheromones act and pest populations behave.

Several comprehensive reviews on pheromones have appeared during the last few years [2–7]. However, most of these have dealt with the chemistry or reception of pheromones, and the discussions of pheromone biology have been quite basic in nature. This review is not intended to be comprehensive, but rather to complement other chapters in this volume by examining more applied aspects of pheromone biology. Examples will be used to highlight biological and behavioral phenomena that might affect the potential use of pheromones in pest management systems for insects, with special emphasis on forest pests.

2. Biological Function

Insect pheromones can be classified most conveniently according to their function in

(1) orientation of the insect towards or away from a location,
(2) signaling danger, or
(3) providing information for social interaction.

Attempts to use pheromones in pest management systems have been essentially limited to those regulating orientation behavior, in particular sex pheromones, aggregation pheromones, and response inhibitors. Other types of orientation pheromones such as trail marking substances may also have potential as control agents, but as yet there are no definitive reports on their application.

2.1. Sex Pheromones

The primary function of sex pheromones is to bring males and females of a species together for the purpose of mating. This may involve orientation of one sex towards the other, excitation, and other short range behavioral reactions. The original idea that a single compound is responsible for sex attraction in each insect has long been disproved. Most insects appear to depend on blends of two or more components for maximum behavioral response. Until recently, bombykol, the first sex attractant to be identified [8], was believed to be an exception. But two additional compounds have since been isolated from the silkworm moth. The (E, E) stereoisomer occurs along with the (E, Z) form and exhibits some excitatory activity, although higher concentrations are required than for the (E, Z) isomer [9]. The (E, Z) aldehyde, bombykal, was also identified from female pheromone glands, and appears to have an inhibitory effect on the response to bombykol [10]. The exact function of each component of the pheromone mixture is not clear, particularly since synthetic samples of bombykol used in behavioral studies have contained 3–5% of the (E, E) isomer as an impurity [9].

Observations on pheromones of numerous Lepidopteran species have emphasized the importance of second or third components in eliciting complete response, as measured by the end effect of insect catch at traps [11]. However, little is known about the role of each component in the total sequence of behavioral events. Baker et al. [12] have indicated that one component may be responsible for long range attraction and another acts as a landing or short range orientation stimulus for a tortricid. Separate stimulation and orientation responses have also been noted for different components of the pheromone system of the Indian meal moth and the almond moth [13], and Cardé et al. [14] have shown that the components of the oriental fruit moth attractant have different behavioral roles. Further experimentation with ultra-pure synthetic samples will be necessary to determine whether this phenomenon may be more widespread throughout the Lepidoptera and possibly other insect orders. It is perhaps unfortunate that the term "pheromone" is now being used to include mixtures of compounds involved in the total orientation behavior of insects. Future studies may reveal that each component is a pheromone in its own right, responsible for a specific behavioral response in a complex communication system.

In many insects, another type of pheromone comes into play after both sexes have been brought together by sex attractants or other means. The so-called aphrodisiac pheromones [15] are involved in a sometimes complex courtship behavior pattern which leads to copulation. In the Lepidoptera, these aphrodisiacs are released by the males from hair-pencils or other specialized eversible organs and result in arresting the female or inducing an acceptance posture. Although many of these volatile secretions have been identified, no field application has been demonstrated. However, it is important to realize that such factors are involved in the successful mating of populations in the field. The possibility exists that attempts to disrupt mating by means of area saturation with a sex attractant may be thwarted by the existence of short range chemical and visual cues.

2.2. Aggregation Pheromones

The attack of host trees by most Scolytidae is mediated by population aggregating pheromones which are emitted by one or both sexes and which attract both sexes to the source. The function of aggregation pheromones is to signal the suitability of a host which would provide a temporary habitat [16] or to recruit large numbers of beetles to overwhelm the natural resistance of healthy trees [17]. In general, the pheromone is released by the sex initiating the attack, viz. the female of monogamous species and the male of polygamous species [18]. However, in some cases the attraction may be complemented by a compound produced predominantly by the opposite sex [19–21].

In contrast to the sex pheromones of the Lepidoptera, many bark beetles appear to aggregate in response to a single compound. For example, *Ips sexdentatus* [22] and *I. duplicatus* [23] are attracted by ipsdienol alone and the southern pine beetle, *Dendroctonus frontalis*, in North and Central America depends on frontalin alone [24]. On the other hand, multiple component attractants do exist, as in the case of *Ips paraconfusus* [25], where three terpene alcohols are required. When both sexes contribute to the attraction, as in *Dendroctonus brevicomis*, and *D. adjunctus*, the component produced by one sex draws more of the opposite sex, so that approximately equal numbers of males and females arrive at the host tree [19, 21].

The aggregation of many bark beetles depends on the presence of host volatiles. As the beetles penetrate the phloem and cambium of a coniferous tree, oleoresin flows from the wounds, and terpene components of the resin complement the pheromone to trigger a mass attack. The most effective terpene is usually the most abundant volatile component of the preferred host's resin [26–32]. For example, the pheromone of *Dendroctonus ponderosae* attacking western white pine in the USA is *trans*-verbenol, which is only active in the presence of resin, and the most effective terpene is α-pinene [28]. A host factor is also a necessary part of the attractant for the elm beetle, *Scolytus multistriatus*, and the active compound was identified as α-cubebene [33]. These host components may serve to guarantee host recognition by the insects. It appears likely that the pheromones are responsible for initial orientation of beetles whereas the host volatiles elicit close range attraction or arrestment.

The involvement of host volatiles may well be responsible for the failure of attempts to disrupt communication in *Dendroctonus frontalis*. When the attractant mixture of frontalin and α-pinene was applied from the air to an active beetle infestation, other volatiles emanating from previously attacked trees appeared to induce landing of the beetles [34]. However, this effect may be used to advantage in pest management efforts. The attractant may be used to contain populations within an infestation. If beetles are induced to reattack infested trees, overcrowding and subsequent mortality may be achieved [35].

2.3. Inhibitors

Compounds that interfere with the response of an insect to its attractant have been found in many pheromone systems. The mechanism of this interference is not generally known, but the effect is to reduce the number of insects arriving at an attractant source. Although the terminology may not accurately reflect the mechanism, these compounds are conveniently referred to as inhibitors.

Certain compounds present in pheromone glands have proved to be inhibitory to the insect producing them. The attraction of male gypsy moths to disparlure is inhibited by a hydrocarbon precursor which is also present in gland extracts [36], and spruce budworm glands contain a substance that inhibits the response of this species to its sex attractant in the forest [37]. However, it is not known whether these compounds are actually released by the moths under natural conditions. Recent studies on the silkworm moth have demonstrated that an inhibitor is in fact released by the female moths [10]. It is difficult to speculate on the possible function of such an inhibitor in the communication system of a moth, but Kaissling and Kasang [10] suggest that the inhibition might serve to ensure that the first female to hatch is not courted by every male in her vicinity.

Natural inhibitors are released by some of the bark beetles as part of a complex pheromone system. In the genus *Dendroctonus*, females initiate the attack of standing trees and emit an aggregation pheromone. Males arriving at the tree may release an additional pheromone to provide a highly attractive combination. However, the males of some species are also capable of releasing an inhibitor during the mass attack. Both the attractants and inhibitors are found in the hindguts of these beetles [38, 39] and are released by defecation. Two inhibitors appear to be produced by *Dendroctonus frontalis* males. Verbenone, the major volatile compound in the hindgut [38], reduces the response of males at relatively low concentrations, and inhibits both sexes at higher concentrations [29]. *endo*-Brevicomin is also produced by *D. frontalis* males and has a strong negative effect on the attraction of both sexes to frontalin and α-pinene [40]. Verbenone in *D. brevicomis* males appears to be an inhibitor for this species [30, 41]. A compound identified from *D. pseudotsugae* as 3-methyl-2-cyclohexen-1-one (MCH) was first thought to be an attractant [42], but later proved to be a powerful inhibitor [43], and *trans*-verbenol also reduces the number of beetles responding to attractive mixtures containing frontalin [43].

The release and function of inhibitors of *Dendroctonus* species have been subjects of some confusion in the field of bark beetle pheromones. Rudinsky et al. [44] suggested that MCH is released by *D. pseudotsugae* females in response to stridulation by the males which join them in their galleries. However, MCH was found in much higher concentrations in males themselves [45], so that most of the inhibitor emanating from the entrance holes undoubtedly comes from the males. Stridulation by bark beetles certainly plays some role in close range behavior [46, 47], but evidence is lacking to support the argument that sonic signals can cause release of pheromone by the opposite sex.

The function of inhibition in *Dendroctonus* is likely to be a form of population regulation. Low concentrations may alter the sex ratio of beetles arriving at an infested tree, whereas high concentrations signal full occupancy of the tree [30]. Another explanation might be that inhibitors function in a close range spacing mechanism on the tree. Since these beetles are monogamous, it is important that approximately equal numbers of each sex are guided to each tree. After landing, the females must space themselves sufficiently to make room for their galleries, and the males must have some means of recognizing whether a female has already been joined by a male. Claims of multiple functions of inhibitory pheromones in *Dendroctonus* [48, 49] i.e. as inhibitors and attractants, have been refuted [50]. Experiments with dilutions of these compounds in ethanol led to the conclusion that they could act as attractants at extremely low concentrations. However the bioassay used actually measured an arrestant effect, which can be obtained with ethanol alone. Ethanol can even have a synergistic effect on the positive response of *D. pseudotsugae* to its aggregation pheromone [50].

Studies on behavioral responses of bark beetles to inhibitors are providing a wealth of information on possible control techniques. The concept of population dispersal is particularly promising in the case of bark beetles since these insects depend on aggregation for their survival. Recent experiments using slow-release formulations of inhibitors for *D. frontalis* demonstrated a significant reduction of beetle landing on host trees [51]. Combinations of *endo*-brevicomin and verbenone were more effective than either inhibitor alone [52], and further results indicate that populations may in fact be reduced, even if specific trees cannot be protected [53].

Many synthetic compounds inhibit the response of insects to their natural pheromones. The structures of these inhibitors often closely resemble those of the pheromones, and the molecules may in fact block pheromone receptor sites [54]. The idea of using such synthetic inhibitors in field control programs is quite appealing, particularly since many of the compounds are relatively inexpensive or simple to synthesize. However, one attempt to disrupt communication of the oriental fruit moth with a known inhibitor apparently backfired [55]. No males were trapped in the immediate vicinity of the inhibitor, but traps baited with the attractant alone caught significantly more males when inhibitor was being released.

2.4. Allelochemicals

The pheromones of many insect species may also have some behavioral effect on other insects, either to the advantage of the emitter (allomones) or of the receiver (kairomones). These effects are an integral result of the evolution of pheromones in a complex ecosystem. In the Lepidoptera, sympatric species using a mutual pheromone have developed the need for additional compounds to ensure their reproductive isolation [11]. These secondary compounds may be inhibitory to closely related species and are thus of mutual benefit. It is remarkable however, that these pheromonal compounds responsible for species integrity may only be active in a narrow range of concentrations [11]. Other compounds released along with the pheromone of a moth may have no known function in the attractant of that moth, but inhibit the response of other insects. Separation of the nun moth and the gypsy moth in Europe is accomplished in part by means of the chirality of the pheromone. The gypsy moth responds only to the (+)-enantiomer whereas the nun moth will respond to a racemic mixture [56], and the (−)-form is inhibitory to the gypsy moth [57]. In this case, (−)-disparlure may be considered to be an allomone.

The exploitation of pheromones by parasites or predators appears to be quite widespread, as exemplified by many parasitic hymenoptera. Larvae of the flour moth, *Anagasta kuhniella*, use a pheromone to achieve population regulation through dispersal and physiological effects [58]. The ichneumonid parasite, *Venturia canescens* has utilized this mandibular gland secretion as a signal for ovipositional probing behavior [58]. Other parasites use the sex pheromone or aggregation pheromone of their hosts as orientation stimuli. Observations on bark beetles have demonstrated that the predacious brachyceran fly, *Medetera bistriata*, responds to attractant produced by *Ips grandicollis* or *Dendroctonus frontalis* [59]. When the synthetic pheromone, frontalin, was tested in combination with α-pinene, a high proportion of *M. bistriata* females responded, indicating a clear kairomonal effect. Other bark beetle pheromones are known to guide coleopteran predators to their prey. The clerid, *Enoclerus lecontei*, is attracted to the three component pheromone system of *Ips paraconfusus* [60] and to the natural attractant of *I. pini* [61]. *Thanasimus dubius* and *T. undatulus* respond to frontalin released by *Dendroctonus frontalis* [62] and *D. rufipennis* [63], respectively, and the trogositid, *Temnochila chlorodia*, is attracted by the *exo*-brevicomin of *Dendroctonus brevicomis* [64, 65]. Although some of these predators are finely tuned to specific compounds, most are less discriminatory, often responding to conifer terpenes as well as to the pheromones [65]. Whenever bark beetle pheromones in combination with terpenes are released in the field, insect predators can be expected to respond.

2.5. Communication Systems

The fact that multiple component mixtures may be involved in chemical communication of insects is now well established. However, few attempts have been made to analyze the behavioral responses to individual components and determine the sequence of events throughout the period of adult activity. Although they are among the most complex, the chemical communication systems of the bark beetles are probably the best understood.

The bark beetles may be roughly divided into two groups, i.e. primary and secondary pests, depending on whether they attack healthy trees or those which have been weakened in some way. In the first case, the trees are marked with "contact" pheromones to signal a mass attack, whereas the "frass pheromones" of the secondary species are produced only upon feeding, to indicate the suitability of a host tree for colonization [66]. Of course, many beetles may be considered to be intermediate, and as a general rule they depend on a combination of both types of pheromones. Most *Dendroctonus* species in North and Central America belong to the aggressive group. Usually the pheromones are already present in the emergent adults so that they can be released immediately upon landing. Otherwise, the attractants are produced as soon as the attacking beetles start to feed and come in contact with host resin. Since *Dendroctonus* spp. are monogamous, some mechanism usually exists to equalize the numbers of each sex arriving on the tree [30]. Also, a host terpene is generally necessary for full response, thus providing assurance that the selected tree is sufficiently rigorous that it will not dry out before the brood can develop. For example, *D. brevicomis* females arrive at their host trees with small amounts of frontalin, *endo*- and *exo*-brevicomin, and *trans*-verbenol in their hindguts [67, 68]. No definite function has been demonstrated for *trans*-verbenol or *endo*-brevicomin in this species, but the frontalin and *exo*-brevicomin attract both sexes of in-flight beetles. As the females feed and construct their galleries in the phloem, larger quantities of *exo*-brevicomin are produced, attracting predominantly males [69]. Myrcene or 3-carene from the exuding resin greatly increases the number of beetles landing [68, 70], and the arriving males release copious quantities of frontalin, which attracts more females. As the colonization progresses and the tree is overwhelmed, resin exudation ceases, and verbenone released by the males inhibits further landing, so that the attack shifts to neighboring trees. Other *Dendroctonus* systems may be equally complicated, but a single pheromone appears to be responsible for the attraction of *D. ponderosae* and for *D. frontalis*. However, inhibitors released by the males are still operative [29, 71] and other means of regulating sex ratios come into play.

The majority of *Ips* bark beetles are secondary pests, colonizing trees which have been felled or weakened in some way. Here the pheromones may be single compounds, as in the case of *I. grandicollis* [72], or multiple component systems such as that of *I. paraconfusus* [25]. In either case, feeding is necessary for production and release of the pheromones. It is not known how the initial selection of a host tree occurs. It may be random trial and error by "pioneer beetles" [73], or some chemical cue may emanate from trees under stress.

Some of the European Ips are relatively aggressive in their behavior, often infesting standing trees. Both *I. typographus* and *I. cembrae* have relatively complicated pheromone systems which include isomers of methyl butenol not yet detected in American species [74, 75]. The exact function of these butenols is not known, but experiments with different trapping systems may throw some light on their significance.

The involvement of host tree volatiles in the aggregation of certain bark beetles has introduced a complication in the pheromone terminology. Silverstein [76] has chosen to include host terpenes such as myrcene as "pheromone components". However, these terpenes are really food plant attractants or arrestants. We prefer to treat the individual components of a chemical communication system as separate entities, each providing its own information. Even if distinct functions for all the compounds are not known at this time, future research might better define the roles of individual components in a communication system.

3. Biosynthesis and Release

Despite the vast number of pheromones that have been identified during the last 20 years, remarkably little is known about their biosynthetic origin in insects. The Lepidoptera in general use relatively long-chain molecules with different functional groups as their pheromones, suggesting that they could be linked to fatty acid metabolism in some way. Other insect orders rely on a wide variety of compounds ranging from monoterpene alcohols to long chain hydrocarbons. In the case of phytophagous insects, host plants can play a major role in orientation, either by attracting both sexes to a beneficial location for mating or by affecting pheromone production in some way. A recent suggestion that pheromones may generally be plant components which are sequestered by lepidopterous larvae [77] was proven to be unfounded for the oak leaf roller [78]. The terpene alcohol, linalool, which has been implicated in the pheromone systems of two bark beetles [79, 80], is also a component of the host tree resin [81]. Secondary compounds produced by

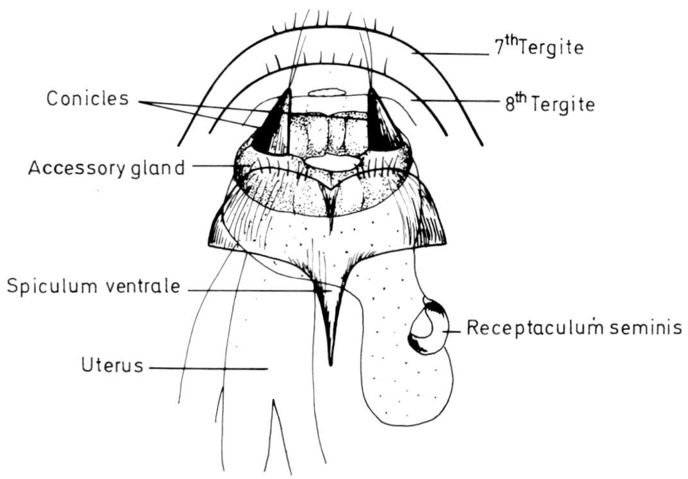

Fig. 1. Accessory glands of *Scolytus scolytus* females (from Grüne [87])

plants are believed to have evolved as defense mechanisms against insects and other organisms. However, many insects have developed the ability to use these compounds for host recognition and may even modify them to produce pheromones [82, 83].

Pheromones are often released from glandular structures. However, they may also be incorporated in the integument or are excreted by defecation. Modified intersegmental membranes or cuticular tissue may serve as glands for the release of sex attractants in many Lepidoptera and some Coleoptera. The most elaborate structures appear to be the scent-disseminating accessories which male Lepidoptera use in close-range courtship behavior. For example, male monarch butterflies (Danaidae) use hair pencils to dust female antennae with a dihydropyrrolizine aphrodisiac as a prerequisite for mating [84].

Male and female bark beetles release the aggregation pheromones from their hindguts, but in females of *Scolytus* a glandular structure is also involved [85–87] (Fig. 1).

3.1. Stimulation of Production

The production and release of pheromones by many insects can be influenced by chemical constituents of the host plant. The polyphemus moth, *Antheraea polyphemus*, depends on perception of *trans*-2-hexenal emanating from oak leaves for release of its sex pheromone, and other volatiles from non-host trees can mask this perception [88, 89]. Several scolytid genera depend on the presence of terpene hydrocarbons in their host trees for the production of terpene alcohols which they utilize as pheromones. Simple exposure of beetles to vapor of the monoterpene precursors results in pheromone biosynthesis by some species [82, 66] whereas other beetles require feeding before pheromones are produced [83]. The mechanism by which bark beetles control the oxidation of terpenes to pheromonal alcohols has been elucidated in certain species. The involvement of juvenile

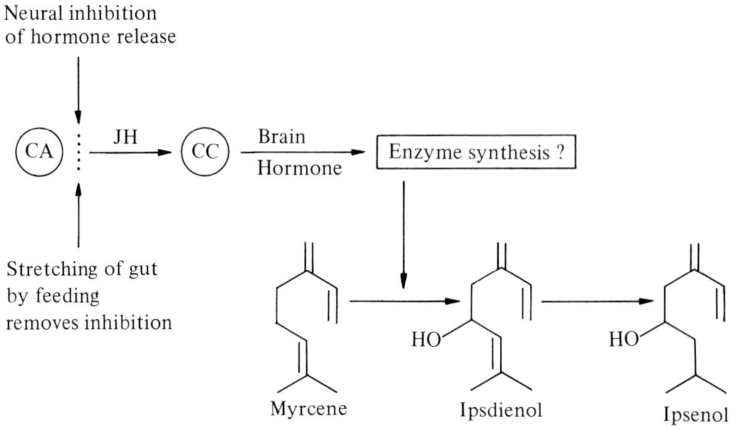

Fig. 2. Generalized scheme for the neural-hormonal control of pheromone biosynthesis proposed for males of the bark beetle, *Ips paraconfusus* (from Hughes and Renwick [91])

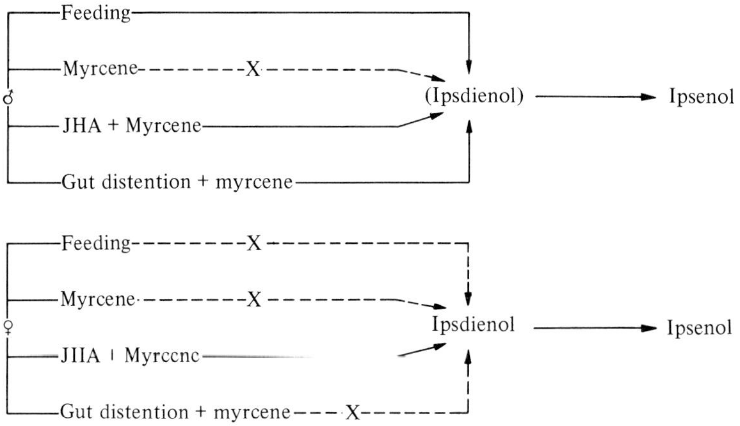

Fig. 3. Variation in pheromone biosynthesis in response to feeding, juvenile hormone analog (JHA) treatment, gut distention by air injection, and myrcene exposure among ♂ and ♀ *Pityokteines* (adapted from Harring [92])

hormone (JH) in pheromone production was suggested in the case of *Ips paraconfusus* [90] and has since been confirmed and clarified [91]. The feeding requirement of this beetle for production of ipsdienol and ipsenol can be by-passed by topical treatment of males with JH before exposure to myrcene. Gland transplantations and decapitation experiments have suggested that the stimulation occurs indirectly (Fig. 2). Distention of the gut, which can be accomplished by air injection, is believed to induce neural stimulation of the corpora allata. The resulting JH output causes the release of brain hormone, which appears to control the oxidation of myrcene to ipsdienol and ipsenol, possibly by controlling synthesis of specific enzymes. Another component of the pheromone complex, *cis*-verbenol, is produced in quantity only in the presence of exogenous α-pinene.

Pheromone biosynthesis and release therefore depend on two limiting factors, i.e. actual feeding (gut distention), and exposure to the monoterpene precursors. These requirements ensure the selection of host material that is relatively fresh, but weak enough to allow easy feeding.

The same principle appears to apply to species of *Pityokteines* which attack white fir in Europe. JH treatment and myrcene exposure lead to the biosynthesis of ipsdienol and ipsenol in both sexes (Fig. 3). However, female beetles, which are not responsible for pheromone release in nature, are not affected by gut distention, whereas males are stimulated to produce pheromone [92].

In other beetles such as *Tenebrio molitor* [93] and certain bark beetles, treatment with JH induces pheromone production directly, without feeding. *Dendroctonus brevicomis* produces large quantities of *exo*-brevicomin after JH treatment, in the absence of any exogenous source of precursor [94]. Here distention of the gut has no apparent effect, but ingestion of or contact with a polar substance(s) in the host phloem provides the stimulus for pheromone production. The use of JH to induce pheromone biosynthesis has proved to be extremely valuable in producing enough material for identification of a pheromone. Treatment of several thousand "Kupferstecher" (*Pityogenes chalcographus*) resulted in the isolation of sufficient compound from these tiny beetles for identification of the pheromone, chalcogran [95, 86, 96].

The oxidation of terpenes by bark beetles attacking conifers appears to be quite widespread. Many products may be formed from each individual terpene [97–99], and each has the potential to function as a pheromone. However, a considerable degree of sex-specific and species-specific control exists. Also, the enantiomeric composition of a compound can affect its utilization as a precursor. The stereo-selective conversion of the α-pinene enantiomers into the geometrical isomers of verbenol [100] has provided a novel approach to determine the optical activity of α-pinene from various sources [101].

The relationship between monoterpene hydrocarbons of conifers and the pheromones of their bark beetle pests is now well established, but the mechanisms involved remain to be explored. The involvement of microflora in the hindguts of beetles has been suggested, since bacteria isolated from *Ips paraconfusus* were shown to ferment α-pinene into numerous derivatives including the verbenols and verbenone [102]. However, no success has been reported in isolating a microorganism that will oxidize myrcene to ipsdienol and ipsenol, and it is unlikely that the microflora of the different sexes and species could vary enough to explain the high degree of specificity of pheromone production. Harring [92] has found that terpenes ingested along with the food of *Pityokteines* pass through the alimentary canal essentially unaltered, so the action of micro-organisms must be extremely limited. On the other hand, terpene vapors entering the beetles through the integument are rapidly converted to relatively large quantities of the specific oxidation products. It appears likely, then, that bark beetles have specialized enzyme systems to transform hydrocarbons into alcohols and other oxidation products. These systems may have evolved as detoxification mechanisms and were later utilized to provide a sophisticated form of chemical communication involving hormonal control and stereospecific selectivity.

3.2. Rhythm of Release

Most insects respond to pheromones in a well-defined diurnal pattern that appears to depend on the photoperiod for temporal placement. The time of dispersal and search is most likely to coincide with atmospheric conditions most favorable for perception and response. When the photoperiod also serves to control circadian rhythms in pheromone release, energy expenditure by the insects is minimized and the risk of predation is reduced.

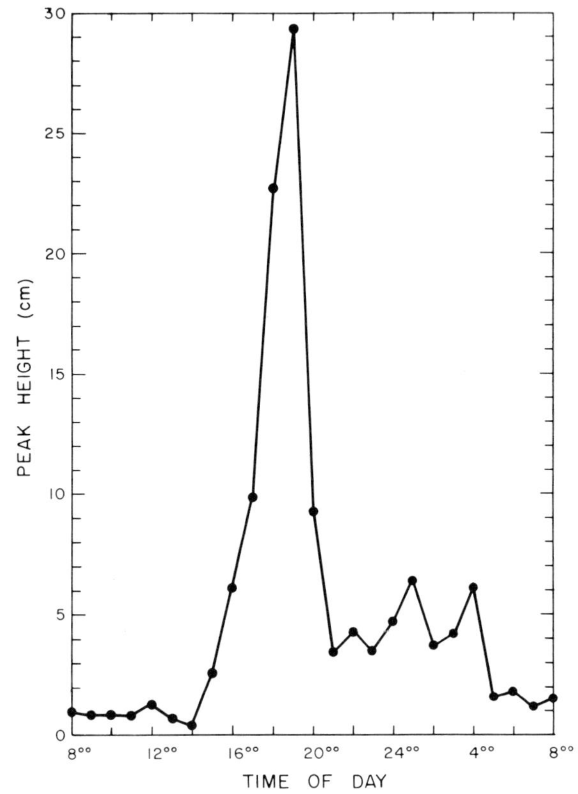

Fig. 4. Temporal variation in pheromone content of ♂ *Polygraphus poligraphus* hindguts, as measured by 4-terpineol peak on gas chromatograms (from Leuze [110])

Diurnal rhythm in the release of pheromone is particularly characteristic in insects where pheromone glands are involved. In many lepidopterous species, increased male activity coincides with the "calling" period of the female [103, 104]. When the same chemical signals are used by sympatric species having overlapping seasonal flight periods, the circadian rhythm of release and response may play an important role in reproductive isolation, as in the case of *Lymantria* [105, 106].

Distinct circadian rhythms have been demonstrated for the dermestid beetles. However, the active periods of *Trogoderma glabrum* and *T. inclusum* (mid-afternoon) overlap, and species separation is achieved through utilization of different isomers of the pheromone, trogodermal [107, 108]. Male-female encounters of another species, *T. variabile*, are maximized at dawn so that interspecific interaction is avoided [109].

Circadian rhythm is not entirely restricted to pheromone release from glands. The timing of release and response in the bark beetle *Polygraphus poligraphus* is also circadian in nature (Fig. 4). The polygamous male of this species attracts both sexes for a brief period only during the evening hours. The temporal placement is at least partially regulated by photoperiod, which affects response as well as causing a dramatic increase in defecation at dusk [110]. As is the case for most scolytids, the pheromonal components are contained in the fecal pellets. Circadian rhythm in other scolytid beetles has been observed in terms of aggregation behavior [111], but not in pheromone release. Continuous defecation and expulsion of feces with the "frass" from the galleries probably provides continued marking of host material. The time of emergence and/or changes in atmospheric conditions are most likely responsible for setting diurnal patterns of aggregation that are already strongly moderated by temperature. However, calling behavior observed in *Scolytus* [86], *Trypodendron* and *Platypus* [112] suggest that circadian release mechanisms may be more common among scolytids and the closely related platypodids.

4. Bioassay Methods

The study of insect pheromones is dependent on appropriate bioassays for demonstration of response, for chemical isolation, and evaluation of synthetic chemicals. The type of bioassay required in each case depends on the type of information sought. Many authors have stressed the need for bioassays that can be used to define the functions of individual components of pheromonal mixtures. Kennedy [113] has pointed out the limitations of most laboratory bioassays and has suggested techniques for better analysis of behavioral responses.

Although it would be ideal if bioassays could be devised to explain mechanisms of response of all insects to their pheromones, agricultural and forest entomologists have to make decisions based on information provided by existing bioassays. With this in mind, some discussion of present methods and the extent of the information obtained might serve as a guide for more applied studies.

4.1. Laboratory Bioassays

The ultimate demonstration of biological activity of pheromones must be performed in the field. However, laboratory bioassays are indispensable in gathering information leading to the final test in the field. Most identifications of pheromones have involved laboratory monitoring of the isolation steps. The sex pheromones of the Lepidoptera have been readily detected by directing an air stream towards the antennae of males and observing the excitation reaction. The orientation behavior of other insects such as the Coleoptera has been followed by means of olfactometers using Y tubes, choice chambers or arenas, where pedestrian insects choose between two or more odor sources. Attempts to more closely mimic natural conditions have made use of flight mills and wind tunnels.

One advantage of laboratory bioassays is that minute quantities of candidate compounds or isolated fractions can be tested. Usually results can be obtained in a relatively short time, so that many repetitions can easily be performed. In addition, observations on specific reactions of the insects to test materials might be possible, thus providing some information on mechanisms of orientation. The disadvantage, however, is that assays using stationary or walking insects often detect only a part of the orientation behavior. From the point of view of using pheromones for control of pest populations, it is essential to know the factors responsible for guiding flying insects to their mates or to host plants. Laboratory tests are most likely to reflect behavioral patterns *after* the insects have landed. In many cases, active compounds have been isolated and identified, but additional compounds are necessary for field response. For example, the attractant of *Ips paraconfusus* was found to be a combination of two compounds in the laboratory [25], but three compounds were necessary in the field [60].

4.2. Electrophysiology

Since the discovery that an electrical potential can be detected when an insect's antenna is exposed to an olfactory stimulus [114, 115], the electroantennogram (EAG) technique has provided a unique bioassay system for the detection of pheromones. This technique has been used for the characterization of many lepidopterous pheromones by screening synthetic compounds of different chain lengths and functional groups [116]. Although the value of this system is now well documented, and EAGs are now obtained on a routine basis from Coleoptera as well as Lepidoptera [54], the limitations are more serious than for other bioassay techniques. The major disadvantage is that EAG response does not indicate how an active compound might affect the behavior of the insect. A positive reaction may be obtained with inhibitors and host plant volatiles as well as attractants [117].

4.3. Field Bioassays

Attempts to utilize behavior-modifying chemicals in pest management systems are usually aimed at populations of flying insects. In order to fully evaluate the effect of pheromones and other chemicals as attractants or inhibitors, field bioassays are essential. Chemical communication systems are generally observed or detected in the field, and identification of the active components can only be confirmed by duplication of the natural conditions. Nevertheless, field bioassays also have their limitations, and the experimental design can have a profound effect on the type of information obtained.

Most bioassay systems for Lepidoptera consist of some kind of sticky trap. This may be a simple rigid sheet of transparent or opaque material such as plexiglass or cardboard, where the insects merely have to make contact to be caught. Other traps have been designed so that the pheromone source and the

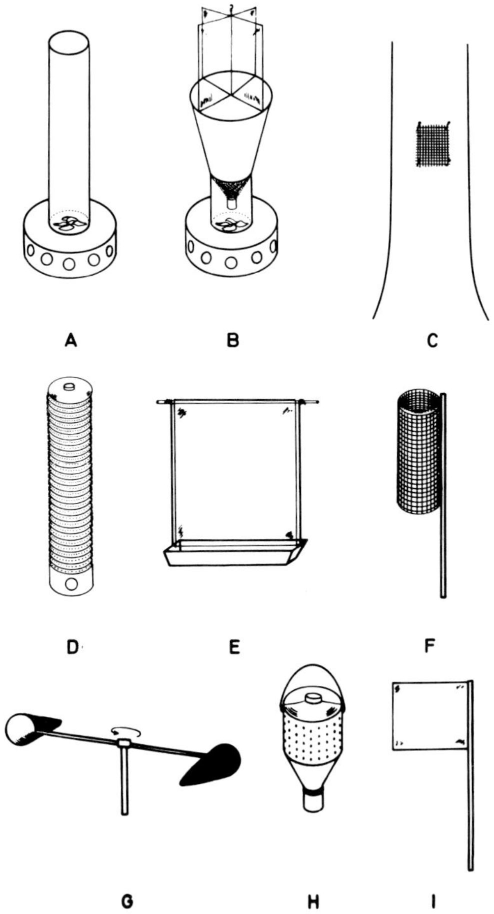

Fig. 5A–I. Examples of trap designs used in bark beetle bioassays. **A** Tree-simulating sleeve olfactometer, **B** funnel olfactometer, **C** hardware cloth sticky screen, **D** perforated cylinder, **E** window trap, **F** hardware cloth cylinder, **G** rotary net, **H** bucket trap, **I** transparent sticky trap

sticky surface are on the inside of an open container. In this case, an additional chemical stimulus may be necessary for responding insects to enter the trap [14]. Thus components of the total pheromone complex may be missed when different trap designs are used.

Probably the greatest number of different systems for assaying pheromone activity have been developed for bark beetles (Fig. 5). These devices range from simple sticky traps to elaborate fan-driven, tree-stimulating olfactometers. The choice of system depends on the behavioral characteristics of the insect species as well as the nature of the information required. Fan-driven funnel olfactometers [111] are convenient for quick bioassays during the flight period of *Ips* species, but for *Dendroctonus* species, a visual stimulus is important, and the tree-simulating sleeve olfactometer [118] is more effective. Window traps [119] are widely used to test bark beetle pheromones and provide results different from those with other traps. Beetles responding to "long-range" attractants fly into a transparent barrier and fall into a water trough. The presence of a "short-range" stimulus is not necessarily required. Other traps such as perforated cylinders [120] require a total complement of compounds for complete response, which includes landing and crawling into one of the holes. In addition to providing more information about the role of individual chemicals in beetle communication systems, such bioassays often reveal the importance of physical stimuli in orientation behavior.

In field as well as laboratory bioassays, the concentration of pheromone perceived by the insects may be critical. When the concentration at any location reaches the level of a natural source of attractant, it is likely that behavior is affected by other stimuli. In many of the Scolytidae, high concentrations of pheromone appear to signal landing [17, 121, 122], and a reduction in flight speed has been observed in other insects [123–125]. Some Lepidoptera males may also use visual cues to locate a female after a certain concentration has been reached [126]. It is clear that concentration effects can produce anomalous results in bioassays, and Roelofs [127] has used a "threshold hypothesis" to explain responses of moths which depend on critical ratios of isomeric compounds. Attraction of male moths can be expected within certain limits of concentration of each component.

Other factors affecting bioassay results in the field include atmospheric conditions [128, 129], circadian rhythms, and seasonal variations in behavior. Obviously, emergence and flight of insects in the field depend on temperature and humidity, but wind may also play an important role. Insects that respond over considerable distances depend on air movement for pheromone molecules to reach them. However, if the wind is too strong, flight may not occur. Also, the elevation at which the insects fly may vary with wind velocity [130]. Wind direction can have a profound effect on responses to pheromones, depending on the original location of the insects. For example, all the bark beetles attracted to a pheromone source may be emerging from a single brood tree, so that traps downwind from this tree will be less effective, as measured by beetle catch.

The timing of field bioassays can be important in compiling the most data from a single insect generation or limited period of activity. Circadian rhythms in response to pheromones may be just as common as rhythms of release. Many bark beetles have characteristic flight patterns, with a small peak in the morning and maximum response in late afternoon or evening [131]. In extreme cases, the flight may be limited to a 2 h period [110]. Most researchers choose to conduct assays over a period of several days so that the results are not affected by rhythms in response. However, such rhythms can be used to advantage. A considerable amount of information can be obtained in a short time by changing baits or positions of traps during the periods of highest insect activity.

The choice of bioassay technique can also be influenced by seasonal factors. Experiments with *Ips typographus* revealed that window traps are superior to perforated cylinder traps in early spring, but no difference occurs in summer [132]. The perforated cylinder traps require that the beetles crawl into the available holes. The behavior of *I. typographus* early in the season appears to involve frequent landing without drilling into the tree, whereas later in the season, less flying time is required to use up the available fat reserves [133]. Seasonal differences in mass attack behavior of *Dendroctonus frontalis* have also been observed [134], thus suggesting the need for different behavioral assay techniques.

Differences in behavior between males and females responding to an aggregation pheromone can have a profound effect on assay results. Traps baited with frontalin and α-pinene have consistently caught more males than females of *Dendroctonus frontalis* [29]. Certain trap designs have caught almost exclusively males [161]. This phenomenon has been explained to a large extent by the effect of trap size and concentration of pheromone [162]. Males apparently orient closer to the source of pheromone, whereas females are induced to land upon reaching a certain threshold concentration [147, 162].

Spacing of traps in a field bioassay system may have a dramatic effect on results. If sources of attractant are in close proximity, interaction may occur. On the other hand, if they are too far apart, the population density of insects is likely to be quite different in each area. Ideally, traps should be separated by a distance slightly greater than the effective response distance of the insect. The distances over which insects respond to pheromones is largely a matter of speculation. Experiments with wind tunnels would suggest that moths are capable of remaining within an odor plume for a long period of time. However, conditions in the field produce pockets of odor which bear little resemblance to plumes observed in the laboratory [129]. The question is whether insects are capable of following a broken trail to the source, or if they accidentally encounter the plume in close proximity to the source. Recent work with Scolytidae would suggest that the distances from which beetles can be attracted have been greatly overestimated. Fares et al. [128] have used a model to suggest that *Dendroctonus frontalis* actively responds from a distance of 5–10 m. Experiments with *Trypodendron lineatum* in Europe have resulted in similar estimates [135]. These relatively short distances of response might explain some of the extreme positional effects obtained in field bioassays. The placement of traps for survey and control of field populations must, therefore, take into account the natural movement of insects throughout their ecosystem.

5. Genetic Factors

The development of resistance to chemical insecticides in insects has raised the question whether similar resistance might be expected when pheromones are utilized in pest management systems. Although the danger is not believed to be as great in the case of pheromones, since the insects depend on the messages for their survival and reproduction, the possibility definitely exists. Several examples have been reported where problems have arisen when pheromones are applied to a pest population, and an understanding of the factors that could limit the successful use of pheromones in essential.

5.1. Alternative Systems

The isolation and identification of pheromones has almost invariably involved characterization of an active compound found in a gland, cuticle, hindgut, or frass of the insects. In some cases, this may be the major constituent of active extracts, and minor components having doubtful activity have been ignored. However, the importance of secondary compounds in insect communication systems is becoming more and more obvious as small percentages of isomers or other compounds are found to be necessary for maximum activity. Trace chemicals have recently been shown to play an important role in the sex attraction of *Heliothis zea* and *H. virescens* [136], species which had previously proved difficult to attract with single compounds or binary mixtures [137]. In other Lepidoptera, additional compounds that have no apparent effect may be present in the pheromone gland [9] and may be released along with the pheromone. When such mixtures of compounds are present, individual insects are likely to show varying degrees of response to each compound. It is therefore possible that the widespread application of pheromones in the field could exert selective pressure on a population to switch to another compound or combination of compounds for the most effective communication.

Many bark beetle pheromones are oxidation products of monoterpenes present in the oleoresin of their hosts. This oxidation process is thought to be a detoxification system which the beetles have used to their advantage [82]. The utilization of specific products has resulted in a high degree of species integrity, and host preferences have developed through the involvement of terpene hydrocarbons in the attraction mechanism. However, the host factors in these aggregation systems are replacable to some extent [29, 30]. This would suggest that a bark beetle species could easily evolve a preference for an alternate host having a different terpene composition. The oxidation of terpenes is never completely selective, so that a variety of alcohols is always produced upon contact with resin [82, 97, 98]. Species specificity is likely to involve differences in sensitivity at the receptor level to a greater extent than differences in ability to produce particular compounds. Each of these terpene alcohols, therefore, must be considered to be a potential pheromone.

5.2. Population Variations

The idea that all insects within a species use the same pheromone system is no longer acceptable. Recent work on the European corn borer has revealed that populations in Europe and the USA use different isomeric mixtures of 11-tetradecenyl acetate [138, 139], and hydrids of the two strains produce and respond to an intermediate mixture according to simple Mendelian inheritance [140]. Populations of the bark beetle, *Ips pini* in western and eastern USA show a distinct preference for the pheromone bouquet associated with their own frass [141, 142], and differences in enantiomeric composition of the ipsdienol produced by the two populations have been reported [143]. Differences in production of *endo*-brevicomin by *Dendroctonus frontalis* males in Texas and Honduras were found, but no significant differences in response have been detected [144]. However, distinct differences in the response of the elm beetle, *Scolytus multistriatus*, in Europe and USA have been demonstrated. North America populations depend on a tripartite mixture containing (−)-α-multistriatin [145], whereas

European populations appear to use the $(-)$-δ-isomer of multistriatin [146]. Differences in pheromone production by laboratory strains of insects may also be expected.

5.3. Taxonomy

Problems in the classification of closely related insect species can lead to confusion in the study of their pheromones. When geographically separate populations of *Dendroctonus frontalis* in North and Central America were compared, chemical analyses as well as biological observations demonstrated considerable differences [144, 147]. Beetles collected in Guatemala and Mexico were subsequently shown to be distinct species [148]. All three species use frontalin as the major aggregation pheromone, but different compounds are probably involved in regulating population density [147]. Studies on the pheromone system of *D. jeffreyi*, which was once considered to be synonymous with *D. ponderosae* [149], revealed a completely different aggregation pheromone [150]. This species has apparently adapted to use 1-heptanol, an oxidation product of n-heptane in its host (*Pinus jeffreyi*), whereas *D. ponderosae* responds to *trans*-verbenol, an oxidation product of α-pinene [28] which is abundant in most pines other than *P. jeffreyi*.

6. Case Histories

6.1. Dendroctonus frontalis

The southern pine beetle, *Dendroctonus frontalis*, is one of the most destructive bark beetles in North and Central America. Attempts to control this pest using conventional chemical insecticides have been largely unsuccessful. Research on the chemical communication of *D. frontalis* has done much to explain the problems of past efforts and has resulted in new hope for a practical pest management system [151–153].

Soon after the discovery of a population aggregating pheromone in the southern pine beetle [154], the potential use of the attractant to manipulate populations was demonstrated [155]. Behavioral studies revealed a strong preference for vertical objects baited with attractive material, so a tree-simulating olfactometer was developed for subsequent bioassays [156]. The isolation of the pheromone was aided by the fact that material collected from the western pine beetle, *D. brevicomis*, was also attractive to *D. frontalis* [154]. The most abundant compounds in both species were identified as *trans*-verbenol in females and verbenone in males [38]. Gas chromatographic analyses of both sexes of several related species, along with bioassays of crushed beetles, led to the isolation of the key pheromonal compound [157]. This application of differential diagnosis [158] indicated *D. brevicomis* males as the best source of the compound, which was then identified as 1,5-dimethyl-6,8-dioxabicyclo[3.2.1]octane (frontalin) [24].

The full activity of frontalin is dependent on the presence of a host factor, α-pinene [159, 29]. However, this may be replaced by *trans*-verbenol or, to some

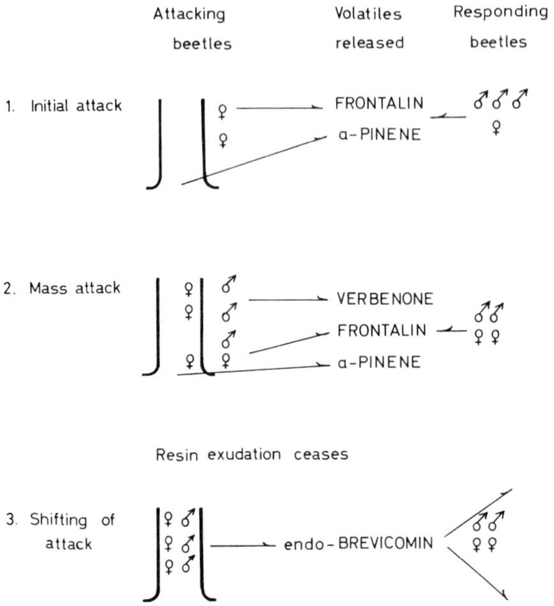

Fig. 6. Mechanism of colonization of pine trees by *Dendroctonus frontalis* (from Vité and Francke [18])

degree, by other terpene hydrocarbons [29]. The combination of frontalin and α-pinene attracts a predominance of males to the pheromone source; but the addition of verbenone appears to equalize the sex ratio of arriving beetles. High concentrations of verbenone inhibit the response of both sexes [29, 30]. *endo*-Brevicomin is also produced by males of *D. frontalis* and has a strong inhibitory effect on response to the attractant [40]. The sequence of events in the colonization of a tree appears to consist of three phases (Fig. 6): initial attack, mass attack, and shifting of the attack. During the initial attack, frontalin released by females along with α-pinene from exuding resin attracts both sexes; but more males than females land on the tree. Verbenone released by the males moderates the response of other males so that equal numbers of both sexes arrive at the tree. As the colonization progresses, the resin flow of the tree ceases, and *endo*-brevicomin from the males probably acts along with verbenone to signal the shift to a fresh tree.

Tests with mixtures containing frontalin have demonstrated a strong response of the clerid predator, *Thanasimus dubius* [62], which is normally abundant on trees undergoing attack. Efforts to use the pheromone in control programs therefore must take into account this kairomonal effect. Traps have been designed to exclude the clerids [161], and the use of trap trees and silvicultural methods which do not harm the predators have been proposed [153, 160].

The application of inhibitors appears to offer a promising approach for moderating the damage inflicted by the southern pine beetle. The effectiveness of *endo*-brevicomin and verbenone has recently been confirmed [51, 52], and the possibility of reducing brood densities in attacked trees has been demonstrated

[53]. Attempts to disrupt communication by the aerial application of rice soaked with Frontalure (frontalin + α-pinene) resulted in the beetles landing on previously attacked trees [34]. This phenomenon may be used to advantage, however. If beetles can be induced to reattack infested trees, overcrowding would occur and survival would probably be drastically reduced. The combined use of attractants and inhibitors may well prove to be the best strategy for combating this pest.

6.2. Trypodendron (=Xyloterus) lineatum

This ambrosia beetle attacks felled coniferous trees and logs throughout the northern hemisphere. Its galleries penetrate the sapwood, resulting in considerable losses through degradation of the lumber, along with the creation of serious quarantine problems [163]. Present conditions in the lumber market and forestry practices actually favor increases in pest populations [164, 165]. In Europe, *T. lineatum* is considered to be the most serious economic pest of all the bark beetles [166].

The aggregation pheromone produced by females of this species was among the first to be studied [167–172], but the identification of "lineatin" as 3,3,7-trimethyl-2,9-dioxatricyclo(3.3.1.$0^{4,7}$) nonane was accomplished only recently [173, 174]. The chemical synthesis [174, 175] has opened the door to field applications [174, 135, 176], which have demonstrated the effectiveness of the pheromone. However, a male-produced inhibitor [177–179] remains to be identified.

Primary and secondary host odors [180, 181] "synergize" the response of *T. lineatum* to its aggregation pheromone. Each component is responsible for a specific step in the sequence of events leading to colonization [182]. Lineatin released by the female entering a new host attracts in-flight beetles; α-pinene, which is common to the coniferous hosts, affords host recognition. At the same time α-pinene repels *T. domesticum*, a sympatric pest of hardwoods in Europe which also responds to lineatin [183]. The actual invasion into the sapwood occurs in response to ethanol [184, 182] which is formed by anaerobic fermentation in tree trunks suitable for breeding (Fig. 7).

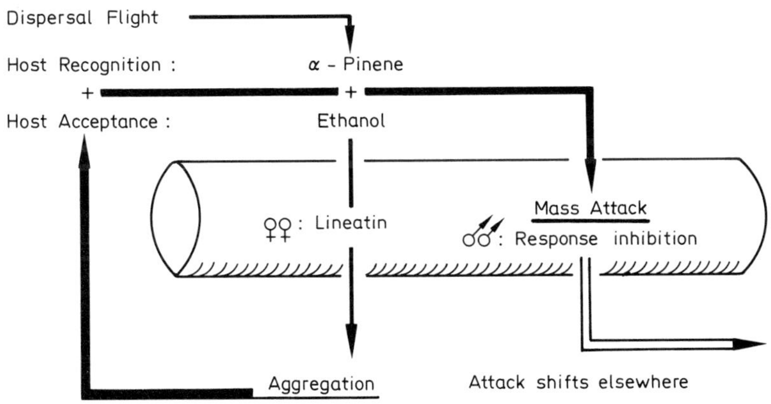

Fig. 7. Mechanism of attack of coniferous logs by *Trypodendron lineatum* (from Vité and Bakke [182])

Visual stimuli induce landing by beetles responding to the attractants [182], so that perforated cylinders simulating tree trunks make highly efficient trapping devices when baited with the mixture of lineatin, α-pinene and ethanol [182]. However, the complete protection of log piles is still thought to be impossible without treatment with insecticide and/or repellent chemicals.

6.3. Ips typographus

The spruce engraver beetle ("Buchdrucker") has posed a major threat to European forestry for several centuries, during which time Norway spruce (*Picea abies*) has been intensively cultivated [185]. Beetle epidemics tend to occur in mature spruce stands that have been damaged by drought, storms or ice. However, the risk has been minimized through silvicultural methods combined with "trap tree" systems [160]. It now appears that the effective but costly trap tree technique used in surveillance and control of this pest is being replaced and improved upon by pheromone trapping [186, 187].

The aggregation pheromone of *I. typographus* is released by male beetles by defecation as they excavate "nuptial chambers" in the inner bark of a new host tree. The attractant consists of two synergistic components, 2-methyl-3-buten-2-ol and (*S*)-*cis*-verbenol [66, 74, 188, 189], which are also produced to some extent by female beetles [190]. Response to the pheromone combination seems to be enhanced somewhat by the male-specific ipsdienol [191, 192], but partially inhibited by ipsenol, another male-specific compound. Some of these compounds also attract clerid predators of the beetle [193].

Methylbutenol is already present in mature beetles during their dispersal flight, but the biosynthesis of *cis*-verbenol requires exposure to (−)-α-pinene in the host tree's oleoresin [190]. This requirement may explain the beetle's tendency to colonize standing trees rather than relatively dry host material with little or no resin flow. The oxidation of α-pinene by *I. typographus* is quite stereospecific. Exposure of the beetles to (+)-α-pinene results in production of *trans*-verbenol, whereas (−)-α-pinene is converted into *cis*-verbenol. Thus the successful attack of a potential host tree may depend to some extent on the enantiomeric composition of its α-pinene [190, 101]. Female beetles which re-emerge from a crowded host to construct their galleries in another tree ("Geschwisterbrut") also produce pheromone, so that they are capable of marking the new host for colonization.

Research and development projects sponsored by the Norwegian and West German governments have led to the rapid development of marketable pheromone formulations and trapping devices [186, 187]. Despite this progress on practical applications, however, a number of basic questions in the pheromone biology of the Norway spruce engraver remain unanswered. The exact roles played by pheromone components and host odors in the beetle's communication system are not yet clear. Conflicting field observations on beetle response to synthetic attractants may be explained by the chiral properties of test substances and/or dosage-response patterns. For example, electrophysiological studies have indicated higher receptor potentials for (−)-α-pinene and (−)-(*R*)-ipsdienol than for their antipodes [194]. Also, seasonal differences in pheromone response behavior of *I.*

typographus populations have been suggested. During spring, the pest populations tend to disperse and are less likely to be caught by trapping devices which are highly effective later in the summer [132].

6.4. Lymantria (=Porthetria)

In Europe, this genus comprises two sympatric pest species [195] which share "disparlure" (*cis*-7,8-epoxy-2-methyloctadecane) [196, 105, 197] as a sex attractant. However, reproductive isolation is achieved by a combination of anatomical and behavioral characteristics [106]. The most notable difference is found in the response of the two species to the disparlure enantiomers [198, 199]. Under field conditions, males of *L. dispar* respond to (7R, 8S)-(+)-disparlure [200] in appreciable numbers only in the absence of the (7S, 8R)-(−)-enantiomer. On the other hand, males of *L. monacha* respond equally well to (+)-disparlure as to the racemic mixture.

The importance of appropriate formulations of disparlure has been emphasized for the success of large scale field trials underway in North America [201, 202]. Recent research on odor perception and behavior of *L. dispar* males in response to disparlure enantiomers [203–208] and analogs [209, 210] appear to be particularly relevant. It is now known that the antagonistic effect of the (−)-enantiomer makes the racemate far less effective than (+)-disparlure for luring males in the field [211]. This phenomenon might suggest the potential use of the inhibitory enantiomer as an agent for "communication disruption" or "disorientation". After two decades of painstaking research, it appears that the structure and configuration of the gypsy moth sex attractant are now established [212].

7. Conclusions

The apparent failure of pheromones to meet their expectations as pest control agents can be traced in most cases to incomplete knowledge of the biological factors involved. Two decades of intensive research on the chemistry of insect pheromones has resulted in the identification of countless biologically active compounds, but the exact function of many of these chemicals remains unclear. Individual compounds or blends of compounds may be responsible for sex attraction, aggregation, or other behavioral responses. Other components of a system may inhibit these responses, and dual functions may even be assigned to certain compounds. In many cases, a complex system of chemical communication is involved in the selection of a host plant, finding a mate, adjusting sex ratios, and controlling population densities on limited resources.

The key to studying behavioral responses of insects to their pheromones lies in the use of appropriate bioassays. Each system inevitably has distinct limitations, and results may be affected by phenomena such as Circadian rhythms or seasonal variations in behavior. Information on competing sources of natural pheromone may be important, and a knowledge of the factors governing production of pheromones by insect pests might provide a means of predicting the danger of outbreaks.

After all the biological facts have been considered, the outlook for pheromones in pest management no longer looks so bleak. However, successful results must always be regarded with caution. The effects of pheromones on nontarget organisms, variations in pest populations, and the possibility of "resistance" are among the dangers to be considered. It appears that the field application of pheromones will benefit from continued research into the biology, behavior and ecology of the pest populations as well as from efforts in the important field of "odor meteorology".

Acknowledgement

We are deeply indebted to Dr. P. R. Hughes of the Boyce Thompson Institute for his critical review of the first draft of this chapter.

8. References

1. Levinson, H.Z.: Z. angew. Entom. *84*, 1 (1977)
2. Anderson, J.F., Kaya, H.K. (eds.): Perspectives in forest entomology. Academic Press. New York, London (1976)
3. Beroza, M. (ed.): ACS Symposium on pest management with insect sex attractants. Plenum Press. New York & London (1976)
4. Birch, M.C. (ed.): Pheromones. North Holland Publishing Co. Amsterdam (1974)
5. Payne, T.L. in: ACS Symposium on the chemistry of neurohormones and neurotransmission. Plenum Press (1978)
6. Sebeok, T.A. (ed.): How animals communicate. Indiana University Press (1978)
7. Shorey, H.H., McKelvey, J.J., Jr. (eds.): Chemical control of insect behavior: theory and application. John Wiley, New York (1977)
8. Butenandt, A. et al.: Z. Naturforsch. *14b*, 283–284 (1959)
9. Kasang, G., Schneider, D., Schäfer, W.: Naturwissenschaften *65*, 337 (1978)
10. Kaissling, K.E. et al.: Naturwissenschaften *65*, 382 (1978)
11. Roelofs, W.L., Cardé, R.T. in: Pheromones (M.C. Birch, ed.). North Holland Publishing Co., Amsterdam, pp. 96–114 (1974)
12. Baker, T.C., Cardé, R.T., Roelofs, W.L.: J. Chem. Ecol. *2*, 333 (1976)
13. Brady, U.E. et al.: Science *171*, 802 (1971)
14. Cardé, R.T., Baker, T.C., Roelofs, W.L.: Nature *253*, 348 (1975)
15. Birch, M.C. in: Pheromones (M.C. Birch, ed.). North Holland Publishing Co., Amsterdam. pp. 115–134 (1974)
16. Atkins, M.D.: Can. Entom. *98*, 285 (1966)
17. Vité, J.P., Pitman, G.B.: Proc. XIV Intern. Union Forest Res. Org. Munich, Sec. *24*, 683 (1967)
18. Vité, J.P., Francke, W.: Naturwissenschaften *63*, 550 (1976)
19. Vité, J.P., Pitman, G.B.: J. Insect Physiol. *15*, 1617 (1969)
20. Bedard, W.D., Silverstein, R.M., Wood, D.L.: Science *167*, 1638 (1970)
21. Hughes, P.R., Renwick, J.A.A., Vité, J.P.: Environ. Entom. *5*, 1165 (1976)
22. Vité, J.P., Bakke, A., Hughes, P.R.: Naturwissenschaften *61*, 365 (1974)
23. Bakke, A.: Norw. J. Entom. *22*, 67 (1975)
24. Kinzer, G.W. et al.: Nature *221*, 477 (1969)
25. Silverstein, R.M., Rodin, J.O., Wood, D.L.: Science *154*, 509 (1966)
26. Vité, J.P., Renwick, J.A.A.: Contrib. Boyce Thompson Inst. *24*, 61 (1968)
27. Vité, J.P., Pitman, G.B.: Can. Entom. *101*, 113 (1969)
28. Pitman, G.B., Vité, J.P.: ibid. *101*, 143 (1969)
29. Renwick, J.A.A., Vité, J.P.: Nature *224*, 1222 (1969)
30. Renwick, J.A.A., Vité, J.P.: Contrib. Boyce Thompson Inst. *24*, 283 (1970)
31. Pitman, G.B.: Science *166*, 905 (1969)

32. Bedard, Silverstein, Wood: Science *167*, 1638 (1970)
33. Pearce, G.T. et al.: J. Chem. Ecol. *1*, 115 (1975)
34. Vité, J.P., Hughes, P.R., Renwick, J.A.A.: Naturwissenschaften *63*, 44 (1969)
35. Payne, T.L., personal communication
36. Cardé, R.T., Roelofs, W.L., Doane, C.C.: Nature (London) *241*, 474 (1973)
37. Sanders, C.J., Lucuile, G.S.: Can. Dept. Envir. Bi-mon. Res. Notes *28*, 10 (1972)
38. Renwick, J.A.A.: Contrib. Boyce Thompson Inst. *23*, 355 (1967)
39. Vité, J.P., Renwick, J.A.A.: Contrib. Boyce Thompson Inst. *24*, 61 (1968)
40. Vité, J.P., Renwick, J.A.A.: Naturwissenschaften *58*, 418 (1971)
41. Wood, D.L. in: H.F. van Emden (ed.), Insect Plant Relationships. Blackwell Sci. Pub., London p. 101–117 (1972)
42. Kinzer, G.W. et al.: J. Econ. Entom. *64*, 970 (1971)
43. Rudinsky, J.A. et al.: Can. Entom. *104*, 815 (1972)
44. Rudinsky, J.A. et al.: Environ. Entom. *2*, 505 (1973)
45. Pitman, G.B., Vité, J.P.: Environ. Entom. *3*, 886 (1974)
46. Rudinsky, J.A., Michael, R.R.: J. Insect Physiol. *19*, 689 (1973)
47. Barr, B.A.: Can. Entom. *101*, 636 (1969)
48. Rudinsky, J.A.: Environ. Entom. *2*, 579 (1973)
49. Rudinsky, J.A.: ibid. *2*, 511 (1973)
50. Pitman, G.B., Hedden, R.L., Gara, R.I.: Z. angew. Entom. *78*, 203 (1974)
51. Payne, T.L., Coster, J.E., Johnson, P.C.: J. Chem. Ecol. *33*, 133 (1977)
52. Payne, T.L. et al.: Environ. Entom. *7*, 578 (1978)
53. Richerson, J.V., Payne, T.L.: Environ. Entom. *8*, 360 (1979)
54. Payne, T.L. in: Pheromones (M.C. Birch, ed.). North Holland Publishing Co., Amsterdam. pp. 35–61 (1974)
55. Rothschild, G.H.L.: Entomol. Exp. & Appl. *17*, 294 (1974)
56. Klimetzek, D. et al.: Naturwissenschaften *63*, 581 (1976)
57. Vité, J.P. et al.: ibid. *63*, 582 (1976)
58. Corbet, S.A.: Nature *232*, 481 (1971)
59. Williamson, D.L.: Ann. Entomol. Soc. Am. *64*, 586 (1971)
60. Wood, D.L. et al.: Science *159*, 1373 (1968)
61. Lanier, G.N. et al.: Can. Entom. *104*, 1917 (1972)
62. Vité, J.P., Williamson, D.L.: J. Insect Physiol. *16*, 233 (1970)
63. Dyer, E.D.A.: Can. J. Forest Res. *3*, 486 (1973)
64. Bedard, W.D. et al.: Science *164*, 1284 (1969)
65. Pitman, G.B., Vité, J.P.: J. Econ. Entom. *64*, 402 (1971)
66. Vité, J.P., Bakke, A., Renwick, J.A.A.: Can. Entom. *104*, 1967 (1972)
67. Silverstein, R.M. et al.: Science *159*, 889 (1968)
68. Pitman, G.B. et al.: J. Insect Physiol. *15*, 363 (1969)
69. Vité, J.P., Pitman, G.B.: J. Insect Physiol. *15*, 1617 (1969)
70. Pitman, G.B.: Science *166*, 905 (1969)
71. Pitman, G.B., personal communication
72. Vité, J.P., Renwick, J.A.A.: J. Insect Phys. *17*, 1699 (1971)
73. Borden, J.H. in: Pheromones (M.C. Birch, ed.). North Holland Publishing Co., Amsterdam. pp. 135–160 (1974)
74. Bakke, A., Frøyen, P., Skattebøl, L.: Naturwissenschaften *64*, 98 (1977)
75. Stoakley, J.T. et al.: Z. angew. Entom. *86*, 174 (1978)
76. Silverstein, R.M. in: Shorey & McKelvey (eds.), Chemical Control of Insect Behavior. Wiley-Interscience. New York. pp. 231–251 (1977)
77. Hendry, L.B. et al.: Science *188*, 59 (1975)
78. Miller, J.R. et al.: Science *192*, 140 (1976)
79. Young, J. C. et al.: J. Insect. Physiol. *19*, 1615 (1973)
80. Birch, M.C., Wood, D.L.: J. Chem. Ecol. *1*, 101 (1975)
81. Brümmer, U.: Thesis, Inst. Org. Chem. Hamburg Univ. (1978)
82. Hughes, P.R.: Z. angew. Entom. *73*, 294 (1973)
83. Hughes, P.R.: J. Insect Physiol. *20*, 1271 (1974)
84. Boppré, M.: Biuz *7*, 161 (1977)

85. Gore, W.E. et al.: J. Chem. Ecol. *3*, 429 (1977)
86. Gerken, B.: Thesis, Freiburg Univ. (1978)
87. Grüne, S.: Handb. z. Best. europäischen Borkenkäfer, M. & H. Schaper, Hannover, 182 pp. (1979)
88. Riddiford, L.M.: Science *158*, 139 (1967)
89. Riddiford, L.M., Williams, C.M.: Science *155*, 589 (1967)
90. Borden, J.H., Nair, K.K., Slater, C.E.: Science *166*, 1626 (1969)
91. Hughes, P.R., Renwick, J.A.A.: Physiol. Entom. *2*, 117 (1977)
92. Harring, C.M.: Z. angew. Entom. *85*, 281 (1978)
93. Menon, M.: Ann. Entom. Soc. Am. *69*, 457 (1976)
94. Hughes, P.R., Renwick, J.A.A.: Physiol. Entom. *2*, 289 (1977)
95. Gerken, B., Hughes, P.: Z. angew. Entom. *82*, 108 (1976)
96. Francke, W. et al.: Naturwissenschaften *64*, 590 (1977)
97. Renwick, J.A.A., Hughes, P.R., Ty, T.Dej.: J. Insect Physiol. *19*, 1735 (1973)
98. Renwick, J.A.A. et al.: J. Insect Physiol. *22*, 725 (1976)
99. Hughes, P.R.: Naturwissenschaften *60*, 261 (1973)
100. Renwick, J.A.A., Hughes, P.R., Krull, I.S.: Science *191*, 199 (1976)
101. Klimetzek, D., Lulev, J., Francke, W., in manuscript
102. Brand, J.M. et al.: Nature *245*, 136 (1975)
103. Shorey, H.H.: Ann. Rev. Entom. *18*, 349 (1973)
104. Schröter, H.: Inaug. Dissert. Univ. Freiburg (1976)
105. Schönherr, J.: Z. angew. Entom. *71*, 260 (1972)
106. Schneider, D. et al.: Oecologia *14*, 19 (1974)
107. Vick, K.W., Drummond, P.C., Coffelt, J.A.: Ann. Entom. Soc. Am. *66*, 1001 (1973)
108. Hammack, K., Burkholder, W.E.: J. Insect Physiol. *22*, 385 (1976)
109. Shapas, T.J., Burkholder, W.E.: J. Chem. Ecol. *4*, 409 (1978)
110. Leuze, C.: Dissert. Univ. Freiburg (in preparation)
111. Vité, J.P., Gara, R.I.: Contrib. Boyce Thompson Inst. *21*, 251 (1962)
112. Madrid, F., Vité, J.P., Renwick, J.A.A.: Z. angew. Entom. *72*, 73 (1972)
113. Kennedy, J.S. in: Chemical control of insect behavior (H.H.Shorey & J.J.McKelvey, Jr. eds.). Wiley Interscience, New York. 231–251 (1977)
114. Schneider, D.: Industrie Elektronik (Hamburg) *5*, 3 (1955)
115. Schneider, D.: Z. Vergl. Physiol. *40*, 8 (1957)
116. Roelofs, W. et al.: Science *174*, 297 (1971)
117. Payne, T.L.: J. Chem. Ecol. *1*, 233 (1975)
118. Vité, J.P., Pitman, G.B.: Canad. Entom. *101*, 113 (1969)
119. Chapman, J.A.: Canad. Entom. *98*, 50 (1966)
120. Vité, J.P.: Allg. Forstz. *33*, 428 (1978)
121. Borden, J.H., Bennett, R.B.: J. Econ. Entom. *62*, 782 (1969)
122. Bennett, R.B., Borden, J.H.: Ann. Entom. Soc. Am. *64*, 1273 (1971)
123. Casida, J.E., Coppel, H.C., Watanabe, T.: J. Econ. Entom. *56*, 18 (1963)
124. Traynier, R.M.M.: Can. Entom. *100*, 5 (1968)
125. Farkas, S.R., Shorey, H.H. in: Pheromones (M.C.Birch, ed.). North Holland Publishing Co. Amsterdam. pp. 81–95 (1974)
126. Shorey, H.H., Gaston, L.K.: Ann. Entom. Soc. Am. *63*, 829 (1970)
127. Roelofs, W.L.: J. Chem. Ecol. *4*, 685 (1978)
128. Fares, Y. et al.: J. Theor. Biol. (in press)
129. Chapman, J.A.: Can. Entom. *99*, 1132 (1967)
130. Kaae, R.S., Shorey, H.H.: Ann. Entom. Soc. Am. *65*, 436 (1972)
131. Vité, J.P., Gara, R.I., Scheller, H.D. v.: Contrib. Boyce Thompson Inst. *22*, 461 (1964)
132. Klimetzek, D., Vité, J.P.: Allgem. Forstz. *33*, 1446 (1978)
133. Atkins, M.D.: Z. angew. Entom. *78*, 209 (1975)
134. Coster, J.E. et al.: J. Georgia Entom. Soc. *12*, 204 (1977)
135. Klimetzek, D., Vité, J.P., Mori, K.: Z. angew. Entom., in press
136. Klun, J.A. et al.: Science *204*, 1328 (1979)
137. Tumlinson, J.H. et al.: J. Chem. Ecol. *1*, 203 (1975)
138. Klun, J.A. et al.: Environ. Entom. *4*, 891 (1975)
139. Cardé, R.T. et al.: ibid. *4*, 413 (1975)

140. Klun, J.A., Maina, S.: ibid. *8*, 423 (1979)
141. Lanier, G.N. et al.: Canad. Entom. *104*, 1917 (1972)
142. Piston, J.J., Lanier, G.N.: ibid. *106*, 247 (1974)
143. Stewart, T.E.: Master's Thesis SUNY College of Environmental Science and Forestry, Syracuse, NY (1975)
144. Vité, J.P. et al.: Z. angew. Entom. *75*, 422 (1974)
145. Pearce, G.T. et al.: J. Chem. Ecol. *1*, 115 (1975)
146. Gerken, B. et al.: Naturwissenschaften *65*, 110 (1978)
147. Renwick, J.A.A., Hughes, P.R., Vité, J.P.: J. Insect Physiol. *21*, 1097 (1975)
148. Wood, S.L.: Great Basin Naturalist *34*, 277 (1974)
149. Wood, S.L.: ibid. *23*, 1 (1963)
150. Renwick, J.A.A., Pitman, G.B.: Environ. Entom. *8*, 40 (1979)
151. Vité, J.P.: Texas Industry *36*, 14 (1968)
152. Vité, J.P.: Allg. Forstz. *25*, 615 (1970)
153. Vité, J.P.: Contrib. Boyce Thompson Inst. *24*, 343 (1970)
154. Vité, J.P., Gara, R.I., von Scheller, H.D.: ibid. *22*, 461 (1964)
155. Gara, R.I., Vité, J.P., Cramer, H.H.: ibid. *23*, 55 (1965)
156. Gara, R.I.: *23*, 349 (1967)
157. Renwick, J.A.A., Vité, J.P.: ibid. *24*, 65 (1968)
158. Vité, J.P., Renwick, J.A.A.: ibid. *24*, 323 (1970)
159. Vité, J.P., Renwick, J.A.A.: ibid. *24*, 61 (1968)
160. Vité, J.P.: Proc. Tall Timbers Conf. Ecol. Anim. Control Habitat Management *3*, 155 (1971)
161. Moser, J.C., Browne, L.E.: J. Chem. Ecol. *4*, 1 (1978)
162. Hughes, P.R.: Z. angew. Entom. *80*, 280 (1976)
163. Nijholt, W.W.: Environm. Canada, Forest Serv. BC-X121; 35 pp. (1979)
164. Nijholt, W.W.: Forest Serv., Pac. For. Res. Cen. Rep. BC-P-25 (1978)
165. Vité, J.P.: Holz-Zentralblatt *105*, 1507 (1979)
166. Uusvaara, O., Loyttyniemi, K.: Folia Forestalia *231*, 1 (1975)
167. Rudinsky, J.A., Daterman, G.E.: Can. Entom. *96*, 1339 (1964)
168. Chapman, J.A.: Canad. Entom. *98*, 50 (1966)
169. Castek, K.L., Barbour, J.F., Rudinsky, J.A.: J. Econ. Entom. *60*, 658 (1967)
170. Borden, J.H., Brownlee, R.G., Silverstein, R.M.: Canad. Entom. *100*, 629 (1968)
171. Francke, W., Heemann, V.: Z. angew. Entom. *75*, 67 (1974)
172. Francke, W., Heemann, V., Heyns, K.: Z. Naturforsch. *29c*, 243 (1974)
173. MacConnell, J.G. et al.: Chem. Ecol. *3*, 549 (1977)
174. Borden, J.H., Handley, J.R., Johnston, B.D., MacConnell, J.G., Silverstein, R.M., Slessor, K.N., Swigar, A.A., Wong, D.T.W.: J. Chem. Ecol. (in press)
175. Mori, K., Sasaki, M.: Tetrahedron Lett. *15*, 1329 (1979)
176. Vité, J.P. et al.: Environm. Entom. (in press)
177. Nijholt, W.W.: Canad. Entom. *105*, 583 (1973)
178. Nihholt, W.W.: ibid. *102*, 894 (1970)
179. Borden, J.H.: Canad. J. Zool. *52*, 533 (1973)
180. Bauer, J., Vité, J.P.: Naturwissenschaften *62*, 539 (1975)
181. Nijholt, W.W., Schönherr, J.: Canad. Forest Serv., Bi-mon. Res. Notes *32*, 31 (1976)
182. Vité, J.P., Bakke, A.: Naturwissenschaften (in press)
183. Kerck, K.: ibid. *65*, 542 (1978)
184. Moeck, H.A.: Canad. Entom. *102*, 985 (1970)
185. Schwerdtfeger, F.: Schriftenr. Forstl. Fak. Univ. Göttingen *13/14*, 135 pp. (1955)
186. Bakke, A., Saether, T.: Skogeieren *65*, 10 (1978)
187. Vité, J.P.: Holz-Zentralblatt *105*, 809 (1979)
188. Krawielitzki, S. et al.: Z. angew. Entom. *83*, 300 (1977)
189. Sauerwein, P., Vité, J.P.: Mitt. Dtsch. Ges. Allg. angew. Entom. *1*, 189 (1978)
190. Hackstein, E., Vité, J.P.: ibid. *1*, 182 (1978)
191. Vité, J.P.: Allg. Forstz. *33*, 428 (1978)
192. Klimetzek, D., Sauerwein, P.: Allg. Forst.- u. Jagdztg. (in press)
193. Bakke, A., Kvamme, T.: Nord. J. Entom. *25*, 41 (1978)
194. Dickens, J.C.: Entom. exp. appl. *24*, 136 (1978)

195. Wellenstein, G., Schwenke, W. in W. Schwenke: Die Forstschädlinge Mitteleuropas Vol. *3*, 334–368. P. Parey, Hamburg u. Berlin (1978)
196. Bierl, B.A., Beroza, M., Collier, C.W.: Science *7*, 815 (1970)
197. Bierl, B.A. et al.: Z. Naturforsch. *30c*, 672 (1975)
198. Vité, J.P. et al.: Naturwissenschaften *63*, 582 (1976)
199. Klimetzek, D. et al.: ibid. *63*, 581 (1976)
200. Mori, K., Takigawa, T., Matsui, M.: Tetrahedron Lett. *44*, 3953 (1976)
201. Beroza, M., Knipling, E.F.: Science *177*, 19 (1972)
202. Cameron, E.A.: Bull. Entom. Soc. Am. *19*, 15 (1973)
203. Plimmer, J.R. et al.: Envir. Entom. *6*, 518 (1977)
204. Cardé, R.T. et al.: ibid. *6*, 768 (1977)
205. Cardé, R.T., Doane, C.C., Farnum, D.G.: Environ. Entom. *7*, 815 (1978)
206. Farnum, D.G. et al.: Tetrahedron Lett. *46*, 4009 (1977)
207. Miller, J.R., Roelofs, W.L.: Environ. Entom. *7*, 42 (1978)
208. Miller, J.R., Mori, K., Roelofs, W.L.: J. Insect Physiol. *23*, 1447 (1977)
209. Schneider, D. et al.: J. Comp. Physiol. *113*, 1 (1977)
210. Roelofs, W.L., Cardé, R.T.: Ann. Rev. Entom. *22*, 377 (1977)
211. Cardé, R.T., Hagaman, T.E.: Environ. Entom. *8*, 475 (1979)
212. Eiter, K. in: R. Wegler (ed.). Chemie Pflanzenschutz- u. Schädlingsbekämpfungsmittel. *1*, 497 (1970)

Chemistry of Insect Pheromones

Chemical Structure, Isolation and Identification, Synthesis and Use in Plant Protection

H. J. Bestmann and O. Vostrowsky

Institute of Organic Chemistry, University Erlangen-Nürnberg, D-8520 Erlangen

Contents

1. Insect Behavior Controlling Chemicals in Plant Protection

In order to avoid environmental damage caused by the uncontrolled use of pesticides and because of strict legal controls on both their use and production [1] recently new methods of pest control, based on biological "insecticides" and therefore largely non-poisonous and harmless to the environment, have been chosen. These new techniques, which by the mid-70's had advanced from the experimental stage in universities and research establishments to applied tests in agriculture, horticulture, fruit farming and forestry, make use of chemicals with which the behavior of insects can be manipulated and controlled [2–4], thus offering new possibilities of insect control.

This class of compounds includes the many synthetic *insect attractants* [5], *repellants* [5] and the so-called "*disruptants*". *Insect attractants*, which have so far been used for insect detection, were (among others) the first non-toxic alternatives in modern plant protection. An infestation may be revealed by capturing pests in treated traps and appropriate precautions taken. This class of substances includes chemicals such as eugenol, methyleugenol, and a large number of esters such as phenethyl propionate, heptyl butyrate, 2,4-hexadienyl butyrate, octyl butyrate and the like. The extermination of the Mediterranean fruit fly, *Ceratitis capitata*, in Florida by means of siglure (sec-butyl 6-methyl-3-cyclohexenecarboxylate) is a significant demonstration of the importance of insect attractants in pest control [6].

Compounds such as deet (N,N-diethyl-m-toluamide), pyrethrine, malathion, methoxychlor, imides, alcohols, esters, ethers and polyfunctional compounds such as 1,3-diols, amide esters, hydroxy esters and diesters belong to the class of *insect repellants*, which are usually discovered, like attractants, *via* "screening programs" on synthetic chemicals. They are used, above all for protection against disease carrying insects such as flies, mosquitos, lice and ticks.

"*Disruptants*" prevent mating, and will be discussed with pheromone inhibitors in the section of pheromones.

As well as these synthetic compounds there is a series of naturally occuring substances which are known as messenger substances (*semiochemicals* [7]). These enable chemical communication between individual insects [8–11] and can therefore also be used for insect manipulation. One class of these compounds, which are active in the producing organism, are the *hormones*. Of primary interest for plant protection are the insect *juvenile hormones* (larval hormones), which control insect moulting. The first of these compounds to be isolated was Cecropia-juvenile hormone from the abdomen of the male cecropia moth *Hyalophora cecropia* [12]. Apart from this and subsequently identified homologs, there are also naturally-occurring compounds such as juvabion ("paper factor"), a cyclic sesquiterpene ester from the North American balsam fir, and a few farnesol derivatives which show juvenile hormone activity. A variety of synthetic substances exceed, however, the activity of natural juvenile hormones, and may play an important role in plant protection in the future.

Among the *semiochemicals* which affect individuals other than the producing organism are those which have an effect on other species (interspecific effect, *allelochemicals* [13]) and are divided into those which bring an adaptive advantage to the receiver (*kairomones* [14]) or to the emitter (*allomones* [14]). Prey-scents, host-scents and phagostimulants are kairomones, whereas allomones are divided into those with antagonistic effect, such as defensive secretions, repellants, warning substances, feeding deterrents, antibiotics and poisons, and those with mutualistic effects, such as flower scents, attractants between symbionts, etc.

Intra-specific *semiochemicals*, i.e. those which are sent and received by individuals of the same species, are called *pheromones* [15]. They cause in the receiving individual organism either an immediate behavior-change (*releaser pheromones* [16]), or they change its endocrinic state, and therewith the receptiveness for later signals (*primer pheromones* [16]). The first group, which is discussed below, includes for insects the sex attractants of female butterflies, the various odor substances of male moths, sex and aggregation pheromones of beetles and cockroaches, alarm substances and trail pheromones of ants, trail substances of termites, the signal substances of social insects, and oviposition stimulants. In a chemical-relevant sense we arrange in this order also substances which influence this type of chemical communication, e.g. *pheromone inhibitors*, *synergists* and *disruptants*, even when they are totally synthetic.

The main emphasis in pheromone research is concentrated with insects, many of which are dangerous plant pests. Their aggregation and above all their sex pheromones provide a nontoxic means [17] of behavior manipulation and species-specific controlling.

There are several techniques for the use of sex pheromones in pest control [2, 3, 18–30]. We shall discuss them by using moth sex attractants as an example:

the volatile attractants are emitted from abdominal glands of the female moths by means of hair tufts, which serve as evaporators. The attractants diffuse over large distances, are received by the olfactory sensors in the antennae of male moths of the same species, enabling the male to find the female, and hence reproduction. This effect is used to lure male moths to sticky traps treated with either synthetic pheromones or the extract of female abdominal glands. It is known, however, that, especially for flight active insects, over 90% of the male insects must be exterminated in order to produce an observable population depression in the next generation. This direct *capture technique (mass trapping)* is therefore only of limited use in pest control.

Considerably more promising is the so-called *monitor* or *prognosis technique*. In this case, as in the capture technique, baited traps, which are now commercially available, are exposed and the trapping rate checked at short intervals. In this way the population density, infestation threshold and flight periods of the insects, which may vary considerably depending upon, for instance, weather conditions, can be determined. The spraying can then be timed exactly and the dose limited to the minimum necessary. With this technology, which is already in use in Europe and overseas, the use of conventional insecticides, and therefore the toxic load on the environment, can be sharply reduced. Some sprayings which were previously necessary in order to be safe can now be totally omitted.

Another possibility for the use of pheromones in plant protection is the *confusion technique*. In this method the infested area is "flooded" with formulated, synthetic pheromone in order to hinder the orientation of the male insect to the female (sex attractants are normally already effective in microgram doses).

The use of pheromone inhibitors and disruptant substances is a very promising method of pest control. An inhibitor is defined as a substance which, when exposed together with a pheromone or beside a pheromone-baited trap, decreases the flight rate towards the odor source or reduces it to zero. Even a species-characteristic sex attractant can act as an inhibitor in high doses. For instance, (Z)-

5-decenyl acetate, the sex pheromone of the turnip moth *Agrotis segetum*, is attractive with as little as 5×10^{-4} μg/trap bait loadings, but is less effective in 1 μg/trap concentrations [31].

A *disruptant* is a compound which, when released in the open air, prevents intersexual transmission of chemical information. This means that the orientation of the males to the females can be hindered by the use of pheromone inhibitors or disruptants, and propagation of the species thus prevented. An essential advantage of this method lies in the isomer purity of the compounds used, for which less requirements are necessary than for the highly specific pheromones (cf. p. 63, multicomponent pheromone complexes).

2. The Chemical Structure of Pheromones

If one attempts to organize insect pheromones into groups according to their chemical structure only the *sex pheromones* of the female Lepidoptera, the *sex attractants*, form a separate homogenous group. They are almost exclusively mono- or poly-olefinic acetates, alcohols or aldehydes, and have been discussed in several reviews [18, 26, 32–40]. The characteristic odors of male moths, the *aggregation* and *sex pheromones* of beetles and cockroaches, the *trail scents* of ants and termites and the *warning secretions* and *signal substances* of social insects on the other hand form structurally inhomogenous classes consisting of aliphatic, alicyclic and aromatic hydrocarbons, alcohols, aldehydes, ketones, esters, etc.

The *sex attractants* of Lepidoptera are only detected by the males. The females usually show no behavioral response, and in electrophysiological tests (electroantennogram, EAG [41]) the attractants have no effect on the receptors of female antennae, except in very high concentrations. They are effective in submicrogram amounts for the male antennae and can be used to investigate the effect of systematic structure variations on electrophysiological activity and hence to help elucidate the mechanism for the reception of insect pheromones [42–45] (cf. Chap. 3).

Male pheromones of butterflies and moths may, as already mentioned, be aromatic compounds such as benzaldehyde, benzyl alcohol, phenyl ethanol and phenyl acetaldehyde, aliphatic aldehydes, alcohols, acids, ketones or terpenoid compounds (cf. Table 1). A male odor substance which stands out because of its chemical structure is 2,3-dihydro-7-methyl-1*H*-pyrrolizin-1-one, found as the main component in danaid butterflies [46–48]. This metabolic product from an alkaloid, which is present in the insects food [49, 50], induces readiness for mating in females [51, 52], and gives high EAG responses in both sexes [53, 54].

The *aggregation pheromones* of the bark beetles (Scolytidae), which are secreted by the "pioneer beetles" and lead the mass flight toward suitable host trees (secondary attraction), consist mostly of terpene alcohols and bicyclic ketals which are produced in supplementary glands of the hindgut *via* metabolism of host material (e.g. vegetable terpenes). The widespread sex specific pheromones of other beetles (Coleoptera) are a chemically heterogeneous class of compounds consisting of compounds like phenol, valeric acid, unsaturated acids, carboxylic esters, alcohols and cyclohexane derivatives (see Table 1).

Table 1. Chemical structure, occurence and behavioral release of insect pheromones and pheromone analogs (® = trade mark, f = female, m = male, ins = insect, host = host plant, sexatt = sex attractant, disrupt = disruptant, inhib = inhibitor, synerg = synergist, aphrod = aphrodisiacum, aggr = aggregation pheromone, att = attractant, sexpher = sexual pheromone, trail = trail pheromone, alarm = alarm pheromone, pher = pheromone, stim = stimulant, ovipos = oviposition stimulant; [a] more than one compound necessary for optimum activity or present in the pheromone complex).

Name, trivial name and trade mark®	Isol. from	Insect species affected	Behavior release	Ref.
2.1. Sex Attractants of Female Butterflies and Moths, and Related Compounds				
a) Monounsaturated Acetates				
Z-5-decenyl acetate		Agrotis fucosa[a]	sexatt	[77]
	f	A. segetum	sexatt	[31]
		Depressaria pulcherimella	sexatt	[78]
		Semioscopis anellanella	sexatt	[78]
E-5-decenyl acetate	f	Anarsia lineatella[a]	sexatt	[79]
Z-7-decenyl acetate		Agrotis fucosa[a]	sexatt	[77]
		Battaristis spec.	sexatt	[80]
		Lobesia aelopa	sexatt	[81, 82]
Z-8-undecenyl acetate		Gaesa aparsella	sexatt	[82]
Z-5-dodecenyl acetate		Autoplusia egena	sexatt	[83]
		Chionodes fuscomaculella	sexatt	[80]
		Euxoa ochrogaster[a]	sexatt	[84]
		E. pleuritica[a]	sexatt	[85]
		Kawabea razawskii	sexatt	[82]
		Scrobipalpa atriplicella	sexatt	[80]
Z-7-dodecenyl acetate	f	Agrotis ypsilon[a]	sexatt	[850]
		Anagrapha falcifera	sexatt	[86]
		Argyrogramma basigera	sexatt	[87]
		A. verruca	sexatt	[87]
		Autographa ampla	sexatt	[80]
		A. biloba	sexatt	[80]
		A. californica	sexatt disrupt	[843, 88, 89]
		A. precationis	sexatt	[80]
		Chrysaspidis contexta	sexatt	[86]
		Conistra grisescens	sexatt	[82]
		Cosmopterix fulminella	sexatt	[82]
		Crymodes devestator[a]	sexatt	[90]
		Epinotia zandana	sexatt	[80]
		Euxoa pleuritica	sexatt	[85]
		Harpypterix xylostella[a]	sexatt	[80]
		Heliothis virescens	disrupt	[88]
		H. zea	disrupt	[88]
		Micardia argentata	sexatt	[82]
		Nippoptilia isikii	sexatt	[81, 82]
		N. vitis	sexatt	[82]
		Orthogonica sera	sexatt	[82]
		Orthosia odiosa	sexatt	[82]
		Pectinophora gossypiella	disrupt	[88]
		Phlyctaenia terreaiis	sexatt	[80]
		Plusia aeroides	sexatt	[86]
	f	Pseudoplusia includens	sexatt disrupt	[88, 91, 92]
		Pterophorus tenuidactylus	sexatt	[80]
		Rhyacionia buoliana	inhib	[93]

Table 1 (continued)

Name, trivial name and trade mark®	Isol. from	Insect species affected	Behavior release	Ref.
		R. neomexicana[a]	sexatt	[94]
		Rhynchaglaea fuscipennis	sexatt	[82]
		Spodoptera dolichos	inhib	[95]
		S. frugiperda	inhib	[96]
		S. ornithogalli	disrupt	[88]
		Sterrha imbecilla	sexatt	[82]
	f	Trichoplusia ni	sexatt disrupt	[88, 92, 97, 98]
		T. oxygramma	sexatt disrupt	[92, 99]
E-7-dodecenyl acetate	f	Cryptophlebia (= Argyroploce)[+] leucotreta	sexatt	[100, 101]
		Micardia argentata	sexatt	[82]
		Rhyacionia neomexicana[a]	sexatt	[94]
Z-8-dodecenyl acetate orfralure, orframone®, funemone®		Alleima loefflingiana	sexatt	[78]
		Aphania infida	sexatt	[80]
		Cnephasia chrysantheana	sexatt	[78]
		Cnephasia alternella	sexatt	[102]
		Cnephasiella incertana	sexatt	[78]
	f	Cryptophlebia (= Argyroploce) leucotreta[a]	sexatt	[62, 103]
		C. ombrodelta	sexatt	[81, 82]
		Epiblema desertana	sexatt	[80]
		E. foenella	sexatt	[82]
		E. scudderiana	sexatt	[102]
		Euxoa ochrogaster[a]	sexatt	[84]
		Grapholitha funebrana[a]	sexatt	[78, 104, 105]
		G. inopinata	sexatt	[82]
		G. janthiana[a]	sexatt	[105]
		G. lobarzewskii[a]	sexatt	[105]
	f	G. molesta[a]	sexatt disrupt	[105–108]
		G. prunivora[a]	sexatt	[106]
		G. tenebrosa	sexatt	[102]
		Gretchena bolliana	sexatt	[109]
		Hedia nubiferana	inhib	[110]
		Heliophobus reticulana	sexatt	[78]
		Laspeyresia aurana	sexatt	[78]
		L. pomonella	inhib	[111]
		Pammene albugina	sexatt	[102]
		P. argyrana	sexatt	[78, 102]
		P. fasciana	sexatt	[102]
		P. inquilana	sexatt	[78, 102]
		P. nemorosa	sexatt	[81, 82]
		Pseudexentera maracana	sexatt	[80]
E-8-dodecenyl acetate	f	Cryptophlebia leucotreta[a]	sexatt	[62, 111]
		Ecdytolopha insiticiana	sexatt	[80]
		Grapholitha janthiana[a]	sexatt	[105]
		G. lobarzewskii[a]	sexatt	[105]
		G. funebrana[a]	sexatt	[105]
		G. molesta[a]	sexatt inhib	[105–107, 112]
		G. packardi	sexatt	[106]

[+] Later corrected into E-8-dodecenyl acetate [62, 111]

Table 1 (continued)

Name, trivial name and trade mark®	Isol. from	Insect species affected	Behavior release	Ref.
		G. prunivora[a]	sexatt inhib	[112]
		Hedia chionosema[a]	sexatt	[80]
		Pammene argyrana	sexatt	[78]
		P. inquilina	sexatt	[78]
		Strophedra nitidana	sexatt	[82]
Z-9-dodecenyl acetate	f	Cnephasia pumicana[a]	sexatt	[113]
grapemone®		Cosmopterix spec.	sexatt	[81]
		Enarmonia formosana[a]	sexatt	[78, 121]
		Episimus argutanus	sexatt	[80]
		Eucosma sonomana[a]	sexatt	[842]
	f	Eupoecilia ambiguella[a]	sexatt	[114, 115]
	f	Homona magnanima[a]	sexatt	[849]
		Leucania loreyi	sexatt	[82]
		Pammene inquilina	sexatt	[78]
		Paralobesia viteana	sexatt disrupt	[116, 117]
		Rhopobota naevana	sexatt	[82]
		Rhyacionia buoliana	inhib	[118, 119]
		R. zozana[a]	sexatt	[842]
		Spodoptera frugiperda	sexatt disrupt	[96, 120]
E-9-dodecenyl acetate		Caradrina morpheus	sexatt	[110]
	f	Cnephasia pumicana[a]	sexatt	[113]
		Dichrorampha spec.[a]	sexatt	[80]
		Dichrorampha plumbana	sexatt	[78]
	f	Diparopsis castanea[a]	sexatt	[122]
		Enarmonia formosana[a]	sexatt	[78, 121]
		Eucosma sonomana[a]	sexatt	[842]
	f	Homona coffearia[a]	sexatt	[126]
		Loxostege chortalis	sexatt	[80]
		Nemophora panzerella	sexatt	[78]
		Pammene argyrana	sexatt	[78]
		P. inquilina	sexatt	[78]
		P. splendidulana	sexatt	[78]
	f	Rhyacionia buoliana	sexatt	[118, 123]
		R. neomexicana[a]	sexatt	[94]
		R. zozana[a]	sexatt	[842]
		Sparganothis pilleriana[a]	sexatt	[124]
		Zeiraphera diniana	sexatt	[125]
Z-10-dodecenyl acetate		Argyrotaenia quadrifasciana	sexatt	[127]
		Cosmopterix victor	sexatt	[82]
		Hedia atropunctana	sexatt	[78]
		H. nubiferana	inhib	[110]
		Incurvata muscalella	sexatt	[78]
		Laspeyresia pomonella	inhibit	[110]
E-10-dodecenyl acetate		Argyroploce aurofasciana	sexatt	[110]
		Cydia nigricana[a]	sexatt	[197]
		Hedia nubiferana	inhib	[110]
		Laspeyresia pomonella	inhib	[110]
	f	Lithocolletis blancardella	sexatt	[128]
11-dodecenyl acetate	f	Diparopsis castanea[a]	synerg	[122]
	f	Homona magnanima[a]	sexatt	[849]

Table 1 (continued)

Name, trivial name and trade mark®	Isol. from	Insect species affected	Behavior release	Ref.
Z-8-tridecenyl acetate		Chrysochrista spec.	sexatt	[82]
E-8-tridecenyl acetate		Chrysochrista spec.	sexatt	[82]
Z-10-tridecenyl acetate		Cosmopterix victor	sexatt	[82]
		Lithocolletis watanabei	sexatt	[82]
Z-11-tridecenyl acetate		Argyrotaenia velutinana	sexatt	[86]
Z-5-tetradecenyl acetate	f	Agrotis exclamationis[a]	sexatt	[129]
		Euxoa auxiliaris[a]	sexatt	[130]
E-5-tetradecenyl acetate		Rhynchopacha spec.	sexatt	[131, 132]
Z-7-tetradecenyl acetate		Adoxophyes orana	inhib	[133]
	f	Amathes c-nigrum	sexatt	[825, 78, 80, 82, 134]
		Lacinipolea lorea	sexatt	[80]
		Orthosia paromoae	sexatt	[82]
E-7-tetradecenyl acetate		Amathes c-nigrum	sexatt	[80]
		Brachima arotraea	sexatt	[82]
		Euxoa auxiliaris[a]	sexatt	[130]
		Zeiraphera diniana	sexatt inhib	[135]
Z-8-tetradecenyl acetate		Spilonota ocellana	sexatt	[110]
Z-9-tetradecenyl acetate prodenialure A	f	Agrotis ypsilon[a]	sexatt	[850]
		Apotomis auricristana	sexatt	[81]
	f	Adoxophyes fasciata[a]	sexatt	[71]
	f	Adoxophyes orana[a]	sexatt	[71, 72, 74, 136]
	f	Agrotis exclamationis[a]	sexatt	[129]
		Amphipoea interoceanica	sexatt	[80]
		Anagasta kuehniella	inhib	[137]
		Aphania auricristana	sexatt	[82]
		Axylis putris	sexatt	[141]
		Bryotropha similis	sexatt	[138]
		Cacoecimorpha pronuba[a]	sexatt	[136]
	f	Cadra cautella[a]	inhib sexpher	[137, 139]
		Clepsis spectrana[a]	sexatt	[140]
		Cucillia chamomillae	sexatt	[78]
		C. intermedia	sexatt	[80]
		Diarsia deparca	sexatt	[82]
		Eurois occulta[a]	sexatt	[142, 143]
		Hermonassa cecilia	sexatt	[81, 82]
		Leucania commoides[a]	sexatt	[144]
		L. phragmitidicola	sexatt	[80]
		Lucania linda[a]	sexatt	[145]
		Mamestra configurata	sexatt	[143]
		Nemapogon apicisignatellus	sexatt	[80]
		Nemophora panzerella	sexatt	[78]
		Ochropleura triangularis	sexatt	[82]
	f	Ostrinia nubilalis[a]	sexatt	[146]
	f	Pandemis heparana[a]	sexatt	[846, 136]
		P. limitata[a]	sexatt	[145, 147]
		P. pyrusana[a]	sexatt	[145, 148]

Table 1 (continued)

Name, trivial name and trade mark®	Isol. from	Insect species affected	Behavior release	Ref.
		Panolis flammea	sexatt	[78, 149, 150]
		Peridroma saucia[a]	sexatt	[141, 151]
		Plodia interpunctuella	inhib	[137]
		Pyreferra citromba	sexatt	[80]
		Sesamia cretica[a]	sexatt	[152]
		Spodoptera cilium[a]	sexatt	[153]
		S. eridiana[a]	sexatt	[75]
	f	S. exempta[a]	sexatt	[154]
	f	S. frugiperda[a]	sexatt	[823]
	f	S. littoralis[a]	sexatt	[122]
	f	S. litura[a]	sexatt	[155]
		S. triturata[a]	sexatt	[156]
E-9-tetradecenyl acetate		Adoxophyes orana[a]	inhib	[157]
		Argyrotaenia velutinana	inhib	[158]
		Bryotropha spec.	sexatt	[71]
		B. terella	sexatt	[110]
		Loxostege neobliteralis	sexatt	[80]
		Polia grandis	sexatt	[80]
		Spodoptera triturata[a]	sexatt	[156]
Z-10-tetradecenyl acetate		Adoxophyes orana	inhib	[133]
		Apotomis corticana	sexatt	[110]
	f	Archips semiferanus[+]	sexatt	[159, 160]
		Endothenia carbonana	sexatt	[110]
		E. benausopis	sexatt	[82]
		Exartema spec.	sexatt	[127]
		Lithocolletis orientalis	sexatt	[82]
Z-11-tetradecenyl acetate	f	Adoxophyes fasciata[a]	sexatt	[71]
riblure, redlamone®	f	A. orana[a]	sexatt	[72, 74, 78, 136]
		Aethes smeathmannia	sexatt	[78]
	f	Archippus breviplicanus[a]	sexatt	[161]
	f	Archips argyrospilus[a]	sexatt inhib	[162]
		A. ingentanus	sexatt	[82]
		A. longicellanus	sexatt	[81]
		A. peratrus	sexatt	[82]
		A. piccanus similis	sexatt	[82]
	f	A. podanus[a]	sexatt	[62, 78, 163]
		A. rosanus[a]	sexatt	[164]
	f	A. semiferanus[a]	sexatt	[159, 160]
		A. semistructus	sexatt	[82]
		A. xylostaena	sexatt	[78]
	f	Argyrotaenia citrana	synerg	[165]
		A. pulchellana	sexatt	[166]
	f	A. velutinana[a]	sexatt disrupt	[117, 167, 168]
		Cacoecimorpha pronuba[a]	sexatt	[136, 169]
		Ceramica picta	sexatt	[80]
		Chimabacche fagella	sexatt	[78]
	f	Choristoneura rosaceana	sexatt	[170]
		Clepsis melaleucana[a]	sexatt	[80]

[+] Later corrected into Z-11-tetradecenyl acetate [160]

Table 1 (continued)

Name, trivial name and trade mark®	Isol. from	Insect species affected	Behavior release	Ref.
	f	C. spectrana[a]	sexatt	[140]
		Cucullia chamomillae[a]	sexatt	[78]
	f	Epichoristodes acerbella	sexatt	[175]
		Eucosma sonomana	inhib	[842]
	f	Homona magnanima[a]	sexatt	[849]
		Hoshinoa longicellanus	sexatt	[82]
		Hyponomeutha mallinellus	sexatt	[78]
		Laspeyresia aurana	sexatt	[78]
		Lucania linda[a]	sexatt	[145]
	f	Ostrinia nubilalis[a]	sexatt	[146, 171, 172, 173]
		O. obumbratalis[a]	sexatt	[174]
	f	Pandemis heparana[a]	sexatt	[846, 136]
		P. limitata[a]	sexatt disrupt	[145, 147]
		P. pyrusana[a]	sexatt	[145, 148]
		Panolis flammea	synerg	[149]
		Phalonidia minimana	sexatt	[82]
		Platynota idaeusalis	inhib	[176]
	f	P. stultana[a]	sexatt inhib	[177]
		Ptycholoma lecheanum[a]	sexatt	[78]
		Pyraustra purpuralis[a]	sexatt	[169]
		Thyris muculata	sexatt	[80]
		Tortrix viridana	sexatt	[78, 150]
		Xenomicta cupreifera	sexatt	[82]
		Yponomeuta padellus-malinellus	sexatt	[110]
		Y. yanagawanus	sexatt	[82]
		Zygaenia transalpina	sexatt	[178]
E-11-tetradecenyl acetate		Adoxophyes orana	inhib	[157]
		Alleima loefflingiana	sexatt	[78]
	f	Archippus breviplicanus[a]	sexatt	[161]
	f	Archips argyrospilus[a]	sexatt inhib	[162]
	f	A. podanus[a]	sexatt	[62, 78, 163]
	f	A. semiferanus[a]	sexatt	[160]
	f	Argyrotaenia velutinana[a]	sexatt inhib disrupt	[158, 168, 172, 179]
		Cacoecimorpha pronuba[a]	sexatt	[169]
		Choristoneura fumiferana	inhib	[176]
		C. rosaceana	inhib	[170]
		C. viridis	sexatt	[180]
		Croesia bergmannia	sexatt	[78]
		C. holmiana	sexatt	[78, 110]
		Dichomeris liguella	sexatt	[80]
		Evergestis forficalis	sexatt	[78]
		Gypsonoma dealbana	sexatt	[78]
		Hysterosia cartwrightana	sexatt	[181]
	f	Loxostege sticticalis	sexatt	[181]
		Lymnaecia phragmitella	sexatt	[181]
	f	Ostrinia nubilalis[a]	sexatt inhib disrupt	[80, 146, 172–174, 179]
		O. umbratalis[a]	sexatt	[174]

Table 1 (continued)

Name, trivial name and trade mark®	Isol. from	Insect species affected	Behavior release	Ref.
		Phalonia spec.	sexatt	[80]
	f	Platynota idaeusalis[a]	synerg	[176, 182, 183]
	f	P. stultana[a]	sexatt inhib	[177]
		Pyraustra ochosalis	sexatt	[80]
		P. purpuralis	sexatt	[169]
		Sathrobrota spec.	sexatt	[81]
		Sparganothis albicaudana	sexatt	[80]
		S. sulfureana	sexatt	[80]
		Spodoptera littoralis[a]	sexatt	[122]
	f	S. litura[a]	sexatt	[155]
		Zeiraphera diniana	sexatt	[125, 135]
E-5-pentadecenyl acetate		Gelechia sestertiella	sexatt	[132]
Z-9-pentadecenyl acetate		Acrobasis rufibimbalis	sexatt	[82]
Z-7-hexadecenyl acetate hexalure		Autographa californica	disrupt	[88]
		Euxoa tessellata	sexatt	[80]
		E. messoria[a]	sexatt	[833]
		Heliothis virescens	disrupt	[88]
		Pectinophora gossypiella	sexatt disrupt	[88, 184, 185]
		Sitotroga cerealella	sexatt	[186]
		Spodoptera ornithogalli	disrupt	[88]
		Tortricodes tortricellus	sexatt	[78]
		Trichoplusia ni	disrupt	[88]
Z-10-hexadecenyl acetate		Mamestra configurata	sexatt	[187]
Z-11-hexadecenyl acetate		Amphipoea velata	sexatt	[80]
		Amphipyra monolitha	sexatt	[81, 82]
		Argyrestia spec.	sexatt	[78]
		Argyrotaenia velutinana	inhib	[158]
		Conistra unimacula	sexatt	[82]
		Crymodes devestator[a]	sexatt	[90, 143]
		Eurois occulta[a]	sexatt	[142, 143]
		Euxoa messoria[a]	sexatt	[833]
	f	Leucania commoides[a]	sexatt	[143, 144]
	f	Mamestra brassicae[a]	sexatt	[141, 188]
	f	M. configurata[a]	sexatt	[143, 189]
		M. illoba	sexatt	[82]
	f	M. oleracea[a]	sexatt	[78, 141]
		Morrisonia confusa	sexatt	[80]
		Orthosia ella	sexatt	[82]
		Pareromene exosectella	sexatt	[82]
		Peridroma saucia[a]	sexatt	[141, 151]
		Plutella xylostella[a]	sexatt	[192]
	f	Pseudaletia unipuncta[a]	sexatt	[834]
		Pseudorthodes crenulata	sexatt	[80]
		P. vecors	sexatt	[80]
	f	Scotogramma trifolii[a]	sexatt	[80, 143, 190, 191]
		Telorta divergens	sexatt	[81, 82]
		T. edentata	sexatt	[82]
		Valeroides viridimacula	sexatt	[82]
E-11-hexadecenyl acetate		Brachmia macroscopa	sexatt	[649]
		Diasemia litterata	sexatt	[82]

Table 1 (continued)

Name, trivial name and trade mark®	Isol. from	Insect species affected	Behavior release	Ref.
		Epiblema moza	sexatt	[81, 82]
		Telorta edentata	sexatt	[81, 82]

b) Polyunsaturated Acetates

Name, trivial name and trade mark®	Isol. from	Insect species affected	Behavior release	Ref.
3,7-dimethyl-1,6-octadien-3-yl acetate linalyl acetate		Pectinophora gossypiella	synerg	[193]
E-7, Z-9-dodecadienyl acetate	f	Lobesia botrana	sexatt	[194, 195]
		Sterrha biselata	sexatt	[196]
E-7, E-10-dodecadienyl acetate		Hedia atropunctana	sexatt	[78]
E-8, E-10-dodecadienyl acetate		Laspeyresia nigricana[a]	sexatt	[197]
		L. pomonella	inhib disrupt	[198, 199]
		L. servillana	sexatt	[78]
	f	Rhyacionia frustrana	sexatt	[200]
		R. rigidana	sexatt	[201]
Z-9, 11-dodecadienyl acetate		Dichrorampha accuminata	sexatt	[78]
		D. gueneeana	sexatt	[78]
		Eurhypara coronata	sexatt	[78]
E-9, 11-dodecadienyl acetate	f	Diparopsis castanea[a]	sexatt	[122, 202]
E-4, Z-7-tridecadienyl acetate	f	Phthorimea operculella[a]	sexatt	[62, 203–205]
Z-7, Z-10-tridecadienyl acetate		Trichoplusia ni	sexatt	[206]
Z-7, Z-11-tridecadienyl acetate		Phthorimea operculella	sexatt inhib	[206]
E-9, Z-11-tridecadienyl acetat		Nemophora panzerella	sexatt	[78]
Z-3, E-5-tetradecadienyl acetate		Prionoxystus robiniae[a]	sexatt	[151, 207–209]
E-3, E-5-tetradecadienyl acetate		Accosus centerensis	sexatt	[207]
Z-4, E-8-tetradecadienyl acetate		Borkhausenia schefferella	sexatt	[210]
Z-7, Z-11-tetradecadienyl acetate		Conistra vaccinii	sexatt	[78, 210]
Z-9, E-11-tetradecadienyl acetate	f	Spodoptera littoralis[a]	sexatt inhib	[122, 211]
litlure	f	S. litura[a]	sexatt inhib	[155, 212]
Z-9, Z-12-tetradecadienyl acetate		Plodia interpunctuella	inhib	[213]
Z-9, E-12-tetradecadienyl acetate		Adoxophyes fasciata	synerg	[214]
litlure, prodenialure B	f	Anagasta kuehniella	sexatt	[215, 216]
	f	Cadra cautella	sexatt inhib	[98, 217–219]
	f	C. figulella	sexatt	[222]
		Elasmopalpus lignosellus	disrupt	[824]
	f	Ephestia elutella	sexatt	[223]
		Ephestiodes infimella	sexatt	[220]
		Myelois crimbrumella	sexatt	[78]
		Orthosia angustipennis	sexatt	[82]
		Panolis flammea	sexatt	[78, 150]
	f	Plodia interpunctuella	sexatt disrupt	[98, 217, 218, 221]
		Pseudoplusia includens	inhib	[95]
		Rusina deprevata	sexatt	[81, 82]
		Spodoptera cilium[a]	sexatt	[153]
		S. dolichos	sexatt	[95]
	f	S. eridiana[a]	sexatt	[75]
	f	S. exempta[a]	sexatt	[154]

Table 1 (continued)

Name, trivial name and trade mark®	Isol. from	Insect species affected	Behavior release	Ref.
		S. exigua[a]	disrupt sexatt	[506, 30, 224]
		S. frugiperda	inhib disrupt	[96]
	f	S. littoralis[a]	sexatt	[211, 212]
	f	S. litura[a]	sexatt	[155, 212]
		Telorta pacifica	sexatt	[82]
		Trichoplusia ni	inhib	[224]
Z-6, Z-11-hexadecadienyl acetate		Eudia pavonia	sexatt	[78]
E-6, Z-11-hexadecadienyl acetate	f	Antheraea pernyi[a]	sexatt	[225]
	f	Antheraea polyphemus[a]	sexatt	[225, 226]
Z-7, Z-11-hexadecadienyl acetate	f	Euxoa messoria[a]	sexatt	[833]
gossyplure	f	Pectinophora gossypiella[a]	sexatt disrupt	[24, 227, 228]
Z-7, E-11-hexadecadienyl acetate	f	Pectinophora gossypiella[a]	sexatt disrupt	[24, 227, 228]
gossyplure, angoulure	f	Sitotroga cerealella	sexatt	[229]
E-7, Z-11-hexadecadienyl acetate		Pectinophora gossypiella	inhib	[228]
E-7, E-11-hexadecadienyl acetate		Pectinophora gossypiella	inhib	[228]
		Podosesia syringae	sexatt	[230]
Z-3, Z-13-octadecadienyl acetate		Aegeria tibialis[a]	sexatt	[233]
		Podosesia aureocincta[a]	sexatt	[230]
		P. syringae	sexatt	[230]
	f	Sanninoidea exitiosa[a]	sexatt disrupt	[231, 232]
		Synanthedon hector[a]	sexatt	[234]
		S. pictipes	inhib disrupt	[231, 232]
E-3, Z-13-octadecadienyl acetate		Podosesia aureocincta[a]	sexatt	[230]
		Sanninoidea exitiosa[a]	disrupt	[232]
		Synanthedon hector[a]	sexatt	[234]
	f	S. pictipes[a]	sexatt disrupt	[231, 232]
		S. bibionipennis[a]	sexatt	[235]
E-4, Z-7, Z-10-tridecatrienyl acetate	f	Phthorimea operculella[a]	sexatt	[62, 204, 205, 236]

c) Monounsaturated Alcohols

E-5-decen-1-ol	f	Anarsia lineatella	sexatt	[79]
		Tricimba cincta	ovipos att	[237]
E-5-dodecen-1-ol		Tricimba cincta	sexatt	[132, 238]
Z-7-dodecen-1-ol		Autographa californica[a]	sexatt	[843]
		Dicentria semirufescens	sexatt	[80]
		Euxoa pleuritica[a]	sexatt	[85]
		Exartema spec.	sexatt	[239]
		Harpipteryx xylostella auct.[a]	sexatt	[80]
		Protobratha leucostola	sexatt	[82]
		Pseudoplusia includens	inhib	[240]
		Raphia frater	sexatt	[80, 239]
		Trichoplusia ni	inhib	[241]
E-7-dodecen-1-ol		Laspeyresia youngana[a]	sexatt	[242]
Z-8-dodecen-1-ol		Celypha striana	sexatt	[78, 110]
		Epiblema scudderiana	sexatt	[80]
		Grapholitha molesta	inhib	[234]
		Hedia nubiferana	inhib	[110]

Table 1 (continued)

Name, trivial name and trade mark®	Isol. from	Insect species affected	Behavior release	Ref.
		Pammene argyrana	sexatt	[78]
		P. rhediella	sexatt	[78]
		P. splendidulana	sexatt	[78]
E-8-dodecen-1-ol		Hedia chionosema[a]	sexatt	[80]
E-9-dodecen-1-ol	f	Barbara colfaxiana	sexatt	[242]
		Dichrorampha spec.	sexatt	[80]
		Sparganothis pilleriana[a]	sexatt	[124]
E-10-dodecen-1-ol		Laspeyresia nigricana[a]	sexatt	[197]
Z-9-tridecen-1-ol		Pammene rhediella	sexatt	[78]
		P. splendidula	sexatt	[78]
Z-11-tridecen-1-ol		Argyrotaenia velutinana	inhib	[158]
E-5-tetradecen-1-ol		Gelechia spec.	sexatt	[238]
		Rhynchopacha spec.	sexatt	[238]
Z-8-tetradecen-1-ol		Lithocolletis pygmaea	sexatt	[82]
Z-9-tetradecen-1-ol		Athetis dissimilis	sexatt	[82]
	f	Homoeosoma electellum[a]	sexatt	[836]
		Sesamia cretica[a]	sexatt	[152]
E-9-tetradecen-1-ol		Epiblema plaquifera	sexatt	[82]
Z-11-tetradecen-1-ol		Aethes smeathmannia[a]	sexatt	[78]
	f	Archips argyrospilus	inhib	[162]
		A. rosanus[a]	sexatt	[164]
		A. sorbana[a]	sexatt	[78]
		Argyrotaenia velutinana	inhib	[158]
		Choristoneura fractivittana	sexatt	[80, 244]
		Clepsis melaleucana[a]	sexatt	[80]
		Eupoecilia kobeana	sexatt	[82]
		Nedra ramulosa	sexatt	[86]
	f	Pandemis heparana[a]	sexatt	[846]
		Platynota idaeusalis	inhib	[183]
	f	P. stultana[a]	synerg inhib	[177, 245]
		Ptycholoma circumclusana	sexatt	[82]
		P. lecheanum[a]	sexatt	[78]
		Sparganothis niveana	sexatt	[80, 86]
		Thyris usitata	sexatt	[82]
		Zygaenia transalpina	sexatt	[178]
E-11-tetradecen-1-ol	f	Archips argyrospilus[a]	inhib	[162]
		A. sorbana[a]	sexatt	[78]
	f	Choristoneura fumiferana[a]	inhib	[176, 246]
	f	Platynota stultana[a]	synerg inhib	[177, 245]
	f	P. idaeusalis[a]	sexatt	[183]
E-5-pentadecen-1-ol		Gelechia spec.	sexatt	[238]
		Rhynchopacha spec.	sexatt	[238]
Z-7-hexadecen-1-ol		Pectinophora gossypiella	inhib	[247]
Z-11-hexadecen-1-ol		Chrysoteuchia topiaria	inhib	[837]
		Eupsilia quadrilinea	sexatt	[82]
		E. satellitia	sexatt	[78]
	f	Heliothis virescens[a]	sexatt	[839]
		Leucania commoides	sexatt	[144]

Table 1 (continued)

Name, trivial name and trade mark®	Isol. from	Insect species affected	Behavior release	Ref.
	f	Mamestra oleracea[a]	sexatt	[141]
	f	Scotogramma trifolii	sexatt	[190, 191]
	f	Pseudaletia unipuncta[a]	sexatt	[834]
E-11-hexadecen-1-ol		Eupsilia quadrilinea	sexatt	[82]
d) Polyunsaturated Alcohols				
E-8, E-10-dodecadien-1-ol codlemone, codlelure		Cnephasiella incertana	sexatt	[78]
		Eucosma campoliliana	sexatt	[78]
		E. nigromaculana	sexatt	[102]
		Hedia nubiferana	sexatt	[78, 110]
		H. ochroleucana	sexatt	[251]
	f	Laspeyresia pomonella	sexatt disrupt	[78, 195, 248, 249, 252]
Z-9, E-12-tetradecadien-1-ol		Cadra cautella	inhib	[253, 250]
	f	Homoeosoma electellum[a]	sexatt	[836]
		Plodia interpunctuella[a]	sexpher	[98, 250]
		Rhycitinae spec.	sexatt	[82]
		Rhynchaglaea scitula	sexatt	[82]
E-10, E-12-hexadecadien-1-ol	f	Bombyx mori[a]	sexpher	[254]
E-10, Z-12-hexadecadien-1-ol bombykol	f	Bombyx mori[a]	sexatt	[255, 256]
Z-3, Z-13-octadecadien-1-ol		Aegeria tibialis[a]	sexatt	[233]
		Podosesia aureocincta[a]	sexatt	[230]
		Synanthedon sequoia	sexatt	[235]
E-3, Z-13-octadecadien-1-ol		Synanthedon bibionipennis[a]	sexatt	[235]
e) Other Pheromone Structures				
8-propoxyoctan-1-ol		Grapholitha molesta	synerg	[257]
undecan-1-ol		Laspeyresia pomonella	inhib	[198]
dodecan-1-ol	f	Archips argyrospilus[a]	synerg	[162]
		Grapholitha molesta	synerg	[112, 257, 258]
	f	Homona coffearia[a]	sexatt	[126]
Z-9-tetradecenyl formate		Heliothis virescens	disrupt	[92]
		H. zea	disrupt	[92]
Z-11-tetradecenyl formate		Argyrotaenia velutinana	inhib	[158]
8-butyloxyoctyl acetate		Adoxophyes orana	inhib	[133]
nonyl acetate		Paralobesia viteana	synerg	[116]
11-methoxyundecyl acetate		Adoxophyes orana	inhib	[132]
11-ethoxyundecyl acetate		A. orana	inhib	[133]
11-propoxyundecyl acetate		A. orana	inhib	[133]
dodecyl acetate	f	Archips argyrospilus	synerg	[162]
	f	Argyrotaenia velutinana[a]	synerg	[158, 168]
		Choristoneura rosaceana	inhib	[170]
	f	Diparopsis castanea[a]	sexatt	[122]

Table 1 (continued)

Name, trivial name and trade mark®	Isol. from	Insect species affected	Behavior release	Ref.
		Grapholitha molesta	inhib	[112, 257]
		G. prunivora	inhib	[112]
	f	Homona coffearia[a]	sexatt	[126]
		Paralobesia viteana	synerg	[116]
		Pectinophora gossypiella	inhib	[193, 259]
	f	Rhyacionia buoliana	inhib	[118]
		Eupoecilia ambiguella[a]	synerg	[114]
9-dodecynyl acetate		Adoxophyes orana	inhib	[260]
tridecyl acetate		Argyrotaenia velutinana	inhib	[158]
9-tridecynyl acetate		Adoxophyes orana	inhib	[260]
tetradecyl acetate	f	Ostrinia nubilalis[a]	sexatt	[146]
	f	Pectinophora gossypiella[a]	inhib	[228, 261]
		Spodoptera littoralis[a]	sexatt	[122]
9-tetradecynyl acetate		Adoxophyes orana	inhib	[260]
Z-9, E-12-tetradecadienyl propionate		Lamprothripa lactaria	sexatt	[82]
11-tetradecynyl acetate		Adoxophyes orana	inhib	[260]
		Argyrotaenia velutinana	inhib disrupt	[158, 179]
		Ostrinia nubilalis	inhib disrupt	[179, 262]
Z-10, 11-epoxytetradecenyl acetate		Argyrotaenia velutinana	inhib	[158]
9-pentadecynyl acetate		Adoxophyes orana	inhib	[260]
hexadecyl acetate	f	Pseudaletia unipuncta[a]	sexatt	[834]
E-8,E-10-dodecadienyl propionate		Laspeyresia pomonella	inhib	[263]
Z-5-decenyl isovalerianate	f	Nudaurelia cytherea	sexatt	[264]
Z-7-tetradecenal	f	Prays citri	sexatt	[265]
	f	P. oleae	sexatt	[851, 829]
Z-9-tetradecenal	f	Heliothis virescens[a]	sexatt	[266, 267]
		Prays oleae	inhib	[851]
Z-11-tetradecenal	f	Argyrotaenia citrana	sexatt	[165]
		Cenopsis acerivorana[a]	sexatt	[244]
		Choristoneura conflictana	sexatt	[244, 268]
		C. fractivittana[a]	sexatt	[244]
		C. fumiferana[a]	sexatt	[244]
		Croesia semipurana[a]	sexatt	[244]
E-11-tetradecenal		Acleris emergana	sexatt	[239]
soolure		Cenopsis acerivorana[a]	sexatt	[244]
		Choristoneura biennis	sexatt	[180]
	f	C. fumiferana[a]	sexatt	[244, 269, 270]
		C. occidentalis	sexatt	[269]
		Croesia semipurana[a]	sexatt	[244]
tetradecanal	f	Heliothis virescens[a]	sexatt	[839]
Z-7-hexadecenal	f	Heliothis zea[a]	sexatt	[839]
	f	H. virescens[a]	sexatt	[839]
Z-9-hexadecanal	f	Heliothis virescens[a]	sexatt	[839]
	f	H. zea[a]	sexatt	[839]
Z-11-hexadecenal		Argyrestia ephipella	sexatt	[78]
		Argyrotaenia citrana	inhib	[165]
	f	Chilo suppressalis[a]	sexatt	[214, 271, 272]
		Chrysoteuchia topiaria	sexatt	[837]
		Crymodes devestator	sexatt	[90]
	f	Heliothis armigera	sexatt	[273]
	f	H. virescens[a]	sexatt	[266, 267]

Table 1 (continued)

Name, trivial name and trade mark®	Isol. from	Insect species affected	Behavior release	Ref.
	f	H. zea	inhib	[266, 274]
		Plutella xylostella[a]	sexatt	[192]
		Xylena formosa	sexatt	[82]
Z-13-hexadecenal		Chrysoteuchia topiaria	inhib	[837]
hexadecanal	f	Heliothis virescens[a]	sexatt	[839]
	f	H. zea[a]	sexatt	[839]
Z-13-octadecenal	f	Chilo suppressalis	sexatt	[271, 272, 275]
		Chrysoteuchia topiaria	inhib	[837]
E-6, Z-11-hexadecadienal	f	Antherea pernyi[a]	sexatt	[255]
	f	A. polyphemus[a]	sexatt	[226, 225]
E-10, Z-12-hexadecadienal bombykal	f	Bombyx mori[a]	sexpher	[256]
		Manduca sexta	sexatt	[276]
E-8, E-10-dodecadienyl ethyl ether		Laspeyresia pomonella	inhib	[263]
E-8, E-10-dodecadienyl propyl ether		L. pomonella	inhib	[263]
Z-7,8-epoxy-2-methyloctadecane[+] disparlure	f	Lymantria dispar	sexatt disrupt	[277, 278, 279]
		L. dispar japonica	sexatt	[280]
		L. fumida	sexatt	[280]
	f	L. monacha	sexatt	[281, 282]
		L. obfuscata	sexatt	[283]
2-methylheptadecane	f	Holomelina spec.	sexatt	[284]
	f	Isia isabella	sexatt	[284]
2-methyl-Z-7-octadecene	f	Lymantria dispar	inhib	[285–287]
Z-8, 9-epoxy-(3S)-3-methylnonadecane		Lymantria dispar	sexatt	[288]
Z-6-heneicosen-11-one		Orgya antiqua	sexatt	[78]
	f	O. pseudotsugata	sexatt	[289, 290]
1,Z-6-heneicosadien-11-one	f	O. pseudotsugata	sexatt	[291]
E-3, Z-6-heneicosadien-11-one	f	O. pseudotsugata	sexatt	[291]
E-7-eicosen-11-one	f	Carposina nipponensis[a]	sexatt	[292]
Z-9-nonadecen-11-one	f	Carposina nipponensis[a]	sexatt	[292]

2.2. Pheromones of Male Lepidoptera

octanal	m	Amauris ochlea[a]	sexpher	[293]
nonanal	m	A. ochlea[a]	sexpher	[293]
	m	Galleria mellonella[a]	sexatt	[297, 294, 295]
undecanal	m	Achroia grisella[a]	sexatt	[296]
	m	Galleria mellonella[a]	sexatt	[294, 295, 297]
Z-11-octadecenal	m	Achroia grisella[a]	sexatt	[296]
2-phenylethan-1-ol	m	Agrocola helvola	sexatt	[298]
	m	Mamestra brassicae[a]	sexatt	[298]
	m	M. configurata	sexatt	[468]
	m	M. persicariae[a]	sexatt	[298, 299]
	m	Peridroma saucia	sexpher	[301]
	m	Phalaenoides glycicae[a]	pher	[830]
	m	Polia nebulosa[a]	sexatt	[299]

[+] For the biological activity of enantiomers of chiral sex attractants c.f. part 5.1.h) "Stereoselective Syntheses of Chiral Moth Pheromones", p. 123.

Table 1 (continued)

Name, trivial name and trade mark®	Isol. from	Insect species affected	Behavior release	Ref.
	m	Polia tincta[a]	sexatt	[298]
	m	Trichoplusia ni	sexatt	[300]
benzaldehyde	m	Apamea monoglypha	pher	[469]
	m	Leucania coniger[a]	pher	[299]
	m	L. impura[a]	pher	[469, 299]
	m	L. pallens[a]	pher	[299]
	m	Leucania spec.	sexatt	[299]
	m	Mamestra persicariae[a]	sexatt	[298, 299]
	m	Melanchra spec.	aphrod	[827]
	m	Morrisonia spec.	sexpher	
	m	Persectania spec.	sexpher	[827]
	m	Phlogophora meticulosa	pher	[469]
	m	Polia nebulosa[a]	sexpher	[299]
	m	Pseudaleta separata	sexatt	[302]
benzyl alcohol	m	Mamestra brassicae	sexatt	[298]
	m	M. persicariae[a]	sexatt	[299, 298]
	m	Polia nebulosa[a]	sexatt	[299, 298]
phenylacetaldehyde	m	Mamestra persicariae[a]	sexatt	[298]
	m	Polia tincta[a]	sexatt	[298]
pinocarvone	m	Apamea monoglypha	sexatt aphrod	[299]
6-methyl-5-hepten-2-one	m	Phlogophora meticulosa[a]	sexpher	[299]
6-methyl-5-hepten-2-ol	m	P. meticulosa[a]	sexpher	[299]
n-butyric acid	m	Bapta temerata	sexpher	
iso-butyric acid	m	Leucania spec.[a]	sexpher	[299]
2-methylbutyric acid	m	Phlogophora meticulosa	sexpher	[299]
3-hexenoic acid	m	Amauris ochlea[a]	sexpher	[293]
2,3-dihydro-7-methyl-1*H*-pyrro-lizin-1-one	m	A. albimaculata	sexpher	[303]
	m	A. echeria	sexpher	[303]
	m	A. niavius	sexpher	[303]
	m	A. ochlea[a]	sexpher	[293, 303]
	m	Danaus affinis affinis[a]	sexpher	[304]
	m	D. chrysippus chrysippus	sexpher	[305]
	m	D. chrysippus dorippus	sexpher	[303]
	m	D. chrysippus petilea[a]	sexpher	[307, 306]
	m	D. gilippus berenice[a]	aphrod	[48, 51]
	m	D. hamatus hamatus	sexpher	[304, 306]
	m	D. limniace petiverana	sexpher	[303]
	m	D. affinis albistriga[a]	sexpher	[307]
	m	D. hamatus moderatus[a]	sexpher	[307]
	m	Lycorea ceres ceres	sexpher	[46, 47]
7-hydroxy-6,7-dihydro-5*H*-pyrrolizin-1-carboxaldehyde	m	Danaus hamatus hamatus	sexpher	[304, 306]
	m	D. pumilis hebridesius	sexpher	[307]
		Euploea lewinii lilyboea	sexpher	[307]
		E. nemertes	sexpher	[307]
		E. sylvester sylvester	sexpher	[304]
	m	Utethesia lotrix[a]	sexpher	[308]
		U. pulchelloides	sexpher	[308]
6,7-dihydro-5*H*-pyrrolizin-1-carboxaldehyde	m	Danaus affinis affinis[a]	sexpher	[304]
	m	D. affinis albistriga[a]	sexpher	[307]
	m	Utethesia lotrix[a]	sexpher	[308]
hexadecyl acetate	m	Lycorea ceres ceres[a]	pher	[46]

Table 1 (continued)

Name, trivial name and trade mark®	Isol. from	Insect species affected	Behavior release	Ref.
Z-11-octadecenyl acetate	m	Lycorea ceres ceres[a]	pher	[46]
2-phenylethyl acetate	m	Phalaenoides glycicae[a]	pher	[830]
methyl salicylate	m	Amauris ochlea[a]	pher	[293]
eugenol	m	A. ochlea[a]	pher	[293]
cis-jasmon	m	A. ochlea[a]	pher	[293]
3,7-dimethyldeca-E,E-2,6-dien--1,10-diol	m	Danaus gilippus berenice	pher	[48]
3,7-dimethyl-10-hydroxy-E,E-2,6--decadienoic acid	m	D. plexippus[a]	pher	[309]
3,7-dimethyloct-E-2-en-1,8-diol	m	D. chrysippus chrysippus	pher	[305]
3,7-dimethyl-E,E-2,6-decadien-1,10-dioic acid	m	D. plexippus[a]	pher	[310]
3,7-dimethyl-2,E-6-octadienal geranial	m	Pieris napi[a]	pher	[311]
3,7-dimethyl-2,Z-6-octadienal neral	m	P. napi[a]	pher	[311]
3,4-dimethoxyacetophenone	m	Amauris niavius	pher	[303]
	m	A. tartaraea	pher	[303]

2.3. Pheromones of the Bark Beetles (Scolytidae)

Name, trivial name and trade mark®	Isol. from	Insect species affected	Behavior release	Ref.
(−)-2-methyl-6-methylene-7-octen--4-ol ipsenol	m	Ips accuminatus	aggr	[312]
	m	I. confusus[a]	aggr	[313–315]
	m	I. cribricollis	aggr	[312, 316]
	ins	I. grandicollis	aggr	[316, 317]
	m	I. latidens	aggr	[316]
	m	I. montanus	aggr	[316]
	m	I. paraconfusus[a]	aggr	[315, 316]
		I. pini	inhib	[318]
	m	Orthotomicus latidens	aggr sexatt	[320]
	m	Pityokteinis curvidens	aggr	[319, 317]
(+)-2-methyl-6-methylene-2,7-octadien-4-ol ipsdienol	m	Ips accuminatus	aggr	[312]
	m	I. avulvus	aggr	[312]
	m	I. bonanseai	aggr	[312]
	m	I. confusus[a]	aggr	[313–315]
	m	I. calligraphus[a]	aggr	[739, 321]
	m	I. cribricollis	aggr	[312]
	m	I. duplicatus	aggr	[317]
	m	I. integer	aggr	[312]
	m	I. interstitialis[a]	aggr	[317]
	m	I. knausi	aggr	[312]
	m	I. paraconfusus[a]	aggr	[315]
	m	I. sexdentatus	aggr	[317, 322]
	m	I. typographus	aggr	[323]
cis-(+)-2-pinen-4-ol cis-(+)-verbenol	m	Ips calligraphus[a]	aggr	[739, 321]
	m	I. confusus[a]	aggr	[313–315]
	m	I. paraconfusus[a]	aggr	[315]
	m	I. typographus	aggr	[323]

Table 1 (continued)

Name, trivial name and trade mark®	Isol. from	Insect species affected	Behavior release	Ref.
trans-2-pinen-4-ol	f	Dendroctonus adjunctus	sexatt	[324, 325]
trans-verbenol, pondelure	f	D. brevicomis	aggr	[326–329]
	f	D. frontalis	aggr	[326–329]
	ins	D. jeffreyi	aggr	[326–329]
	f	D. ponderosa[a]	aggr	[326–329]
	f	D. pseudotsugae	aggr inhib	[326–329]
	m	Ips typographus	aggr	[323]
exo-7-ethyl-5-methyl-6,8-dioxa-	f	Dendroctonus brevicomis	aggr	[330, 331]
bicyclo[3.2.1]octane,	ins	D. jeffreyi	aggr	[326]
exo-brevicomin	f	D. ponderosa	aggr	[326]
		Temnochila chlorodia [Trogosit.]	aggr kair	[332]
1,5-dimethyl-6,8-dioxabibyclo-	f	Dendroctonus adjunctus	sexatt	[324, 325]
[3.2.1]octane	m	D. brevicomis	aggr sexatt	[333]
frontalin, douglure	f	D. frontalis[a]	aggr sexatt	[317, 330, 331]
	f	D. pseudotsugae[a]	aggr	[334–336]
		Thanasimus bistriata [Clerid.]	aggr kair	[337]
		T. dubius [Clerid.]	aggr kair	[338]
		T. undatulus [Clerid.]	aggr kair	[335]
3-methyl-2-cyclohexen-1-one	f	Dendroctonus pseudotsugae	aggr inhib	[334, 339, 340,
		D. rufipennis	inhib disrupt	341, 342]
ethanol		Dendroctonus pseudotsugae	synerg	[343]
4-methylheptan-3-ol	f	Scolytus multistriatus[a]	aggr	[344]
		Scolytus scolytus[a]	aggr	[374, 345]
6-methylhept-5-en-2-ol		Gnathotrichus sulcatus	aggr	[346]
sulcatol		Platypus flavicornis[a]	aggr	[347]
3,7-dimethyl-1,6-octadien- -3-ol linalool	f	Ips paraconfusus	inhib	[318, 348]
3-methyl-2-cyclohexen-1-ol seudenol	f	Dendroctonus pseudotsugae[a]	synerg pher	[336, 343]
3-methylcyclohexanone		D. pseudotsugae	inhib	[349]
2-isopropyl-5-methylcyclohexanone menthone		D. pseudotsugae	inhib	[349]
2-isopropylidene-5-methyl-5-cyclo- hexen-1-one piperitone		D. pseudotsugae	inhib	[349]
5-ethyl-2,4-dimethyl-6,8-dioxabi-	f	Scolytus multistriatus[a]	aggr	[344, 350]
cyclo[3.2.1]octane multistriatin		S. scolytus[a]	aggr	[345]
1-methoxy-2-phenylbenzene o-phenylanisol		Dendroctonus rufipennis	aggr	[351]
2-methylene-3,3-dimethylbicyclo- [2.1.1]heptane camphene	host	D. pseudotsugae	synerg	[335]
2,6,6-trimethylbicyclo[3.1.1]-	host	D. frontalis	synerg	[352, 353]
hept-2-ene α-pinene	host	D. ponderosae	synerg	[353]
	host	D. pseudotsugae	synerg	[354]
3-methyl-3-cyclohexen-1-ol	ins	D. pseudotsugae	pher	[355]
endo-7-ethyl-5-methyl-6,8-dioxa-	f	D. brevicomis	pher	[356]
bicyclo[3.2.1]octane	m	D. frontalis	pher inhib	[356]
endo-brevicomin	m	D. ponderosae	inhib	[357]

Table 1 (continued)

Name, trivial name and trade mark®	Isol. from	Insect species affected	Behavior release	Ref.
2,6,6-trimethylbicyclo[3.1.1]-heptan-3-one pinocarvone	ins	D. brevicomis	pher	[358, 359]
	m	D. frontalis	pher	[358]
E-6,6-dimethyl-2-methylenebicyclo-[3.1.1]heptan-3-ol E-pinocarveol	ins	D. brevicomis	pher	[358, 359]
6,6-dimethyl-2-hydroxymethylbi-cyclo[3.1.1]hept-2-ene myrtenol	ins	D. brevicomis	pher	[359, 358]
	ins	D. frontalis	pher	[358]
2,6,6-trimethylbicyclo[3.1.1]-hept-2-ene 3-carene	ins	Blastophagus piniperda	pher	[360]
	host	Dendroctonus brevicomis[a]	synerg	[353]
	host	D. frontalis[a]	synerg	[353]
7-methyl-3-methylene-1,6-octa-diene myrcene	ins	Blastophagus piniperda[a]	synerg	[361]
	host	Dendroctonus brevicomis[a]	synerg	[361]
3a,3b,4,5,6,7-hexahydro-4α-isopro-pyl-3,7β-dimethyl-1*H*-cyclo-penta[1.3]cyclo[1.2]benzene α-cubebene	host	Scolytus multistriatus[a]	synerg	[344]
	host	S. scolytus[a]	synerg	[374, 345]
2,6,6-trimethylbicyclo[3.2.1]-hept-2-en-4-one verbenone	m	Dendroctonus brevicomis	stim inhib	[352, 329]
	m	D. frontalis	stim	[329]
	f	D. jeffreyi	pher	[326]
	ins	Blastophagus piniperda[a]	pher	[360]
3-hydroxy-3-methylbutan-2-one	ins	Trypodendron domesticus	sexatt aggr	[362]
	ins	T. lineatum	sexatt	[362]
	ins	T. signatus	sexatt	[362]
3-methylbutan-1-ol	m	Platypus flavicornis[a]	aggr	[347]
hexan-1-ol	m	P. flavicornis[a]	aggr	[347]
2-phenylethanol	m	Dendroctonus brevicomis	pher	[363]
	m	Ips paraconfusus	pher	[363]
3,3,7-trimethyl-2,9-dioxatricyclo-[3.3.1.$0^{4;7}$]nonane lineatin	w	Trypodendrum lineatum	aggr	[364]
2-ethyl-1,6-dioxaspiro[4.4]nonane chalcograne	ins	Pityogenes chalcographus	aggr	[365]
2-methylbut-3-en-1-ol		Ips cembrae	aggr	[848]
	m	I. typographus	aggr	[740]
1-heptanol	f	Dendroctonus jeffreyi	aggr att	[841]
2-heptanol	f	D. jeffreyi	aggr att	[841]

2.4. Sex and Aggregation Pheromones of Other Beetle Families

(+)(Z)-2-isopropenyl-1-methylcyclo-butanethanol, grandlure I, grandisol	m	Anthonomus grandis[a]	sexatt aggr disrupt	[366, 367]
		Rhabdoscelus obscurus[a]	sexatt	[368]
(Z)-3,3-dimethyl-$\Delta^{1,\beta}$-cyclohexan-ethanol, grandlure II	m	Anthonomus grandis[a]	sexatt aggr disrupt	[366, 367]
	f	Curculio caryae	sexatt	[845]
		Rhabdoscelus obscurus[a]	sexatt	[368]

Table 1 (continued)

Name, trivial name and trade mark®	Isol. from	Insect species affected	Behavior release	Ref.
(E)-3,3-dimethyl-$\Delta^{1,\beta}$-cyclohexan-ethanol, grandlure II	m	Anthonomus grandis[a]	sexatt aggr disrupt	[366]
		Rhabdoscelus obscurus[a]	sexatt	[368]
(Z)-3,3-dimethyl-$\Delta^{1,\alpha}$-cyclohexan-acetaldehyde, grandlure III	m	Anthonomus grandis[a]	sexatt aggr disrupt	[366, 367]
(E)-3,3-dimethyl-$\Delta^{1,\alpha}$-cyclohexan-acetaldehyde, grandlure IV	m	A. grandis[a]	sexatt	[366, 367]
valeric acid	f	Limonius californicus	att	[410]
methyl (–) (E)-2,4,5-tetradeca-trienoate	m	Acanthoscelides obtectus	sexatt	[369, 370]
(–)-14-methyl-(Z)-8-hexadecen-1-ol		Trogoderma granarium	sexatt	[371]
	f	T. inclusum	sexatt	[372]
14-methyl-(E)-8-hexadecen-1-ol	f	T. glabrum	sexatt	[372]
methyl (–)-14-methyl-(Z)-8-hexa-decenoate	f	T. inclusum	sexatt	[373]
methyl 14-methyl-(E)-8-hexadecen-oate	f	Trogoderma glabrum	sexatt	[372]
E-3,Z-5-tetradecadienoic acid megatomic acid		Attagenus elongatulus	sexatt	[375]
	f	A. megatoma	sexatt	[376]
	f	A. picens	sexatt	[377]
phenol	f	Costelytra zealandica	att	[378]
phenethyl propionate		Popillia japonica	att	[379]
Z-5-(1-decenyl)dihydro-2(3*H*)--furanone	f	P. japonica	sexatt	[380]
1-pentadecene	ins	Tribolium confusum	sexpher	[381]
hexadecane	ins	T. confusum	sexpher	[381]
-heptadecene	ins	T. confusum	sexpher	[381]
eptadecadiene	ins	T. confusum	sexpher	[381]
utyl E-2, E-4-hexadienoate butyl sorbate		Amphimallon majalis	att	[382]
nethyl Z-3-decenoate		Anthrenus flavipes	sexatt	[383]
-3-decenoic acid	f	A. flavipes	sexatt	[383]
ıethyl Z-7-hexadecenoate	f	Trogoderma glabrum	sexatt	[372]
thyl 3-isobutyl-2,2-dimethyl-cyclopropanecarboxylate ethyl dihydrochrysanthemoate chrislure		Oryctes rhinoceros	att	[384]
thyl 2,2-dimethyl-3-(2-methylpro-penyl)-cyclopropanecarboxylate ethyl chrysanthemoate		O. rhinoceros	att	[385]
ethyl cyclohexylpropionate		Popillia japonica	att	[386]
ropyl 1,4-benzodioxan-2-car-boxylate, amlure		Amphimallon majalis	att	[387]
exanoic acid	f	Trogoderma glabrum	sexatt	[372]
	ins	Limonius canus	sexpher	[388]

Table 1 (continued)

Name, trivial name and trade mark®	Isol. from	Insect species affected	Behavior release	Ref.
14-methyl-Z-8-hexadecenal	f	Trogoderma inclusum	sexatt	[389]
trogodermal	f	T. variabile	sexatt	[389, 390]
14-methyl-E-8-hexadecenal trogodermal	f	T. glabrum	sexatt	[372]
4-hydroxyhexanoic lactone	ins	T. glabrum	sexpher	[372]
4,6,6-trimethylbicyclo [3.1.1]hept--3-en-2-one, (−)-verbenone		Hyalotrupes bajulus	ovipos	[391]
p-cymenol		H. bajulus	ovipos	[391]
2,3-dihydro-2,3,5-trimethyl-6-(1--methyl-2-oxobutyl)-4*H*-pyran-4-one	f	Stegobium paniceum	sexpher	[392]
(±)-des-N-morphinane		Oryctes rhinoceros	att	[466]
cyclohexyl phenylacetate		Trichodes ornatus	att	[467]
2-fluorophenol		Costelytra zealandica	sexatt	[835]
4-fluorophenol		C. zealandica	sexatt	[835]
3-fluorophenol		C. zealandica	inhib	[835]
isopropyl Z-9-dodecenoate	m	Dermestes maculatus[a]	pher	[832]
isopropyl Z-9-tetradecenoate	m	D. maculatus[a]	pher	[832]
isopropyl Z-9-hexadecenoate	m	D. maculatus[a]	pher	[832]
isopropyl Z-9-octadecenoate	m	D. maculatus[a]	pher	[832]

2.5. Sex and Aggregation Pheromones of Cockroaches (Blattidae, Orthoptera)

propyl cyclohexylacetate		Blatella germanica	att	[393, 394]
(Z-1, E-5)-1,10(14)-diepoxy-4-(15)--5-germacradien-9-one Periplanone B	f	Periplaneta americana	sexatt	[62, 395, 396]
Periplanone A	f	P. americana	sexatt	[62, 395, 396]
29-hydroxy-3,11-dimethyl-2-nonacosanone		Blatella germanica	sexatt	[397, 398]
3,11-dimethyl-2-nonacosanone	f	B. germanica	sexatt	[397, 399, 400]

2.6. Alarm Substances of Ants (Formicidae, Hymenoptera)

The alarm and alerting substances of ants are chemical compounds released in the nest region in cases of disturbance, causing locomotoric activities, aggression behavior and further typical behavioral reactions of worker insects. The type of behavior release depends on distinct pheromone concentrations; these chemicals can beside alarm activity also show attractancy in low concentrations, as well as repellency in higher doses. They represent a chemical-structurally heterogenous class of compounds, often posess polyfunctional activities and were secreted by the mandibular gland, the supraanal gland, Dufour gland, hindgut or poison gland, dependent on the ant species. Ants alarm pheromones, which do not have great importance for an insect control, will be listed here according to their chemical structure and reported only with the reference to a few literature reviews [63, 64, 401–409].

Table 1 (continued)

Name, trivial name and trade mark®	Isol. from	Insect species affected	Behavior release	Ref.

alkanes: nonane, decane, undecane, dodecane, tridecane, pentadecane, heptadecane, octadecane, nonadecane, docosane, 3-methylundecane, 5-methylundecane

alkenes: undecene, dodecene, tridecene, pentadecene, heptadecadiene, Z-8-heptadecene, nonadecene, heneicosene, tricosene

saturated ketones: 4-methylhexan-3-one, heptan-2-one, 2-methylheptan-4-one, 4-methylhexan-2-one, 2-methylheptan-3-one, 2-methylcyclopentanone, Z-1-acetyl-2-methylcyclopentane, octan-3-one, 6-methyloctan-3-one, nonan-3-one, decan-3-one, undecan-3-one, tridecan-2-one, pentadecan-2-one, heptadecan-2-one, nonadecan-2-one, tetradecan-3-one, hexadecan-3-one

unsaturated ketones: 6-methyl-5-hepten-2-one, 4,6-dimethyl-E-4-octen-3-one

saturated alcohols: 4-methylheptan-3-ol, octan-3-ol, nonan-3-ol, decanol, undecanol, dodecanol, hexadecanol

unsaturated alcohols: 2,6-dimethylhept-5-en-1-ol, tetradecenol

aldehydes: 2-hexanal, 2,6-dimethylhept-5-enal, 2-butyloct-2-enal, acetaldehyde

acids: acetic acid, propionic acid, isovaleric acid, 4-hydroxyhexadec-9-enoic acid, 4-hydroxyoctadec-9-enoic acid, formic acid, 4,6-dimethylhex-E-2-enoic acid

esters: decyl acetate, undecyl acetate, dodecyl acetate, tetradecyl acetate, hexadecyl acetate, octadecyl acetate, 4-hydroxyoctadec-9-enolid, methyl 6-methylsalicylate

terpenoid compounds: citral, citronellol, geraniol, farnesene, 2,3-dihydrofarnesal, geranylgeraniol, geranylcitronellal, farnesal, perillene, farnesyl acetate, geranylgeraniol, citronellal, α-pinene, β-pinene, limonene, myrcene

heterocyclic compounds: alkylpyrazines

2.7. Trail Pheromones of Ants and Termites (Formicidae and Isoptera)

methyl 4-methylpyrrol-2-carboxylate	ins	Atta cephalotes	trail	[411]
attalure	ins	A. texana	trail	[65, 66]
12-isopropenyl-1,5,9-trimethyl-cyclotetradeca-1,5,9-triene neocembrene A (nasutene)	ins	Nasutitermes spec.	trail	[412, 413]
Z-3,Z-6,E-8-dodecatrien-1-ol		Coptotermes formosanus	trail	[414]
		Leucotermes speratus	trail	[414]
	ins	Reticulitermes flavipes	trail	[414–416]
		R. hesperus	trail	[414]
		R. santonensis	trail	[417]
		R. tibialis	trail	[414]
	m	R. virginicus	trail	[414, 415]
$C_{15}H_{26}O$ dihydroagarofurane		R. santonensis	trail	[418]
$C_{12}H_{18}O$ tricycloekasantalal		R. santonensis	trail	[418]
Z-3-hexadecen-1-ol	ins	Calotermes flavicollis	trail	[424]
	ins	Microcerotermes edentatus	trail	[424]
4-phenyl-Z-3-buten-1-ol		Coptotermes formosanus	trail	[414]
		Reticulitermes speratus	trail	[414]
Z-3,Z-6-dodecadien-1-ol		Coptotermes formosanus	trail	[414]
		Reticulitermes speratus	trail	[414]
hexanoic acid	ins	Lasius fuliginosus	trail	[422, 423]
	ins	Zootermopsis nevadensis	trail	[419–421]

Table 1 (continued)

Name, trivial name and trade mark®	Isol. from	Insect species affected	Behavior release	Ref.
heptanoic acid	ins	Lasius fuliginosus	trail	[422, 423]
octanoic acid	ins	L. fuliginosus	trail	[422, 423]
nonanoic acid	ins	L. fuliginosus	trail	[422, 423]
decanoic acid	ins	L. fuliginosus	trail	[422, 423]
methyl 4-chloropyrrol-2-carboxylate		Atta texana	trail	[425]
methyl 4-bromopyrrol-2-carboxylate		A. texana	trail	[425]
3,7,11-trimethyl-E-2,E-6,10-dodecatrien-1-ol farnesol	ins	Zootermopsis nevadensis	trail	[421]
5-methyl-3-butyloctahydroindolizine	ins	Monomorium pharaonis	trail	[426]
3,4,7,11-tetramethyltrideca-E-6,Z-10-dienal faranal	ins	M. pharaonis	trail	[427]

2.8. Sex Attractants of Flies (Diptera)

Z-9-tricosene, muscalure, muscamone®	f	Musca domestica[a]	sexatt	[428, 429]
Z-9-heneicosene	f	M. domestica[a]	sexatt	[429]
Z-9-pentacosene		Fannia canicularis	pher	[432, 831]
Z-14-nonacosene	ins	M. autumnalis	sexatt	[430, 431]
Z-13-nonacosene	ins	M. autumnalis	sexatt	[430, 431]
Z-13-heptacosene	ins	M. autumnalis	sexatt	[430, 431]
Z-11-hentriacontene		Fannia pusio F. femoralis	pher pher	[831] [831]
1,Z-7,Z-13-pentacosatriene	ins	Stomoxys calcitrans	pher	[433]
methyl E-6-nonenoate	m	Ceratitis capitata[a]	sexatt	[434, 435]
E-6-nonen-1-ol	m	C. capitata[a]	sexatt	[435, 434]
Z-6-nonen-1-ol	f	Dacus oleae[a]	sexatt	[464]
p-cymene	f	D. oleae	sexatt	[464]
nonacosane	ins	Musca autumnalis	inhib	[430]
heptacosane	ins	M. autumnalis	inhib	[430]
Z-11-octadecenyl acetate	m	Drosphila melanogaster	pher	[436]
15,19-dimethylheptatriacontane	f	Glossina morsitans	sexpher	[437]
17,21-dimethylheptatriacontane	f	G. morsitans	sexpher	[437]
15,19,23-trimethylheptatriacontane	f	G. morsitans	sexpher	[437]
methyl eugenol		Dacus zonatus	sexatt	[840]

2.9. Aphids and Scales (Homoptera)

Z-3-methyl-6-isopropenyl-3,9-decadienyl acetate	ins	Aonidiella aurantii	sexatt	[438, 439]

Table 1 (continued)

Name, trivial name and trade mark®	Isol. from	Insect species affected	Behavior release	Ref.
3-methyl-6-isopropenyl-9-decenyl acetate	ins	A. aurantii	sexatt	[438, 439]
β-farnesene	ins	Metapolophonium dirhodum	alarm	[440]
	ins	Rhopalosiphum padi	alarm	[440]
1,5-dimethyl-8-(1-methylethenyl)--1,4-cyclodecadiene germacrene A	ins	Therioaphis maculata	alarm	[441]
	ins	T. riehmi	alarm	[441]

2.10. True Bugs (Heteroptera)

Name, trivial name and trade mark®	Isol. from	Insect species affected	Behavior release	Ref.
E-2-hexanal		Cimex lectularius	alarm	[442]
E-2-octenal		C. lectularius	alarm	[442]
vanillin	m	Eurygaster integriceps	sexpher	[443, 462]
ethyl acrylate	m	E. integriceps[a]	sexpher	[462]
hexyl butyrate	ins	Lygus linolearis	pher	[463]
E-2-hexenyl butyrate	ins	L. linolearis	pher	[463]

2.11. Bees, Wasps, Bumble Bees, Sawflies (Hymenoptera)

Name, trivial name and trade mark®	Isol. from	Insect species affected	Behavior release	Ref.
isopentyl acetate	ins	Apis mellifera	alarm	[444]
2-decenyl acetate	ins	A. dorsata	alarm	[445]
	ins	A. florea	alarm	[445]
9-oxo-E-2-decenoic acid queen substance	f	A. cerana	att	[446, 447]
	f	A. dorsata	att	[446, 447]
	f	A. florea	att	[446]
	f	A. mellifera	att	[448–451]
10-hydroxy-E-2-decenoic acid	f	A. mellifera	sexatt	[448–451]
ll-trans-farnesyl hexanoate		Andrena spec.	pher	[452]
eranyl octanoate		A. spec.	pher	[452]
2,4-hexadienyl butyrate		Vespula spec.	att	[453]
etrahydro-6-undecyl-2H-pyran-2-one	f	Vespa orientalis	att	[454, 455]
−)-3,7,11-trimethyl-E-6,10-dode-cadien-1-ol, terrestrol	m	Bombus terrestris	sexpher	[456–458]
,7-dimethylpentadecan-2-ol	ins	Neodiprion sertifer	sexpher	[459]
,7-dimethylpentadec-2-yl acetate	m	N. lecontei	sexpher	[459]
		N. pinetum	sexatt	[844]
,7-dimethylpentadec-2-yl propionate		N. lecontei	sexpher	[459]
-2-methyl-5-hydroxyhexanoic lactone	m	Xylocopa hirutissima	sexatt	[460]
nalool	ins	Colletes thoracicus[a]	aggr	[461]
	ins	C. inequalis[a]	aggr	[461]
	ins	C. validus[a]	aggr	[461]
ral	ins	C. thoracicus[a]	aggr	[461]
	ins	C. inequalis[a]	aggr	[461]
	ins	C. validus[a]	aggr	[461]

Table 1 (continued)

Name, trivial name and trade mark®	Isol. from	Insect species affected	Behavior release	Ref.
geranial	ins	C. thoracicus[a]	aggr	[461]
	ins	C. inequalis[a]	aggr	[461]
	ins	C. validus[a]	aggr	[461]
methyl-1,6-dioxaspiro-4,5-decane	ins	Paravespula vulgaris	pher	[465]
geranyl acetate	f	Sceliphron caementarium[a]	aggr	[828]
2-decen-1-ol	f	Sceliphron caementarium[a]	aggr	[828]
Z-citral	ins	Apis mellifera[a]	pher	[838]
E-citral	ins	A. mellifera[a]	pher	[838]
nerol	ins	A. mellifera[a]	pher	[838]
geraniol	ins	A. mellifera[a]	pher	[838]
nerolic acid	ins	A. mellifera[a]	pher	[838]
geranic acid	ins	A. mellifera[a]	pher	[838]
E,E-farnesol	ins	A. mellifera[a]	pher	[838]

Cockroach (Blattidae) *sex pheromones* control enticement and mating behavior of the males [55, 56], and a successful mating is often dependent on the reception of the pheromones from the male tergal gland secretion (seducin) [57–59]. The chemical structures, or partial structures, of the pheromones of the American cockroach have been elucidated only recently [60–62].

Among the ant *alarm pheromones*, which may be produced in mandibular glands, supraanal glands, hindgut, Dufour glands or poison glands are found aliphatic compounds in the form of hydrocarbons, terpenoid aldehydes and alcohols, and above all a large number of ketones, as well as two sulphur containing compounds (Table 1). They cause aggressive behavior, alerting or alarm reactions, but may have attractive or repellant effects in other concentrations [63, 64].

Trail substances are used by many ant species to mark the way from food source to nest. One such, 4-methyl-pyrrol-2-carboxylic acid methylester, has been isolated in 600pg quantities ($\triangleq 6 \cdot 10^{-10}$ g) from the poison gland of the myrmicine *Atta texana* [65, 66]. Trail substances identified for termites include a diterpene hydrocarbon, sesquiterpenes, a dodecatrienol and hexanoic acid (cf. literature cited in Table 1).

Chemical signals for honey bees come from secretions from the mandibular glands, the Nassanoff glands and the lining of the sting pouch. They perform multiple functions, depending upon the situation in which they are released or received. They include among others the queen substance, 9-oxo-(E)-2-decenoic acid, 10-hydroxy-(E)-2-decenoic acid, 2-heptanone, isopentyl acetate, geraniol, geranic acid, citral and geranial (Table 1).

Table 1 presents the chemical names according to nomenclature, trivial names and trade marks as well as the natural occurence and the insect species, from which the pheromone has been isolated. Furthermore the kind of behavioral release and literature references are indicated.

3. Physiological and Behavioral Activity of Pheromones

3.1. Electrophysiology

Investigations of the physiological activity of pheromones were particularly carried out with butterflies and moths. After being secreted by the females, the sex attractant molecules attain, by convection, the male antennae which are closely occupied with sensory hairs, thus representing an air filter system. These hairs, containing the dendrits of the sensory cells, have pores into which the pheromone molecules may diffuse (Fig. 1). They arrive at the pore vessels from which tubuli lead directly to the membrane of the dendrit.

Biological membranes represent an electrical condensor and at the same time a variable resistor. Usually they possess a so-called resting potential. Impinging on the dendritic membrane, the pheromone molecules open, in a hitherto unknown mechanism, ionic channels, into which a few thousands of ions can penetrate for a short period and change the membrane potential. This change of potential, called the slow receptor potential, extends electrotonically over the membrane surface and is converted into signals (nerve impulses). The generation of these signals is due to a periodical change of the potential; they are transmitted to the central nervous system to produce a behavioral release.

The slow receptor potentials as well as the elementary potentials can be measured by means of microcapillary electrodes (single – cell recording).

In our studies on the relationships between the molecular structure of pheromones and biological activity we used the electroantennogram technique (EAG technique) developed by D. Schneider [41] and schematically shown in Fig. 2. The first microelectrode (c) is implanted into the base of the antenna stem and the second one either into the tip of the latter or into an antennal branch. Then an

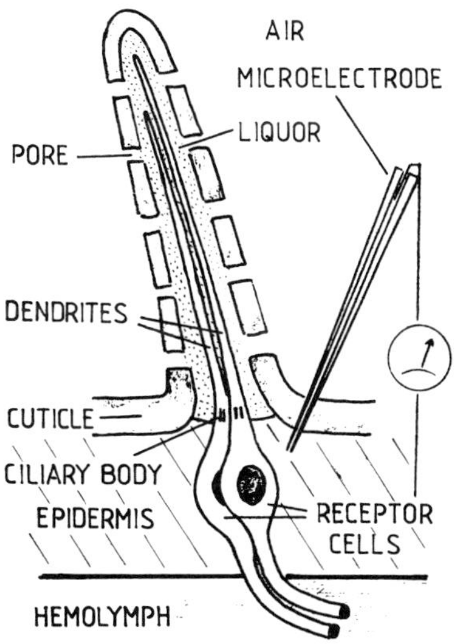

Fig. 1. Schematic longitudinal section of an olfactory hair, according to D. Schneider [852]

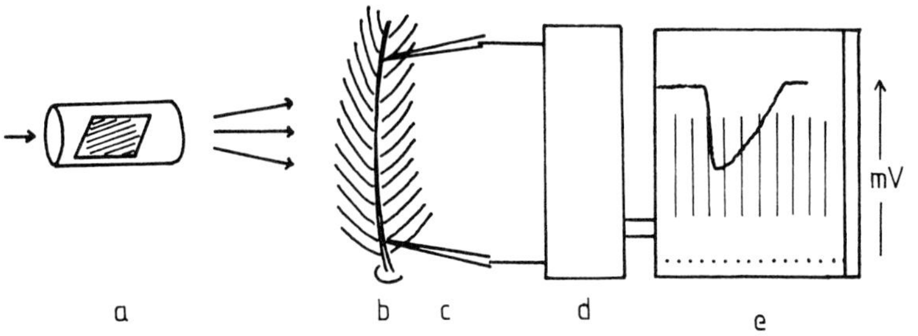

Fig. 2a–e. Scheme for measuring an electroantennogram (EAG). **a** Stimulus source, **b** antenna, **c** electrodes, **d** amplifier, **e** recorder

air stream of defined intensity and duration is blown over a stimulus source (a), containing a filter paper soaked with a certain amount of a test substance (10^{-3} to 10 μg of pheromone or structural variation).
This stream of air carries the desorbed molecules to the antenna; from there they penetrate into the dendritic membrane as described in Fig. 1 and open the ionic channels. Using the arrangement of instruments (d, e) shown in Fig. 2, the sum of many slow receptor potentials is registered ("summated response").

3.2. Structure-Activity Relationships

Using the EAG technique, the physiological activities of various compounds can be compared due to the different amplitudes of their sum potentials released. These investigations led to a series of structure-activity rules for *Lepidoptera* [42–45], which are highly important for the understanding of the mechanism of pheromone reception, the structure elucidation of these signal substances (cf. Chap. 4.3, p. 69), and the synthesis of new chemicals with pheromone activity.

Both elongation and shortening of the carbon chain of a naturally occurring pheromone of the general formula *I*

$$CH_3{-}(CH_2)_n{-}CH{=}CH{-}(CH_2)_m{-}O{-}COCH_3$$

I

by one methylene group results in a 3 to 30 fold decrease of EAG activity. Alteration of the chain length by two methylene groups causes a loss of activity of a factor 10 to 100. With even greater chain length variations of molecules, elongation produces a more drastic decrease in activity than shortening.

Structural variations in the molecular section – $(CH_2)_n$ – give rise to a more pronounced loss of activity in the EAG test than the same variation of the – $(CH_2)_m$ – molecular segment. If, for instance, the carbon chain of a pheromone is elongated by two methylene groups, to elicit maximum EAG response, the double bond must be shifted by two positions such that, with respect to the original – $(CH_2)_n$ – structure, e.g. the terminal alkyl moiety becomes identical with that of the original signal substance.

Altering the geometry of the double bond of the olefinic pheromone *I*, the opposite isomer always gives rise to an approximately 3 fold decrease of physiological activity. The dependence of the EAG amplitude on the functional end group of *I* results in another structure-activity relationship.

EAG studies with noctuid and tortricid moths using alkyl branched pheromone analogs [44] gave the following results:

The receptors on the antennae of various moth species, maximally responding to pheromones of type *I*, can be classified according to the number n of the methylene groups of the terminal alkyl moiety of their speciesown semiochemical. Receptors for stimulus molecules with equal number n and different m respond to identical alterations (e.g. branching) in the – $(CH_2)_n$ – section almost with the same decrease of activity. Variation of the m-segment can induce qualitative gradations. This is valid for receptors maximally responding to pheromones *I* with n = 1, 3 or n = 5.

Elongating the n part of a pheromone molecule with n = 1 (compound *II*) for two methylene groups (*III*) results in a decrease of biological activity by a factor of 30 to 300. Introducing a methyl group into the *1*-position with respect to the double bond in the elongated molecule (compound *IV*) leads to an increase of activity by a factor of 3 compared to the unbranched compound *III* (cf. Table 2, Part A).

The activating influence of a methyl group in the α-position within the n segment of elongated pheromone analogs is the more pronounced in the comparison of the activities of compounds *V* and *VI* (Table 2, Part A). Offering receptors, maximally responding to (Z)-9-dodecenyl acetate (*II*, m=8), molecule *V* – elongated by four CH_2 groups – leads to a drop of the EAG activity by a factor of 300 to 1000. On the other hand, the α-methyl branched pheromone-variation product *VI* shows a decrease of activity by a factor of only 3, compared to the speciesown pheromone *II* (m = 8).

From these results can be concluded, that the receptor recognizes a defined conformation of the n part in a rather flexible way. If the chain length of this molecular section exceeds a certain value, no correct fitting in the receptor region can result. The "hanging over" part of the stimulus molecule could interact with other regions of the membrane, thus blocking and preventing the opening of ionic channels. Molecules *IV* and *VI*, however, assume a conformation which enables them to fit the methyl group dynamically almost in the receptor region. This is requested for the same group of the speciesown pheromone *II*. In this case, the longer chain end will occupy other membrane areas where the interactions do not markedly effect the change of potential [493].

Similar results could be obtained in investigations with moths maximally responding to pheromone molecules *III* (identical with *I*, n = 3). Introducing a methyl group in the α-position relative to the double bond lowers the activity of the molecule by a factor of 10 to 30. Replacing the methyl group by an ethyl group (*VII*) has only little influence on activity. For m = 6 and 10 only small changes of activity are found in further elongation of the side chain by a CH_2 group (*VIII*) whereas with m = 8 an increase of activity by a factor 10 is observed for the propyl group, in contrast to compounds *IV* and *VII* (Table 2, Part B).

If the propyl group, which is essential for an exact fitting in the corresponding receptor region, is retained in these molecules, only small activity loss can be

Table 2. Relative electroantennogram efficiencies of alkyl branched pheromone analogs of lepidoptera. Activity of pheromone ≙ 100. (The comparisons of activities are related to the stimulus source loadings, giving the same response potentials as the pheromone of the species)

Compound		Relative activity		
Part A: Measured on receptor types, maximally responding to compounds *I* with $n = 1$				
			$m = 8$	$m = 10$
II	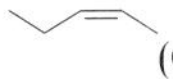 $(CH_2)_m$—O—$COCH_3$	—	100	100
III	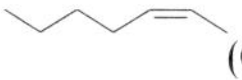 $(CH_2)_m$—O—$COCH_3$	—	3	0.3
IV	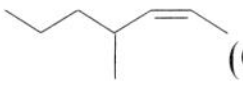 $(CH_2)_m$—O—$COCH_3$	—	10	1
V	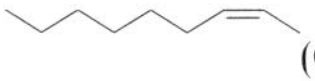 $(CH_2)_m$—O—$COCH_3$	—	0.1	—
VI	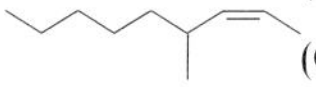 $(CH_2)_m$—O—$COCH_3$	—	30	—
Part B: Measured on receptor types, maximally responding to compounds *I* with $n = 3$				
		$m = 6$	$m = 8$	$m = 10$
III	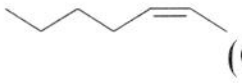 $(CH_2)_m$—O—$COCH_3$	100	100	100
IV	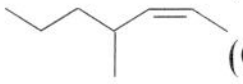 $(CH_2)_m$—O—$COCH_3$	3	3	3
VII	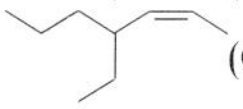 $(CH_2)_m$—O—$COCH_3$	3	3	3
VIII	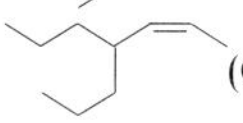 $(CH_2)_m$—O—$COCH_3$	3	30	3
IX	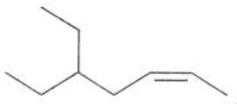 $(CH_2)_m$—O—$COCH_3$	3	0.3	1
X	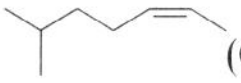 $(CH_2)_m$—O—$COCH_3$	10	10	10

observed when introducing an α-positioned side chain, independent of the size of the latter.

Of interest is the fact, that symmetrical branching in α-, β-, and γ-position relative to the double bond in *III* causes different effects. The ethyl group in the β-position only is responsible for the greatest loss of activity (compound *IX*) followed by α-propyl branching (*VIII*). γ-Methyl branching results (compound *X*) in a decrease of activity by a factor 10, as compared to *III*. Molecular models reveal that the ethyl group in the β-position has a more pronounced effect on the formation of distinct conformations than a propyl group in the α- or a methyl group in the γ-position.

All these structure-activity investigations indicate that the release of stimulus at the membrane by interactions between molecule and receptor region is neither based on a rigid "key-lock principle" nor on a "three-point interaction". We rather assume that the pheromone molecule is flexibly inserted into the receptor region as schematically shown in Fig. 3.

It can be assumed that the receptor region contains some elements of chirality and that the insertion of the multiple prochiral pheromone molecule with its individual

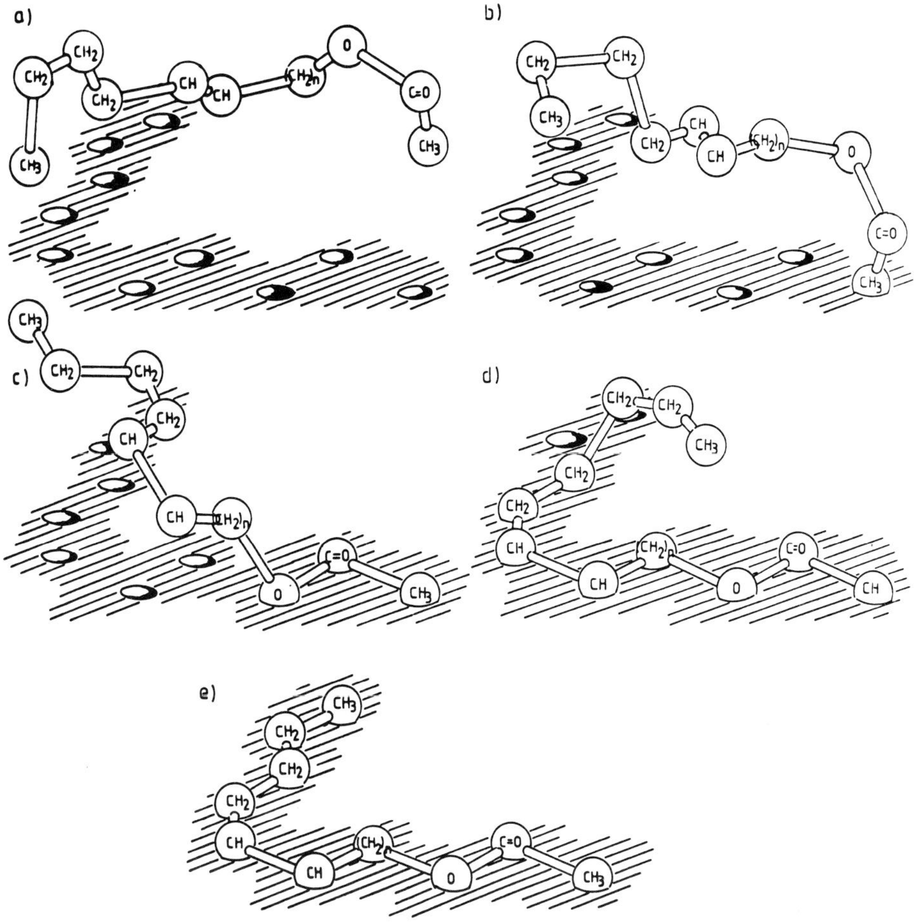

Fig. 3a–e. Schematic presentation of the flexible insertion of a pheromone molecule into the receptor region. **a** Approach of the molecule; **b** first contact between pheromone and receptor; **c** and **d** dynamic insertion into the receptor; **e** final phase; synchronously occurring changes of the receptor region are not considered

sections (e.g. A or B in *I*, Fig. 4) can occur only from one of the two enantiotopic sides of each center of prochirality. This implies that the signal molecule has to be embedded into a definite conformation. During this process conformational changes can also be expected within the "receiver part". If this concept is true receptors, maximally responding to achiral pheromone molecules *I*, should distinguish between enantiomers of chiral pheromone analogs *XI* because of the occurrence of additional diastereoisomeric interactions, and because only one of both enantiomers can adopt the conformation of section A of the signal substance required by the receptor region.

To verify this postulate we synthesized the following optically active pheromone analogs (Fig. 4) [648] and compared their EAG activities with those of the corresponding unbranched achiral pheromones.

$$CH_3{-}(CH_2)_n{-}CH{=}CH{-}(CH_2)_m{-}O{-}COCH_3 \qquad I$$
$$\longleftarrow A \longrightarrow \qquad \longleftarrow B \longrightarrow$$

$$CH_3{-}(CH_2)_{n-1}{-}\overset{*}{C}H(R){-}CH{=}CH{-}(CH_2)_m{-}O{-}COCH_3 \qquad XI \quad R = \text{alkyl}$$

$$CH_3{-}\overset{*}{C}H(C_2H_5){-}CH{=}CH{-}(CH_2)_{10}{-}O{-}COCH_3 \qquad XII$$

$$CH_3{-}\overset{*}{C}H(C_2H_5){-}CH{=}CH{-}CH{=}CH{-}(CH_2)_8{-}CH_2OH \qquad XIII$$

$$CH_3{-}\overset{*}{C}H(C_2H_5){-}CH{=}CH{-}CH{=}CH{-}(CH_2)_8{-}CHO \qquad XIV$$

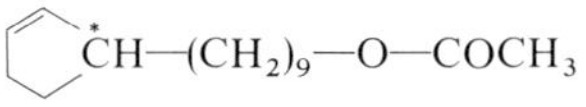

XV[a]

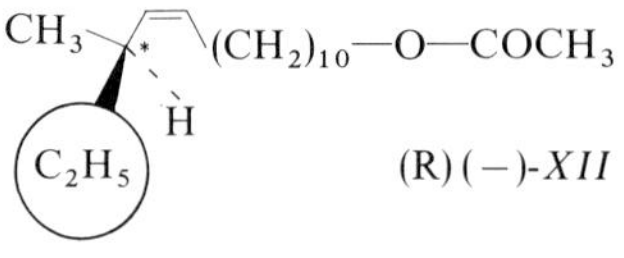

(R)(−)-*XII*

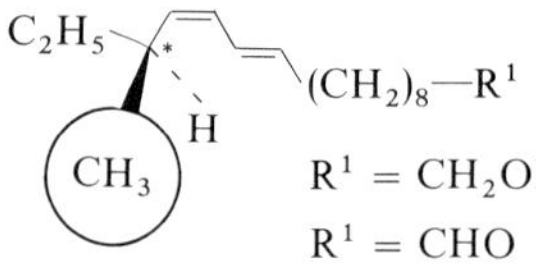

$R^1 = CH_2OH$ *XIII*

$R^1 = CHO$ *XIV*

(S)(+)-*XIII* or *XIV*

[a]Compound *XV* was given to our disposal by O. L. Chapman (USA)

Fig. 4. Optically active lepidoptera pheromone analogs [648]

In contrast to molecule *XII*, molecule *XV* does not exhibit free conformational adjustment of its section A, because the conformation is determined by the ring structure. Furthermore, the center of chirality in *XV* is in the B part (Fig. 4).

The relative EAG activities listed in Table 3 reveal that the receptors can distinguish between the enantiomers in all cases investigated. In Ostrinia nubilalis and Argyrotaenia velutinana, which both use (Z)-11-tetradecenyl acetate (*II*, m = 10) as their pheromone molecule but belong to different moth families, the (R)-enantiomer of *XII* is more active than the (S)-isomer by a factor of 100. Also, the enantiomers of the cyclic pheromone analog *XV* are clearly distinguished by the receptors (by a factor of 10 to 30). The activity of the racemic mixture in all cases is intermediate between those of the chiral antipodes.

Furthermore, Table 3 shows that for both moth species the differences in activity between (R)-*XII*, containing a flexible terminal alkyl moiety, and the rigid (S)-*XV* are of the order of 30 and 100 fold activity, respectively. Compared with pheromone *II*, only a 10 fold loss of activity has been reported with (R)-*XII*, whereas with (S)-*XV*, this loss is 300 to 1000 fold. This reveals the importance of conformational mobility.

Table 3. Relative EAG activities of chiral pheromone analogs [648]. Activity of the naturally occurring pheromone ≙ 100

Compound no.	Moth species		Compound no.	Moth species
	O. nubilalis	A. velutinana		Bombyx mori
II	100	100	*XVI*	100
(*R*)(−)−*XII*	10	10	(*S*)(+)−*XIII*	10
(*R*, *S*) −*XII*	1	1	(*R*, *S*) −*XIII*	3–10
(*S*)(+)−*XII*	0.1	0.1	(*R*)(−)−*XIII*	3
(*S*)(−)−*XV*	0.3	0.1	*XVII*	100
(*R*, *S*) −*XV*	0.03	0.03	(*S*)(+)−*XIV*	10
(*R*)(+)−*XV*	0.01	0.01	(*R*, *S*) −*XIV*	3–10
			(*R*)(−)−*XIV*	3

With Bombyx mori, which utilizes (E)-10,(Z)-12-hexadecadien-1-ol (bombykol, *XVI*) and (E)-10,(Z)-12-hexadecadienal (bombykal, *XVII*) as pheromones, also a discrimination of the (S)- and (R)-enantiomer could be observed both with *XIII* and *XIV*. In both cases, the (S)-enantiomer was about three times more active than the (R)-form, the activity of the racemate lying in between.

All these results support the assumptions and ideas on the dynamics of pheromone molecule interactions and chiral receptor regions. We furthermore assume that to the receptor region, involved in the interactions with the signal molecule, also belong proteins as well as lipid regions. Thus, we use the term "receptor region" instead of the usual termini "receptor" or "receptor molecule". Hence, structural variations in different sections of pheromone molecules can result in different and varying amplitudes of the receptor and elementary potentials.

3.3. Multicomponent Pheromone Complexes

The intersexual transmission of information in insects is generally produced by multicomponent pheromone complexes consisting of several stimulus molecules [67–70]. Each of these molecules stimulates a specific receptor cell on the antennae of the "receiver". Thus, in the males of the silkworm moth *Bombyx mori* there exist receptors only responding to (E)-10,(Z)-12-hexadecadien-1-ol (bombykol), and others only stimulated by the corresponding aldehyde (bombykal) [537]. This means that in this case the interaction of the polar functional group with a group-specific "receptor region" is almost essential for an answer of the nerve cell. The effectiveness of the pheromone, however, can also definitely be determined from other sections of the molecule.

The chemical communication system of pheromones is therefore comparable with other information systems which use different symbols or signs for the transmission of information (e.g. letters in languages, point and dash in the morse code, the signs zero and one in the computer language, or three basic colors in color television). The information system itself represents an analog system.

There are often only slight differences in the chemical structure of the components of a pheromone complex. Thus, some moth species make use of a mixture of the (Z)- and (E)-isomers of an olefinic acetate while others utilize an olefinic alcohol and the corresponding aldehyde.

The few investigated cases show that for each molecular species of a pheromone complex exist specific receptor cells on the male antennae (Fig. 5). Each receptor cell is connected with the central nervous system by a separate nerve fibre which transmits nerve impulses when the cell is stimulated by a pheromone compound. The rate of impulses in each fibre depends on the amount of stimulating molecules. Therefore, a certain mixture of pheromone components produces a characteristic ratio of impulses among the different receptor cell types (impulse modulation). Only the impulse pattern induced by the species-specific pheromone complex leads to a release of a behavioral response. If the proportions of the pheromone components are varied, the impulse pattern of the receptor types also will be changed. This results in a "false" information, causing either a wrong behavior mode or a disruption of information transmission (Fig. 5).

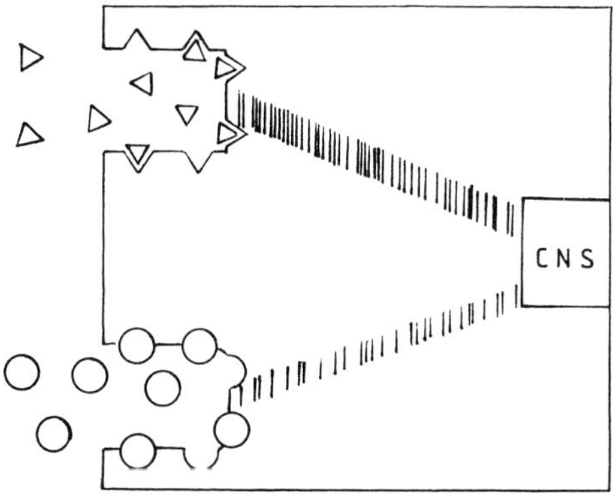

Fig. 5. Schematic description of the mechanism of the reaction of two different molecules of a pheromone complex with their specific receptor cells, and of the signal transmission to the central nervous system

Changing the proportions of the single components of a pheromone complex thus yields numerous possibilities of variations for chemical information. The effectivity and species specificity of a pheromone complex are governed by the ratio of the quantitative composition of its components. A variation of this ratio allows exchange of species-specific information using two molecules, even if related species intimately co-exist. This is demonstrated by the pheromone of the tortricids *Adoxophyes fasciata* and *A. orana*, identified by Tamaki [71, 72], Ritter [73], and Meijer [74]. The two moth species produce (Z)-9-tetradecenyl acetate as the main component of the pheromone and (Z)-11-tetradecenyl acetate as the minor component in the ratios 1.8 : 1 and 3.4 : 1, respectively. In this case, the individual compounds reveal no response to both species in behavior tests, and only the correct ratio of compounds for the individual species results in full attractivity of the natural pheromone. Multicomponent pheromones may also release the individual sections of a behavioral pattern in their correct order. It appears that for the noctuid moth *Prodenia eridiana* (Z)-9,(E)-12-tetradecadienyl acetate provides the remote orientation stimulus whereas the second component of the pheromone complex, (Z)-9-tetradecenyl acetate, causes excitation of the males [75, 76].

The change of the composition of the complex can also result in a wrong information and in a variation of the behavior released. Thus, the transition from an (*E*)- to a (*Z*)-olefin in the electrophysiological test (EAG) only causes a loss of activity by a factor of 3 [43]. It may, however, entail a release of inverse behavior [119].

Heliothis virescens females produce (Z)-11-hexadecenal as an attractive component of their sex pheromone [266, 267], *H. zea* females the same compound as an inhibitor [266, 274]. Exchanging the last CH_2 group by an oxygen atom (i.e. structural variation from the aldehyde to the formate of the same chain length CH_3-$(CH_2)_3$-CH=CH-$(CH_2)_8$-O-CHO) yields a substance with almost the same EAG activity as the pheromone of the two species [43, 853]. This compound, first assumed to be a sex attractant [853], later turned out to be a disruptant [92] which is successfully used in field tests.

3.4. Olfactometer Experiments and Field Tests

Behavior controlling chemicals like sex attractants can be tested in field experiments as well as in laboratory in a suitable olfactometer apparatus, thus providing qualitative and quantitative information about the biological activity of a sample. These olfactometers [488] are scent analyzers and usually consist of several chambers or tubes, which can be loaded with test substances, being connected with empty chambers or tubes. The test insects may be required to make a "yes or no" decision by allowing them to have access to individual chambers in which the compounds are released; thus, the form of the behavior release (e.g. attractive, inhibiting or repellent effect, escaping movements, locomotive activity, wing flutter reaction, etc.) may be observed. Similar results can be obtained from a Y-tube olfactometer in which the test animals are released in a Y-shaped glass tube and, moving against an air stream, can choose between the arm leading to the scent source and that leading away from it.

The single-tube olfactometer [489] has led to special technical developments in olfactometry. Test insects, e.g. butterflies, may be released in a number of parallel tubes under the same wind conditions in a climate- and light/dark-controlled room, and their behavior observed with respect to the experimental scent sources. With this arrangement the effects of different quantities of scent, mixtures and of individual components on the activity can be observed [490]. Using a time lapse television camera, the experiments can be monitored by recording on a magnetic tape, and evaluated with a video recorder.

The success of field trials, which are most efficient in the determination of the biological activity of test chemicals, is affected by many parameters such as age, maturity, sex, and vitality of the test insects, temperature, humidity, time of day and season, circadian rhythm and the light-dark cycle of the animals, brightness and air movement in the test area, concentration of the samples, impurities in the air and by the way in which the test substances are exposed. Field tests can therefore actually only be carried out by entomologists, require exact compliance with certain test conditions and are often difficult to reproduce.

4. Isolation and Structure Elucidation of Pheromones

4.1. General

For the structure elucidation of (E)-10,(Z)-12-hexadecadien-1-ol (bombykol), the sex attractant of the silkworm moth *Bombyx mori*, Butenandt et al. [255] managed to isolate 6 mg of an active compound from the abdomina extract of more than 500,000 female moths. Although this material and the work involved are no longer representative because of modern pheromone analytical techniques, the studies illustrate the main difficulty of determining pheromone structures, i.e. the detection of minute quantities of biologically active materials among a large excess of body material or impurities. The concentration of the sex attractants in female moths is of the order of a few ng per insect ($1\ ng = 10^{-9}\ g$); the concentration of a trail substance in an Australian termite has been estimated as 1:4,000,000 [412]. This substance-to-impurity ratio, which governs the necessity of performing separation steps, presents extraordinary difficulties, despite the sensitivity of modern analytical techniques such as liquid chromatography, gas chromatography and mass spectrometry. These difficulties have, however, led to the development of many new analytical techniques. These have effectively been utilized in the entire field of trace analysis. The analytical problems in pheromone chemistry are, however, not confined to the isolation and structure determination of natural pheromones or of behavior controlling substances but also cover the analysis of synthetic materials. The determination of impurities, residues, chemical degradation products, vaporization rates as well as the constant analytical control of periphery material such as insect traps, glue, etc. are necessary in order to avoid prejudicing an estimate of the effectiveness of the active compound.

In this chapter the details of special analytical techniques will not be discussed, but several examples which illustrate individual steps in the analysis are used, indicating the range and limits of the methods. A knowledge of general analytical practice is assumed.

The isolation and structure determination of a pheromone [470] (Fig. 6), such as the sex attractant from butterflies, can be divided into isolation and prepurification steps, such as extraction, filtration and lipid precipitation, and the actual fractionation operations such as gel permeation chromatography (GPC), thin layer chromatography (TLC), column chromatography, preparative gas chromatography (prep. GC) and high performance liquid chromatography (HPLC). Then comes the actual identification stage and the structure proof, in which gas chromatography (GC-FID), reaction gas chromatography, high-resolution gas chromatography using glass capillary columns ["$(GC)^2$"], gas chromatography coupled with mass spectrometry (GC-MS), mass chromatography (MC), mass fragmentography (GC-MID, SID, MF) and gas chromatography coupled to an electroantennogram detector (GC-EAD) may be used. The usual techniques of mass spectrometry, n.m.r., IR and UV spectroscopy and "wet" chemical analysis are also used in pheromone structure determination. In order to decide between biologically active and inactive fractions the above mentioned

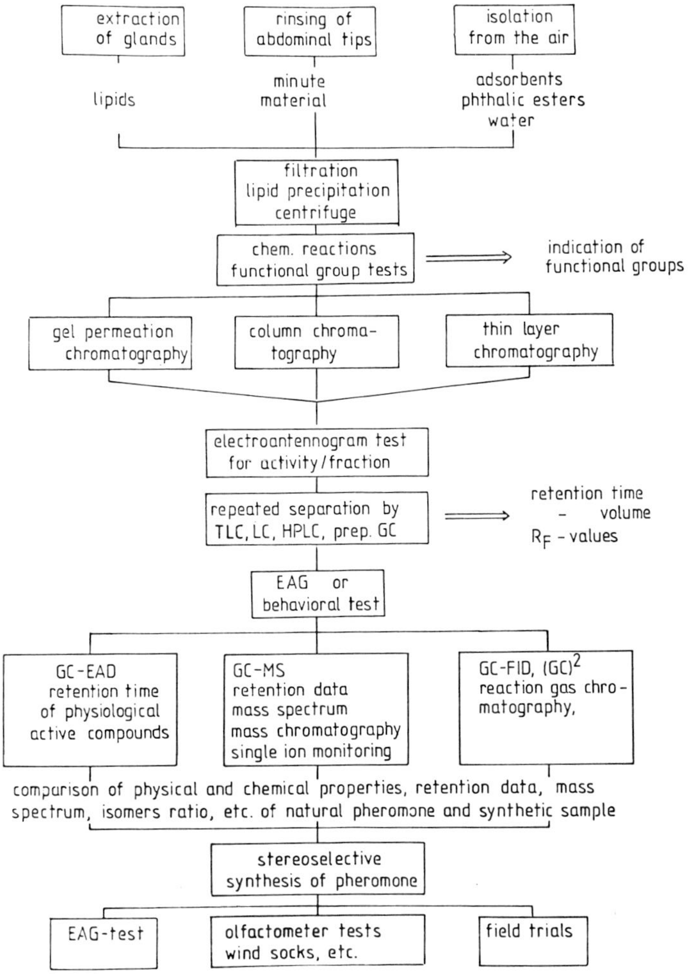

Fig. 6. General scheme for the isolation of moths sex pheromones [470]

EAG-test is employed as a biological detector between the individual separation steps to select the active fractions only for further purification.

The final proof for a pheromone structure is the subsequent synthesis and comparison of the physical properties, retention data, molecular spectra, isomer ratio and chemical reactivity of the synthetic pheromone with those of its natural counterpart, as well as determination of its effectivity in laboratory and field trials. Further information on the identification of behavior controlling agents is given in review articles by Jacobson [18], Shorey [471], Beroza [35, 472], Silverstein [473], Bedoukian [474] and Ställberg-Stenhagen [475]. The various analytical methods are covered in the specialist literature and the references cited therein [476–487].

4.2. The Bioassay

Biological tests for behavior controlling chemicals rely mostly on the observation of stereotyped behavioral responses of the insects to specific substances, e.g. pheromones or attractants, or on the measurement of the amplitude of receptor potentials during sensory stimulation (electrophysiological measurements). These tests serve to determine the activity of the pheromone extract and to distinguish between active and inactive fractions of separation in the structure determination.

Field tests are affected by many experimental details and conditions, can only be carried out by entomologists, are very difficult to reproduce, and thus not often are used as the biological test during pheromone identification (cf. Chap. 3.4, p. 65). The analysis of insect pheromones requires, however, a test which can easily be carried out in the laboratory, which the analyzing chemist can use whenever necessary, and which gives quantitative and qualitative informations of a sample. These conditions are fulfilled by scent analyzers like chamber or tube olfactometers [488]. (For the different types of olfactometers see p. 65.)

Another test for a pheromone like a sex attractant of female butterflies or moths relies on the wing fluttering reaction, a typical behavioral answer of male moths to a female pheromone by wing vibration and anemotactic orientation reactions, is particularly easy to be carried out. The classical form of this test, as it was used to identify bombykol [255], consists of a glass rod which is dipped into the test solution and then held close to the male antennae, the scent organs of the moth. A positive result is taken as the onset of the flutter reaction in at least 50% of the test insects.

The most commonly used technique for determining pheromone activity in test solutions is the electrophysiological measurement of the slow cell potentials of the antenna sensilla during excitement by scents [41, 491], as it was already discussed in Chap. 3.1. By means of microelectrodes the excitation potentials of different test solutions and fractions are amplified, recorded, and the form and size of the amplitudes compared. In this way many excitants can be tested and the activity of a series of test substances graded using a single insect, especially because the response amplitude is largely independent on the environmental factors which influence behavior tests [492].

A further advantage of EAGs, apart from the simplicity and economy in the use of test insects, is the fact that each compound of a multicomponent pheromone complex (cf. p. 63) causes a reaction, corresponding to its receptory activity. In a behavioral test, however, the presence of minor components, which may be synergists, inhibitors or essential pheromone ingredients, can considerably heighten or reduce activity of sex attractants and therefore decrease the reliability of the results of behavior tests. Often only the exact isomer ratio is responsible for the full attractant activity, whereas the individual components are fully inactive [71–74, 107, 172]. Furthermore often one percent of the opposite geometrical isomer of the pheromone seriously can reduce its attractancy [205]. This in fact means, that more or less every separating operation and fractionation of pheromone complexes and controlling by means of behavioral biotests is inadmissible, because a minor component, necessary for full efficacy could have been lost.

4.3. Pheromone Screening

The beginning of a pheromone isolation and structure elucidation usually commences with a "pheromone screening". For the identification of female sex attractants, traps, with a great number of synthetic chemicals and their mixtures, are mounted outside or in open wind socks in the infestation areas of the corresponding moth, and the capture rate of the traps is compared with untreated control traps. This method does not lead conclusively to the discovery of the pheromone structure of the insect species investigated, but in many cases yields the knowledge of attractive substances, which can be used the same as the naturally occurring pheromones for the control of butterflies and moths.

A pheromone screening by means of the EAG technique in many cases gives already concrete hints for the molecular structure of the requested compound. Thus systematical synthetic variations of the structure of known pheromones result in defined reductions of efficacy of the corresponding EAG-amplitudes compared with the amplitude of the speciesown pheromone or the most effective compound.

Knowing these structure-activity relationships, as they were already discussed in Chap. 3.2 on p. 58, the corresponding "response spectrum" of a lepidopterous species over a number of test compounds already gives indications with reference to the chain length, the number, position and geometry of double bonds and the functional group of the sex attractant. The identity of two response spectra of two different species consequently leads to the identity of the sex attractants of both insects.

4.4. Pheromone Sources

For the isolation of unknown pheromone systems it seems useful to separate the insect species according to sexes, to extract different separate parts of the insect body and to examine the extracts by means of bioassays. Although in a few cases for the isolation of sex attractants produced from one sex only also active extracts have been produced without separation of sexes [215, 218], the extraction of separate parts of the body is recommended to find out the site of production of the pheromone.

The sex attractants of female butterflies and moths (*lepidoptera*) are produced in glands, which are differentiations of intersegmental membranes. In the family of psychidae areas of gland cells are found in segments of the meso- and metathorax running into particular channels (for a morphological review of pheromone glands of lepidoptera cf. Barth [494]). The sites of pheromone producing glands of male moths exhibit a great morphological variety in contrast to the females pheromone glands: they are reported for different segments of the front- or the rear wing and the three pairs of legs, for the first to the ninth abdominal segment, the genital appendix and the meso- and metathorax [494], and are mostly provided with auxiliary apparatus like hair brushes and hair pencils. Thus often abdominal scent organs consisting of pairy hair pencils and/or hair brushes are found in the family of the noctuid moths. In the males of the pyralid family these are for example wing glands, in the danaids the male pheromones were produced in protrudible pairy brush organs of the abdomen end.

The sex and aggregational pheromones of the bark beetles (*Scolytidae*) are formed in appendix glands of the hindgut by metabolisation of host material, and the emission of pheromone is linked to the reception of food. In the family of the curculionid beetles, pheromones have been isolated from the feces [495], female elaterid beetles produce them in abdominal glands, dermestid females in intersegmental glands of abdominal sternits. From female *Costelytra zealandica (Scarabaeidae)* phenol was isolated from the abdomen as the main component of the sex pheromone [378]. The aphrodisiac odor secret of the male melid beetle *Cerocoma schäfferi*, known because of its scent fanning, should be mentioned, which is produced in glands of the mandible and the third antennal link, and is fanned with the hairy front legs during mating towards the female [496–498]. The main location of cockroach pheromone production seems to be the digestive tract [62, 499].

The alarm pheromones of ants are generated from the mandibular glands, the supraanal gland, and the hindgut; the trail substances are found in the hindgut, the poison gland, the Dufour, the Pavan and tibial glands. The trail substances of termites were found in sternal glands localized in the fifth abdominal segment. The chemical signal substances of the honey bee were isolated from mandibular, Nassanoff gland and from the lining of the sting pouch.

Apart from the direct isolation of a pheromone from the corresponding sections of the insect body, other pheromone sources can be noted. Tumlinson et al. isolated the pheromones of the boll weevil [366, 495], Persoons the pheromones of roaches [62] and Carlson the sex pheromones of flies [428] from the feces of the insects, pheromones of bark beetles were obtained from frass and boring dust of the beetles [313, 331, 346, 500]. The trail substances of ants were isolated as well from parts of the insects [411] or whole animals [411], those of termites from a fungus infected wood [415], a copulation stimulant was found in oak leaves [501], in the absence of which *Antheraea polyphemus (Lepidoptera)* is not able to copulate. A larval attractant and an oviposition stimulant was found in the peel of apples [502], methyl eugenol, which is used as synthetic attractant for male orient fruit flies, is found in the blossoms of the golden shower *Cassia fistula* [503].

Of crucial importance concerning the yields of the primary isolation of a pheromone is the physiological state and the state of development of the animals. For many insects the proper time of optimum pheromone production is three to four days after eclosion, whereas Steinbrecht established the highest pheromone content in the silkworm moth *Bombyx mori* immediately after eclosion [504]. A dependence of pheromone production from the time of the day could also be observed [505], and it should be of advantage to extract the insects to a distinct point of time during their light-dark rhythm. Thus the extract of female beet army worm abdomina, prepared shortly before the end of dark period were more active than those, prepared three hours after beginning of light phase [506]. Because the pheromone production will be finished or diminished after copulation usually virgin and uncopulated insects are used for the isolation.

4.5. Isolation of Pheromones

As it can be seen also from Fig. 6, for isolation of pheromones generally exist four procedures, the extraction of the gland segments namely, the rinsing of glands, the destillation of the volatile compounds, and the isolation of the odor substances from the surrounding air.

a) Extraction

The extraction of the gland segments is by far the most usual method to win active material; the abdominal tips can be collected in solvent or extracted hot. For the destruction of cellular tissue the animal material can be grind or mercerated in a cell tissue grinder. Shocking of the probe in liquid nitrogen facilitates the grinding procedure, and often the substrate had been homogenized by ultra sonic and thus the cells distroyed.

The solvents used can be decisive for the quality of the extract. Thus for the isolation of sex attractants among others ether, methylene chloride, carbon disulfide, methyl formate, benzene, acetone, methanol and ethanol were tested as solvents, whereby ether and methylene chloride mostly resulted in a greater amount of active material than the others. Nevertheless it is recommended to chose a rather unefficient solvent, which extracts just enough of the pheromone desired, but reduces the accumulation of other impurities. Decisive for the choice of solvent moreover is a low boiling point and simple purification procedures. For methylene chloride it is recommended to wash it acid free before use.

b) Rinsing of the Gland Region

Rinsing the separated abdominal ends of female moths Sower et al. could demonstrate gas chromatographically the sex attractants of the Angoumois grain moth *S. cerealella*, the Indian meal moth *P. interpunctuella*, the almond moth *C. cautella*, the cabbage looper *T. ni*, and the noctuid moth *P. includens* with 20–30 females each only [98].

From washing solutions of the male beetle *Aconthoscelides obtectus* Horler isolated (−)(E)-2,4,5-tetradecatrienoic acid methyl ester [369], and Henzell and Lowe phenol from ether washings of female *Costelytra zealandica* abdomina [378].

c) Distillation

The steam distillation of suspensions or homogenates of insect parts often was used for the isolation of pheromones [264], and the distillate subsequently or continuously extracted. The odorous substances of the Dufour gland of the ant *Myrmica rubra* were isolated by vacuum distillation in a short-path distillation apparatus [507], the volatile components of different glands of ants and bumble bees by vacuum distillation into deep cooled open or into closed glass spirals with cooled ends [64, 508–511]. By "sweep-codistillation" with nitrogen from the injection port of a gas chromatograph Buser and Arn isolated the sex attractants of *L. botrana* and *L. pomonella* [195].

d) Isolation from the Air

Because the most of the referred pheromone compounds are volatile, often the isolation from the air, surrounding the insects, was preferred to extraction, thus often saving further purification steps. Moreover the pheromone emission can be taken from the insect containers over longer periods by ventilating them with air and thereby obtaining a greater amount of pheromone than with extraction or washing procedures, which represent "snapshots" of the pheromonal state of the insect only.

Several working groups dealing with the isolation of sex attractants of roaches and moths, condensed the volatile substances from the air, perstreaming the insect containers, by cooling to −78 °C [294, 512–514], or the whole air together with the odor substances by cooling to −193 °C [515] and subsequent controlled evaporation of the liquidified air. Both methods have the disadvantage of condensing aside the desired compounds also a great amount of water into the cool traps, so that the condensates must be worked up by salting out and solvent extraction procedures.

An improvement of this technique is the separation of the volatile compounds from the air by adsorption. In this case the air, streaming out from the insect containers is directed over adsorption materials like silica gel, GC column material, or over Porapak Q^{TM} (p-divinylbenzene-vinylbenzene copolymer) conditioned by extraction and heating [516], and resolves the adsorbate by solvent extraction from the support (cf. isolation of bark beetle pheromones [339, 357, 358] and the sex attractant of *Choristoneura fumiferana* [270]).

The adsorption of volatile compounds in a static system also had been used for the isolation of pheromones and odorous substances. The filter paper layers of insect containers often served as pheromone sources [62, 517–519], as well as parts of insects or plants were kept in contact with a GC-support material, loaded with a silicium high vacuum grease, and this adsorber material after a few days used in a GC-precolumn for GC–MS analyses [520].

Deep Temperature Precipitation of Lipids

For the separation of lipid material acetonic extracts or methylene chloride extracts treated with acetone were cooled down according usual methods from −20 °C [521] to −70 °C [522], and the lipids precipitated removed by decantation, filtration or centrifugation.

Wet Chemical Reactions – Group Tests

By means of wet chemical reactions for the analysis of functional groups and for the determination of the degree of the unsaturation in connection with a simple bioassay like electroantennogram or wing flutter test in most cases first concrete

Table 4. Functional group tests and biological activity determination of reaction products by means of bioassay

Functional group of pheromone	Chemical reaction	Activity
Double bond	Hydrogenolysis[a], ozonolyses, $KMnO_4$, perbenzoic	–
	acid, bromination	–
	Reaction of bromination products with Zn	+
Ester	Hydrolysis, $LiAlH_4$ reduction,	–
acetate	Acetylation of hydrolysis product in the neutral fraction	+
methylester	Esterification of the acidic hydrolysis products with diazomethane	+
Carboxylic acid	Extraction with alkali, esterification with diazomethane	–
	Acidification of alkali extracts, hydrolysis of methylated products	+
	$LiAlH_4$ or $NaBH_4$ reduction, ZnO_2 – reaction loop[a]	–
	oxidation of reduction products with CrO_3/H_2SO_4	+
Alcohols	Acetylation and subsequent hydrolysis	– +
	boric acid reaction loop[a]	–
	Phthalic acid anhydride and subsequent hydrolysis	– +
	phenyl isocyanate, naphthyl isocyanate	–
Aldehyde	$LiAlH_4$ or $NaBH_4$, semicarbazide, 2,4-dinitrophenyl-hydrazone, hydroxylamine, benzidine subtraction loop[a]	–
	o-dianisidine subtraction loop[a]	–
	Girard-T – or Girard-P-reagent	–
	Subsequent acidification with H_2SO_4	+
Acetal	Hydrolysis with mineral acids,	–
Epoxide	Treatment with dilute acid, $LiAlH_4$, periodic acid	–
	phosphoric acid reaction loop[a]	–
Amine	Extraction with acids	–
	Neutralizing the acidic layer	+

[a] Used in the form of GC-loops in reaction gas chromatography (cf. p. 79)

hints of the chemical structure of an insect pheromone can be obtained directly from the crude extract. Thereby the biological activity of a substance can be destroyed by reactions carried out on a micro scale and often by the subsequent "back reaction" reobtained again [472].

Table 4 shows a series of simple test reactions with which the presence or absence of certain classes of compounds in crude pheromone extracts can be proved. At the same time it has to be considered strictly that the crude material also contains a great number of other compounds and impurities, which can falsify or mask the proper result. The same these investigations must be secured by control tests with each of the chemicals solely, to exclude a possible influence of test chemicals onto the test insects. These group tests can be carried out as well in the form of a tüpfel reaction with the corresponding fraction of a thin layer chromatographic separation directly on the plate, or in reaction loops during gas chromatographic separations [523] (cf. p. 90).

The technique of the reaction on the thin layer plate and the subtraction of substances in reaction loops in the gas chromatography, respectively, has been used for the identification of the sex attractant of the female gypsy moth *Lymantria dispar* (Lepidoptera) [277, 524]. Thereby male moths served during the GC-separation as the GC-detector, showing characteristic copulational movements when the column effluent was directed over them. A particularly beautiful example for a partial structure elucidation of a butterfly sex attractant by means of wet chemical reactions demonstrated Tomida and Ishii [525] in their investigation of the pheromone of the Eri-silkworm moth *Philosamia cynthia ricini*. The authors compared parallely and in cross tests the flutter reaction of the silkworm *Bombyx mori* and *P. cynthia* each versus their extracts and corresponding reaction solutions, and partially identified the sex attractant as a conjugated-unsaturated carbonyl compound [525].

4.6. Chromatographic Purification and Fractionation for the Identification of Pheromones

Numerous techniques of adsorption as well as of partition chromatography and their combinations have been used for the purification and fractionation of pheromone extracts, whereby in most cases beside the separation also already first indications like retention volumina, retention times and r_F-values for the chemical structure of the compounds can be obtained.

a) Gel Permeation Chromatography (GPC)

The gel permeation chromatography, also called steric exclusion chromatography, gel filtration chromatography or gel chromatography [526] is based on different speed of permeation of different molecules through pores of swollen gels. Thereby the smaller molecules penetrate through the pores of the gel, and are retained maximally and eluted last. Medium sized molecules will penetrate only a few pores and therefore will leave the gel column sooner, whereas molecules of a defined size or larger will pass the column without retardation with the same streaming velocity as the eluent. In the contrary to the other chromatographical separations in the case of the gel chromatography thus the great molecules are eluted first from the column, the smaller following later according to their molecule size (Fig. 7a). This method is generally suitable for separations of compounds of molecular weights from 1,000 to 2,000,000 (Fig. 7b), but is also applicable for the prepurification of crude extracts and for the separation of high molecular impurities. Nevertheless demonstrates Fig. 7c, that suitable gels and the successive connection of several columns with this chromatographical technique separations can also be achieved for substances of molecular weights between 100 to 300.

The gel permeation chromatography offers the advantage of symmetrical peak shapes, of short and predictible times of elution, and no necessity for gradient elution. On the other hand it has the disadvantage of a simultaneously competing adsorption chromatography, which must be avoided strictly, and the limited

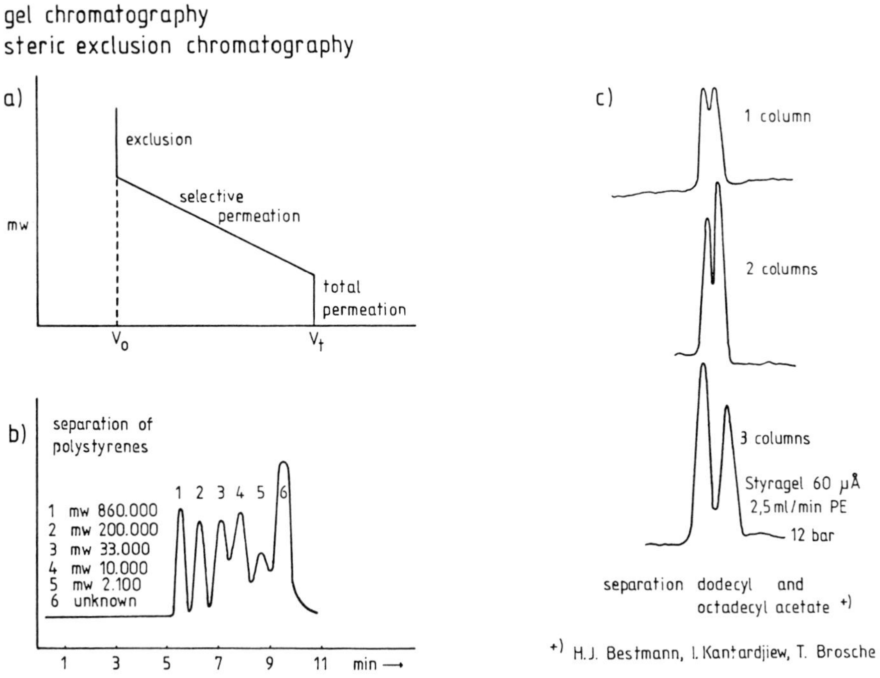

Fig. 7. Gel permeation chromatography

loading capacity of the column material. Because of this a more preparative chromatography of this kind becomes difficult and often results in detection problems.

The materials used for gel chromatography are swollen gels like e.g. sephadex, polystyrenes, polyester and polyamine gels, silica gels and porous glass pearls. While the weak and medium hard gels swell differently in solvents and therefore make the gel columns unpermeable, the harder gels like polystyrenes, silica gels and porous glasses resist high pressure, high temperatures, show equal behavior of swelling to organic or aqueous solvents, and hence can be used for a chromatography under higher pressure.

The relative young method of gel chromatography nowadays has been already used often for the prepurification of extracts in isolation and identification studies of insect pheromones, a few examples will be given here. Thus Klun et al. [146] used Styragel 100 µÅ for the isolation of the sex attractant of the European corn borer *Ostrinia nubilalis*, achieving separations of compounds of molecular weights between 235 and 275, i.e. acetates with alcohol moieties of C_{13} to C_{15}, and the separation of tripalmitin and higher molecular impurities [146]. Kochansky et al. used BioBeads SX-2 as gel for the structure determination of the sex attractants of *Antheraea polyphemus* [226] and *O. nubilalis* [173], Persoons Sephadex LH 20 for the isolation of the sex pheromone of the cockroach *Periplaneta americana* [62], Tumlinson and Heath Poragel 60A (37–75 µm) for the isolation of the pheromone of the tobacco budworm *Heliothis virescens* [527].

b) Liquid Chromatography under Normal Pressure

One of the first steps in the purification of crude pheromone extracts and the concentration of active material in most cases is a liquid chromatographic separation on adsorbent filled columns. For these cases usually simple glass columns with adsorbents like silica gel, aluminium oxide, Florasil® (silica gel and magnesium oxide) in ratios of 100:1 relative to the mixture to be separated are used. Using eluents of increasing polarity, e.g. according to the method of Hirsch and Ahrens [528] crude mixtures can be separated according to the polarities of each of the components. Thus with increasing portions of diethyl ether in hexane a separation according to the classes of compounds can already be achieved, the compounds usually eluting from the column in the order saturated hydrocarbons, olefins, ethers, esters, ketones, aldehydes, alcohols and carboxylic acids. Thus the determination of the elution volume of the biologically active compound by means of the EAG technique already gives a first hint on the polarity and thus on the structure of the pheromone.

With this method the active components of the sex pheromone of the pink bollworm *Pectinophora gossypiella* (Lepidoptera), which later turned out to be two isomeric hexadecadienyl acetates, were found in the 5% ether/hexane and the following 15% ether/hexane fraction, using silica gel for the separation [228].

Figure 8 gives an example of the efficacy determination of the column separation of the attractant of the noctuid moth *Mamestra brassicae* (Lepidoptera) using the EAG-test, the attractant (Z)-11-hexadecenyl acetate found in the 20% ether/pentane fraction [188]. Buser and Arn determined E-8,E-10-dodecadien-1-ol, the pheromone of the codling moth *Laspeyresia pomonella* by mass fragmentography (cf. p. 85), and E-7,Z-9-dodecadienyl acetate, the pheromone of the

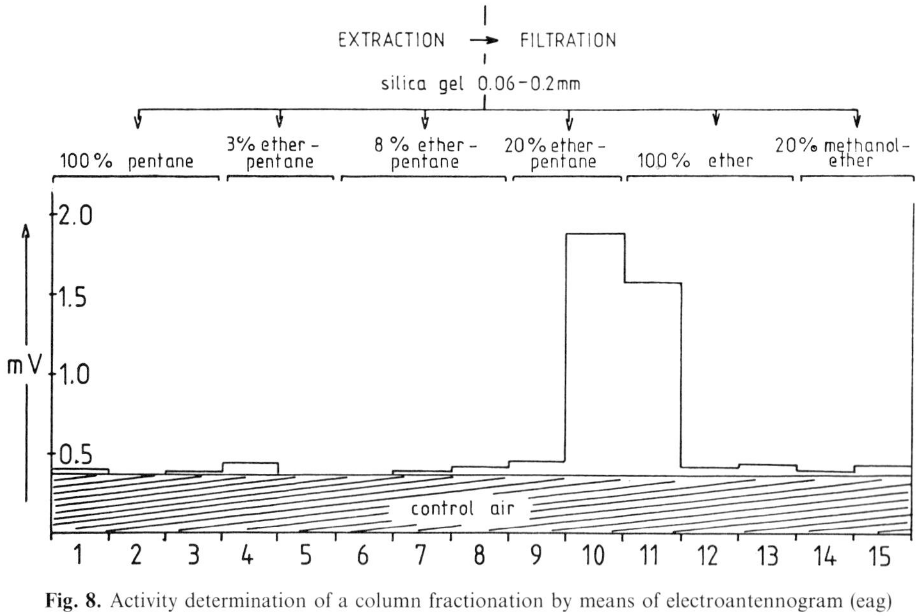

Fig. 8. Activity determination of a column fractionation by means of electroantennogram (eag)

tortricid *Lobesia botrana* under similar conditions in the 8% ether/pentane fraction [195]. Similar conditions were chosen for the isolation of the sex attractants of *Antheraea polyphemus* [226], *Diparopsis castanea* [122], *Lymantria monacha* [282], *Sitotroga cerealella* [229], *Synanthedon pictipes* [231], *Sanninoidaea exitiosa* [231] and *Carposina nipponensis* [165].

Silver nitrate impregnated silica gel or Florosil®-columns were applied for the separation of (Z)- and (E)-olefins. This technique of argentation chromatography (usually 20–25% $AgNO_3$, Adsorbil CABN®, Hi-Florisil-Ag®) makes possible the separations of olefins according their number of double bonds, the (E)-isomer being eluted sooner than the (Z)-isomer. Although this chromatographic technique seldom was used for the isolation or purification of olefinic pheromones, it posesses great importance for the preparative isomer purification of synthetic sex attractants, in particular when the concentration of the opposite geometrical isomer becomes critically for the efficacy of a sex attractant.

c) High Performance Liquid Chromatography/High Pressure Liquid Chromatography/HPLC

The conventional techniques of column chromatography, the effectivity of which were unquestionable and which found a broad field of application, in the last years more and more were displaced by the high performance liquid chromatography under high pressures (HPLC). This technique gains the effectivity of gaschromatographical separations, yields fast analysis times and allows automatical instrumentation [529]. While in the classical form of the liquid chromatography adsorbtion materials with particle size of 150–200 μ, columns with about 1000 theoretical plates/m and an injection volume of a few millilitres lead to separations with ±30% of reproducibility, with HPLC techniques, using particle size of 5–50 μ and columns with 50,000 plates/m and a few microlitres injection volume the good reproducibility of ±1% and because of the low detector cell volume a high detection sensitivity can be obtained. Beside the adsorbtion chromatography, using silica gels and aluminium oxides as packing materials, in analogy to gas chromatography exists a method of partition chromatography between two liquid phases.

The stationary liquid phases used are thermically and chemically stable, and any solvent can be used as the mobile phase. Thereby two different kind of partition chromatography are possible: the normal phase technique namely, the separation principle represented by polar bound phases and unpolar mobile phases, and unpolar bound stationary phases (reversed phase technique) with polar mobile phases like methanol, acetonitrile and water. Applying the normal phase technique, the less polar compounds are eluted first, in the reversed phase technique the most polar substance first leaves the separation system.

For the detection of insect pheromones in fractions obtained by HPLC-technique considerable difficulties arise. Thus refractometer detectors for lepidopteran pheromones like the mono-unsaturated acetates possess only detection limits of approximately 1 μg. In the case of the more sensitive UV-detectors because of the absorbtion of the acetate carbonyl group only those with extremely low range of wave length are applicable. Often it is recommended to derivatize certain classes of compounds specifically in the crude mixture and thus facilitating and making possible the separation, detection and identification. For instance, in the crude extract of female lepidoptera, using aldehydes as pheromones, those can be derivatized to dinitrophenyl hydrazones and separated by HPLC-chromatography [530]. The DNP-derivatives of different aldehydes are easily separatable by HPLC-reversed phase technique [531], are because of their dinitrophenyl group already in the ng-range detectable [530, 531], and can be easily identified by mass spectrometrical methods because of their specific fragment ions (cf. p. 86).

Tumlinson et al. separated alkenyl acetates into their (E)- and (Z)-isomers using silver nitrate impregnated silica gel columns (1–11 µ), benzene for elution and refractometer detection [231, 532, 533], with this method also separating alcadienyl acetates into their isomers sufficiently [533]. For the separation of isomer olefins nowadays HPLC-columns with kationic exchange resins are commercially available, being used by Houx [534] for the separation of geometrical isomers of alkenyl acetates with attractant activity, and Persoons et al. [103] for the isolation of (Z)- and (E)-8-dodecenyl acetate, the sex attractant of the false codling moth *Cryptophlebia leucotreta*. An interesting method of (Z)- and (E)-isomer separation lies in the application of π-complex forming metall salts like silver nitrate and similars in the mobile phase by using the reversed phase technique. Therewith isomeric olefin hydrocarbons could be separated on chemically bound RP-8 phase simply using isopropanol/water/$AgNO_3$ as the eluent system [535].

By means of HPLC-techniques the sex attractants of *Phthorimaea operculella* [203], *Ostrinia nubilalis* (New York tribe) [173] and *Ostrinia nubilalis* (west coast) [146] could be isolated, and synthetical (Z)-3,(E)-5-tetradecadienyl acetate, the pheromone of *Pryonoxystus robiniae* [208] separated from its (E),(E)-isomer [527].

With Adsorbil-2-ADN (20% $AgNO_3$-silica gel) columns from synthetical (E)-3,(Z)-13-octadecadienyl acetate, the attractant of *Synanthedon pictipes* [234], the (Z),(Z)-isomer can be separated, as well as the separation of the (E)-alkenal pheromones from *Heliothis virescens* succeds [527].

d) Thin Layer Chromatography

The thin layer chromatography [480, 484] is because of its small material demand, the high separation achievement, the good reproducibility and the simple and fast procedure often used for isolation and purification. Color and spray reagents, vapours of chemicals (J_2, Br_2) or UV-light (in connection with fluorescent indicators) are used for the detection of the adsorbed compounds. Identification reactions for the presence or absence of certain functional groups can be carried out directly on the thin layer plates [480, 484] (cf. p. 73).

The method of high performance thin layer chromatography (HPTLC) [536], developed in the last years, gains because of new techniques of charging the mobile phases, activation of the layer and improvements of the probe dosation an extremely high precision in chromatographic separations, fast analysis times and a better reproducibility than the elder standard thin layer chromatography methods.

By means of thin layer chromatographical separations Kasang et al. [256] isolated bombykal, a second component of the silkworm moth *Bombyx mori* sex attractant [537]; the pheromone of the saturnid moth *Antheraea polyphemus*, (E)-6,(Z)-11-hexadecadienyl acetate and (E)-6,(Z)-11-hexadecadienal was isolated using silica gel/silver nitrate [226]. Similar methods were applied for the identification of (Z)-10-tetradecenyl acetate from females of *Archips semiferanus* [538], (E)-5-dodecenyl acetate from *Anarsia lineatella* [79] and the sex attractants from *Argyrotaenia velutinana* [168] and *Ostrinia nubilalis* [173].

A broad field of application of argentation chromatography lies in the determination of configuration of double bonds, comparing the r_F-values of unknown compounds with those of (Z)- and (E)-isomers of structurally related substances (e.g. Bierl et al. [277], Carlson et al. [428], Roelofs et al. [248], Kochansky et al. [226]). Prasad reported, that he obtained better separations of certain terpene olefins using a silver perchlorate impregnation of the thin layer plates instead silver nitrate [539].

For the investigation of metabolites of bombykol, the sex attractant of female *Bombyx mori*, Kasang and Weiss applied tritium-marked bombykol onto silkworm antennae, and isolated by means of thin layer chromatography and argentation chromatography the different metabolic products [540].

e) Gas Chromatography

Similarily to the former described liquid chromatography gas chromatographic techniques can also be applied in many phases of isolation and purification of insect pheromones. E.g. can the GC-column effluent be splitted after the separation and one part collected, or condensed directly after passing a thermal conductivity detector. For the condensation or collection of the separated volatiles, exhausting from the oven, practically colled glass capillaries [62, 158, 163, 440, 541, 542], capillary tubes with a temperature gradient [76, 313, 348, 411, 527, 543], stationary phases for gas chromatography [544, 545, 546], glass wool [547] powdered graphite [548] are used or the substances introduced into cooled solvents [101, 264, 287].

With these microscale preparative gas chromatography among others the sex attractants of *Antheraea polyphemus* [226] and *A. pernyi* [225], *Chilo suppressalis* [271], *Diparopsis castanea* [122], *Cryptophlebia leucotreta* [103], *Pseudoplusia includens* [91] and *Laspeyresia pomonella* [549] have been isolated.

4.7. Identification and Structure Elucidation

The usual methods of molecular spectroscopic and chromatographic identification, determination of physico-chemical data and the comparison with synthetic compounds (cf. Fig. 6) can be used for the identification and elucidation of pheromone structures. The purification and fractionation procedures already show first indications of functional groups and of partial structural features of the substance. The knowledge of polarity, approximate molecule size, behavior with solvents and adsorbtion materials, with the techniques described below, allows for the final identification.

The minute amount of substance available for pheromone analyses often affords the elaboration of special micromethods and gave rise to ingenious innovations and improvements of already known techniques in the field of general analysis of natural materials [472, 527, 550].

Molecular Spectroscopic Methods

a) Ultrared Spectroscopy

The importance of the UR-(IR)-spectroscopy for the determination of functional groups in unknown molecule structures also can be demonstrated in the field of pheromone analysis. By means of micro-UR-cells, micro-KBr-disks and stream condensers, spectra can be determined in the µg-scale [551]. Using computer accumulation and fourier transform equipment [552, 847], spectra can be obtained in the submicrogram level. Riley et al. used the infrared spectroscopy for the identification of the trail pheromone of the termite *Atta cephalotes*, methyl 4-methylpyrrolcarboxylate, by taking spectra of 25 µg samples [411]. Persoons used 7–10 µg samples for the identification of the cockroach pheromone periplanone B [62] and the attractants of *Archips podana* [163] and *C. leucotreta* [103], Tumlinson et al. similar dimensions for those of the moths *Synanthedon pictipes* and *S. exitiosa* [231].

Beside the diagnostic IR-absorptions (e.g. for the characterization of C=O, C–O, C=C, O H, N–H, C–C and other bonds) the strong $=C-H_{(out\ of\ plane)}$-deformation vibration of the trans-standing hydrogen atoms of (E)-olefins ($\sim 965\ cm^{-1}$) are used for the determination of configuration of olefinic pheromones or for checking the stereochemical purity of synthetic compounds. Thus knowing the spectra of two stereochemically unique isomeric compounds like (Z)- and (E)-alkenyl acetates, the amount of (Z)-isomer can be estimated fairly well by comparing a certain absorption band relative to the $960\ cm^{-1}$ signal (so-called base line method [553]). The corresponding $=C-H_{(out\ of\ plane)}$-absorption of (Z)-olefins ($\sim 720\ cm^{-1}$) can not be used for an investigation like this, since it represents only a weak absorption and is superimposed by the antisymmetrical polymethylene rocking vibration [$-(CH_2)_n-$, for $n \geqq 4$, $720\ cm^{-1}$].

b) UV-Spectroscopy

Ultraviolett spectroscopy had found application for the identification of insect pheromones only in those cases, when an UV-absorbing chromophore is present in the molecular structure investigated [62]. On the other hand highly sensitive UV-detectors for the recognition of UV-absorbing molecules in chromatography, especially in the HPLC, are used in the purification and fractionation of pheromones.

c) Nuclear Magnetic Resonance Spectroscopy, NMR-Spectroscopy

The nuclear magnetic resonance spectroscopy in the form of ^{1}H-proton or ^{13}C-carbon resonance spectroscopy nowadays certainly is the most widely used analytical technique in organic chemistry.

Ritter et al. detected two different methyl groups and three methyne groups adjacent to nitrogen and thus the structure of the trail pheromone of the pharao ant *Monomorium pharaonis* [426, 427]. Byrne et al. determined two structural elements in the beetle pheromone sulcatol [346], and Persoons the structure of periplanone B with 1–2 µg samples [62].

The ^{13}C-nuclear magnetic resonance spectroscopy gives signals which correspond to the order of binding and to the chemical and magnetical environment of each carbon atom. This is the most suitable method for structure determination and above all for determination of the actual configuration and geometry of the substance analyzed.

Until recently the smallest sample size with which reliable proton-NMR-spectra could be obtained, was approximately 100 µg. Using special micro-NMR-sample tubes and the CAT-technique (computer accumulation of several hundreds of spectra, thus improving the signal-noise ratio) good spectra could be obtained in the 50–100 µg scale. Only in recent years has the technique of taking spectra considerable improved and spectra of 2 µg samples (e.g. in 75 µl solvent) with a data system (pulse-Fourier-transform technique) are reported in the literature [527]. ^{13}C-Spectroscopy can be achieved with the aid of the naturally occuring 1.1% isotope ^{13}C. Because of this much more sample is needed, and spectra from samples in milligram scale can be determined. Nishida et al. [399] determined the structure of a cockroach pheromone (36 mg sample) by means of ^{13}C-spectroscopy (cf. Table 1).

For the synthesis of pheromones and the investigation of structure and purity of synthetic insect sex attractants both methods (^{1}H and ^{13}C) are applied to a rather great extent. Thus for the investigation of the geometry of olefinic double bonds in alkenols, alkenyl acetates and alkenoic acid esters ^{13}C-spectroscopy is suitable. The differently substituted carbon atoms of a double bond in a sex attractant of the general formula $CH_3(CH_2)_nCH{=}CH(CH_2)_m$-OAc gives in a proton noise decoupled spectrum (e.g. 100 MHz, PFT) two singlet signals, the low field shift of which depends on the configuration of the double bond. E.g. show the corresponding olefinic carbon atoms of methyl oleate and methyl elaidinate resonance frequencies differing by approx. 0.485 ppm, and inspite of the different relaxation times can be used for a rough isomer ratio estimation [621].

Disselnköter et al. observed by means of ^{13}C-NMR spectroscopy the shift of each olefinic carbon atom of the synthesis intermediates in their preparation of the four isomeric 7,11-hexadecadienyl acetates and could assure the configuration of their end products [554].

d) Mass Spectroscopy

Ordinary Techniques

The most frequent analytic technique in pheromone chemistry and in the analysis of natural substances is mass spectroscopy, especially in combination with a gas chromatographic separation and data system. The mass spectrometers used are magnetic field instruments as well as quadrupole or multipole instruments. The simple quadrupole spectrometers generally produce spectra with lower resolution, with discrimination of higher ion masses, and of smaller mass range than the sector field instruments. But they are often used because their construction is principally simpler and hence cheaper. By using a circulating electric field instead of the magnetic field, they don't show hysteresis effects, and thus can scan faster compared to a magnetic sector instrument. In combination with gas chromatographs they are especially useful in the application of fast separating capillary columns.

From 10 ng of substance a complete mass spectrum can be obtained, with the molecular weight and the chemical composition of the compounds investigated. In the case of electron impact ionization (EI-spectra) the vaporized substances are bombarded with electrons and thus ionized. The resulting, more or less instable molecular ions collapse and give neutral radicals and positive ions, characteristic for the substance analyzed.

Only decay products of the analyzed compound and no direct information about the molecular weight is obtained, when the molecular ion is unstable or short lived because of thermical or energetical reasons. In these cases chemical ionization (CI-spectra) [555, 556] is used. A reagent gas which is present in excess in the ion source, is ionized by electron bombardement, and the thus ionized gas can abstract or add a proton from or to the investigated substance, depending on its basicity. The "quasi molecular ion" is now stable under this ion source conditions, and a spectrum of the molecular ion with a small portion of decay products is obtained.

The field ionization and field desorption mass spectroscopy (FI and FD technique) cause ionization by means of high electric fields and is especially useful for evidence of unstable or non-volatile compounds. In the case of the FD spectroscopy the test sample is coated on an ionization wire, and no actual vaporization and hence no thermically stressing, no electron bombardement and minute collisions of molecules appear. Thus the molecular ions even of unstable and hardly volatile substances can be observed. However, this form of mass spectrometry is not suitable for coupling with the gas chromatograph.

In the combination of a gas chromatograph and mass spectrometer (GC-MS combination) [557, 558], the substance mixture investigated first is chromatographically separated because of different retention times and then subsequently introduced into the ion source of the mass spectrometer according to the foregoing separation. For GC-MS combinations non-polar stationary GC-phases are mainly used which show less "bleeding" and hence giving less contamination of the ion source and a lower background in the spectra. By using WCOT-capillary columns, the GC-exit can be connected directly to the mass spectrometer by a platinum iridium capillary as interface. In the case of packed columns which could because of their high carrier gas flow disturb the high vacuum, a separator is needed for a partial removing of the carrier gas. From any form of GC-MS combinations the mass spectra of each of the compounds as well as their retention times can be obtained.

Only a data system, connected to a GC-MS combination [559], actually enables a perfect evaluation of all structure information obtained from a mass spectrometric analysis. The spectra, especially when using capillary columns, are obtained so fast, that only continous recording (e.g. a few seconds per spectrum), storage and subsequent elaboration is possible. At the same time the data system directs the mass spectrometer and regulates scanning parameters like mass range, integration times, etc. By summation of each single ion intensity of each spectrum the computer reconstructs a chromatogram from which any spectrum wanted can be recalled. Automatically the retention time is registered, the area of each signal can be determined, and the single ion intensities be measured. A great simplification for the interpretation of spectra is the possibility of subtraction of two different spectra and hence the subtraction of the spectrum background. Often the spectra can be coded, stored in a library, and at a later point of time recalled and used for comparison.

Mass Chromatography

A special technique in computerized GC-mass spectrometry is called mass chromatography. With it the computer locates the presence and the intensity of any wanted ion from all spectra recorded (approx. 200–500 spectra per run), and reflects them graphically as a "mass chromatogram". This mathematical manipulation shows the temporal partition of a distinct ion over a whole chromatographical separation and is a great facilitation of work in GC-MS analysis. Since all ions, generated from the same compound, must appear at the same time (i.e. the retention time of the compound), the signals in each mass chromatogram must coincide temporally (cf. Fig. 9). In order to look for a certain compound or class of compounds, one enters their specific ions into the computer. The temporal coincidence at the retention time of the desired compound ascertains the presence of a substance which generates the wanted ions, having the retention time of the chemical compound which is sought.

Thus, e.g. for the investigation of saturated hydrocarbons the series of chain fragment ions like 43^+ (C_3H_7), 57^+ (C_4H_9), 71^+ (C_5H_{11}), 85^+ (C_6H_{13}), etc. can be represented in form of mass chromatograms. Many compounds like acetates, alcohols, carboxylic acids, carboxylic acid methylesters, ethylesters, ketones, etc. fragmentate in a mass spectrometer because of energetic reasons (e.g. McLafferty rearrangement) or because of thermal reasons with the loss of uncharged and saturated molecules (e.g. acetic acid from acetates and carboxylic acids, water from alcohols, methyl and ethyl acetate from the corresponding methyl and ethyl esters, acetone from aliphatic ketones, etc.), forming an unsaturated hydrocarbon ion (Fig. 9). Recording these hydrocarbon fragments and simultanously ions like 43^+ (CH_3CO^+ from acetates, ketones), 60^+ (CH_3COOH^+ from acetates, carboxylic acids), 61^+ ($CH_3COOH_2^+$ from acetates), 74^+ ($CH_3COOCH_3^+$ from methyl-

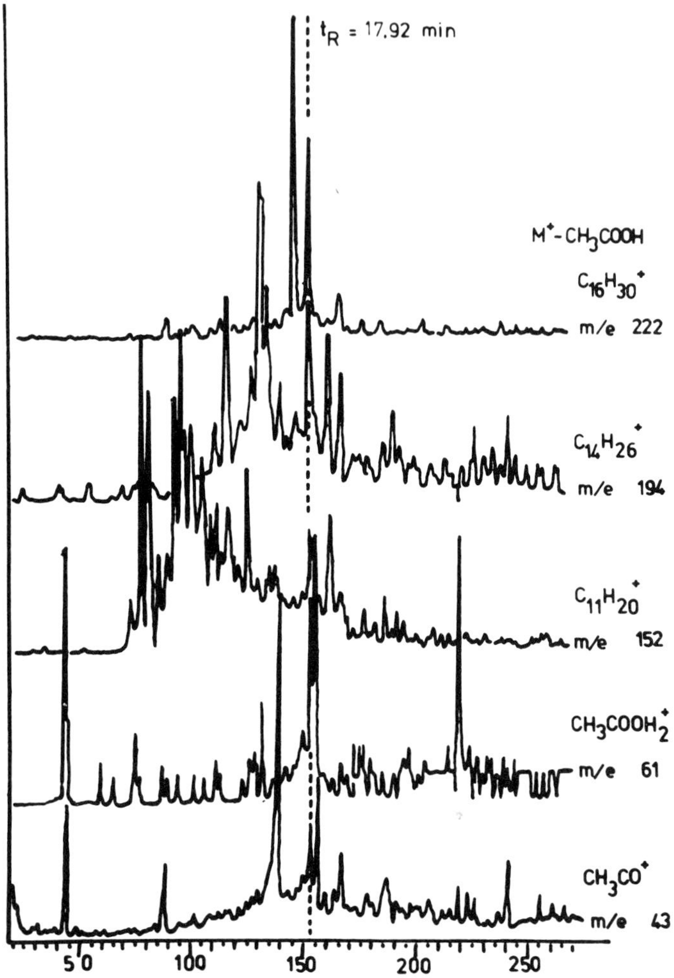

Fig. 9. Mass chromatograms for the analysis of monounsaturated acetates in *M. brassicae* [188]

esters), 88^+ ($CH_3COOC_2H_5^+$ from ethylesters), 58^+ ($CH_3COCH_3^+$ from aliphatic ketones) and additional characteristic fragment ions in most cases represents sufficient proof for the presence of the corresponding compound.

In identifying the sex attractant of the noctuid moth *Mamestra brassicae* [188] the mass chromatograms of the ions CH_3CO^+ (m/e=43), $CH_3COOH_2^+$ (m/e=61) and the chain fragment ions $C_{16}H_{30}^+$ (m/e=222), $C_{14}H_{26}^+$ (m/e=194), $C_{12}H_{22}^+$ (m/e=166), $C_{11}H_{20}^+$ (m/e=152) were registered superimposed and gave the temporally coincidence in spectrum No. 153 at the retention time of 17.92 min for (Z)-11-hexadecenyl acetate (Fig. 9).
This technique often allows a clear separation of unresolved GC-signals into two or more distinct compounds because of different ion intensities along the GC-profile. This was used e.g. by Beroza et al. for (E)-8,(E)-10-dodecadien-1-ol (codlemone, codlelure) in a poorly purified extract of female codling moths

Laspeyresia pomonella by observing the intensity partition of the ion m/e = 182 (codlemone molecular ion) within an unresolved GC-signal [249]. A subtraction of a mass spectrum of the GC-peak flank from the spectrum with highest m/e = 182 – intensity yielded a well recognizable spectrum of the dodecadienol with the corresponding retention time [249].

A further advantage of the mass chromatography technique is the elimination of irrelevant signals from a chromatogram. Such apparent impurities from the solvent, from support or adsorbtion materials, from injector septa, etc., are confusing and must be eliminated. Thus it seems practical to investigate the whole chromatogram for the presence of ions with m/e = 149 (base peak of phthalic acid esters), 73 and 147 [$(CH_3)_3Si^+$ and $(CH_3)_3Si-O=Si(CH_3)_2^+$], or 215 and 220 (di-tert-butyl p-kresol as stabilizer), and to eliminate the corresponding irrelevant GC-signals.

Mass Fragmentography

In contrast to mass chromatography, in which actually complete spectra are recorded and single ions mathematically manipulated, mass fragmentography, or single ion or multiple ion detection, represents a certain scanning technique [560]. Thus the computer directs the mass spectrometer and registers only one or few characteristic ions of the substance analyzed. Since the time for the registration of the other ions now can be saved, the repetitive scanning of the spectrometer can be increased considerably. As a consequence the integration times for these few ions can be extended drastically (normal scan mode approx. 1–5 ms/ion mass, in mass fragmentography to 100 ms and more), thus improving the signal/noise ratio and consequently the sensitivity of the instrument into the subnanogram range. But now no complete mass spectrum is obtained; only the appearance of one or more characteristic ions at the retention time and with the relative ion intensities of the wanted compound.

By using this special technique of GC-MS analysis, Buser and Arn demonstrated the presence of (E)-7,(Z)-9-dodecadienyl acetate in the female pheromone of the European grapevine moth *Lobesia botrana* and (E)-8,(E)-10-dodecadienol in the codling moth *Laspeyresia pomonella* [195], and Hendry et al. the pheromone components of the oak leaf roller *Archips semiferanus* [159, 561].

Other Techniques of Mass Spectroscopy

Apart from taking the mass spectra of actual pheromones the spectra of reaction products have been obtained for identification. In these cases the hydrogenolysis of double bonds [62, 277, 369, 412, 428, 508, 527, 558, 562] and the degradative hydrogenation for the determination of the carbon skeleton [331] of the compound investigated were the methods mostly used.

Reactions for the analysis of double bond positions, which can be carried out mass spectrometrically, are of great importance in pheromone chemistry. Thus for the determination of positions of double bonds of mono- or multiple unsaturated acetates, known as moth attractants, the pheromone extracts are often ozonized on a microgram scale and the ozone cleavage products (alkanals and ω-

formylalkyl acetates) analyzed by means of gas chromatography, GC-MS analysis or mass fragmentography [62, 91, 118, 122, 163, 173, 217, 236, 472, 563–566] (cf. Fig. 10).

The stereospecific epoxidation with m-chloroperbenzoic acid (MCPB) of olefins into epoxides can be used for the determination of double bond positions by GC-MS analysis. By using electron impact ionization [567] and chemical ionization [568] the position of the epoxide oxirane ring in the carbon chain and thus the original position of the double bond can be analyzed. In the investigation of the double bond positions of (Z)-7, (E)-11-hexadecadienyl acetate, the attractant of the Angoumois grain moth *Sitotroga cerealella*, the epoxidation reaction could be stopped at the stage of the monoepoxides, and the configuration of the monoepoxides determined by NMR- and chemical ionization MS-data.

Figure 10 illustrates two further interesting reactions which can be used for mass spectroscopic identification of olefinic double bonds: oxidation with osmium tetroxide yields a 1,2-glykol, which can be converted into the bis-trimethylsilyl compound [569, 570]. The reaction of the glykol with phenylboric acid yields the cyclic boronate, which gives characteristic MS-fragments at the original double bond position [569, 570] (Fig. 10).

A possibility for evidence of unsaturated aliphatic aldehydes, which are known as moth sex pheromones and give only unspecific mass spectra, is the derivatization to 2,4-dinitrophenyl hydrazones. These can be separated by HPLC-technique and because of their minute volatility analyzed by direct inlet into the mass spectrometer. The excellent molecular ions, because of the aromatic dinitrophenyl group, and the characteristic ion m/e = 204 from the McLafferty rearrangement give spectra, which are recognizable down to the nanogram level.

e) Gas Chromatographic Techniques

General

The retention time of a compound, i.e. the elution time from the gas chromatographic column, which is determined on polar and non-polar separation columns, gives, by comparison with the retention times of known test substances, considerable information with respect to polarity, molecular size and the presence of functional groups in the unknown compound. The introduction of Kovats-retention indices [571, 572], which represent the net retention time t'_R (t'_R = total retention time t_R minus gas hold-up time t_t) relatively with the net retention time t'_R of an n-alkane, expressed by the number of carbon atoms times 100 (e.g. hexane = 600, heptane = 700, etc.), considerably improves this analytic technique. The number of carbon atoms under isothermal oven conditions is proportional to the logarithm of the net retention time t'_R and gives a straight line, graphically drawn, from which the retention index of each substance can be taken. Since the log t_R's of a homologous series of compounds like alcohols, acetates, esters, etc. also give a straight line, it can be shown, that the retention index is the sum of increments, derived from the chain lenght and the functional group. As a consequence, the determination of retention index of an unknown compound on several columns gives information on the functional group.

Since butterfly pheromones are mostly longchain unsaturated compounds with one single functional group, it is possible to conclude the structure of the compound from retention indices. Many similar index systems with their reference substances, which refer to each analysis theme, have been elaborated, e.g. for straight chain fatty acid esters [573] and for steroids [574]. The structural proof of chemically complicated pheromones, which relies on the use of retention data and comparising chromatography with known test samples, however, is often not sufficient. Thus Young could demonstrate, that the retention times of two terpenoid alcohols, linalool and 2 methyl-6-methylene-7-octen-4-ol namely, on three of five different columns used had identical t_R-values and even by means of cochromatography could not be distinguished [348].

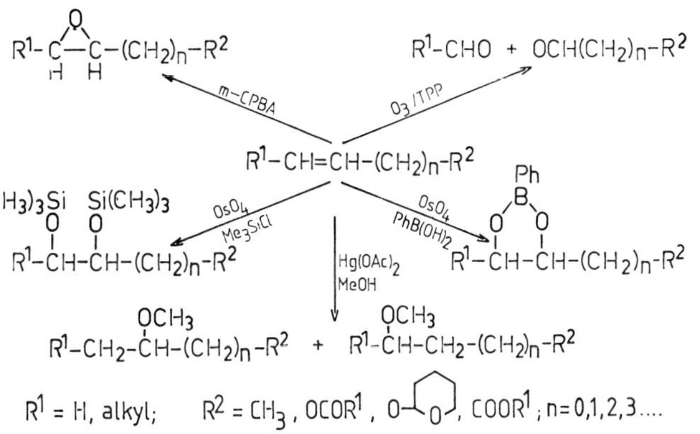

Fig. 10. Gas chromatographical and mass spectroscopical determination of double bond position

The analysis of (Z)- and (E)-isomers of olefins by means of gas chromatography presently offers considerable difficulties, and mostly is restricted to certain GC-conditions and substances. Thus polar liquid phases which were treated with silver nitrate [575, 576] give good separation results, but allow only temperatures up to 65 °C and hence are applicable only for compounds with approx. 6–8 C-atoms. The use of thallium nitrate and polyethylene glycol as the stationary phase, for the separation of terpenoid olefins [577], allows column temperatures of 100–130 °C. Separation on packed columns with derivatization of the double bond gives better resolution. The cis-stereospecific epoxidation with m-chloroperbenzoic acid (m-CPBA) [578] converts (Z)-olefins into (Z)-epoxides, and (E)-olefins into (E)-epoxides, respectively, and is excellent for gas chromatographic separations which use polar stationary phases [579].

In recent times GC-phases are available which enable the separation of positional isomers and geometrical isomers of fatty acid esters and alkenyl acetates used in the form of packed columns. Thus e.g. Silar 10 C separates isomeric tetradecenyl acetates with retention time quotients of approximately 1.015 to 1.030 for the positional, and 1.06 to 1.15 for geometrical isomers. The relative retention volumina of (E)- and (Z)-isomers on highly polar phases like diethyleneglykol succinate (DEGS) come to approx. 1.01, and hence these phases can be used in the form of wall coated open tubular capillary columns directly for the determination of isomer amount of olefins because of their separation capability [554, 580, 581].

Capillary Gas Chromatography

Capillary gas chromatography [582] uses columns of an inner diameter of <1 mm. These are separated into two types: the micropacked columns and the actual capillary columns, like the WCOT- and SCOT-types of columns. In the case of WCOT-(wall coated open tubular)-columns the glass or metal surface of the column tube is coated with a film of the liquid phase; in the SCOT-(support coated open tubular)-columns the layer of support is on the tube wall and is coated itself with separation liquid. These columns, which are open inside the tube and give no restriction for the carrier gas, possess a high permeability and hence can be produced in long form. The advantage compared to classical columns, which generally have a separation number *TZ* between 3 and 5 per m, lies in the high resolution (approximately 500–5000 theoretical plates/m) and in separation numbers of 25 to 50 for column lengths of about 25 m. Since the capillary columns are easily overloaded (maximum injection volume 0.1–1 μl, the maximum substance amount 2–300 ng per component) and are run with low carrier gas flows (0.5–4 ml N_2/min, 1–10 ml H_2 or He/min), the inject port normally has a splitting device which also causes a loss of analysis substance. This means in pheromone analysis, which represents a trace analysis, that special injection techniques [583] must be applied to guarantee working without a loss.

Because of their high resolution capability capillary columns are used primarily for the determination of configuration of olefinic attractants and pheromones by comparing retention times with those of test compounds of known geometry.

Biological Detectors in Gas Chromatography

Since the analysis of insect pheromones represents an analysis of behavior releasing and physiologically active compounds, combinations between gas chromatographical separation procedures and biological test methods have been elaborated. Here the insect itself serves as the detector for the differentiation between biological active and inactive fractions of the GC-separation.

In the investigation of the pheromone system of the silkworm moth *Bombyx mori* (Lepid.) Bayer and Anders directed the GC-column effluent of an extract of female moths over small cages containing male moths of the same species. With the starting wing flutter reaction the authors could detect beside the already known bombykol two more behavior releasing compounds [584], none of them could be identical with bombykal, a second component of the pheromone of these moths [256, 537].

A rather simple determination of the retention time of sex attractants is the collection of the compounds, which come from the separating column, by condensation into cooled glass capillaries. These can be used afterwards either directly as stimulus sources for the electroantennogram test (EAG), or the condensates with a few µl solvent washed out from the tubes and the solutions used directly for EAG tests.

In the determination of the pheromone components of the saturnids *Antheraea pernyi* and *A. harti* (Lepid.) an interesting variant of this technique in the form of single cell recording of pheromone receptor cells was applied [225]. First hints from EAG-GC-combination indicated the presence of (E)-6,(Z)-11-hexadecadienal and (E)-6,(Z)-11-hexadecadienyl acetate, the same attractant components which were found in *A. polyphemus* [226], however with an other mixture ratio [225]. Thus the single GC-fractions for the evidence of the aldehyde were tested versus the aldehyde sensory cell of *A. pernyi* and *A. harti*, which specifically responded to that compound. Since the acetate sensory cell of the antennae of both species is not very sensitive towards the hexadecadienyl acetate, the evidence for the presence of the acetate was given with the acetate cell of an *A. polyphemus* antenna, which is more sensitive for this compound than the latter [225]. With the amplitudes and the frequency pattern of single cell responses from single GC-fractions one possesses a sensitive and highly specific detector for the analysis of insect pheromones by means of gas chromatography.

A synchronous registration of a GC-analysis by using an EAG detector was developed by Moorhouse et al. [585], who diluted the column eluate of a gas chromatographical separation of the pheromone extract of the red bollworm moth *Diparopsis castanea* with nitrogen, the flow was intercepted to avoid EAG-base line drifting (10 s intercept), and blown over a male antenna of the same species [585]. Accordingly the sex attractants of *Spodoptera exempta* [154, 586] and *Ephestia cautella* [587] (Lepidoptera) were identified.

The synchronous registration of an EAG- and a FID-detector in capillary gas chromatography succeeded Arn et al. [588]. The eluted compounds were split after separation and one part analyzed with a flame ionization detector; the second part was analyzed with a separate insect antenna. In contrast to the previous method which avoided adaption of the sensory cells by intercepting the air flow, the exhausting gases were humidified with wet air, mixed and cooled, and thus could be directed for even longer periods over the antenna without drying or adaption.

With this technique (EAD-GC) in 0.1 female equivalent of an extract of *Lobesia botrana* (E)-7,(Z)-9-dodecadienyl acetate, the pheromone, could be detected. The analysis of seven females of *Spilonota ocellana* only yielded (Z)-8-tetradecenyl acetate as the sex attractant of this moth [110].

Solid Injection Technique

Gas chromatographic solid injection techniques include the analysis of compounds enclosed in glass ampoules or metal capsules, that can be inserted directly into the injection port of the GC-oven. Thus pheromone glands or distinct parts of the insect bodies can be directly introduced in the gas chromatograph. In contrast to the usual chromatographical separations which use solutions, the effect of dilution is lost. No additional impurities can come from the solvent, hence there is no danger of overlapping peaks because of solvent peak tailing.

By using solid injection techniques Morgan and Wadhams investigated the content of ant Dufour glands [507], and Descoins et al. studied the sex attractants of the noctuid moths *Mamestra brassicae* and *M. oleracea* [141].

Reaction Gas Chromatography

Gas chromatography in conjunction with chemical reactions gives the retention times of the substances investigated and information about their chemical structures [482, 589]. By means of simple chemical reactions, performed as well before the separation of the analysis mixture as after its separation, a signal of a certain compound can be eliminated or a new signal can arise. In a similar manner the biological activity of a GC-fraction can increase or decrease.

The most widely used chemical reaction in gas chromatography is the catalytic hydrogenolysis, where the catalyst is mounted in a small on-line reactor in front or behind the column, with hydrogen as the carrier gas [590, 591].

Since many pheromones are unsaturated compounds, this method often has been used for the identification of these substances (cf. the isolation of the sex attractants of the housefly *Musca domestica* (Diptera) [428], of the summerfruit tortrix *Adoxophyes orana* [74], both peachtree borers *Synanthedon pictipes* and *Sannoidea exitiosa* (Lepidoptera) [231], of the bark beetle *Anthonomus grandis* (Coleoptera) [367], and the pheromone of the cockroach *Periplaneta americana* (Blattidae) [62]).

The GC-hydrogenation (mostly with 1% palladium on the support material, 150–250 °C) usually takes place quantitatively and fast enough, so that the retention times of the saturated compounds generated can still be reliably determined. This technique was used for the identification of numerous different substances like unsaturated esters, ethers, nitriles, ketones, amines, amides and alcohols, and the cleavage of functional groups like halogens, mercaptans and aldehydes.

Before the compounds are separated by gas chromatography, they might be treated with hydrogen at elevated temperature (200–300 °C) in a catalytic reactor (length approx. 20 cm), which completely reduces the compounds to their hydrocarbon skeletons [590, 592, 593].

Thus the structure of the bark beetle pheromone brevicomin could be determined from the final hydrogenation product, n-nonane, and additional spectral data [331]. Similarily the structure determination of the pheromone of the smaller European elm bark beetle *Scolytus multistriatus* has been elucidated [594].

A logical adjunct to the "carbon skeleton chromatography" is the Kovats retention index system (cf. p. 86). The retention index of the hydrocarbon obtained by hydrogenolysis can be compared and added to the index increments of different functional groups. The sum is then compared with the original, undegraded substance. Thus the desired structure of the compound and its functional group can be determined [589].

The use of reaction or subtraction loops, connected to the GC-gas ways, can employ chemicals for specific reactions. This enables group analysis even for minute analytical samples. If a signal disappears after passing a ZnO-loop, the presence of a carboxylic acid is indicated, while with boric acid the signals of primary and secondary alcohols are eliminated. Benzidine removes carbonyl compounds, o-dianisidine aldehydes, and phosphoric acid epoxides. With lithium aluminium hydride or boron hydride substances which contain oxygen functional groups can be subtracted. By means of sodium boron hydride the difference between ketones and aldehydes can be determined; by means of sodium trimethoxy boron hydride ketones can be distinguished from esters [595].

5. Synthesis of Insect Pheromones

The increasing interests in the behavioral release of insect pheromones and in their application to environmental plant protection has led to a great number of new modified synthesis procedures for the preparation of these compounds. The high species specificity of pheromones often can be achieved only by a distinct ratio of mixture of positional, geometrical or optical isomers. Because of this fact, special synthetic methods with high stereochemical purity and selectivity are required for pheromone synthesis. The following chapter deals with special stereoselective and specific syntheses of pheromones, by trying to assemble them according to reaction chemical points of view and insect genus (Reviews for the synthesis of insect pheromones cf. [18, 36, 596–603]).

5.1. Butterflies and Moths – Lepidoptera

The female sex attractants of butterflies and moths are, as already several times mentioned, mostly mono- or polyunsaturated alcohols, their acetates or aldehydes [18, 26, 32–40] (see Table 1). Because of the rather simple chemical structures of these compounds more than only structural factors are needed to guarantee the isolation mechanism between the different species and the maintainance of species specificity. Thus, e.g. in the family of the tortricid moths *(Tortricidae)* this specificity can be obtained by employing different blends of compounds as pheromones, varying the functional end group (acetate, alcohol, aldehyde), the position of the double bond, the configuration, the number of double bonds and/or the chain length of the pheromone molecule [127, 604, 605]. As an example, (Z)-8-dodecenyl acetate, the female sex attractant of the oriental fruit moth *Grapholitha molesta* shows optimum attractancy to males of the same species in the presence of approx. 7% of the (E)-isomer, for males of *G. prunivora* however with an amount of 2.2% [107, 112, 606]. A tribe of the European corn borer *Ostrinia nubilalis* and the tortricid *Argyrotaenia velutinana*, both having (Z)-11-tetradecenyl acetate as the main pheromone component, are weakly attracted by the sole (Z)-isomer, whereas the attractivity is significantly enhanced by addition of 3 and 6% of (E)-isomer, respectively [168, 172, 607].

Because of these facts a strictly defined isomer ratio or substance ratio is necessary for optimum attractivity of moth sex attractants. Moreover a high degree of stereochemical purity and unequivocality of each synthesis step is required. For this reason many new and stereoselective synthesis procedures for the preparation of pheromone molecules were developed.

a) Wittig Olefination and Related Methods

The conversion of alkylidene triphenylphosphoranes with carbonyl compounds, known as Wittig reaction, gives olefins by generation of a double bond [608]. Although this reaction in its original form allowed poor sterical control of the geometry of the double bond, it was modified to give a highly stereoselective synthesis method of (Z)-olefin preparation by treating aliphatic, unstable alkylidene triphenyl phosphoranes and aliphatic aldehydes in non-polar solvents i) in the absence of lithium salts ("saltfree conditions") [609], ii) in polar, aprotic solvents like dimethyl formamide [610], dimethyl sulfoxide [611, 612], or iii) hexamethyl phosphoric acid triamide (HMPA) [611–614]. The formation of the phosphorus ylides from the corresponding phosphonium salts with sodium bis(trimethylsilyl)amide ("silazid-technique") and the subsequent carbonyl olefination at $-78\,°C$ affords (Z)-alkenes with a (Z)-isomer amount of $\geqq 98\%$ [615, 616].

For a correct understanding of the stereochemistry of the Wittig reaction more recent results concerning its reaction mechanism will be given here [617]: A phosphorus ylide *1* and an aldehyde *2* form an oxaphosphetane *3* by a [2+2]-cycloaddition, the ring substituents R^2 and R^3 of *3* are in the cis-configuration. Subsequently the ylide carbon atom goes to the apical position of the trigonal bipyramid of the pentacoordinated phosphorus compound *4* by a ligand rearrangement process (pseudorotation). According to the rules which govern the attack and leaving of nucleophiles during the formation of tetra- and pentacoordinated phosphorus compounds [618], the C–P-bond can break while betain *5* is formed. If R^2 is an electron donating substituent and the phosphorus ligands R^1 are phenyl groups, the life time of *5* is very short, triphenylphosphine oxide is rapidly eliminated and (Z)-olefins *7* are formed (Scheme 1).

However, if R^2 is an electron accepting group (e.g. $COOCH_3$), the life time of *5* is significantly prolonged. Now a rotation can take place around the C–C-axis, which gives rise to the thermodynamically more stable conformational betain isomer *6*, thus resulting in the formation of (E)-olefins *8* (cf. p. 103) (Scheme 1).

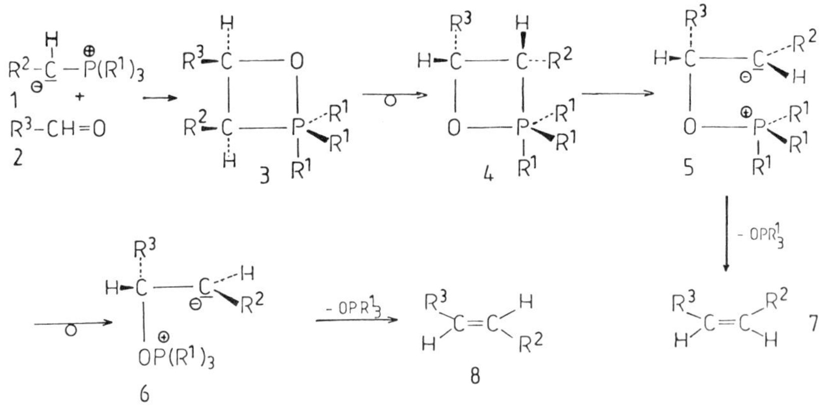

Scheme 1

(Z)-Monoolefinic Alcohols, Acetates, Esters, and Aldehydes

For the preparation of mono-(Z)-unsaturated butterfly sex attractants four different approaches exist, namely

a) the reaction of ω-substituted aldehydes *9* with alkylidene triphenylphosphoranes *10* to (Z)-alkenoic acid esters, (Z)-alkenyl acetates and (Z)-alkenyl ethers *11*, respectively,

b) reacting ω-substituted alkylidene triphenylphosphoranes *12* with simple alkanals *13* afords the same olefines *11*,

c) condensation of terminally unsaturated alkenals *14* and alkylidene triphenylphosphoranes *10* obtaining 1,(Z)-m-alkadienes *15*, and

d) (Z)-stereoselective carbonyl olefination of terminally unsaturated alkylidene triphenylphosphoranes *16* and alkanals *13* yielding 1,(Z)-m-alkadienes *15*. The introduction of functional end groups into the alkadienes *15* [pathway c) and d)] can be achieved by hydroboration and subsequent oxidation (Scheme 2).

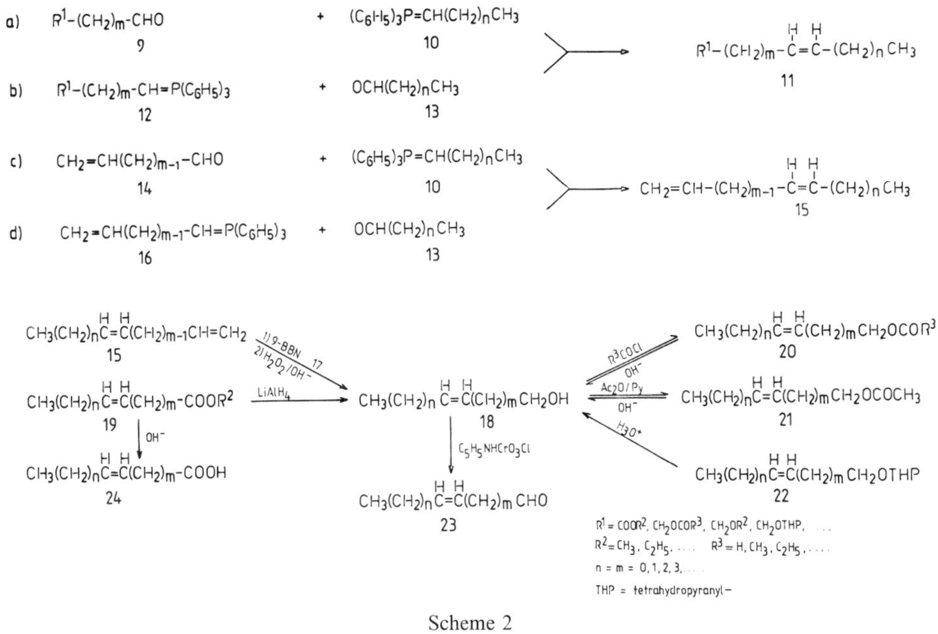

Scheme 2

The second part of Scheme 2 presents the consecutive reactions and the converting reactions to the particular Wittig reactions within this class of compounds. Thus, the 1, (Z)-m-alkadienes *15* resulting from the olefin generating reactions can be hydroborated regioselectively with 9-borabicyclo[3.3.1]nonane (9-BBN *17*) and subsequently oxidized to the (Z)-alkenols *18*. *18* can also be obtained by reduction of the (Z)-alkenoic acid esters *19* or by hydrolysis of the (Z)-alkenyl acetates *21*, alkenyl carboxylates *20*, and the (Z)-alkenyl tetrahydropyranyl ethers *22*. Oxidation of alkenols *18* with pyridinium chlorochromate (PCC) yields the (Z)-alkenals *23*, hydrolysis of *19* the (Z)-unsaturated fatty acids *24*.

By means of the Wittig synthesis (Z)-sex attractants can be prepared in the form of a "unitized construction system" [602]. Using this construction system Bestmann and coworkers prepared numerous (Z)-7-, (Z)-9-, and (Z)-11-unsaturated pheromones and analogues like alkenyl formates, acetates, propionates, butyrates, isobutyrates, alkenoic acid esters, alkenols, alkenals, alkenyl ethers, alkenyl methyl ketones etc. [44, 493, 614, 616] and used them for investigations of structure-activity relationships of lepidopterous pheromones [42–45].

Reacting 3-ethoxycarbonyl propylidene triphenylphosphorane with alkenals according the "silazid-technique" yields 1-substituted (Z)-4-alkenes [602, 619, 620], the same reaction with ethyl 4-formylbutanoate, obtained from ozone cleavage of cyclopentyl ethylether, leads to the corresponding (Z)-5-alkenes [602, 619].

The olefination of 4-pentenal leads to 1,(Z)-5-alkadienes in an elegant access, which can be hydroborated regiospecifically to (Z)-5-alken-1-ols [31, 602, 619]. The same reaction with 5-formylpentanoic acid esters, 6-formylhexanoic esters and 7-formylheptanoic esters yields 1-substituted (Z)-6-alkenes [602, 621], (Z)-7-alkenes [44, 602] and (Z)-8-alkenes [602, 621], respectively.

The ozone cleavage of derivatives of oleic acid gives rise to the corresponding ω-substituted nonanals, of which the stereoselective Wittig reaction gives numerous pheromones and pheromone analogues [44, 493, 602, 614, 615]. The ozonolysis product of methyl 10-undecenoate gives (Z)-10-unsaturated compounds [602, 621]. For the synthesis of (Z)-11-sex attractants ozonolysis products of vaccenic acid derivatives are available, as is 10-undecenal for the "alkadiene approach" [493, 616]. From thus prepared 1,(Z)-11-alkadienes (Z)-11-sex pheromones can easily be obtained. The same reaction sequence with the ozone cleavage product of erucic acid or brassidic acid esters yields (Z)-13-unsaturated insect attractants [622].

By stereoselective Wittig reaction

(Z)-5-decenyl acetate	[31],
(Z)-7-decenyl acetate	[621],
(Z)-7-dodecenyl acetate	[44, 623],
(Z)-8-dodecenyl acetate	[621],
(Z)-9-dodecenyl acetate	[614, 615],
(Z)-10-dodecenyl acetate	[602, 621],
(Z)-5-tetradecenyl acetate	[129],
(Z)-7-tetradecenyl acetate	[44, 134, 619, 623],
(Z)-8-tetradecenyl acetate	[619],
(Z)-9-tetradecenyl acetate	[609, 614, 615],
(Z)-10-tetradecenyl acetate	[602, 621],
(Z)-11-tetradecenyl acetate	[616],
(Z)-7-hexadecenyl acetate	[44],
(Z)-11-hexadecenyl acetate	[188, 616],
(Z)-13-octadecenyl acetate	[622],
(Z)-7-dodecenol	[44, 623],
(Z)-8-dodecenol	[621],
(Z)-11-tridecenol	[616, 621],
(Z)-7-tetradecenol	[623],
(Z)-9-tetradecenol	[614],
(Z)-11-tetradecenol	[616],

(Z)-7-hexadecenol	[44],
(Z)-11-hexadecenol	[188, 616],
(Z)-13-octadecenol	[622],
(Z)-9-tetradecenal	[614],
(Z)-11-tetradecenal	[616],
(Z)-11-hexadecenal	[616],
(Z)-13-octadecenal	[622],
(Z)-9-tetradecenyl formate	[614, 615],
(Z)-11-tetradecenyl formate	[616],
(Z)-5-decenyl 2-methylbutyrate	[619]

and other compounds have been synthesized. Most of the listed compounds are synthetic attractants, female sex pheromones, pheromone inhibitors or disruptants (see Table 1).

Bisunsaturated Lepidopterous Sex Attractants

Among the butterfly sex pheromones are a great number of double-unsaturated compounds, in which the positions and configurations of the double bonds considerably vary, in no apparent regular relationship among different Lepidopteran species. In their synthesis, the generation of (Z)-double bonds by means of Wittig reaction generally does not seem to raise difficulties.

For the preparation of bisolefins with one double bond of (E)-geometry this binding can either be formed by (E)-selective carbonyl olefination with phosphoranes having an electronsubtracting group at the ylide carbon atom (e.g. $\text{Ø}_3P{=}CHCOOR$, $\text{Ø}_3P{=}CHCHO$) and combined with the (Z)-Wittig synthesis or other olefin generating reactions. Many of these syntheses are characterized by a lower stereoselectivity than preparations of monounsaturated compounds. This makes the enrichment or the separation of distinct isomers a necessity.

In the synthesis of conjugated-unsaturated attractants with (E),(Z)-geometry like (E)-10, (Z)-12-hexadecadienol (bombykol, *30*), a pheromone component of the silkworm moth *Bombyx mori*, Bestmann and coworkers [624] olefinated methyl 9-formylnonanoate *25* with formylmethylene triphenylphosphorane *26* and reacted the so formed α, β-unsaturated aldehyde *27* (Z)-selectively to (E)-10,(Z)-12-hexadecadienoic acid methylester *29*. The reduction of *29* yielded (E)-10,(Z)-12-hexadecadienol (bombykol, *30*), and the oxidation of which (E)-10,(Z)-12-hexadecadienal (bombykal, *31*), the second component of the silkworm moth pheromone [537]. The isomer distribution was 89–95% (E),(Z)-, 3–6% (Z),(Z)-, 2–4% (E),(E)-, and $\leqq$1% (Z),(E)-form [624] (Scheme 3).

Following the same reaction sequence also (E)-9, (Z)-11-tetradecadienyl acetate, (E)-9,(Z)-11-pentadecadienyl acetate, (E)-9,(Z)-11-hexadecadienyl acetate, (E)-10, (Z)-12-pentadecadienyl acetate, (E)-10, (Z)-12-heptadecadienyl acetate and the corresponding alcohols could be obtained [624]. Analogous conversions of 10-undecenal led *via* “alkadiene approach” to (E)-11, (Z)-13-hexadecadienol, (E)-11, (Z)-13-heptadecadienol, (E)-11, (Z)-13-octadecadienol and the appropriate acetates [624].

Truscheit and Eiter [625] prepared bombykol *30* by the Wittig reaction of 11-formyl-(E)-10-undecenyl acetate *32*, and separated the small amount of (E), (E)-isomer in form of its urea inclusion compound. “Exchanging” the reaction partners of the Wittig reaction, i.e. the condensation of (Z)-2-hexenylidene triphenylphosphorane *33* and aldehyde *34*, formed a mixture of the (Z),(Z)- and (E),(Z)-ester *35*, which could have been resolved in form of the carboxylic acids *36* by fractionated recrystallization [625]. This example demonstrated, that the (Z)-stereoselectivity of the Wittig reaction is maintained when reacting conjugated-unsaturated aldehydes and basic ylides, whereas the same reaction between allylic ylides and alkanals yields olefins with approximately 50% (E)-isomer of the newly generated double bond [625] (Scheme 3).

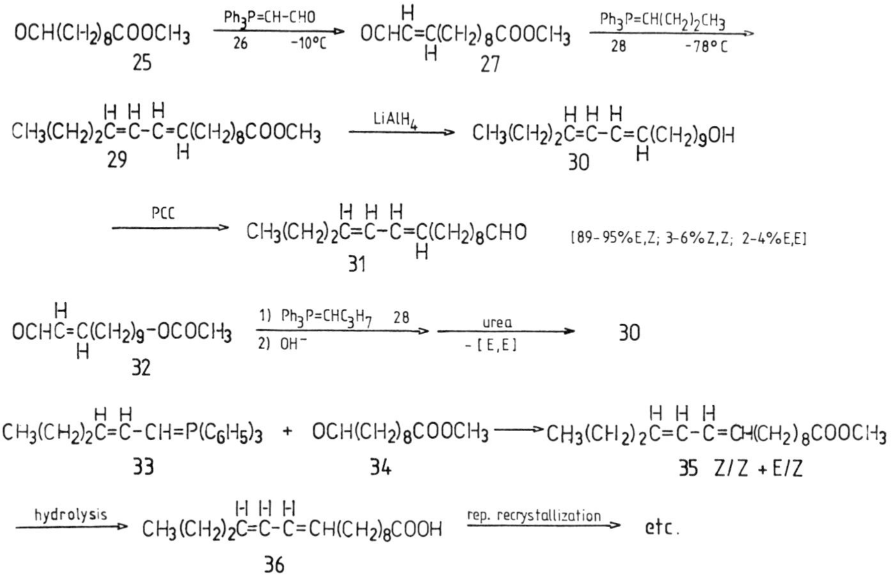

Scheme 3

Bombykol was the first time prepared by Butenandt using Wittig reaction of 2-hexynylidene triphenylphosphorane and 9-formyl nonanoic ester. The resulting (Z)- resp. (E)-10-hexadecen-12-ynoic acid esters were resolved by urea, the (E)-compound hydrogenolyzed with Lindlar catalyst and reduced with lithium aluminium hydride. Further purification *via* the urea inclusion compound and low temperature crystallization gave (E)-10,(Z)-12-hexadecadienol *30* [626, 627].

Several methods based on the Wittig reaction have been used for the synthesis of (Z)-9,(E)-11-tetradecadienyl acetate *40*, the main component of the pheromone of *Spodoptera littoralis* and *S. litura*. The condensation of (E)-2-pentenal *37* with the ylide of the THP-alkyl triphenylphosphoniumsalt *38* or the ylide of 9-acetoxynonyl triphenylphosphonium bromide *39* yielded acetate *40* with an (Z),(E)-isomer amount of 80–95% [621, 628–630] depending on the reaction conditions. The small amount of the (E),(E)-isomer can be precipitated from the mixture with tetracyanoethylene as the Diels-Alder adduct [629]. The reversed procedure, i.e. reacting (E)-2-pentenylidene triphenylphosphorane *41* and 8-formyloctyl acetate *42*, gave isomeric mixtures between 40:60 and 60:40% of the (Z),(E)- and the (E),(E)-form and required subsequent isomer separation. *40* could also be obtained by the reaction of (E)-2-pentenal *37* and 9-hydroxynonyl triphenylphosphonium bromide *43* with butyl lithium *44* as the base in HMPA and subsequent acetylation [631] (Scheme 4).

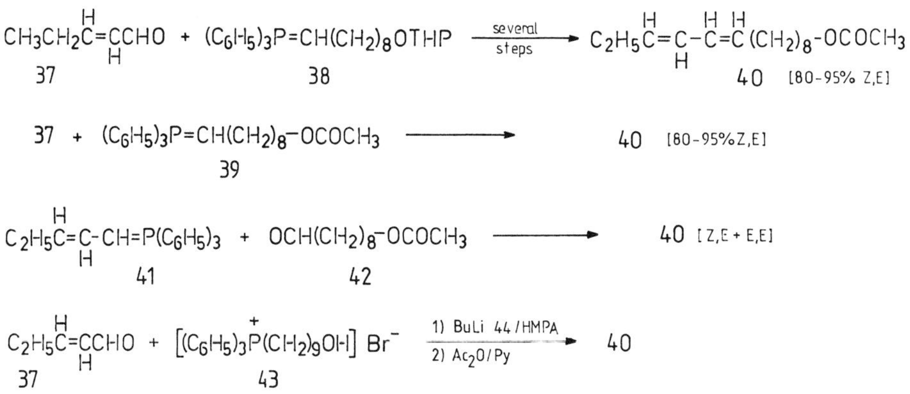

Scheme 4

(E)-7,(Z)-9-Dodecadienyl acetate *54* is the sex attractant of the European grape berry moth *Lobesia botrana*, a dangerous vine pest of Southern Europe. Originally Roelofs et al. [194] prepared *54* by Wittig reaction of diethyl 2-(cyclohexylimino)-ethyl phosphonate *48* with methyl 6-formylhexanoate *47*, which was formed by pyridine-N-oxide oxidation from methyl 7-bromoheptanoate *45*. A second Wittig condensation of the aldehyde *49* formed with the ylide *50* gave the dienic ester *51*. *51* was hydrolyzed to *52* and subsequently reduced, while the dienic alcohol *53* was esterified to *54* (~88% (E),(Z)-isomer after prep. GC-purification) [194] (Scheme 5). In principle a similar synthesis for *54* involves the use of formylmethylene triphenylphosphorane and diethyl ethoxycarbonylmethyl phosphonate. The second olefination step should then be performed under "saltfree" conditions [621, 630, 632].

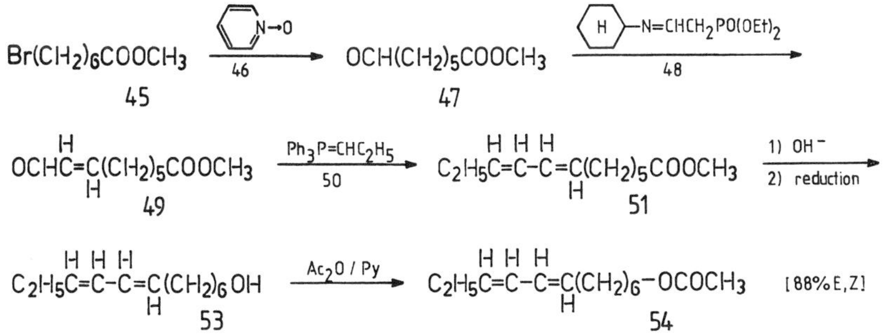

Scheme 5

The female of the codling moth *Laspeyresia pomonella* uses (E)-8,(E)-10-dodecadienol (codlemone®, *60*) as the attractive component of its sex pheromone system (cf. Table 1). This compound was first synthesized by Roelofs et al., as shown in Scheme 6, by Wittig reaction of (E)-2-butenyl triphenylphosphonium bromide *55* and methyl 7-formylheptanoate *56*. The isomeric esters *57* were hydrolyzed and the acid *58* was purified by crystallization and then reduced to the dienols *59*. The isomeric mixture was isomerized with iodine/sun light and an enriched (E),(E)-dienol *60* obtained. For the isolation of pure dienol *60* the crude product was acetylated, the acetate *61* separated into isomers by means of $AgNO_3$-impregnated thin layer chromatography plates (cf. p. 79), and again hydrolyzed to *60* [98% (E),(E)] [248] (Scheme 6). A modified method of reacting (E)-2-butenylidene phosphorane *62* with ethyl 7-formylheptanoate *63* in the presence of lithium bromide yielded a mixture of diene esters *64* (~45% E, E), and their reduction gave the corresponding dienols *59*. The esters *61* formed by acetylation could again be separated into their isomers by silver nitrate/silica gel – liquid chromatography (95% E, E) [633] (Scheme 6). The conversion of (E)-2-butenal *65* with 8-THP-octylidene phosphorane *66* afforded predominantly the (Z)-8, (E)-10-isomer *67* with 77% yield.

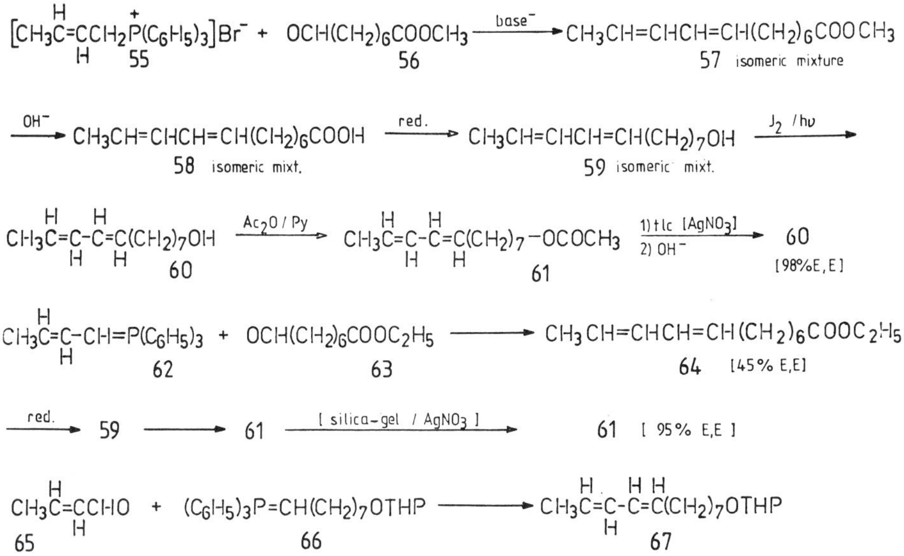

Scheme 6

Vig et al. have described a synthesis of codlemone *60* using a condensation according to the phosphonate method [634]. The metallated phosphonate *68* was olefinated (E)-selectively with the THP-aldehyde *69* giving rise to the (E), (E)-alkadienoic ester *70*. The terminal ethoxycarbonyl group was reduced, mesylated, and the mesylate *71* reduced and hydrolyzed to *60* (Scheme 7). The authors do not document, however, the steric composition of their synthesis product [634].

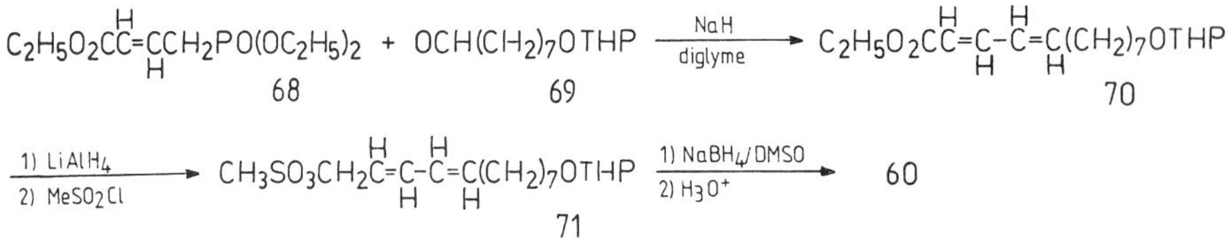

Scheme 7

Isomers and derivatives of codlemone can also be obtained with fairly good stereochemical purity by a Wittig reaction [635]. (Z)-Carbonyl olefination of butenal *65* with the ylide *72* of 8-acetoxyoctyl triphenylphosphonium bromide yielded (Z)-8,(E)-10-dodecadienyl acetate *73* and (Z)-8,(E)-10-dodecadienol *74*. Starting with 1,7-octadiene *75* one can reach 7-octenal *76* by monohydroboration and oxidation. This can be olefinated with phosphorane *26* giving (E)-2,9-decadienal *77* and subsequently with *78* (Z)-stereoselectively to 1,(E)-8, (Z)-10-dodecatriene *79*. The specific hydroboration of the terminal double bond led to (E)-8, (Z)-10-dodecadienol *80*, which gave (E)-8, (Z)-10-dodecadienyl acetate *81* upon being acetylated (isomer ratio 85% E,Z; 14% E,E; 1% Z,E) [635] (Scheme 8).

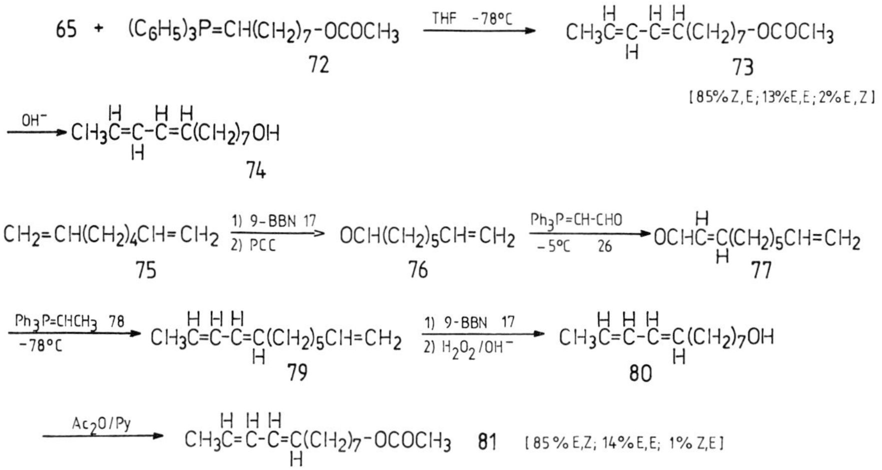

Scheme 8

(E)-9,11-Dodecadienyl acetate *85* is the sex attractant of the red bollworm *Diparopsis castanea*, a cotton pest of South Africa. Witting reactions of the allylidene phosphorane *82* and 8-formyloctyl acetate *42* afforded E:Z-isomer ratios of *83* from 4:1 to 3:7 depending on the solvent polarity. Isomeric resolution by silver nitrate/silica gel caused difficulties, but by formation of sulfolen *84* and its stereospecific thermolysis the (E)-isomer *85* could be isolated in +99% stereochemical purity. Reacting the mixture with tetracyanoethylene *86* the (E)-form *85* could be separated and thus the (Z)-isomer (Z)-9,11-dodecadienyl acetate *87* be obtained [636] (Scheme 9). The conversion of *42* with formylmethylene phosphorane *26* and a second Wittig reaction with methylene triphenylphosphorane *89* gave a significant higher stereoselectivity with 96% (E)-isomer for the synthesis of *85* [630] (Scheme 9).

(Z)-3,(E)-5-Tetradecadienyl acetate *92* recently was proposed as the main component of the pheromone of *Prionoxystus robiniae* on the basis of electroantennogram responses, and could be prepared by a Wittig reaction of (E)-2-undecenal *90* and the ylide *91*. Thereafter the (Z)-3,(E)-5-isomer *92* could be separated in a purity of 95% from the distillation residue (98% (E)-3, (E)-5-form) by means of a spinning ribbon column [209] (Scheme 9).

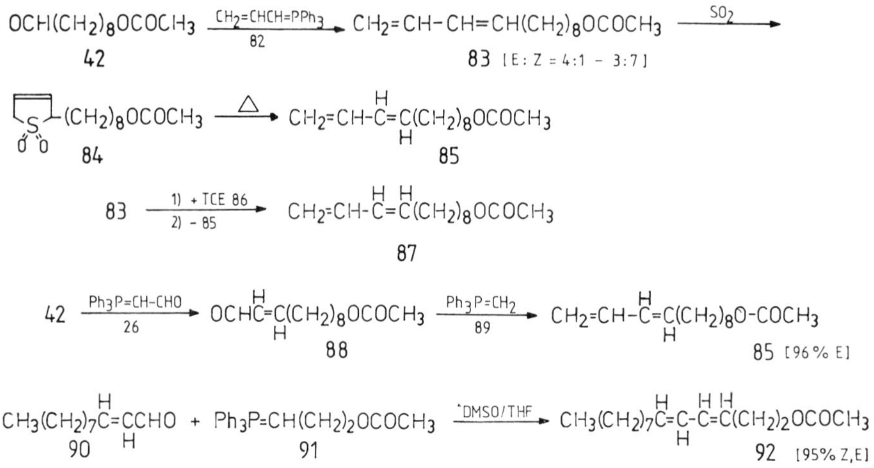

Scheme 9

(Z)-9,(E)-12-Tetradecadienyl acetate *97*, a pheromone substance of the stored products pests *C. cautella*, *P. interpunctuella*, *S. eridiana*, *A. kühniella*, *E. elutella*, *S. exigua*, *C. figulella* and *S. litura*, having "methylene interrupted" double bonds has been synthesized, among other routes, by carbonyl olefination. Starting from cyclopropylethanol *93* with Julia synthesis [637] one gets (E)-3-pentenyl bromide *94*. The phosphonium salt *95* of compound *94* can generate the ylide *96* when reacted with potassium/HMPA [638] or sodium/DMSO [587], and olefinated with 8-formyloctyl acetate *42* (5–9% E,E-isomer content) [587, 638] (Scheme 10). The synthesis product could be obtained isomer free by high performance liquid chromatography on a $AgNO_3$-loaded florisil column.

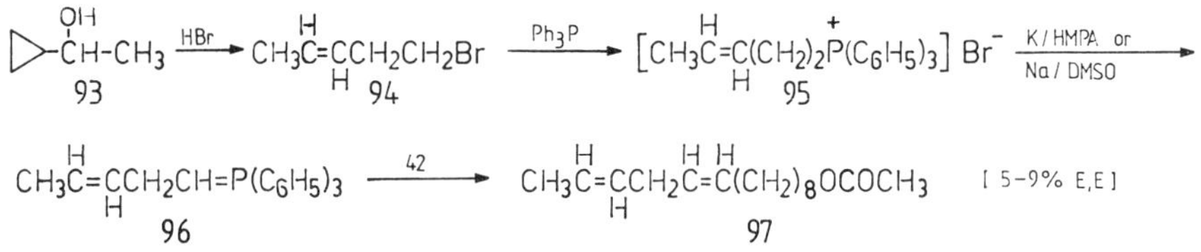

Scheme 10

(Z)-7, (Z)-11-Hexadecadienyl acetate *104* and (Z)-7, (E)-11-hexadecadienyl acetate *105* make up in equal proportions the sex attractant of the pink bollworm *Pectinophora gossypiella*. The (Z)-7,(E)-11-isomer *105* is the sole attractant of the grain moth *Sitotroga cerealella* (cf. Table 1), while the other isomers represent pheromone inhibitors. Sonnet synthesized the (Z),(Z)-isomer *104* by alkylation of the lithium acetylide *98* with 1-bromo-3-chloropropane *99*. Then he formed the phosphoniumsalt *100* and olefinated the corresponding ylide *101* with pentanal *102*. Subsequent acetylation led to (Z)-11-hexadec-7-ynyl acetate *103* (Z : E = 96 : 4) [639]. The (Z)-hydrogenolysis of the triple bond in *103* gave rise to (Z)-7,(Z)-11-hexadecadienyl acetate *104*. The author was not able to prepare the corresponding (E)-isomer *105* by means of Wittig reaction [639] following the method of Schlosser [640] (Scheme 11).

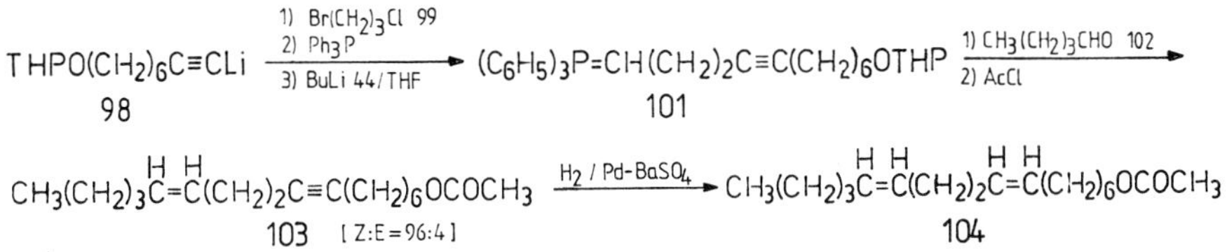

Scheme 11

Bestmann et al. [620] prepared the isomers *104* and *105* each *via* carbonyl olefination according to their "silazid technique" [615], and condensed (Z)-4-nonenylidene triphenylphosphorane *109*, and (E)-4-nonenylidene triphenylphosphorane *115*, respectively, with ethyl 6-formylhexanoate *110*. The bromides *108* and *114* necessary for the ylide formation could be obtained by Wittig synthesis of pentanal *102* and 3-ethoxycarbonylpropylidene triphenylphosphorane *106* or by Crombie reaction of 2-butyl-3-chlorotetrahydropyrane *113*. The dienic acid esters *111* and *116* were reduced to alkadienols *112* and *117* and subsequently esterified to acetates *104* and *105* (*104*: 94.5% Z,Z-amount; *105*: 83.7% Z,E-form) [620].

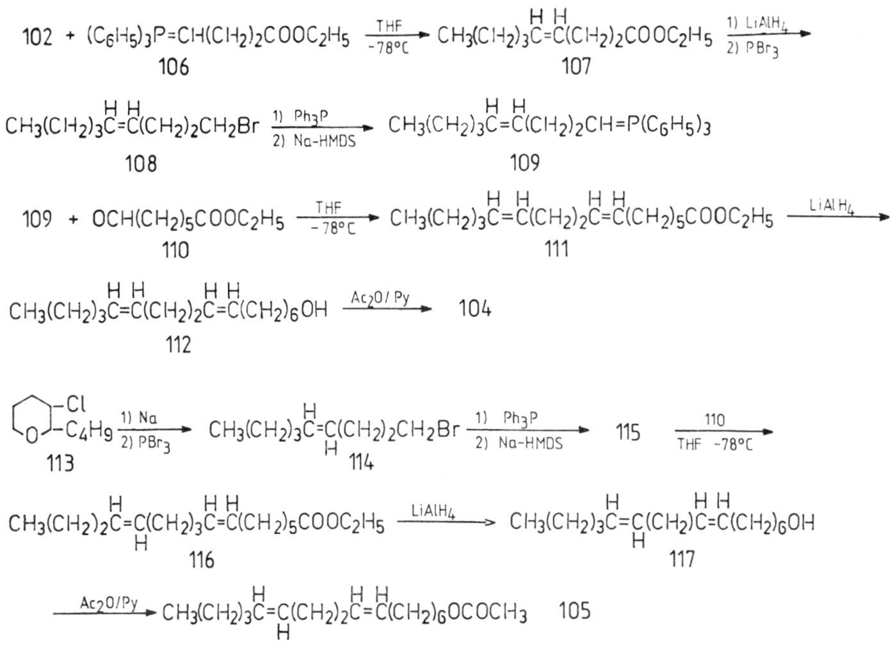

Scheme 12

An elegant method, directly leading to a 1:1 mixture of the acetates *104* and *105* (gossyplure) was developed by Anderson and Henrick [641], who started the synthesis with (Z)-1,(Z)-5-cyclooctadiene *118*. The selective, unsymmetrical cleavage of one of the two double bonds *via* the epoxide *119* and the ketol *120* led to a synthon *121*, the double bond of which was already purely (Z)-configurated because of the original geometry of the cyclooctadiene starting material.

Stereocontrolled Wittig reaction in presence of LiBr and ethanol yielded a 1 : 1 mixture of the (Z),(E)- and (Z),(Z)-esters *123*, which could have been coupled together with the three carbon unit *124*. Hydrolysis and acetylation gave the both pheromone acetates, *104* and *105* [641] (Scheme 13).

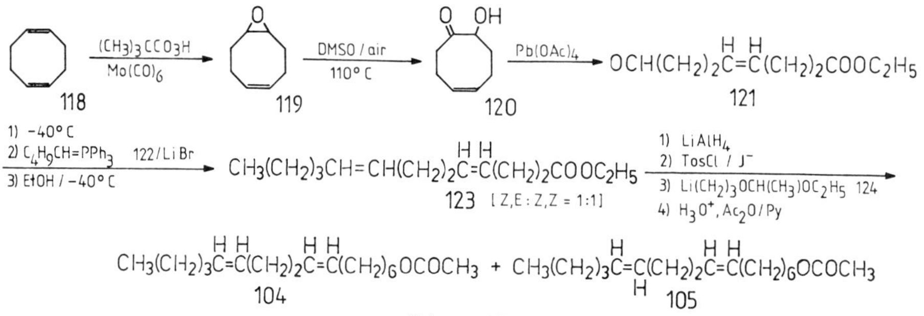

Scheme 13

For the synthesis of single *105* Hammond and Descoins also used a Wittig reaction [642]. Starting with 1-hepten-3-ol *125* they obtained after a few steps the phosphonium salt *115*, which could be olefinated with the aldehyde *110*, available from suberic acid monoethylester *128*. Reduction and subsequent acetylation gave pheromone *105* according Scheme 14 [642].

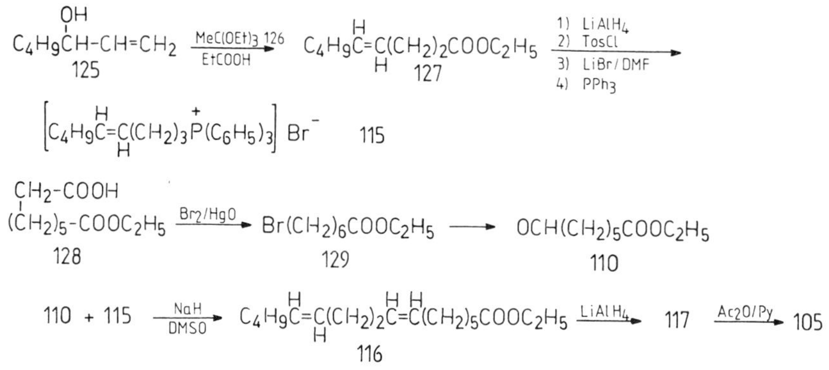

Scheme 14

Other Structures

(Z)-7,8-Epoxy-2-methyloctadecane *136* (disparlure, disparmone®) can be easily obtained as a racemic mixture (cf. p. 123) from 2-methyl-(Z)-7-octadecene *134* by stereoselective epoxidation with m-chloroperbenzoic acid *135* (m-CPBA). The olefin, which at the same time was proved to be a precursor of biosynthesis [285], was prepared by Wittig reactions in many cases [277, 287, 643]. Thus the Grignard reagent of isoamyl bromide *130* can be reacted with oxetane *131* giving isooctyl triphenylphosphonium bromide *132*, and this olefinated with undecanal *133* giving 2-methyl-(Z)-7-octadecene *134*. The (Z)-stereospecific epoxidation yields disparlure *136*, the pheromone of the gypsy moth *Lymantria dispar* [643] (Scheme 15). (For the synthesis of optically active disparlure and its biological effect refer to p. 123).

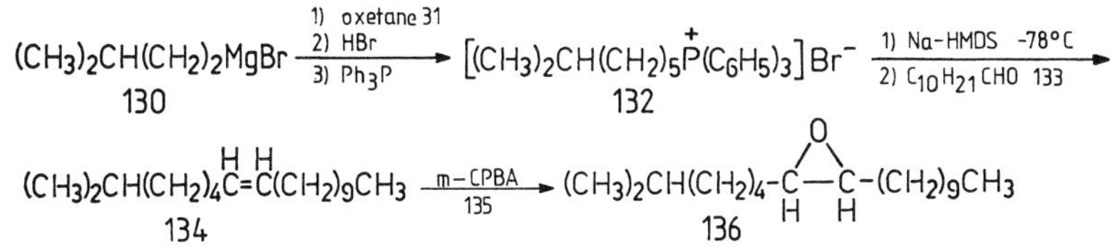

Scheme 15

(Z)-6-Heneicosen-11-one *143*, the sex attractant of the Douglas fir tussock moth *Orgya pseudotsugata*, was prepared by Fetizon and Lazare using a combination of Wittig reaction and Grignard condensation [644]. 2-Hydroxytetrahydropyrane *137* is converted under (Z)-selective olefination conditions with hexylidene triphenylphosphorane *138* to (Z)-5-undecen-1-ol *139*, which is oxidized to the corresponding aldehyde *140*. The Grignard reaction of *140* and decyl magnesium bromide *141* gave (Z)-6-heneicosen-11-ol *142*, which could be oxidized to the attractant *143* using pyridinium chlorochromate [644] (Scheme 16).

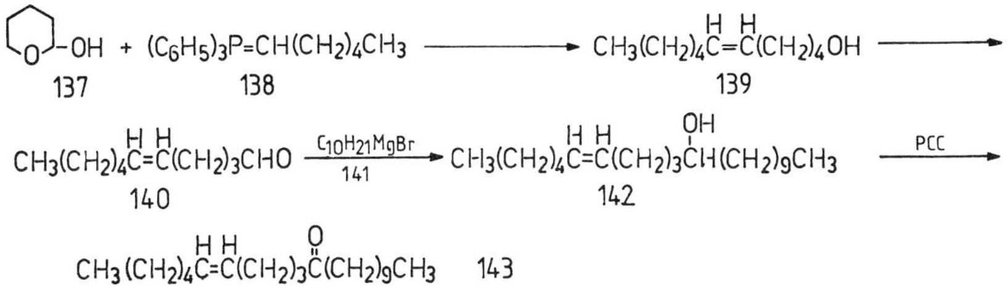

Scheme 16

(E)-Olefinic Moth Sex Pheromones by Wittig Reaction

Limited because of the short lifetime of the betain intermediate *5* [617], as it was discussed on p. 92 (cf. Scheme 1), alkylidene triphenylphosphoranes are principally inappropriate for Wittig synthesis of (E)-olefins. Only one (E)-selective olefin generating reaction, based on β-oxidoalkylidene triphenylphosphoranes as intermediates, was reported to yield olefinic hydrocarbons with 98% (E)-isomer content [640]. Nevertheless, this variation of carbonyl olefination has not been used successfully in pheromone synthesis till now (see [639]). In fact, often (E)-isomers have been obtained by preparation of the corresponding (Z)-olefins and subsequent stereomutation, using HNO_2 [645], irridiation by UV-light, and thiols and diphenyl phosphine [646, 649], or treating the corresponding epoxide with diphenyl phosphine lithium and methyl iodide [647, 643].

Only phosphoranes with electron withdrawing substituents in α-position, like $(C_6H_5)_3P{=}CH{-}CHO$, $(C_6H_5)_3P{=}CH{-}COOR$, $(C_6H_5)_3P{=}CH{-}CN$, etc., are able to generate (E)-olefins with a stereoselectivity of more than 95%. Thus the condensation of formylmethylene triphenylphosphorane *26* with aliphatic aldehydes often has been used for the formation of (E)-α,β-unsaturated aldehydes in pheromone synthesis (see Schemes 3, 9).

Another reaction of comparible importance in (E)-olefin synthesis is the Wittig reaction of e.g. $(C_6H_5)_3P{=}CH{-}COOCH_3$ *144* with alkanals, resulting in methyl

(E)-2-alkenoates. These α,β-unsaturated esters, which can be reduced to (E)-2-alken-1-ols, converted into (E)-2-alkenyl halogenides, the corresponding phosphonium salts, etc., hence are important intermediates for the preparation of bisolefinic sex attractants (see [630]).

b) The "Acetylenic Approach" to Pheromone Synthesis
Alkenols and Alkenyl Acetates

The partial hydrogenolysis of triple bonds by means of palladium catalysts or homogenous hydrogenation using sodium in liquid ammonia seems to be the most universal reaction for the synthesis of (Z)- or (E)-olefinic sex attractants. The (Z)-reduction of triple bonds with Lindler catalyst (Pd–$CaCO_3$–$Pb(OAc)_2$) in presence of quinoline usually gives (Z)-olefins in good yields [650, 651]. The (E)-isomer content is about 1.5–5%, and about 10% at higher reaction temperature. In some cases this method is inferior to stereoselective Wittig syntheses. However, new investigations demonstrate, that Lindlar hydrogenations at −30 to −10 degrees in pentane, hexane, or hexane/tetrahydrofurane, afford (Z)-olefins with 99.5% (Z)-content. The remaining 0.5% (E)-isomer are formed during the reaction and not by consequent stereomutation [652]. Other alternatives for the partial hydrogenation of alkynes to (Z)-alkenes are represented by catalysts like palladium/barium sulfate [653], "P-2 nickel" [654], other nickel catalysts [655] and sterically hindered hydroboration reagents like disiamyl borane [656, 657].

The reduction of acetylenes to (E)-olefins is usually performed using sodium or lithium in liquid ammonia [658, 659] and affords a high (E)-stereoselectivity with very small or undetectable traces of (Z)-isomers [660, 581]. In spite of this good selectivity the reaction gives rather poor yields with alkynes having chain lengths of more than C_{13} [659, 660]. For these cases the corresponding reduction of the alkynes using lithium aluminium hydride in tetrahydrofurane and ether should be preferred [660, 661]. On the other hand, surprisingly, using this method neither β-alkynols nor γ-alkynols could have been converted to the corresponding (E)-alkenols [662]. Lower and higher molecular (E)-β- or (E)-γ-alkenols have been obtained by reduction using a great excess of aluminate in a mixture of diglyme/tetrahydrofurane, as was demonstrated in the synthesis of (E)-7-dodecenyl acetate [663].

Scheme 17 pertains to the general preparation of monounsaturated sex attractants *via* "acetylenic approach", and the synthesis of (Z)-9- and (E)-9-dodecenyl acetate *155* and *159*, (Z)-9- and (E)-9-dodecen-1-ol *156* and *160*, and (Z)-9- and (E)-9-dodecenal *157* and *161*. 1,8-Octanediol *145* is converted *via* the monobromo compound *146* to the ω-bromoalkyl tetrahyropyranylether *148*, and with it lithium acetylide *149* alkylated to the decynyl tetrahydropyranylether *150*. A second alkylation with ethylbromide *151* affords 9-dodecynyl-THP-ether *153*, which either can be converted to alkynyl acetate *154* and (Z)-stereoselectively hydrogenated to (Z)-9-dodecenyl acetate *155*, or partially hydrogenated to (E)-9-dodecenyl THP-ether *158* and subsequently acetylated to (E)-9-dodecenyl acetate *159*. Hydrolysis of either acetates or THP-ethers, respectively, affords (Z)-9-dodecenol *156* and (E)-9-dodecenol *160*, which can be oxidized to the corresponding dodecenals *157* and *161*, respectively (Scheme 17).

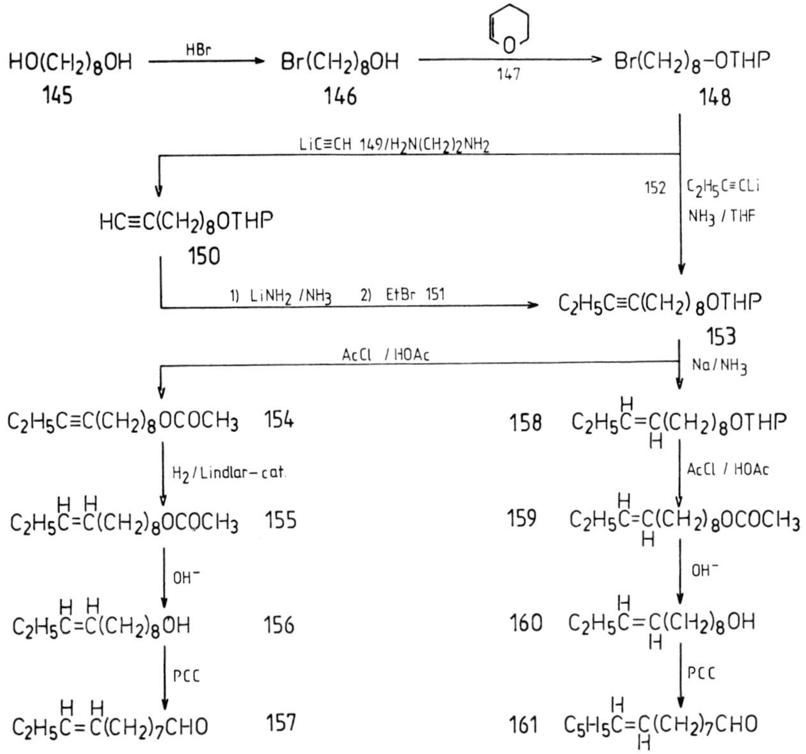

Scheme 17

For the synthesis of the ω-bromoalkanols, e.g. *146*, a continously extraction method with HBr and heptane, petrol ether, toluene, etc., for the bromination of polymethylenglykols [826, 596, 664, 665] is practically used. 5-Halopentan-1-ol can be obtained in good yields by treating tetrahydropyrane with acetyl chloride or bromide and zink chloride, with subsequent hydrolysis of the haloacetate [666].

The alkylation of acetylides, e.g. *149*, *150*, and *152*, with primary halogenides [667, 668] has been investigated because of the poor solubility of long chain alkyl halides and the metal acetylides (especially the sodium salts) in liquid ammonia and in many other solvents. Often an excess of sodium amide is used for the deprotonation of the alkynes; alkyl chlorides normally give only poor yields. Polar solvents like dimethyl sulfoxide (DMSO) [667, 669, 670], hexamethyl phosphoric acid amide (HMPA) [671–676], dimethyl formamide (DMF) [672, 677, 678], DMF and xylene [672], or DMF and tetrahydrofurane (THF) [673], are preferred. For simple laboratory scale syntheses the lithium acetylides were generated best with butyl lithium *44* in THF or HMPA/THF at -20 to 0 degrees [671, 675, 676, 679] or with $LiNH_2$ in THF at 65 °C [596]. DMSO is best used only as reaction solvent for alkylations with the commercially available lithium acetylide ethylenediamine complex [680–683], because ordinary alkylation of long chain lithium or sodium alkynides gives low yields in DMSO [684], since the metallation of the sulfoxide to the methyl sulfinylmethid anion prevails a competitive reaction. 1-Alkynes can already be isomerized at room temperature to

2-alkynes using an excess lithium acetylide in HMPA or HMPA/THF [670, 685]. The same is possible with longer heating with sodium amide in DMSO [686].

An advantage of the acetylenic approach for pheromones is the possibility of purifying the alkynols by low temperature recrystallization from pentane (−20 °C), because their melting points generally lie in the region of room temperature.

Alkadienols, Alkadienyl Acetates, Alkadienals and Other Structures

The preparation of (Z)- and (E)-monounsaturated butterfly attractants via the acetylenic route in most cases represents the most suitable synthesis technique. However for the preparation of double unsaturated pheromones, especially with different geometries of double bonds, either the combination of acetylene coupling reactions with other olefin generating reactions is needed, or a strict separation of the different hydrogenolysis steps. Numerous different methods by means of partial acetylene hydrogenations have been elaborated for the synthesis of conjugated-unsaturated attractants with (E),(Z)-configuration, e.g. (E)-7,(Z)-9-dodecadienyl acetate, the sex pheromone of the grape berry moth *Lobesia botrana* (Scheme 18–20).

Labovitz et al. [687] heated 1-hepten-4-yn-3-ol *164* with trimethyl orthoacetate *165* obtaining methyl 4-nonen-6-ynoate *166* with an isomer ratio of E:Z=4:1. *166* was converted over several steps into the bromoalkenyne *167*. This was coupled together with lithium 3-(1-ethoxy)ethoxypropyl *124*, and the reaction product *168* hydrolyzed to the alkynol *169*. The crude product *169* was recrystallized at −35 °C from pentane and (E)-7-dodecen-9-yn-1-ol *169* was obtained in a purity of 99%. The corresponding trimethylsilyl derivative *171* was selectively hydroborated with bis(1,2-dimethylpropyl)borane *172* (disiamyl borane) and the protecting group hydrolytically splitted off. The resulting dienol *173* was recrystallized at −50 °C from pentane and acetylated using acetic anhydride/pyridine to the pheromone *174* with a gaschromatographically determined isomer ratio of 97.5% E,Z and 1.8% E,E [687].

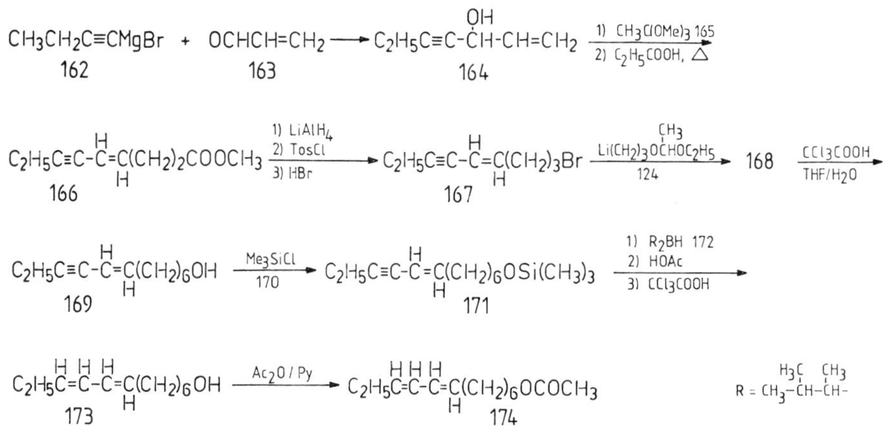

Scheme 18

Negishi [688] obtained by "acetylene zipp reaction" [689] 7-octynyl acetate *176* from 2-octyn-1-ol *175*, the former could be converted to (E)-7-dodecen-9-ynyl acetate *177* with borane *172* and lithium butynide *152*. Partial hydrogenation of the triple bond of *177* with powdered zinc in propanol or by hydroboration afforded the attractant *174* (Scheme 19) [688].

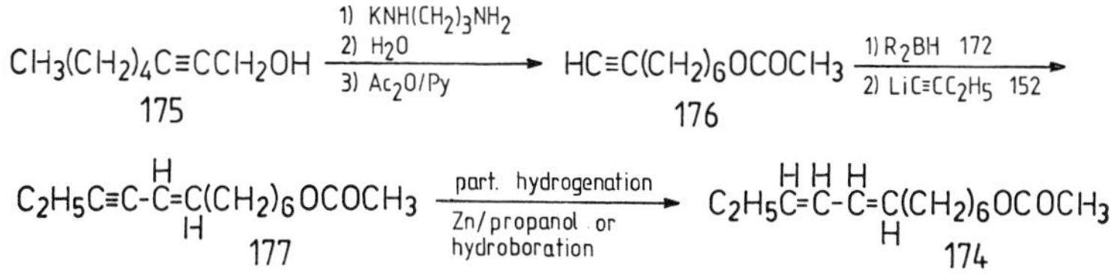

Scheme 19

Descoins and Samain also synthesized the pheromone *174 via* alkynol *169* as a synthesis intermediate [690]. They modified the Julia rearrangement of 1-cyclopropyl-2-pentyn-1-ol *178* [691], which usually gives (Z)-3-octen-5-ynyl bromide *182* with only 67% (Z)-amount, by increasing the sterical hindrance of the ring opening reaction by the triple bond by formation of complex *180* with dicobalt octacarbonyl *179*. Treating the complex *180* with bromic acid and zinc bromide and subsequent oxidizing with iron-III-nitrate yielded the (E)-homoallylic bromide *182* with a stereoselectivity of 99% [690]. *182* was copper-I-calatyzed coupled with the Grignard reagent *183*, hydrolyzed to alkynol *169* (96% E-isomer), recrystallized to a +99% product, and at −20 °C partially hydrogenated to the alkadienol *173* using a special Lindlar catalyst (92% E, Z; 8% E, E). The acetylation of the gaschromatographically purified dienol *173* gave once more (E)-7, (Z)-9-dodecadienyl acetate *174* (Scheme 20).

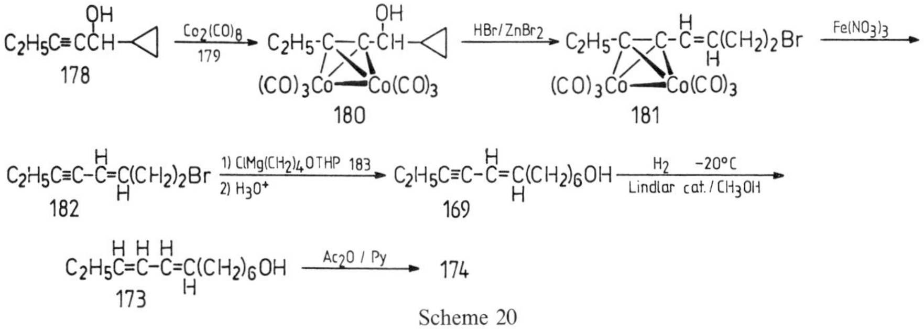

Scheme 20

(Z)-9, (E)-11-Tetradecadienyl acetate *40* (a component of the pheromone complex of the cotton pest *Spodoptera littoralis* and the cereals pest *S. litura* and another example for a conjugated E,Z-system), was synthesized by Tamaki et al. [212] *via* acetylenic route as a mixture with the Z,Z-isomer (Scheme 21). By preparative gas chromatography *40* could be isolated followingly [212].

Carney and Baum [652] synthesized *40 via* (E)-3-hexen-1-yne *191*, which could be obtained from 1-hexyn-4-ol, *190*, the Grignard product of *188* and *189* [692, 693]. Alkylation with *148*, trimethylsilylation, hydroboration with disiamyl borane *172*,

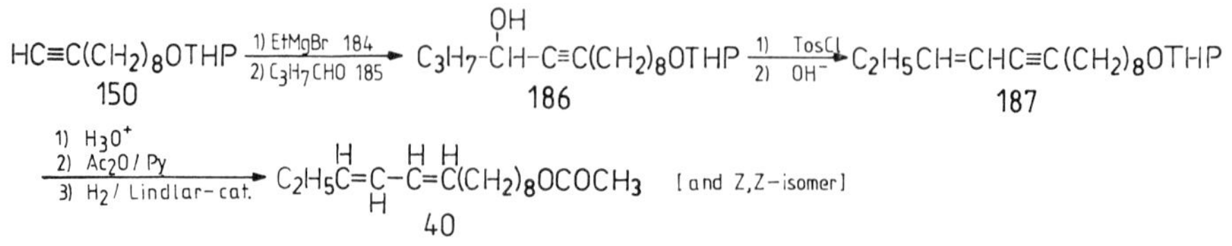

Scheme 21

hydrolysis and acetylation afforded the sex attractant *40*, as shown in Scheme 22. The (E)-11-tetradecen-9-yn-1-ol *192* was purified by crystallization from pentane (mp. 11 °C), the alcohol function protected (*193*), and the alkadienol *194* obtained by hydroboration and hydrolysis, again purified by low temperature recrystallization from pentane (−60 °C). The following acetylation afforded (Z)-9,(E)-11-tetradecadienyl acetate *40*, containing 0.5% E,E-isomer [652].

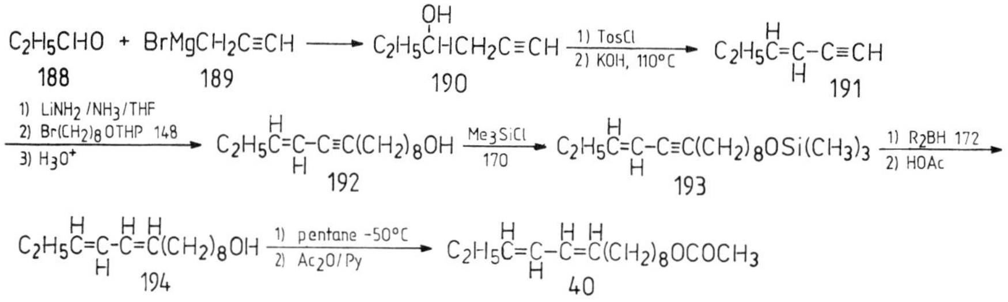

Scheme 22

Using the same sequence of synthesis steps and starting from (Z)-3-hexen-1-yne the (Z)-9, (Z)-11-isomer was prepared. Isomerization of (Z)-9, (Z)-11-tetradecadienol (mp. 17 °C) with thiophenol [694] and repeated crystallizations from pentane yielded the corresponding E,E-alcohol and the (E)-9,(E)-11-tetradecadienyl acetate in 99.4% purity [652].

(E)-10, (Z)-12-Hexadecadienol *30* (bombykol), the sex attractant pheromone of the silk worm moth *Bombyx mori*, was among many other ways synthesized by combination of Wittig olefination and acetylene synthesis (cf. p. 96) [626, 627]. An alternative route was the condensation of epoxide *195* with lithium acetylide *196* to the alkynol *197*. Elimination of the resulting tosylate, and resolution *via* the urea inclusion compound led to pure (E)-alkynol *198*, the hydrogenolysis of which by means of Lindlar catalyst yielded bombykol *30* [626, 627] (Scheme 23).

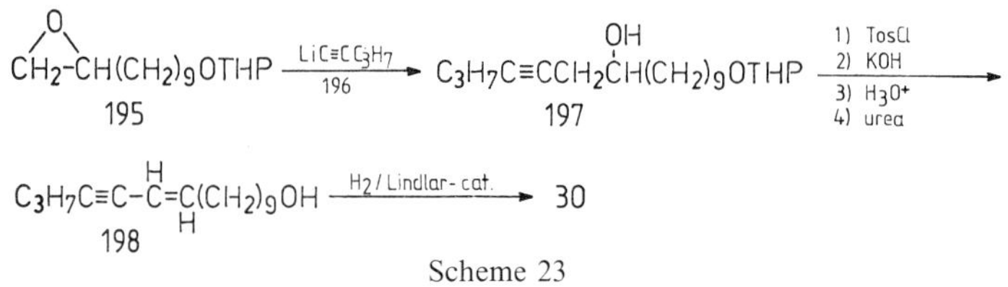

Scheme 23

Truscheit and Eiter reported a similar synthesis involving the conversion of the alkynol ester *200*, which is derived from *199* and *25*, into the chloride *201*. The dehalogenation, reduction, etherification and alkylation of this compound afforded the enynol-THP-ether *204*, and further hydrolysis and partial hydrogenation gave (E)-10, (Z)-12-hexadecadien-1-ol *30* [625].

This could be separated at the stage of the (E)-enynol *198* from the E-isomer by crystallization of the urea inclosure compound, and the final product *30* obtained by repeated crystallization from petrol ether purified [625] (Scheme 24).

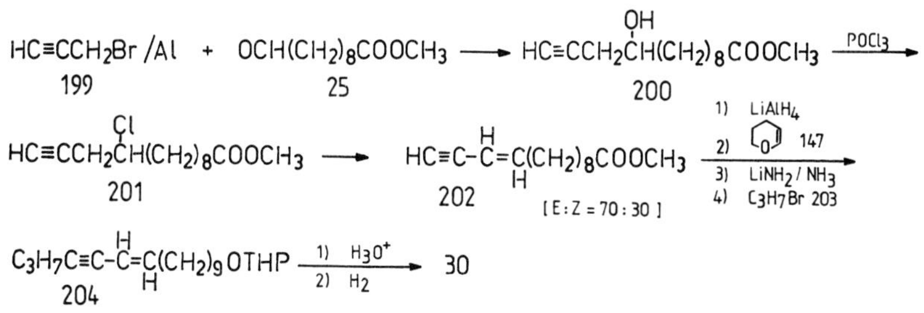

Scheme 24

Negishi et al. [695] synthesized bombykol *30* using their method for stereoselective synthesis of conjugated (E)-enynes. Addition of disiamylborane *172* to the alkynol derivative *205* and subsequent alkynylation and treatment with iodine yielded the alkenynyl-trimethylsilyl ether *209*, which could be hydroborated affording *30* (Scheme 25). This hydroboration-protonolysis for the conversion of conjugated enynes to dienes seems to be superior than the partial hydrogenolysis.

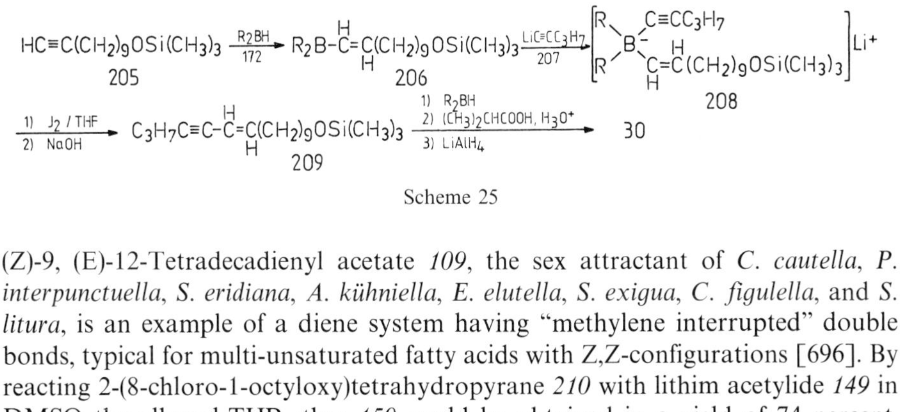

Scheme 25

(Z)-9, (E)-12-Tetradecadienyl acetate *109*, the sex attractant of *C. cautella*, *P. interpunctuella*, *S. eridiana*, *A. kühniella*, *E. elutella*, *S. exigua*, *C. figulella*, and *S. litura*, is an example of a diene system having "methylene interrupted" double bonds, typical for multi-unsaturated fatty acids with Z,Z-configurations [696]. By reacting 2-(8-chloro-1-octyloxy)tetrahydropyrane *210* with lithim acetylide *149* in DMSO the alkynyl-THP-ether *150* could be obtained in a yield of 74 percent. Further reaction with (E)-2-butenyl bromide *211* gave (E)-12-tetradecen-9-inyl-THP-ether *212* (29% yield) [75]. The partial hydrogenation to *213* and subsequent acetylation to *109* (94% yield) followed the usual methods (Scheme 26) [75].

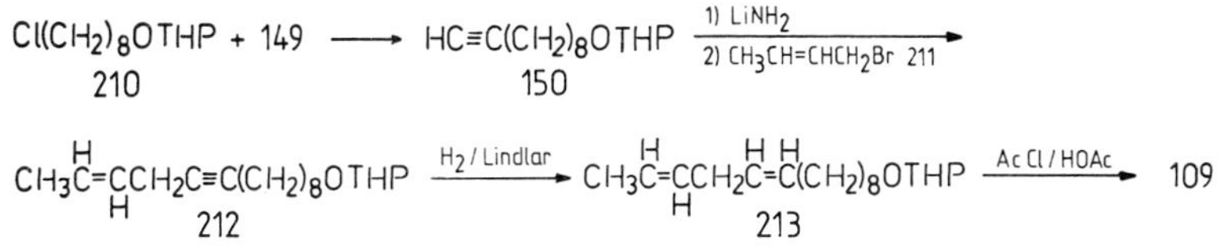

Scheme 26

Fumaki synthesized all four geometrical isomers of 9,12-tetradecadienyl acetate, in doing so the preparation of the (Z)-9, (E)-12-isomer was similar to the one shown in Scheme 26 [697].

Su et al. [698] started their synthesis of *109* with 1,9-decadiyne *214*, which was converted to acetal *215* [699] and mixed with crotyl bromide *211* yielding *216* (59% yield) after 16 hours heating in dioxane. The alternative reaction of the Grignard reagent of *215* with *211* in THF in the presence of copper salts gave *216* after 36 hours heating in only 29% yield. Hydrolysis, reduction, partial hydrogenation using 5% palladium on barium sulphate and subsequent acetylation yielded acetate *109*, which was purified gaschromatographically [698] (Scheme 27).

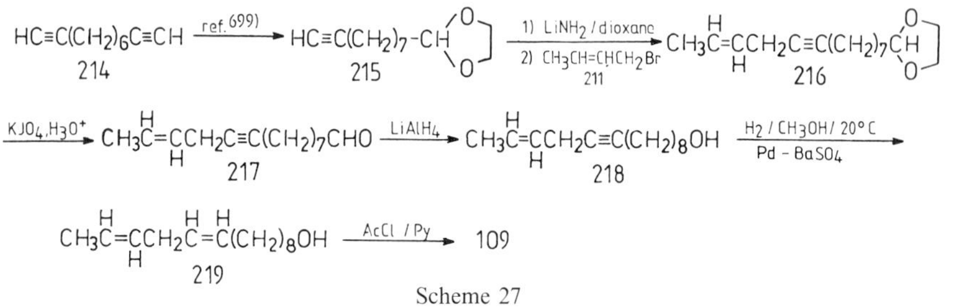

Scheme 27

In the synthesis of (E)-4, (Z)-7-tridecadienyl acetate *229*, the sex pheromone of the potato tuberworm moth *Phthorimaea operculella*, 4-pentyn-1-ol *220* was converted to the unsymmetrical acetal *222* and the corresponding lithium compound reacted with paraformaldehyde *223* to *224*. Stereoselective reduction and chlorination according the method of Collington and Meyers [700] afforded (E)-alkenyl chloride *225*, which could be coupled together with heptynyl magnesium bromide *226*, in presence of catalytic amounts of cuprous chloride, to (E)-4-tridecen-7-ynyl ether *227*. Hydrolysis with trichloroacetic acid in aquaous tetrahydrofurane and low temperature crystallization gave rise to enynol *228* (98.8% GC-analysis), which could be partially hydrogenated and subsequently acetylated [203] (17% total yield from pentynol *220*, 98% GC-purity) (Scheme 28).

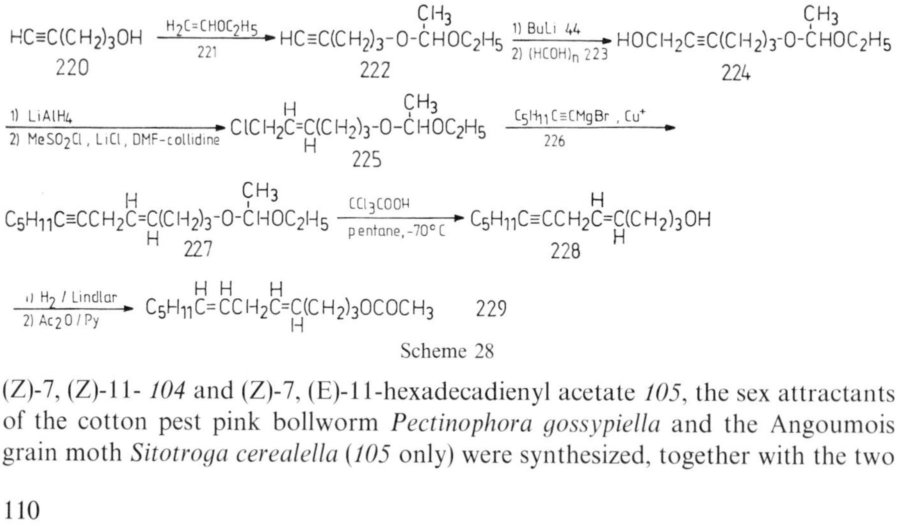

Scheme 28

(Z)-7, (Z)-11- *104* and (Z)-7, (E)-11-hexadecadienyl acetate *105*, the sex attractants of the cotton pest pink bollworm *Pectinophora gossypiella* and the Angoumois grain moth *Sitotroga cerealella* (*105* only) were synthesized, together with the two other isomers, by Su and Mahany [701] *via* the acetylenic route, starting with 3-octyn-1-ol *232* and 2-(6-chloro-1-hexyloxy)tetrahydropyrane. A group of chemists from Bayer AG (Leverkusen, Germany) improved this method to a synthesis with total yields between 31 and 34 percents (Scheme 29) [554]. Therein 3-octyn-1-ol *232*, available from 1-hexyne lithium *230* and ethylene oxide *231*, converted into the bromides *233* and *238* (PBr_3 in dioxane/pyridine), and these coupled together with the lithium salt *98* to *234* and *239* and reacted to the acetates *235* and *240*, respectively. The (E)-configuration of *237* and *241* was obtained by homogeneous hydrogenolysis (sodium in ammonia), the (Z)-geometry of *105* and *104* by selective hydrogenation over palladium/calcium carbonate [701] or by means of modified Lindlar catalyst in petrol ether [554]. (Z)-7,(Z)-11-Hexadecadienyl acetate *104* could thus be prepared in a purity of 95.7% (2.4% Z,E and 1.9% E,Z), and (Z)-7, (E)-11-hexadecadienyl acetate *105* in 97.1% (1.4% Z,Z and 1.5% E,E; gas chromatographically determined with DEGS-WCOT capillary columns) [554] (Scheme 29).

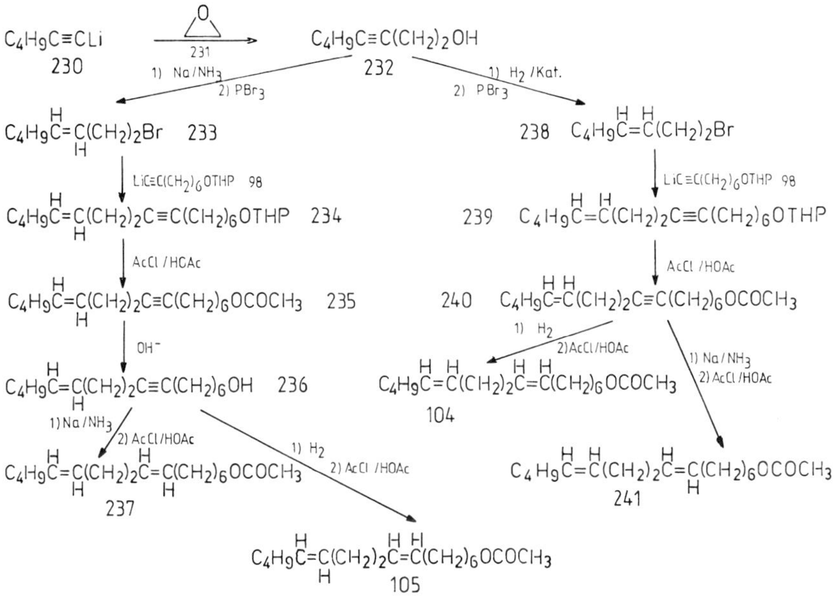

Scheme 29

Mori, Tominaga and Matsui synthesized both the acetates *104* and *105* by reacting propargyl magnesium bromide *189* with 1-bromohept-2-yne *242* and (E)-2-heptenyl bromide *246*, respectively, and subsequent alkylation of the enynes *243* and *247*, respectively, with 2-(6-bromohexyloxy)tetrahydropyrane *244* [702, 703]. The resulting THP-ethers *245* and *234* were hydrolyzed, partially hydrogenated (P-2 nickel catalysis [654] for the (Z)-product) and acetylated to *104* and *105* (total yield for Z, Z was 32%, for Z, E 47%) [702, 703] (Scheme 30). Bierl et al. [228] prepared the (Z)-7, (E)-11-hexadecadienyl acetate *105* starting with 1,5-hexadiyne *248*. Monoalkylation of the mono lithium salt with butyl bromide *249* gave 1,5-decadiyne *243* in 40% yield. By reducing the corresponding lithium salt, using sodium in ammonia, they obtained (E)-5-decen-1-yne *247* and followed Scheme 30 for the remaining synthesis steps [228] (Scheme 30).

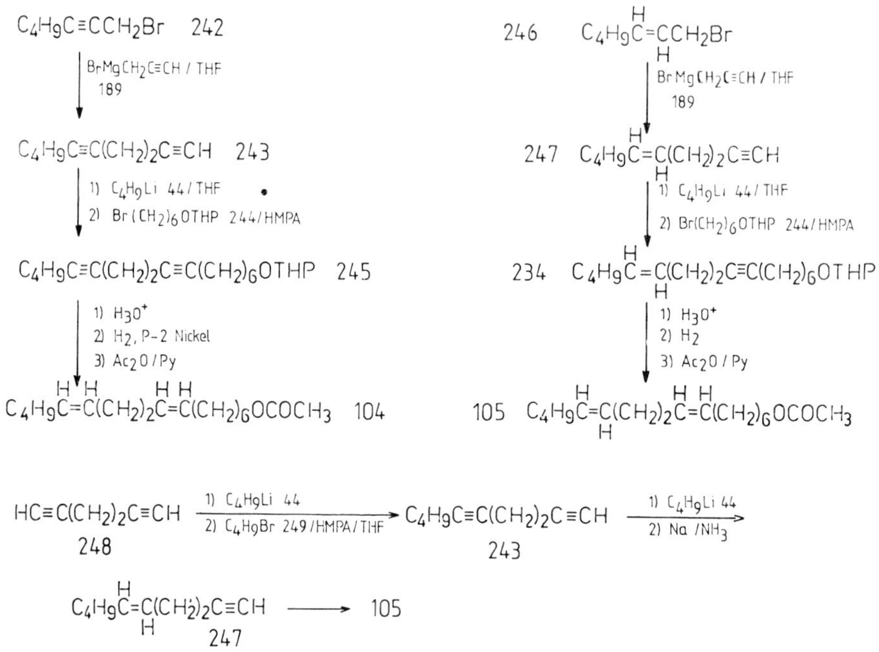

Scheme 30

Using a combination of an acetylenic synthesis for the double bond in position 7 and a Wittig olefination for the other one, Sonnet obtained (Z)-7,(Z)-11-hexadecadienyl acetate *104* in a purity of 93% (about 3.5% Z,E and 3.3% E,Z-isomer) [639] (cf. p. 100), Wittig reactions). Hammond and Descoins obtained the lithium derivative of 5-decen-1-yne *247* from 3-butyn-1-ol. This could be alkylated with a 6-iodohexyl ether, the ether protecting group split off and the enynol acetylated. The resulting (E)-11-hexadecen-7-ynyl acetate was hydrogenated by means of Lindlar catalyst yielding *105* [642].

The four geometrical isomers of 3,13-octadecadienyl acetate, of which (E)-3, (E)-13-octadecadienyl acetate *257* is the pheromone of the lesser peach tree borer *Synanthedon pictipes*, and (Z)-3, (Z)-13-octadecadienyl acetate *256* the pheromone of the peach tree borer *Sanninoidea exitiosa* (cf. Table 1), were synthesized according Scheme 31 [231]. The (E), (Z)-isomer *261* could be obtained by sodium ammonia hydrogenation of the enyne *260* and subsequent preparative GC-separation containing 3–5% each of the other isomers.

(E)-6, (Z)-11-Hexadecadienyl acetate *267* and (E)-6, (Z)-11-hexadecadienal *268*, the pheromone components of the saturnid moths *Antheraea pernyi* [225] and *Antheraea polyphemus* [226], were synthesized by alkylation of 4-pentynyl tetrahydropyranyl ether *262*, conversion to (Z)-4-nonenyl bromide *108*, and further alkylation with the alkynyl-THP-ether *265*. (Z)-11-Hexadecen-6-ynyl ether *266*, which resulted in low yields, was (E)-partially hydrogenated and acetylated to the

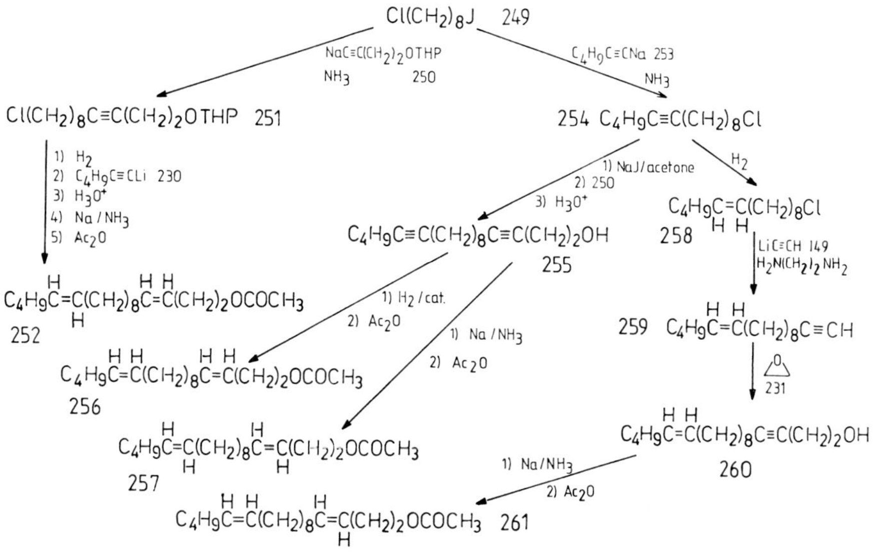

Scheme 31

alkadienyl acetate *267*. Hydrolysis and oxidation with chromic oxide and pyridine afforded (E)-6,(Z)-11-hexadecadienal *268*, the second pheromone component of the above mentioned saturnids [226] (Scheme 32).

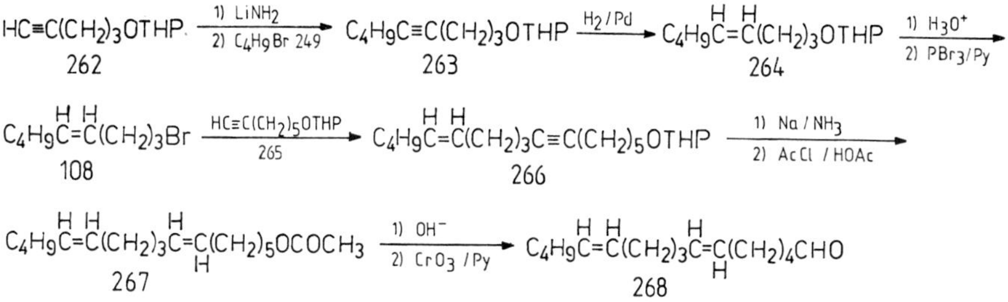

Scheme 32

c) Reduction of Allyl Diethyl Phosphonates

The (E)-stereoselective reduction of γ-alkylsubstituted allyl diethylphosphonates with lithium aluminiumhydride, as it was described by Kondo et al. [704] for the synthesis of (E)-11-tetradecenal *273*, the sex pheromone of the spruce budworm *Choristoneura fumiferana* (cf. Table 1), is an alternative for the (Z)-stereoselective Wittig reaction based on a phosphorus organic compound. In this work metallated 2-butenyl diethylphosphonate *269* was alkylated with 2-(10-bromodecyloxy)tetrahydropyrane *270*. The reduction of the resulting phosphonate *271* using lithium aluminium hydride yielded an (E)-olefin *272* with shifted double bond, the geometry of the original allyl phosphonate *269* having no influence to the stereoselectivity of the reaction [704] (Scheme 33).

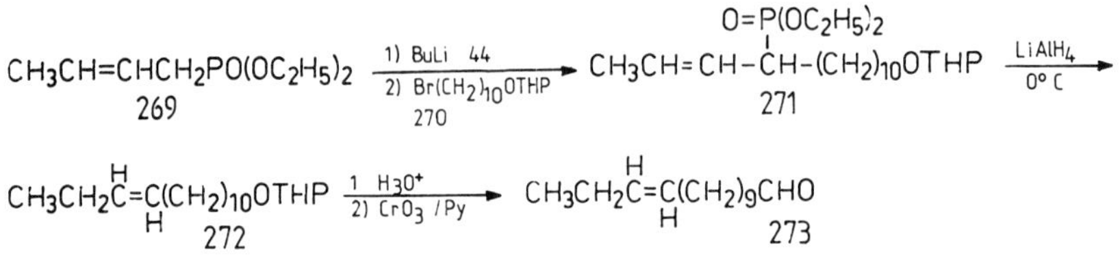

Scheme 33

Using slightly altered experimental conditions Bestmann et al. [705] prepared with this method (E)-6-nonenyl acetate, (E)-6-nonenol, (E)-8-dodecenyl acetate, (E)-8-dodecenol, (E)-9-dodecenyl acetate, (E)-9-dodecenol, (E)-7-tetradecenyl acetate, (E)-7-tetradecenol, (E)-11-tetradecenyl acetate, (E)-11-tetradecenol, (E)-9-hexadecenyl acetate, and (E)-9-hexadecenol (for physiological activities cf. Table 1) [705].

An interesting variant of this reaction is its application on conjugated-unsaturated allyl phosphonates, e.g. *276* in Scheme 34. They also can be metallated and alkylated, but in the subsequent reduction only the first double bond is shifted to the lower position. Using this method, which allows a general access to "methylene interrupted" double bonds with (E), (E)-geometry, a working group of the University Erlangen [706] synthesized (E)-7, (E)-10-dodecadienol *279a*, (E)-9, (E)-12-tetradecadienol *279b*, (E)-11,(E)-14-hexadecadienol *279c*, (E)-7,(E)-10-dodecadienyl acetate *280a*, (E)-9, (E)-12-tetradecadienyl acetate *280b*, and (E)-11, (E)-14-hexadecadienyl acetate *280c*, containing 85–86.5% E,E-isomers, 1.5–2% Z,E-, 12–13% E,Z- and $\leqq$1% Z,Z-isomers (gaschromatographically determined, 100 m DEGS steel capillary columns) [706] (Scheme 34).

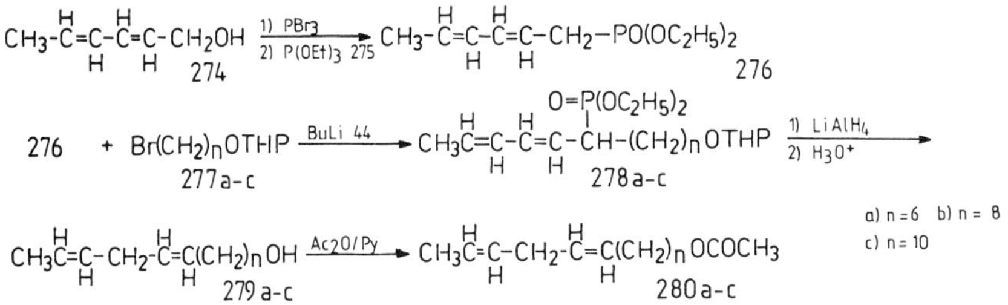

Scheme 34

d) Coupling Reactions with Organometallic Reagents

Coupling reactions leading to a new carbon carbon bond, which does not present a part of the ene- or diene-system, has the advantage that one can start from synthons already possessing the double bonds of desired configuration. Thus, derivatives of (E)-2,(E)-4-hexadien-1-ol *247* (sorbyl alcohol) with its conjugated E,E-double bonds were predestinate starting compounds for the synthesis of codlemone, (E)-8,(E)-10-dodecadien-1-ol *60*, the sex attractant of the codling moth *Carpocapsa pomonella* (Table 1). The (E)-2,(E)-4-hexadienyl bromide *281* (sorbyl bromide) was coupled in several cases with the Grignard compound *282* giving the THP-ether *283*, which was hydrolyzed to the alkadienol *60* [263, 707, 708]

(Scheme 35). Using the same method Mori was able to obtain (E)-9,11-dodecadienyl acetate *85* [708], a pheromone component of the African cotton pest *Diparopsis castanea* (Table 1).

A method superior to this halogenide coupling reaction is the corresponding coupling of sorbyl acetate *285* with organometallic compound *282* in presence of dilithium tetrachlorocuprate according the technique of Fouquet and Schlosser [709], as it was used by Samain and Descoins [710] and Bestmann and coworkers [635] for the preparation of *60*. The latter started with sorbic acid *284*, which could be converted over several steps to sorbyl acetate *285*. This was reacted over a copper catalyst with the Grignard reagent *282* and the resulting ether *283* hydrolyzed to the dienol *60*. Oxidation using pyridinium chlorochromate (PCC) yielded (E)-8,(E)-10-dodecadienal *286*, and acetylation gave (E)-8,(E)-10-dodecadienyl acetate *61*, containing 4% E,Z- and 1% Z,E-isomer amount [635] (Scheme 35).

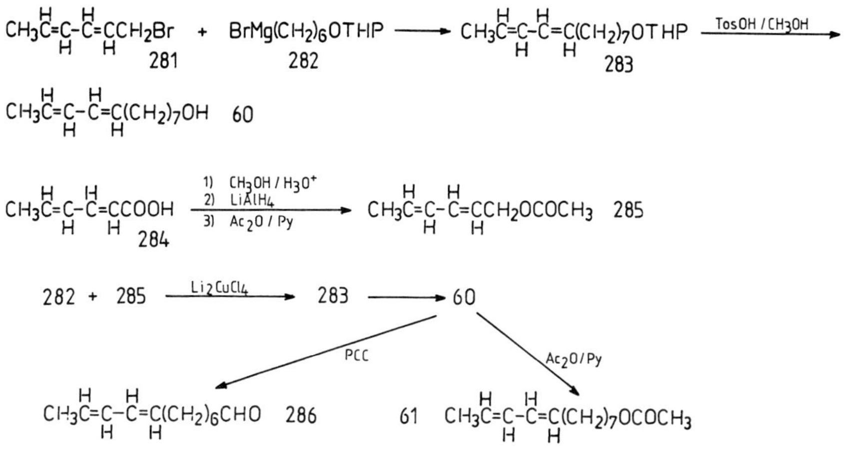

Scheme 35

In the same paper the authors described the synthesis of (E)-10,(E)-12-tetradecadien-1-ol, (E)-12,(E)-14-hexadecadienol, (E)-10,(E)-12-tetradecadienyl acetate, and (E)-12,(E)-14-hexadecadienyl acetate, and using Wittig reactions the corresponding (Z),(E)- and (E),(Z)-isomers (cf. p. 99) [635]. In the case of codlemone *60* a purity of +99% of E,E-form easily can be obtained by simple crystallization from pentane at −5 °C (mp. 29 °C).

Coupling a Grignard reagent under CuJ catalysis conditions allowed Mandai et al. [711] the synthesis of the *Diparopsis* pheromone *85* using an industrially available starting material. The catalytical telomerization of butadiene *287* with phenol *288* yielded 2,7-octadienyl phenyl ether *289*, which could be converted into alkenol *290* using 9-borabicyclononane *17* (9-BBN), and this into iodide *291*. By coupling *291* with Grignard compound *183* in presence of CuJ/bipyridyl they obtained 12-phenoxy-10-dodecenyl tetrahydropyranyl ether *292*, which could easily be hydrolyzed and acetylated to *293*. The elimination of phenol gave (E)-9,11-dodecadienyl acetate *85* with about equal amounts of (Z)-9- and (E)-9-isomers [711] (Scheme 36).

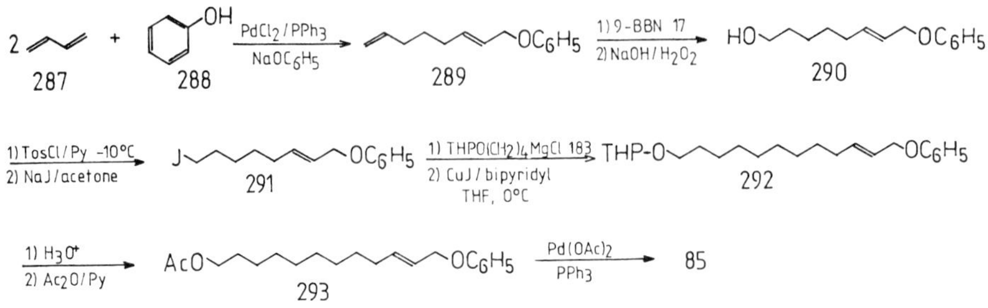

Scheme 36

Coupling reactions with allylic compounds of the kind which were used for the synthesis of *60* create the possibility of allylic rearrangement and thus of the generation of rearranged and undesired products. Additionally it is necessary to overcome the "homocoupling" of each of the reaction partners as a competing reaction. So monoolefinic allyl halides together with nonallylic organometallic reagents generally give mixtures of olefin isomers, and even though the cross coupling product from the primary allylic compound predominates, one can find considerable amounts of dienic products from the "homocoupling" of the allyl fragment [712]. Thus Henrick and Garcia could demonstrate, that treatment of sorbyl alcohol *274* with thionyl chloride and tributyl amine in ether yields a mixture of isomeric chlorides *294a–c*. The coupling of this isomer mixture with Grignard reagent *282* in ether/tetrahydrofurane with subsequent hydrolysis afforded a complex reaction mixture with about 75% of the desired E, E-isomer *60*, as well as other isomeric dienols and tetraene compounds besides 1,12-dodecanediol [713]. In the same way, variations of the halogenation conditions according to Storck et al. [714], Collington and Meyers [700] and the use of CCl_4/HMPA in ether [715] gave rise to mixtures of *294a–c* in the synthesis of sorbyl chloride from the dienol *274*. The reaction of sorbyl alcohol *274* and phosphoric tribromide (0.4 mol equivalents) in ether at −40 °C yielded a mixture of bromides *281a and b* and a complex dienol mixture at the further preparation of codlemone [713], the same with the analoguous reaction of sorbyl bromide with the lithiumorganic compound in presence of dilithium tetrachlorocuprate, and the coupling with sorbyl mesylate [713], respectively. Only the usage of esters of sorbyl alcohol, e.g. sorbyl acetate *285* in presence of Li_2CuCl_4 gave reaction products in sufficient isomer purity, fairly good yields and only little formation of competition coupling products (Scheme 37).

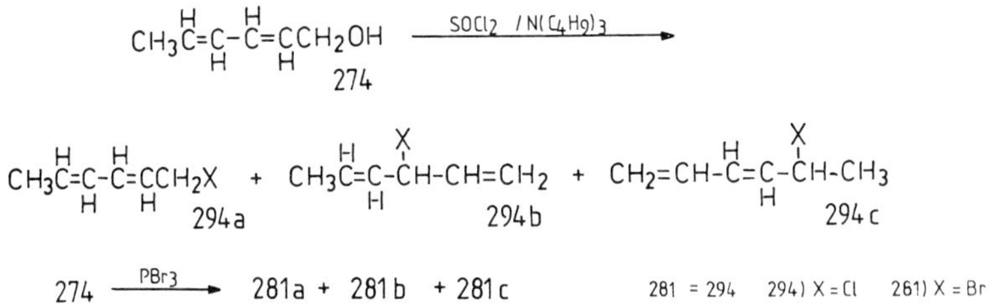

Scheme 37

Another version of coupling avoiding any allylic reaction intermediates is shown in Scheme 38 in the synthesis of *60*. By Julia rearrangement [637] of cyclopropyl propenylcarbinol *295* Descoins and Henrick obtained the homoallylic bromide *296* containing about 10% of (Z)-3,(E)-5-isomer, which could have been coupled with Grignard reagent *297* and a catalytic amount of Li_2CuCl_4 giving *283*, and then hydrolyzed to *60* [716] (79% E,E). The alternative coupling, namely the Grignard product of homoallylic bromide *298* with ω-bromoalkyl tetrahydropyranyl ether *299* gave on the other hand a reaction product distribution of *283* with only 66% of E,E-isomer [716] (Scheme 38).

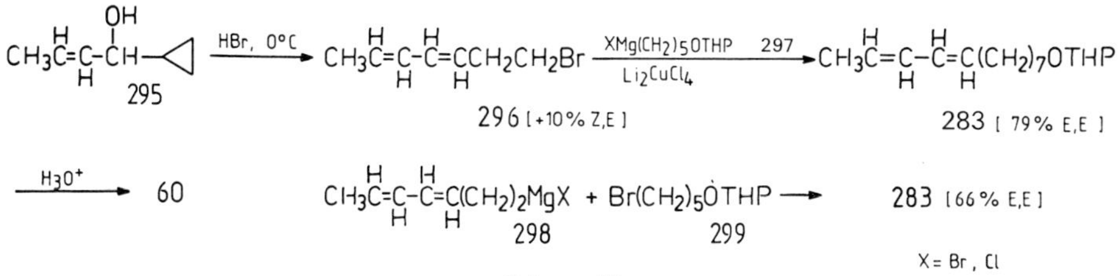

Scheme 38

Beside the above described building methods for butterfly pheromones by means of a coupling reaction with Grignard compounds or the analoguous lithium organyls a great part of similar syntheses has been carried out using triple-unsaturated reaction synthons, e.g. lithium acetylides, magnesium halogenides of alkynes and substituted propargylic derivatives. Because these methods terminate into steps of partial hydrogenations of the triple bond into (Z)- or (E)-double bonds, they have been discussed already in Chap. 5.1b) ("acetylenic approach", p. 104).

Three synthesis examples of the application of organometallic compounds for the preparation of a lepidoptera pheromone and terminating in acetylene hydrogenations are demonstrated by Smith et al. [289], Mori and coworkers [717], and Kocienski and Cernigliaro [718] in their syntheses of (Z)-6-heneicosen-11-one *143*, the sex attractant of the Douglas tussock moth *Orgya pseudotsugata* (cf. Table 1), a fir pest living in the western parts of North America. The first deprotonated 2-decyl-1,3-dithiane *300* according to Seebach [719] with butyl lithium *44* and alkylated the lithium derivative with 4-decynyl bromide *301* to the disubstituted dithiane *302*, which was hydrolyzed and (Z)-stereoselectively hydrogenated with P-2 nickel to pheromone *143* [289] (Scheme 39).

Mori's synthesis of *143* starts with a Grignard-like alternative, the conversion of 3-ethoxycyclohex-2-enone *303*, epoxidation and Eschenmoser cleavage [720] of the resulting epoxy-ketone *306* giving alkynone *307*, the ketale *308* of which could be alkylated in the usual manner, selectively hydrogenated and hydrolyzed affording *143* [717] (Scheme 39).

The same cleavage of epoxy-cycloalkanones was used by Kocienski and Cernigliaro [718], who synthesized 2-pentyl-3-decyl-2,3-epoxycyclohexanone *312* in two steps from the enol ether *310* and converted it by means of p-toluene sulfonyl hydrazide into the alkynone *313*. Lindlar catalyzed hydrogenolysis gave *143* in 58% total yield [718] (Scheme 39).

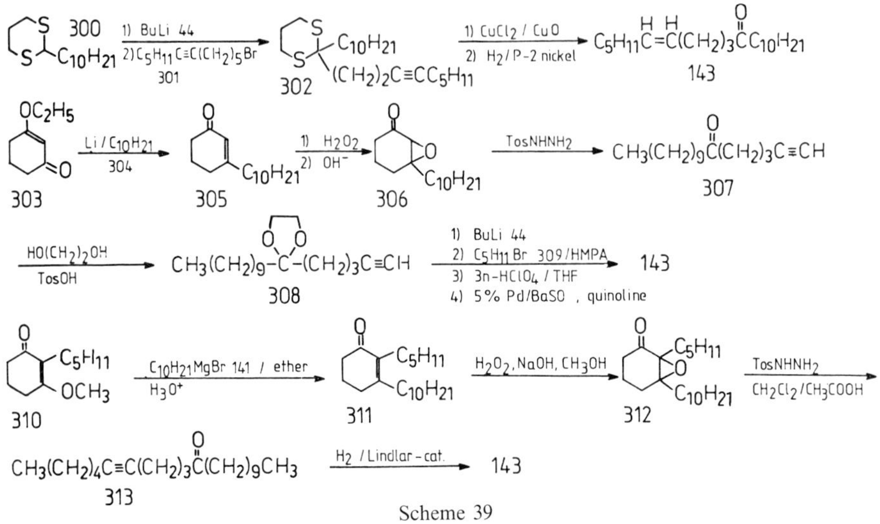

Scheme 39

The possibility of application of organocopper compounds for the synthesis of pheromones was demonstrated by Normant et al. in their preparation of (E)-10,(Z)-12-hexadecadienol (*30*, bombykol) [721]. Therein the propargylic alcohol *318*, obtained from propyl magnesium bromide *314*, cuprous bromide and 3-bromo-2-propynyl trimethylsilylether *317*, was reduced to the E,Z-alkadienol *319* in high yields (95%). The resulting acetate *320* was subsequently coupled with Grignard reagent *321* in the presence of catalytic amounts of Cu-(I)-salts and hydrolyzed to bombykol *30* [721] (Scheme 40).

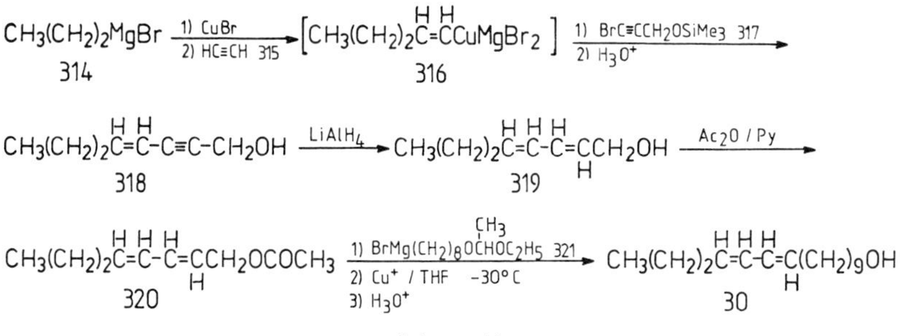

Scheme 40

The authors could extend the same principle of synthesis for the preparation of 1,4- and 1,5-dienes, respectively, by reacting allylic epoxides like *324* with vinylic and allylic organo-copper reagents. Thus the reaction of n-$(C_5H_{11})_2CuLi$ *322* and acetylene *315* led to (Z)-heptenyl cuprate *323*, which gave alcohol *325* upon reacting with butadiene oxide *324*. The reaction of crude *325* and N-chlorosuccinimide gave the dienyl chloride *326*, which could be alkylated to the alkadienic ester *327*. Reduction and subsequent acetylation led to (E)-4,(Z)-7-tridecadienyl acetate *229*, a component of the sex pheromone of *Phthorimaea operculella* (E,Z/Z,Z = 96/4) [722] (Scheme 41).

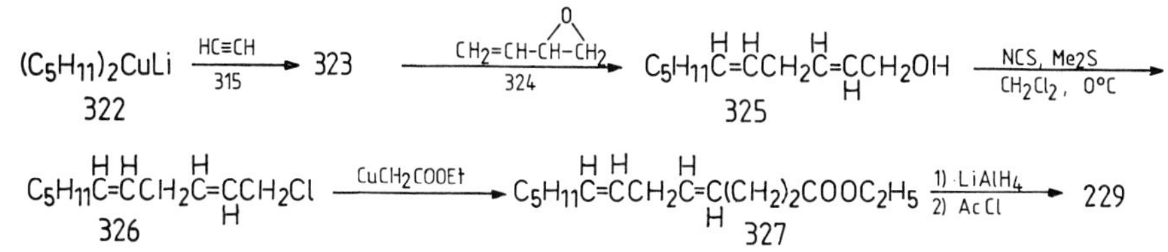

Scheme 41

Beside lithium-, magnesium- and copperorganyls for controlled generation of C–C bonds, organoboron compounds have also been used in pheromone synthesis. Negishi et al. [695] reported a highly stereoselective synthesis of conjugated enynes by addition of disiamylborane *172* onto alkynes and lithium acetylides by formation of 1-alkynyl-1-alkenyl dialkylboronate complexes. Those complexes (e.g. *208*) can be decomposed by iodine and sodium hydroxide obtaining (E)-enynes (e.g. *209*), thus yielding starting compounds for the synthesis of bombykol *via* acetylene hydrogenations [695] (cf. Scheme 25).

A simple synthesis of racemic disparlure *136* by means of coupling reactions has been described by Russian chemists very recently [723]. Starting with 1,5-cyclooctadiene *118* the dibromo compound *328* could be obtained, and then coupled with the appropriate Grignard compounds *329* and *330* to (Z)-olefin *134* and subsequently oxidized to the epoxy-attractant *136* [723] (Scheme 42).

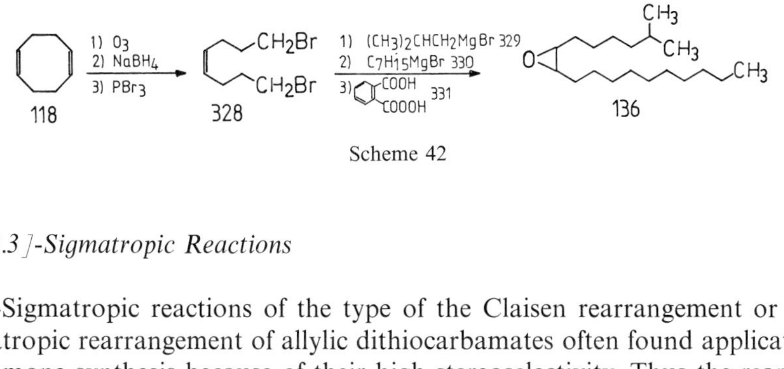

Scheme 42

e) [3.3]-Sigmatropic Reactions

[3.3]-Sigmatropic reactions of the type of the Claisen rearrangement or of the sigmatropic rearrangement of allylic dithiocarbamates often found application in pheromone synthesis because of their high stereoselectivity. Thus the rearrangement of the enol ether derivative *334* of acetate *332* gave 10-mesitoyloxy-4,8-dimethyldeca-(E)-4,(E)-8-dienoic acid *335* with 98% stereoselectivity, which easily could be reduced to 3,7-dimethyldeca-(E)-4, (E)-8-dien-1,10-diol *336* [724], a component of the pheromone of male *Danaus gillipus berenice* (cf. Table 1, p. 48) (Scheme 43).

Heating of allylic alcohol *337* with trimethyl orthoacetate *165* in the presence of propionic acid, a version of the Claisen rearrangement, was used for the preparation of the methyl alkenynoate *166*, a synthesis intermediate for (E)-7,(Z)-9-dodecadienyl acetate *54*, the sex pheromone of the grape berry moth *Lobesia botrana* [687]. The resulting orthoester *338* loses methanol by the generation of the ketene acetale *339*, which rearranges to *166*. Nevertheless is the stereoselectivity of this reaction sequence about 80% only [687] (see also Part 5.1b), p. 106 and Scheme 18) (Scheme 44).

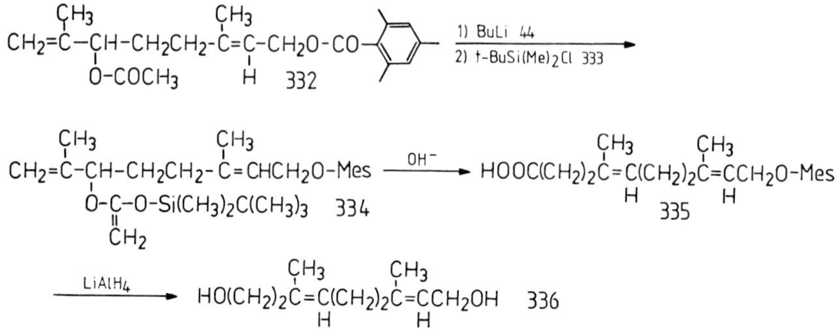

Scheme 43

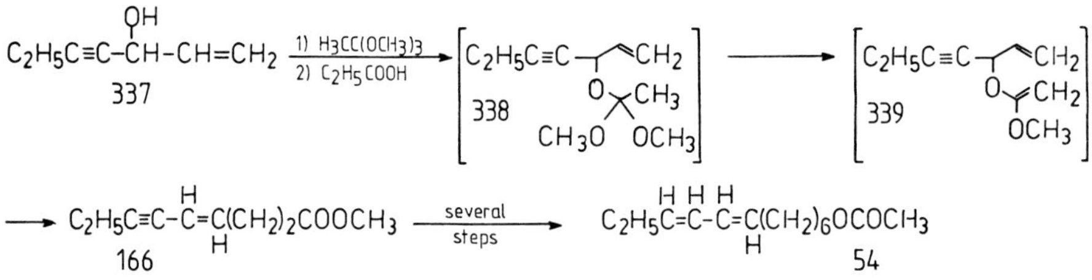

Scheme 44

The method of sigmatropic rearrangement of allylic dithiocarbamates, published by Hayashi and Midorikowa [725], represents an interesting synthesis for (E)-alkenole pheromones.

By alkylation of N,N-dimethyl dithiocarbamates *340* can be obtained and by heating this the rearrangement product *341*, which can be alkylated a second time with the halo-THP-ether *343* yielding 1-(ω-tetrahydropyranyloxyalkyl)-alkenyl dithiocarbamate *344*. Removing of the dithiocarbamate group resulted in the alkenyl-THP-ether *345* which could be converted into the (E)-alkenyl acetate, as it was demonstrated by the synthesis of (E)-6-nonenyl acetate and (E)-7-dodecenyl acetate [725] (Scheme 45).

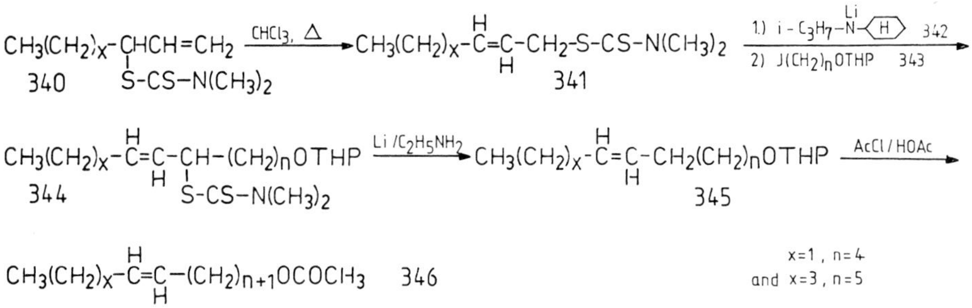

Scheme 45

Samain and Descoins investigated the stereoselectivity of Claisen rearrangements of secondary enynols and synthesized bombykol *30* and bombykal *31*, the pheromone components of the silkworm moth [256], by rearrangement of the tert-butyl dimethylsilyl ether *350* of the enol of acetate *348* [726]. 1-Octen-4-yn-3-

ol *347* was acetylated and by means of cyclohexyl isopropyl amide lithium *342* and tert-butyl dimethylsilyl chloride *333* converted *via* the unisolated intermediates *350* and *351* into the (E)-4-decen-6-yn-1-ol *352*. By means of TosCl and LiBr the authors got the enynyl bromide *354* in 95% yield, which could be coupled with the ω-chloroalkyl-THP-ether *355* to *356*, partially hydrogenolyzed with diisoamylborane *172*, and subsequently hydrolyzed to (E)-10,(Z)-12-hexadecadien-1-ol *30* (bombykol). Acetylation resulted in a bombykol acetate *357* containing 96% E, Z-, 1.6% Z,Z-, and 2.4% of Z,E-isomer amount, oxidation of 30 by pyridinium chlorochromate (PCC) yielded bombykal *31* [726] (Scheme 46).

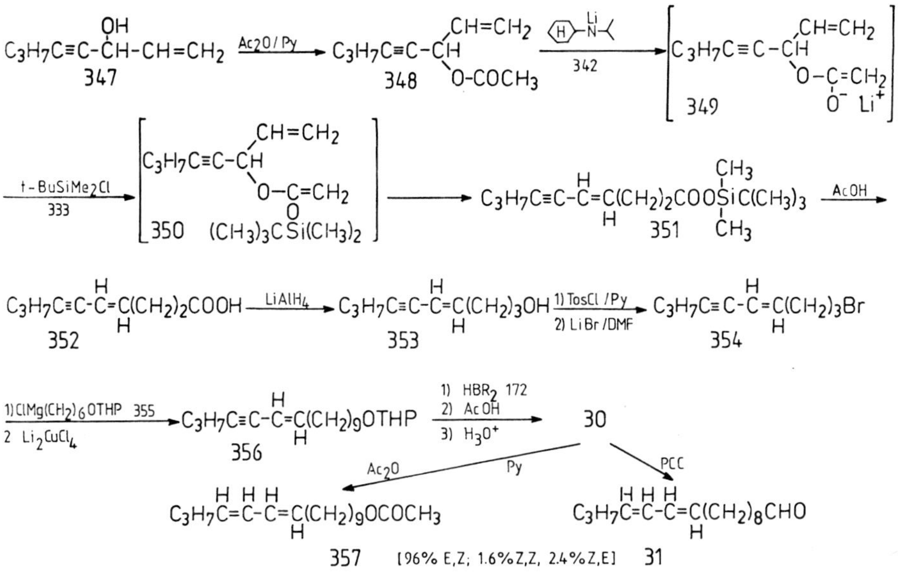

Scheme 46

f) Transition Metal Catalyzed Metathesis Reaction

The homogenous, metal catalyzed metathesis olefin synthesis [727] often found use in the preparation of insect sex attractants [728–730] since this method offers the possibility of starting from cheap petrochemicals. Of course, this reaction, which can be carried out easily in great scale being a onepot reaction, does not have any stereoselectivity, so that the resulting Z, E-isomer mixtures must be resolved by conventional techniques.

Because in the case of the crossed metathesis reaction from two unsymmetrically substituted olefins four products were produced, it is preferable to use 1-alkenes, so that the generated ethylene can be removed easily (Scheme 47) from the reaction mixture. Thus by reacting 1-tetradecene *364* and 4-methylpent-1-ene *365* with a catalyst of dichlorodinitrosyl bis(triphenylphosphine)molybdanum aluminium chloride 2-methyl-4-heptadecene *366* could be obtained, and be reduced to 2-methylheptadecane *367*, the sex attractant of *Holomelina* species [728] (Scheme 47).

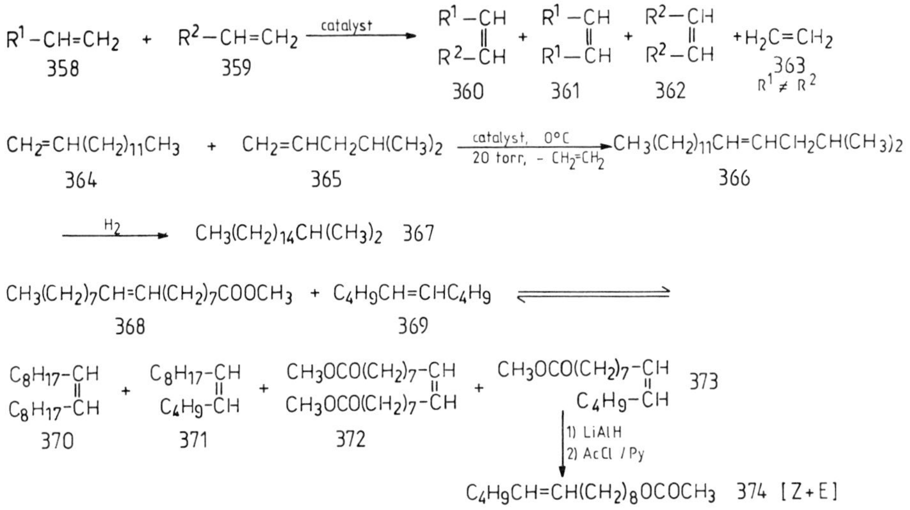

Scheme 47

Analoguously to the metathesis reaction with unsaturated fatty acid esters [731] recently also sex pheromones with functional groups like -OCOR, -COOR, etc. could be prepared [732], using the same reaction. Thus the crossed metathesis reaction of oleic acid methylester *368* and 5-decene *369* with a catalyst system consisting of wolfram hexachloride and tetramethyl tin resulted in methyl 9-tetradecenoate *373* in 8% yield, which easily could have been converted into the (E)- and (Z)-isomeric mixture of 9-tetradecenyl acetate *374* [732] (Scheme 47).

In the same way Nakamura and coworkers [733] succeeded in the synthesis of 9-tetradecenyl acetate *374* directly by using the metathesis reaction of oleyl acetate and 5-decene [733] (for the use of this reaction type for the synthesis of flies pheromones see the corresponding chapters).

g) Electrochemical Procedures

The anodic oxidation of two different carboxylic acids, the mixed Kolbe electrolysis, gives a synthetic approach towards (Z)-unsaturated pheromones. Seidel et al. [734] obtained (Z)-7-dodecenoic acid methylester *379* from the mixed electrolysis of (Z)-4-nonenoic acid *377* together with glutaric acid methyl monoester *378*. *379* was a synthesis intermediate of (Z)-7-dodecenyl acetate *382*, the sex attractant of the female cabbage looper *Trichoplusia ni*. The authors condensed the ylide of the 3-carboxypropyl triphenylphosphonium bromide *376*, which resulted from γ-butyrolactone *375*, with pentanal *102* to (Z)-4-nonenoic acid *377*, which was subjected coelectrolysis together with the monoester *378*. In addition to the desired methylester *379* also the symmetrical dimer (Z)-5,(Z)-11-hexadecadiene *380* and suberic acid dimethylester *381* were produced, whereby the dependence of the reaction yield on the electrolysis current density and on the ratio of *377*:*378* was studied. The methylester *379* was isolated according to usual methods and converted into (Z)-7-dodecenyl acetate *382* (70–80% (Z)-isomer content) [734] (Scheme 48).

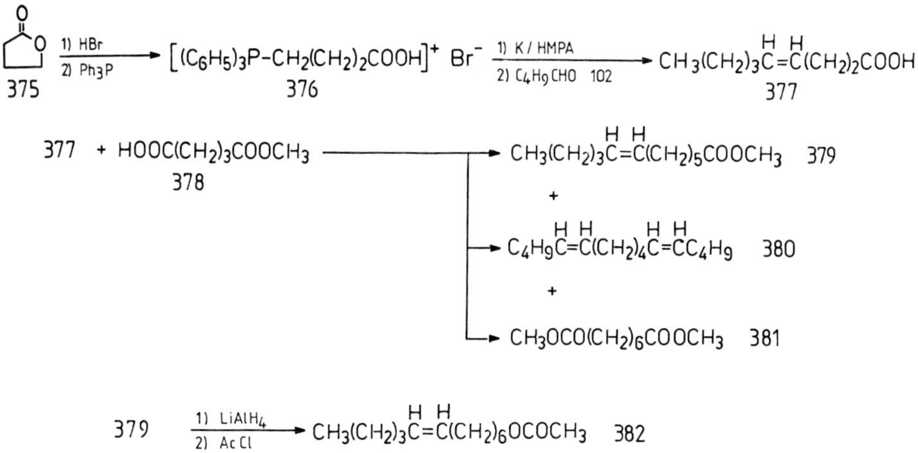

Scheme 48

h) Stereoselective Syntheses of Chiral Lepidoptera Pheromones

While one was assuming until a few years ago that insect pheromones consist each of single compounds, today the proof of the existence of multi-component pheromone complexes has often been brought forth. And only the exact natural substance ratios of the single components, which often are chemically-structurally related compounds like geometrical isomers, homologous compounds or substances with only different functional groups, can effect an optimum behavioral release. Because of this not only the type of compounds but also their ratio of mixture can represent an isolation mechanism and prevent cross attraction and thus the intermixture of pheromone-related insect species.

Therefore it was obvious, that in the case of the natural occurence of chiral substances as sex attractants only one enantiomer of these compounds or a distinct ratio of both of the optical antipodes determines the species specific behavioral answer. Because the amount of pheromone, as it is isolated for example from female moths, normally lies in the nanogram scale ($1 ng = 10^{-9} g$), it is not possible to measure the kind of optical rotation of naturally occuring pheromones with current methods or to determine their absolute configuration. Therefore only the stereoselective syntheses of both of the enantiomers of a chiral pheromone and the subsequent biological test of the single compounds, the racemic mixture and defined isomer mixtures can provide evidence about the configuration and the composition of chiral pheromones.

(Z)-7,8-Epoxy-2-methyloctadecane *136* (disparlure) is till today the only known chiral female sex attractant occuring in moths, and a great number of syntheses of the racemic product had been published (cf. p. 102). The first selective preparation of the both enantiomers was reported in 1974 by Iwaki et al. [735], who started the synthesis with (S)-glutamic acid, whereby the end products were each contaminated with the other antipode [735]. Later Mori and coworkers synthesized (7R, 8S)-disparlure, which was known to be the biologically more active of the two enantiomers from the former work [735], in a stereochemical purity of +98%, starting from (2S,3S)-tartaric acid *383* [736] (Scheme 49). The resulting tosylate *384* was coupled together with lithium bis(3-methylbutyl)cuprate *385* to *386*, and the latter converted into the hydroxylactone *387* (56% yield). The corresponding THP-ether *388* was reduced by diisobutyl aluminium hydride to *389* and olefinated with phosphorane *390* to *391*. The subsequent hydrogenation of the (Z,E)-isomer mixture *391*, tosylation and cleavage of the protecting group yielded the tosyl alcohol *392* in 47% yield, which could be converted by ring closure reaction into (7R,8S)-*136* $[\alpha]_D^{23}$: $+0.8 \pm 0.1°$ (CCl_4) [736] (Scheme 49). The other isomer (7S,8R)-disparlure (7S,8R)-*136* was prepared in a similar way starting with *384*, but with the sequence of the alkylation steps correspondingly changed.

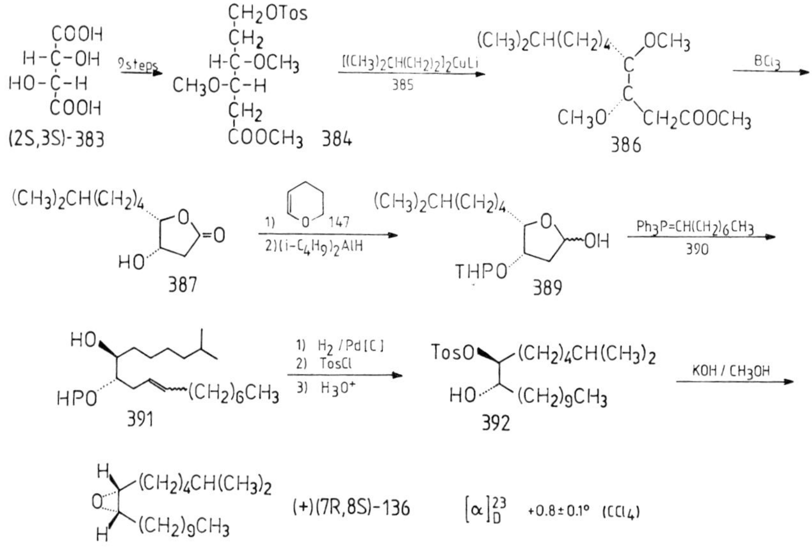

Scheme 49

Recently (+)- and (−)-disparlure *136* was synthesized by a group of chemists of Michigan State University [737], starting from 1-menthyl p-toluenesulfinate *393*. In their procedure diastereomerically pure 1-menthyl p-toluenesulfinate *393* was reduced to sulfoxide *395* by means of 6-methylheptyl magnesium bromide *394* (65% yield), and this converted into sulfoxide *397*. The corresponding lithium salt was substituted with undecanal *133* and the mixture resolved into the diastereomers. The isomer *398* was reduced with lithium aluminiumhydride to *399*, methylated with trimethyl oxonium tetrafluoroborate *400*, and from the sulfonium salt *401* (+)(7R,8S)-disparlure *136* was obtained [737] (Scheme 50).

The other diastereomer of *398*, which could be separated by chromatography, resulted by the same reaction sequences in the formation of (−)-trans-disparlure *136*; and by exchanging the two alkylating steps using undecyl magnesium bromide and 6-methylheptanal, respectively, the (+)-trans-epoxide *136* and (−)-cis-disparlure (7S,8R)-epoxy-2-methyloctadecane *136* could be obtained from the same scheme (Scheme 50).

Trapping experiments and luring tests with the gypsy moth *L. dispar* showed a greater attractancy of (+)(7R,8S)-disparlure *136* compared to the (−)(7S,8R)-

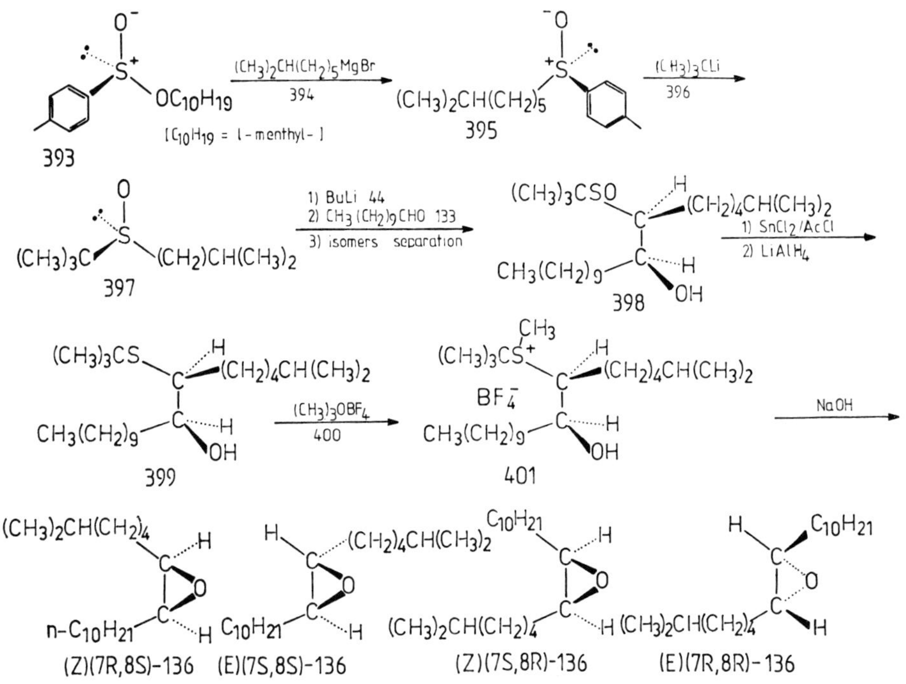

Scheme 50

product [738, 739] and were in accordance with former laboratory test results [735]. But while the addition of (7S,8R)-disparlure to (+)-*136* significantly reduced the male behavioral answers of *L. dispar* under natural conditions, the same addition as consequently the racemic mixture (1:1) had no reducing effect for males of the nun moth *L. monacha* [738, 739].

5.2. Beetles – Coleoptera

The pheromones of beetles (Coleoptera), the greatest order in the class of insects, are a structurally inhomogenous class of compounds, produced either by the males or by the males and females of a species. Especially in the family of the bark beetles (Scolytidae) pheromones were objects of numerous investigations because of the economical importance of these pest insects [317, 323, 740]. The beetle pheromones are, with a few exceptions, almost all chiral compounds (cf. Table 1), and thus we want to restrict the description of syntheses to stereoselective preparations of optically active members of this class of substances.

a) Scolytidae – Bark Beetles

The main pheromone components of the bark beetle *Ips paraconfusus* are the three optically active compounds (+)-cis-verbenol *403*, (−)-2-methyl-6-methylene-7-octen-4-ol *413* (ipsenol), and (−)-2-methyl-6-methylene-2,7-octadien-4-ol *421* (ipsdienol), which were furthermore identified as the pheromones of many other species (Table 1). Optically pure (+)-cis-verbenol *403* had been synthesized as the first chiral bark beetle pheromone by reduction of (−)-verbenone *402* [740] (Scheme 51).

The determination of the absolute configuration of *413* as (−)(4S)-*413* succeeded only recently by a stereoselective synthesis with (S)-leucine *404* serving as starting material [741, 742]. The resulting leucic acid was esterified, the alcohol group protected as THP-ether, und the acid reduced and converted into the tosylate *406*. The cleavage of the protecting group and subsequent treatment with alkali yielded the optically active epoxide *407*. Malonic ester condensation of *407* followed by hydrolysis afforded lactone *408*, which was converted with formaldehyde and diethylamine into (S)-*409*. The exomethylene group of *409* could have been protected by selenoid formation, while the carbonyl function was reduced to the lactol. Wittig reaction of *412* with methylene triphenylphosphorane *89* and cleavage of the protecting group afforded (S)-*413* ($[\alpha]_D^{24} = 16.5$ in ethanol) [741, 742] (Scheme 51).

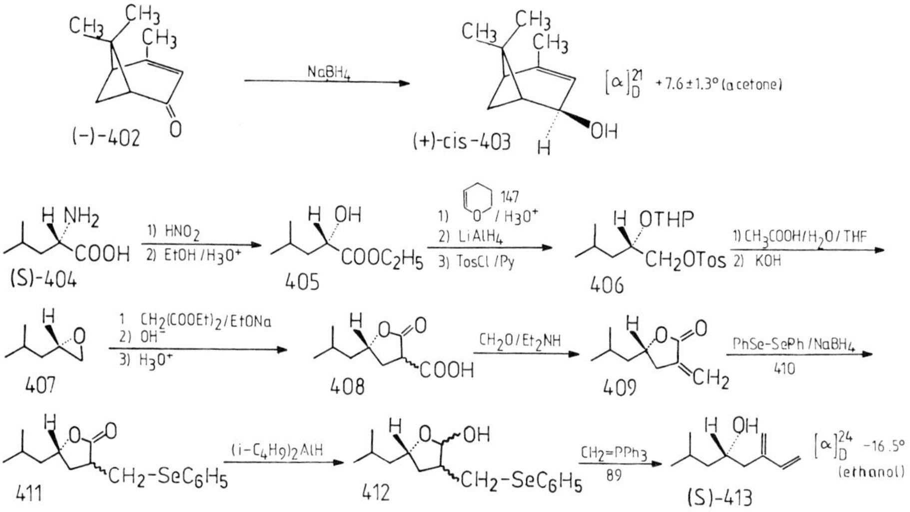

Scheme 51

Ips grandicollis responds solely to (S)-*413* as an aggregation attractant, while the (R)-isomer, which was deduced from (R)-leucine [742], was almost inactive [743].

Ipsenol *413*, ipsdienol *421*, cis-verbenol *403* and trans-verbenol *414* were found also in the pheromone of the forest pest *Ips typographus* (cf. Table 1). While already numerous syntheses of racemic ipsdienol *421* were reported [744, 745, 746], only in 1976 the preparation of (−)(R)-*421* was published by Mori [747], thereby allowing the absolute configuration of the pheromone to be determined as (+)(S)-*421*.

The acetonide of the (+)(R)-glyceraldehyde *415*, which was derived from D-mannit, served as the starting material. *415* was olefinated with the ylide of the phosphonium salt *416* and the resulting olefin *417* converted into the tertiary alcohol *418* by oxymercuration. The acetonide protecting group was reacted to the monotosylate *419* and the latter by treatment with alkali to the epoxide *420*, which by means of the same reaction sequence as for *413* in Scheme 51 afforded optically active (R)-*421* [747] (Scheme 52).

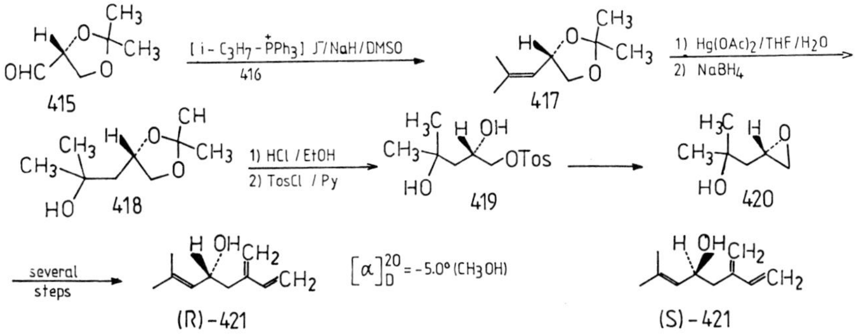

Scheme 52

Ipsdienol of unknown configuration is also found in the population pheromone of *Ips sexdentatus* (Table 1).

The absolute configuration of natural trans-verbenol *414*, which was found in the hindguts of *Dendroctonus frontalis*, *D. brevicomes*, *D. ponderosae* and in many other bark beetle species (cf. p. 49, Table 1), has not been proven unequivocally til today. Pure (+)-trans-verbenol *414* was synthesized from commercially available (+)(1R, 5R)-2-pinene *422* by oxidation using lead tetraacetate. The resulting enol acetate *423* was transesterified to the ester *424* of 3-acetoxyetienic acid, repeatedly recrystallized und converted with lithium aluminium hydride to optically pure (+)-trans-verbenol *414* [748] (Scheme 53). The analogous sequence of synthetic steps, starting from (1S, 5S)-pinene, afforded (−)-*414* [748].

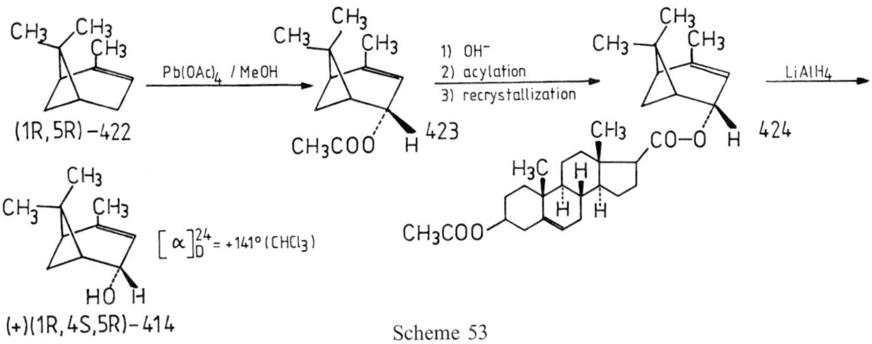

Scheme 53

α-Multistriatin is one of the three components of the aggregation pheromone of the bark beetle *Scolytus multistriatus*, a dangerous elm pest of Europe and the United States, and was structurally elucidated as 2,4-dimethyl-5-ethyl-6,8-dioxabicyclo[3.2.1]octane *431* with (1S,2R,4S,5R)-configuration [350], the synthesis [749] starting with (R)-glyceraldehyde acetonide *415*, which, *via* the secondary alcohol *425* and the oxidation product *426*, afforded olefin *428* in high optical purity by Wittig olefination. Hydroboration, tosylation and Finckelstein reaction gave the iodomethyl compound *429*, which was reacted with the magnesium salt of cyclohexylimine to *430*. Heating of the crude *430* in presence of HCl resulted in a mixture of the four possible multistriatin isomers, the pure (−)-multistriatin being isolated by preparative gas chromatography [749] and therefore the configuration of the naturally occuring substance being established (Scheme 54).

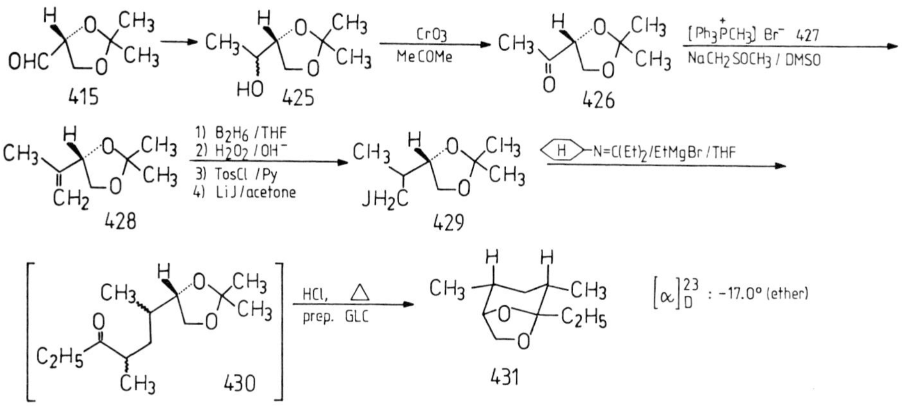

Scheme 54

A synthesis of racemic *431* [750], reported four years ago, could possibly also be used for the preparation of optically active forms of multistriatin, because the racemic synthesis intermediate 2′-(2,2-dimethyl-1,3-dioxacyclopent-4-yl)-1-propanol can also be deduced in optically active forms of known configuration [751].

(−)-4-Methylheptan-3-ol *442* and (−)-α-cubebene *443* are the other two components of the pheromone of *Scolytus multistriatus*, the latter being produced as synergist by the host plants. The absolute configuration of (−)-4-methylheptan-3-ol *442* only recently was established by Mori [752], who prepared threo-*440* and erythro-*441* from stereochemically uniform nerol and geraniol, respectively [752]. The naturally occuring isomer, established as threo-form because of GC-, IR-, and NMR-data, was synthesized starting with methyl (R)-citronellate *432*, as it is shown in Scheme 55 [752]. Oxidation of *432* with m-chloroperbenzoic acid afforded epoxide *433*, which was cleaved to the aldehyde *434*. Wolff-Kishner reduction yielded *435*, which could be converted *via* acyl chloride with dimethylamine into the amine *436*, which could be reacted to (R)-3-methylhex-3-ene *437* according the method of Cope [753]. Treatment with m-chloroperbenzoic acid resulted in the diastereomeric epoxides *438*, and subsequent ring opening reaction yielded a mixture of (3R,4R)-*400* and (3S,4R)-4-methylheptan-3-ol *441*, the (3R,4R)-threo-*441*-isomer of which, having a positive angle of rotation of $[\alpha]_D^{23}$: +21°, could be separated gaschromatographically [752]. Because the naturally occuring pheromone on the other hand had threo-configuration and was levorotatory [344], its absolute configuration must have been assigned with (3S, 4S)-*442* [752] (Scheme 55).

Sulcatol, the aggregation pheromone produced by males of the ambrosia beetle *Gnathotrichus sulcatus* consists of 65% (+)(S)- and 35% (−)(R)-6-Methyl-5-hepten-

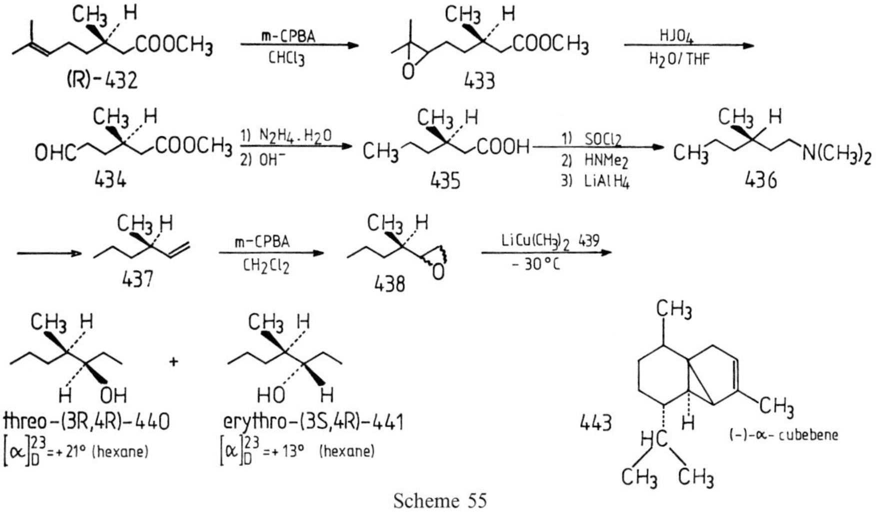

Scheme 55

2-ol *450*, as was determined by NMR analysis of the corresponding (R)-1-methoxy-1-trifluoromethylphenylacetate [346]. The synthesis of both of the optically active antipodes was carried out by Mori, (R)- and (S)-glutamic acid *444*, respectively, being the starting material [754]. Scheme 56 shows by the example of the synthesis of (R)-*450* the conversion of (S)-glutamic acid *444* *via* lactone *445* into the iodo compound *446*, which could be dehalogenized with Raney nickel to *447* and reduced to the lactol *448*. Wittig olefination with the phosphorane of phosphonium salt *449* afforded optically pure (−)(R)-6-methyl-5-hepten-2-ol *450* with $[\alpha]_D^{23}$: − 14.5° (ethanol) [754].

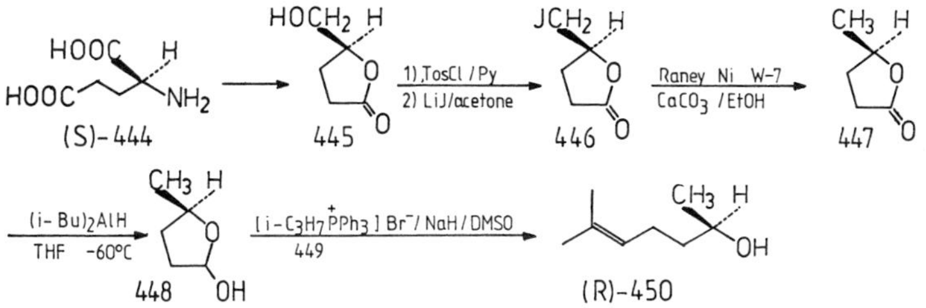

Scheme 56

One from the frass of freshly hatched females of *Dendroctonus brevicomis* isolated component of a population attractant was identified as exo-7-ethyl-5-methyl-6,8-dioxabicyclo[3.2.1]octane *457* (exo-brevicomin, Table 1). In 1974 Mori synthesized the (1R,5S,7R)-*457* as well as the enantiomer (1S,5R,7S)-*457* starting from (2S,3S)-tartaric acid *451* and the enantiomeric tartaric acid, respectively [755] (Scheme 57). The bromide *452* obtained in twelve steps from *451* was condensed with acetoacetic ester *453* to the ketoester *454*, which was *via* *455* oxidized to the diformate *456*. Basic hydrolysis and acidification yielded cyclic ketal *457* with (1R, 5S,7R)-configuration, the analogous reaction sequence starting from (2R,3R)-tartaric acid gave the enantiomeric compound (−)(1S,5R,7S)-*457* [755] (Scheme 57).

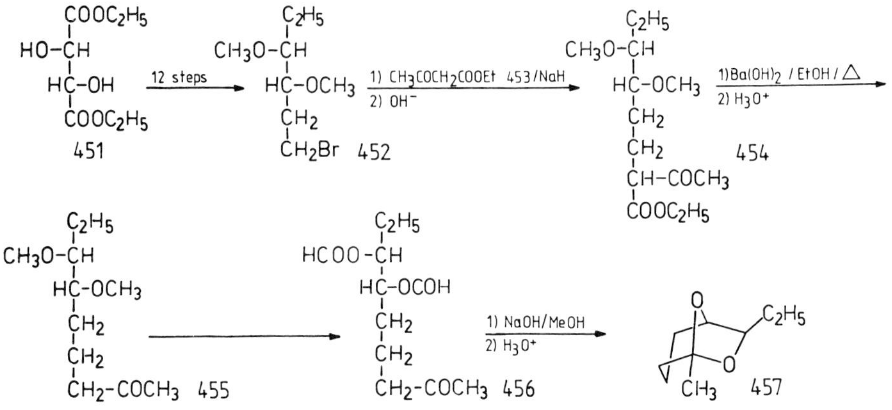

Scheme 57

Apart from a series of published syntheses of racemic exobrevicomin *457* [756–760] the preparation of racemic *457* by means of Kolbe electrolysis [761] should perhaps be pointed out, as it presents a rather simple and short synthesis procedure (Scheme 58). By mixed Kolbe dimerization of (E)-3-hexenoic acid *458* and levulinic acid *459* Knolle and Schäfer [761] obtained beside 2,7-octandione *460* the unsymmetrical coupling products (E)-6-nonen-2-one *461* and 5-ethyl-6-hepten-2-one *462*. cis-Hydroxylation of *462* with osmium tetroxide and subsequent ketalization yielded racemic exo-7-ethyl-5-methyl-6,8-dioxabicyclo-[3.2.1]octane *457* (Scheme 58).

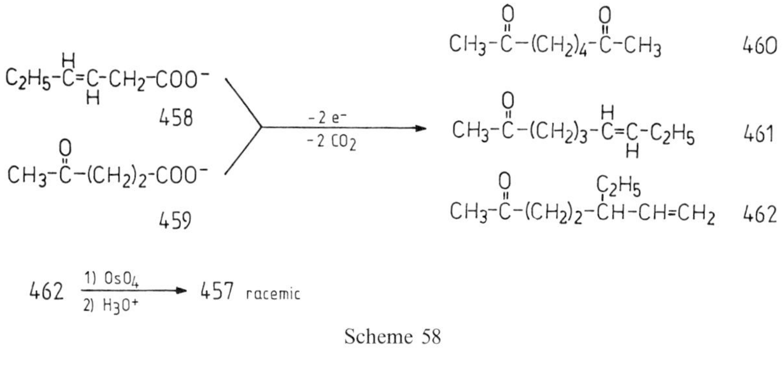

Scheme 58

A further component of the aggregation pheromone of *Dendroctonus brevicomis*, which also has been isolated from females of *D. frontalis*, has been identified as 1,5-dimethyl-6,8-dioxabicyclo[3.2.1]octane *468* (frontalin). Because of the small content of pheromone its absolute configuration has not been elucidated as today, but the levorotatory (−)(1S,5R)-frontalin *468* shows higher biological activity than the other enantiomer.

While several methods of preparation were already known for racemic frontalin [760, 762], Mori [763] synthesized (+)-frontalin *468* and the (−)-enantiomer starting with the lactone *463* of (±)-2-hydroxy-2-methylpentandioic acid. The (R)-lactone *463* and the corresponding (S)-form were obtained by resolution of the racemate with quinine and quinchonine and following crystallization of the resolved acids [763]. The (S)-configuration of (−)-*463* was determined by comparising the CD-spectra with the spectra of (S)-2-hydroxypentan-1,5-diacid lactone. Optical pure (R)-*463* was reduced to triol *464*, converted into acetonide *465* and the corresponding tosylate *466* reacted with sodium cyanide. The reaction of *467* with methyl magnesium iodide and subsequent acidification afforded (1R,5S)-frontalin *468*, $[\alpha]_D^{23}$: + 53.4° (ether) [763] (Scheme 59). A second, recently published synthesis of (1R,5S)- and (1S,5R)-frontalin uses the ketone *470* as starting chemical, which could be obtained in four steps from methyl-γ-D-glycopyranoside *469* by benzylidenation, formation of oxirane, reduction and oxidation [764] (Scheme 59).

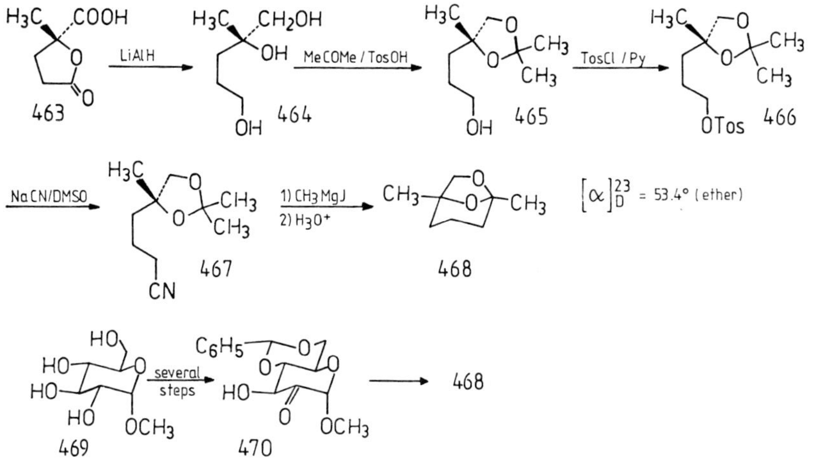

Scheme 59

In addition to the bicyclic ketals brevicomin *457*, multistriatin *431* and frontalin *468* an aggregation pheromone with spiroketal structure could very recently be identified from the *Pityogenes chalcographus* [365]. 2-Ethyl-1,6-dioxaspiro[4.4]-nonane (Chalcogran, *476*) was synthesized by aldol condensation from furfural *471* and butanone *472*, hydrogenolysis to 1-(2-furyl)-3-pentanone *473*, reduction to *474* and cyclisation–hydrogenation (Scheme 60) [765]. The authors prepared not only the pheromone but in four different ways altogether eleven 1,6-dioxaspiro[4.4]nonanes and described their mass spectrometrical fragmentation [765].

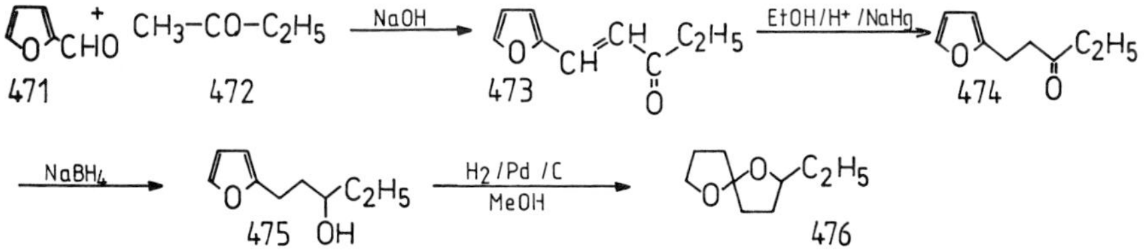

Scheme 60

A synergistically acting pheromone component of the bark beetle *Dendroctonus pseudotsugae* was identified as 3-methyl-2-cyclohexen-1-ol *478* (seudenol)

[336, 343]. The optically active forms of this substance were prepared by Mori et al. [766] by racemate resolution of iodocyclohexenol *477* and reaction with dimethyl lithium cuprate *439*, and thus the configuration of the naturally occuring seudenol *478* was determined (Scheme 61).

Scheme 61

b) Dermestidae

In the dermestid family are many beetle species, known as stored product pests and therefore interesting for a pest control based on pheromones. The pheromone of the stored grain pest *Trogoderma inclusum* consists of four compounds, two of them have been identified as (−)-14-methyl-(Z)-8-hexadecen-1-ol *483* and methyl (−)-14-methyl-(Z)-8-hexadecenoate *484*. Mori synthesized (S)-*483* and (S)-*484*, starting from (S)-2-methylbutan-1-ol [767, 768]. The resulting (S)-5-methylheptyl bromide *479* was coupled with lithium acetylide *480* and the tetrahydropyranyl

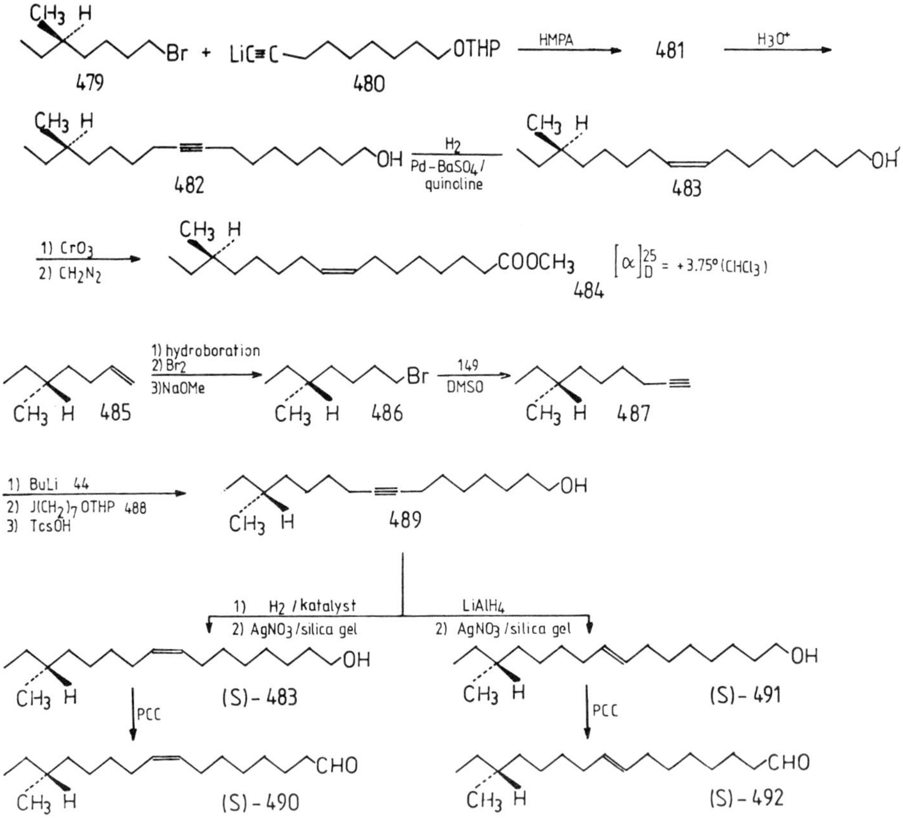

Scheme 62

ether *481* obtained hydrolyzed. Partial hydrogenolysis of the triple bond of *482* afforded (S)-*483* with a positive sign of rotation, the subsequent Jones oxidation and esterification (S)-*484* with $[\alpha]_D^{25} = +3.75°$ ($CHCl_3$) (Scheme 62). Because (S)-*483* and (S)-*484* were dextrorotatory, whereas the natural product was levorotatory, this must posess (R)-configuration [767, 768].

Two stereoisomers of *483* and *484*, namely 14-methyl-(E)-8-hexadecen-1-ol *491* and methyl 14-methyl-(E)-8-hexadecenoate, and 4-hydroxyhexanoic acid lactone *499* was identified from the multicomponent sexual pheromone of female *Trogoderma glabrum* [327]. 14-Methyl-(Z)-8-hexadecenal *490* was reported as the main component of the pheromone of *T. inclusum* and *T. variabile* females, the proof for the existence of 14-methyl-(E)-8-hexadecenal *492* in the pheromone of *T. glabrum* was made [769] and the compound was synthesized as shown in the second part of Scheme 62 [770]. Hydroboration of (S)-5-methyl-1-heptene *485*, obtained from (S)-2-methylbutanol, and treatment of the borane with bromine yielded bromide *486*, which was substituted with the lithium acetylide-ethylene diamine complex to *487* in 55 percent yield. The resulting alkynyl lithium was alkylated with the iodo-THP-ether *488* and the obtained alkynyl-THP-ether hydrolyzed to (S)-*489* (54% yield).

The catalytic hydrogenolysis afforded (Z)-alkenol *483*, which could be oxidized to (S)-14-methyl-(Z)-8-hexadecenal *490* with pyridinium chlorochromate (PCC). The reduction of *489* with $LiAlH_4$ yielded (E)-alkenol *491*, and the PCC-oxidation gave (S)-14-methyl-(E)-8-hexadecenal (S)-*492* [770] (Scheme 62).
The two enantiomers of the 4-hydroxyhexanoic acid lactone were synthesized by Ravid and Silverstein [771] from the commercially available enantiomeric forms of glutamic acid *493*. Diazotation of (S)-*493* led to the (S)-lactonic acid *494*, which was esterified to *495* (Scheme 63). Reduction afforded the lactonic alcohol *496* and the triol *497*, which were chromatographically separated. Tosylation of *496* and coupling with $LiCu(CH_3)_2$ yielded the (R)-4-hydroxyhexanoic acid lactone *499*, $[\alpha]_D^{20}: +42.7°$ (CH_3OH) [771] (Scheme 63). (S)-*499* was prepared from (R)-glutamic acid *493* by the same method [771].

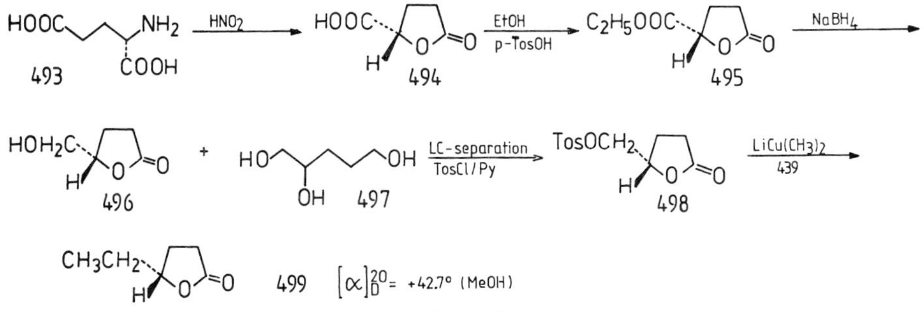

Scheme 63

c) Curculionidae

One of the most dangerous cotton pests of the United States is the curculionid beetle *Anthonomus grandis*, which is responsible for more than three quarters of all insect cotton damage.

(+)-cis-1-Methyl-1-(2-hydroxyethyl)-2-isopropenylcyclobutane *502* (grandisol) is the main component of the four compounds of the pheromone [366, 367], and several stereoselective synthesis procedures of this substance have been already published [772–779]. Of most interest is the method of Billups and coworkers [774], who obtained grandisol *502* in two steps by cyclopentadienyl nickel catalyzed isoprene dimerization *via* diene *501* (12–15% yield) and subsequent hydroboration – oxidation [774] (Scheme 64).

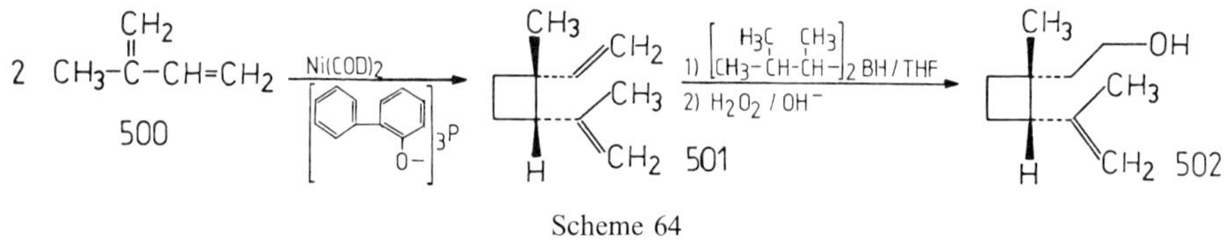

Scheme 64

Hobbs and Magnus synthesized optically active grandisol *502* starting from (−)-β-pinene, and thereby determined the absolute configuration of (+)-*502* with (1R,2S) [780]. The β-pinan-2-ol *503*, deduced from β-pinene, was etherified into *504* and subsequently oxidized to lactone *505*. The reduction with lithium triethoxyaluminium hydride yielded 6,9-dimethyl-8-hydroxy-7-oxatricyclo[$4.3.0.0^{3,9}$] nonane *506*, which was olefinated with phosphorane *89* to *507*. Hydroboration-oxidation with following acetylation and dehydratisation led to a mixture of acetates *509* and *510*, which could be oxidized and hydrogenated to 2-α-H-9-acetoxymethylpinan-4-one *511*. Photolysis and decarbonylation afforded *via* grandisol acetate *512* the (+)(1R,2S)-grandisol *502*, $[\alpha]_D^{21.5}$: +18.5° (hexane) [780] (Scheme 65).

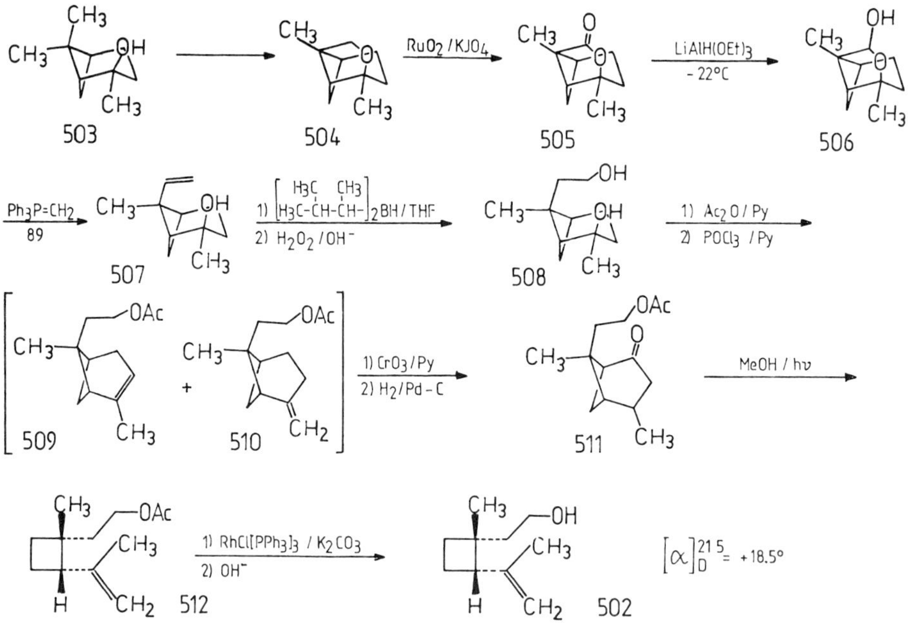

Scheme 65

d) Bruchidae

In the first days after hatching from the pupae the male of *Acanthoscelides obtectus* produces a sex attractant, which has been identified in 1970 as (−)-(E)-2,4,5-tetradecatrienoic acid methyl ester *517* [369]. For a long time only the synthesis of the racemic form of this chiral allenic component of an insect pheromone was known. It was only at the end of 1978 that the synthesis of the corresponding optically active compounds has been published. The first preparation of racemic *517* started with 1-undecin-3-ol *513*, which could be reacted to alkadien-1-ol *514* over several intermediate steps. Manganese dioxide oxidation afforded allenic aldehyde *515*, which could be olefinated with the anion of methoxycarbonylmethyl dimethylphosphonate *516* to methyl (E)-2,4,5-tetradecatrienoate *517* [781] (Scheme 66).

In the synthesis of ester *517*, as it was carried out by Baudouy and Goré [782], the stereo- and regioselective reduction of racemic 4,5-tetradecadien-2-yn-1-ol *518* using lithium methoxyaluminium hydride is the key reaction step. The oxidation of the resulting trienol *519* and esterification yielded also racemic methyl (E)-2,4,5-tetradecatrienoate *517* [782] (Scheme 66).

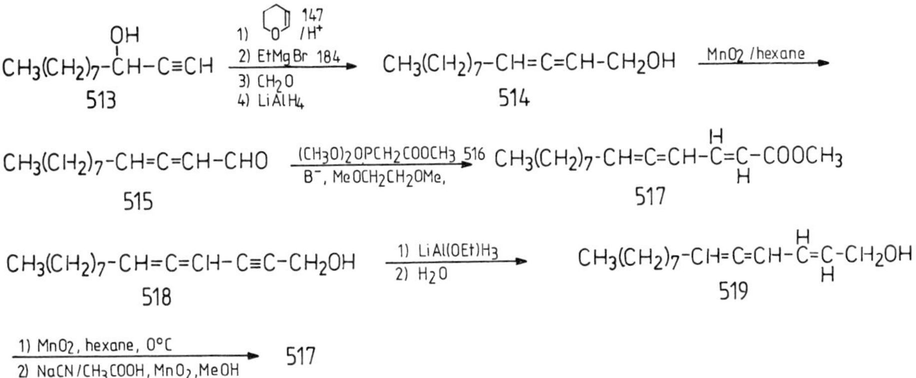

Scheme 66

Two further syntheses of *517* were carried out by means of lithium dialkylcuprate couplings. Descoins et al. [783] startet with the monoacetal *520* of fumaric dialdehyde, which reacted by Grignard reaction to alkenynol *522*. The reaction of the corresponding propargyl acetate *523* with lithium dioctylcuprate *524* gave enallene acetal *525* in 25% yield, which could be hydrolyzed, oxidized and esterified to *517* [783] (Scheme 67). A more simple synthesis, as it was carried out by Michelot and Linstrumelle [784], is based on the preparation of the allenic cuprate *529* and its reaction with methyl propynoate *530* (Scheme 67). They alkylated allene *526* to 1,2-undecadiene *528*, the lithium cuprate of which (*529*) reacted with methyl propynoate *530* to *517* [784] (Scheme 67).

In a more recent work for the preparation of racemic *517* the 1-undecyn-3-ol *513* also served as the starting material [785]. A modified Claisen rearrangement afforded allenic ester *531* in 95% yield, which could be converted via alkadienol *532* into the bromo allenic compound *533*. The carboxylation of the correspond-

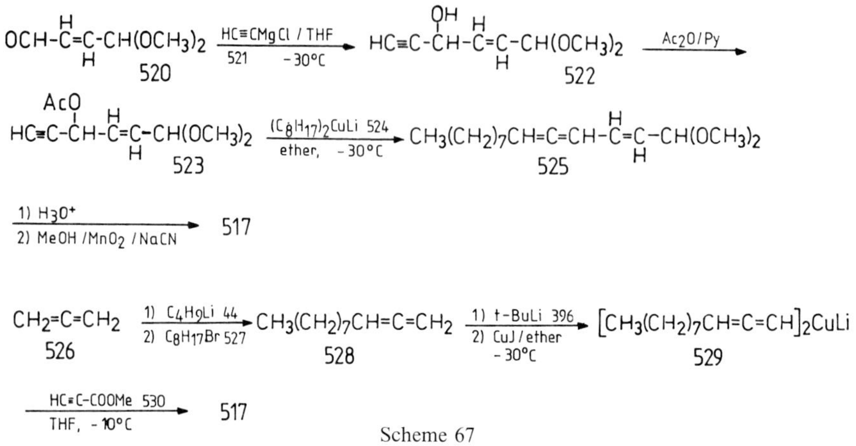

Scheme 67

ing Grignard compound and subsequent esterification gave ester *534*, the lithium enolate of which with diphenyl diselenide *535* and subsequent oxidation of the resulting phenyl selenoester reacted to methyl (E)-2,4,5-tetradecatrienoate *517* [785] (Scheme 68).

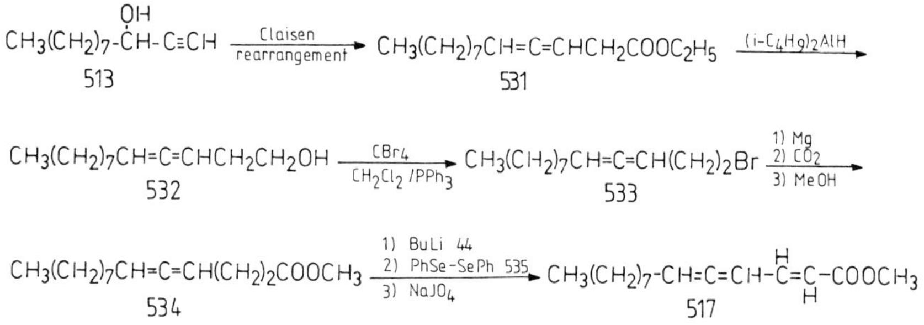

Scheme 68

Only at the end of 1978 did Pirkle and Boeder [786] report the synthesis of both enantiomers of *517*. Cyanopropionic aldehyde *536* reacted with Grignard reagent *537* to the racemic adduct *538*, which could be resolved into its antipodes *via* optically active urethanes. Coupling with dioctyl lithium cuprate *524* generated the isomeric allenes *540*, which could have been converted into the allene olefins *517* [786] (Scheme 69).

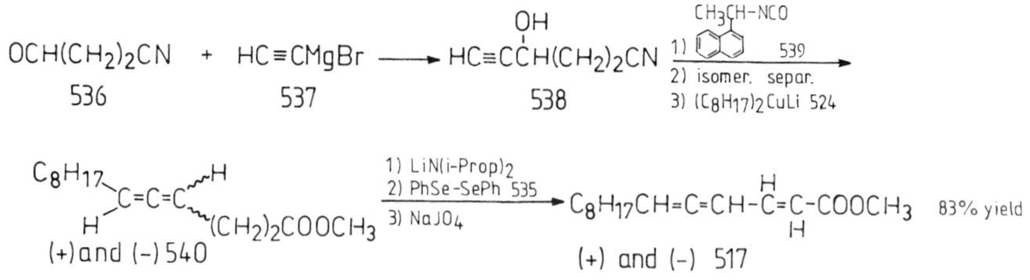

Scheme 69

5.3. Flies – Diptera

Among the known fly pheromones except methyl (E)-6-nonenoate *546* and (E)-6-nonenol *542*, the pheromone components of the mediterranean fruit fly *Ceratitis capitata*, only olefinic hydrocarbons have been isolated as sex attractants. Syntheses for the both (E)-6-compounds have been reported several times and performed using different types of reactions.

Canevet et al. [705] prepared (E)-6-nonenol *542* by $LiAlH_4$ reduction of 1-(5-tetrahydropyranyloxy)-(E)-2-butenyl diethylphosphonate *541*. The phosphonate *269* was prepared by Arbusow reaction from crotyl bromide *211*, the anion generated with n-butyllithium *44*, and alkylated in α-position with 5-bromopentyl tetrahydropyranyl ether *299*. The resulting phosphonate *541* was reduced with a shift of the double bond according the method of Kondo [704] and hydrolyzed to *542* [705] (Scheme 70). (For the reduction of allyl diethylphosphonates cf. p. 113.)

A coupling reaction between (E)-3-hexenyl magnesium bromide *543* and allylic bromide *544* and subsequent regioselective hydroboration of the terminal double bond of alkadiene *545* with 9-borabicyclo[3.3.1]nonane *17* (9-BBN) also yielded (E)-6-nonen-1-ol *542* [787]. Oxidation with pyridinium chlorochromate/silver oxide and esterification with diazomethane easily led to (E)-6-nonenoic acid methylester *546*, the second component of the fruit fly pheromone [788] (Scheme 70).

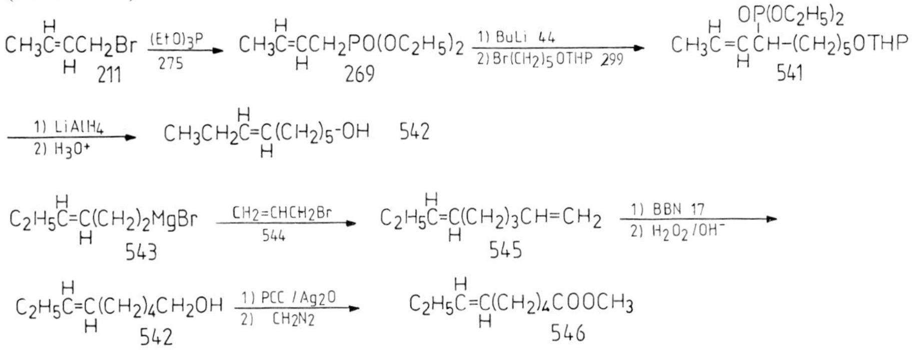

Scheme 70

(E)-6-Nonen-1-ol *542* was obtained by a group of scientists at Boston University from carbonyl metathesis reaction [789]. From propionaldehyde *188* and cyclohexa-1,3-diene *547* the authors obtained the bicyclic oxetane *548 via* photo-Paterno-Büchi cycloaddition, the hydrogenation and pyrolysis of which led to (E)-6-nonenal *550*. Its lithium tetrahydroaluminate reduction gave the pheromone *542*, 95% (E)-isomer [789] (Scheme 71).

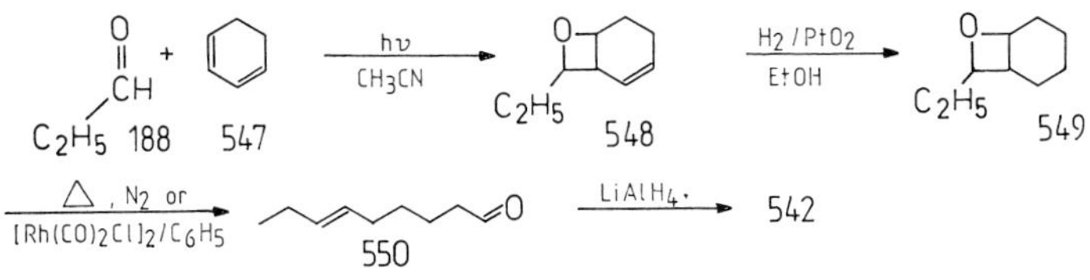

Scheme 71

(E)-6-Nonenyl acetate, which represents an intermediate in the synthesis of pheromone *542*, was prepared by Hayashi and Midorikova [725] with a high degree of stereoselectivity by a [3.3]-sigmatropic rearrangement of allylic dithiocarbamates (cf. 5.1e, p. 120 and Scheme 45).

(Z)-9-Tricosene *553* (muscalure, muscamone®) was isolated as the sex attractant, together with a number of homologous hydrocarbons, from faecal and cuticular lipids of female house flies *Musca domestica* [428]. Mansingh et al. [429] reported, that a 7:3 mixture of *553* and (Z)-9-heneicosene *556* showed a higher attractivity in olfactometric tests than muscalure itself, though this result could not be proved in field trials [790]. (Z)-14-Nonacosene *561* has been identified in cuticular extracts of both sexes of *Musca autumnalis*, as have the less active compounds (Z)-13-nonacosene *562* and (Z)-13-heptacosene *563* [431].

These hydrocarbons, which all possess (Z)-configuration, had been, among other ways, synthesized by selective Wittig reactions [428, 790, 791]. Carlson et al. [428, 790] obtained (Z)-9-tricosene *553* (muscalure) by condensation of nonanal *551* with the ylide of the phosphonium salt *552* in 73% yield and a Z/E-isomer ratio of approximately 85:15, in DMSO with lithium butyl as the base for ylide formation (Scheme 72). Laboratory olfactometer tests showed a higher potency of the (Z)-isomer than for (E)-*553* to male house flies. A reduction of attractivity could not be observed when the (E)-isomer was added to the (Z)-olefin [790].

Bestmann et al. [791] synthesized the (Z)-olefin from tetradecanal *555* by means of a Wittig reaction (Scheme 64) using potassium in HMPA as the base for the generation of the ylide from the phosphonium salt *554* and obtained *553* in 33% yield with only 4% (E)-isomer amount [791].

$$\underset{551}{CH_3(CH_2)_7CHO} + \underset{552}{[(C_6H_5)_3\overset{+}{P}(CH_2)_{13}CH_3]\,Br^-} \xrightarrow[10-20°C]{BuLi\ 44/DMSO} \underset{553\quad 15\%E,\ 73\%\ yield}{CH_3(CH_2)_7\overset{H}{C}{=}\overset{H}{C}(CH_2)_{12}CH_3}$$

$$\underset{554}{[CH_3(CH_2)_8\overset{+}{P}(C_6H_5)_3]\,Br^-} + \underset{555}{OCH(CH_2)_{12}CH_3} \xrightarrow{K\ /\ HMPA} 553\quad 4\%\ E,\ 32\%\ yield$$

Scheme 72

Using the same method, i.e. stereoselective Wittig reaction with potassium in HMPA, (Z)-9-heneicosene *556* could be prepared in 37% yield (4% (E)-isomer amount) [791], similarily the condensation of tetradecanal *555* with pentadecylidene triphenylphosphorane *557* in HMPA/THF resulted in (Z)-14-nonacosene *561* (70% yield) with 94–96% (Z)-amount, as could be estimated from the intensity of the 960 cm^{-1} IR-band [430].

Muscalure *553* was furthermore obtained from partial hydrogenation of the corresponding acetylenic compound [792, 793]. The alkylation of the lithium salt of 1-pentadecyne *559* with bromooctane *527* resulted in the formation of 9-tricosyne *560* in high yields in diglyme [793] or HMPA/THF [792], the hydrogenolysis of which by means of Lindlar catalyst [793] or 5% palladium on barium sulfate [792], respectively, afforded *553* with 95% (Z)-isomer amount (cf. p. 104).

Scheme 73 shows, how (Z)-9-tricosene *553* is produced in 85% yield from erucic acid *564* [792]. The latter was reacted to the methyl ketone *565* and this reduced to the alkene *553* according to Huang-Minlon (89% yield) [792]. A similar method was the conversion of oleic acid *566* with lithium pentyl *567* to the ketone *568*, which gave *553* in a 77% total yield, using Huang-Minlon conditions [794] (Scheme 73).

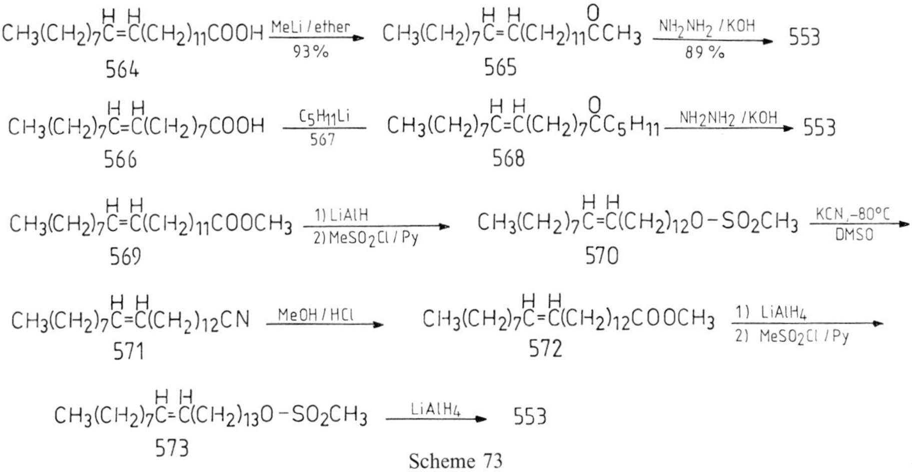

Scheme 73

The second part of Scheme 73 shows a multistep synthesis of *553*, starting with erucic acid methylester *569*. Chain elongation of the (Z)-13-docosenyl methane-sulfonate *570* with KCN and subsequent methanolysis, reduction of the ester *572*, mesylation and reduction of the sulfonate *573* resulted in *553* with about 80% yield [795]. Though this synthesis produces good yields in all steps, it seems to be unapplicable for a muscalure preparation on a technical scale because of its length.

The homogenous, transition metal catalyzed metathesis reaction [727] (cf. 5.1f, p. 121) often found use for the synthesis of simple olefinic diptera pheromones. But the isomeric mixtures formed must be resolved into their components. (Z)-9-Tricosene *553* (muscalure) was prepared from 1-decene *574* and 1-pentadecene *575* in 26% yield [728], and from 2-hexadecene *576* and 9-octadecene *577* in 11% yield [730]. The metathesis olefin synthesis of 1-tetradecene *578* and 2-hexadecene *576*, or 2-octadecene *579* yielded 13-heptacosene *563* and 13-nonacosene *562*, respectively [730], the same reaction with 1-hexadecene *580* and 2-hexadecene *576* gave 14-nonacosene *561* [730] the third pheromone component of *Musca autumnalis*.

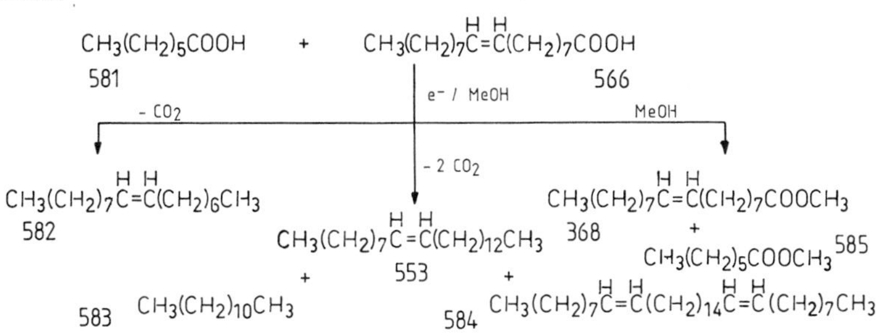

Scheme 74

The mixed Kolbe electrolysis of heptanoic acid *581* and oleic acid *566* led in a single step synthesis to the house fly pheromone muscalure *553*. The electrolysis of *581* and *566* in methanol resulted in the formation of 14% muscalure *553*, 20% dodecane *583*, 7% (Z)-9,(Z)-25-tetratriacontadiene *584*, 2% (Z)-8-heptadecene *582* and the methylated products methyl heptanoate *585* (16%) and oleic acid methylester *368* (68%) in the neutral fraction [796] (Scheme 74). Because the esters *585* and *368* are easily to hydrolyse, they therefore can serve as starting material again, the actual yield of *553* approximately being 40%.

5.4. Hymenoptera

The order of Hymenoptera comprises a great number of state forming insects, which regulate their social order and their life in colonies by very complex pheromone systems.

For (E)-9-oxodec-2-enoic acid *594*, the "queen substance", originated from the secretions of the honey bee, of which a number of syntheses are known, the more recent ones will be reviewed in here. Bestmann et al. [797] started from pimelic acid *586*, which could be converted over several steps into the phosphorane *587* and hydrolyzed into ketoacid *588*. *588* was reduced *via* acid chloride *589* and thiolic ester *590* to aldehyde *591*, and the latter olefinated with methoxycarbonylmethylene triphenylphosphorane *592* to 9-oxodecenoic ester *593*. Careful hydrolysis led to the "queen substance" *594* [797] (Scheme 75).

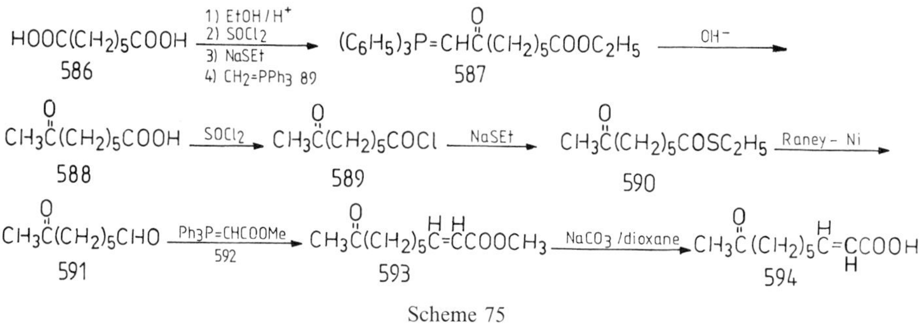

Scheme 75

Trost and Salzmann [798] started the synthesis with commercially available acelaic acid methylester *595*, which could be converted to 9-oxodecanoate *597*. *597* could be converted into methyl 2-methylthio-9-oxodecanoate *599* and oxidized to the ester *593*. Hydrolysis again yielded pheromone *594* [798] (Scheme 76).

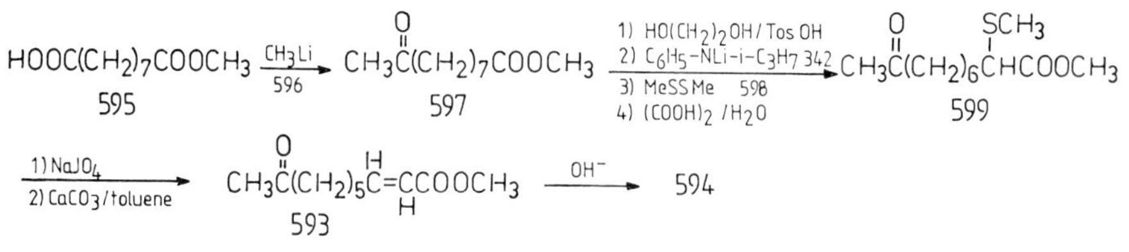

Scheme 76

Tsuji and coworkers [799] reported a simple synthesis of *594* from butadiene *286* and malonic ester *600* using palladium catalysis. The octadienyl malonic ester *601* was oxidized to *602*, hydrogenated, and with diphenyl diselenide *535* converted into the selenoester *604*. Oxidation and hydrolysis resulted again in *594* [799] (Scheme 77).

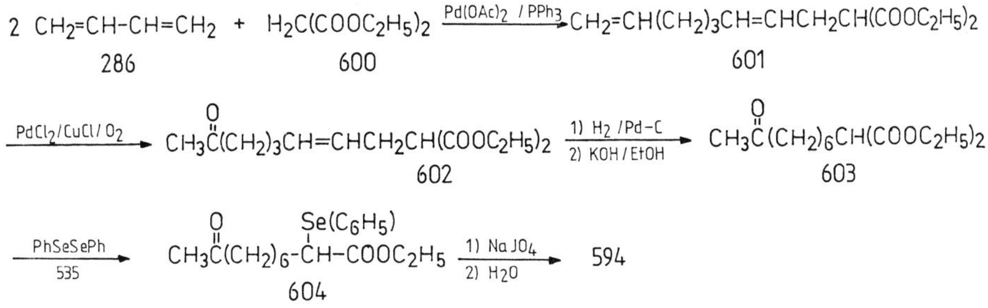

Scheme 77

(E)-10-Hydroxydec-2-enoic acid *608* ("royal jelly acid") was found in the brood food of the Weisel cells ("gelée royale") of the honey bees and synthesized from suberic acid monoester *128* [797]. The hydroxyacid *605* deduced was acetylated, converted into the thiolester *606* and with Raney nickel and methoxycarbonylmethylene triphenylphosphorane *592* olefinated to *607*. The α,β-unsaturated ester *607* was then deacetylated and hydrolyzed to acid *608* [797] (Scheme 78).

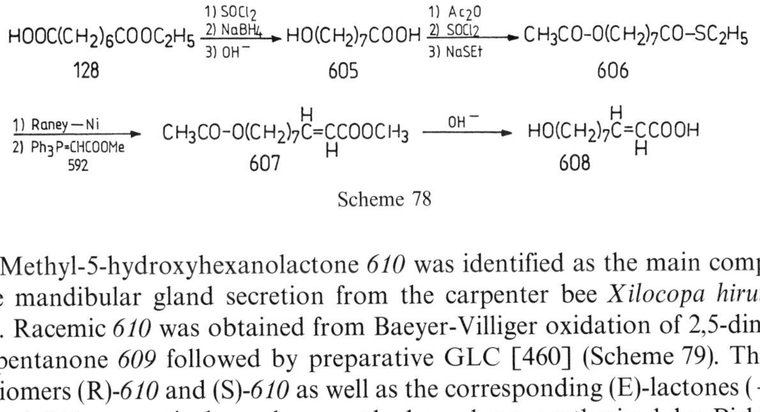

Scheme 78

(Z)-2-Methyl-5-hydroxyhexanolactone *610* was identified as the main component of the mandibular gland secretion from the carpenter bee *Xilocopa hirutissima* [460]. Racemic *610* was obtained from Baeyer-Villiger oxidation of 2,5-dimethylcyclopentanone *609* followed by preparative GLC [460] (Scheme 79). The both enantiomers (R)-*610* and (S)-*610* as well as the corresponding (E)-lactones (+)-*616* and (−)-*616*, respectively, only recently have been synthesized by Pirkle and Adams [800], starting from racemic 5-cyanopentan-2-ol *611*. This could be reacted with (R)(−)-1-(1-naphthyl)-ethyl isocyanate *612*, resolved into the diastereomeric cyanocarbamates *613* by HPLC separation, and by silanolysis with trichlorosilane again converted into the enantiomeric starting substances (S)-*611* and (R)-*611* (Scheme 79). Hydrolysis and subsequent lactonization gave the corresponding enantiomers of δ-methyl-δ-valerolactone *614*. Alkylation of *614* with lithium diisopropylamide and methyl iodide yielded a 1:1-mixture of each the (Z)- and (E)-lactones *610* and *616*, which could be separated gaschromatographically [800] (Scheme 79).

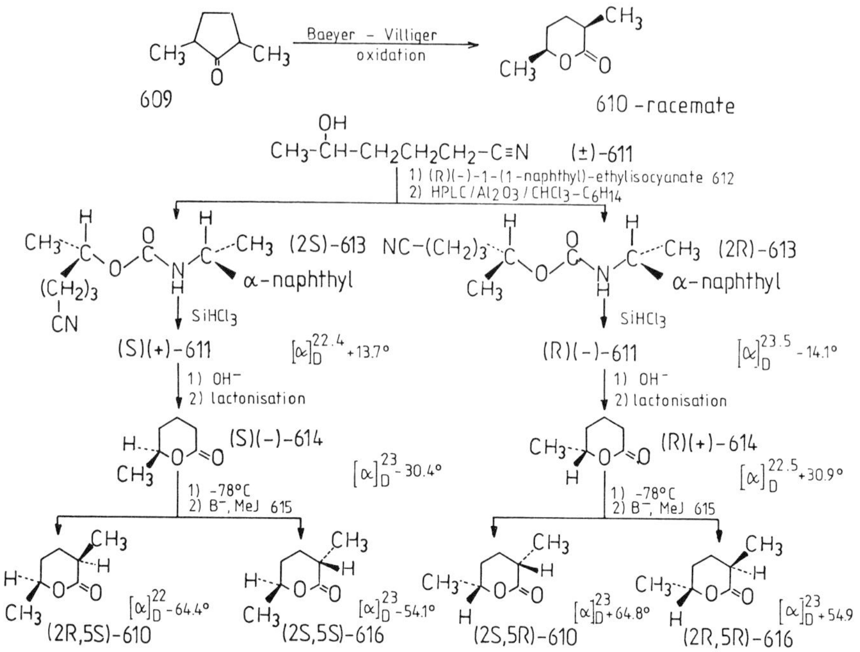

Scheme 79

Another chiral lactone *621* was found in the mandibular glands of the orient hornet *Vespa orientalis* [454], which possesses the function of a "queen substance". The (R)(+)- and (S)(−)-enantiomer of *621* was obtained from (R)(+)- and (S)(−)-1,2-epoxytridecane *617* and the dianion of propiolic acid *620* and subsequent reduction [801]. The enantiomeric epoxides *617* could be prepared by racemate resolution of 1-dimethylaminotridecan-2-ol *618* with dibenzoyl tartaric acid and Hofmann elimination [801] (Scheme 80).

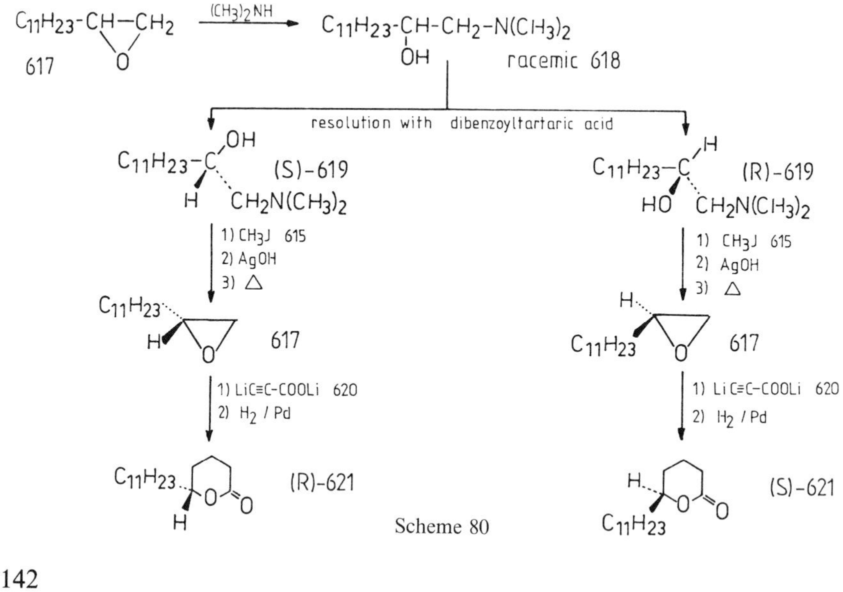

Scheme 80

The main component of the marking substance of the male bumble bee *Bombus terrestris* was identified as (E)(−)-3,7,11-trimethyl-6,10-dodecadien-1-ol *627* (terrestrol) [456–458]. Terrestrol was synthesized by a stereospecific pathway starting from geraniol *623* [457]. The deduced homogeranic acid *624* was chain elongated by Kolbe electrolysis with enantiomeric 3-methylglutaric acid monomethylester *625* to *626*. The reduction of the enantiomers *626* yielded terrestrol (S)(−)-*627* and the enantiomer (R)(+)-*627* [457] (Scheme 81).

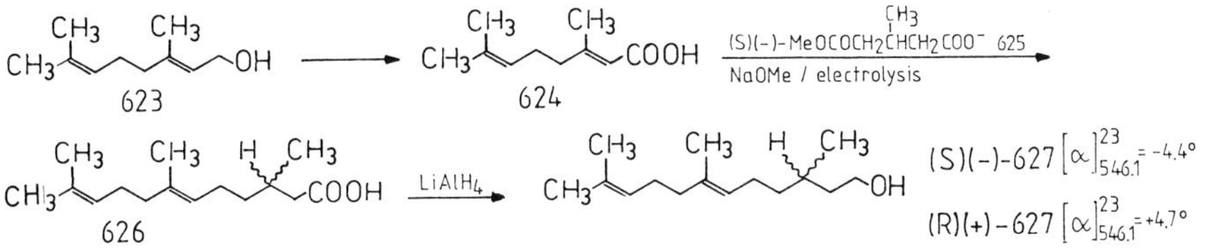

Scheme 81

The specificity of the pheromone receptors and the behavioral responses to the enantiomers of chiral pheromones was proven for the alarm pheromone of the leaf cutting ants *Atta texana* and *A. cephalotes* [802, 803] and of the myrmicine *Pogonomyrmex barbatus* [804]. Only the (S)(+)-isomer, (+)-4-methylheptan-3-one *630* is synthesized by the two Atta species [802, 803], whereas the optical activity of the *P. barbatus* compound still is unknown. *630* and a mixture of the both diastereomers of the corresponding alcohol *631* are components of the alarm substance of *Neoponera villosa* [805]. Both enantiomers, (S)-*630* and (R)-*630* were prepared stereospecifically from optical pure 2-methylpentanoic acid *628* in 78% yield [806] (Scheme 82). Comparing both compounds, the (S)-4-methylheptan-3-one *630* showed an approximately 400 fold higher activity towards *A. texana* than the corresponding (R)-isomer [806].

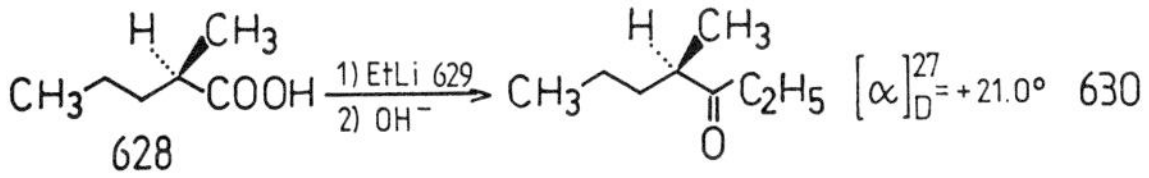

Scheme 82

Manicone, (E)-4,6-dimethyloct-4-en-3-one *634*, and 4-methylhexan-3-one *639* were isolated as alarm pheromones from the mandibular glands of *Manica mutica* and *M. bradley* [409], the optical activity of it not known. Racemic *634* was prepared by the selective dehydration of the 3-hydroxyester *632 via* the 3-alanoxy-ester [807] (Scheme 83).

Only recently racemic *634* was synthesized by Kocienski et al. [808] from 2,4-dimethyl-3-oxohexanoic acid methylester, who used a reaction, which allows the introduction of an α,β-unsaturated carbonyl unit by the reaction of 4-chloro-2-pyrazolin-5-one *635* with sodium hydroxide [809] (Scheme 83). (S)-4-Methylhexan-3-one *639* was obtained from a substitution reaction of (S)-2-methylbutanoyl chloride *638* with ethyl magnesium bromide [810] (Scheme 83). (S)(+)-Manicone *634* and the (R)-enantiomer have been prepared recently in good yields by the titanium-IV-chloride catalyzed reaction of 3-trimethylsiloxypent-2-ene with (S) and (R) 2-methylbutan-1-ol, respectively [811].

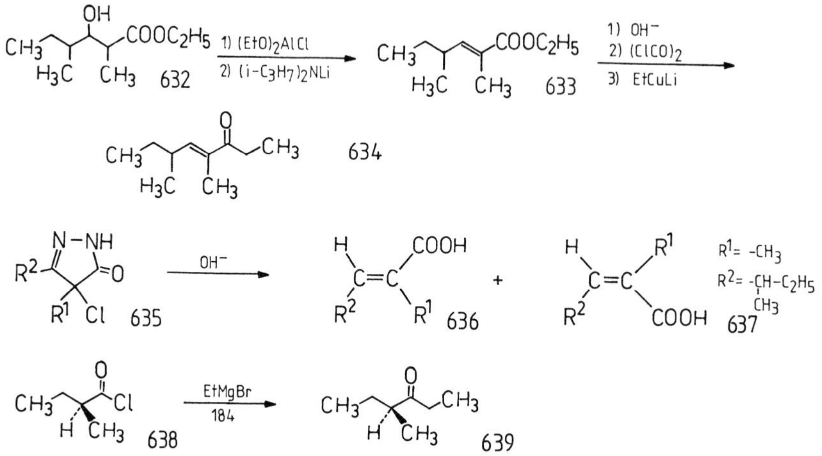

Scheme 83

The trail pheromone of the pharao ant *Monomorium pharaonis* had been identified as (6E, 10Z)-3,4,7,11-tetramethyl-6,10-tridecatrienal (faranal, cf. Table 1) [427], the structures of the chiral centres, however, have not been determined until today. A stereospecific synthesis of this compound has not yet been reported.

Esters of erythro-3,7-dimethylpentadecan-2-ol *646* are the main components of the pheromone of several species of pine saw flies (*Diprionidae*) [459]; the configuration of C-7 is not yet known. *646* has been synthesized by Magnussum [812] starting from commercially available 2,3-dimethylcyclohexanol *640*. The higher boiling fraction of *640* was converted into the ketone *641* by a Jones oxidation and oxidized according to Bayer-Villiger to the lactone *642*. The reaction with lithium octyl *643* yielded the hydroxyketone *644*, which was olefinated with *89* and subsequently catalytically hydrogenated to *646* in a total yield of 25% [812] (Scheme 84).

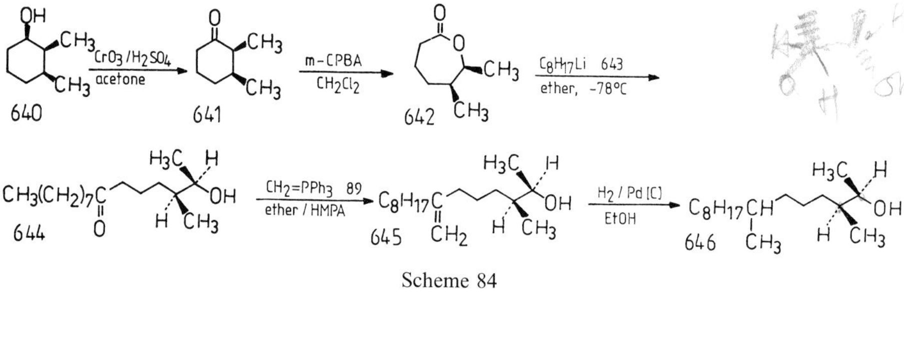

Scheme 84

5.5. Pheromones of Isoptera

The diterpene hydrocarbon cembrene *651* was isolated from the termite *Nasutitermes exitiosus* and its structure elucidated [412]. This compound was synthesized in its racemic form from *trans, trans*-geranyllinalool *647*, which was converted into the *trans, trans, trans*-geranylgeranyl phenyl sulphide with subsequent epoxidation

648. Intramolecular cyclization in the presence of butyl lithium *44* and 1,4-diazabicyclo[2.2.2]octane *649* afforded terpene alcohol *650*, which could be hydrated and desulfurized to *651* [412] (Scheme 85).

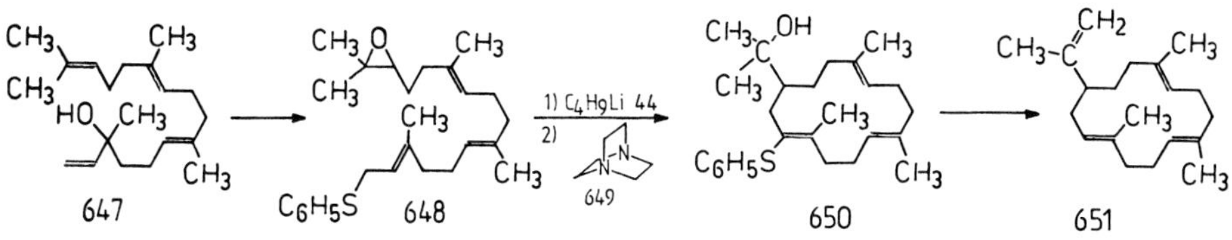

Scheme 85

Furthermore racemic *651* had been prepared by nickel carbonyl catalyzed coupling reaction of terminal allylic bromides [813] in a four step synthesis from *trans*-geranylgeranic acid [814].

The levorotatory enantiomer of 4,11-epoxy-(Z)-eudesmane *657*, a tricyclic sesquiterpene, was isolated from the defensive secretion of the African termite *Amitermes evuncifer* [815] and in a rather unambiguous reaction sequence synthesized from (−)-carvone *652* [816]. The *epi-γ*-eudesmol *656* was obtained from *653* by ringclosure reaction with 1-chloropentan-3-one *654* as the key reaction step. Oxymercuration and subsequent reduction of the organomercury adduct yielded a mixture of *657* and *658* (ratio 3:1) [816] (Scheme 86).

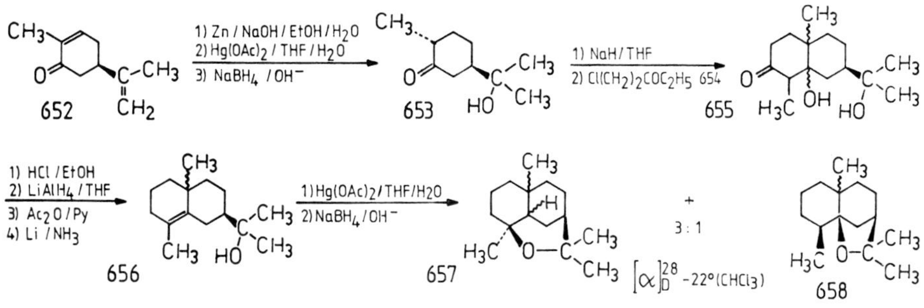

Scheme 86

5.6. Synthesis of Cockroach Pheromones (Blattaria)

The female German cockroach *Blatella germanica* secretes a pheromone, which can be received by the males by antennal contact. It causes wing fluttering and enables them to recognize the opposite sex. One of the both pheromone components isolated has been identified as 3,11-dimethylnonacosan-2-one *662* [399, 400]; its complete stereochemistry still unknown.

Until today five syntheses of *662* have been described [400, 817–820], of which the method of Schwarz and coworkers [818] seems to be very interesting to us. The anion of acylmethylene phosphorane *660* was alkylated with halogenide *659* to the stable ylide *661*, which could be hydrolyzed to the pheromone *662* [818] (Scheme 87).

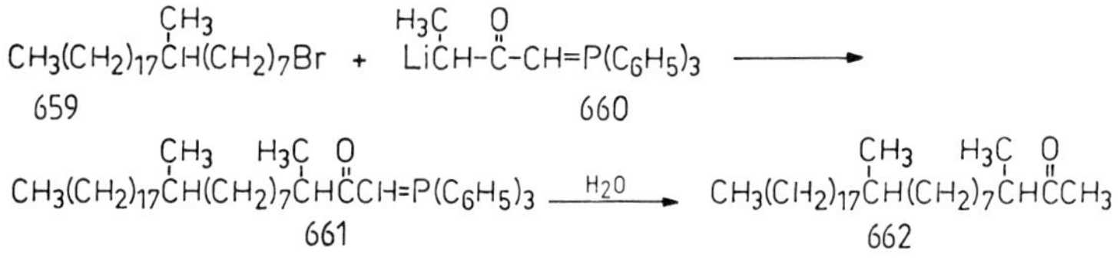

Scheme 87

In a more recent synthesis of *622* [819] the preparation of ketone *665* from S-arylthioester *663* and lithium dialkylcuprate *664* was the key reaction step. Ketoacetale *665* and phosphor ylide *89* gave methylene acetal *666*, which was hydrogenated and hydrolyzed to *667*. The corresponding bromide *668* was coupled with 3-(1-ethoxy)ethoxy-2-methylbutyllithium *669* to *670*, the hydrolysis of which yielded *671*. The subsequent Jones oxidation afforded racemic *662* in 74% yield [819] (Scheme 88).

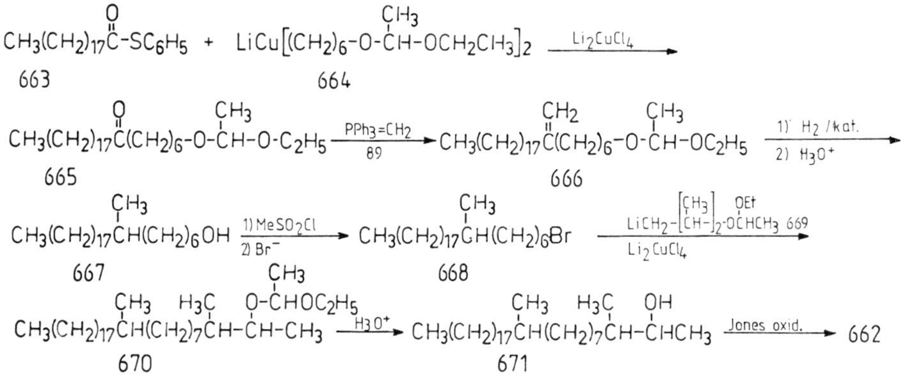

Scheme 88

The application of an oxy-Cope reaction of 4-ethynyl-4-methylhexa-1,5-dien-3-ol *674* to aldehyde *676* was reported by Place et al. [820] in their synthesis of *662*. Thereby 5-chloro-3-methyl-3-penten-1-yne *672* was the starting compound, which was reacted via the steps *674*, *676*, *678*, and *679* to the pheromone *662* [820] (Scheme 89).

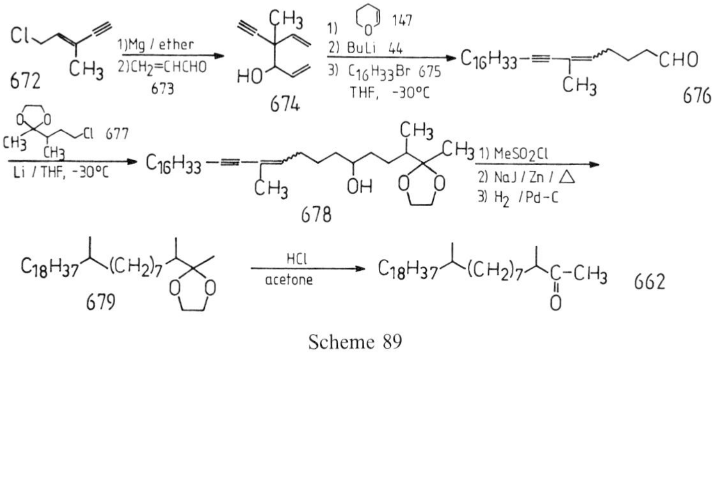

Scheme 89

The second substance which causes wing fluttering of *B. germanica* had been identified as 3,11-dimethyl-29-hydroxynonacosan-2-one *686* [398]. This compound shows positive optical rotation. The racemate was synthesized starting with 9-bromo-3-methylnonan-2-one *680*, which could be obtained from 2-methylacetoacetate [397]. *680* was converted into acetal *681* and by means of a second acetoacetate synthesis the ketone acetal *683* formed. Grignard reaction with *684* led to THP-ether *685*, hydrolysis, cleavage of water and hydrogenation to the pheromone 29-hydroxy-3,11-dimethylnonacosan-2-one *686* [398] (Scheme 90).

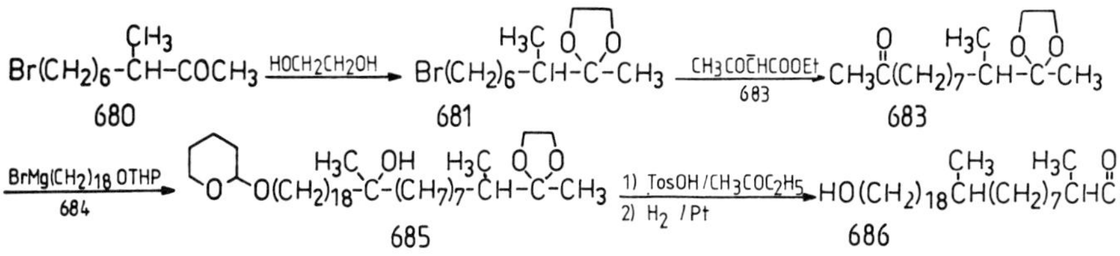

Scheme 90

Another preparation of *686* was reported by Burgstahler et al. in 1977 [821]. 4-Benzyloxy-1-bromobutane *687* was reacted with the lithium acetylide/ethylenediamine complex to *688*, a second alkylation with 1,12-dibromododecane *689* gave *690* in 82% yield. The corresponding triphenylphosphorane and 9-bromononan-2-one *691* yielded a Z/E-isomeric mixture of olefin *692*, the alkylation of methylacetoacetate *693* the ketoester *694*. Hydrogenation, hydrolysis and decarboxylation gave rise to *686* [821] (Scheme 91).

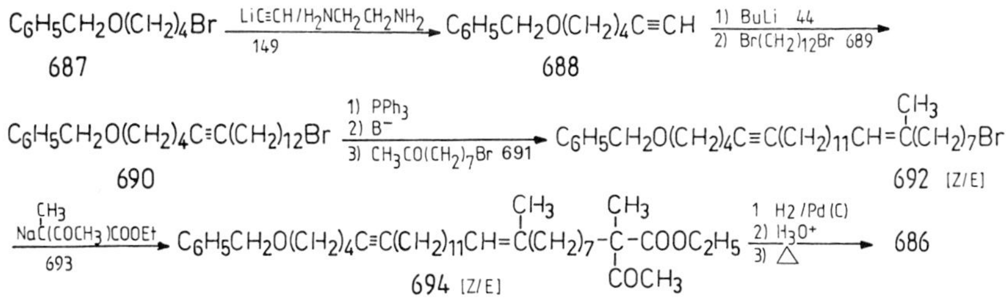

Scheme 91

The first synthesis of (Z)-1,(E)-5-1,10(14)-diepoxy-4(15)-5-germacradien-9-one *705* (periplanone B), the pheromone of the American cockroach, has been presented by Still at a meeting of the American Chemical Society in 1978 [822]. In the course of this synthesis the cyclodecadienone *700* was the key intermediate, available from the cyclohexenone derivate *695*. Aldolization, acetylation and addition of Me_3SnLi and Me_3SiCl gave the β-stannyl enol silylether *696*. Coupling with lithium dimethylcuprate and oxidative workup afforded *697*, ring expansion to *700* required two more steps and an oxy-Cope rearrangement. *700* was protected, then epoxidized, and converted to the bisepoxide *704*. Subsequent oxidation with pyridinium chlorochromate (PCC) yielded the pheromone *705* of the American cockroach *Periplaneta americana* [822] (Scheme 92).

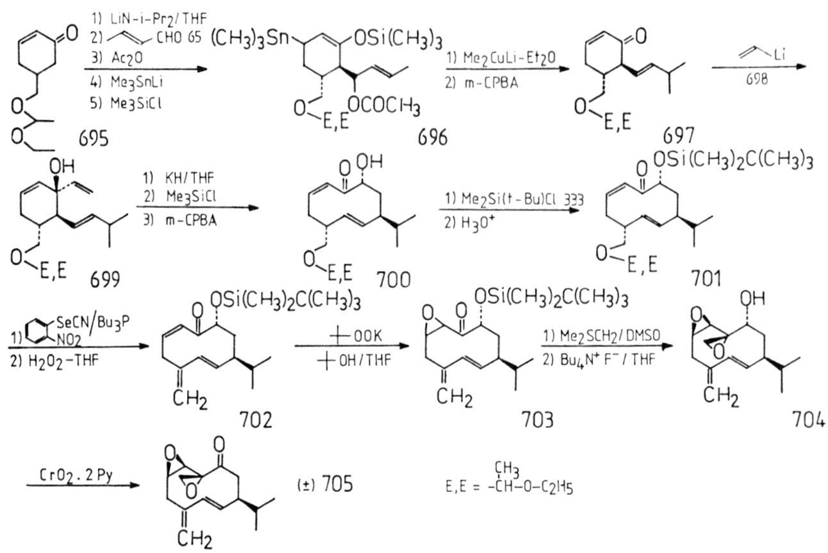

Scheme 92

6. Acknowledgement

The authors thank Drs. T. Clark (Erlangen), W. Kaska and P. Stein (USA) for assistance in the translation of the manuscript.

7. References

1. Wolff, J.: in: Government Regulations, Pheromone Analysis, Additional Pesticides (G. Zweig and J. Sherma, eds.), p. 21, Academic Press, New York 1976
2. Control of Insect Behavior by Natural Products (D. L. Wood, R. M. Silverstein, M. Nakajima, eds.), Academic Press, New York 1970
3. Chemicals Controlling Insect Behavior (M. Beroza, ed.), Academic Press, New York 1970
4. Chemical Control of Insect Behavior (H.H. Shorey, J.J. McKelvey Jr., eds.), Wiley-Interscience, New York 1977
5. Beroza, M.: in: Chemicals Controlling Insect Behavior, [3], p. 145
6. Steiner, L.F. et al.: J. Econ. Entomol. *54*, 30 (1961)
7. Law, J.H., Regnier, F.E.: Ann. Rev. Biochem. *40*, 533 (1971)
8. Hendry, L.B. et al.: Rec. Adv. Phytochem. *10*, 351 (1976)
9. Von Endt, D.W.: Diss. Abstr. Int. B *37* (9), 4474 (1977)
10. Animal Communication by Pheromones (H.H. Shorey, ed.), Academic Press, New York 1976
11. Schneider, D.: in: Sensory Physiology and Behavior (R. Galun, P. Hillman, I. Parnas, R. Werman, eds.), p. 173, Plenum Publishing Corp., New York 1975
12. Röller, H. et al.: Angew. Chem. *79*, 190 (1967); Angew. Chem. Intern. Ed. Engl. *6*, 179 (1967)
13. Whittaker, R.H., Feeny, P.P.: Science *171*, 757 (1971)
14. Brown, W.L., Eisner, T., Whittaker, R.H.: Bio Science *20*, 21 (1970)
15. Karlson, P., Lüscher, M.: Naturwissenschaften *46*, 63 (1959)
16. Wilson, E.O., Bossert, W.H.: Rec. Progr. Hormone Res. *19*, 673 (1963)
17. Beroza, M. et al.: Toxic. Appl. Pharmacol. *31*, 421 (1975)
18. Insect Sex Pheromones (M. Jacobson, ed.), Academic Press, New York 1972

19. Insecticides of the Future (M. Jacobson, ed.), M. Dekker, New York 1975
20. Roelofs, W.L.: Envir. Lett. *8*, 41 (1975)
21. Cardé, R.T.: Envir. Health Perspect. *14*, 133 (1976)
22. Roelofs, W.L., Cardé, R.T.: Ann. Rev. Entomol. *22*, 277 (1977)
23. Tumlinson, J.H., Mitchell, E.R., Chambers, D.L.: in: Pest Management with Insect Sex Attractants (M. Beroza, ed.), ACS Symp. Ser. Nr. *23*, p. 53, Amer. Chem. Soc., Washington 1976
24. Shorey, H.H., Gaston, L.K., Kaae, R.S.: ibid. p. 67 (1976)
25. Beroza, M.: ibid., p. 99 (1976)
26. Pheromones (M.C. Birch, ed.), North Holland-American Elsevier, 1974
27. Slater, G.A.: Progr. Farmer, Jan. *1975*, 8a (1975)
28. Clements, A.N.: Sittingborne Res. Centre (1975)
29. Vité, J.P.: Z. angew. Entomol. *77*, 325 (1975)
30. Mitchell, E.R.: Bio Science *25*, 493 (1975)
31. Bestmann, H.J. et al.: Angew. Chem. *90*, 815 (1978); Angew. Chem. Intern. Ed. Engl. *17*, 768 (1978)
32. Karlson, P., Schneider, D.: Naturwissenschaften *60*, 113 (1973)
33. Priesner, E.: Fortschr. Zool. *22*, 49 (1973)
34. Inscoe, M.N., Beroza, M.: in: Pest Management with Insect Sex Attractants, [23], p. 145 (1976)
35. Beroza, M.: Toxicol. Envir. Chem. Rev. *1*, 109 (1972)
36. McConnell, J.G., Silverstein, R.M.: Angew. Chem. *85*, 647 (1973); Angew. Chem. Intern. Ed. Engl. *12*, 644 (1973)
37. Evans, D.A., Green, C.L.: Chem. Soc. Rev. *2*, 75 (1973)
38. Mayer, M.S., McLaughlin, J.R.: An Annotated Compendium of Insect Sex Pheromones, Florida Agric. Exp. Station Monograph Series Nr. 6 (1975)
39. Weatherston, J., Percy, J.: Endevour *1*, 83 (1977)
40. Eiter, K.: in: Chemie der Pflanzenschutz- und Schädlingsbekämpfungsmittel (R. Wegler, ed.), Vol. *1*, p. 497, Springer-Verlag, Berlin-Heidelberg-New York 1970
41. Schneider, D.: Z. vergl. Physiol. *40*, 8 (1957)
42. Vostrowsky, O., Bestmann, H.J., Priesner, E.: Nachr. Chem. Techn. *21*, 501 (1973)
43. Priesner, E., Jacobson, M., Bestmann, H.J.: Z. Naturforsch. *30c*, 283 (1975)
44. Priesner, E. et al.: Z. Naturforsch. *32c*, 979 (1977)
45. Bestmann, H.J.: Z. angew. Entomol. *82*, 110 (1976)
46. Meinwald, J., Meinwald, Y.C.: J. Amer. Chem. Soc. *88*, 1305 (1966)
47. Meinwald, J. et al.: Science *151*, 583 (1966)
48. Meinwald, J., Meinwald, Y.C., Mazzochi, P.H.: Science *164*, 1174 (1969)
49. Schneider, D. et al.: J. comp. Physiol. *97*, 245 (1975)
50. Edgar, J.A., Culvenor, C.C.J.: Experientia *31*, 393 (1975)
51. Pliske, T., Eisner, T.: Science *164*, 1170 (1969)
52. Brower, L.P., Brower, J.V.Z., Cranston, F.P.: Zoologica (N.Y.) *50*, 1 (1965)
53. Schneider, D., Seibt, U.: Science *164*, 1173 (1969)
54. Seibt, U., Schneider, D., Eisner, T.: Z. Tierpsychol. *31*, 513 (1973)
55. McCluskey, R., Wright, C.G., Yamamoto, R.T.: J. Econ. Entomol. *62*, 1465 (1969)
56. Barth, R.H.: Z. Tierpsychol. *27*, 722 (1970)
57. Roth, L.M.: Ann. Entomol. Soc. Amer. *61*, 176 (1968)
58. Roth, L.M., Barth, R.: J. Insect Physiol. *10*, 965 (1964)
59. Roth, L.M., Dateo, G.P.: J. Insect Physiol. *12*, 255 (1966)
60. Persoons, C.J., Ritter, F.J., Lichtendon, W.J.: Proc. Kon. Ned. Akad. Wetensch. Amsterdam, C *77*, 201 (1974)
61. Persoons, C.J. et al.: Tetrahedron Lett. *1976*, 2055
62. Persoons, C.J.: Thesis University Wageningen (Netherlands), 1977
63. Ritter, F.J. et al.: in: Crop Protection Agents (Mc Farlane, ed.), p. 201, Academic Press, London
64. Bergström, G., Löfquist, J.: J. Insect Physiol. *14*, 995 (1968); *16*, 2353 (1970); and in: Pesticide Chemistry, Vol. 3: Chemical Releasers in Insects, (A.S. Tahori, ed.), p. 195, Gordon and Breach, London 1971
65. Tumlinson, J.H. et al.: Nature *234*, 348 (1971)
66. Tumlinson, J.H. et al.: J. Insect Physiol. *18*, 809 (1972)
67. Bestmann, H.J.: Conf. Dtsch. Ges. Fettwiss. Würzburg (Germany), 20.–23. 9. 1977

68. Anonymous: Chem. Eng. News. March 20., p. 23 (1978)
69. Silverstein, R.M., Young, J.C.: in: Pest Management with Insect Sex Attractants, [23], p. 1 (1976)
70. Bestmann, H.J.: Mitt. Dtsch. Ges. Allgem. Angew. Entomol. *1*, 147 (1978)
71. Tamaki, Y. et al.: Appl. Entomol. Zool. *6*, 139 (1971)
72. Tamaki, Y. et al.: Kontuy *39*, 338 (1971)
73. Ritter, F.J.: Medd. Fak. Landbouwwet. Gent *36*, 874 (1971)
74. Meijer, G.M. et al.: Science *175*, 1469 (1972)
75. Jacobson, M. et al.: Science *170*, 542 (1970)
76. Redfern, R.E. et al.: J. Econ. Entomol. *64*, 1570 (1971)
77. Wakamura, S.: Appl. Entomol. Zool. *13*, 290 (1978)
78. Thomas, W. et al.: Entomol. Exp. Appl., in press (1980)
79. Roelofs, W.L. et al.: Envir. Entomol. *4*, 580 (1975)
80. Roelofs, W.L., Comeau, A.: in: Chemical Releasers in Insects (A.S. Tahori, ed.), Proc. IUPAC 2nd Intern. Congr. Pest. Chem., Tel Aviv, Israel, Vol. 3, p. 91, Gordon and Breach, New York and London 1971
81. Ando, T. et al.: Agr. Biol. Chem. *39*, 1163 (1975)
82. Ando, T. et al.: ibid. *41*, 1485 (1977)
83. Kaae, R.S. et al.: Ann. Entomol. Soc. Amer. *66*, 444 (1973)
84. Struble, D.L., Swailes, G.E.: Envir. Entomol. 7, 384 (1978)
85. Steck, W. et al.: Can. Entomol. *110*, 775 (1978)
86. Roelofs, W.L., Comeau, A.: J. Econ. Entomol. *63*, 969 (1970)
87. McLaughlin, J.R., Shorey, H.H.: in press, cited in [38]
88. Kaae, R.S. et al.: Envir. Entomol. *1*, 651 (1972)
89. Kaae, R.S., Shorey, H.H., Gaston, L.K.: Science *179*, 487 (1973)
90. Steck, W. et al.: Envir. Entomol. *6*, 270 (1977)
91. Tumlinson, J.H. et al.: Envir. Entomol. *1*, 466 (1972)
92. Mitchell, E.R., Jacobson, M., Baumhover, A.H.: Envir. Entomol. *4*, 577 (1975)
93. Daterman, G.E., Daves jr., G.D., Jacobson, M.: Envir. Entomol. *1*, 382 (1972)
94. Jacobson, M., Jennings, D.T.: J. Envir. Sci. Health, Part *A*, *13*, 429 (1978)
95. Mitchell, E.R., Tumlinson, J.H.: Ann. Entomol. Soc. Amer. *66*, 917 (1973)
96. Mitchell, E.R. et al.: Envir. Entomol. *3*, 778 (1974)
97. Berger, R.S.: Ann. Entomol. Soc. Amer. *59*, 767 (1966)
98. Sower, L.L., Coffelt, J.A., Vick, K.W.: J. Econ. Entomol. *66*, 1220 (1973)
99. Mitchell, E.R. et al.: Envir. Entomol. *1*, 365 (1972)
100. Read, J.S., Warren, F.L., Hewitt, P.H.: Chem. Commun. *1968*, 792
101. Read, J.S. et al.: J. Insect Physiol. *20*, 441 (1974)
102. Chambon, H.P., d'Anguilar, J.: Ann. Zool.-Ecol. Animal *6*, 423 (1974)
103. Persoons, C.J., Ritter, F.J., Nooijen, W.J.: J. Chem. Ecol. in press
104. Granges, J., Baggiolini, M.: Rev. Suisse Vitic. Arboric. *3*, 93 (1971)
105. Biwer, G.: Recherche *9*, 799 (1978)
106. Roelofs, W.L., Comeau, A., Selle, R.: Nature *224*, 723 (1969)
107. Beroza, M., Muschik, G.M., Gentry, C.R.: Nature (New Biol.) *244*, 149 (1973)
108. Gentry, C.R. et al.: J. Econ. Entomol. *67*, 607 (1974)
109. Gentry, C.R., Beroza, M., Blythe, J.L.: Envir. Entomol. *4*, 227 (1975)
110. Arn, H. et al.: Experientia *30*, 1142 (1974)
111. Persoons, C.J. et al.: Med. Fac. Landbouwwetensch. Reijksuniv. Gent *41*, 937 (1976)
112. Roelofs, W.L., Cardé, R.: Envir. Entomol. *3*, 586 (1974)
113. Biwer, G. et al.: Ann. Zool. Ecol. Anim. *10*, 129 (1977)
114. Arn, H., Rauscher, S.: 1st OILB-Meeting, Rothamsted (England), October 1977
115. Arn, H. et al.: Z. Naturforsch. *31c*, 499 (1976)
116. Roelofs, W.L. et al.: J. Insect. Physiol. *17*, 2235 (1971)
117. Taschenberg, E.F., Cardé, R.T., Roelofs, W.L.: Envir. Entomol. *3*, 239 (1974)
118. Smith, R.G. et al.: J. Insect Physiol. *20*, 661 (1974)
119. Lange, R., Hoffmann, D.: Naturwissenschaften *59*, 217 (1972)
120. Sekul, A.A., Sparks, A.N.: USDA Techn. Bull., in press, cited in [34]
121. Minks, A.K. et al.: Entomol. Exp. Appl. *19*, 301 (1976)
122. Nesbitt, B.F. et al.: Nature (New Biol.) *244*, 208 (1973)

123. Preiss, R.: Z. angew. Entomol. *83*, 311 (1977)
124. Saglio, P. et al.: C. R. Hebd. Seances Acad. Sci., Ser. D *284*, 2007 (1977)
125. Baltensweiler, W. et al.: Mitt. Schweiz. Entomol. Ges. *51*, 133 (1978)
126. Kochansky, J.P., Roelofs, W.L., Sivapalan, P.: J. Chem. Ecol. *4*, 623 (1978)
127. Comeau, A., Roelofs, W.L.: Entomol. Exp. Appl. *16*, 191 (1973)
128. Roelofs, W.L., Reissig, W.H., Weires, R.W.: Envir. Entomol. *6*, 373 (1977)
129. Bestmann, H.J. et al.: Tetrahedron Lett. *1980*, 747
130. Struble, D.L., Swailes, G.E.: Envir. Entomol. *6*, 719 (1977)
131. Subchev, M.A., Kuznetzov, N.V.: Dokl. Bulg. Akad. Nauk *27*, 993 (1974)
132. Subchev, M., Myrsina, R.A.: Nauk Pr.-Ukr. Sil's'kogospod. Akad. *161*, 69 (1976)
133. Voerman, S., Minks, A.K.: Envir. Entomol. *2*, 751 (1973)
134. Bestmann, H.J. et al.: Tetrahedron Lett. *1979*, 497
135. Roelofs, W.L. et al.: Experientia *27*, 1438 (1971)
136. Milaire, H.G.: Pom. Franc. *17*, 13 (1975)
137. Brady, U.E.: J. Ga. Entomol. Soc. *4*, 41 (1969)
138. Roelofs, W.L., Comeau, A.: Science *165*, 398 (1969)
139. Brady, U.E.: Life Sci. *13*, 227 (1973)
140. Minks, A.K. et al.: Science *180*, 1073 (1973)
141. Descoins, C. et al.: C. R. Acad. Sci. Ser. D *286*, 77 (1978)
142. Steck, W.F. et al.: Envir. Entomol. *5*, 523 (1976)
143. Steck, W.F., Underhill, E.W., Chisholm, M.D.: J. Chem. Ecol. *3*, 603 (1977)
144. Struble, D.L. et al.: Canad. Entomol. *109*, 1393 (1977)
145. Roelofs, W.L. et al.: Envir. Entomol. *5*, 649 (1976)
146. Klun, J.A., Junk, G.A.: J. Chem. Ecol. *3*, 447 (1977)
147. Roelofs, W.L. et al.: in: Pest Management with Insect Sex Attractants, (M. Beroza, ed.), p. 77, [23] (1976)
148. Roelofs, W.L., Lagier, R.F., Hoyt, S.C.: Envir. Entomol. *6*, 353 (1977)
149. Priesner, E. et al.: Z. Naturforsch. *33c*, 1000 (1978)
150. Knauf, W. et al.: Z. angew. Entomol. *88*, 307 (1979)
151. Struble, D.L. et al.: Envir. Entomol. *5*, 988 (1976)
152. Arsura, E. et al.: Experientia *33*, 1423 (1977)
153. Campion, D.G.: Misc. Rept. No. *4*, Centre Overseas Res., London, 1972
154. Beevor, P.S. et al.: Experientia *31*, 22 (1975)
155. Tamaki, Y. et al.: Jap. J. Appl. Entomol. Zool. *20*, 81 (1976)
156. Balir, B.W., Tannock, J.: Rhod. J. Agr. Res. *16*, 221 (1978)
157. Minks, A.K., Voerman, S.: Entomol. Exp. Appl. *16*, 541 (1973)
158. Roelofs, W.L., Comeau, A.: J. Insect Physiol. *17*, 435 (1971)
159. Hendry, L.B. et al.: Science *187*, 355 (1975)
160. Hindenlang, M., Wichmann, J.K.: Science *195*, 86 (1977)
161. Sugie, H., Yaginuma, K., Tamaki, Y.: Appl. Entomol. Zool. *12*, 69 (1977)
162. Roelofs, W.L. et al.: Envir. Entomol. *3*, 747 (1974)
163. Persoons, C.J. et al.: J. Insect. Physiol. *20*, 1181 (1974)
164. Roelofs, W.L. et al.: Envir. Entomol. *5*, 362 (1976)
165. Hill, A.S. et al.: J. Chem. Ecol. *1*, 215 (1975)
166. Maini, S.: Inform. Fitopatol. *23* (9), 11 (1973)
167. Roelofs, W.L., Arn, H.: Nature *219*, 513 (1968)
168. Roelofs, W.L., Hill, A., Cardé, R.: J. Chem. Ecol. *1*, 83 (1975)
169. Aglade, P.: Rev. Zool. Agr. Pathol. Veg. *73*, 37 (1974)
170. Roelofs, W.L., Tette, J.P.: Nature *226*, 1172 (1970)
171. Klun, J.A., Brindley, T.A.: J. Econ. Entomol. *63*, 779 (1970)
172. Klun, J.A. et al.: Science *181*, 661 (1973)
173. Kochansky, J. et al.: J. Chem. Ecol. *1*, 225 (1975)
174. Klun, J.A., Robinson, J.F.: Ann. Entomol. Soc. Amer. *65*, 1337 (1972)
175. Rossi, R. et al.: 3rd OILB-Meeting; Wädenswil (Switzerland), Oct. 1979
176. Sanders, C.J., Bartell, R.J., Roelofs, W.L.: Can. Forestry Serv. Bi-monthly Res. Notes *28*, 9 (1972)
177. Hill, A.S., Roelofs, W.L.: J. Chem. Ecol. *1*, 91 (1975)
178. Benz, G., v. Salis, G.: Experientia *29*, 729 (1973)

179. Klun, J.A. et al.: Envir. Entomol. *4*, 871 (1975)
180. Sanders, C.J. et al.: Can. Entomol. *106*, 157 (1974)
181. Struble, D.L., Lilly, E.C.: Can. Entomol. *109*, 261 (1977)
182. Bode, W.M., Asquith, D., Tette, J.P.: J. Econ. Entomol. *66*, 1129 (1973)
183. Hill, A.S. et al.: Envir. Entomol. *3*, 249 (1974)
184. Green, N., Jacobson, M., Keller, J.C.: Experientia *25*, 682 (1969)
185. McLaughlin, J.R. et al.: Envir. Entomol. *1*, 645 (1972)
186. Sower, L.L., Vick, K.W., Long, J.S.: Ann. Entomol. Soc. Amer. *66*, 184 (1973)
187. Struble, D.L. et al.: Can. Entomol. *107*, 355 (1975)
188. Bestmann, H.J. et al.: Tetrahedron Lett. *1978*, 605
189. Chisholm, M.D. et al.: Can. Entomol. *107*, 361 (1975)
190. Struble, D.L., Swailes, G.E.: Envir. Entomol. *4*, 632 (1975)
191. Underhill, E.W., Steck, W.F., Chisholm, M.D.: Envir. Entomol. *5*, 307 (1976)
192. Yen-Hsin Chow, Chin Ling Hsu, Yu Mei Lin, K'o Hsueh Fa Chan Yuch K'an *6*, 651 (1978)
193. Neumark, S., Jacobson, M., Teich, I.: Envir. Lett. *7*, 21 (1974)
194. Roelofs, W.L. et al.: Mitt. Schweiz. Entomol. Ges. *46*, 71 (1973)
195. Buser, H.R., Arn, H.: J. Chromatogr. *106*, 83 (1975)
196. Biwer, G. et al.: C. R. Hebd. Seances Acad. Sci. Ser. D *280*, 1469 (1975)
197. Wall, C., Greenway, A.R., Burt, P.E.: Physiol. Entomol. *1*, 151 (1976)
198. Hathaway, D.O. et al.: Envir. Entomol. *3*, 522 (1974)
199. Moffitt, H.R., unpublished results, cited in: Pest Management with Insect Sex Attractants, [23], p. 155 (1976)
200. Berisford, C.W.: Envir. Entomol. *6*, 245 (1977)
201. Hill, A.S. et al.: Envir. Entomol. *5*, 959 (1976)
202. Nesbitt, B.F. et al.: J. Insect Physiol. *21*, 1091 (1975)
203. Roelofs, W.L. et al.: Life Sci. *17*, 699 (1975)
204. Persoons, C.J. et al.: Med. Fak. Landbouww. Reijksuniv. Gent *41*, 945 (1976)
205. Voerman, S., Minks, A.K., Persoons, C.J.: Potato Res. *20*, 123 (1977)
206. Fouda, H.G., Seiber, J.N., Bacon, O.G.: J. Econ. Entomol. *68*, 423 (1975)
207. Doolittle, R.E., unpublished, cited in: Pest Management with Insect Sex Attractants, [23], p. 160 (1976)
208. Doolittle, R.E. et al.: Symp. Insect Chem. ACS 25th Southeastern Reg. Meet. Charleston, S.C., Nov. 7.–9., (1973)
209. Doolittle, R.E. et al.: J. Chem. Ecol. *2*, 399 (1976)
210. Knauf, W., Bestmann, H.J., Koschatzky, K.H.: unpublished
211. Tamaki, Y., Yushima, T.: J. Insect Physiol. *20*, 1005 (1974)
212. Tamaki, Y., Noguchi, H., Yushima, T.: Appl. Entomol. Zool. *8*, 200 (1973)
213. Vick, K.W., Sower, L.L.: J. Econ. Entomol. *66*, 1258 (1973)
214. Negishi, T., Ishiwatari, T., Asano, S.: Appl. Ent. Zool. *12*, 178 (1977)
215. Kuwahara, Y. et al.: Agr. Biol. Chem. *35*, 447 (1971)
216. Dahm, K.H. et al.: Life Sci. *10* (II), 531 (1971)
217. Brady, U.E. et al.: Science *171*, 802 (1971)
218. Kuwahara, Y. et al.: Science *171*, 801 (1971)
219. Takahashi, S. et al.: Botuy-Kagaku *37*, 56 (1972)
220. Adler, V.E. et al.: J. Econ. Entomol. *69*, 706 (1976)
221. Sower, L.L., Turner, W.K., Fish, J.C.: J. Chem. Ecol. *1*, 335 (1975)
222. Brady, U.E., Daley, R.C.: Ann. Entomol. Soc. Amer. *65*, 1356 (1972)
223. Brady, U.E., Nordlund, D.A.: Life Sci. *10* (II), 797 (1971)
224. Mitchell, E.R., unpublished, cited in [23], p. 161 (1976)
225. Bestmann, H.J., Vostrowsky, O., Brosche, T., Platz, H., Kasang, G., Kaißling, K.E.: in preparation
226. Kochansky, J. et al.: J. Insect Physiol. *21*, 1977 (1975)
227. Hummel, H.E. et al.: Science *181*, 873 (1973)
228. Bierl, B.A. et al.: J. Econ. Entomol. *67*, 211 (1974)
229. Vick, K.W. et al.: Experientia *30*, 17 (1974)
230. Nielsen, D.G., Purrington, F.F.: Envir. Entomol. *7*, 708 (1978)
231. Tumlinson, J.H. et al.: Science *185*, 614 (1974)
232. McLaughlin, J.R. et al.: J. Chem. Ecol. *2*, 73 (1976)

233. Underhill, E.W. et al.: Can. Entomol. *110*, 495 (1978)
234. Yagimuna, K. et al.: Appl. Entomol. Zool. *11*, 266 (1976)
235. Nielsen, D.G. et al.: Envir. Entomol. 7, 544 (1978)
236. Yamaoka, R., Fukami, H., Ishii, S.: Agr. Biol. Chem. *40*, 1971 (1976)
237. S'bchev, M.A., Myrsina, R.A.: Nauk Pr.-Ukr. Sil's'kogospod. Akad. *200*, 52 (1977)
238. Myrsina, R.A. et al.: Fiziol. Akt. Veshchestva *9*, 68 (1977)
239. Weatherston, J., Davidson, L.M., Simonini, D.: Can. Entomol. *106*, 781 (1974)
240. McLaughlin, J.R. et al.: Envir. Entomol. *3*, 677 (1974)
241. Tumlinson, J.H. et al.: Envir. Entomol. *1*, 354 (1972)
242. Weatherston, J. et al.: Experientia *33*, 723 (1977)
243. Beroza, M. et al.: J. Econ. Entomol. *66*, 1307 (1973)
244. Weatherston, J. et al.: J. Chem. Ecol. *4*, 543 (1978)
245. Baker, J.L. et al.: Envir. Entomol. *4*, 90 (1975)
246. Weatherston, J., Maclean, W.: Can. Entomol. *106*, 281 (1974)
247. Neumark, S., Teich, J., Green, N.: Envir. Lett. *5*, 1 (1973)
248. Roelofs, W.L. et al.: Science *174*, 297 (1971)
249. Beroza, M., Bierl, B.A., Moffitt, H.R.: Science *183*, 89 (1974)
250. Sower, L.L., Vick, K.W., Tumlinson, J.H.: Envir. Entomol. *3*, 120 (1974)
251. Descoins, C., Samain, D.: personal communication, cited in G. Decodts, G. Dressaire, Y. Langlois, Synthesis *1979*, 510
252. Moffit, H.R.: unpublished, cited in [23], p. 149 (1976)
253. Kuwahara, Y., Casida, J.E.: Agr. Biol. Chem. *37*, 681 (1973)
254. Kasang, G., Schneider, D., Schäfer, W.: Naturwissenschaften *65*, 337 (1978)
255. Butenandt, A. et al.: Z. Naturforsch. *14b*, 283 (1959)
256. Kasang, G. et al.: Angew. Chem. *90*, 74 (1978); Angew. Chem. Intern. Ed. Engl. *17*, 60 (1978)
257. Roelofs, W.L., Cardé, R.T., Tette, J.P.: Envir. Entomol. *2*, 252 (1973)
258. Cardé, R.T., Baker, T.C., Roelofs, W.L.: Nature *253*, 348 (1975)
259. Staten, R.T. et al.: J. Econ. Entomol. *66*, 1263 (1973)
260. Voerman, S., Minks, A.K., Houx, N.W.H.: Envir. Entomol. *3*, 701 (1974)
261. Beroza, M., Staten, R.T., Bierl, B.A.: J. Econ. Entomol. *64*, 580 (1971)
262. Klun, J.A., Robinson, J.F.: Ann. Entomol. Soc. Amer. *64*, 1083 (1971)
263. George, D.A. et al.: Envir. Entomol. *4*, 606 (1975)
264. Henderson, H.E. et al.: J. Insect Physiol. *19*, 1257 (1973)
265. Nesbitt, B.F. et al.: Insect Biochem. 7, 355 (1977)
266. Roelofs, W.L. et al.: Life Sci. *14*, 1555 (1974)
267. Tumlinson, J.H. et al.: J. Chem. Ecol. *1*, 203 (1975)
268. Weatherston, J., Percy, J.E., MacDonald, L.M.: Experientia *32*, 178 (1976)
269. Weatherston, J. et al.: Can. Entomol. *103*, 1741 (1971)
270. Weatherston, J., Descoins, C., Grant, G.G.: C. R. Hebd. Seances Acad. Sci. Ser. D *281*, 1111 (1975)
271. Nesbitt, B.F. et al.: J. Insect Physiol. *21*, 1883 (1975)
272. Ohta, K. et al.: Agr. Biol. Chem. *40*, 1897 (1976)
273. Piccardi, P. et al.: J. Insect Physiol. *23*, 1443 (1977)
274. Sekul, A.A. et al.: J. Econ. Entomol. *68*, 603 (1975)
275. Tatsuki, S. et al.: Botuy-Kagaku *42*, 1 (1977)
276. Starratt, A.N. et al.: Z. Naturforsch. *34c*, 9 (1979)
277. Bierl, B.A., Beroza, M., Collier, C.W.: Science *170*, 87 (1970)
278. Beroza, M. et al.: J. Econ. Entomol. *67*, 659 (1974)
279. Cameron, E.A. et al.: Science *183*, 972 (1974)
280. Beroza, M. et al.: Envir. Entomol. *2*, 966 (1973)
281. Schönherr, J.: Z. angew. Entomol. *71*, 260 (1972)
282. Bierl, B.A. et al.: Z. Naturforsch. *30c*, 672 (1975)
283. Beroza, M., Punjabi, A.A., Bierl, B.A.: J. Econ. Entomol. *66*, 1215 (1973)
284. Roelofs, W.L., Cardé, R.: Science *171*, 648 (1971)
285. Cardé, R.T. et al.: Nature *241*, 474 (1973)
286. Kasang, G., Schneider, D., Beroza, M.: Naturwissenschaften *61*, 130 (1974)
287. Bierl, B.A., Beroza, M., Collier, W.A.: J. Econ. Entomol. *65*, 659 (1972)
288. Niccoli, A., Conti, C., Rossi, R.: Chim. Ind. *58*, 12. Dez. (1976)

289. Smith, R.G., Daterman, G.E., Daves jr., G.D.: Science *188*, 63 (1975)
290. Daterman, G.E. et al.: Envir. Entomol. *5*, 1187 (1976)
291. Smith, L.M. et al.: J. Org. Chem. *43*, 2361 (1978)
292. Tamaki, Y., Honma, K., Kawasaki, K.: Appl. Entomol. Zool. *12*, 60 (1977)
293. Petty, R.L. et al.: Experientia *33*, 1324 (1977)
294. Leyrer, R.L., Monroe, R.E.: J. Insect Physiol. *19*, 2267 (1973)
295. Finn, W.E., Payne, T.L.: Southwest Entomol. *2*, 62 (1977)
296. Dahm, K.H. et al.: Naturwissenschaften *58*, 265 (1971)
297. Röller, H. et al.: Acta Entomol. Bohemoslov. *65*, 208 (1968)
298. Bestmann, H.J., Vostrowsky, O., Platz, H.: Experientia *33*, 874 (1977)
299. Aplin, R.T., Birch, M.C.: Experientia *26*, 1193 (1970)
300. Jacobson, M. et al.: Experientia *32*, 964 (1976)
301. Birch, M.C., Grant, G.G., Brady, U.E.: Ann. Entomol. Soc. Amer. *69*, 491 (1976)
302. Clearwater, J.W.: J. Insect Physiol. *18*, 781 (1972)
303. Meinwald, J. et al.: Experientia *30*, 721 (1974)
304. Edgar, J.A., Culvenor, C.C.J., Smith, L.W.: Experientia *27*, 761 (1971)
305. Meinwald, J. et al.: Tetrahedron Lett. *1971*, 3485
306. Edgar, J.A., Culvenor, C.C.J.: Nature *248*, 614 (1974)
307. Edgar, J.A., Culvenor, C.C.J., Robinson, G.S.: J. Austr. Entomol. Soc. *12*, 144 (1973)
308. Culvenor, C.C.J., Edgar, J.A.: Experientia *28*, 627 (1973)
309. Meinwald, J. et al.: Tetrahedron Lett. *1968*, 4893
310. Meinwald, J. et al.: Chem. Commun. *1969*, 86
311. Bergström, G., Lundgren, L.: Zoon Suppl. *1*, 67 (1973)
312. Vité, J.P., Bakke, A., Renwick, J.A.A.: Can. Entomol. *104*, 1967 (1972)
313. Young, J.C., Silverstein, R.M., Birch, M.C.: J. Insect Physiol. *19*, 2273 (1973)
314. Silverstein, R.M. et al.: Tetrahedron Lett. *1966*, 1929
315. Silverstein, R.M., Rodin, R.M., Wood, D.L.: Science *154*, 509 (1966)
316. Vité, J.P., Renwick, J.A.A.: J. Insect Physiol. *17*, 1699 (1971)
317. Vité, J.P., Francke, W.: Naturwissenschaften *63*, 550 (1976)
318. Birch, M.C., Wood, D.L.: J. Chem. Ecol. *1*, 101 (1975)
319. Harring, C.M., Vité, J.P., Hughes, R.P.: Naturwissenschaften *62*, 488 (1975)
320. Wood, D.L. et al.: Nature *215*, 206 (1967)
321. Renwick, J.A.A., Vité, J.P.: J. Insect Physiol. *18*, 1215 (1972)
322. Vité, J.P., Bakke, A., Hughes, P.R.: Naturwissenschaften *61*, 365 (1974)
323. Bakke, A.: Naturwissenschaften *63*, 92 (1976)
324. McNew, G.L.: Contrib. Boyce Thompson Instit. *24*, 251 (1970)
325. Vité, J.P., Renwick, J.A.A.: Contrib. Boyce Thompson Instit. *24*, 323 (1970)
326. Pitman, G.B. et al.: J. Insect Physiol. *15*, 363 (1969)
327. Pitman, G.B.: J. Econ. Entomol. *64*, 426 (1971)
328. Pitman, G.B., Vité, J.P.: Canad. Entomol. *101*, 143 (1969)
329. Renwick, J.A.A.: Contr. Boyce Thompson Instit. *23*, 355 (1967)
330. Kinzer, G.W. et al.: Nature *221*, 477 (1969)
331. Silverstein, R.M. et al.: Science *159*, 889 (1968)
332. Pitman, G.B., Vité, J.P.: J. Econ. Entomol. *64*, 402 (1971)
333. Vité, J.P., Pitman, G.B.: J. Econ. Entomol. *63*, 1132 (1970)
334. Kinzer, G.W. et al.: J. Econ. Entomol. 64, 970 (1971)
335. Pitman, G.B., Vité, J.P.: Ann. Entomol. Soc. Amer. *63*, 661 (1970)
336. Vité, J.P. et al.: Naturwissenschaften *59*, 469 (1972)
337. Williamson, D.L.: Ann. Entomol. Soc. Amer. *64*, 586 (1971)
338. Vité, J.P., Williamson, D.L.: J. Insect Physiol. *16*, 233 (1970)
339. Rudinsky, J.A. et al.: Envir. Entomol. *2*, 505 (1973)
340. Rudinsky, J.A.: Envir. Entomol. *2*, 579 (1973)
341. Rudinsky, J.A. et al.: Z. angew. Entomol. *75*, 254 (1974)
342. Kline, L.N. et al.: Can. Entomol. *106*, 485 (1974)
343. Pitman, G.B., Hedden, R.L., Gara, R.J.: Z. angew. Entomol. *78*, 203 (1975)
344. Pearce, G.T. et al.: J. Chem. Ecol. *1*, 115 (1975)
345. Blight, M.M. et al.: Experientia *33*, 845 (1977)

346. Byrne, K.J. et al.: J. Insect Physiol. *20*, 1895 (1974)
347. Renwick, J.A.A., Vité, J.P., Billings, R.F.: Naturwissenschaften *64*, 226 (1977)
348. Young, J.C. et al.: J. Insect Physiol. *19*, 1615 (1973)
349. Rudinsky, J.A., Kline, L.N., Diekman, J.D.: J. Econ. Entomol. *68*, 527 (1975)
350. Gore, W.E., Pearce, G.T., Silverstein, R.M.: J. Org. Chem. *40*, 1705 (1975)
351. Wright, R.H., Chapman, J.A., Dyer, E.D.A.: Can. For. Serv. Bi-Mon. Res. Notes *30*, 10 (1974)
352. Renwick, J.A.A., Vité, J.P.: Nature *224*, 1222 (1969)
353. Renwick, J.A.A., Vité, J.P.: Contr. Boyce Thompson Institute *24*, 283 (1970)
354. Knopf, J.E.A., Pitman, G.B.: J. Econ. Entomol. *65*, 723 (1972)
355. Libbey, L.M. et al.: J. Insect Physiol. *22*, 871 (1976)
356. Vité, J.P., Renwick, J.A.A.: Naturwissenschaften *58*, 418 (1971)
357. Rudinsky, J.A. et al.: Envir. Entomol. *3*, 90 (1974)
358. Libbey, L.M. et al.: J. Insect Physiol. *20*, 1667 (1974)
359. Hughes, P.R.: Thesis Univers. California, Davis (1973)
360. Francke, W., Heemann, V.: Z. angew. Entomol. *75*, 67 (1974)
361. Bedard, W.D. et al.: Science *164*, 1284 (1969)
362. Francke, W., Heemann, V., Heyns, K.: Z. Naturforsch. *29c*, 243 (1974)
363. Renwick, J.A.A., Vité, J.P., Pitman, G.B.: Naturwissenschaften *63*, 198 (1976)
364. McConnell, J.G. et al.: J. Chem. Ecol. *3*, 549 (1977)
365. Francke, W. et al.: Naturwissenschaften *64*, 590 (1977)
366. Tumlinson, J.H. et al.: Science *166*, 1010 (1969)
367. Tumlinson, J.H. et al.: J. Org. Chem. *36*, 2616 (1971)
368. Chang, V.C.S., Curtis, G.A.: Envir. Entomol. *1*, 476 (1972)
369. Horler, D.F.: J. Chem. Soc. (C), *1970*, 859
370. Halstead, D.G.H.: J. Stored Prod. Res. *9*, 109 (1973)
371. Levinson, H.Z., Levinson, A.R.: Naturwissenschaften *61*, 685 (1974)
372. Yarger, R.G., Silverstein, R.M., Burkholder, W.E.: J. Chem. Ecol. *1*, 323 (1975)
373. Rodin, J.O. et al.: Science *165*, 904 (1969)
374. Blight, M.M. et al.: Experientia *34*, 1119 (1978)
375. Barak, A.V., Burkholder, W.E.: Proc. North Centr. Branch Entomol. Soc. Amer. *28*, 175 (1973)
376. Silverstein, R.M. et al.: Science *157*, 85 (1967)
377. Anonymous: Chem. Eng. News *45*, 35 (1967)
378. Henzell, R.F., Lowe, M.D.: Science *168*, 1005 (1970)
379. McGovern, T.P. et al.: J. Econ. Entomol. *63*, 1727 (1970)
380. Tumlinson, J.H. et al.: Science *197*, 789 (1977)
381. Keville, R., Kannowski, P.B.: J. Insect Physiol. *21*, 81 (1975)
382. Tashiro, H. et al.: J. Econ. Entomol. *57*, 230 (1964)
383. Fukui, H. et al.: Tetrahedron Lett. *1974*, 3563
384. Barber, I.A. et al.: J. Econ. Entomol. *64*, 1041 (1971)
385. Maddison, P.A., Beroza, M., McGovern, T.P.: J. Econ. Entomol. *66*, 591 (1973)
386. McGovern, T.P. et al.: J. Econ. Entomol. *63*, 276 (1970)
387. McGovern, T.P. et al.: J. Econ. Entomol. *63*, 168 (1970)
388. Butler, L.I. et al.: Envir. Entomol. *4*, 229 (1975)
389. Silverstein, R.M., Young, J.C.: in Pest Management with Insect Sex Attractants, [23], p. 29, (1976)
390. Cross, H.J. et al.: J. Chem. Ecol. *3*, 115 (1977)
391. Higgs, M.D., Evans, D.A.: Experientia *34*, 46 (1978)
392. Kuwahara, Y. et al.: Tetrahedron *34*, 1769 (1978)
393. Sugawara, R., Kurihara, S., Muto, T.: J. Insect Physiol. *21*, 957 (1975)
394. Sugawara, R., Tominaga, Y., Suzuki, T.: Insect Biochem. *7*, 438 (1977)
395. Persoons, C.J., Ritter, F.J.: 1st OILB-Meet. Rothamsted (England), Oct. 1977
396. Talman, E. et al.: Israel J. Chem. *17*, 227 (1978)
397. Nishida, R. et al.: Agr. Biol. Chem. *40*, 1407 (1976)
398. Nishida, R.: J. Chem. Ecol. *2*, 449 (1976)
399. Nishida, R., Fukami, H., Ishii, S.: Experientia *30*, 978 (1974)
400. Nishida, R., Fukami, H., Ishii, S.: Appl. Entomol. *10*, 10 (1975)
401. Cavill, G.W.K., Williams, P.J.: J. Insect Physiol. *13*, 1097 (1967)

402. Casnati, G., Ricca, A., Pavan, M.: Chem. Ind. (Milano) *49*, 57 (1967)
403. Blum, M.S., Padovani, F., Amante, E.: Comp. Biochem. Physiol. *26*, 291 (1968)
404. McGurk, D.J. et al.: J. Insect Physiol. *12*, 1435 (1966)
405. Moser, J.C., Brownlee, R.C., Silverstein, R.M.: J. Insect Physiol. *14*, 529 (1968)
406. Crewe, R.M., Blum, M.S.: J. Insect Physiol. *18*, 31 (1972)
407. Blum, M.S., Brand, J.: Amer. Zool. *12*, 553 (1972)
408. Wheeler, J.W., Blum, M.S.: Science *182*, 501 (1973)
409. Fales, H.M. et al.: J. Insect Physiol. *18*, 1077 (1972)
410. Jacobson, M., Lilly, E.C., Harding, C.: Science *159*, 208 (1968)
411. Riley, R.G. et al.: J. Insect Physiol. *20*, 651 (1974)
412. Moore, B.P.: Nature *211*, 746 (1966)
413. Birch, A.J. et al.: J. Chem. Soc., Perkin Trans. *1*, 2653 (1972)
414. Matsumura, F., Jewitt, D.M., Coppel, H.C.: J. Econ. Entomol. *65*, 600 (1972)
415. Matsumura, F., Coppel, H.C., Tai, A.: Nature *219*, 963 (1968)
416. Tai, A., Matsumura, F., Coppel, H.C.: J. Org. Chem. *34*, 2180 (1969)
417. Ritter, F.J., Coenen-Saraber, C.M.A.: Entomol. Exp. Appl. *12*, 611 (1969)
418. Ritter, F.J., van Oosten, A.M., Persoons, C.J.: 7 Intern. Congr. Int. Union Study Soc. Insects (1973)
419. Hummel, H.: Ins. Sociaux *15*, 213 (1968)
420. Hummel, H., Karlson, P.: Hoppe-Seylers's Z. Physiol. Chemie *349*, 725 (1968)
421. Karlson, P., Lüscher, M., Hummel, H.: J. Insect Physiol. *14*, 1763 (1968)
422. Huwyler, S., Grob, K., Viskonti, M.: Helv. Chim. Acta *56*, 976 (1973)
423. Huwyler, S., Grob, K., Viskonti, M.: J. Insect Physiol. *21*, 299 (1975)
424. Verron, H., Barbier, M.: C.R. Acad. Sci. D *254*, 4089 (1962)
425. Sonnet, P.E., Moser, J.C.: Envir. Entomol. *2*, 851 (1973)
426. Ritter, F.J. et al.: Experientia *29*, 530 (1973)
427. Ritter, F.J. et al.: Tetrahedron Lett. *1977*, 2617
428. Carlson, D.A. et al.: Science *174*, 76 (1971)
429. Mansingh, A. et al.: Can. Entomol. *104*, 1963 (1972)
430. Uebel, E.C. et al.: J. Chem. Ecol. *1*, 195 (1975)
431. Sonnet, P.E., Uebel, E.C., Miller, R.W.: Envir. Entomol. *4*, 761 (1975)
432. Uebel, E.C. et al.: J. Chem. Ecol. *3*, 269 (1977)
433. Sonnet, P.E., Uebel, E.C., Miller, R.W.: J. Chem. Ecol. *3*, 251 (1977)
434. Ohinata, K. et al.: J. Envir. Sci. Health, part A, *12*, 67 (1977)
435. Jacobson, M. et al.: J. Med. Chem. *16*, 248 (1973)
436. Brieger, G., Butterworth, F.: Science *167*, 1262 (1967)
437. Carlson, D.A., Langley, P.A., Huyton, P.: Science *201*, 750 (1977)
438. Anonymous: Chem. Eng. News *54* (50), 5 (1976)
439. Roelofs, W.L. et al.: J. Chem. Ecol. *4*, 211 (1978)
440. Wientjens, W.H.J.M., Lakwijk, A.C., van der Masel, T.: Experientia *29*, 658 (1973)
441. Bowers, W.S. et al.: Science *196*, 680 (1977)
442. Levinson, H.Z. et al.: J. Insect Physiol. *20*, 1231 (1974)
443. Ubik, K. et al.: Naturwissenschaften *62*, 348 (1975)
444. Boch, R., Shearer, D.A., Stone, B.C.: Nature *195*, 1018 (1962)
445. Veith, H.J., Weiss, J., Koeniger, N.: Experientia *34*, 423 (1978)
446. Sannasi, A., Rajulu, G.S.: Life Sci. *10* (II), 195 (1971)
447. Shearer, D.A. et al.: J. Insect Physiol. *16*, 1437 (1970)
448. Gary, N.E.: Science *136*, 773 (1962)
449. Butler, C.G.: J. Entomol. *A46*, 1 (1971)
450. Pain, J.: Bee World *54*, 11 (1973)
451. Butler, C.G., Fairey, E.M.: J. Apic. Res. *3*, 65 (1964)
452. Tengö, J., Bergström, G.: J. Chem. Ecol. *1*, 253 (1975)
453. Davis, H.G. et al.: J. Med. Entomol. *4*, 275 (1967)
454. Ikan, R. et al.: J. Insect Physiol. *15*, 1709 (1969)
455. Ishay, J.: Insectes Soc. *20*, 243 (1973)
456. Ställberg-Stenhagen, S.: Acta Chem. Scand. *24*, 358 (1970)
457. Ahlquist, L., Ställberg-Stenhagen, S.: Acta Chem. Scand. *25*, 1685 (1971)

458. Stenhagen, E.: Ark. Kemi *28*, 453 (1968)
459. Jewett, D.M., Matsumura, F., Coppel, H.C.: Science *192*, 51 (1976)
460. Wheeler, J.W. et al.: Tetrahedron Lett. *1976*, 4029
461. Hefetz, A., Batra, S.W.T., Blum, M.S.: Experientia *35*, 319 (1979)
462. Vrkoc, J. et al.: Acta Entomol. Bohemoslov. *74*, 205 (1977)
463. Gueldner, R.C., Parrot, W.L.: Insect Biochem. *8*, 389 (1978)
464. Rossi, R. et al.: 3 OILB-Meet. Wädenswil (Switzerland), Oct. 1979; and R. Rossi, A. Carpita, G. Vita, Gazz. Chim. Ital. *108*, 709 (1978)
465. Francke, W., Hindorf, G., Reith, W.: Angew. Chem. *90*, 915 (1978)
466. Van der Meer, R.K. et al.: Envir. Entomol. *8*, 6 (1979)
467. Davis, H.G., Eves, J.D., McDonough, L.M.: Envir. Entomol. *8*, 147 (1979)
468. Clearwater, J.R.: Comp. Biochem. Physiol. *50 B*, 77 (1975)
469. Aplin, R.T., Birch, M.C.: Nature *217*, 1167 (1968)
470. Vostrowsky, O., Bestmann, H.J.: Mitt. Dtsch. Ges. Allgem. Angew. Entomol. *1*, 152 (1978)
471. Shorey, H.H., Gaston, L.K., Jefferson, R.N.: Adv. Pest Contr. Res. *8*, 57 (1968)
472. Beroza, M.: J. Chrom. Science *13*, 314 (1975)
473. Silverstein, R.M.: in Control of Insect Behaviour by Natural Products (D. L. Wood, R. M. Silverstein, H. Nakajima, eds.), p. 285, Academic Press, New York 1970
474. Bedoukian, P.Z.: Adv. Chemoreception *1*, 19 (1970)
475. Ställberg-Stenhagen, S., Stenhagen, E., Bergström, G.: Zoon Suppl. *1*, 77 (1973)
476. Kovats, E.Sz. (ed.): Column Chromatography; Fifth Intern. Symp. on Separation Methods, Lausanne 1969, Sauerländer AG, Aarau, Switzerland 1970
477. Ettre, L.S., McFadden, W.H. (eds.): Ancillary Techniques of Gas Chromatography, Wiley (Interscience), New York 1969
478. Budzikiewicz, H., Djerassi, C., Williams, D.H.: Mass Spectrometry of Organic Compounds, Holden-Day, San Francisko/Calif., 1967
479. Karger, B.L., Snyder, L.R., Horvath, C.: An Introduction to Separation Science, Wiley (Interscience), New York 1973
480. Kirchner, J.G.: Thin-layer Chromatography, Vol. XII, Technique of Organ. Chem., Wiley (Interscience), New York 1971
481. Kirkland, J.J. (ed.): Modern Practice of Liquid Chromatography, Wiley (Interscience), New York 1971
482. Sigga, S. (ed.): Instrumental Methods of Organic Functional Group Analysis, Wiley (Interscience), New York 1972
483. Silverstein, R.M., Bassler, G.C., Morrill, T.C.: Spectrometric Identification of Organic Compounds, 3 ed., Wiley, New York 1974
484. Stahl, E.: Dünnschicht-Chromatographie, 2 ed., Springer-Verlag, Berlin-Heidelberg-New York 1976
485. Stenhagen, E., Abrahamson, S., McLafferty, F.W. (eds.): Atlas of Mass Spectral Data, Wiley (Interscience), New York 1969
486. Stenhagen, E., Abrahamson, S., McLafferty, F.W. (eds.): Registry of Mass Spectral Data, Vol. 1–4, Wiley (Interscience), New York 1974
487. Bentley, K.W., Kirby, B.W. (eds.): Elucidation of Organic Structure by Physical and Chemical Methods, 2 ed., Vol. 1–3, Wiley (Interscience), New York 1972
488. Moulton, D.G., Turk, A., Johnston jr., J.W.: in Methods in Olfactory Research, Academic Press 1975
489. Flaschenträger, B., Sayed Amin, El, Jarczyk, H.J.: Mikrochim. Acta *1957* (3–4), 385
490. Knauf, W., Vostrowsky, O., Bestmann, H.J.: in preparation (1980)
491. Schneider, D.: Science *163*, 1031 (1969)
492. Payne, T.L., Shorey, H.H., Gaston, L.K.: J. Insect Physiol. *16*, 1043 (1970)
493. Bestmann, H.J., Rösel, P., Vostrowsky, O.: Liebigs Ann. Chem. *1979*, 1189
494. Barth, R.: Rio de Janeiro: Parque Nacional do Itatiaio Bol. 7, 1 (1960)
495. Tumlinson, J.H. et al.: J. Econ. Entomol. *61*, 470 (1968)
496. Matthes, D.: Zool. Anz. Suppl. *33*, 316 (1969)
497. Matthes, D.: Umschau *1970*, 112
498. Matthes, D.: Entomol. Blätter *66*, 33 (1970)
499. Bodenstein, W.G.: Ann. Entomol. Soc. Amer. *63*, 336 (1970)

500. Peacock, J.W. et al.: Envir. Entomol. *2*, 355 (1973)
501. Riddiford, L.M.: Science *158*, 139 (1967)
502. Wearing, C.H., Hutchins, R.F.N.: J. Insect Physiol. *19*, 1251 (1973)
503. Kawano, Y., Mitchell, W.C., Matsumoto, H.: J. Econ. Entomol. *61*, 986 (1968)
504. Steinbrecht, R.A.: Z. vergl. Physiol. *48*, 341 (1964)
505. Coffelt, J.A., Sower, L.L., Vick, K.W.: Envir. Entomol. *7*, 502 (1978)
506. Brady, U.E., Ganyard jr., M.C.: Ann. Entomol. Soc. Amer. *65*, 898 (1972)
507. Morgan, E.D., Wadhaus, L.J.: J. Insect Physiol. *18*, 1125 (1972)
508. Bergström, G. et al.: Ark. Kemi *28*, 453 (1968)
509. Andersson, C.O. et al.: Ark. Kemi *26*, 191 (1967)
510. Ställberg-Stenhagen, S., Stenhagen, E.: Sv. Kem. Tidskr. *80*, 178 (1968)
511. Bergström, G., Kullenberg, B., Ställberg-Stenhagen, S.: Chem. Scripta *4*, 174 (1973)
512. Yamamoto, R.: J. Econ. Entomol. *56*, 119 (1963)
513. Jones, P.A., Coppel, H.C., Casida, J.E.: J. Econ. Entomol. *58*, 465 (1965)
514. Jacobson, M. et al.: J. Med. Chem. *16*, 248 (1973)
515. Browne, L.E., Birch, M.C., Wood, D.L.: J. Insect Physiol. *20*, 183 (1974)
516. Byrne, K.J. et al.: J. Chem. Ecol. *1*, 1 (1975)
517. Bell, W.J., Burns, R.E., Barth, R.H.: Behav. Biol. *10*, 419 (1974)
518. Nordlund, D.A., Brady, U.E.: Envir. Entomol. *3*, 797 (1974)
519. Wharton, D.R.A. et al.: Science *137*, 1062 (1962)
520. Bergström, G.: Chem. Scripta *4*, 135 (1973)
521. Solomon, J.D., Doolittle, R.E., Beroza, M.: Ann. Entomol. Soc. Amer. *65*, 1058 (1972)
522. Klun, J.A.: J. Econ. Entomol. *61*, 484 (1968)
523. Beroza, M.: in Column Chromatography; 5 Intern. Symp. on Separation Methods, Lausanne 1969 (E. Sz. Kovats, ed.), Chimia, Suppl. p. 92, Säuerländer AG, Aarau, Switzerland (1970)
524. Bierl, B.A., Beroza, M., Ashton, W.T.: Mikrochim. Acta *1969*, 637
525. Tomida, I., Ishii, S.: Appl. Ent. Zool. *3*, 103 (1968)
526. Determann, H.: Gel Chromatography, Springer-Verlag, Berlin-Heidelberg-New York 1969
527. Tumlinson, J.H., Heath, R.R.: J. Chem. Ecol. *2*, 87 (1976)
528. Hirsch, J., Ahrens jr., E.H.: J. Biol. Chem. *233*, 311 (1958)
529. Engelhardt, H.: Hochdruck-Flüssigkeits-Chromatographie, 2ed., Springer-Verlag, Berlin-Heidelberg-New York 1977
530. Bestmann, H.J., Brosche, Th., Vostrowsky, O.: in preparation
531. Selim, S.: J. Chromatogr. *136*, 271 (1977)
532. Heath, R.R. et al.: J. Chrom. Science *15*, 10 (1977)
533. Heath, R.R. et al.: ibid. *13*, 380 (1975)
534. Houx, N.W.H., Voerman, S., Jongen, W.M.F.: J. Chromatogr. *96*, 25 (1975)
535. Schomburg, G., Zegarski, K.: J. Chromatogr. *114*, 174 (1975)
536. Zlatkis, A., Kaiser, R.E.: in HPTLC, High Performance Thin Layer Chromatography, J. Chromatography Library, Vol. *9*, Elsevier, Amsterdam, Netherlands 1977
537. Kaißling, K.E. et al.: Naturwissenschaften *65*, 382 (1978)
538. Hendry, L.B. et al.: Experientia *30*, 886 (1974)
539. Prasad, R.S., Gupta, A.S., Dev, S.: J. Chromatogr. *92*, 450 (1974)
540. Kasang, G., Weiss, N.: J. Chromatogr. *92*, 401 (1974)
541. Meijer, G.M. et al.: Science *175*, 1469 (1972)
542. Burson, K.R., Kenner, C.T.: J. Chrom. Sci. *7*, 63 (1969)
543. Brownlee, R.G., Silverstein, R.M.: Anal. Chem. *40*, 2077 (1968)
544. Hornstein, I., Crowe, P.F.: Anal. Chem. *34*, 1354 (1962)
545. Beroza, M.: J. Gaschromatogr. *2*, 330 (1964)
546. Bierl, B.A., Beroza, M., Ruth, J.M.: J. Gaschromatogr. *6*, 286 (1968)
547. Houghton, E.: J. Chromatogr. *90*, 57 (1974)
548. Duffield, R.M., Blum, M.S., Brand, J.M.: Ann. Entomol. Soc. Amer. *67*, 821 (1974)
549. McDonough, L.M. et al.: J. Econ. Entomol. *62*, 62 (1969)
550. Morgan, E.D., Tyler, R.C.: J. Chromatogr. *134*, 174 (1977)
551. Price, G.O., Sunos, E.C., Williams, J.F.: Anal. Chem. *39*, 138 (1967)
552. King, S.T.: J. Agr. Food Chem. *21*, 526 (1973)

553. Heigl, J.J., Bell, M.F., White, J.H.: Anal. Chem. *19*, 293 (1947);
Brügel, W.: in Einführung in die Ultrarotspektroskopie, p. 227–264, D. Steinkopf-Verlag, Darmstadt, Germany 1954
554. Disselnköter, H. et al.: Tetrahedron *32*, 1591 (1976)
555. Field, F.H.: Acc. Chem. Res. *1*, 42 (1968)
556. Munson, M.S., Field, F.H.: J. Amer. Chem. Soc. *84*, 4697 (1962)
557. Henneberg, D.: Z. Anal. Chem. *183*, 12 (1961)
558. Henneberg, D., Henrichs, U., Schomburg, G.: J. Chromatogr. *112*, 343 (1975)
559. Henneberg, D. et al.: Angew. Chem. *84*, 381 (1972); Intern. Ed. Engl. *11*, 357 (1972)
560. Gordon, E., Frigerio, A.: J. Chromatogr. *73*, 401 (1972)
561. Hendry, L.B., Hindelang, D.M.: Finnigan Spectra *5*, No. 1 (1975)
562. Issenberg, P., Kobayashi, A., Mysliwy, T.J.: J. Agr. Food Chem. *17*, 1377 (1969)
563. Beroza, M., Bierl, B.A.: Mikrochim. Acta *1969*, 720
564. Beroza, M., Bierl, B.A.: Anal. Chem. *39*, 1131 (1967)
565. Beroza, M., Bierl, B.A.: Anal. Chem. *38*, 1976 (1966)
566. Moore, B.P., Brown, W.V.: J. Chromatogr. *60*, 157 (1971)
567. Bierl, B.A., Beroza, M.: J. Amer. Oil Chem. Soc. *51*, 466 (1974)
568. Tumlinson, J.H., Heath, R.R., Doolittle, R.E.: Anal. Chem. *46*, 1309 (1974)
569. Blum, W., Richter, W.J.: Helv. Chim. Acta *57*, 1744 (1974)
570. Blum, W., Richter, W.J.: Tetrahedron Lett. *1973*, 835
571. Ettre, L.S.: Anal. Chem. *36*, 31 (1964)
572. Kovats, E.: Adv. Chromatogr. *1*, 229 (1965)
573. Miwa, T.K. et al.: Anal. Chem. *32*, 1739 (1960)
574. Vanden Heuvel, W.J.A., Horning, E.C.: Biochem. Biophys. Acta *64*, 416 (1962)
575. Bednas, M.E., Russel, D.S.: Can. J. Chem. *36*, 1272 (1958)
576. Muhs, M.A., Weiss, F.T.: J. Amer. Chem. Soc. *84*, 4697 (1962)
577. Banthorpe, D.V., Gatford, C., Hollebone, B.R.: J. Gaschromatography *6*, 852 (1959)
578. Swern, D.: in: Organic Reactions (R. Adams, ed.), Vol. 7, 386, J. Wiley and Sons, New York 1953
579. McDonough, L.M., George, D.A.: J. Chrom. Sci. *8*, 158 (1970)
580. Lipsky, S.R., Londowne, R.A., Lovelock, J.E.: Anal. Chem. *31*, 852 (1959)
581. Warthen, D., Green, N.: J. Amer. Oil Chem. Soc. *46*, 191 (1969)
582. Novotny, M., Zlatkis, A.: Chromatogr. Rev. *14*, 1 (1971)
583. Grob, K.: Chromatographia *8*, 423 (1975)
584. Bayer, E., Anders, F.: Naturwissenschaften *46*, 380 (1959)
585. Moorhouse, J.E. et al.: Nature *223*, 1174 (1969)
586. Beevor, P.S. et al.: Bull. Entomol. Res. *62*, 439 (1973)
587. Read, J.S., Beevor, P.S.: J. Stored Prod. Res. *12*, 55 (1976)
588. Arn, H., Städler, E., Rauscher, S.: Z. Naturforsch. *30c*, 722 (1975)
589. Beroza, M., Inscoe, M.N.: in Ancillary Techniques of Gas Chromatography (L.S. Ettre, W.H. McFadden, eds.), p. 89, Wiley (Interscience), New York, 1969
590. Beroza, M., Sarmiento, R.: Anal. Chem. *35*, 1353 (1963)
591. Beroza, M., Sarmiento, R.: Anal. Chem. *38*, 1042 (1966)
592. Beroza, M.: Anal. Chem. *34*, 1801 (1962)
593. Beroza, M., Sarmiento, R.: Anal. Chem. *36*, 1744 (1964)
594. Anonymous: Chem. Eng. News *52*, (39), 16 (1974)
595. Regnier, F.E., Huang, J.C.: J. Chromatogr. Sci. *8*, 267 (1970)
596. Henrick, C.A.: Tetrahedron *33*, 1845 (1977)
597. Rossi, R.: Synthesis *1977*, 817
598. Rossi, R.: Synthesis *1978*, 413
599. Pyatnova, Y.B., Ivanov, L.L., Kyskina, A.S.: Usp. Khim. *38*, 126 (1969)
600. Eiter, K.: Fortschr. Chem. Org. Naturstoffe *28*, 204 (1970)
601. Katzenellenbogen, J.A.: Science *194*, 139 (1976)
602. Bestmann, H.J., Vostrowsky, O.: Chem. Phys. Lipids *24*, 335 (1979)
603. Jacobson, M.: in Control of Insect Behavior by Natural Products (D.L. Wood, R.M. Silverstein, M. Nakajima, eds.), p. 111, Academic Press, New York 1970
604. Roelofs, W.L., Cardé, R.T.: in Pheromones (M.C. Birch, ed.), p. 96, Elsevier, New York 1974
605. Cardé, R.T. et al.: J. Chem. Ecol. *3*, 71 (1977)

606. Cardé, R.T., Baker, T.C., Roelofs, W.L.: J. Chem. Ecol. *1*, 475 (1975)
607. Baker, T.C., Cardé, R.T., Roelofs, W.L.: J. Chem. Ecol. *2*, 333 (1976)
608. Wittig, G., Geißler, G.: Liebigs Ann. Chem. *580*, 44 (1953)
609. Bestmann, H.J., Range, P., Kunstmann, R.: Chem. Ber. *104*, 65 (1971)
610. Petragnani, N., Schill, G.: Chem. Ber. *97*, 3293 (1964)
611. Kovaleva, A.S. et al.: Zh. Org. Khim. *8*, 2613 (1972)
612. Kovaleva, A.S. et al.: Zh. Org. Khim. *10*, 696 (1974)
613. Bestmann, H.J., Stransky, W.: Synthesis *1974*, 798
614. Bestmann, H.J. et al.: Chem. Ber. *108*, 3582 (1974)
615. Bestmann, H.J., Stransky, W., Vostrowsky, O.: Chem. Ber. *109*, 1694 (1976)
616. Bestmann, H.J. et al.: Chem. Ber. *111*, 248 (1978)
617. Bestmann, H.J.: Pure Appl. Chem. *51*, 515 (1979); Bestmann, H.J., Downey, W., Geibel, K., Ugi, I., Marquarding, D., Kamal, R.: in preparation
618. Marquarding, D. et al.: Angew. Chem. *85*, 99 (1973), Intern. Ed. *12*, 91 (1973)
619. Bestmann, H.J., Koschatzky, K.H.: unpublished
620. Bestmann, H.J. et al.: Tetrahedron Lett. *1976*, 353
621. Bestmann, H.J. et al.: unpublished
622. Bestmann, H.J., Wax, R., Vostrowsky, O.: Chem. Ber. *112*, 3740 (1979)
623. Bestmann, H.J., Koschatzky, K.H., Vostrowsky, O.: Chem. Ber. *112*, 1923 (1979)
624. Bestmann, H.J. et al.: Tetrahedron Lett. *1977*, 121
625. Truscheit, E., Eiter, K.: Liebigs Ann. Chem. *658*, 65 (1962)
626. Butenandt, A., Hecker, E.: Angew. Chem. *73*, 349 (1961)
627. Butenandt, A. et al.: Liebigs Ann. Chem. *658*, 39 (1962)
628. Hall, D.R. et al.: Chem. Ind. (London) *1975*, 216
629. Goto, G. et al.: Chem. Lett. *1975*, 103
630. Bestmann, H.J., Süß, J., Vostrowsky, O.: Tetrahedron Lett. *1979*, 2467
631. Neumark, S. et al.: Envir. Letters *10*, 97 (1975)
632. Henrick, C.A., Geigel, M.A., Willy, W.E.: Zoecon Corp., unpublished results, cited in [596]
633. Descoins, C., Henrick, C.A.: unpublished, cited in [596]
634. Vig, O.P., Gauba, A.L., Gupta, K.C.: J. Indian Chem. Soc. *52*, 541 (1975)
635. Bestmann, H.J., Süß, J., Vostrowsky, O.: Tetrahedron Lett. *1978*, 3329
636. Nesbitt, B.F. et al.: Tetrahedron Lett. *1973*, 4669
637. Julia, M., Julia, S., Guegan, R.: Bull. Soc. Chim. France *1960*, 1072
638. Bestmann, H.J., Vostrowsky, O., Plenchette, A.: Tetrahedron Lett. *1974*, 779
639. Sonnet, E.: J. Org. Chem. *39*, 3793 (1974)
640. Schlosser, M., Christmann, K.F.: Angew. Chem. *76*, 683 (1964)
641. Anderson, R.J., Henrick, C.A.: J. Amer. Chem. Soc. *97*, 4327 (1975)
642. Hammond, A., Descoins, C.: Bull. Soc. Chim. France *1978*, 299
643. Bestmann, H.J., Vostrowsky, O., Stransky, W.: Chem. Ber. *109*, 3375 (1976)
644. Fetizon, M., Lazare, C.: J. Chem. Soc. *1978*, 842
645. Litchfield, C. et al.: J. Amer. Oil Chem. Soc. *42*, 73 (1965)
646. Sgoutas, D.S., Kummerov, F.A.: Lipids *4*, 283 (1969)
647. Vedejs, E., Fuchs, P.: J. Amer. Chem. Soc. *95*, 822 (1976)
648. Bestmann, H.J. et al.: Angew. Chem. *92*, 492 (1980); Intern. Ed.
649. Hirano, C., Muramoto, H., Horiike, M.: Naturwissenschaften *63*, 439 (1976)
650. Gutmann, H., Lindlar, H.: in Chemistry of Acetylenes, (H.G. Viehe, ed.), Marcel Dekker, New York, p. 355, 1969
651. Lindlar, H., Dubuis, R.: Org. Synth. *46*, 89 (1966)
652. Carney, R.L., Baum, J.W.: unpublished results, cited in [596]
653. Cram, D.J., Allinger, N.D.: J. Amer. Chem. Soc. *78*, 2518 (1956)
654. Brown, C.A., Ahuja, V.K.: Chem. Commun. *1973*, 553
655. Marvell, E.N., Li, T.: Synthesis *1973*, 457
656. Brown, H.C., Zweifel, G.: J. Amer. Chem. Soc. *83*, 3834 (1961)
657. Holan, G., O'Keefe, D.F.: Tetrahedron Lett. *1973*, 673
658. Campbell, K.N., Normant, H.: Bull. Soc. Chim. France *1965*, 1872
659. Gunstone, D.F., Jie, M.L.K.: Chem. Phys. Lipids *4*, 1 (1970)
660. Warthen jr., J.D., Jacobson, M.: Synthesis *1973*, 616

661. Slaugh, L.H.: Tetrahedron *22*, 1741 (1966)
662. Garanti, L. et al.: Gazz. Chim. Ital. *106*, 187 (1976)
663. Rossi, R., Carpita, A.: Synthesis *1977*, 561
664. Perrine, T.D.: J. Org. Chem. *18*, 1356 (1953)
665. Campbell, K.N., Sommers, A.H.: Org. Synth. Coll. Vol. *3*, 446 (1955)
666. Synerholm, M.E.: J. Amer. Chem. Soc. *69*, 2851 (1947); Olsen, S., Gundersen, J., Finsnes, E.: Acta Chem. Scand. *6*, 641 (1952)
667. Brandsma, L.: in Preparative Acetylenic Chemistry, Elsevier Publ. Comp., New York 1971
668. Ziegenbein, W.: in Chemistry of Acetylenes (H.G. Viehe, ed.), p. 169, Marcel Dekker, New York 1969
669. Kriz, J., Benes, M.J., Peska, J.: Tetrahedron Lett. *1965*, 2881
670. Lewis, R.G., Gustafson, D.H., Ermann, W.F.: ibid. *1976*, 401
671. Hendry, L.B., Korzeniowski, S.H.: Science *195*, 86 (1977)
672. Rutledge, T.F.: J. Org. Chem. *24*, 840 (1959)
673. Normant, J.F.: Bull. Soc. Chim. France *1965*, 859
674. Cuvigny, T., Normant, H.: ibid. *1965*, 1872
675. Schwarz, M., Waters, R.M.: Synthesis *1972*, 567
676. Brattesani, D.N., Heathcock, C.H.: Synth. Commun. *3*, 245 (1973)
677. Jenny, E.F., Meier, K.D.: Angew. Chem. *71*, 245 (1959)
678. Fray, G.I. et al.: Tetrahedron *15*, 18 (1961)
679. Beckmann, C. et al.: Synthesis *1975*, 423
680. Dear, R.E.A., Pattison, F.L.M.: J. Amer. Chem. Soc. *85*, 622 (1963)
681. Smith, W.N., Beumel jr., O.F.: Synthesis *1974*, 441
682. Warthen, D.: J. Med. Chem. *11*, 371 (1968)
683. Warthen, D., Jacobson, M.: J. Med. Chem. *11*, 373 (1968)
684. Smith, W.N., Kuehn, E.D.: J. Org. Chem. *38*, 3588 (1973)
685. Tyman, J.H.P.: Synth. Commun. *5*, 21 (1975)
686. Klein, J., Gurfinkel, E.: Tetrahedron *26*, 2127 (1970)
687. Labovitz, J.N., Henrick, C.A., Corbin, V.L.: Tetrahedron Lett. *1975*, 4209
688. Negishi, E., Abramovitch, A.: Tetrahedron Lett. *1977*, 411
689. Brown, C.A., Yamashita, A.: J. Amer. Chem. Soc. *97*, 891 (1975)
690. Descoins, C., Samain, D.: Tetrahedron Lett. *1976*, 745
691. Julia, M., Descoins, C., Risse, C.: Tetrahedron Suppl. *8* (II), 443 (1966)
692. Petrov, A.A., Porfiryeva, Y.I., Semenov, G.I.: J. Gen. Chem. USSR *27*, 1258 (1957)
693. Eglington, G., Whiting, M.C.: J. Chem. Soc. *1950*, 3650
694. Henrick, C.A. et al.: J. Org. Chem. *40*, 1 (1957)
695. Negishi, E., Lew, G., Yoshida, T.: J. Chem. Soc. Chem. Commun. *1973*, 874
696. Kunau, W.H.: Chem. Phys. Lipids *11*, 254 (1973)
697. Fukami, H.: in: Environmental Toxicology of Pesticides (F. Matsumura, G.M. Boush, T. Misato, eds.), p. 571, Academic Press, New York 1972
698. Su, H.C.F., Mahany, P.G., Brady, U.E.: J. Econ. Entomol. *66*, 845 (1973)
699. Walborsky, W.M., Greenway, A.R., Burt, P.E.: Physiol. Entomol. *1*, 151 (1976)
700. Collington, E.W., Meyers, A.I.: J. Org. Chem. *36*, 3044 (1971)
701. Su, H.C.F., Mahany, P.G.: J. Econ. Entomol. *67*, 319 (1974)
702. Mori, K., Tominaga, M., Matsui, M.: Agric. Biol. Chem. *38*, 1551 (1974)
703. Mori, K., Tominaga, M., Matsui, M.: Tetrahedron *31*, 1846 (1975)
704. Kondo, K., Negishi, A., Tunemoto, D.: Angew. Chem. *86*, 415 (1974); Intern. Ed. *13*, 407 (1974)
705. Canevet, J.C. et al.: Chem. Ber. *113*, 1115 (1980)
706. Bestmann, H.J., Süß, J., Vostrowsky, O.: Tetrahedron Lett. *1978*, 245
707. Butt, B.A. et al.: J. Econ. Entomol. *67*, 37 (1974)
708. Mori, K.: Tetrahedron *30*, 3807 (1974)
709. Fouquet, G., Schlosser, M.: Angew. Chem. *86*, 50 (1974); Intern. Ed. *13*, 82 (1974)
710. Samains, D., Descoins, C.: Synthesis *1978*, 388
711. Mandai, T. et al.: Tetrahedron *35*, 309 (1979)
712. De Wolfe, R.H., Young, W.G.: Chem. Rev. *56*, 753 (1956)
713. Henrick, C.A., Garcia, B.A.: unpublished, cited in [596]
714. Stork, G., Grieco, P.A., Gregson, M.: Org. Synth. *54*, 68 (1974)

715. Downie, I.M., Lee, J.B., Matough, M.F.S.: J. Chem. Soc. Chem. Commun. *1968*, 1350
716. Descoins, C., Henrick, C.A.: Tetrahedron Lett. *1972*, 2999
717. Mori, K., Uchida, U., Matsui, M.: Tetrahedron *33*, 385 (1977)
718. Kocienski, P.J., Cernigliaro, C.J.: J. Org. Chem. *41*, 2927 (1976)
719. Seebach, D.: Angew. Chem. *81*, 690 (1969); Intern. Ed. *8*, 639 (1969)
720. Schreiber, J. et al.: Helv. Chim. Acta *50*, 2101 (1967)
721. Normant, F., Commercon, A., Villieras, J.: Tetrahedron Lett. *1975*, 1465
722. Alexakis, A., Cahiez, G., Normant, J.F.: Tetrahedron Lett. *1978*, 2027
723. Tolstikov, G.A. et al.: Tetrahedron Lett. *1978*, 1857
724. Katzenellenbogen, J.A., Cristy, K.J.: J. Org. Chem. *39*, 3315 (1974)
725. Hayashi, T., Midorikowa, H.: Synthesis *1975*, 100
726. Samain, D., Descoins, C.: in press
727. Calderon, N. et al.: J. Amer. Chem. Soc. *90*, 4133 (1968)
728. Rossi, R.: Chim. Ind. (Milano) *57*, 243 (1975)
729. Küpper, F.W., Streck, R.: Chem. Ztg. *99*, 464 (1975)
730. Küpper, F.W., Streck, R.: Z. Naturforsch. *31b*, 1256 (1976)
731. van Dam, P.B., Mittelmejer, M.C., Boelhouwer, C.: J. Chem. Soc. Chem. Commun. *1972*, 1221
732. Rossi, R.: Atti del X. Convegno Nazionale di Chimica Organica, Ferrara, 3.–7. Oct. 1977
733. Nakamura, R. et al.: Chem. Lett. *1976*, 253
734. Seidel, W., Knolle, J., Schäfer, H.J.: Chem. Ber. *110*, 3544 (1977)
735. Iwaki, S. et al.: J. Amer. Chem. Soc. *96*, 7842 (1974)
736. Mori, K., Takigawa, T., Matsui, M.: Tetrahedron Lett. *1976*, 3953
737. Farnum, D.G. et al.: Tetrahedron Lett. *1977*, 4009
738. Klimetzek, D. et al.: Naturwissenschaften *63*, 581 (1976)
739. Vité, J.P. et al.: Naturwissenschaften *63*, 582 (1976)
740. Bakke, A., Froyen, P., Skattebøl, L.: Naturwissenschaften *64*, 98 (1977)
741. Mori, K.: Tetrahedron Lett. *1975*, 2187
742. Mori, K.: Tetrahedron *32*, 1101 (1976)
743. Vité, J.P. et al.: Naturwissenschaften *63*, 582 (1976)
744. Riley, R.C. et al.: Chem. *39*, 1957 (1974)
745. Mori, K.: Agric. Biol. Chem. *38*, 2054 (1974)
746. Garbers, C.F., Scott, F.: Tetrahedron Lett. *1976*, 1625
747. Mori, K.: Tetrahedron Lett. *1976*, 1609
748. Mori, K.: Agric. Biol. Chem. *40*, 415 (1976)
749. Mori, K.: Tetrahedron *32*, 1979 (1976)
750. Elliot, W.J., Fried, J.: J. Org. Chem. *41*, 2475 (1976)
751. Elliot, W.J., Fried, J.: J. Org. Chem. *41*, 2469 (1976)
752. Mori, K.: Tetrahedron *33*, 289 (1977)
753. Cope, A.C., Ciganek, E.: Org. Synth. Coll. Vol. *4*, 612 (1963)
754. Mori, K.: Tetrahedron *31*, 3011 (1975)
755. Mori, K.: Tetrahedron *30*, 4223 (1974)
756. Wasserman, H.H., Barber, E.H.: J. Amer. Chem. Soc. *91*, 3674 (1969)
757. Lipkowitz, K.B., Mundy, B.P., Geeseman, D.: Synth. Commun. *3*, 453 (1973)
758. Chaquin, P., Morizur, J.P., Kossanyi, K.: J. Amer. Chem. Soc. *99*, 903 (1977)
759. Cooke, J.L., Williams, H.J., Natarajan, S.: J. Org. Chem. *42*, 2380 (1977)
760. Mundy, B.P., Otzenberger, R.D., De Barnardis, A.R.: J. Org. Chem. *26*, 2390 (1971)
761. Knolle, J., Schäfer, H.J.: Angew. Chem. *87*, 777 (1975); Intern. Ed. *14*, 758 (1975)
762. D'Silva, T.D.J., Peck, D.W.: J. Org. Chem. *37*, 1828 (1972)
763. Mori, K.: Tetrahedron *31*, 1381 (1975)
764. Hicks, D.R., Fraser-Reid, B.: Chem. Commun. *1976*, 869
765. Francke, W., Reith, W.: Liebigs Ann. Chem. *1979*, 1
766. Mori, K. et al.: Tetrahedron *34*, 1901 (1978)
767. Mori, K.: Tetrahedron *30*, 3817 (1974)
768. Mori, K.: Tetrahedron Lett. *1973*, 3869
769. Cross, J.H. et al.: J. Chem. Ecol. *2*, 457 (1976)
770. Rossi, R., Carpita, A.: Tetrahedron *33*, 2447 (1977)
771. Ravid, U., Silverstein, R.M.: Tetrahedron Lett. *1977*, 423

772. Zurflüh, R. et al.: J. Amer. Chem. Soc. *92*, 425 (1970)
773. Gueldner, R.C., Thompson, A.C., Hedin, P.A.: J. Org. Chem. *37*, 1854 (1972)
774. Billups, W.E., Cross, J.H., Smith, C.V.: J. Amer. Chem. Soc. *95*, 3438 (1973)
775. Stork, G., Cohen, J.F.: J. Amer. Chem. Soc. *96*, 5272 (1974)
776. Babler, J.H.: Tetrahedron Lett. *1975*, 2045
777. Cargill, R.L., Wright, B.A.: J. Org. Chem. *40*, 120 (1975)
778. Trost, B.M., Keeley, D.E.: J. Org. Chem. *40*, 2013 (1975)
779. Ayer, W.A., Browne, L.M.: Can. J. Chem. *52*, 1352 (1974)
780. Hobbs, P.D., Magnus, P.D.: J. Amer. Chem. Soc. *98*, 4594 (1976)
781. Landor, P.D., Landor, S., Musaka, S.: Chem. Commun. *1971*, 1638
782. Baudouy, R., Goré, J.: Synthesis *1974*, 573
783. Descoins, C., Henrick, C.A., Siddal, J.B.: Tetrahedron Lett. *1972*, 3777
784. Michelot, D., Linstrumelle, G.: Tetrahedron Lett. *1976*, 275
785. Kocienski, P.J., Cernigliaro, G., Feldstein, G.: J. Org. Chem. *42*, 353 (1977)
786. Pirkle, W.H., Boeder, C.W.: J. Org. Chem. *43*, 2091 (1978)
787. Rossi, R.: Chim. Ind. (Milan) *59*, 591 (1977)
788. Rossi, R.: unpublished results, cited in [597]
789. Jones, G., Acquadro, M.A., Carmody, M.A.: Chem. Commun. *1975*, 206
790. Carlson, D.A. et al.: J. Agric. Food Chem. *22*, 194 (1974)
791. Bestmann, H.J., Vostrowsky, O., Platz, H.: Chem. Ztg. *98*, 161 (1974)
792. Cargill, R.L., Rosenblum, M.G.: J. Org. Chem. *37*, 3971 (1972)
793. Eiter, K.: Naturwissenschaften *59*, 468 (1972)
794. Ho, T.L., Wong, C.M.: Can. J. Chem. *52*, 1923 (1974)
795. Richter, I., Mangold, H.K.: Chem. Phys. Lipids *11*, 210 (1973)
796. Gribble, G.W., Sanstead, J.K., Sullivan, J.W.: Chem. Commun. *1973*, 735
797. Bestmann, H.J., Kunstmann, R., Schulz, H.: Liebigs Ann. Chem. *699*, 33 (1966)
798. Trost, B.M., Salzmann, N.T.: J. Org. Chem. *40*, 148 (1975)
799. Tsuji, J., Masaoka, K., Takahashi, T.: Tetrahedron Lett. *1977*, 2267
800. Pirkle, W.H., Adams, P.E.: J. Org. Chem. *43*, 378 (1978)
801. Cooke, J.L., Richon, A.B.: J. Org. Chem. *41*, 3516 (1976)
802. Riley, R.G., Silverstein, R.M., Moser, J.C.: Science *183*, 760 (1973)
803. Riley, R.G., Silverstein, R.M., Moser, J.C.: J. Insect Physiol. *20*, 1629 (1974)
804. Benthuysen, J.L., Blum, M.S.: J. Ga. Entomol. Soc. *9*, 235 (1974); B. A. *59*, 61232 (1975)
805. Duffield, R.M., Blum, M.S.: Ann. Entomol. Soc. Amer. *66*, 1357 (1973)
806. Riley, R.G., Silverstein, R.M.: Tetrahedron *30*, 1171 (1974)
807. Katzenellenbogen, J.A., Utawanit, T.: J. Amer. Chem. Soc. *96*, 6153 (1974)
808. Kocienski, P.J., Ansell, J., Ostrow, R.W.: J. Org. Chem. *41*, 3625 (1976)
809. Carpino, L.A.: J. Amer. Chem. Soc. *80*, 601 (1958)
810. Pieroni, O. et al.: J. Polymer Sci. (C) *22*, 993 (1969)
811. Banno, K., Mukaiyama, T.: Chem. Lett. *1976*, 279
812. Magnussum, G.: Tetrahedron Lett. *1977*, 2713
813. Kodana, M., Matsuki, Y., Ito, S.: Tetrahedron Lett. *1975*, 3065
814. Dauben, W.G. et al.: J. Amer. Chem. Soc. *97*, 4973 (1975)
815. Wadhams, L.J., Baker, R., Houx, P.E.: Tetrahedron Lett. *1974*, 1967
816. Baker, R., Evans, D.A., McDowell, P.G.: Chem. Commun. *1977*, 111
817. Burgsthaler, W. et al.: J. Org. Chem. *40*, 3456 (1975)
818. Schwarz, M., Oliver, J.E., Sonnet, P.E.: J. Org. Chem. *40*, 2410 (1975)
819. Rosenblum, L.D., Anderson, R.J., Henrick, C.A.: Tetrahedron Lett. *1976*, 419
820. Place, P., Roumestant, M.L., Goré, J.: Tetrahedron *34*, 1931 (1978)
821. Burgsthaler, A.W. et al.: J. Org. Chem. *42*, 566 (1977)
822. Still, C.: 175 Nat. Meet. Amer. Chem. Soc. March 13, Anaheim, Calif. (1978); Still, W.C.: J. Amer. Chem. Soc. *101*, 2493 (1979)
823. Sekul, A.A., Sparks, A.N.: J. Econ. Entomol. *60*, 1270 (1967)
824. Mitchell, E.R.: J. Ga. Entomol. Soc. *11*, 211 (1976)
825. Fujimura, T.: Jap. J. Appl. Ent. Zool. *20*, 133 (1976)
826. Pattison, F.L.M., Stothers, J.B., Woolford, R.G.: J. Amer. Chem. Soc. *78*, 2255 (1956)
827. Clearwater, J.R.: M. Sc. Thesis, Massey Univers. Palmerston, N. Z. (1971)

828. Hefetz, A., Batra, S.W.T.: Experientia *35*, 1138 (1979)
829. Campion, D.G. et al.: Experientia *35*, 1146 (1979)
830. Edgar, J.A., Cockrum, P.A., Carrodus, B.B.: Experientia *35*, 861 (1979)
831. Uebel, E.C. et al.: Fla. Entomol. *61*, 139 (1978)
832. Levinson, H.Z. et al.: Naturwissenschaften *65*, 543 (1978)
833. Struble, D.L., Swailes, G.E., Ayre, G.E.: Can. Entomol. *109*, 975 (1977)
834. McDonough, L.M., Kamm, J.A., Bierl-Leonhardt, B.A.: J. Chem. Ecol., in press
835. Osborne, G.O., Chapman, R.B., Boyd, J.F.: Envir. Entomol. *8*, 610 (1979)
836. Underhill, E.W., Arthur, A.P., Chisholm, M.D., Steck, W.F.: Envir. Entomol. *8*, 740 (1979)
837. Kamm, J.A., McDonough, L.M.: Envir. Entomol. *8*, 773 (1979); J. Chem. Ecol. *5*, 211 (1979)
838. Williams, I.H.: 3 OILB-Meet. Wädenswil/Horgen (Switzerland), Oct. 1979
839. Klun, J.A. et al.: Science *204*, 1328 (1979)
840. Qureshi, Z.A. et al.: Pak. J. Sci. Ind. Res. *19*, 22 (1976)
841. Renwick, J.A.A., Pitman, G.B.: Envir. Entomol. *8*, 40 (1979)
842. Sower, L.L. et al.: ibid. *8*, 265 (1979)
843. Steck, W.F. et al.: ibid. *8*, 373 (1979)
844. Kraemer, M. et al.: ibid. *8*, 519 (1979)
845. Hedin, P.A. et al.: ibid. *8*, 521 (1979)
846. Frerot, B., Gallois, M., Einhorn, J.: C. R. Hebd. Seances Acad. Sci. Ser. D *288*, 1611 (1979)
847. Cournoyer, R., Shearer, J.C., Anderson, D.H.: Anal. Chem. *49*, 2275 (1977)
848. Stoakley, J.T. et al.: Z. angew. Entomol. *86*, 174 (1978)
849. Noguchi, H., Tamaki, Y., Yushima, T.: Appl. Entomol. Zool. *14*, 225 (1979)
850. Hill, A.S. et al.: J. Chem. Ecol. *5*, 439 (1979)
851. Renon, M. et al.: C. R. Hebd. Seances Acad. Sci. Ser. D *288*, 1559 (1979)
852. Schneider, D.: Sci. American *231*, 28 (1974)
853. Anonymous: Chem. Eng. News *50* (49), 19 (1972)

Die praktische Verwendung von Insektenpheromonen

M. Boneß

BAYER Pflanzenschutz Anwendungstechnik, Biologische Entwicklung
D-5090 Leverkusen

Inhalt

1. Einleitung

Pheromone sind definitionsgemäß Stoffe, die der chemischen Kommunikation unter Artgenossen dienen. Damit bieten sie zugleich vielfältige Möglichkeiten zur Manipulation von Schädlingen und Nützlingen; das gilt sinngemäß auch für Einzelkomponenten, Derivate, Mimetica und Antagonisten. Unter ihnen greifen die im angelsächsischen Schrifttum als „primer pheromones" zusammengefaßten Substanzen in die Individualentwicklung ein und führen zur Kastendifferenzierung oder zur Entstehung gregärer Phasen; sie wurden bisher kaum auf ihre Anwendungsmöglichkeiten untersucht und können hier deshalb übergangen werden. Die dann verbleibenden „releaser pheromones" lösen Verhaltensreaktionen aus und sind sowohl funktionell wie chemisch von großer Vielfalt. Gemeinsam sind ihnen jedoch gewisse Grundzüge, die bei jeder Anwendung für praktische Zwecke berücksichtigt werden müssen. Anders als Insektizide führen sie nicht unmittelbar zur Abtötung nach peroraler oder perkutaner Aufnahme oder Einatmung, sondern erregen oder unterbinden Verhaltensweisen nach dem Durchlaufen eines vielstufigen Weges: Abgabe auf ein Substrat oder in ein Medium (meist Luft); Ausbreitung in und mit diesem; Perzeption durch Sinnesorgane, Verarbeitung im Zentralnervensystem, und schließlich eine spezifische Reaktion. Viel mehr als bei Insektiziden müssen daher exogene und endogene Faktoren berücksichtigt werden, die in diese Prozesse eingreifen: einerseits Dampfdruck und Stabilität des betreffenden Pheromons, Wind und sonstige Wetterfaktoren; andererseits Tagesrhythmus und Reaktionsbereitschaft, Lauf- und Flugaktivität und Konkurrenzfaktoren. Die vielfach extreme Artspezifität erlaubt einerseits ein gezieltes Eingreifen und erschwert andererseits die gleichzeitige Erfassung oder Bekämpfung gemeinsam auftretender Arten. Spezielle Probleme bei der Prüfung und Anwendung von Pheromonen für die Praxis ergeben sich aus ihrer flüchtigen Natur: Die Grenzen zwischen „behandelten" und „unbehandelten" Arealen sind fließend und veränderlich, und zur Klärung vieler Fragen wie zur Erzielung einer optimalen Wirkung werden deshalb oft recht große Flächen benötigt.

Der erfolgversprechende Gebrauch von Pheromonen setzt voraus, daß der Anwender sich mit diesen Gegebenheiten vertraut macht und aus ihnen praktikable Maßnahmen ableitet.

2. Sexuallockstoffe und Aggregationspheromone

Die weitaus meisten anwendungsorientierten Publikationen befassen sich mit Pheromonen, die – zum Teil neben weiteren Funktionen – anlockend wirken, sei es als Sexualduftstoffe auf das andere Geschlecht oder als Aggregationspheromone auf Artgenossen ganz allgemein. Nicht immer ist eine ganz eindeutige Trennung zwischen diesen beiden Gruppen möglich, und aus der anziehenden Wirkungsweise ergeben sich weitgehend gleichartige Anwendungsformen, so daß eine gemeinsame Behandlung angebracht ist.

2.1. Pheromone als Indikatoren

Die Ausnutzung attraktiver Pheromone für Indikatorzwecke ist die naheliegenste und am längsten bekannte Form, mit der derartige Effekte praktisch genutzt wurden und werden. Durch die schon seit Jahrzehnten praktizierte Verwendung von gekäfigten Tieren oder Drüsenextrakten einiger Arten (insbesondere des Schwammspinners, *Lymantria dispar*) wurde dieses Verfahren bereits lange vor der Aera synthetischer Pheromone in vielen Details abgeklärt [1]. Dabei geht es um die Feststellung des Vorkommens und der Ausbreitung, etwa im Rahmen von Quarantänemaßnahmen oder bei neu eingeschleppten Arten, sowie um die Ermittlung von Termin und Stärke des Auftretens als Entscheidungshilfe für die

Durchführung von Gegenmaßnahmen im Rahmen eines gezielten oder integrierten Pflanzenschutzes [2, 3]. Hierfür sind Pheromone anderen Methoden wie Lichtfallen, Schlüpfproben oder visuellen Kontrollen oft überlegen in Bezug auf Effektivität, Spezifität, Arbeitsaufwand und Witterungsabhängigkeit, insbesondere bei schwachem Auftreten; die Ergebnisse laufen jedoch nicht immer mit denen traditioneller Methoden parallel.

Wie bei allen Verfahren, die nicht eindeutig auf eine Fläche bezogen werden können, lassen sich die erhaltenen Werte nicht allgemeinverbindlich in Befallsprognosen umsetzen; als spezifische Eigenschaft dieser Technik ist zu berücksichtigen, daß die Ausbeuten mit zunehmender Populationsdichte meist nur unterproportional anzusteigen pflegen infolge der Konkurrenz mit den zahlreicher werdenden Artgenossen, die ebenfalls Pheromone aussenden [4–6]. In seltenen Fällen kann dies bis zur Verschiebung der Fangmaxima führen [7]. Auch der Zuflug aus der Umgebung kann selbst bei Arten mit nur geringer Wander- und Ausbreitungstendenz erheblich ins Gewicht fallen [5, 8], wobei zwischen den Geschlechtern große Unterschiede auftreten können [9]. Nur ausnahmsweise allerdings gelingt es überhaupt nicht, eine verwertbare Korrelation zwischen den Fängen in Pheromonfallen auf der einen Seite und Zeitpunkt und Stärke des Befalls auf der anderen zu ermitteln [10]. Reproduzierbare Ergebnisse lassen sich nur erzielen durch eine permanente Abgabe des Pheromons, die während der ganzen Überwachungszeit qualitativ und quantitativ gleich bleibt. Die überragende Wichtigkeit der Reinheit und – bei Beteiligung mehrerer Komponenten – der korrekten Zusammensetzung des Duftstoffes braucht hier nicht im Detail erörtert zu werden. Die quantitative Seite ist für die meisten Arten von geringerer Bedeutung; vielfach steigt das Fangergebnis über einen weiten Bereich mit der Aufwandmenge an. Immerhin gibt es auch Fälle, bei denen eine positive Reaktion im Nahbereich an eine auch nach oben relativ eng begrenzte Konzentrationsspanne gebunden ist, vor allem bei einigen Pyraliden [11, 12]. Die meisten Pheromone sind so flüchtig, daß sie ohne Verwendung verdunstungshemmender Zusätze nur wenige Tage attraktiv bleiben. Da die Erhöhung der Aufwandmenge die Wirkungsdauer nur mäßig verlängert [13, 14], führt dieser Weg nicht weiter; auch die Ausbringung auf Sand, Zellstoff und ähnlichen Substraten oder in einfachen Lösungsmitteln wie Hexan, Benzol, Xylol erleichtert zwar die Handhabung der üblicherweise im Milligrammbereich liegenden Wirkstoffmengen, verlängert die Dauerwirkung aber kaum.

In einigen Fällen haben sich Lösungsmittel mit geringem Dampfdruck bewährt, zum Beispiel Trioctanoin [15] oder pflanzliche Öle [16], die besten Resultate jedoch werden in der Regel erzielt durch Einschluß oder Einarbeitung in makromolekulare Materialien. Viele Pheromone lösen sich gut in Kautschuk und werden aus ihm nur allmählich wieder freigesetzt [17], für andere eignen sich besser Kapseln, Beutel oder Schläuche aus Polyaethylen [18, 19]; auch für die Einarbeitung in feste Wachse, Polystyrol-Latices und imprägnierte Diatomeenerde liegen positive Ergebnisse vor [20]. Erstarrende Plastisole [21] und Mehrschichten-Folien aus Materialien mit unterschiedlichen Eigenschaften [22–24] lassen breite Variationsmöglichkeiten zu und kommen den Erfordernissen einer rationellen Fertigung entgegen. Bei weniger stabilen Verbindungen kann auch ein Wirkungsverlust durch chemische Veränderungen eintreten [25], und einige

Pheromone sind auch recht lichtempfindlich [26], so daß bei ihrer Aufbereitung ein Schutz gegen derartige Einflüsse erforderlich ist.

Von Pheromonen angelockte Insekten bleiben meist nur für kurze Zeit in der Nähe der Duftquelle; sie müssen deshalb gefangen oder auf andere Weise festgehalten werden. Hierfür finden Leimfolien wegen ihrer einfachen Handhabung besonders breite Anwendung. Bei Verwendung von Polybutenen und ähnlichen Substanzen bleiben sie viele Wochen fängig, sofern nicht durch Verschmutzung oder zu reichlichen Anflug ihre Kapazitäten vorzeitig erschöpft werden. Wenig flüchtige, konservierende Flüssigkeiten, eventuell mit einem Zusatz von Netzmitteln, sind auch einem anhaltend starken Andrang gewachsen [27, 28]; von Nachteil ist ihre erhöhte Anfälligkeit gegen Störungen und der höhere Arbeitsaufwand für ihre Hantierung. Kombinationen mit elektrischen Fanggittern sind zwar sehr effektiv [29], lassen sich aber ohne überhöhtem Aufwand nur in der Nähe von Gebäuden einsetzen.

Bei manchen Arten ist es möglich, unter Ausnutzung gewisser Verhaltensweisen wie Lichtreaktion oder Deckungsbedürfnis Lebendfallen mit Reuseneffekten zu verwenden und auf jedes weitere Fangmittel zu verzichten [30–32].

Mit Kontaktinsektiziden lassen sich allenfalls bei der Verwendung in Fallen für einzelne Arten befriedigende Resultate erzielen; im Allgemeinen erfolgt der Anflug zu flüchtig, um die Aufnahme der erforderlichen Wirkstoffmenge sicherzustellen. Fraßgifte haben sich zwar in Ködern vielfach bewährt, sind aber zur Kombination mit Pheromonen in den meisten Fällen ungeeignet. Nur Atemgifte wie Dichlorvos lassen sich breiter einsetzen unter der Voraussetzung, daß die Fallenkonstruktion den Luftaustausch in Grenzen hält und eingedrungene Tiere am vorzeitigen Entkommen hindert [33, 34]; außerdem dürfen keine zu starken Repellent-Effekte auftreten [35].

Eine geradezu uferlose Zahl von Veröffentlichungen befaßt sich mit der Gestaltung und Plazierung von Fallen. In empirischen Vergleichen führen unterschiedliche Formen oft zu sehr stark voneinander abweichenden Resultaten, wobei die jeweiligen Ergebnisse sich kaum auf andere Arten übertragen lassen [36–38]. Eine direkte Beobachtung des hierfür verantwortlichen Anflug- und Einflugverhaltens ist nur bei tagaktiven Tieren leicht durchführbar [39]; für die weitaus zahlreicheren, vorwiegend oder ausschließlich bei Nacht aktiven Arten sind derartige Untersuchungen schwierig und nur bedingt möglich [40–42]. Dabei können im Nahbereich erhebliche Komplikationen durch störende Reaktionen auftreten [40]. Farben sind nur bei einigen Tagtieren von wesentlicher Bedeutung [43, 44], während für viele Baumbewohner senkrechte „Stamm"-Strukturen eine große Rolle spielen [45, 46]. In etlichen Fällen muß auch die Höhe der Fallen den Fluggewohnheiten angepaßt werden: Es gibt neben vielen weniger spezialisierten Arten solche, die sich bevorzugt in Bodennähe, zwischen den Pflanzen, über der Pflanzendecke oder im Kronenraum aufhalten.

Temperaturaturschwankungen und mäßige Niederschläge haben nur einen relativ geringen Einfluß auf die Leistungen von Pheromon-Fallen, bei völlig unbewegter Luft oder zu starkem Wind dagegen erfolgt ein starker Wirkungsabfall [6, 47, 48]. Nicht selten unterscheiden sich verschiedene, im Laufe des Jahres aufeinander folgende Generationen aus noch weitgehend unbekannten Gründen erheblich in der Intensität ihrer Reaktionen auf Pheromone [49]. Ein großer

Vorteil der Pheromon-Fallen im Vergleich zu Licht- und Köderfallen liegt in ihrer gewöhnlich sehr ausgeprägten Spezifität. Komplikationen bei der Auswertung können jedoch auftreten, wenn in ihrem Einzugsbereich gleichzeitig mehrere Angehörige einer Gattung oder Gattungsgruppe auftreten, die auf das gleiche Pheromon ansprechen, wie im Falle von diversen Grapholitha- und Pammene-Arten bei Lockstoffen mit der Hauptkomponente cis-8-Dodecenylacetat [50, 51]. Seltener werden systematisch einander fernstehende, aber oberflächlich ähnlich aussehende Formen vom gleichen Duftstoff angezogen, so zum Beispiel die Tortricide Adoxophyes orana und die Noctuide Oligia fasciuncula [52]. Vermutlich werden durch das permanente Angebot einer relativ hohen Duftstoffkonzentration einerseits gleichsinnig reagierende Arten mit tageszeitlich verschieden gelegenen Aktivitätsphasen zusammengefaßt und andererseits zusätzliche Differenzierungsmechanismen im Nahbereich überspielt.

Gelegentlich können auch unterschiedliche Pheromone in einer Falle ohne Wirkungsverlust kombiniert werden, sofern es dadurch weder zu einer Verfälschung der artspezifischen Duftreize kommt noch antagonistische Effekte auftreten [53–55]. Eine gemeinsame Ausbringung von Pheromonen und Fraßlockstoffen [56] oder eine Zugabe zu Lichtfallen [57] kann ebenfalls von Vorteil sein.

2.2. Bekämpfung von Schädlingen im Anlockverfahren

Prinzipiell können Insekten mit anlockenden Pheromonen auch dezimiert oder angereichert werden. Dabei gelten die gleichen Regeln und Abhängigkeiten wie für den Gebrauch zu Indikatorzwecken – abgesehen davon, daß es sich erübrigt, auf den Erhaltungszustand und die Identifizierbarkeit der gefangenen Tiere Rücksicht zu nehmen. Stattdessen gewinnen Kostenfragen und Probleme der praxisgerechten, rationellen Durchführung an Gewicht. Da sich die Lockwirkung in der Regel auf adulte Insekten beschränkt, die Schäden aber zumeist durch ihre Nachkommen im Jugendstadium verursacht werden, müssen derartige Maßnahmen fast immer vorbeugend erfolgen. Die Anzahlen gefangener Tiere und selbst deren prozentualer Anteil an der Gesamtpopulation sind kein ausreichendes Maß für den angestrebten Erfolg: Entscheidend ist allein die Wirkung auf die Nachkommenschaft. Nur Exemplare, die vor der Ausübung ihrer Fortpflanzungsfunktionen ausgeschaltet werden, sind hierfür von Belang. Faktoren wie Lebensdauer, natürliche Mortalität, Reifungszeit, Nachkommenzahl, Kopulationshäufigkeit, Wettbewerb mit der vorhandenen Population und Generationenfolge beeinflussen die für eine erfolgreiche Bekämpfung erforderlichen Wirkungsgrade. Um die rechnerische Durcharbeitung solcher Populationsmodelle haben sich Knipling und McGuire sowie Nakamura und Oyama bemüht [58, 59]. Die Übertragung dieser Überlegungen auf ein konkretes Objekt, den Schwammspinner *Lymantria dispar*, durch Beroza und Knipling [60] führte zu dem Schluß, daß zur allmählichen Verminderung einer in Zunahme begriffenen Population Fangquoten im Bereich von 90–95% erforderlich sind. Dieser Wert dürfte zwar für Objekte mit günstigeren biologischen Eigenschaften etwas niedriger liegen, setzt aber den Einsatzmöglichkeiten dieses Verfahrens insbesondere bei einem starken Ausgangsbefall Grenzen. Um so hohe Quoten zu erreichen, muß sich die zu schützende Fläche möglichst vollständig und pausenlos im Bereich einer Pheromon-Konzentration befinden, die der von Artgenossen ausgehenden Lockwirkung überlegen ist.

Unter optimalen Bedingungen ermittelte Grenzwerte und maximale Reichweiten sind als Grundlage für die in praxi erforderlichen Dosierungen und Fallenabstände nicht geeignet. Das gilt besonders für Arten, bei denen die Mehrheit der Individuen sich weitgehend ortsstet verhält und nur über geringe Distanzen abwandert. Bei den vielen nur für Männchen attraktiven Sexualduftstoffen kann der Bekämpfungserfolg, besonders auf kleinen Flächen, außerdem durch die Zuwanderung bereits begatteter Weibchen aus der Umgebung beeinträchtigt werden. Dies ist vermutlich einer der Gründe die dazu führen, daß der Erfolg einer Pheromon-Anwendung mit der Größe und Geschlossenheit der behandelten Fläche meist besser wird; daneben nimmt gleichzeitig der relative Anteil der Randzonen ab, die je nach Windrichtung zeitweilig nur unvollkommen erfaßt werden. Bei zu starkem Ausgangsbefall kann es vorteilhaft sein, diesen durch Anwendung von Insektiziden oder anderen Maßnahmen zunächst auf ein Niveau zu senken, bei dem ein befriedigender Wirkungsgrad zu erreichen ist. Wo primär befallsfreie Flächen vorliegen, deren Besiedlung aus der Umgebung her erfolgt – sei es aus Winterlagern, von abgeernteten Flächen, aus anderen Kulturen oder im Zuge weiträumiger Wanderungen –, kann ein Ring von Fallen um das betreffende Feld oder in seinen Randzonen ausreichen, um einen schädigenden Befall zu verhindern oder wenigstens zu verzögern [61]. In den Fällen, wo die von Pheromonen, meist im Zusammenwirken mit pflanzlichen Lockstoffen, angezogenen Tiere auf ihren Wirtspflanzen zum Fraß oder Einbohren verbleiben, können Fangbäume oder zu diesem Zweck angesäte oder angepflanzte Pflanzenbestände an die Stelle von Fallen treten. Diese werden entweder mit geeigneten Insektiziden behandelt, mitsamt den angelockten Schädlingen und ihrer Brut abgetötet, untergepflügt oder auf andere Weise beseitigt. Wo derartige Verfahren bereits früher im Gebrauch waren, konnten sie durch die zusätzliche Anwendung von Pheromonen zum Teil erheblich verbessert werden [62–64].

Fangergebnisse allein sagen noch nichts über die Wirksamkeit einer Pheromon-Anwendung aus, und für die anschließend ermittelte Stärke des Befalls durch die folgende Generation wiederum sind in der Regel bereits weitere Faktoren mit verantwortlich; deshalb haben sich spezielle Methoden eingebürgert, um zu quantitativen Werten für die erzielten Wirkungsgrade zu kommen. Hierher gehören Modellversuche mit ausgesetzten Tieren außerhalb der Flugzeit oder in befallsfreien Gebieten [65], Freilassung und Wiederfang markierter Tiere innerhalb einer natürlichen Population [66–68], und schließlich das Aussetzen unbegatteter Weibchen mit anschließender Kontrolle der Kopulationsrate im Vergleich zu unbeeinflußten Parzellen [69].

Bei der praktischen Erprobung des Anlockverfahrens hat sich bei den meisten Arten durchaus eine signifikante Reduktion des Schadauftretens erreichen lassen. Die Wirkungsgrade anderer Bekämpfungsmaßnahmen wurden jedoch oft nicht erreicht oder erforderten einen Material- und Arbeitsaufwand, der mit diesen nicht konkurrieren konnte. Weitere Probleme können sich besonders auf intensiv bewirtschafteten, großflächigen Feldkulturen daraus ergeben, daß andere Bearbeitungsvorgänge mit dem Fallengebrauch kollidieren. Der Abwurf von großen

Mengen einfacher Kleinstfallen würde diese Schwierigkeiten möglicherweise umgehen, ist aber aus anderen Gründen problematisch [60].

2.3. Störung der chemischen Kommunikation

Ein gänzlich anderer Weg zur Nutzbarmachung von Pheromonen wurde erstmals 1960 von Beroza vorgeschlagen [70], nämlich die Störung der chemischen Kommunikation durch eine permanente „Überflutung" des Geländes mit dem Duftstoff. Über die physiologischen Grundlagen dieser auch als „Konfusionsverfahren" oder „Desorientierungsmethode" bezeichneten Technik herrscht noch keine völlige Klarheit: Neben der Unmöglichkeit, unter diesen Bedingungen natürliche chemische Reizquellen zu lokalisieren, können Ermüdungserscheinungen bei zeitlich begrenzter Reaktionsbereitschaft, Anhebungen der Schwellenwerte („habituation") und Fehlleitungen zu „falschen" Zielen beteiligt sein [71–73].

In einer großen Zahl von Versuchen wurde seither gezeigt, daß dieses zunächst rein spekulativ geäußerte Konzept tatsächlich realisierbar ist, allerdings mit sehr unterschiedlichem Erfolg. Die Gründe für diese Differenzen liegen nur selten so offen zutage wie beim Vergleich der fast ausschließlich bei Dunkelheit aktiven Nonne, *Lymantria monacha*, mit dem tagaktiven Schwammspinner, *Lymantria dispar*, bei dem ein hochentwickeltes optisches Suchverhalten weitgehend an die Stelle der chemischen Orientierung treten kann [74].

Nach allen bisher vorliegenden Erfahrungen wird mit jenen synthetischen Substanzen oder Kombinationen, die möglichst genau den arteigenen Pheromonen entsprechen, nicht nur der stärkste Anlockeffekt erzielt, sondern auch die intensivste Desorientierung. (Ob für diesen Anwendungsbereich auch die gleichen hohen Anforderungen an Reinheit und Einheitlichkeit der Produkte gestellt werden müssen wie für eine optimale Lockwirkung, ist bislang kaum untersucht worden.) Sofern nur einzelne Komponenten eines mehrgliederigen Pheromon-Systems ausgebracht werden, sind zumindest höhere Dosierungen erforderlich [32, 75]; das gilt noch mehr für manche „Mimetica", die den jeweiligen Pheromonen strukturell nahestehen und entsprechende, aber schwächere Reaktionen auslösen [75, 76]. Arten, deren Pheromon-Systeme sich in einzelnen Bestandteilen überschneiden oder ähneln, können unter Umständen durch Ausbringung nur einer Substanz zugleich betroffen werden [75, 77].

Als „Hemmstoffe" oder „Antipheromone" werden in der Literatur oft Substanzen vorgestellt, die sich negativ auf die Fangleistung von Pheromon-Fallen auswirken, wenn sie den Lockstoffen zugesetzt oder in ihrer unmittelbaren Umgebung ausgebracht werden.

Über die Grundlagen dieser Wirkung ist damit noch nichts ausgesagt. Versuche, sie zur Erzielung von Desorientierungseffekten heranzuziehen, schlugen zumeist völlig fehl, und in manchen Fällen wurde sogar das Gegenteil erreicht, nämlich eine Intensivierung der Reaktion auf lockende Artgenossen oder synthetische Pheromone [78, 79]. Als wenigstens teilweise erfolgreich erwies sich die Verwendung der trans-Homologen des aus cis-Isomeren bestehenden Pheromons von *Adoxophyes orana* [80], sowie des Acetats vom alkoholischen Pheromon des Apfelwicklers, *Carpocapsa pomonella* [81]. Positive Ergebnisse liegen außerdem für Antipheromone von Borkenkäfern vor, mit denen sich die Besiedlung von Wirtsbäumen oder geschlagenen Stämmen eindämmen ließ – so für Methylcyclohexenon bei *Dendroctonus pseudotsugae* [82, 83] und für Brevicomin-Isomere und Verbenon bei *Dendroctonus frontalis* [84].

Kaum etwas ist bekannt über die Konzentrationen von Pheromonen, die während der Aktivitätsperiode eines Schädlings in seinem Lebensraum auftreten; sie dürften jedoch selbst bei individuenreichen Populationen recht bescheiden sein und sich also mit synthetischen Präparaten leicht überbieten lassen [85].

Alle Angaben über die für den gewünschten Effekt erforderlichen Aufwandmengen und – seltener – über die damit erzielten Konzentrationen, die sich in der Literatur finden, wurden rein empirisch ermittelt; sie bewegen sich zwischen wenigen Gramm pro Hektar und bis zu hundertfach höheren Dosierungen, insbesondere bei nur mäßig wirksamen Mimetica [76, 86–88]. Noch mehr als beim Anlockverfahren muß dabei eine langanhaltende und möglichst gleichbleibende Abgabe angestrebt werden. Hierfür haben sich drei Methoden als die zur Zeit aussichtsreichsten herauskristallisiert: Das Versprühen von Mikrokapseln mit Hüllsubstanzen passender Durchlässigkeit [65, 69, 89, 90], das Ausbringen pheromongefüllter kapillarer Hohlfasern [86, 91], und schließlich die Verteilung relativ weniger Verdampfer mit hoher Einzeldosis im Bestand [90, 92], wobei jedoch den maximalen Abständen Grenzen gesetzt sind [32]. Bei lang auseinandergezogenen Flugzeiten oder mehreren, sich überlappenden Generationen ist es, unabhängig von der benutzten Methode, vorteilhafter, mehrmals geringe Aufwandmengen auszubringen als nur einmal oder in großen zeitlichen Abständen hohe Dosen [93].

Während beim Gebrauch von Fallen die Anzahl gefangener Tiere als erstes Maß der Wirksamkeit dienen kann, läßt sich der Erfolg einer auf Störung der Kommunikation angelegten Maßnahme nicht unmittelbar fassen. Die objektivste Methode zur Ermittlung dieser Störwirkung ist die bereits erwähnte Exposition unbegatteter Weibchen [69, 89, 94]. Da dieses Verfahren arbeitsaufwendig und störanfällig ist, hat sich die Verwendung von Pheromonfallen als Indikator für die Beeinträchtigung der chemischen Orientierungsleistung eingebürgert. Beim Vergleich beider Techniken laufen die Ergebnisse in den meisten Fällen parallel; bei einzelnen Arten auftretende grobe Diskrepanzen müssen als Hinweis auf zusätzliche Möglichkeiten der Partnersuche auf anderer Basis gelten.

Wie alle sonstigen Formen des Pheromon-Gebrauchs ist auch das Konfusionsverfahren abhängig vom Wetter und anderen Umwelteinflüssen, nur lassen sich deren Auswirkungen hier schwieriger erfassen. Hierher gehört zunächst der Einfluß des Windes auf die Verdunstungsrate, die Ausbreitung und Verdriftung [95, 96], sowie eine wiederholt beobachtete Abhängigkeit von der Temperatur [52, 80, 95]: Möglicherweise geht bei hohen Temperaturen ein zunehmender Anteil des Pheromons durch Konvektion verloren, so daß die Wirkung nachläßt. Auf der anderen Seite kann sich in Wäldern und ähnlich strukturierten Kulturen (Obst, Oliven etc.) eine Anreicherung des Pheromons in Bodennähe herausbilden, was bei vorwiegend im Kronenraum aktiven Schädlingen von Nachteil sein kann [97].

Zusätze von Haftmitteln, sowie faserige und bandförmige Trägermaterialien, die sich im Laubwerk verfangen, können hier Abhilfe bringen; sie sind auch in niederen Feldkulturen von Nutzen, um Wirkungsverluste durch Einwaschung in den Boden nach Niederschlägen oder Beregnung zu verhindern.

2.4. Lockstoffe als Hilfsmittel für andere Verfahren

Bei allen Methoden zur direkten Bekämpfung oder mittelbaren Manipulation von Schädlingen ist es notwendig, einen möglichst hohen Anteil der Population mit den jeweiligen Agentien in Berührung zu bringen. Es liegt nahe, die starke und spezifische Anlockwirkung vieler Pheromone hierfür auszunutzen. So können Aggregationspheromone herangezogen werden, um die Repellentwirkung mancher Insektizide zu überwinden [98]. Im Rahmen der biologischen Schädlingsbekämpfung könnten sie die Aufnahme von Insektenpathogenen fördern [99, 100], beim Einsatz von Chemosterilantien deren Anwendung in freilebenden Populationen ermöglichen. Eine praktische Erprobung dieser Vorschläge steht jedoch noch aus.

3. Anwendungsmöglichkeiten sonstiger Pheromone

Neben den anlockenden Sexualduftstoffen und Aggregationslockstoffen finden sich andere Pheromone, die eine vorwiegend abstoßende Wirkung entfalten. Hierher gehören Markierungssubstanzen, etwa bei Fruchtfliegen, die eine nochmalige Eiablage in bereits belegte Früchte unterbinden [101]. Prinzipiell können sie für die Verhinderung eines Schadenfalls genutzt werden [102]. Alarmpheromone veranlassen gesellig lebende Insekten zu Flucht- oder Fallreaktionen, so zum Beispiel bei Blattläusen [103]; Vorschläge zur Ausnutzung dieses Effekts haben bislang nur spekulativen Charakter. Das gilt in gleicher Weise für die entsprechenden Pheromone sozialer und subsozialer Insekten, und ebenso für die vielerlei Stoffe, die in deren Zusammenleben andere Funktionen erfüllen [104]. Vor allem die Spurfolgepheromone pflanzenschädigender Termiten und Ameisen legen Bemühungen um ihren Gebrauch nahe [105, 106]. Auch die von der Brut ausgehenden Reize wurden in ersten Tastversuchen auf ihre Verwendbarkeit überprüft [107]. Sogenannte „Aphrodisiaca" oder „Arrestantien", die bei der Paarung im Nahbereich wirksam werden, könnten möglicherweise ebenfalls zur Manipulation des Verhaltens eingesetzt werden [108, 109]. Wie bei allen anderen hier aufgeführten Pheromonen existieren dafür bestenfalls erste Ansätze, obwohl es sich zum Teil um relativ einfache und leicht zugängliche Chemikalien handelt.

4. Umweltprobleme und Nebenwirkungen

Auch wenn Pheromone fast durchweg nur in geringen Mengen angewendet werden und nur ausnahmsweise direkt mit dem Erntegut in Berührung kommen, muß eine gefahrlose Handhabung ebenso sichergestellt werden wie bei anderen Umweltchemikalien. In bezug auf akute Toxizität dürfte es dabei kaum Schwierigkeiten geben [110]; über ihre Rückstände ist bis jetzt kaum gearbeitet worden, weil wegen der Flüchtigkeit fast aller Duftstoffe hierzu wenig Anlaß besteht [111]. Daten über Abbau und Transport durch Umweltfaktoren finden sich zerstreut in vielen Arbeiten; eine umfassendere Untersuchung am Beispiel des Gossyplure berücksichtigt die Beständigkeit in Wasser und Boden, Fragen der Lichtstabilität, der eventuellen Einwaschung ins Erdreich, und die wichtigsten Abbauprodukte [112]. Auch unerwünschte biologische Effekte sind bei der Anwendung von Pheromonen denkbar; an erster Stelle steht hier eine eventuelle Verstärkung des Befalls im Umkreis von Lockstoffquellen. Sie ist immer dann möglich, wenn nicht nur Männchen angelockt werden und das Konzentrationsgefälle die Ansprüche der zuwandernden Tiere schon in einigem Abstand von der Falle oder dem Fangbaum erfüllt. In der Praxis gibt es vor allem bei Borkenkäfern und beim Baumwollkapselkäfer Schwierigkeiten mit diesem Phänomen [34, 113]. Andererseits können Pheromone zugleich als Kairomone für ihre natürlichen Gegenspieler fungieren und ihnen den Weg zu ihrer Beute oder ihrem Wirt anzeigen. Eine Diskussion des „Kairomon"-Begriffs und der Möglichkeiten, die sich hier insbesondere für die biologische Schädlingsbekämpfung eröffnen [114], würde den Rahmen dieses Beitrags überschreiten. Zwar dienen vorwiegend solche Reize als „Kairomone", die für die chemische Kommunikation innerhalb der Art keine Rolle spielen – zum Beispiel Geruchsstoffe von der Oberfläche der Cuticula oder der Flügelschuppen [115], doch nicht selten auch Pheromone. Werden diese in Fallen, auf insektizidbehandelten Fangbäumen oder zur Vernichtung bestimmten Parzellen geboten, können dadurch auch Nutzarthropoden in großem Umfang dezimiert werden. Das gilt besonders für manche Feinde von Borkenkäfern, die auf deren Aggregationslockstoffe ebenfalls ansprechen [116, 117], doch muß hierauf auch bei Sexualpheromonen und während der Eiablage abgegebenen Markierungsgerüchen geachtet werden [118, 119]. Wieweit auch beim Konfusionsverfahren Nützlinge positiv oder negativ beeinflußt werden, wurde noch nicht untersucht.

5. Entwicklung und Stand der Pheromon-Anwendung

5.1. Forst und Gehölze

Im Forst wurde schon lange vor der Identifizierung der ersten Pheromone mit den Lockstoffen wichtiger Schädlinge erfolgreich gearbeitet: Einerseits mit Weibchen und Drüsenextrakten von *Lymantria dispar* [1], andererseits – noch ohne klare Kenntnis der Zusammenhänge – mit den Aggregationspheromonen von Borkenkäfern, die wesentlich zur Wirksamkeit von Fangbäumen beitrugen. Diese beiden Gruppen spielen auch in der neueren Entwicklung die Hauptrollen [120]. Aufbauend auf den Erfahrungen mit gekäfigten Faltern und Extrakten wurde *Lymantria dispar* zum „klassischen" Objekt für die Entwicklung der praxisorientierten Anwendung von Pheromonen und diente für viele Aspekte als Modell. Das gilt für Aufbereitung und Ausbringung des Wirkstoffs ebenso wie für Methoden zur Erfolgskontrolle, Überlegungen zur Populationsdynamik und Analysen des Anflugverhaltens [60, 65, 69, 72, 89]. Dabei stellte sich heraus, daß sich Fallen mit Disparlure hervorragend zur Überwachung eignen; auch der Fang großer Individuenzahlen ist möglich, wenn die Reaktionen im Nahbereich bei der Gestaltung der Fallen berücksichtigt werden [121]. Sowohl durch Anlockung wie auch Desorientierung lassen sich Bekämpfungserfolge erzielen; die Wirkungsgrade befriedigen jedoch nur bei schwachem bis mäßigem Auftreten. *Lymantria monacha*, die Nonne, kann ebenfalls mit Fallen gut überwacht werden [37], und für eine Bekämpfung mit Hilfe von Pheromonen liegen aussichtsreiche Ansätze vor [33, 74]. Entsprechendes gilt auch für einen weiteren Vertreter der Lymantriidae, *Orgyia pseudotsugata* [122], und die *Tortriciden Choristoneura fumiferana* [123, 124] und *Rhyacionia buoliana* [78, 125].

Weite Waldgebiete werden permanent durch Borkenkäfer gefährdet. Umfangreiche Untersuchungen an *Dendroctonus*-Arten zeigten, daß ihr Auftreten durch eine komplexe Abfolge von Reizen und Reaktionen gesteuert wird, in der Inhaltsstoffe der Wirtspflanze den Initialbefall auslösen.

Verschiedene Pheromone locken in der Folge jene Massen an, die erforderlich sind, um die Abwehrkräfte des Baumes zu überwinden, und beugen andererseits einer übermäßigen Befallsdichte vor. Bei allen Unterschieden im Detail gilt dieses Schema prinzipiell auch für Angehörige anderer Gattungen [126]. Für den Einsatz von Borkenkäferpheromonen im Forstschutz schaffen diese Abhängigkeitsverhältnisse Voraussetzungen, die von denjenigen für andere Schädlingsgruppen erheblich abweichen [127, 128]. Das Anlocken und Fangen großer Mengen von Käfern macht keine Schwierigkeiten [129, 130]. Daraus resultiert jedoch nicht unbedingt auch schon eine praktisch bedeutsame Verminderung des Befalls [131]. Erfolge wurden mit dieser Methode bisher erzielt gegen *Ips typographus* in Skandinavien [132] und *Gnathotrichus sulcatus* auf frisch geschlagenem Holz im Bereich von Sägewerken [133]. Das Desorientierungskonzept versagte in der Praxis [134]. Aussichtsreicher erscheinen modifizierte Fangbaumverfahren, bei denen durch Pheromonabgaben die Attraktivität von gefällten Stämmen oder Baumstümpfen verstärkt oder verlängert wird [62, 135] oder die Käfer zur Annahme gesunder, stehender Bäume veranlaßt werden. Die Ausschaltung der angelockten Tiere samt Brut erfolgt entweder durch Fällen und Entrinden, mit Insektiziden [62] oder durch Abtötung der Stämme, etwa mit Hilfe von Kakodylsäure [136]. Eine andere Möglichkeit besteht darin, durch Anti-Aggregationspheromone, die allein oder in Kombination mit den Pheromonen geboten werden, das Zustandekommen eines erfolgreichen Befalls zu verhindern [82, 137, 138].

5.2. Obst, Wein, Oliven

Obstanlagen und Bestände sonstiger baum- und strauchförmiger Kulturpflanzen entsprechen in ihrer Struktur weitgehend dem Forst, unterscheiden sich von ihm aber in der Intensität der Bewirtschaftung und damit in den tolerierbaren Schadensschwellen. Diese liegen insbesondere bei Fruchtschäden recht niedrig – so auch für den Apfelwickler, *Carpocapsa pomonella*, der im Kernobstbau in den meisten Regionen eine Schlüsselstellung einnimmt. Pheromonfänge als Bezugsbasis für Bekämpfungsmaßnahmen haben sich vielerorts gut eingeführt [5, 9, 36, 139]. Bekämpfungsmaßnahmen mit Fallen verliefen teils befriedigend [140, 141], teils enttäuschten sie selbst bei sehr hohem Aufwand [142, 143]. Neuere Veröffentlichungen über Desorientierungsverfahren sind vorwiegend positiv [88, 144], auch solche mit Anti-Pheromonen [81].

Von zahlreichen weiteren, regional auf Kernobst schädigenden Tortriciden sind die Pheromone bekannt und werden mit Erfolg zur Überwachung benutzt [29, 145, 146]. Bekämpfungserfolge mit verschiedenen Techniken wurden erzielt gegen *Argyrotaenia velutinana* [147], *Archips podana* [35] und *Adoxophyes orana* [3, 80]. In die engste Verwandtschaft der letztgenannten Art gehört der wichtige Teeschädling *Adoxophyes fasciata* [148].

Auf Steinobst nehmen *Grapholitha*-Arten die Stelle des Apfelwicklers ein, insbesondere *Gr. molesta, funebrana* und *prunivora*. Ihre Pheromone sind einander sehr ähnlich, wodurch bei gemeinsamem Vorkommen die Überwachung erschwert [50, 51], eine Bekämpfung dagegen vereinfacht wird. Massenfänge brachten keinen Erfolg, während das Konfusionsverfahren bei schwachem bis mäßigem Auftreten funktionierte [87, 144, 149]. Auch die Pheromone einiger Pfirsichschädlinge werden bereits in der Praxis eingesetzt, so von der *Gelechiide Anarsia lineatella* [2, 150] und den *Sesiiden Synanthedon pictipes* und *Sanninoidea exitiosa* [151]. Hinzu kommt in jüngster Zeit die *Pyralide Amyelois transitella* auf Mandeln [152].

Im Weinbau erwies sich die Desorientierungstechnik als wirksam gegen *Argyrotaenia velutinana* [153] und *Paralobesia viteana* [154] in Nordamerika, während sie bei den europäischen Arten *Clysia (Eupoecilia) ambiguella* und *Lobesia botrana* noch nicht befriedigte [155, 96]. Für *Prays citri* auf Agrumen und *Prays oleae* auf Oliven wurden Pheromone erst vor kurzem aufgeklärt und befinden sich noch in der Erprobung [156, 157]. Pheromone anderer Insektengruppen sind in einigen Fällen ebenfalls bereits bekannt – so für Fruchtfliegen und Schildläuse –, wurden bislang jedoch noch nicht in die Praxis eingeführt.

5.3. Getreide

Im Gegensatz zu den vielfältigen Aktivitäten im Forst und Obstbau spielt der Einsatz von Pheromonen in Getreide und anderen Kulturgräsern bis heute nur eine bescheidene Rolle. Daran sind gewiß auch die für ihre Anwendung ungünstigen Verhältnisse mit Schuld, wie sie in einjährigen, niedrigen und meist großflächigen Kulturen herrschen. Das seit langem bekannte und intensiv untersuchte Pheromon des Maisstengelbohrers, *Ostrinia nubilalis*, macht selbst bei der Verwendung für Überwachungszwecke Schwierigkeiten [7, 25, 49], und die Aussichten auf einen Einsatz zur Bekämpfung werden nach einer Analyse des Paarungsverhaltens skeptisch beurteilt [158]. Für den Reisstengelbohrer *Chilo suppressalis* sehen die ersten Resultate günstiger aus [11].

5.4. Baumwolle

Der Baumwollkapselkäfer *Anthonomus grandis* war als eingeschleppter Großschädling in den Vereinigten Staaten Objekt intensiver Bemühungen zur Eindämmung und Ausrottung, wobei auch die Pheromonabgabe lebender Käfer in großem Umfang genutzt wurde. In diesem Rahmen fanden auch die synthetischen Pheromone sehr bald und in großer Breite den Weg in die Praxis [159, 45]. Schwerpunkt der Bemühungen war und ist eine möglichst weitgehende Ausschaltung der zuwandernden Käfer zu Beginn der Vegetationsperiode – entweder mit Fallen um die Felder oder in ihrem Inneren [61, 34], oder durch Anreicherung der Käfer auf Fangflächen oder in Teilen der Felder, die davor oder danach mit Insektiziden behandelt werden [63]. Als alleinige Abwehrmaßnahme reichen Fallen nicht aus [160], doch sind sie ein wichtiges Glied bei der Entwicklung integrierter Bekämpfungssysteme. Unbefriedigend waren die Ergebnisse von Desorientierungsversuchen, vermutlich infolge des starken Anteils von Wirtspflanzengerüchen und Farbreaktionen an der Orientierung [161].

Auch bei der *Gelechiide Pectinophora gossypiella*, die ebenfalls in den Baumwollkapseln lebt, förderten Ausbreitungstendenzen und Quarantänemaßnahmen das Interesse am Pheromoneinsatz [159]. Dafür wurde zunächst das empirisch gefundene Mimeticum „Hexalure“ verwandt [27, 76, 77], ehe das weitaus wirksamere wahre Pheromon „Gossyplure“ verfügbar war. Dieses führte bei Verwendung von Fallen zu eindeutigen Befallsreduktionen [52, 162, 163]. Auch das Desorientierungsverfahren war auf großen Flächen erfolgreich [93]. Mit neuartigen Formulierungen (Hohlfasern, bzw. Mehrschichtenfolien in Flockenform) ließ sich der zunächst reichlich hohe Arbeitsaufwand erheblich vermindern [86, 164], ein entsprechendes Handelspräparat ist inzwischen erhältlich [165].

Diparopsis castanea, eine in Teilen Afrikas ebenfalls in Baumwollkapseln auftretende Noctuide, ist mit Pheromonfallen nur unzureichend zu überwachen [10] und wird erst durch relativ hohe Dosierungen bei der Partnersuche gestört [166]. Weitere Baumwollschädlinge treten auch in anderen Kulturen auf und werden deshalb im nächsten Abschnitt mit behandelt.

5.5. Sonstiger Acker- und Gartenbau

Die Gattung *Spodoptera* (unter Einschluß von *Prodenia* und *Laphygma*) enthält mehrere Arten, die als Blattfresser große Schäden in verschiedenen Kulturen anrichten. Ihre Pheromone sind inzwischen weitgehend bekannt und wurden an vielen Stellen getestet [167]. Obwohl es sich um polyphage Arten mit starker Tendenz zu lokalen und weiträumigen Wanderungen handelt, wurden beachtliche Erfolge erzielt, vor allem gegen die einander sehr nahestehenden, von Südeuropa bis Ostasien verbreiteten Arten *Sp. littoralis* und *litura* [32]. Außer der Flugüberwachung ist auch eine Bekämpfung mit Fallen möglich [68], Reusenfallen sind im Handel [31]. Für eine wirksame Desorientierung bestehen ebenfalls gute Aussichten [73, 168]. Gearbeitet wird auch an Fang und Konfusion von *Spodoptera frugiperda* und *exigua* [38, 169].

Zahlreiche Veröffentlichungen befassen sich mit *Trichoplusia ni* und weiteren, ähnlich reagierenden Plusiinen. Dabei liegt der Schwerpunkt bei Anlockungs- und Fangverfahren [40, 48, 54, 56, 57] und den Grundlagen der Desorientierung [92, 95]. Ihre Bewährung unter echten Praxisverhältnissen steht jedoch noch aus. Im Falle der wirtschaftlich sehr bedeutenden und mit Insektiziden nur schwer bekämpfbaren Heliothis-Arten (*H. zea*, *virescens* und *armigera*) wurden Fortschritte lange Zeit durch Schwierigkeiten bei der Aufklärung ihrer Pheromone behindert. Dementsprechend ist sowohl die Entwicklung von Anlockverfahren [16, 170] wie von Konfusionstechniken [169, 171] noch im Fluß. Dies gilt in verstärktem Maße für etliche weitere Noctuiden aus den Gattungen *Agrotis*, *Mamestra*, *Leucania*, *Sesamia* und deren Verwandtschaft, für die erst in jüngster Zeit Pheromone oder Mimetica gefunden wurden [172–175].

Für den wichtigsten Schädling im Erbsenanbau, den Erbsenwickler, *Cydia nigricana*, wurde ein Überwachungssystem auf Pheromonbasis entwickelt [41, 176]. Die Kartoffelmotte, *Phthorimae operculella*, läßt sich ebenfalls gut mit Fallen erfassen [177]. Aus anderen Insektenordnungen wäre allenfalls der Käfer *Costelytra zealandica* zu erwähnen, dessen Männchen von Phenol angezogen werden und damit die Basis für eine gezielte Bekämpfung auf Grünland liefern [178].

5.6. Vorratsschutz

Auf den ersten Blick bietet sich der Vorratsschutz geradezu an für den Einsatz von Pheromonen: Ihre Anwendung kann auf geschlossene Räume oder engbegrenzte Lokalitäten beschränkt werden; andere Verfahren sind in vielen Fällen nur schwierig durchführbar, zu aufwendig oder unerwünscht, und für mehrere wichtige Schädlingsgruppen sind Lockstoffe seit Jahren vorhanden. Trotzdem waren die Erfolge bisher bescheiden [179–181]. Fallengebrauch zur Überwachung ist möglich und nützlich – so für die *Phycitinae* (*Ephestia*, *Cadra*, *Plodia*) und für *Sitotroga cerealella* [182, 183]. Weitergehende Maßnahmen waren höchstens partiell erfolgreich [184, 185]. Beim Auftreten der Kartoffelmotte im Lager lassen sich ebenfalls Pheromonfallen verwenden [186]. Vorschläge zur Kombination von Lockstoffen mit anderen Agentien sind noch zu erproben [99, 100, 187].

5.7. Hygiene, Medizin und Veterinärmedizin

Auch von Insekten mit hygienischer Bedeutung, von Ektoparasiten des Menschen und seiner Haustiere wurden bereits Pheromone verschiedener Art gefunden, jedoch erst in wenigen Fällen zu ihrer Bekämpfung herangezogen. „Muscalure", ein Sexuallockstoff der Stubenfliege, dient zur Steigerung der Wirksamkeit von Insektiziden [188, 189]. Bei Schaben und Zecken befinden sich vergleichbare Bemühungen noch im Anfangsstadium [98, 190], und der praktische Wert des „Monomorins" der Pharaoameise *Monomorium pharaonis*, wird nicht hoch eingeschätzt [191]. Veröffentlichungen über Anwendungserfolge mit den verfügbaren Pheromonen von Stechmücken, blutsaugenden Fliegen und Wanzen fehlen noch völlig.

6. Tendenzen, Probleme, Aussichten und Grenzen

Pheromone wurden bereits für eine große Zahl von teilweise sehr bedeutenden Schadinsekten ganz oder in den Grundzügen aufgeklärt und synthetisch zugänglich gemacht [192], und die Kenntnisse auf diesen Gebieten wachsen weiterhin schnell. Für einige Pheromone wichtiger Schädlinge gibt es schon technisch brauchbare Synthesen. Es hat sich aber in jüngster Zeit gezeigt, daß eine optimale

Anlockwirkung nicht immer durch nur eine chemisch wohldefinierte Verbindung erzielt werden kann, sondern daß ein komplizierterer, oft noch gar nicht ganz übersehbarer Pheromon-Komplex notwendig ist. Im Gegensatz zu den anfangs gehegten Hoffnungen, daß mit der weiteren Erforschung der Pheromone der Einsatz einfacher würde, ergeben sich jetzt zusätzliche Komplikationen (vgl. den vorangehenden Beitrag "Biology of Pheromones" by J.P. Vité and J. A. A. Renwick). Auf Probleme der Lichtsensibilität von Pheromonen sei nur hingewiesen.

Nur die Sexualpheromone von Lepidopteren-Weibchen und die Aggregationspheromone einiger Käferfamilien wurden bisher in größerem Umfang auf ihre Anwendungsmöglichkeiten hin untersucht. Es ist zu erwarten, daß auch Pheromone mit anderen Funktionen und von anderen Insektenordnungen für die Praxis nutzbar gemacht werden können, etwa Spurfolge- und Markierungs-Pheromone oder solche Stoffe, die das Beziehungsgefüge sozialer Insekten regulieren, sowie Pheromone von einigen Dipteren, Hymenopteren, Wanzen, Schildläusen und Termiten. Für andere Gruppen allerdings sind die Chancen hierfür nur gering (so bei Blattläusen, Psylliden, Zikaden, Tagfaltern, Heuschrecken, Thysanopteren, Milben), und auch innerhalb der für einen Pheromon-Einsatz prinzipiell geeigneten Kategorien ist von Art zu Art mit sehr unterschiedlichen Erfolgsaussichten zu rechnen. Die an einem Objekt gewonnenen Erkenntnisse lassen sich dabei selbst auf nahe Verwandte nur sehr bedingt übertragen.

Unter den verschiedenen Einsatzmöglichkeiten hat sich die Verwendung von Pheromonen für Überwachungs- und Prognosezwecke bereits einen festen Platz gesichert. Sie bedarf zwar vielfach noch einer genaueren Durcharbeitung, aber kaum noch wesentlicher Veränderungen. Die Eignung von Pheromonen zur Bekämpfung oder Regulation der Populationen von Schadinsekten konnte ebenfalls in zahlreichen Untersuchungen bewiesen werden, jedoch gelang der Übergang vom Versuchsstadium in die Praxis erst in wenigen Fällen. Immerhin brachten die letzten Jahre nach einer Phase weitgehender Stagnation beschleunigte Fortschritte. Im Falle der Anlockverfahren beruhen diese auf detaillierten Analysen des Verhaltens der zu bekämpfenden Arten und dessen Berücksichtigung bei der Ausgestaltung von spezifischen Fallen- und Fangpflanzentechniken [193]. In kleinräumigen und arbeitsintensiven Kulturen und bei hinreichend stark reagierenden Arten dürften damit durchaus sinnvolle und wirtschaftlich konkurrenzfähige Einsatzmöglichkeiten zu finden sein. Insgesamt jedoch hat sich das Interesse der meisten Stellen mehr und mehr auf Bemühungen um die Unterbrechung der chemischen Kommunikation verlagert; hier haben in jüngster Zeit neuartige, besser auf die Eigenschaften und die Wirkungsweise dieser flüchtigen Substanzen zugeschnittene Fomulierungen (Mikrokapseln, Hohlfasern, Mehrschichtenfolien) wesentliche Fortschritte gebracht. Sie führten sowohl zu einer besseren Ausnutzung und längeren, gleichmäßigeren Freisetzung (und dadurch geringeren Aufwandmenge) der Pheromone wie zu einer Vereinfachung der Ausbringung. Soweit vom Gewicht der pheromonalen Einflüsse auf das Verhalten der jeweiligen Art her überhaupt eine Basis für derartige Eingriffe gegeben ist, liegen damit die wichtigsten Voraussetzungen für die Einführung von Pheromonen in die Praxis vor. Erforderlich sind noch Bemühungen um die Steigerung der Wirkungsgrade und der Zuverlässigkeit solcher Verfahren, vertiefte

Kenntnisse über ihre Abhängigkeit von abiotischen und biotischen Faktoren (Witterung, Entwicklungszyklus, Mikroklima, Populationsdichte) [193] und vor allem die reibungslose Eingliederung in den Gesamtkomplex der Bewirtschaftungsmaßnahmen mit einem Minimum an Material- und Arbeitskosten. Die Forschungs- und Entwicklungsarbeiten auf den hier skizzierten Gebieten erfordern einen erheblichen personellen und materiellen Aufwand und die Kooperation von Fachleuten recht unterschiedlicher Arbeitsrichtungen innerhalb einer entsprechend gut ausgebauten Institution.

Wichtig ist auch die Organisation und Koordination solcher Maßnahmen, die im Rahmen eines einzelnen Betriebes keine befriedigende Wirksamkeit erwarten lassen. Während die Verwendung von Pheromonen zu Prognosezwecken nach der Ausarbeitung der erforderlichen Grundlagen im Bedarfsfalle durchaus in die Hand des einzelnen Land- oder Forstwirts gelegt werden kann, dürfte dies beim Einsatz zu Bekämpfungszwecken nur in Ausnahmefällen möglich sein. In der Regel werden hierfür alle Anbauer unter der Anleitung durch einen leistungsfähigen Beratungsdienst zusammenwirken müssen.

In bestimmten Fällen könnte eine längerfristige prophylaktische Pheromon-Anwendung sinnvoll sein, insbesondere in großflächigen, extensiven Kulturen mit hoher Schadensschwelle (Wälder, Grünland). Unbefriedigend ist noch der Stand unserer Kenntnisse über den Gebrauch von Pheromonen in Vorratslagern, Ställen, Gewächshäusern und sonstigen Gebäuden, wie überhaupt im engeren Umkreis des Menschen. Notwendig und erfolgversprechend sind weiterhin Arbeiten über ihre Kombination mit anderen chemischen, physikalischen und biologischen Reizen (Fraßlockstoffe, Licht und Farben, Deckungsbedürfnis) und mit anderen Bekämpfungstechniken (Insektizide, Sterilisantien, Pathogene). Eine stürmische Entwicklung ist in Anbetracht der sehr komplexen Voraussetzungen und Abhängigkeiten auf chemischem, biologischem, meterologischem, technischem und wirtschaftlichem Gebiet auch in Zukunft kaum zu erwarten; solide Fundamente für einen stetigen Fortschritt dürften aber nunmehr, 20 Jahre nach der Aufklärung des ersten Insekten-Pheromons, gegeben sein.

7. Literatur

1. Holbrook, R.F., Beroza, M., Burgess, E.D.: J. Econ. Entomol. *53*, 751 (1960)
2. Milaire, H.-G.: Defense des vegetaux *30*, 177, 6 (1976)
3. Minks, A.K.: Biological aspects of the use of pheromones in integrated control with particular reference to the summer fruit tortrix moth, Adoxophyes orana. OILB/SROB 5e Symposium Bolzano Lutte Integree en vergers (1975)
4. Howell, J.F.: Envir. Entomol. *3*, 803 (1974)
5. Mani, E., Wildbolz, T.: Schweiz. Z. Obst- u. Weinbau *111*, 351 (1975)
6. Neuffer, G.: Anz. Schädlingsk. Pflanzen-Umweltschutz *47*, 116 (1974)
7. Oloumi-Sadeghi, H., Showers, W.B., Reed, G.L.: J. Econ. Entomol. *38*, 663 (1975)
8. Mani, E., Wildbolz, T.: Z. angew. Entomol. *83*, 161 (1977)
9. Myburgh, A.C. et al.: Proc. 1st Congr. Entom. Soc. Southern Africa, p. 99 (1974)
10. Marks, R.J.: Bull. entomol. Res. *67*, 575 (1977)
11. Beevor, P.S. et al.: Bull. entomol. Res. *67*, 439 (1977)
12. Klun, J.A., Robinson, J.F.: Ann. entom. Soc. Amer. *65*, 1337 (1972)
13. Mc Kibben, G.H. et al.: J. Econ. Entomol. *65*, 1512 (1972)
14. Mc Kibben, G.H., Johnson, W.L.: Bull. Ent. Soc. Amer. *22*, 323 (1976)

15. Beroza, M. et al.: J. Econ. Entomol. *64*, 1499 (1971)
16. Hendricks, D.E., Hartstack, A.W., Shaver, T.N.: J. Chem. Ecol. *3*, 497 (1977)
17. Flint, H.M. et al.: Envir. Entomol. *7*, 57 (1978)
18. Glass, H.E. et al.: J. Econ. Entomol. *63*, 370 (1970)
19. Kuhr, R.J., Comeau, A., Roelofs, W.L.: Envir. Entomol. *1*, 625 (1972)
20. Plimmer, J.R. et al.: Proc. Controlled Release Pesticide Symp. Akron, Ohio, 3.29–3.39, 1976
21. Fitzgerald, T.D. et al.: Envir. Entomol. *2*, 607 (1973)
22. Beroza, M. et al.: Envir. Entomol. *3*, 926 (1974)
23. Hardee, D.D., Mc Kibben, G.H., Huddleston, P.M.: J. Econ. Entomol. *68*, 477 (1975)
24. Kohorn, H. v., Kydonieus, A.F.: DBP 2356155 v. 9. 11. 1973
25. Mc Leod, D.G.R., Starratt, A.N.: Canad. Entomol. *110*, 51 (1978)
26. Goto, G., Matsuoka, Y., Hirago, K.: Chem. Lett. *1974*, 1975
27. Neumark, S., Teich, I.: J. Econ. Entomol. *66*, 298 (1973)
28. Boness, M.: Pflanzensch. Nachr. Bayer *28*, 155 (1975)
29. Mitchell, E.R. et al.: Envir. Entomol. *1*, 365 (1972)
30. Granett, J.: J. Econ. Entomol. *66*, 359 (1973)
31. Anon: Japan Pestic. Inform. No. 31, 21 (1977)
32. Kamano, S., Wakamura, S., Oyama, M.: Proc. Symp. Insect Pheromones and their Applications, Nagaoka and Tokyo, p. 135, Dec. 8–11, 1976
33. Boness, M., Schulze, W., Skatulla, U.: Anz. Schädlingskde. Pflanzen-Umweltschutz *47*, 119 (1974)
34. Hardee, D.D. et al.: J. Econ. Entomol. *68*, 502 (1975)
35. Boness, M.: Z. angew. Entomol. *82*, 104 (1976)
36. Charmillot, P.J. et al.: Schweiz. Landwirtsch. Forsch. *14*, 57 (1975)
37. Shuhravý, V., Hochmuth, R.: Anz. Schädlingskde., Pflanzen-Umweltschutz *49*, 55 (1976)
38. Tingle, F.C., Mitchell, E.R.: J. Econ. Entomol. *68*, 613 (1975)
39. Mastro, V.C., Richerson, J.V., Cameron, E.A.: Environ. Entomol. *6*, 128 (1977)
40. Kilinen, R.G., Ost, R.W.: J. Econ. Entomol. *64*, 310 (1971)
41. Lewis, T. et al.: Ann. Appl. Biol. *80*, 257 (1975)
42. Murlis, J., Bettany, B.W.: Nature *268*, 433 (1977)
43. Cross, W.H.: Ann. Rev. Entomol. *18*, 17 (1973)
44. Leggett, J.E., Cross, W.H.: Envir. Entomol. *7*, 4 (1978)
45. Cardé, R.T. et al.: Envir. Entomol. *6*, 765 (1977)
46. Kerck, K.: Naturwissenschaften *59*, 423 (1972)
47. Marks, R.J.: Bull. ent. Res. *67*, 243 (1977)
48. Sower, L.L., Kaae, R.S., Shorey, H.H.: Ann. Entomol. Soc. Amer. *66*, 1121 (1973)
49. Roelofs, W.L. et al.: Envir. Entomol. *1*, 606 (1972)
50. Alford, D.V.: Bull. ent. Res. *68*, 97 (1978)
51. Chambon, J.P., d'Aguilar, J.: Ann. Zool.-Ecol. Animale *6*, 423 (1974)
52. Boness, M.: Anz. Schädlingskde., Pflanzen-Umweltschutz *51*, 161 (1978)
53. Coudriet, D.L., Henneberry, T.J.: J. Econ. Entomol. *69*, 603 (1976)
54. Hendricks, D.E., Hartstack, A.W., Raulston, J.R.: Envir. Entomol. *6*, 556 (1977)
55. Marks, R.J.: Bull. ent. Res. *66*, 267 (1976)
56. Creighton, C.S., McFadden, T.L., Cuthbert, E.R.: J. Econ. Entomol. *66*, 114 (1973)
57. Debolt, J.W., Jay, D.L., Ost, R.W.: J. Econ. Entomol. *68*, 186 (1975)
58. Knipling, E.F., McGuire, J.U., jr.: U.S.D.A. Agric. Inform. Bull. No. 308, p. 20 (1966)
59. Nakamura, K., Oyama, M.: Appl. Entomol. Zool. *13*, 176 (1978)
60. Beroza, M., Knipling, E.F.: Science *177*, 19 (1972)
61. Hardee, D.D., Lindig, O.H., Davich, T.B.: J. Econ. Entomol. *64*, 928 (1971)
62. McLean, J.A., Borden, J.H.: Canad. Entom. *109*, 675 (1977)
63. Scott, W.P. et al.: J. Econ. Entomol. *67*, 281 (1974)
64. Lloyd, E.P. et al.: J. Econ. Entomol. *65*, 1144 (1972)
65. Beroza, M. et al.: Envir. Entomol. *2*, 1051 (1973)
66. Charmillot, P.-J.: Mitt. Schweiz. Entomol. Ges. *50*, 37 (1977)
67. Flint, H.M. et al.: J. Econ. Entomol. *67*, 738 (1974)
68. Hirano, Ch.: Jap. J. appl. Entom. Zool. *21*, 216 (1977)
69. Cameron, E.A.: Bull. Ent. Soc. Amer. *19*, 15 (1973)
70. Beroza, M.: Agric. Chem. *15*, 37 (1960)

71. Roelofs, W.: Environ. Lett. *8*, 41 (1975)
72. Richerson, J.V.: J. Chem. Ecol. *3*, 291 (1977)
73. Oyama, M., Wakamura, S.: Appl. Entomol. Zool. *12*, 372 (1977)
74. Boness, M.: Rep. Inform. 8. Internat. Plant Prot. Congr. Moscow, Section V, 41–47 (1975)
75. Roelofs, W.: Proc. Symp. Insect Pheromones and Applications, Nagaoka and Tokyo, p. 123–134, Dec. 8–11, 1976
76. Shorey, H.H., Kaae, R.S., Gaston, L.K.: J. Econ. Entomol. *67*, 347 (1974)
77. Kaae, R.S. et al.: Envir. Entomol. *3*, 87 (1974)
78. Daterman, G.E., Daves, G.D. jr., Smith, R.G.: Envir. Entomol. *4*, 944 (1975)
79. Shorey, H.H.: Proc. Symp. Insect Pheromones and Applications, Nagaoka and Tokyo, p. 97–108, Dec. 8–11, 1976
80. Minks, A.K., Voerman, S., Klun, J.A.: Entom. Exp. Appl. *20*, 163 (1976)
81. Hathaway, D.O. et al.: Envir. Entomol. *8*, 318 (1979)
82. Furniss, M.M. et al.: Canad. Entom. *109*, 1063 (1977)
83. St. Clair, A., Goulding, R.L., Rudinsky, J.A.: Z. angew. Entomol. *83*, 297 (1977)
84. Richerson, J.V., Payne, T.L.: Envir. Entomol. *8*, 360 (1979)
85. Browne, L.E. et al.: J. Chem. Ecol. *5*, 397 (1979)
86. Gaston, L.K. et al.: Science *196*, 904 (1977)
87. Arn, H. et al.: Entom. exp. appl. *19*, 139 (1976)
88. Audemard, H., Beauvais, F., Descoins, C.: Rev. Zool. Agric. Pathol. Veget. *76*, 15 (1977)
89. Beroza, M. et al.: Envir. Entomol. *4*, 705 (1975)
90. Campion, D.G.: Pestic. Sci *1976/7*, 636
91. Campion, D.G., Lester, R., Nesbitt, B.F.: Pestic. Sci. *1978/9*, 434
92. Farkas, S.R., Shorey, H.H., Gaston, L.K.: Envir. Entomol. *3*, 876 (1974)
93. Shorey, H.H.: Summ. Proc. Beltwide Cotton Prod.-Mechaniz. Conf. 24, 1975
94. Hathaway, D.O.: USDA Sci. Educat. Admin. AAT-W-6 (1979)
95. Shorey, H.H. et al.: Envir. Entomol. *1*, 641 (1972)
96. Roehrich, R., Carles, J.P., Tresor, C.: Rev. Zool. Agric. Pathol. Veget. *76*, 25 (1977)
97. Plimmer, J.R., Caro, J.H., Freeman, H.P.: J. Econ. Entomol. *71*, 155 (1978)
98. Rust, M.K., Reierson, D.A.: J. Econ. Entomol. *70*, 34 (1977)
99. Burkholder, W.E., Boush, G.M.: EPPO Bull. *4*, 455 (1974)
100. Shapas, J., Burkholder, W.E., Boush, G.M.: J. Econ. Entomol. *70*, 469 (1977)
101. Prokopy, R.J., Reissig, W.H., Moericke, V.: Entomol. Exp. Appl. *20*, 170 (1976)
102. Katsoyannis, B.I., Boller, E.F.: Envir. Entomol. *5*, 151 (1976)
103. Bowers, W.S. et al.: Science *177*, 1121 (1972)
104. Blum, M.S.: Proc. Symp. Insect Pheromones and Applications, Nagaoka and Tokyo, p. 49–60, Dec. 8–11, 1976
105. Robinson, S.W., Cherrett, J.M.: Proc. Symp. Pheromones Def. Secret. Soc. Insects, Dijon, 91 (1974)
106. Ritter, F.J. et al.: Symp. Evaluation of Biolog. Activity, Wageningen, 16.–18. April 1975
107. Robinson, S.W., Cherrett, J.M.: Bull. entom. Res. *63*, 519 (1974)
108. Anonym: Chem. Engng. News *54*, H. 23, p. 14 (1976)
109. Hirai, K., Shorey, H.H., Gaston, L.K.: Science *202*, 644 (1978)
110. Beroza, M. et al.: Toxicol. Appl. Pharmacol. *31*, 421 (1975)
111. Knipling, E.F.: Envir. Health Persp. *14*, 145 (1976)
112. Henson, R.D.: Envir. Entomol. *6*, 821 (1977)
113. Vité, J.P.: Contrib. Boyce-Thompson Inst. *24*, 343 (1970)
114. Lewis, W.J. et al.: Behav. Biol. *16*, 267 (1976)
115. Gross, H.B., Jr. et al.: J. Chem. Ecol. *1*, 431 (1975)
116. Pitman, G.B., Vité, J.P.: J. Econ. Entomol. *64*, 402 (1971)
117. Kennedy, B.H.: Bull. Entom. Soc. Amer. *25*, 116 (1979)
118. Hrdý, I., Sedivý, J.: Acta Entomol. Bohemoslov. *76*, 59 (1979)
119. Prokopy, R.J., Webster, R.P.: J. Chem. Ecol. *4*, 481 (1977)
120. Vité, J.P.: Biolog. unserer Zeit *8*, 112 (1978)
121. Granett, J.: J. Econ. Entomol. *66*, 359 (1973)
122. Sower, L.L., Daterman, G.E.: Envir. Entomol. *6*, 889 (1977)
123. Sanders, C.J.: Canad. Entom. *110*, 43 (1978)

124. Sanders, C.J.: Envir. Entomol. *5*, 868 (1976)
125. Daterman, G.E.: USDA Forest Serv., Pacific Northwest Forest Range Exp. Sta. 180 (1974)
126. Wood, D.L.: Pheromones of Bark Beetles, in: Control of Insect Behavior by Natural Products (ed. D.L. Wood, R.M. Silverstein, M. Nakajima) p. 301. New York and London (1970)
127. Bakke, A.: OEEP/EPPO Bull. No. 9, 5 (1973)
128. Vité, J.P.: Z. angew. Entomol. *77*, 325 (1975)
129. Browne, L.E.: J. Chem. Ecol. *4*, 261 (1978)
130. Cuthbert, R.A., Peacock, J.W., Cannon, W.N., Jr.: J. Chem. Ecol. *3*, 527 (1977)
131. Pitman, G.B.: Envir. Entomol. *2*, 109 (1973)
132. O'Sullivan, D.A.: Chem. Engng. News. *57*, No. 31, 10 (1979)
133. McLean, J.A., Borden, J.H.: J. Econ. Entomol. *72*, 165 (1979)
134. Vité, J.P., Hughes, P.R., Renwick, J.A.A.: Naturwissenschaften *63*, 44 (1976)
135. Bühler, H., Günther, G.: Allgem. Forstzeitschr. H. 5, p. 108 (1979)
136. Copony, J.A., Morris, C.L.: J. Econ. Entomol. *65*, 754 (1972)
137. Rudinsky, J.A. et al.: Z. angew. Entomol. *75*, 254 (1974)
138. Hedden, R.L., Pitman, G.B.: J. Econ. Entomol. *71*, 633 (1978)
139. Riedl, H., Croft, B.A., Howitt, A.J.: Canad. Entomol. *108*, 449 (1976)
140. MacLellan, C.R.: Canad. Entomol. *108*, 1037 (1976)
141. Madsen, H.F., Vakenti, J.M., Peters, F.E.: J. Econ. Entomol. *69*, 597 (1976)
142. Proverbs, M.D., Logan, D.M., Newton, J.R.: Canad. Entomol. *107*, 1265 (1975)
143. Charmillot, P.J., Baggiolini, M.: Schweiz. Landw. Forsch. *14*, 57 (1975)
144. Cardé, R.T., Baker, T.C., Castrovillo, P.J.: Entom. Exp. Appl. *22*, 280 (1977)
145. Trammel, K.: Use and field application of pheromones in orchard pest management programs, in: Environmental Quality and Safety, (ed. Coulston, F., Korte, F.), Vol. 5, p. 69, Stuttgart (1973)
146. Alford, D.V. et al.: Ann. Appl. Biol. *91*, 165 (1979)
147. Cardé, R.T., Trammel, K., Roelofs, W.L.: Envir. Entomol. *4*, 448 (1975)
148. Tamaki, Y., Ishiwatari, T., Osakabe, M.: Jap. J. appl. Entom. Zool. *19*, 187 (1975)
149. Mani, E. et al.: Mitt. Schweiz. Entomol. Ges. *51*, 307 (1978)
150. Rice, R.E., Jones, R.A.: J. Econ. Entomol. *68*, 358 (1975)
151. McLaughlin, J.R. et al.: J. Chem. Ecol. *2*, 73 (1976)
152. Anonym: Chem. Engng. News *57*, H. 41, 8 (1979)
153. Taschenberg, E.F., Roelofs, W.L.: Envir. Entomol. *7*, 103 (1978)
154. Taschenberg, E.F., Roelofs, W.L.: Envir. Entomol. *5*, 688 (1976)
155. Rauscher, S., Arn, H.: Entom. Exp. Appl. *25*, 16 (1979)
156. Sternlicht, M. et al.: Phytoparasitica *6*, 101 (1978)
157. Renon, M. et al.: C. R. Hebd. Seances Acad. Sci. Ser. D *288*, 1559 (1979)
158. Showers, W.B. et al.: Envir. Entomol. *5*, 1099 (1976)
159. Knipling, E.F.: Ginners J. and Yearbook, p. 23 (1971)
160. Taft, H.M., Hopkins, A.R.: J. Econ. Entomol. *71*, 598 (1978)
161. Huddleston, P.M., Mitchell, E.B., Wilson, N.M.: J. Econ. Entomol. *70*, 83 (1977)
162. Flint, H.M. et al.: J. Econ. Entomol. *69*, 535 (1976)
163. Huber, R.T., Moore, L., Hoffmann, M.P.: J. Econ. Entomol. *72*, 222 (1979)
164. Flint, H.M.: Briefl. Mitteilung 1979
165. Anonym: Chem. Week *122*, H. 24, 33 (1978)
166. Marks, R.J.: Bull. Entom. Res. *66*, 279 (1976)
167. Campion, D.G.: Meded. Fac. Landbauww. Gent *40*, 283 (1975)
168. Campion, D.G. et al.: Bull. Entom. Res. *66*, 335 (1976)
169. Mitchell, E.R., Baumhover, A.H., Jacobson, M.: Envir. Entomol. *5*, 484 (1976)
170. Mitchell, E.R., Tumlinson, J.H., Baumhover, A.H.: J. Chem. Ecol. *4*, 709 (1978)
171. Hendricks, D.E.: Envir. Entomol. *5*, 978 (1976)
172. Steck, W. et al.: Canad. Entom. *111*, 91 (1979)
173. Arsura, E. et al.: Experientia *33*, 1423 (1977)
174. Struble, D.L., Swailes, G.E.: Canad. Entom. *110*, 769 (1978)
175. Takahashi, S. et al.: Jap. J. Appl. Entom. Zool. *23*, 78 (1979)
176. Macaulay, E.D.M.: Plant Path. *26*, 179 (1977)
177. Thal, J.: Gesunde Pflanze *29*, 277 (1977)
178. Henzell, R.F., Lauren, D.R.: N. Z. J. Agric. Res. *20*, 75 (1977)

179. Burkholder, W.E.: Proc. Symp. Insect Pheromones and Applications, Nagaoka and Tokyo, p. 111, Dec. 8.–11. 1976
180. Haines, C.P.: Pestic. Sci. *7*, 647 (1976)
181. Weidner, H.: Z. Pflanzenkrankh. Pflanzenschutz *85*, 617 (1978)
182. Chow, Y.S., Yang, L.C., Lee, M.Y.: Plant Protect. Bull. *17*, 418 (1975)
183. Vick, K.W. et al.: J. Econ. Entomol. *72*, 245 (1979)
184. Reichmuth, Ch. et al.: Z. angew. Entomol. *85*, 205 (1978)
185. Brady, U.E. et al.: Envir. Entomol. *4*, 441 (1975)
186. Thal, J.: Anz. Schädlingskde., Pflanzenschutz, Umweltschutz *52*, 44 (1979)
187. Levinson, H.Z., Levinson, A.R.: Z. angew. Entomol. *84*, 337 (1977)
188. Carlson, D.A., Beroza, M.: Envir. Entomol. *2*, 555 (1973)
189. Morgan, P.B., Gilbert, J.H., Fye, R.L.: Fla. Entomol. *57*, 136 (1974)
190. Rechav, Y., Whitehead, G.B.: J. Econ. Entomol. *71*, 149 (1978)
191. Edwards, J.P., Pinniger, D.B.: Ann. Appl. Biol. *89*, 395 (1978)
192. Bestmann, H.J.: Chemie der Pheromone, in: Chemie der Pflanzenschutz- und Schädlingsbekämpfungsmittel, Bd. 6 (Herausg. Wegler, R.), Springer-Verlag, Heidelberg 1980
193. Renwick, J.A.A., Vité, J.P.: Biology of Pheromones, in: Chemie der Pflanzenschutz- und Schädlingsbekämpfungsmittel, Bd. 6 (Herausg. Wegler, R.), Springer-Verlag, Heidelberg 1980

The Use of Juvenoids in Insect Pest Management

J. P. Edwards

Ministry of Agriculture, Fisheries and Food, Slough Laboratory, Slough, Berks., U.K.

J. J. Menn

Zoecon Corporation, Palo Alto, CA 94304, USA

Contents

1. Introduction

The realisation that insect moulting and metamorphosis were controlled by hormones circulating in the haemolymph led to the suggestion that such chemicals might be used as "third generation pesticides" [256]. Subsequent research identified a large number of physiological and behavioural processes controlled by the moulting hormones (ecdysteroids) and the juvenile hormones (JHs). The determination of the chemical structure of naturally occurring JH in the moth

Table 1. Juvenile hormones and representative juvenoids[a]

Code	Structure	Other designation	Ref.
1	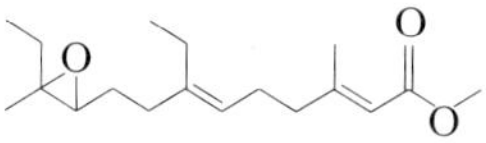	JH-I	[189, 190]
2	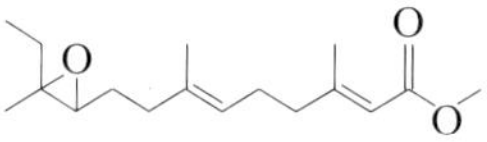	JH-II	[142]
3	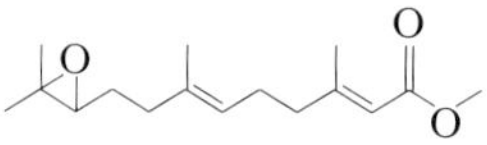	JH-III	[26, 104]
4	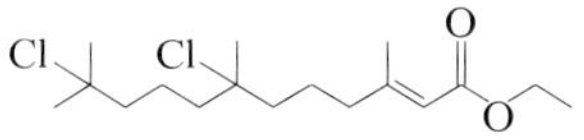	Mixture of hydro-chlorinated farnesoates	[117]
5	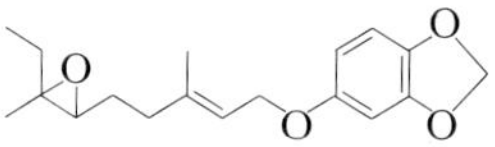	Hoffmann-La Roche RO 20-3600 Bowers 2B	[24, 25]
6	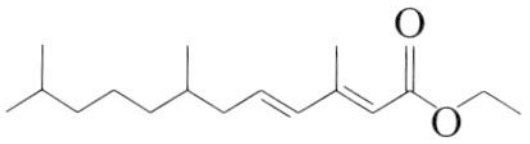	Hydroprene Altozar® IGR Zoecon ZR-512	[88, 89]
7	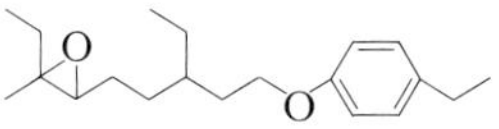	Epofenonane Hoffmann-La Roche RO 10-3108	[85]
8	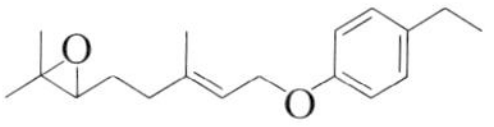	Stauffer R-20458	[170, 171]
9	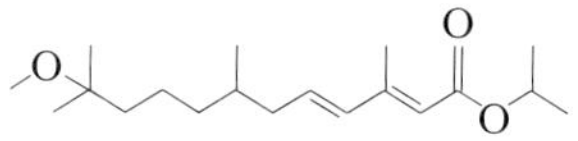	Methoprene Altosid® IGR Zoecon ZR-515 Kabat® IGR Manta® IGR Apex™ IGR	[88, 89]
10	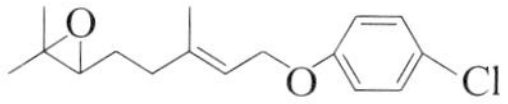	Epoxy geranyl chlorophenyl ether	[107, 113]
11	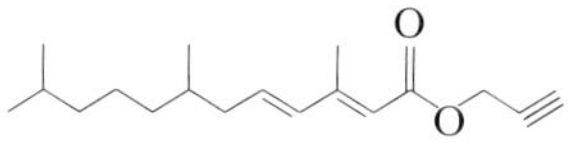	Kinoprene Zoecon ZR-777 Enstar® IGR	[88, 156, 226]
12	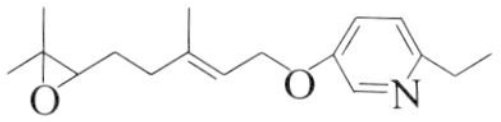	AI 3-70644	[111, 219]
13	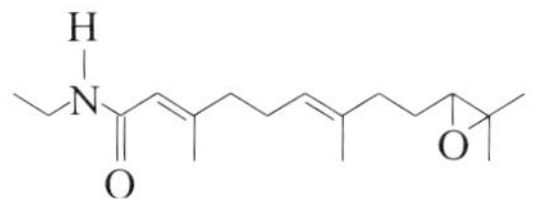	Terpenoid amide	[48]

[a] The juvenoids are listed in the order of discussion in the chapter

Table 1 (continued)

Code	Structure	Other designation	Ref.
14	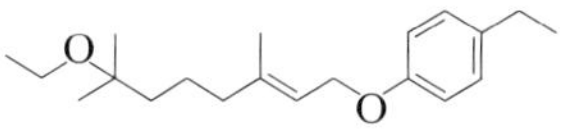	JH-25	[88, 195]
15	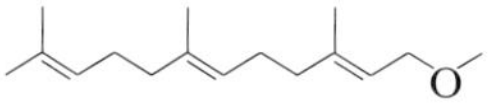	Farnesyl methyl ether	[199]

Hyalophora cecropia – JH-I (1, Table 1)[1] [190] and the subsequent identification of two other naturally occurring juvenile hormones [104, 142] – JH II (2) and III (3) – gave impetus to the systematic synthesis of chemical analogues with potential usage as pesticides. For convenience, synthetic chemicals with juvenile hormone activity are collectively termed "juvenoids" [81] to distinguish them from other chemicals with insect growth or development regulating properties. This chapter gives a brief account of some experimental and practical attempts to apply juvenoids in pest management. In addition, we have attempted to assess the future use of juvenoids and examined the philosophy of pest management strategies employing such chemicals.

2. The Role of Endogenous Juvenile Hormones

In all insects so far examined, the JHs are the humoral factors responsible for the maintenance of the larval state, i.e., the regulation of metamorphosis. The hormones are produced by paired, retrocerebral glands, the corpora allata, which have direct nervous connection to the brain.

The first systematic study of the morphology of the corpora allata was made by Nalbert [153] who proposed that their function was endocrine. Wigglesworth [255] in his classic parabiosis experiments, recognised the metamorphosis-preventing nature of the secretions from the corpora allata of the Hemipteran bug *Rhodnius prolixus* and demonstrated that removal of the corpora allata from nymphs caused premature moulting to the adult stage. Later work showed that the implantation of activity secreting corpora allata into final-instar nymphs prevented metamorphosis and led to the production of supernumerary giant nymphs [253]. The effect of the corpora allata hormone (JH) on the maturation of eggs was also first demonstrated by Wigglesworth [255] who recorded the absolute dependence upon the hormone for the ripening of eggs in the ovaries of *R. prolixus*. Wigglesworth [254] also demonstrated that the hormone was necessary for the functioning of the male accessory glands in *R. prolixus* and Novak [161] demonstrated that the hormone responsible for the retention of larval characteristics – the JH – was the same as the factor responsible for egg maturation and accessory gland development.

1 All compounds mentioned in the text are listed in Table 1 with a sequential code

Despite the intensive research in recent years on the physiological and biochemical effects of juvenile hormones and their chemical analogues, the precise mechanism by which these hormones regulate metamorphosis and reproduction remains largely obscure. It has become clear that the physiological effects of allatectomy or hormone administration in one insect order or genus, are not necessarily manifest in closely related insects. Even today, our knowledge of the titre and physiological action of endogenous JH is limited. For example, it is not clear if all three naturally occurring JHs are found in all insects or whether additional JHs exist in insects not yet studied. Certainly, in the Lepidoptera, the evidence points to the possibility that JH I (1) and/or JH II (2) are the hormones responsible for regulating metamorphosis, whereas JH III (3) may play a purely gonadotropic role. Our knowledge in this area is limited by the fact that most identifications of naturally occurring juvenile hormones have been only in adult insects. Notwithstanding these difficulties, it is possible to outline the probable action of the juvenile hormones in insect development and reproduction. In general, we may assume that JH is present in relatively high titre at certain stages in the haemolymph of larval (or nymphal) insects. During the majority of the larval stage, the presence of JH serves to maintain the insect in the larval stage, i.e., to prevent metamorphosis. At some point during the final larval instar, the haemolymph titre of JH drops and this signals the organism to proceed with pupal, and subsequently, with adult development. However, it is possible that, because of the fundamental differences in development between holometabolous and hemimetabolous insects, the precise way in which JH regulates development in these major groups of insects is different. At some point during the late pupal or early adult stage, the corpora allata resume secretion of JH. In these instances, the presence of JH in the adult acts as a mechanism for inducing ovarian development and/or vitellinogenesis in females and spermatogenesis and accessory gland development in males. The gonadotropic action of juvenile hormone is more apparent in those insects which produce cyclic batches of eggs, e.g., cockroaches, earwigs, and locusts where egg production is apparently linked with cyclical changes in the activity of the corpora allata [35, 248]. Apart from various general effects of JH on intermediary metabolism, it seems that the hormone also plays a role in diapause, pigmentation and color polymorphism, phase and caste determination, pheromone production and some aspects of behaviour. Clearly then, the juvenile hormones influence a wide range of important physiological processes in both the developing and the mature insect.

3. Effects of Exogenous Application of Juvenoids

Insect juvenile hormones and synthetic analogues have several effects when applied to insects at appropriate times during their development. Such applications are usually made topically, by injection, or by incorporation into the food medium. In most cases, application of juvenoids results in the retention of some juvenile characteristics manifest as developmental retardation or disruption or inhibition of the formation or function of the reproductive system.

3.1. Morphogenetic Effects

Because the juvenile hormones are intimately involved in the programming and regulation of post-embryonic development, it is not surprising that artificial application of hormones typically disrupts development. Application of juvenoids to larval insects during the final instar gives effects which may range from the formation of perfect supernumary larvae (which often attain twice the size of normal final-instar larvae) through insects which have both larval and pupal or pupal and adult features, to almost perfect adults which have some minor deformity. Giant larvae are usually only produced by the application of relatively high doses of juvenoids or in instances where the application is made at a specific critical time or when exposure is continuous (i.e., during exposure to treated food). Lower doses tend to allow successful pupation with subsequent disruption of adult development leading to the formation of pupal-adult intermediates or adults with varying degrees of deformity. Generally speaking, insects are sensitive to these morphogenetic effects of juvenoids only during the final larval and early pupal stages. However, some Hemipteran groups appear to be sensitive in the penultimate larval instar.

3.2. Ovicidal Effects and Disruption of Embryogenesis

Application of juvenoids to the eggs of some species can result in an ovicidal effect due to the disruption of embryogenesis. However, these effects can be confused with the direct toxic or suffocating action of the application solvent or of the oily juvenoid itself [47]. True disruption of embryogenesis can be recognised by the fact that it occurs at a distinct point during embryogenesis (at or before blastokinesis). From a practical point of view it is perhaps unfortunate that, in those species which are susceptible to the ovicidal action of juvenoids, the effect is often only apparent when eggs are treated at a very early stage in embryonic development, i.e., within the gravid female or soon after oviposition. Moreover, those chemicals which exhibit ovicidal activity against the eggs of certain species are often most effective by topical application or as a vapour. Application to the substrates are generally much less effective. In addition to direct ovicidal effects, Riddiford [186] has reported that treatment of the eggs of *Hyalophora cecropia* can lead to delayed morphogenetic effects in subsequent post-embryonic development. However, delayed effects resulting from treatment of eggs with juvenoids need further confirmation in other species.

3.3. Effects on the Reproductive System and Fecundity

In only relatively few species are adult insects affected by juvenoid treatment. Such effects are usually manifest as the disruption of the reproductive capacity of the female insect. The mechanism responsible is not fully understood and the matter is made especially puzzling since there is good evidence that JH is gonadotropic in several species [3, 115, 175, and others]. Reduced fertility in those female insects which are sensitive may result from a direct physiological action of excess juvenoid on the JH-regulated reproductive system or by direct ovicidal activity on the eggs inside the gravid female. Again, it seems that to achieve significant effects upon the reproductive system, it is often necessary to treat the insect soon after adult emergence (i.e., at the start of reproductive activity). In some species, treatment of the pupal insect or the pharate adult may also reduce reproductive capacity [140]. However, the interference with reproduction resulting from juvenoid application may sometimes arise from a decrease in mating ability and may be secondary rather than direct action on the reproductive system. Such effects have been implicated in the beetle *Tribolium castaneum* where the retention of vestigial pupal urogomphi in otherwise normal adults may impair or prevent copulation [54]. Similarly, the known ability of juvenoids to interfere with the production or secretion of sex pheromones [99] and with behavioural reponses to such pheromones might also reduce fecundity. Although JH has been implicated in the proper development and functioning of the male reproductive system [22, 70, and others] effects on male reproductive capacity are rare except in some Diptera where normal rotation of male genitalia is impaired [223]. In the social Hymenoptera, particularly ants, treatment with juvenoids may cause reversible or irreversible sterility in queens [59, 238]. In such case, the phenomenon appears to be dose dependent and may result from a saturation of the endocrine system normally responsible for the regulation of oogenesis, vitellogenesis or accessory gland function.

3.4. Effects on Diapause, Polymorphism and Behaviour

Several authors have reported the close relationship between endogenous JH titre and the onset or termination of diapause in insects [44, 205]. It appears that larval diapause is associated with a prolonged high titre of JH in the haemolymph and adult reproductive diapause with depressed JH titre. Thus, application of juvenoids to diapausing larvae does not usually result in any significant change in this state. However, juvenoid treatment of adult insects in reproductive diapause often leads to the break of this condition (e.g. [107]). From the practical viewpoint, juvenoids might be used to break adult diapause in a population at a time inappropriate for the development, feeding or reproduction of the ensuing generation. It is noteworthy that in one race of silkworms, application of juvenoids to adult females changed the voltinism of the larval offspring [150].

In recent years, it has become apparent that a large number of caste, phase and colour polymorphisms are directly regulated by the endocrine system. In those insects where this is the case, application of juvenoids may change the caste bias, for example, in Hymenoptera [172, 258] and Isoptera [95], or may change the phase of treated individuals, e.g. locusts [103]. In some species, application of relatively high levels of juvenoid results in marked colour changes, for example, treated nymphs of the cockroach *Blattella germanica* are unusually dark [51]. Although the colour change reported in this instance would be unlikely to be of any practical value, colour changes may be more practically significant in insects which depend upon precise colour matching with background as a defence mechanism or those in which colour plays an important role in sex recognition.

In summary, application of juvenoids is aimed at disruption of the development, reproduction, behaviour and general physiology of treated insects and this is the basis for considering their use as pest control agents. In the following section we will review some of the recent studies aimed at evaluating juvenoids as control agents for insect pest species.

4. The Use of Juvenoids Against Pest Species

4.1. Pests of Field Crops, Silviculture, and Horticulture

This category is the major target for crop-protecting chemicals. Therefore, it is not surprising that the first attempts to control insects by disrupting development with juvenoids were made in this area [138, 225]. The most important agricultural pest species are found in the orders Lepidoptera and Hemiptera with the balance largely made up with representatives of the Coleoptera, Diptera and various minor orders. The following is a brief documentation of attempts at practical application of juvenoids in the plant protection area.

Several authors have evaluated the potential of juvenoids as control agents for a variety of lepidopterous pests of field crops and forestry. Richmond [185] used Law's mixture (4) and Compound 5 in laboratory studies against the western spruce budworm *Chloristoneura fumiferana* and produced a variety of developmental effects with decreased egg production and hatch. Outram [169] reported that pupae of this species were sensitive to juvenoid treatment only up to two days old, and later [168] that treatment of this insect adversely affected the development of three hymenopteran parasites. On the other hand, Smilowitz et al. [217] demonstrated that hydroprene (6) was reasonably active against the cabbage looper *Trichloplusia ni* and did not affect the development of the hymenopteran parasite *Hyposoter exiguae*. Early experiments with *Tortrix viridana* [163] indicated the control potential of juvenoids against this species and, more recently, Schooneveld et al. [204] obtained complete control of *Adoxophyes orana*, the summer fruit tortrix, with RO-10-3108 (7) sprayed on foliage at 0.1% a.i. Moreover, this study demonstrated the good field stability of this chemical which remained biologically active on leaves for up to 3 months after treatment. It also showed no deleterious effects on two hymenopteran parasites *Colpoclypeus florus* and *Metiorus ictericus* although another parasite, *Teleutia striata* showed some possible adverse reaction to the treatment. Schmid et al. [200] obtained good results in practical trials with Compound 7 against populations of *Adoxophyes orana*, the rose tortrix *Archips rosanus* and the winter moth *Operophthera brumata* on fruit trees. In addition, they reported that control of populations of the aphid *Dysaphis* sp. could be obtained by an extra spray treatment carried out when the fruit was at the early green cluster stage. Several authors have reported rather poor activity of juvenoids against the tobacco budworm *Heliothis virescens* (e.g. [12]). Vinson [241] reported the adverse effects of juvenoid treatment on development of endoparasites and Guerra et al. [82] tested five juvenoids against this species in field cages and reported slightly increased larval development time and reduced fecundity in the F_2 generation. Furthermore, two of the juvenoids used in this study produced some phytotoxicity although none apparently affected the development of the endoparasite *Trichogramma* sp. El-Guindy and Bishara [66] observed total inhibition of mating in the American bollworm *Heliothis armigera* when pre-pupae were treated with 1 μg of the juvenoid R-20458 (8). Sehnal et al. [207] found methoprene (9) to be the most effective of 32 juvenoids tested against a number of lepidopterous larvae. Ovicidal action was recorded when freshly laid eggs of the Egyptian cotton leafworm *Spodoptera littoralis* and *Mamestra brassicae* were treated with 0.1 % methoprene (9). Inhibition of metamorphosis was obtained in *Autographa (Plusia) gama* at 0.05 μg methoprene/larva; in *Spodoptera littoralis* at 0.8 μg/larva; *Mamestra brassicae* at 6 μg/larva and *Scotia (Agrotis) ypsilon* at 50 μg/larva; thus demonstrating a considerable range of susceptibility to the juvenoid within members of the family Noctuidae. These authors also suggest that the juvenile hormone may be involved in the change from gregarious to solitary behaviour exhibited by some Noctuid larvae [207]. Abdallah et al. [1] reported juvenoid-induced reduction in fecundity in adult females of *Spodoptera littoralis* and delayed effects during subsequent larval development when eggs were treated. These latter effects were probably artefacts since they were not apparent when larvae were reared on another diet. In the same species, El-Guindy et al. [65] found resistance to Compound 8 in a carbamate-resistant strain and reported that pre-treatment of larvae with chlorpyrifos during the 4th instar increased the tolerance of 6th instar larvae to the juvenoid in both susceptible and resistant strains. The authors suggested that this increased tolerance may have resulted from the induction of detoxifying enzymes by the insecticide. In a field trial, Skuhravy [211] demonstrated the efficacy of 0.1% hydroprene (6) spray treatments in reducing field populations of the larch case bearer *Coleophora laricella*, and Jobin and Retnakaran [102] obtained good control of *Lambdina fiscellaria* (the hemlock looper) with methoprene (9) applied by air. Varjas and Sehnal [239] applied Compound 10 in field trials against the fall webworm *Hyphantria cunea* and achieved an 80% reduction in the number of pupae formed. However, they also reported the production of many giant larvae, but several parasites appeared unaffected. Against the Pyralid *Ostrinia nubilalis* (the European corn borer) Lewis et al. [125] claimed that four juvenoids tested under field conditions were ineffective and that liquid formulations gave some phytotoxicity. Cawich et al. [36] obtained poor results against the pink boll worm *Pectinophora gossypiella* with R-20458 (8), although more recently it has been suggested that this species and other members of the family Gelechiidae are much less sensitive to the morphogenetic effects of juvenoids than other lepitopterous species (Stockel and Edwards, in press). Additionally, Cawich et al. [36] report that diapausing larvae of *P. gossypiella* required higher doses of compound 8 than non-diapausing larvae to give the same morphogenetic response. Gelbic and Sehnal [78] reported that the codling moth *Cydia pomonella* was perhaps a suitable target for juvenoid-based control when they produced reduced fertility and inhibition of embryogenesis as well as morphogenetic effects. Recently, Street [229] obtained ovicidal effects on eggs of the cabbage white butterfly *Pieris brassicae*, particularly if juvenoid treatment occurred on the day of oviposition.

During studies with juvenoids against lepidopterous pests of field crops and forestry, several interesting "side effects" have been recorded. These include the induction of rectal prolapse (and subsequent death) in *Spodoptera litura* at doses lower than those required to produce giant larvae [158], cessation of spinning behaviour in *Pieris brassicae* treated with juvenile hormone [18] and the induction of abnormal development in the maxillae of fifth instar *P. brassicae* larvae which, as a result, were unable to feed and subsequently died of starvation. These atypical effects might provide the basis for practical control of some lepidopterous species. However, this would hinge on the development of more stable and cost/effective juvenoids.

Another important group of agricultural pests, the Hemiptera (Homoptera and Heteroptera) contains the true plant bugs as well as aphids and scale insects. Studies on the control potential of juvenoids against true bugs have been scarce. Both *Oncopeltus fasciatus* and *Pyrrhocoris apterus* have been extensively used for laboratory studies of juvenoid action although neither species is of great economic importance. Reissig and Kamm [183] prevented the emergence of nymphs and terminated reproductive diapause in female *Draeculacephala crassicornis* with foliar sprays of JH 1 and Compound 5 and Lefevre [119] reported that several juvenoids showed activity against the sugar beet lace bug *Piesma quadratum*. Promising results were obtained by Kontev et al. [108] with Compound 10 against *Eurygaster integriceps* where the chemical gave good ovicidal activity, inhibited metamorphosis and terminated adult diapause. A temporary suspension of obligatory adult reproductive diapause was reported in a field strain of *Eurygaster maura* when adults were topically treated with juvenoids [159]. Temporary inhibition of reproductive diapause in a parent generation might well provide the basis for control of F_1 populations which, as a result, would be "out of step" with fluctuating seasonal factors such as temperature or food plants.

The aphids are an important group of plant pests both in the field and in glasshouses. Their complex biology, combining alternation of generations with parthenogenetic viviparity – factors which may be regulated by the endocrine system – make them potential targets for hormone-based pest control. Kuhr and Cleere [114] examined the control potential of juvenoids against eight aphid species including the pea aphid *Acyrthosiphon pisum*, the raspberry aphid *Amphorophora agathonica* and the turnip aphid *Hyadaphis pseudobrassicae* with generally promising results. Tamaki [232] used R-20458 (8) and hydroprene (6) in spray treatments against *Myzus persicae* on infested peach twigs and reduced the production of winged migrants by up to 75%. He suggested that such treatments might be useful in protecting summer host crops although El-Gayar [64] has reported that two juvenoids did adversely affect *Aphidoletes aphidimyza*, an important Dipteran parasite of *M. persicae*. Nassar et al. [156] reported the effect of three juvenoids against the aphid *Schizaphis graminium* in laboratory studies where they obtained developmental effects and sterility. These authors demonstrated the effectiveness of kinoprene (11) which was the most active compound against this species and concluded that it would give effective control. In his review, Staal [225] points out that juvenoid-induced effects on reproduction in the Aphididae can result either from direct effects on the development of the embryo inside the parent or by indirect effects which cause incomplete metamorphosis of the genital pore. Methoprene (9) has been shown to affect the developmental stages of the apterous form of the potato aphid *Macrosiphum euphorbiae* and this chemical was not repellent to the insect [17]. Thanassoulopoulos [234] obtained control of *Aphis fabae* under greenhouse conditions with kinoprene (11) and Hrdy and Zeleny [97] conducted successful trials with juvenoids against *Phorodon humuli* in cage experiments. Gaudchau [77] obtained permanent sterility in *Acyrthosiphon pisum* and *Aphis craccivora* five days after insects were exposed to host plants sprayed with 0.1% kinoprene (11) although (unusually) no direct toxic effects were seen. Della-Giustina and Le Rumeur [52] reported the efficacy of using methoprene (9) in conjunction with the pyrethroid bioresmethrin against the white fly *Trialeurodes vaporariorum* on tomato and cucumber plants. Although most aphid species seem generally sensitive to juvenoids, it is noteworthy that Lema and Poe [122] reported that they had to use 3,000 ppm kinoprene (11) to significantly reduce adult emergence in the aphid *Liriomyza sativae* and that such large doses were also detrimental to the parasite *Opius dimidiatus*.

Several researchers have examined the effects of juvenoids on scales and mealy bugs. Staal et al. [226] showed kinoprene (11) gave high mortality and reduced viable progeny when sprayed on parthenogenetic females of the citrus mealy bug *Planococcus citri*. Vogel et al. [244] obtained a progressive reduction in the number of anal pores following juvenoid treatment of nymphs of the same species and construed this phenomenon as evidence of reversal of metamorphosis. In recent laboratory studies, Hamlen [83] demonstrated that kinoprene (11) and RO-10-3108 (7) were as effective as conventional insecticides in reducing greenhouse populations of the mealy bugs *Pseudococcus longispinus* and *Phenococcus solani*. Against scale insects, Boboye and Carman [23] produced inhibition of development and cessation of growth with several juvenoids in *Aonidiella aurantii*. In this species, the juvenoids were significantly more active against males than females and the authors concluded that the complete reduction of the male population would be an effective basis for control. Asano and Kamei [11] reported that methoprene (9) stimulated oviposition and decreased egg hatch in the wax scale *Ceroplastes pseudoceriferus* and Kozar and Varjas [109] obtained good control of the San Jose scale *Quadraspidiotus perniciosus* on apples and red currants using several juvenoids at 0.1% a.i. In these tests, males were considerably more sensitive than females to the effects of the juvenoids and no adverse effects were apparent in two hymenopteran parasites of this species. In fact, many of these recent experiments have indicated that juvenoids are perhaps ideally suited for use against greenhouse pests particularly as Klein-Koch [106] has reported ovicidal and morphogenetic effects of juvenoids in the red spider mite *Tetranychus urticae* and in the mite *Phytoseiulus persimilis*.

There is relatively little published information on the potential of juvenoids for the control of coleopteran pest species although this group contains a number of important agricultural pests. Walker [245] showed that field applications of Compound 10 reduced egg hatch on the Mexican bean beetle *Epilachna varivestis* by 90% although the compound showed some phytotoxicity. Sehnal and Skuhravy [206] concluded that the Colorado potato beetle, *Leptinotarsa decemlineata* was not a suitable target for hormone-based pest control even though this insect appeared to be very sensitive to juvenoid treatment. These authors observed morphogenetic and ovicidal effects together with a change in burrowing behaviour of treated larvae. The same authors also reported that juvenoid treatments of the adult Colorado potato beetle disrupted reproductive diapause although this effect is more easily obtained if the insects have been in diapause for several months. Moreover, termination of diapause is only temporary in this species if the insects remain in diapause-inducing (short-day) conditions [203]. Topical treatment of 0–24 h-old eggs of the white pine weevil *Pissodes strobi* with Compound 8 reduced hatch and gave some mortality in subsequent post-embryonic development [184]. Neal and Hower [157] reported that the alfalfa weevil *Hypera postica* was largely unaffected by treatments with Compound 5. Novak and Sehnal [164] and Novak et al. [162] tested juvenoids against the spruce bark beetle *Ips typographus* and the large pine weevil *Hylobius abietis*. Morphogenetic effects were produced in both species although the weevil was noticeably less sensitive. Treatment of adult *H. abietis* with several juvenoids resulted in an increase in the numbers of eggs laid together with a reduction in their fertility. These and other studies with members of the Curculionidae (weevils) indicate that this family is much less sensitive to the morphogenetic effects of juvenoids than other members of the order Coleoptera. Some of the possible reasons for the insensitivity of this group will be discussed later in this chapter.

Little has been published with reference to the bioactivity of juvenoids against insect pests of other orders. Fogal and Sullivan [72] tested JH 1 and Compound 8 as foliar sprays against the pine sawfly *Neodiprion sertifer* and reported some ovicidal effects which, they conclude, might be useful in protecting certain ornamentals and Christmas trees from sawfly defoliation. Fytizas and Mourikis [76] obtained inhibition of adult eclosion and ovicidal effects against the olive fly *Dacus oleae* when infested olives were dipped in aqueous solutions of emulsified methoprene (9) and Hamlen and Mead [84] reported control of fungus gnats (*Bradysia* spp.) with kinoprene (11) and methoprene (9) in greenhouse conditions although reductions in population were slower than with conventional insecticides.

Overall, it appears that juvenoids are less effective against pests of agriculture and forestry than they are against some other pest groups. There are several factors which contribute to the relative inefficiency of juvenoids used against pests of field crops. Firstly, in many agricultural pests it is the phytophagous larval stage feeding on plant material which causes damage. Because the larvae are only sensitive to the effects of juvenoids in the final larval instar, the activity of juvenile hormones is limited to the period after which most of the damage has been done. Indeed, in some instances the formation of giant supernumerary larvae resulting from juvenoid treatment merely increases the amount of crop damage [205]. Secondly, most juvenoids are unstable in ultra-violet light and, under field conditions, breakdown of the chemical is often too rapid to permit good control even when used against species which are highly sensitive. The insensitivity of the feeding larva to juvenoids is a fundamental problem which will be impossible to overcome even with the development of highly active and stable juvenoids. The problem of persistence under field conditions, however, may be solved by the rational synthesis of more stable analogues or by the employment of various protective formulations, e.g., microencapsulation. Another consideration pertaining to the use of juvenoids against pests of field crops is that, even in instances where effective control can be obtained, it may be necessary to treat very large areas to preclude reinfestation of the crop by nearby untreated populations. This factor, coupled with the relatively high cost of juvenoids by comparison with conventional "commodity" pesticides, may limit the use of juvenoids as crop protection chemicals on grounds of cost alone. At present, it appears that juvenoids may prove most effective in glasshouses where UV radiation levels are low compared to field conditions. Moreover, progress has already been made in the synthesis of UV-stable juvenoids with the discovery of epofenonane (7) [85]. However, no commercial exploitation of this chemical is as yet apparent. It thus appears, for the reasons mentioned above, that juvenoids will find, at best, only limited use against pests of field crops and forestry. However, should the use of juvenoids in these areas become more efficient, it seems that chemicals like epofenonane (7) are sufficiently selective not to pose a threat to either the development or population density of honey bees [80].

4.2. Pests of Medical and Veterinary Importance

The group of insect pests is dominated by members of the Diptera (flies). Of particular importance are a large number of biting flies which carry debilitating and sometimes fatal diseases of man and his domestic animals. As early as 1964, Lewallen [124] reported the effects of farnesol on mosquito larvae and these studies paved the way for future successful control measures employing juvenoids against this group of insects. Jakob and Schoof [101] tested four species of mosquitoes with twelve juvenoids and concluded that control using juvenoids would be practicable. Lewis and Christenson [126] reported that the juvenoid, Compound 5, gave promising control of *Culex pipiens* in both laboratory and small-scale field tests. Schaefer and Wilder [198] also tested Compound 5 and methoprene (9) against *Aedes nigromaculis* and obtained encouraging results with methoprene (9) applied at 0.125 1b/acre. In 1973, these researchers obtained practical control of *A. nigromaculis* and promising results against *A. melanimon* with aerial application of the slow-release (encapsulated) formulation of methoprene (9) at the rate of 0.025 1b/acre [197] although *Culex tarsalis* was apparently less affected. Schaefer and Dupras [196] demonstrated rapid breakdown of methoprene (9) in water, particularly in sunlight and at high temperature and Miura and Takahashi [147] showed that a range of aquatic micro-organisms including protozoans, annelids and molluscs were unaffected by methoprene (9) and concluded that encapsulated methoprene was safe to use on the field-pasture community. In addition, Quistad et al. found no toxicological or persistence problems with methoprene (9) in a model ecosystem [180]. However, more recently Breaud et al. [29] reported that six applications of methoprene over an eighteen-month period temporarily reduced the populations of 14 aquatic organisms although other species remained stable or increased over the same period. These authors, however, concluded that no species would be eliminated as a result of such treatment and that recovery of all affected populations would eventually take place. Mulla et al. [152] obtained good results with slow-release formulations of methoprene (9) against both susceptible and organophosphorus ester-resistant *Chironimus* spp. in the laboratory, small-plot tests, and in breeding fields, against flies of the genera *Chironimus*, *Procladius* and *Tanytarsus*. Mulla et al. [151] also evaluated several juvenoids (excluding methoprene) (9) against *Anopheles albimanus* and *Culex pipiens* without particularly encouraging results. Hoppe et al. [94] evaluated five juvenoids in small-scale field tests against *C. pipiens*. Of these, R-20458 (8) was the most active and Compound 5 was the most persistent. Hoppe [93] also claimed that granular formulations were more effective than emulsifiable concentrates. Pree and Stewart [177] examined the persistence of encapsulated methoprene (9) in both salt and fresh water and obtained a half-life value of between four and five weeks. Dorn et al. [55] obtained about 70% persistence of intact epofenonane (7) after 4 weeks incubation in polluted water, confirming the remarkable environmental stability of this compound. The discovery that methoprene had no adverse effects on the parasitic nematode *Romanomermis culicivorax* led Finney et al. [71] to suggest that integrated control using the parasite in combination with juvenoid treatments would be feasible especially against *Aedes aegypti*. Self et al. [208] conducted field trials against *Culex quinquefasciatus* where application of 1 ppm to larval habitats in Jakarta reduced the female adult population by about 80% for a period of 2 to 5 weeks after treatment. Against the related species *Culex tarsalis*, Case and Washino [34] reported rather poor results (10–15% reduction in adult emergence) when methoprene (9) was applied at 0.1 1b/acre in Californian rice fields. Naqvi et al. found [154] that methoprene (9) treatment of *A. aegypti* inhibited morphogenesis, embryogenesis and produced sterility. Good results were also obtained against the related *Anapheles stephensi* with 0.001 ppm methoprene (9) [155].

In the field of veterinary and medical entomology, another area in which juvenoids have been assessed as potential control agents is against manure-breeding flies and other Dipteran pests of livestock. Morgan and Labrecque [148] reported marginal sterilising effects of juvenoids in the house fly *Musca domestica* and, more recently, Morgan et al. [149] have demonstrated that 0.01% methoprene (9), incorporated into chicken feed, gave 99% larval mortality when house flies were reared on the resultant manure. Interestingly, direct application of the juvenoid to poultry manure was considerably less effective. Bhaskaran [19a] obtained various morphological abnormalities when compound 4 was applied to pupae of the flesh fly *Sarcophaga bullata* and Sivasubramanian [210] obtained ovicidal effects in the same species when females were treated with juvenoids, particularly if the treatment occurred before ovulation. Wright and Spates [260] showed two juvenoids, including Compound 5 to be effective in producing morphogenetic effects in pupae of the stable fly *Stomoxys calcitrans* and noted that such treatment did not affect the development or oviposition of the Hymenopteran parasite *Muscidifurax raptor*. Wright et al. [261] obtained good results with two juvenoids against the stable fly in small-plot field tests although Bowman et al. [27] had reported relatively rapid breakdown (70–90% in 22 days) of juvenoids in synthetic larval-rearing medium. More recently, Wright and Smalley [259] evaluated 4 juvenoids against the stable fly in field trials and obtained good control by area-application to larval-breeding sites. Methoprene (9) was the most effective chemical tested, requiring only 0.025% a.i. to achieve control. No adverse toxicological effects were found with any of the juvenoids in swine, hamsters or sheep. Harris et al. [87] fed 3 juvenoids to cattle to control the horn fly, *Haematobia irritans* and the stable fly in manure. These authors reported that methoprene (9) was the most active of the compounds tested giving good results at feed-through rates of 100 mg per animal per day against the stable fly and at only 0.7 mg per animal per day against the horn fly. Miller and Uebel [145] tested methoprene (9), R-20458 (8) and Compound 5 against larvae of the house fly and the face fly *Musca autumnalis*. Once again, methoprene (9) gave the best results and a feeding rate of 1 mg/kg/day gave complete control of the face fly and significant control of the house fly. Wasti and Mahadeo [247] reported the effects of R-20458 (8) on the house fly and the blowfly *Phormia regina* and Harris et al. [86] used methoprene-impregnated mineral blocks (diet supplements licked by cattle) to obtain 87% inhibition of development of horn flies and effective control of stable flies. Wright et al. [262] reported that in large-plot field tests, R-20458 (8) gave an overall 83% reduction in adult stable flies when sprayed on the manure in a cattle feed lot. However, these authors also suggested that the efficiency of the treatment was diminished as a result of crust formation on the manure pats preventing the penetration of the juvenoid. Metabolic studies on the degradation of methoprene (9) in cattle [178, 179] have shown that this juvenoid is metabolised to simple natural products, mostly acetate, which were incorporated into fatty acids, lactose and cholesterol. Chamberlain et al. [42] reported that some unchanged methoprene passed out with the faeces following oral administration to steers. Using a slightly different approach, Beadles et al. [15] used methoprene (9) admixed in drinking water (0.34 mg/l) to completely inhibit the formation of adult horn flies in manure and reduced the number of flies in the treated pasture from about 360/animal to about 7/animal. These authors concluded that this technique showed good control potential particularly in those areas where sources of drinking water could be strictly controlled. Ivie et al. [100] conducted an interesting series of experiments to examine the fate of R-20458 (8) following oral and dermal administration to cattle. After dermal application, about 40% of the intact juvenoid remained on the skin after seven days, indicating that dermal application might perhaps be effective for the control of skin or hair-infesting arthropods on livestock animals. Miller et al. [143, 144] used a sustained-release bolus containing methoprene (1% a.i.) to inhibit horn fly development in faeces for a period of 10 to 12 weeks and [144] obtained good control of horn fly with methoprene (9) administered to cattle through drinking water in a large-scale field trial on an Hawaiian island. This group of workers also conducted a successful field trial using the methoprene feed-through technique against the face fly *Musca autumnalis* [146].

Outside the areas of mosquito and manure-breeding fly control, little has been published on the practical use of juvenoids for the control of other veterinary and medical pests. Along similar lines to the work with cattle, Herald and Knapp [90] and Breeden et al. [30] evaluated juvenoids for the control of houseflies in poultry manure. The latter authors reported good control of houseflies when micro-encapsulated methoprene (5 ppm a.i.) was incorporated into chicken feed. Meleney and Roberts reported [137] the effects of juvenoids against the short-nosed cattle louse and the warble fly and Chamberlain and Hopkins [40] tested juvenoids

Table 2. Registered uses of methoprene[a]

Product name	Purpose	Where used	Remarks
ALTOSID® SR-10[b]	Mosquito larvicide	Worldwide	Micro encapsulated formulation
ALTOSID Briquet	Mosquito larvicide	U.S., Canada, Latin America	Sustained release product
ALTOSID 10F	Fly larvicide	Japan	Flowable formulation
ALTOSID CP-10	Manufacturing concentrate for Cattle Mineral Supplements	U.S.	Feed through horn fly larvicide
ALTOSID Bolus	Single Dose Horn Fly larvicide	U.S.	Sustained release in rumen. Investigational (Fall 1979)
Houseplant Mist	Houseplant insecticide/miticide spray	U.S., Canada	Combination with 2 other active ingredients
Pharorid®	Pharaoh Ant Control	U.K.	Bait concentrate formulation
KABAT®	Stored Tobacco Protection	Worldwide	Cigarette beetle control
MANTA®	Silkworms – commercial silk production	Japan	Increases silkworm productivity
APEX	Mushroom Fly control	U.S.	Investigational (Fall 1979)
ALTOSID 5E®	Stored Product Protection (nuts, grains, rodent baits, pet food, etc.)	U.S.	Investigational (Fall 1979)

[a] This information was provided by J. Young, Zoecon Corporation, Palo Alto, CA
[b] First registered use of methoprene – approved June 1973 in US

against the angora goat biting louse and the cattle biting louse [41]. Recently Mansingh and Rawlins [130] reported the anti-gonadotropic effect of juvenoids on the cattle tick *Boophilus microplus* and Solomon and Evans [220] found ovicidal effects when juvenoids were applied to adult female *B. microplus*, *B. decoloratus* and *Amblyomma hebraeum*. Prassert et al. [176] used nasal sprays of methoprene (9) emulsion on artificially infested lambs against the sheep bot fly *Oestrus ovis* and reported 84–100% inhibition of adult emergence with doses ranging from 0.1 to 1.0 mg a.i./lamb. These authors concluded that this could be a viable technique for achieving population control of this species. Denlinger [53] induced abortion of second and third instar larvae from tsetse flies *Glossina morsitans* by topical application of methoprene (9) and, although the aborted larvae failed to pupate, the females produced a viable larva within 10 days of the abortion. Finally, Pinchin et al. [173] found several juvenoids active against *Panstrongylus megistus*, a primary vector of Chaga's disease.

Thus, in recent years, there has been considerable effort aimed at the utilisation of juvenoids to control both water-breeding and manure-breeding flies. Such attempts have frequently met with considerable success, probably because in most cases the larval stages of these pests are innocuous and it is only the adult insect, sometimes even only one sex, which causes injury. Moreover, it was the result of efforts to control pests in these categories which led, in 1973, to the first commercial registration of the juvenoid, methoprene (9) for the control of floodwater mosquitoes in the USA. Indeed, methoprene which has good activity in a variety of insect groups, seems to be particularly active against dipterous insects. To date, of the many juvenoids reported and tested, only methoprene (9) has achieved commercial status. A summary of its world-wide registered uses is presented in Table 2.

4.3. Pests of Stored Products

This category almost exclusively comprises Coleopterous and Lepidopterous species and it was soon realized that pests of stored commodities, particularly grain and grain-products destined for human consumption, would be ideal targets for the use of the essentially non-hazardous and non-persistent juvenoids. Thomas and Bhatnagar-Thomas [235] were the first to report the effect of a juvenoid on 3 stored-products insects, the rust-red flour beetle *Tribolium castaneum*, the drugstore beetle *Stegobium paniceum*, and *Callosobruchus chinensis*. A variety of morphogenetic effects were observed at relatively low doses and, in a later series of experiments, Bhatnagar-Thomas [21] reported a reduction of oviposition of *C. chinensis*, *S. paniceum* and the lesser grain borer, *Rhizopertha dominica* resulting from hormone treatment. Socha and Sehnal [218] reported that juvenoids prevented imaginal differentiation in the yellow mealworm *Tenebrio molitor* at doses as low as 0.01 μg insect and that the maximum sensitivity occurred just after the larval-pupal ecdysis. Walker and Bowers [246] further reported high activity with Compound 5 and analogues in the same insect. Bransby-Williams [28] obtained rather poor results with juvenoids against the maize weevil *Sitophilus zeamais*, although tests against the moth *Ephestia cautella* were more promising. Promising results were also obtained by Nowock [165] against the related *E. kuehniella* with farnesyl methyl ether (15) which, additionally, prevented the fusion and torsion of the testes. Metwally and Landa [140] demonstrated that topical application of juvenoids to pupae of the khapra beetle *Trogoderma granarium* resulted in sterility with extremely small doses (1 picogramme per insect), and morphogenetic effects in this insect and in *Caryedon gonagra* were reported by Metwally and Sehnal [141]. Karnavar [105] demonstrated that juvenoid-treated diapausing larvae of *T. granarium* failed to pupate even when they were exposed to diapause-terminating conditions. The sterilising effects of juvenoids in *T. granarium* appear to result from interference with the normal working of the female repoductive system since juvenoids do not appear to affect spermatogenesis in this species [116]. Recently, Bhatnagar-Thomas [20] suggested the possibility of using juvenoid-impregnated paper strips or chalk granules for the practical control of this species. In tests with Compound 5 against *Ephestia kuehniella*, Hoppe [93] completely prevented adult emergence with 3 and 5 ppm of a juvenoid in grain for a period of four months. This author noted, however, that larvae removed from treated grain after this period were capable of pupation and the formation of viable adults which, even when JH-induced deformity prevented flight, were able to copulate and produce viable eggs. Strong and Diekman [230] conducted an extensive evaluation of 15 juvenoids against a number of storage beetles and moths. In this study, methoprene (9) and hydroprene (6) provided the most effective control. Hydroprene (6) was effective against *Ephestia cautella*, *Plodia interpunctella* (the Indian meal moth), *Tribolium castaneum*, *Trogoderma inclusum* and *T. variabile* at the rate of 5 ppm in the food media and at 10 ppm against *Rhizopertha dominica*. The granary weevil, *Sitophilus granarius* was effectively controlled with 50 ppm hydroprene (6) but the rice weevil, *S. oryzae* was only partially controlled at this concentration. These authors also reported good control of the tobacco beetle, *Lasioderma serricorne*, *R. dominica*, the saw-toothed grain beetle *Oryzaephilus surinamensis*, *O. mercator* and *T. castaneum* with 5 ppm methoprene (9) and of the confused flour beetle, *T. confusum* with 10 ppm methoprene. The latter insect was also effectively controlled by R-20458 (8) at this concentration [171]. Subsequently, these results have been expanded and confirmed by several authors who have evaluated juvenoids against a variety of stored-products pests [127, 128, 136, and others]. Recently Kramer and McGregor [111] tested 6 juvenoids (pyridyl and phenyl ethers) against 5 stored-products beetles and 3 stored-products moths. They found the pyridine juvenoid (12) to be the most active. However, Compound 12 only gave 50% suppression of *S. oryzae* development when grain was treated with 100 ppm. Interestingly, this dose remained ineffective even when combined with 600 ppm of the synergist piperonyl butoxide. However, in grains pre-fabricated from treated flour, they did obtain 100% inhibition of adult development in *S. oryzae* with 100 ppm of the same juvenoid and concluded that the low activity found in ordinary grain resulted from the protective action of the grain shell although they noted that little or no larval development (a rather atypical juvenoid effect) occurred in the artificial grain. Oberlander et al. [166] demonstrated that sub-lethal treatment of *Plodia interpunctella* larvae with juvenoids reduced mating behaviour in adults, not by reducing pheromone production but, they suggest, by direct interference with calling behaviour and the induction of abnormalities in antennal morphology. Indeed, Menon and Nair [139] have shown in *Tenebrio molitor* that, prior to the complete maturation of the ovaries, juvenoid treatment can stimulate pheromone production. With the exception of *Trogoderma*, effects on reproduction seem to be relatively rare in both beetles and moths infesting stored commodities. However, Tan and Mordue [233] have reported that hydrochlorinated juvenoids give significant sterility and reduce oviposition in the moth *Ephestia kuehniella* when applied to female adults. Williams and Amos found [257] no effects on oviposition as a result of juvenoid treatment of *T. castaneum* although Amos et al. [9] reported a reduction in fertility of the same species following treatment with methoprene and hydroprene. However, they did not conclusively demonstrate a direct effect of the treatment on ovaries or testes. Moreover, as De Vries and Brown [54] have observed, reproductive capacity may be reduced in this (and perhaps other) species by the incomplete or disturbed metamorphosis of the reproductive organs or by the juvenoid-induced retention of pupal urogomphi in adults. Strong and Diekman [230] reported no effects on the reproduction of any of the species tested. More recently, it has been shown that the reproductive capacity of both male and female *Sitotroga cerealella* (the angoumois grain moth) is not impaired by sub-lethal applications of methoprene (9) or hydroprene (6) to pupae (Stockel and Edwards, in press). El-Tantawi et al. [67] have however, obtained ovicidal effects with juvenoids in *Callosobruchus maculatus* and have demonstrated that females of this species were more sensitive than males to the morphogenetic effects following treatment of eggs. Moreover, the discovery that juvenoids have both high morphogenetic and ovicidal activity in the cigarette beetle *Lasioderma serricorne* [121, 132] provided important efficacy data in support of subsequent commercial registration of methoprene (Kabat®, Table 2) for the control of this species and the tobacco moth (*Ephestia elutella*) in hogsheads of stored tobacco [120, 131]. Against the rice moth *Corcyra cephalonica* Ramakrishnan and Joshi [181] tested both hydroprene (6) and methoprene (9) and reported that 2 ppm hydroprene (6) on sorghum grains completely inhibited adult emergence whereas 10 ppm methoprene (9) was required to obtain the same effect. These authors also reported that hydroprene (6), remained biologically active for up to 6 months on sorghum. The higher activity of hydroprene (6) versus methoprene (9) demonstrated against the rice moth is also apparent in other stored-products moths, including *Sitotroga cerealella* (Stockel and Edwards, in press). In small-scale practical evaluations Hoppe [92] has demonstrated that the relatively stable juvenoid, epofenonane (7) when applied at 10 ppm to grain remained biologically active against *P. interpunctella*, *E. kuehniella*, *R. dominica* and a susceptible and non-specific malathion-resistant strain of *T. castaneum* for a period of a least one year. In the same experiment, only a 30% reduction in adult production from larvae of *Sitophilus granarius* was achieved. Earlier, Dyte [56] reported about 10-fold cross resistance to JH I in the same resistant strain (CTC 12) of *T. castaneum* although this particular strain appears to be resistant to all xenobiotic chemicals, irrespective of structure or origin, so far tested against it. Moreover, Amos et al. [10] found no evidence of cross-resistance to 9 juvenoids in a malathion-resistant strain of *T. castaneum* and it is also known that a pyrethrum-resistant strain of *S. granarius* is not cross-resistant to JH I [62]. However, Brown et al. [33] reported the induction of 7-fold resistance to the juvenoid hydroprene following laboratory selection of *T. confusum*. Against insecticide-resistant moths Silhacek et al. [209] showed JH I to be more effective in reducing adult emergence in the malathion-resistant strain of *Plodia interpunctella* than in the susceptible strain.

In addition to the reported effects on insects, it is interesting to note that Topinska and Stepien [236] have demonstrated both toxic and morphogenetic effects with 2 juvenoids in the stored-products mite *Tyrophagus putrescentiae*.

The overall conclusion from these studies is that although the majority of stored-product pests are highly susceptible to juvenoids, this is not the case with beetles of the genus *Sitophilus*. Early suggestions as to the reason for the apparent insensitivity of *Sitophilus* spp. were that larvae were protected from the effects of the juvenoid which was unable to penetrate through the grain inside which the larvae develop [92, 111, 227]. This suggestion has also been made in relation to *Sitotroga cerealella*, another relatively insensitive species that develops inside grain. However, Rowlands [191] has demonstrated that three juvenoids rapidly penetrate grain and accumulate largely in the region of the aleurone layer. Moreover, since final-instar larvae are often found in close proximity to the aleurone layer at the time of pupation and therefore of maximum sensitivity to juvenoids (Edwards, unpublished results) and since treatment of last instar larvae or newly-formed pupae removed from grain does not result in greatly increased sensitivity to juvenoids (Edwards, unpublished results, [111]), inhibition of penetration does not appear to be the only reason for the insensitivity of the weevil to juvenoids. Several other factors such as metabolism [62], the presence or absence of gut flora [13] and vapour effect or ventilation [8, 231] have also been implicated. Metabolism, as a determining factor in the sensitivity of insects to juvenoid treatment, will be covered more extensively later in this chapter. However, it has been clearly demonstrated [58] that the age-related sensitivity of *T. castaneum* to JH I can be closely correlated with the activity of juvenoid-metabolising enzymes [63]. Whatever the reason for the insensitivity of *Sitophilus* (and to some extent *Sitotroga* and some Bruchids), it is clear that this problem has been paramount in hindering the otherwise potentially valuable contribution of juvenoids to the control of pests affecting stored products.

4.4. Household, Domestic, and Other Pests

Within this group of pests are found representatives of a wide variety of insect orders. Many species have evolved in close association with man and his domestic environment and some have spread across the world largely as a result of trade between nations [167]. Only a few species cause serious economic damage (e.g., termites) although many are important as "nuisance pests" and potential transmitters of disease. Perhaps the most important group of pests in this category are the cockroaches, including the German cockroach *Blattella germanica*, the oriental cockroach *Blatta orientalis*, the American cockroach *Periplaneta americana* and *P. australasiae*. The effects of juvenoids on cockroaches and the importance of the endocrine system in regulating reproduction in these insects have been reasonably well documented. Bell and Barth [16] reported on the effect of JH on vitellogenin production, yolk deposition, colleterial gland activity and pheromone production in the cockroach *Byrsotria fumigata*. The gonadotropic action of juvenile hormone has been investigated in *B. germanica* [115] as has the role of JH in accessory gland development and function in *P. americana* [22]. Das and Gupta [51] reported a variety of effects when last-instar female nymphs of *B. germanica* were treated with juvenoids. These included morphogenetic effects, impairment of reproductive capability resulting from ovarian disfunction, inability to produce oothecae and failure to respond to normal male courtship behaviour. Earlier, Cruickshank [48] had reported that the incorporation of amide-substituted juvenoids, including Compound 13, into food induced premature ootheca drop and reduced the number of nymphs hatching from successfully retained oothecae in the same species. Riddiford et al. [187] found hydroprene (6) to be effective against small mixed-age populations of *B. germanica*. Admixture of 5 ppm hydroprene (6) in the food material prevented normal metamorphosis in 97% of the population. However, even at 100 ppm hydroprene (6), complete eradication was not obtained after 7 months when some adultoids remained alive although no reproduction occurred. These authors concluded that control of this species with juvenoids was feasible as did Cristodorescu et al. [46] who found 1.0% epofenonane (7) effective against *B. germanica.* A similar conclusion was reached by Edwards [59] who evaluated methoprene (9) against all nymphal stages of *B. germanica* and reported that feeding first, second or third instar nymphs food containing 0.1% methoprene (9) completely prevented reproduction in the subsequent adult stages. Similar results were obtained with methoprene against *Blatta orientalis.* Das and Gupta [50] reported the absence of repellency of two juvenoids (Compounds 5 and 8) at 200 ppm in food material to the cockroach *B. germanica.* Abdel-Gawaad [2] however, reported little success with injections of hydroprene (6) or methoprene (9) in nymphs of *Periplaneta americana.*

Several researchers have investigated the potential of juvenoids against ants. Cupp and O'Neal [49] reported the morphogenetic effects of hydroprene (6) and methoprene (9) on larvae on the fire ants *Solenopsis invicta* and *S. richteri.* Methoprene (9) was effective topically, orally and by trophallaxis. Troisi and Riddiford [238] tested the same two juvenoids against small laboratory colonies of *S. invicta* and reported blocking of larval metamorphosis and the cessation in the production of larval brood. This latter phenomenon they attributed to a combination of ovicidal effects followed by cannibalism of dead or abnormal eggs. In these experiments, they also noted an increase in mortality of worker ants following contact exposure to hydroprene (6). Vinson et al. [243] screened several juvenoids for activity against *S. invicta* and reported high activity with the compounds JH-25 (14), JH I and methoprene (9). It is interesting to note that this is one of the few instances where the naturally-occurring hormone JH I has shown activity comparable to that of the synthetic analogues. Vinson and Robeau [242] fed juvenoids to colonies of *S. invicta* and again reported JH-25 (14) to be the most active compound tested. Activity was noted primarily against the pupal stage and there was a gradual decline in egg production. Death of the colony occurred at doses of 5 mg/colony or higher and these authors also reported a degree of repellency to all juvenoids tested with the exception of JH-25 (14). More recently, further studies have confirmed the high activity of JH-25 (14) against fire ants [14]. Both Troisi and Riddiford [238] and Vinson and Robeau [242] reported changes in the caste bias resulting from juvenoid treatment of fire ant colonies. Such changes, which have been reported in other ant species [118, 172] may be useful in control measures against ants and other social insects, e.g., termites [45, 96]. Wendel and Vinson [250] studied the metabolism and distribution of [14C]-R-20458 (8) fed to colonies of *S. invicta* and reported that, following administration in soybean oil, larvae contained 7% of the total radiocarbon after 36 h. Pupae, queens and eggs received only traces of radiocarbon. They further reported that the half-life of the juvenoid never exceeded two days and suggested the need to develop a (solid) bait that would be fed directly to larvae rather than stored (and thus metabolised) in the worker crop. Edwards [60] and Rupes et al. [193] reported the effects of methoprene (9) on laboratory colonies of the Pharaoh's ant *Monomorium pharaonis.* Incorporation of 0.5% methoprene in ant food resulted in the destruction of brood stages and a cessation of oviposition by queens. Subsequent examination of ovaries from treated queens revealed considerable ovarian atrophy and the absence of oocytes and trophocytes in the ovarioles. In small-scale field trials, bait containing 0.5% methoprene (9) eradicated infestations of this species in small premises [57, 59]. Subsequently, the practical utility of methoprene (9) bait for the control of Pharaoh's ant was demonstrated in a large-scale field trial in a British hospital [61]. Hrdy et al. [98] have reported similar success using juvenoids against the same species in Czechoslovakia. At present, methoprene bait concentrates for Pharaoh's ant control have been registered for use in the United Kingdom (Table 2). Preliminary studies have indicated that methoprene baits are also effective against ants of the genus *Pheidole* (Edwards, unpublished). However, attempts to control some other house-infesting ant species (e.g., the Argentine ant *Iridomyrmex humilis*, *Paratrechina longicornis* and *P. vividula*) with juvenoid baits have not proved successful although the reason for this may be that such species are not readily attracted to simple bait formulations (Edwards, unpublished).

The role of the endocrine system in the regulation of caste in termites has been recently reviewed by several authors [95, 123, 129]. Practical studies against this important pest group have, however, been sparse. The typical response to juvenoid treatment in a number of different species is the development of pre-soldiers or soldiers from pseudergates normally destined to form workers. French and Robinson reported [74] inhibition of feeding and only negligible mortality in *Mastotermes darwiniensis* exposed to juvenoid-treated plywood blocks, and French et al. [75] were unsuccessful in controlling field colonies of this species with wood blocks impregnated with hydroprene (6).

In recent years, a number of publications have reported the effects of juvenoids on a variety of other domestic pests. Matolin and Rohdendorf [133] demonstrated the ovicidal activity of farnesyl methyl ether (15) vapour on eggs of fire brat *Thermobia domestica* and Rohdendorf and Sehnal [188] induced sterility in the same insect with topical application of JH-I. Caussanel reported [35] that the corpora allata of the earwig *Labidura riparia* showed cyclic variation consistent with changes in the reproductive system. However, Srihari et al. [224] obtained no apparent effects on reproduction in this insect although they reported that juvenoid treatment induced morphogenetic effects and the production of giant adults which were reproductively active. Mauchamp [134] obtained morphogenetic and reproductive effects induced by juvenoids in the clothes moth *Tineola bisselliella* and Vinson and Williams [240] have reported ovicidal, developmental and antifertility effects in the human body louse *Pediculus humanus*. Moreover, in recent years, several investigations have indicated that juvenoids may be useful in controlling fleas especially the cat flea *Ctenocephalides felis*. Chamberlain and Becker [39] showed that 1 ppb methoprene (9) was effective in preventing cocoon formation and adult eclosion in the oriental rat flea *Xenopsylla cheopis* confirming the good activity of this juvenoid against fleas. These authors suggest that dusting treatments against dog and cat fleas or pelleted rat-baits containing methoprene (9) might prove effective as a control measure. However, against the tick *Dermacentor variabilis*, a pest of dogs and other domestic animals, McDaniel and Oliver [135] showed that hydroprene (6) and R-20458 (8) produced effects on development but not on female reproductive capacity or male spermatogenesis.

Although house flies and other Dipterans frequently rank high on the list of domestic pest species, their control with juvenoids in the domestic environment is probably not feasible since, usually, only adults are present. However, we have seen that the breeding sites of both flies (cattle-lofts, poultry houses, etc.) and mosquitoes can be successfully treated with juvenoids, thus reducing the adult population capable of infesting domestic premises.

4.5. The Use of Juvenoids in Sericulture

The implication for the utilisation of juvenoids for increasing silk production were quickly realised following the report by Akai et al. [7] that injection of juvenoids into fifth instar silkworm larvae *(Bombyx mori)* produced giant larvae which subsequently spun extra-large cocoons. Nihmura et al. reported [160] similar increases in the weight of silk cocoons when larvae were fed JH I and JH II and Chang et al. [43] obtained similar results with methylene dioxyphenyl-substituted juvenoids together with an increase in egg production by treated females. Subsequently, the juvenoid methoprene (Manta®) was commercially registered in Japan for improving silk yield (Table 2). This is a unique example whereby a juvenoid has been applied to increase the productivity of insects rather than to reduce damage resulting from their overproductivity.

5. Some Factors Influencing the Practical Use of Juvenoids

5.1. Choice of Target Species

In an earlier chapter, it was shown that certain biological factors (e.g. developmental stage) influence the sensitivity of insects to treatment with juvenoids. Furthermore, the limited stability of juvenoids in the environment makes the timing of application a crucial factor in their successful use against certain pest species. In addition to such considerations, several other factors can influence the effectiveness of juvenoids. Suitable target pests have to be carefully identified and the techniques of delivering the active chemical to the insect at the time(s) of maximum sensitivity have to be carefully planned in order to obtain the desired effect(s) on the pest population. Broadly speaking, the ideal target pests are those in which, for reasons already discussed, only the adult stage causes damage. In addition, control measures utilising juvenoids are likely to be most effective in species which are susceptible to both morphogenetic and reproductive effects. Moreover, pest populations which are confined to specific, isolated areas or habitats – thus precluding reinfestation by untreated individuals – may be controlled more effectively than widely-distributed species. In situations where all those conditions can be fulfilled, juvenoids are likely to be highly effective in providing selective management of insect populations. Unfortunately, most pest species attacking agricultural crops do not lend themselves to effective control with juvenoids [138].

5.2. Metabolism, Resistance, and Synergism

Naturally occurring juvenile hormones are relatively unstable compounds both in the environment and in the insect haemolymph. For example, the half-life of exogenously administered JH-I in insect haemolymph is generally less than one hour [213] and the same is true for many juvenoids [6, 251]. As we have seen, earlier, juvenoids are also highly sensitive to breakdown by UV radiation and to hydrolysis in (non-sterile) aqueous media [202]. Whilst the low stability of the juvenoids makes them environmentally acceptable as pesticides, this property does not always allow sufficient persistence for optimum practical usage. Consequently, one of the major requirements for a practically viable compound is effective environmental persistence.

Because the precise regulation of endogenous JH titre is an important factor in controlling development and reproduction in insects, it is not surprising that, in all species examined, there are efficient enzyme systems which rapidly metabolise JHs to inactive products [4, 6, 73, 91, 213, 251, and others]. In most cases, exogenously applied natural JHs are metabolised via two primary pathways to form a 10,11-diol derivative and an acid derivative. The enzymes responsible for these conversions have been designated as JH-epoxide hydrase and JH-esterase respectively [31, 249]. Metabolism by these enzyme systems leads to the formation of the corresponding acid-diol derivative which may be further metabolised and/or conjugated. In *T. castaneum*, the activity of the JH-epoxide hydrase increases during the final larval instar [63] at a time when the endogenous level of JH is rapidly declining. Similar age-related changes in the activity of JH-metabolising enzymes have been found in other species (e.g., the southern armyworm [214, 263]; the tobacco hornworm [249]; the Colorado potato beetle [112]; the southwestern corn borer [32, and others]). It seems probable therefore, that the reduction of endogenous JH titre at critical periods during development results from both the cessation of production of the JH by the corpora allata [69], (Bergot et al., in preparation [19]) and an increase in the activity of the JH-metabolising enzymes. In addition, it is probable that secretion of JH by the corpora allata is continuously regulated by a feed-back mechanism responsive to the titre of hormone in the haemolymph [201]. In some (if not all) insects, endogenous JH appears to be bound to a so-called carrier protein [68, 237, 252]. The carrier protein may have several functions (e.g., to protect the JH molecule from metabolism by non-specific enzymes, to facilitate the transport of the lipophilic JH molecule through an essentially aqueous medium or to deliver the molecule to the site of action in an appropriate molecular configuration). The first of these possibilities seems most reasonable since Sanberg et al. have demonstrated [194] the presence of a JH-specific esterase in *Manduca sexta*, which is present at critical times during development, and which, unlike non-specific enzymes, can metabolise protein-bound JH. A similar system has been suggested in *Leptinotarsa decemlineata* [110]. It is not clear to what extent enzymes which metabolise natural juvenile hormones are active against synthetic juvenoids. However, insects, as do higher animals, possess *m*ixed-*f*unction *o*xidase enzymes (mfo) capable of metabolising a wide range of xenobiotics including juvenoids [4, 251]. Such enzymes may confer on insects the ability to develop populations resistant to conventional insecticides when exposed to appropriate selection pressure. It seems reasonable, therefore, to expect certain strains of insects to be less sensitive to juvenoids either as a result of cross-resistance or by direct selection of resistant strains with juvenoids [79]. Dyte first reported [56] cross-resistance to a juvenoid (JH-I) in a multi-resistant strain of the flour beetle *Tribolium castaneum* and since then, cross-resistance has been reported in *Musca domestica* [37, 38, 174, 192] and in some other species. In addition, resistance to juvenoids has been induced by laboratory selection in *Tribolium confusum* and *Culex pipiens* [33] and in the house fly *M. domestica* [79]. However, as previously indicated, not all insecticide-resistant strains show cross-tolerance to juvenoids and, at present, there are no reports of resistance resulting from field treatments. It thus remains to be seen whether continued practical use of juvenoids against pest species will lead to the selection of resistant strains in the field and, if so, whether such resistance will be sufficiently high to be of practical significance.

The realisation of the involvement of enzymes in rapidly degrading exogenous and endogenous JH and the reports of cross-resistance to juvenoids prompted the attempted use of synergists to increase the activity of juvenoids in both the susceptible and resistant insects. However, all attempts to use synergists containing a methylene dioxyphenyl moiety (MDP) which inhibit mfo activity (e.g., tri-ortho cresyl phosphate) have been notably unsuccessful [5, 25, 182, 221, 222]. Although some authors have reported that synergists like piperonyl butoxide have intrinsic juvenile hormone activity recent studies have indicated that the JH-like effects produced by MDP type synergists are not entirely consistent with those produced by true juvenoids [212].

Although attempts are continually being made to improve the stability of juvenoids both in the environment and in the target insects, recent studies have shown that, in one respect at least, the presence of catabolic enzymes may be used to some advantage. In this instance, endogenous enzymes are presumed to release active juvenoid molecules from (inactive) long chain fatty acid esters – the so-called juvenogens (or pro-juvenoids) – thus allowing the physical and chemical properties of the parent molecule to be tailored to suit the practical situation [215]. However, further evidence of practical success with this approach is needed to justify further attempts to develop viable projuvenoids.

6. Conclusions

In this review we have attempted to present a summary of the novel approach to insect pest control utilizing juvenoid – compounds developed as a consequence of recent basic research findings in insect endocrinology. The many functions of juvenile hormones in insects and their uniqueness to the endocrine system of arthropods has offered the possibility of developing effective pesticides which are relatively non-hazardous to the environment and to non-target organisms including man and his domestic animals. The analysis of recent studies in this area indicates that some of the compounds do have promising utility in a variety of applications, especially those in which only the adult forms have significant pest status and where pest populations are limited to specific areas, e.g. lakes, stored products, domestic institutions, etc. However, we have also seen that, for a variety of hitherto mentioned reasons, practical control of agricultural and forestry pests is probably not feasible. Nevertheless, it is evident that we have not yet fully explored the potential of these "third generation pesticides" which have, in their own way, provided an illustration of the rapid and worthwhile conversion of major breakthroughs in basic research findings into practical control techniques.

7. References

1. Abdallah, M.D., Zaazou, M.H., El-Tantawi, M.: Z. angew. Entom. *78*, 176 (1975)
2. Abdel-Gawaad, A.A.: Z. angew. Entom. *80*, 346 (1976)
3. Adams, T.S.: J. Insect Physiol. *20*, 263 (1974)
4. Agosin, M. et al.: J. Biol. Chem. *254*, 9915 (1979)
5. Ajami, A.M.: J. Insect Physiol. *21*, 1017 (1975)
6. Ajami, A.M., Riddiford, L.M.: J. Insect Physiol. *19*, 635 (1973)
7. Akai, H., Kiguchi, K., Mori, K.: Appl. Ent. Zool. *6*, 218 (1971)
8. Amos, T.G., Williams, P.: Aust. J. Zool. *25*, 201 (1977)
9. Amos, T.G., Williams, P., Semple, R.L.: Experientia *34*, 469 (1978)
10. Amos, T.G. et al.: J. Econ. Entomol. *67*, 474 (1974)
11. Asano, S., Kamei, M.: Botyu Kagaku *42*, 45 (1977)
12. Bagley, R.W., Bauernfeind, J.C.: Field experiences with juvenile hormone mimics. In: Insect juvenile hormones – chemistry and action (Ed. J.J. Menn and M. Beroza). Academic Press, N.Y., pp. 113–151 (1972)
13. Baker, J.E., Lum, P.T.M.: J. Ga Ent. Soc. *11*, 213 (1976)
14. Banks, W.A., Lofgren, C.S., Plumley, J.K.: J. Econ. Entomol. *71*, 75 (1978)
15. Beadles, M.L. et al.: J. Econ. Entomol. *68*, 781 (1975)
16. Bell, W.J., Barth, R.H.: J. Insect Physiol. *16*, 2303 (1970)
17. Benskin, J., Perron, J.M.: Canad. Entom. *105*, 619 (1973)
18. Benz, G.: Experientia *29*, 1437 (1973)
19. Bergot, B.J., Schooley, D.A., Hall, M.S., Judy, K.: In preparation

19a. Bhaskaran, G.: J. exp. Zool. *181*, 127 (1972)

20. Bhatnagar-Thomas, P.L.: J. Food Sci. Technol. *13*, 159 (1976)
21. Bhatnagar-Thomas, P.L.: J. Econ. Entomol. *66*, 277 (1973)
22. Blaine, W.D., Dixon, S.E.: Canad. Ent. *105*, 1275 (1973)
23. Boboye, S.O., Carman, G.E.: J. Econ. Entomol. *68*, 473 (1975)
24. Bowers, W.S.: Science *164*, 323 (1969)
25. Bowers, W.S.: Science *161*, 895 (1968)
26. Bowers, W.S., Thompson, M.J., Uebel, E.: Life Sci. *4*, 2323 (1965)
27. Bowman, M.C., Wright, J.E., Beroza, M.: J. Econ. Entomol. *66*, 301 (1973)
28. Bransby-Williams, W.R.: E. Afr. Agric. For. J. *38*, 170 (1972)
29. Breaud, T.P. et al.: Mosquito News *37*, 704 (1977)
30. Breeden, G.C., Turner, E.C., Beane, W.L.: J. Econ. Entomol. *68*, 451 (1975)
31. Brooks, G.T.: Nature (London) *245*, 382 (1973)
32. Brown, J.J., Chippendale, G.M., Turunen, S.: J. Insect Physiol. *23*, 1255 (1977)
33. Brown, T.M., Devries, D.H., Brown, A.W.: J. Econ. Entomol. *71*, 223 (1978)
34. Case, T.J., Washino, R.K.: Mosquito News *38*, 191 (1978)
35. Caussanel, M.C.: C.R. hebd. Seanc. Acad. Sci. Paris *277*, 1521 (1973)
36. Cawich, A., Crowder, L.A., Watson, T.F.: J. Econ. Entomol. *67*, 173 (1974)
37. Cerf, D.C., Georghiou, G.P.: Pestic. Sci. *5*, 759 (1974)
38. Cerf, D.C., Georghiou, G.P.: Nature (London) *239*, 401 (1972)
39. Chamberlain, W.F., Becker, J.D.: Southwest Entomol. *2*, 179 (1977)
40. Chamberlain, W.F., Hopkins, D.E.: Ann. Entomol. Soc. Am. *63*, 1363 (1970)
41. Chamberlain, W.F., Hopkins, D.E., Gingrich, A.R.: J. Econ. Entomol. *66*, 127 (1973)
42. Chamberlain, W.F. et al.: J. Agric. Food Chem. *23*, 736 (1975)
43. Chang, C.F., Murakoshi, S., Tamura, S.: Agric. Biol. Chem. *36*, 692 (1972)
44. Chippendale, G.M.: An. Rev. Entom. *22*, 121 (1977)
45. Chu, H., Tai, T., Chen, T., King, M.: Acta entomol. Sinica *17*, 161 (1974)

46. Cristodorescu, G., Nosec, I., Tacu, V.: Arch. Roum. Path. Exp. Microbiol. *36*, 73 (1977)
47. Critchley, B.R., Almeida, A.A.: Bull. Entom. Res. *63*, 1 (1973)
48. Cruickshank, P.A.: Bull. Wld. Hlth Org. (Int. Conf.) *44*, 395 (1971)
49. Cupp, E.W., O'Neal, J.: Environ. Entomol. *2*, 191 (1973)
50. Das, Y.T., Gupta, A.P.: Entomologia exp. appl. *20*, 195 (1976)
51. Das, Y.T., Gupta, A.P.: Experientia *30*, 1093 (1974)
52. Della-Giustina, W., le Rumeur, J.Y.: Phytiat. Phytopharm. *26*, 277 (1977)
53. Denlinger, D.L.: Nature *253*, 347 (1975)
54. De Vries, D.H., Brown, T.M.: J. Econ. Entomol. *70*, 272 (1977)
55. Dorn, S. et al.: J. agric. Food Chem. *24*, 637 (1976)
56. Dyte, C.E.: Nature *238*, 48 (1972)
57. Edwards, J.P.: Proc. 8th Internat. Cong. I.U.S.S.I., Wageningen, pp. 81–82 (1977)
58. Edwards, J.P.: J. Stored Prod. Res. *12*, 71 (1976)
59. Edwards, J.P.: Proc. 8th Brit. Insectic. Fungic. Conf., pp. 267–275 (1975)
60. Edwards, J.P.: Bull. Ent. Res. *65*, 75 (1975)
61. Edwards, J.P., Clarke, B.: Internat. Pest Control *20*, 5 (1978)
62. Edwards, J.P., Rowlands, D.G.: Insect Biochem. *8*, 23 (1978)
63. Edwards, J.P., Rowlands, D.G.: Pestic. Biochem. Physiol. *7*, 194 (1977)
64. El-Gayar, F.: Entomophaga *21*, 297 (1976)
65. El-Guindy, M.A., Abdel-Sattar, M.N., El-Asar, M.R.S.: Z. angew. Entom. *84*, 424 (1977)
66. El-Guindy, M.A., Bishara, S.I.: ibid. *81*, 75 (1976)
67. El-Tantawi, M.A. et al.: ibid. *81*, 37 (1976)
68. Emmerich, H., Hartmann, R.: J. Insect Physiol. *19*, 1663 (1973)
69. Fain, M.J., Riddiford, L.M.: Biol. Bull. *149*, 506 (1975)
70. Fernez, H.J.: Experientia *29*, 356 (1973)
71. Finney, J.R. et al.: Mosquito News *37*, 6 (1977)
72. Fogal, W.H., Sullivan, C.R.: Canad. J. Forest Res. *6*, 523 (1976)
73. Fox, P.M., Massare, J.S.: Comp. Biochem. Physiol. (B) *53*, 195 (1976)
74. French, J.R.J., Robinson, P.J.: Z. angew. Entom. *85*, 360 (1978)
75. French, J.R. et al.: ibid. *87*, 239 (1979)
76. Fytizas, E., Mourikis, P.A.: ibid. *84*, 107 (1977)
77. Gaudchau, Von M.: Z. angew. Entom. *83*, 261 (1977)
78. Gelbic, I., Sehnal, F.: Bull. Ent. Res. *63*, 7 (1973)
79. Georghiou, G.P., Lee, S., De Vries, D.H.: J. Econ. Entomol. *71*, 544 (1978)
80. Gerig, L.: Proc. VIIIth Int. Cong. I.U.S.S.I., Wageningen, p. 88 (1977)
81. Gill, S.S. et al.: Preliminary chromatographic studies on the metabolism and photodecomposition products of the juvenoid 1-(4′-ethylphenoxy)-6,7-epoxy-3,7-dimethyl-2-octene. In: Insect juvenile hormones – Chemistry and Action (Menn, J.J. and Beroza, M., Eds.), Academic Press N.Y., pp. 177–189 (1972)
82. Guerra, A.A. et al.: Ann. Ent. Soc. Am. *70*, 771 (1977)
83. Hamlen, R.A.: J. Econ. Entomol. *70*, 211 (1977)
84. Hamlen, R.A., Mead, F.W.: J. Econ. Entomol. *72*, 269 (1979)
85. Handgartner, W.W. et al.: J. Agric. Food Chem. *24*, 169 (1976)
86. Harris, R.L., Chamberlain, W.F., Frazar, E.D.: J. Econ. Entomol. *67*, 384 (1974)
87. Harris, R.L., Frazar, E.D., Younger, R.L.: ibid. *66*, 1099 (1973)
88. Henrick, C.A., Willy, W.E., Staal, G.B.: J. Agric. Food Chem. *24*, 207 (1976)
89. Henrick, C.A., Staal, G.B., Siddall, J.B.: ibid. *21*, 354 (1973)
90. Herald, F., Knapp, F.W.: Proc. N. Cent. Brch Am. Ent. Soc. *28*, 197 (1973)
91. Hooper, G.H.S.: Insect Biochem. *6*, 255 (1976)
92. Hoppe, T.H.: J. Stored Prod. Res. *12*, 205 (1976)
93. Hoppe, T.H.: J. Econ. Entomol. *67*, 789 (1974)
94. Hoppe, T.H., Isler, H., Vogel, W.: Mosquito News *34*, 293 (1974)
95. Hrdy, I.: The influence of juvenile hormone analogues on caste development in termites. In: Phase and caste determination in insects – Endocrine Aspects (Ed. M. Luscher) Pergamon Press, New York, p. 71 (1976)
96. Hrdy, I., Krecek, J.: Insectes Soc. *19*, 105 (1972)
97. Hrdy, I., Zeleny, J.: Acta Ent. Bohemoslov. *70*, 386 (1973)

98. Hrdy, I. et al.: Proc. 8th Int. Congr. Int. Union Study Soc. Insects, Wageningen, pp. 83–84 (1977)
99. Hughes, P.R., Renwick, J.A.A.: Physiol. Entomol. *2*, 289 (1977)
100. Ivie, G.W., Wright, J.E., Smalley, H.E.: J. Agric. Food Chem. *24*, 222 (1976)
101. Jakob, W.L., Schoof, H.F.: Mosquito News *31*, 540 (1971)
102. Jobin, L., Retnakaran, A.: Symp. Entom. Soc. Am., Workshop on insect growth regulators, Las Vegas, USA (1974)
103. Joly, L., Joly, P.: C.R. hebd. Seances Acad. Sci. Ser. D (Paris) *279*, 1007 (1974)
104. Judy, K.J. et al.: Proc. Nat. Acad. Sci. USA *70*, 1509 (1973)
105. Karnavar, G.K.: Indian J. exp. Biol. *11*, 138 (1973)
106. Klein-Koch, C. von: Z. angew. Entom. *82*, 193 (1976)
107. Kontev, C.H. et al.: Acta Entomol. Bohemoslov. *71*, 72 (1974)
108. Kontev, C.H. et al.: ibid. *70*, 377 (1973)
109. Kozar, F., Varjas, L.: Acta Phytopath. Acad. Sci. Hung. 11, 295 (1976)
110. Kramer, K.J., Childs, C.N.: Insect Biochem. 7, 397 (1977)
111. Kramer, K.J., McGregor, H.E.: J. Econ. Entomol. *71*, 132 (1978)
112. Kramer, S.J., De Kort, C.A.: Molec. Cell. Endocrinol. *4*, 43 (1976)
113. Krimer, M.Z. et al.: Zh. Vses. Khim. O-va *21*, 356 (1976)
114. Kuhr, R.J., Cleere, J.S.: J. Econ. Entomol. *66*, 1019 (1973)
115. Kunkel, J.G.: J. Insect Physiol. *19*, 1285 (1973)
116. Landa, V., Metwally, M.M.: Acta Entomol. Bohemoslov. *71*, 145 (1974)
117. Law, J.H., Yuan, C., Williams, C.M.: Proc. Nat. Acad. Sci. Am. *55*, 576 (1966)
118. Ledoux, A.: C.R. hebd. Seances Acad. Sci. Ser. D (Paris) *280*, 569 (1976)
119. Lefevre, M. Von: Z. angew. Entom. *82*, 187 (1976)
120. Lehman, R.M. et al.: Tobacco Workers' Conf. 28th, Orlando, FL 11 p., (1979)
121. Lehman, R.M., Walker, G.B., Manzelli, M.A.: Tobacco Chemists' Res. Conf., Greensboro, N.C., 6 p. (1977)
122. Lema, K.M., Poe, S.L.: Florida Entom. *61*, 67 (1978)
123. Lenz, M.: The dependence of hormone effects in termite caste determination on external factors. In: Phase and Caste determination in insects – Endocrine Aspects (Ed. M. Luscher), Pergamon Press, New York, pp. 73–89 (1976)
124. Lewallen, L.L.: Mosquito News *24*, 43 (1964)
125. Lewis, L.C., Lynch, R.E., Berry, E.C.: J. Econ. Entomol. *66*, 1315 (1973)
126. Lewis, L.F., Christenson, D.M.: Proc. 40th Ann. Conf. Calif. Mosquito Contr. Assoc., 49 (1972)
127. Loschiavo, S.R.: J. Econ. Entomol. *69*, 395 (1976)
128. Loschiavo, S.R.: Manitoba Entomol. *9*, 43 (1975)
129. Lüscher, M.: Evidence for an endocrine control of caste determination in higher termites. In: Phase and caste determination in insects – Endocrine Aspects (Ed. M. Lüscher), Pergamon Press, New York, pp. 91–103 (1976)
130. Mansingh, A., Rawlins, S.C.: Naturwissenschaften *64*, 41 (1977)
131. Manzelli, M.A.: Int. Work. Conf. Stored Prod. Entomol. Proc., 2nd, 432–436 (1978)
132. Manzelli, M.A., Lehman, R.M.: Tobacco Workers' Conference, 27th, 6 p. (1977)
133. Matolin, S., Rohdendorf, E.B.: Acta Entomol. Bohemoslov. *69*, 1 (1972)
134. Mauchamp, B.: An. Zool. Ecol. Anim. *6*, 407 (1974)
135. McDaniel, R.S., Oliver, J.H.: J. Parasitol. *64*, 571 (1978)
136. McGregor, H.E., Kramer, K.J.: J. Econ. Entomol. *68*, 668 (1975)
137. Meleney, W.P., Roberts, I.H.: J. Parasitol. *61*, 956 (1975)
138. Menn, J.J., Pallos, F.M.: Environ. Lett. *8*, 71 (1975)
139. Menon, M.D., Nair, K.K.: Ann. Entom. Soc. Am. *69*, 457 (1976)
140. Metwally, M.M., Landa, V.: Z. angew. Entom. *72*, 97 (1972)
141. Metwally, M.M., Sehnal, F.: Biol. Bull. Mar. Biol. Lab., Woods Hole *144*, 368 (1973)
142. Meyer, A.S., Schneiderman, H.A., Hanzmann, E.: Proc. Nat. Acad. Sci. USA *60*, 853 (1968)
143. Miller, J.A. et al.: J. Econ. Entomol. *70*, 589 (1977)
144. Miller, J.A. et al.: J. Econ. Entomol. *70*, 417 (1977)
145. Miller, R.W., Uebel, E.C.: ibid. *67*, 69 (1974)
146. Miller, R.W., Pickens, L.G., Hunt, L.M.: ibid. *71*, 274 (1978)
147. Miura, T., Takahashi, R.M.: ibid. *66*, 917 (1973)
148. Morgan, P.B., Labrecque, G.C.: ibid. *64*, 1479 (1971)

149. Morgan, P.B. et al.: Canad. Entom. *107*, 413 (1975)
150. Morohoshi, S., Ishida, S., Fugo, H.: Proc. Japan Acad. *48*, 736 (1972)
151. Mulla, M.S., Darwazeh, H.A., Norland, R.L.: J. Econ. Entomol. *67*, 329 (1974)
152. Mulla, M.S. et al.: ibid. *67*, 165 (1974)
153. Nalbert, A.: Z. wiss. Zool. *104*, 181 (1913)
154. Naqvi, S.N.H., Rashid, S., Ashrafi, S.H.: Z. angew. Entom. *80*, 316 (1976)
155. Naqvi, S.N.H. et al.: ibid. *85*, 61 (1978)
156. Nassar, S.G., Staal, G.B., Armanious, N.I.: J. Econ. Entomol. *66*, 847 (1973)
157. Neal, J.W., Hower, A.A.: ibid. *67*, 93 (1974)
158. Negishi, T., Shimada, S., Asano, S.: J. Insect. Physiol. *22*, 1523 (1976)
159. Nemeth, I., Varjas, L.: Acta Phytopath. Acad. Sci. Hung. *11*, 317 (1976)
160. Nihmura, M. et al.: Agric. Biol. Chem. (Japan) *36*, 889 (1972)
161. Novak, V.J.A.: Nature *167*, 132 (1951)
162. Novak, V., Sehnal, F.: Lesnictvi *22*, 87 (1976)
163. Novak, V., Sehnal, F.: Z. angew. Entom. *73*, 312 (1973)
164. Novak, V. et al.: ibid. *80*, 118 (1976)
165. Nowock, J.: Wilhelm Roux, Arch. Entw. Mech. Org. *172*, 303 (1973)
166. Oberlander, H., Sower, L., Silhacek, D.L.: J. Insect Physiol. *21*, 681 (1975)
167. Odhiambo, T.R.: Proc. Int. Congr. Entom. XV. Washington, pp. 52–59 (1976)
168. Outram, I.: Environ. Entomol. *3*, 361 (1974)
169. Outram, I.: J. Econ. Entomol. *66*, 1033 (1973)
170. Pallos, F.M., Menn, J.J.: Synthesis and activity of juvenile hormone analogues. In: Insect juvenile hormones – Chemistry and Action (Menn, J.J., Beroza, M., Eds.), Academic Press, New York, pp. 303–316 (1972)
171. Pallos, F.M. et al.: Nature *232*, 486 (1971)
172. Passera, L., Suzzoni, J.P.: C.R. hebd. Seanc. Acad. Sci., Ser. D (Paris) *286*, 615 (1978)
173. Pinchin, R. et al.: J. Econ. Entomol. *71*, 950 (1977)
174. Plapp, F.W., Vinson, S.B.: Pestic. Biochem. Physiol. *3*, 131 (1973)
175. Postlethwait, J.H., Wiesner, K.: Nature New Biology *244*, 284 (1973)
176. Prassert, V. et al.: J. Econ. Entomol. *68*, 639 (1975)
177. Pree, D.J., Stewart, D.K.R.: Bull. Environ. Contam. Toxicol. 14, 117 (1975)
178. Quistad, G.B., Staiger, L.E., Schooley, D.A.: J. Agric. Food Chem. *4*, 750 (1975)
179. Quistad, G.B. et al.: ibid. *4*, 743 (1975)
180. Quistad, G.B. et al.: Pestic. Biochem. Physiol. *6*, 523 (1976)
181. Ramakrishnan, V., Joshi, N.K.: J. Food Sci. Technol. *14*, 87 (1977)
182. Redfern, R.E., McGovern, T.P., Beroza, M.: J. Econ. Entomol. *63*, 540 (1969)
183. Reissig, W.H., Kamm, J.A.: J. Econ. Entomol. *67*, 181 (1974)
184. Retnakaran, A.: Canad. Entom. *105*, 591 (1973)
185. Richmond, C.E.: J. Econ. Entomol. *65*, 950 (1972)
186. Riddiford, L.M.: Juvenile hormone and insect embryonic development: its potential role as an ovicide. In: Insect Juvenile Hormones – Chemistry and Action (Menn, J.J. and Beroza, M., Eds.), Academic Press, New York, pp. 95–111 (1972)
187. Riddiford, L.M., Ajami, A.M., Boake, C.: J. Econ. Entomol. *68*, 46 (1975)
188. Rohdendorf, E.B., Sehnal, F.: Experientia *28*, 1099 (1972)
189. Röller, H., Dahm, K.H.: Rec. Prog. Horm. Res. *24*, 651 (1968)
190. Röller, H. et al.: Angew. Chem. Int. Ed. *6*, 179 (1967)
191. Rowlands, D.G.: J. Stored Prod. Res. *5*, 413 (1976)
192. Rupes, V. et al.: Entomologia Exp. Appl. *19*, 57 (1976)
193. Rupes, V. et al.: Acta Entomol., Bohemoslav. *75*, 155 (1978)
194. Sanberg, L.L. et al.: Nature *253*, 266 (1975)
195. Sarmiento, R. et al.: Science *179*, 1342 (1973)
196. Schaefer, C.H., Dupras, E.F.: J. Econ. Entomol. *66*, 923 (1973)
197. Schaefer, C.H., Wilder, W.H.: ibid. *66*, 913 (1973)
198. Schaefer, C.H., Wilder, W.H.: ibid. *65*, 1066 (1972)
199. Schmialek, P.: Z. Naturforsch. *18*, 516 (1963)
200. Schmid, A. et al.: Entomologia Exp. Appl. *24*, 65 (1978)

201. Schooley, D.A., Bergot, B.J.: Abstr. 70 Pest. Div., 178th ACS Nat. Meetg., Washington D.C. Sept. 9–14, 1979
202. Schooley, D.A., Quistad, G.B.: in Pesticide and Xenobiotic Metabolism in Aquatic Organisms (Kuhn, M.A.Q., Lech, J.J., Menn, J.J., Eds.) Am. Chem. Soc., Washington, pp. 161–176 (1979)
203. Schooneveld, H., Sanchez, A.O., De Wilde, J.: J. Insect Physiol. *23*, 689 (1977)
204. Schooneveld, H., Molen, J.P. van der, Wiebenga, J.: Entomologia Exp. Appl. *19*, 227 (1976)
205. Sehnal, F.: Endocrinol. Exp. *5*, 29 (1971)
206. Sehnal, F., Skuhravy, V.: Z. angew. Entom. *81*, 401 (1976)
207. Sehnal, F., Metwally, M.M., Gelbic, I.: ibid. *81*, 85 (1976)
208. Self, L.S. et al.: Mosquito News *38*, 74 (1978)
209. Silhacek, D.L., Oberlander, H., Zettler, J.L.: J. Stored Prod. Res. *12*, 201 (1976)
210. Sivasubramanian, P.: Entomologia Exp. Appl. *25*, 165 (1979)
211. Skuhravy, V.: Acta Entomol. Bohemoslav. *70*, 313 (1973)
212. Slade, M., Wilkinson, C.F.: Science *181*, 672 (1973)
213. Slade, M., Zibitt, C.H.: Metabolism of Cecropia juvenile hormone in insects and mammals. In: Insect Juvenile Hormones – Chemistry and Action (Menn, J.J., Beroza, M., Eds.) pp. 155–176. Academic Press, New York (1972)
214. Slade, M., Hetnarski, H.K., Wilkinson, C.F.: J. Insect Physiol. *22*, 619 (1976)
215. Slama, K., Romanuk, M.: Insect Biochem. *6*, 579 (1976)
216. Slama, K., Romanuk, M., Šorm, F.: Insect Hormones and Bioanalogues, Springer-Verlag, Wien, p. 428 (1974)
217. Smilowitz, Z., Martinka, C.A., Jowyk, E.A.: Environ. Entomol. *5*, 1178 (1976)
218. Socha, R., Sehnal, F.: J. Insect Physiol. *18*, 317 (1972)
219. Solli, H. et al.: Pestic. Sci. 7, 503 (1976)
220. Solomon, K.R., Evans, A.A.: J. Med. Entomol. *14*, 433 (1977)
221. Solomon, K.R., Metcalfe, R.L.: Pestic. Biochem. Physiol. *4*, 127 (1974)
222. Solomon, K.R. et al.: Life Sci. *13*, 733 (1973)
223. Spielman, A., Williams, C.M.: Science *154*, 1043 (1966)
224. Srihari, T., Breuzet, M., Caussanel, C.: C.R. hebd. Seanc. Acad. Sci. Ser. D (Paris) *280*, 1107 (1975)
225. Staal, G.B.: An. Rev. Entom. *20*, 417 (1975)
226. Staal, G.B., Nassar, S., Martin, J.W.: J. Econ. Entomol. *66*, 851 (1973)
227. Stockel, J.: Bull. EPPO *6*, 413 (1976)
228. Stockel, J., Edwards, J.P.: J. Stored Prod. Res. (in press)
229. Street, M.L.: Experientia *34*, 544 (1978)
230. Strong, R.G., Diekman, J.: J. Econ. Entomol. *66*, 1167 (1973)
231. Swatonek, F. von: Bodenkultur *26*, 278 (1975)
232. Tamaki, G.: Canad. Entom. *105*, 761 (1973)
233. Tan, K.H., Mordue, W.: Bull. Entom. Res. *67*, 483 (1977)
234. Thanassoulopoulos, A.C.: Z. angew. Entom. *77*, 171 (1974)
235. Thomas, P.J., Bhatnagar-Thomas, P.L.: Nature *219*, 949 (1968)
236. Topinska, J.C., Stepien, Z.: Zesz. Probl. Postep. Nauk Roln. *171*, 269 (1975)
237. Trautman, K.H.: Z. Naturforsch. *276*, 263 (1972)
238. Troisi, S.J., Riddiford, L.M.: Environ. Entomol. *3*, 112 (1974)
239. Varjas, L., Sehnal, F.: Entomologia Exp. Appl. *16*, 115 (1973)
240. Vinson, J.W., Williams, C.M.: Proc. Nat. Acad. Sci. USA *58*, 294 (1967)
241. Vinson, S.B.: J. Econ. Entomol. *67*, 335 (1974)
242. Vinson, S.B., Robeau, R.: ibid. *67*, 584 (1974)
243. Vinson, S.B., Robeau, R., Dzuik, L.: ibid. *67*, 325 (1974)
244. Vogel, W. et al.: Experientia *33*, 1537 (1977)
245. Walker, W.F.: J. Econ. Entomol. *66*, 30 (1973)
246. Walker, W.F., Bowers, W.S.: J. Agric. Food Chem. *21*, 145 (1973)
247. Wasti, S.S., Mahadeo, C.R.: Entomol. Scand. *5*, 56 (1974)
248. Weaver, R.J., Pratt, G.E., Finney, J.R.: Experientia *31*, 597 (1975)
249. Weirich, G., Wren, J., Siddall, J.B.: Insect Biochem. *3*, 397 (1973)
250. Wendel, L.E., Vinson, S.B.: J. Econ. Entomol. *71*, 561 (1978)
251. White, A.F.: Life Sci. *11*, 201 (1972)
252. Whitmore, E., Gilbert, L.I.: J. Insect. Physiol. *18*, 1153 (1972)

253. Wigglesworth, V.B.: J. Exper. Biol. *29*, 620 (1952)
254. Wigglesworth, V.B.: Quart. J. Microsc. Sci. *79*, 91 (1936)
255. Wigglesworth, V.B.: Nature *136*, 338 (1935)
256. Williams, C. M.: Nature *178*, 212 (1956)
257. Williams, P., Amos, T.G.: Austr. J. Zool. *22*, 147 (1974)
258. Wirtz, P., Beetsma, J.: Entomologia Exp. Appl. *15*, 517 (1972)
259. Wright, J.E., Smalley, H.E.: Arch. Environ. Contam. Toxicol. *5*, 191 (1977)
260. Wright, J.E., Spates, G.E.: Science *178*, 1292 (1972)
261. Wright, J.E., Campbell, J.B., Hester, P.: Environ. Entomol. *2*, 69 (1973)
262. Wright, J.E., Campbell, J.B., Oehler, D.D.: J. Econ. Entomol. *67*, 459 (1974)
263. Yang, R.S., Pelliccia, J.G., Wilkinson, C.F.: Biochem. J. *136*, 817 (1973)

Pflanzenschutzpräparate mikrobieller Herkunft

K. Bauer, E. Bischoff, H. v. Hugo

Bayer AG, Sparte Pharma – Institut für Biochemie
D-5600 Wuppertal 1

D. Berg

Bayer AG, Sparte Pflanzenschutz – Chemische Forschung
D-5600 Wuppertal 1

P. Kraus

Bayer AG, Sparte Pflanzenschutz – Anwendungstechnik – Biologische Forschung
D-5090 Leverkusen

Inhalt

Häufig verwendete Abkürzungen

LD_{50}	lethal dose 50% für 50% der Versuchstiere lethale Dosis
TLm	Tolerance Limit medium für 50% der Versuchstiere lethale Konzentration
l	löslich
ll	leicht löslich
nl	nicht löslich
MHK	Minimale Hemmkonzentration
p.o.	per os

1. Einführung

Schädlinge, Pflanzenkrankheiten und Unkräuter reduzieren die potentielle Welternte jährlich um etwa ein Drittel. Neben Kulturmaßnahmen und Resistenzzüchtung ist der chemische Pflanzenschutz ein wesentlicher Faktor bei der Bekämpfung dieser hohen Verluste.

Die Möglichkeiten des chemischen Pflanzenschutzes werden erweitert durch die Anwendung wirksamer Naturstoffe. Dabei wird die Fähigkeit von Mikroorganismen ausgenutzt, Sekundärstoffe zu bilden. Heute werden bereits speziell auf den Pflanzenschutz ausgerichtete Prüf- und Entwicklungsprogramme mit Mikroorganismen durchgeführt. Das Vorgehen dabei unterscheidet sich von dem bei synthetisierten Substanzen üblichen dadurch, daß hier von hochwirksamen Stoffen unbekannter Struktur und Konzentration ausgegangen wird; die chemische Synthese liefert dagegen definierte Substanzen, deren biologische Wirkung geklärt werden muß.

Der Entwicklungsaufwand für ein mikrobielles Pflanzenschutzmittel ist außerordentlich hoch. Nach der Isolierung von Mikroorganismen, deren Sekundärstoffwechsel man noch nicht kennt, werden in ersten Kultivierungs- und Prüfverfahren aktive Stämme ausgewählt. Es folgen Arbeiten zur Fermentation und Reindarstellung der Wirkstoffe. Da der Wirkstoffgehalt bei noch nicht optimierten Stämmen gering sein kann (< 100 mg/l), müssen zur Bereitstellung für breite biologische Prüfungen oft große Fermentationsvolumina (mehrere tausend Liter) aufgearbeitet werden. In dieser Stufe ist die Reproduzierbarkeit sowohl der Wirkstoffbildung als auch der Aufarbeitung nicht in gleichem Maße wie bei chemisch-synthetischen Präparaten gewährleistet. Eine systematische Verfahrensbearbeitung und Stammoptimierung ist daher erforderlich.

Voraussetzung für jede weitere Entwicklung wirksamer Naturstoffe ist ihre Reindarstellung sowie die Aufklärung der Konstitution. Damit sind bereits bis zur Entscheidung über die Brauchbarkeit eines Naturstoffs große finanzielle Aufwendungen notwendig. Um das Risiko zu mindern, sind staatliche Institutionen bereit, sich an solchen Programmen finanziell zu beteiligen [1, 2]. Man verspricht sich von solchen Arbeiten Pflanzenschutzmittel mit neuen Strukturen, die von Wirkungsspektrum, Selektivität und Wirkungsmechanismus her neue Aspekte eröffnen. Gleichzeitig ergeben sich aus eventuellen neuen Strukturen auch Ansätze für semi- und vollsynthetische Arbeiten. Eine Reihe bereits im Handel befindlicher Naturstoffe zeigt, daß diese Vorgehensweise zum Erfolg führen kann. Indessen können auch bei mikrobiell gewonnenen Wirkstoffen gleiche Probleme hinsichtlich Toxikologie und Resistenzentwicklung auftreten wie bei Präparaten aus der chemischen Synthese. Bei der Abbaubarkeit von Naturstoffen ist zu erwarten, daß enzymatisch synthetisierte Stoffe grundsätzlich einem enzymatischen Abbau zugänglich sind. Damit sind sie im Sinne einer geringeren Umweltbelastung positiv zu beurteilen.

In der folgenden Zusammenstellung, die keinen Anspruch auf absolute Vollständigkeit erheben kann, wird die Vielfalt an Strukturen und Wirkungen deutlich, die sich bei der bisherigen Suche nach Pflanzenschutzmitteln aus Mikroorganismen ergeben hat. Eine Vielzahl von Forschungs- und Entwicklungsprodukten sind in ihrer Konstitution noch unbekannt. Aus diesem Grund wurden zur Charakterisierung die physikalisch-chemischen Parameter ausführlich wiedergegeben.

2. Fungizide und Bakterizide

2.1. Einleitung

Industrielle Präparate auf chemisch-synthetischer Grundlage stehen den Landwirten seit dem 20. Jahrhundert zur Verfügung. Ziel der fortlaufenden Entwicklungen war einerseits, wirksamere Substanzen zu finden, die in niedrigeren Aufwandmengen eingesetzt werden können, oder die ein breiteres Wirkungsspektrum aufwiesen. Andererseits galt es, die Toxizität dieser Substanzen gegen Pflanzen und andere Organismen möglichst niedrig zu halten. Trotz intensiver Forschung und Entwicklung betragen auch heute noch die Verluste, die bei Großkulturen durch Pflanzenkrankheiten hervorgerufen werden, jährlich etwa 10–12% der potentiellen Welternte [3].

Obwohl in den USA zwischen 1951 und 1960 jährlich etwa 130 Mill. $ zur Bekämpfung von Pflanzenkrankheiten ausgegeben wurden, belief sich der durch Pflanzenkrankheiten (einschließlich Nematoden) verursachte Schaden auf etwa 4 Mrd. $ [4]. Durch intensive Bewirtschaftung nahmen die krankheitsbedingten Probleme sogar weiter zu. Dies wird durch den weltweit wachsenden Anteil der Fungizide und Bakterizide am chemischen Pflanzenschutz im Zeitraum 1971–1974 deutlich [5].

Früh erkannte man, daß Krankheitserregern durch natürliche Konkurrenz Grenzen gesetzt sind. So konnte Whetzel 1909 zeigen, daß der Pilz *Trichothecium roseum* das Wachstum mehrerer pflanzenpathogener Pilze hemmt [6]. Yarwood versuchte 1932 mit Konidien-Suspensionen von *Ampelomyces quisqualis* den Mehltau an Klee (*Erysiphe polygoni*) zu bekämpfen [7].

Diese Beobachtungen, daß Antagonisten und Hyperparasiten eine spezifische Bekämpfungsmöglichkeit pflanzenpathogener Pilze darstellten, und die Tatsache, daß in den 40er Jahren die Entdeckung der Antibiotika auf dem Humangebiet eine neue Epoche der Chemotherapie einleitete, führten zu einer intensiven Suche nach Wirkstoffen aus Mikroorganismen. Man setzte auch mit Erfolg breitwirksame Antibiotika aus der Humanmedizin im Pflanzenschutz ein [8]. Diese gemeinsame Anwendung bringt jedoch Probleme, besonders auf dem Gebiet der Resistenzentwicklungen, die den medizinischen Einsatz gefährden können [9]. Mit aus diesem Grund ist es in der Bundesrepublik Deutschland nicht erlaubt, im Pflanzenschutz Antibiotika zu benutzen, die auch in der Medizin Verwendung finden [10].

Seit einiger Zeit wird deshalb die Suche nach fungizid und bakterizid wirksamen Stoffwechselprodukten aus Mikroorganismen in speziell auf den Pflanzenschutz ausgerichteten Programmen betrieben. Erste Erfolge waren in Japan zu verzeichnen, wo in Screeningprogrammen jährlich bis 10000 Stämme von Mikroorganismen auf die Bildung aktiver Wirksubstanzen überprüft werden. Mit

®Blasticidin S kam 1961 das erste ausschließlich für die Anwendung in der Landwirtschaft entwickelte Antibiotikum in den Handel und gab damit Anstoß für Folgeprogramme [11]. Dabei erwiesen sich die Actinomyceten als eine besonders fündige Gruppe von Mikroorganismen [12].

Die im ersten Teil dieser Arbeit zusammengestellten Wirkstoffe sind Fungizide und Bakterizide, die eine Zulassung für die Verwendung als Pflanzenschutzmittel erhalten haben und zum größeren Teil Handelspräparate geworden sind. Im zweiten Teil werden Substanzen beschrieben, die gegen phytopathogene Mikroorganismen in Labor-, Gewächshaus- oder Freilandversuchen Wirkung gezeigt haben und somit eine intensivere Bearbeitung rechtfertigen.

Synonyma

Acetylendicarboxamid s. Cellocidin
Actidion s. Cycloheximid
Agrimycin s. Streptomycin
Agri-Strep s. Streptomycin
Albimycin s. Hondamycin
Aminomycin s. Valinomycin
Antibiotic 720-A s. Antimycin A_1
Antipiriculin s. Antimycin A_1
Aquamycin s. Cellocidin
Aureomycin s. Tetracyclin
B-98891 s. Mildiomycin
Cathomycin s. Novobiocin
Cellomate s. Cellocidin
14-Deoxylagosin s. Filipin
FH-3582-A s. Nonactin
Grisetin s. Griseofulvin
Kasumin s. Kasugamycin
KSM s. Kasugamycin
Lenamycin s. Cellocidin
N-329-A s. Nonactin
N-329-B s. Valinomycin
Natamycin s. Pimaricin
Pyrogreen s. Pyrrolnitrin
Shirahagen s. Chloramphenicol
TA-25-M-l s. Nonactin
Terramycin s. Oxytetracyclin
Tricin s. Trichothecin
Validacin s. Validamycin A
Virosin s. Antimycin A_1

Literatur

1. Yamaguchi, T.: Japan Pesticide Information *18*, 5 (1974)
2. Forschung Aktuell: Biotechnologie, Herausgeber: Der Bundesminister für Forschung und Technologie, Umschau-Verlag Frankfurt, S. 150ff. (1978)
3. Cramer, H.H.: Pflanzenschutz und Welternte; Pflanzenschutz-Nachrichten BAYER AG *20*, 1 (1967)
4. Apple, J.L. in: World Food, Pest Losses and the Environment, Kap. 3 (Hrsg. Pimentel, D.) AAAS Selected Symp. 13, Westview Press, Boulder, Colorado 1978
5. Wellmann, R.H.: Ann. Rev. Phytopathol. *15*, 153 (1977)
6. Whetzel, H.H.: Ont. Nat. Sci. Bull. *5*, 3 (1909)
7. Yarwood, C.E.: Phytopathol. *22*, 31 (1932)
8. Dekker, J.: World Rev. Pest Control *10*, 9 (1971)
9. N.N.: Phytopathol. News *12*, 55 (1978)
10. Heitefuß, R.: Pflanzenschutz, G. Thieme Verlag, Stuttgart 1975
11. Misato, T. in: Plant Protection in Japan 1976, Agriculture Asia, Special Issue *10*, 381 (1976)
12. Réhácek, Z.: Folia Microbiol. (Prag) *8*, 228 (1963)

2.2. Handelsprodukte

Blasticidin S
(Nihon Nohyaku Co., Ltd.)

Blasticidin S wurde in Japan als erstes Antibiotikum ausschließlich für die Anwendung auf dem landwirtschaftlichen Sektor entwickelt. Die stark basische Substanz wurde im Kulturfiltrat des Stammes *Streptomyces griseochromogenes* von Takeuchi et al. [1] gefunden. Die Isolierung gelang durch Adsorption an Aktivkohle mit anschließender Chromatographie an stark sauren Ionenaustauschern. Die chemische Struktur haben Yonehara und Otake [2] aufgeklärt.

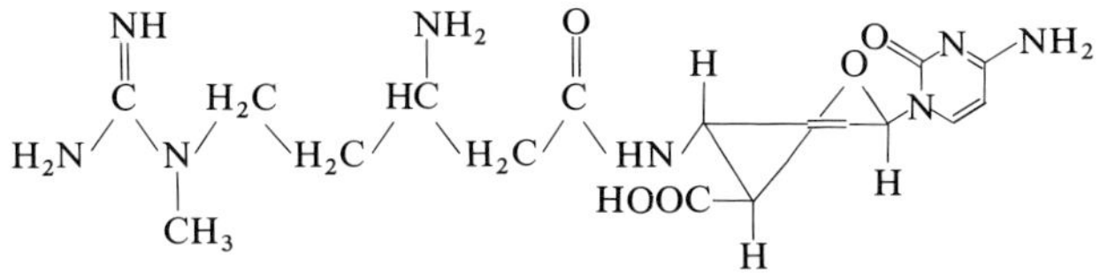

Gegen eine Reihe von Pilzen, Bakterien und Viren [3] wirksam, wird Blasticidin S nur gegen den besonders empfindlichen Pilz *Pyricularia oryzae* an Reis mit Feldkonzentrationen von 10–20 ppm eingesetzt [4]. Die biologische Wirkung wurde 1960 patentiert [5]. Bereits bei niedrigen Konzentrationen (10–20 ppm) kommt es bei zahlreichen Pflanzenarten zu Schäden, deshalb wird im Handelsprodukt das am wenigsten phytotoxische Monobenzyl-aminophenylsulfat verwendet [6]. Die fungitoxische Wirkung beruht auf der Hemmung der Proteinsynthese, wobei die Peptidbildung aus der Amino-acyl-t-RNS gehemmt wird [4]. Als weiterer Wirkungsmechanismus wird die Hemmung der DNS-Synthese angenommen [7]. Aufgrund der systemischen Eigenschaften von Blasticidin S kann es kurativ eingesetzt werden. Die Stabilität in der Pflanze ist gering [8].

Im Labor erzeugte resistente Stämme von *Pyricularia oryzae* zeichneten sich durch geringere Fungizid-Aufnahmeraten aus [9]. Im Freiland hat sich noch kein Fall von Resistenz gezeigt [10].

Auf dem Markt befindet sich Blasticidin S seit 1961. Es wird fast ausschließlich als Staub (0.16%) formuliert eingesetzt. Die Produktionszahlen für 1977 lagen bei 7500 t Präparat.

Neben den phytotoxischen Problemen wirkten sich auch die relativ hohe Toxizität (LD_{50} Maus p.o. 39,5 mg/kg) und Schleimhautreizungen bei der Staubapplikation negativ auf die Marktentwicklung aus [11].

Literatur: 1. Takeuchi, S. et al.: J. Antibiotics Ser. A. *11*, 1 (1958). – 2. Yonehara, H., Otake, N.: Tetrahedron Lett. 3785 (1966). – 3. Hirai, T. et al.: Phytopathology *56*, 1236 (1966). – 4. Huang, K.T., Misato, T.: Rev. Plant Protect. Res. *3*, 12 (1970). – 5. JA 16449 (1960), Nihon Nohyaku Co., Ltd. – 6. Dekker, J.: World Rev. Pest Control *10*, 9 (1971). – 7. Sullia, S.B., Griffin, D.H.: Biochem. Biophys. Acta *475*, 14 (1977). – 8. Misato, T., Yamaguchi, I., Homma, Y. in: Environmental Toxicology of Pesticides (Hrsg. Matsumara, F., Boush, G.M., Misato, T.) 587, Academic Press, New York, London 1972. – 9. Huang, K.T., Misato, T., Asuyama, H.: J. Antibiotics *17*, 71 (1964). – 10. Woodcock, D.: Structure Activity Relationships, in: Systemic Fungicides (Herausg. Marsh, R.W.) 2. Aufl. Longman, London und New York 1977. – 11. Misato, T. in: Environmental Quality and Safety, Global Aspects of Chemistry and Technology as applied to the Environment, Vol. 5, 48 (Hrsg. Coulston, F., Korte, F.) G. Thieme Publ. Stuttgart, Academic Press New York 1976.

Cellocidin

Cellomate (Toa Noyaku Co.; Ihara Chem. Co.; Nihon Nohyaku Co.; Hokko Chem. Ind. Co.)

Cellocidin wird von dem Stamm *Streptomyces chibaensis* produziert [1]. Seine Struktur wurde durch Suzuki und Okuma aufgeklärt [2]. Cellocidin ist identisch mit Aquamycin, Lenamycin und Acetylendicarboxamid.

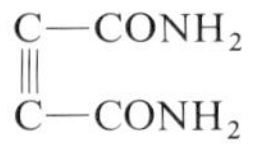

Die Synthese gelingt aus Fumarsäure oder Butandiol [3].

Das Antibiotikum ist mit 100–200 ppm wirksam gegen *Xanthomonas oryzae* an Reis. Da bereits 200 ppm zu Phytotoxizität an Reis führten, wird das 1964 in den Handel gebrachte Präparat nicht mehr produziert [4].

Die bakterizide Wirkung beruht auf einem Eingriff in den Citronensäure-Cyclus durch Hemmung NAD-abhängiger Dehydrogenasen [5]. Die akute Toxizität beträgt 90–125 mg/kg p.o. an der Maus [6].

Literatur: 1. Suzuki, S. et al.: J. Antibiotics *11*, 81 (1958). – 2. Suzuki, S., Okuma, K.: J. Antibiotics *11*, 84 (1958). – 3. Saggiomo, A.J.: J. org. Chem. *22*, 1171 (1957). – 4. Dekker, J.: World Rev. Pest Control *10*, 9 (1971). – 5. Huang, K.T., Misato, T.: Rev. Plant Prot. Res. *3*, 12 (1970). – 6. Woodcock, D.: Toxicological Considerations, in: Systemic Fungicides, Kap. 4, 2. Aufl. (Hrsg. Marsh, R.W.), Longman, London and New York 1977.

Chloramphenicol

®Shirahagen (Sankyo Co., Ltd.)

Chloramphenicol wurde 1947 von Ehrlich et al. im Kulturfiltrat von *Streptomyces venezuelae* entdeckt [1]. Die Struktur von Chloramphenicol wurde von Rebstock et al. durch Abbau und Synthese ermittelt [2]. Heute wird der Wirkstoff ausschließlich totalsynthetisch dargestellt.

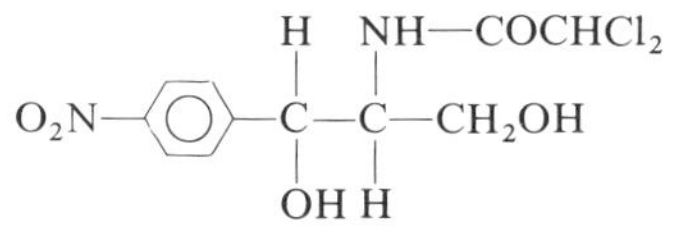

Chloramphenicol kann zur Bekämpfung phytopathogener Bakterien verwendet werden und kam 1964 nach der Zulassung in Japan als Reis-, Obst- und Gemüsebakterizid in den Handel [3]. Wirksam ist die D(–)Threo-Form, die in Mischung mit der L-Form Bestandteil des Handelspräparates Shirahagen ist, das mit 100–200 ppm Wirkstoff ohne Anzeichen von Phytotoxis gegen *Xanthomonas oryzae* an Reis wirksam ist [3]. Obwohl die Substanz ausgeprägt systemisches Verhalten zeigt [4], fand Chloramphenicol keine größere Anwendung (1974 wurden in Japan 10 t produziert [3]). Die Registrierung des für den Pflanzenschutz teuren Präparates lief 1975 aus. Chloramphenicol hemmt die ribosomale Proteinsynthese der 70 S-Einheit prokaryontischer Zellen sowie die mitochondriale Proteinbiosynthese [5].

Literatur: 1. Ehrlich, J. et al.: Science *106*, 417 (1947). – 2. Rebstock, M. et al.: J. Am. Chem. Soc. *71*, 2458 (1949). – 3. Misato, T. in: Plant Protection in Japan 1976, Agriculture Asia, Spec. Issue *10*, 381 (1976). – 4. Brandl, E. in: Die Antibiotica, Bd. II, 816 (Hrsg. Brunner, R., Machek, G.) Vlg. Hans Carl, Nürnberg 1965. – 5. Sisler, H.D.: Ann. Rev. Phytopathol. 7, 311 (1969).

Cycloheximid

®Actidion (Upjohn Co.)
Stereoisomere: z. B. Naramycin B, Isocycloheximid

Cycloheximid wurde 1946 von Whiffen et al. aus dem Kulturfiltrat von *Streptomyces griseus* isoliert [1] und seine Struktur von Leach [2] und Kornfeld et al. [3] beschrieben. Es wurde 1948 durch Upjohn Co. unter der Handelsbezeichnung Actidion als Pflanzenschutzmittel eingeführt.

Da Cycloheximid 4 Asymmetriezentren besitzt, sind zahlreiche Stereoisomere möglich.

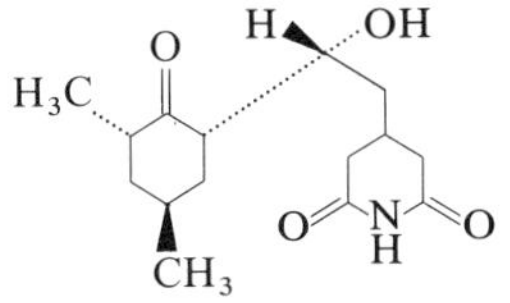

Cycloheximid ist breit wirksam gegen Eukaryonten. Trotz der hohen Wirksamkeit (bereits 1–4 ppm sind gegen Echte Mehltaupilze wirksam) ist die Möglichkeit praktischer Nutzung im Pflanzenschutz durch phytotoxische Nebenwirkungen auf wenige Indikationen beschränkt [4]. In den USA wird Actidion seit 1952 begrenzt gegen pilzliche Rasenkrankheiten genutzt [5]. In Japan wird das Präparat seit 1959 gegen Falsche Mehltaupilze an Zwiebeln und seit 1964 gegen *Guignardia laricini* an japanischer Lärche [6], allerdings jedoch nur in geringen Mengen (1974: 17 t) angewendet [7].

Das Antibiotikum wird von der Wurzel aufgenommen und in der Pflanze akropetal verteilt [8], wo es auch kurativ wirksam wird. Es greift in den pilzlichen Proteinstoffwechsel ein, indem es die Bildung der Peptidbindungen hemmt [9]. Unabhängig davon wurde die Hemmung der DNS-Synthese in Kulturen von *Achlya bisexualis* beobachtet [10]. Der Wirkungsmechanismus, der sich nicht nur gegen Mikroorganismen, sondern auch gegen Pflanzen und Tiere auswirkt, hat LD_{50}-Werte für die Maus p.o. von 133 mg/kg zur Folge.

Literatur: 1. Whiffen, A., et al.: J. Bact. *52*, 610 (1946). – 2. Leach, B.E., et al.: J. Am. Chem. Soc. *69*, 474 (1947). – 3. Kornfeld, E.C., et al.: J. Am. Chem. Soc. *71*, 150 (1949). – 4. Brandl, E., in: Die Antibiotica, Band 2, 816 (Hrsg. Brunner, R., Machek, G.) Verlag Hans Carl, Nürnberg 1965. – 5. Dekker, J., in: Fungicides, Bd. 2, Chemistry and Physiology, 580 (Hrsg. D.C. Torgeson) Academic Press, N.Y., London 1969. – 6. Dekker, J.: World Rev. Pest Control *10*, 9 (1973). – 7. Misato, T., Agriculture Asia, Spec. Issue *10*, 371 (1976). – 8. Fawcett, C.H., Spencer, D.M.: Ann. Rev. Phytopathol. *8*, 403 (1970). – 9. Sisler, H.D.: Ann. Rev. Phytopathology, *7*, 311 (1969). – 10. Sullia, S.B., Griffin, D.H.: Biochem. Biophys. Acta *475*, 14 (1977).

Ezomycine
(Sankyo Co., Ltd.)

Ezomycine werden von einem Streptomyces-Stamm, der *Streptomyces kitazawaensis* ähnlich ist, produziert. Takoaka et al. [1] isolierten einen Antibiotika-Komplex, dem bisher 8 Ezomycine zugeordnet werden konnten [2, 3]. Die Aufklärung der jeweiligen Strukturen gelang Sakata et al. [4, 5]. Die Verbindungen, die zur Gruppe der Pyrimidinnucleoside gehören, sind nach Sakata et al. [6] aufgrund des in den Ezomycinen A_1, B_1, C_1 und D_1 enthaltenen l-Cystathionins fungitoxisch.

Als Beispiel sei die Struktur von Ezomycin A_1 wiedergegeben:

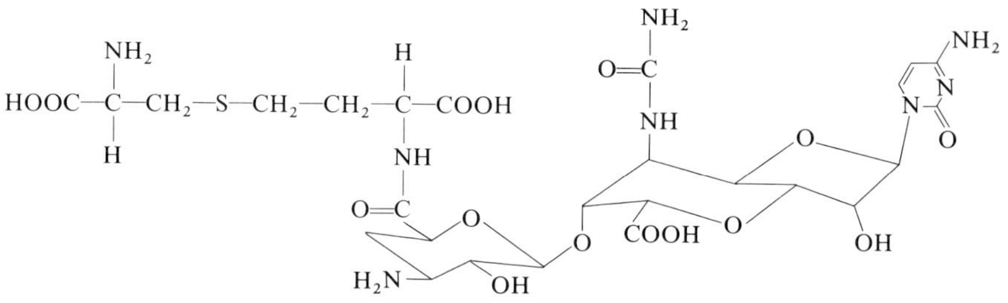

Die Ezomycine gehören zu den Antibiotika, die in Japan bei der Suche nach neuen mikrobiellen Stoffwechselprodukten gefunden worden sind. Auffallend ist das enge Wirkungsspektrum der Substanzen gegen die phytopathogenen Pilze *Sclerotinia sclerotiorum* und *Botrytis cinerea* [2]. Zwar erhielten die Ezomycine 1970 die Zulassung als Pflanzenschutzmittel gegen die Stengelfäule an Bohnen, verursacht durch *Sclerotinia sclerotiorum*, doch haben sie auf dem Markt keine Rolle gespielt [7].

Literatur: 1. JA 515 332, Sankyo Co., Ltd., 1971. – 2. Sakata, K. et al.: Agr. Biol. Chem. *38*, 1883 (1974). – 3. Sakata, K. et al.: Agr. Biol. Chem. *41*, 2027 (1977). – 4. Sakata, K. et al.: Tetrahedron Lett. *49*, 4327 (1974). – 5. Sakata, K. et al.: Tetrahedron Lett. *37*, 3191 (1975). – 6. Sakata, K. et al.: Agr. Biol. Chem. *37*, 697 (1973). – 7. Misato, T.: Agriculture Asia, Spec. Issue *10*, 371 (1976).

Griseofulvin
(Glaxo Laboratories Ltd.)
Grisetin, Grisovit, Murphulvin

Griseofulvin wurde aus dem Mycel des Stammes *Penicillium griseofulvum* durch Oxford et al. 1939 isoliert [1]. Bereits 1946 wurde die breite Wirkung gegen phytopathogene Pilze von Brian et al. beschrieben [2]. Die Struktur wurde von Grove et al. 1951 aufgeklärt [3].

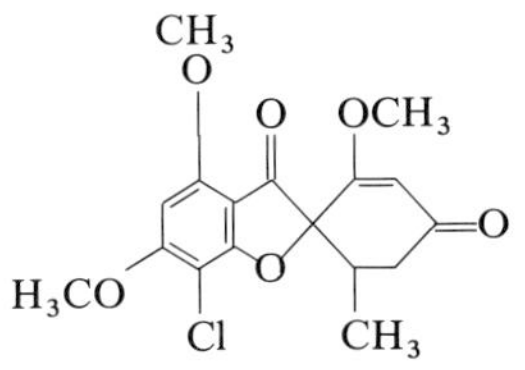

Griseofulvin erhielt 1957 in England die Zulassung gegen *Botrytis cinerea* an Salat [4]. 1959 wurde das Fungizid in Japan zur Bekämpfung von *Sclerotinia mali* im Obstbau und der Fusarium-Welke an Melonen registriert [5]. Daneben wurde auch gute Wirkung gegen Echte Mehltaupilze an Zierpflanzen und Gemüse nachgewiesen [6]. Die Wirkung, die sich in einer Hemmung von Hyphenwachstum und Infektionsorganen (Haustorien) äußert, beruht auf einem Eingriff in Vorgänge bei der Mitose [7]. Eine Wechselwirkung von Griseofulvin und Tubulin, das zum Aufbau des Spindelapparates benötigt wird, weist ebenfalls in diese Richtung [8].

Griseofulvin, das später auch im tiermedizinischen Bereich Anwendung fand, konnte sich als Pflanzenschutzmittel nicht durchsetzen. Obwohl es sich nach Aufnahme über Wurzel und Blatt akropetal verteilt [9] und pflanzenverträglich ist, zeigte es gegenüber den etablierten Fungiziden keine Vorteile. Einer größeren Anwendung stand vor allem der durch hohe Produktionskosten bedingte Preis entgegen.

Der Wirkstoff ist mit einer LD_{50} von 400 mg/kg bei Ratten p.o. wenig toxisch [10].

Literatur: 1. Oxford, A.E. et al.: Biochem. J. *33*, 240 (1939). – 2. Brian, P.W. et al.: Trans. Brit. Mycol. Soc. *29*, 173 (1946). – 3. Grove, J.F. et al.: Chem. and Ind. (London) *11*, 219 (1951). – 4. Brandl, E., in: Die Antibiotica, Bd. 2, 877 (Hrsg. Brunner, R., Machek, G.) Verlag Hans Carl, Nürnberg 1965. – 5. Dekker, J.: World Rev. Pest Control *10*, 9 (1971). – 6. Brian, P.W.: Trans. Brit. Mycol. Soc. *43*, 1 (1960). – 7. Gull, K., Trinci, A.P.J.: Nature *244*, 292 (1973). – 8. Herzog, W., Weber, K.: J. Molec. Biol. *111*, 329 (1977). – 9. Crowdy, S.H., Pramer, D.: Chem. Ind. 160 (1955). – 10. Woodcock, D.: Toxicological Considerations, in: Systemic Fungicides. (Hrsg. Marsh, R.W.) 2. Aufl., Longman, London, New York 1977.

Kasugamycin
KSM, Kasumin (Hokko Chemical Industr. Co., Ltd.)

Die Substanz wurde bei Kultivierung von *Streptomyces kasugaensis* M 338-M 1 1964 von Umezawa et al. gefunden [1], beschrieben [2] und patentiert [3].

Die Isolierung erfolgte aus dem Kulturfiltrat durch Adsorption an Kohle, Elution mit 0,1 N HCl und Chromatographie an stark sauren Ionenaustauschern. Das Antibiotikum wird als kristallines, wasserlösliches Hydrochlorid gewonnen.

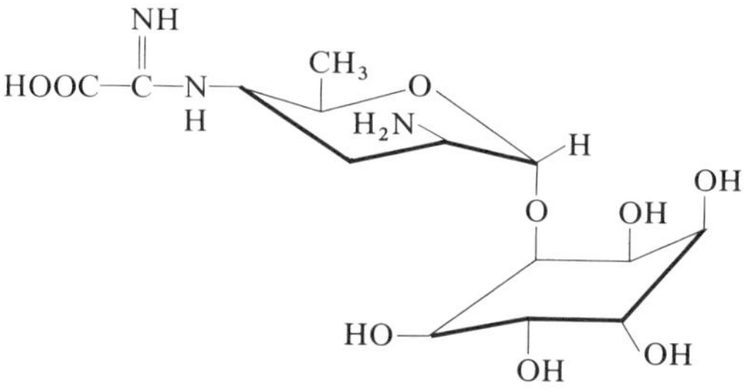

Kasugamycin ist spezifisch wirksam gegen *Pyricularia oryzae*, einen Schadpilz, der den Reisanbau in Asien vor große Probleme stellt. Bereits mit 20 ppm wird *Pyricularia oryzae* an Reis wirksam bekämpft, wobei das Mycelwachstum stärker als die Sporenkeimung gehemmt wird [4].

Die Fungitoxizität von Kasugamycin ergibt sich aus der Hemmung der Proteinsynthese [5], wobei Okuyama et al. [6] fanden, daß bei *Escherichia coli* Kasugamycin die Bindung von Aminoacyl-t-RNS an den Komplex der m-RNS mit den 30 S Ribosomenuntereinheiten stört. Seit 1965 wird Kasugamycin unter der Handelsbezeichnung Kasumin® (Hokko K.K.) in Japan gegen *Pyricularia oryzae* eingesetzt, wobei hauptsächlich eine 0,3%ige Staubformulierung zur Anwendung kommt. Probleme durch resistente *Pyricularia oryzae*-Stämme traten zunehmend in den Distrikten auf, wo ausschließlich Kasugamycin zur Bekämpfung eingesetzt wurde [8]. 1973 wurde deswegen die Anwendung eingestellt. Daraufhin nahmen die resistenten Stämme deutlich ab, so daß danach Kasugamycin – allerdings verstärkt in Kombination mit anderen Fungiziden – wieder eingesetzt wurde [7]. 1977 betrug in Japan die Produktion von Kasugamycin einschließlich der Kombinationspräparate ca. 35000 t. Bei der Anwendung ergeben sich keine Probleme der Phytotoxizität, die Toxizität bei Säugetieren (LD_{50}: 2000 mg/kg p.o. bei der Maus) und Fischen (TLm: 1 mg/ml bei Karpfen) ist niedrig [7].

Literatur: 1. Umezawa, H. et al.: J. Antibiotics Ser. A *18*, 101 (1965). – 2. Ishiyama, T. et al.: J. Antibiotics Ser. A *18*, 115 (1965). – 3. JA 21757, 19.12.1966, Hokko Chemical Industr. Co., Ltd. – 4. Okamoto, H.: Japan Pesticide Information *10*, 66 (1972). – 5. Tanaka, N. et al.: J. Biochem. *60*, 429 (1966). – 6. Okuyama, A. et al.: J. Antibiotics *28*, 903 (1975). – 7. Misato, T., in Coulston, F., Korte, F.: Environmental Quality and Safety. Bd. 5, 48, Academic Press, New York 1976. – 8. Ito, I. et al.: Ann. Phytopath. Soc. Japan *43*, 301 (1977).

Mildiomycin
(Takeda Chem. Ind.)
B 98891

1974 wurde gefunden, daß *Streptoverticillium rimofaciens* B 98891 eine fungizid wirksame Substanz bildet, die sich durch Chromatographie an Amberlite IRC-50 darstellen läßt [1]. Diese Substanz wird seitdem von Takeda Chem. Ind. für die Verwendung im Pflanzenschutz entwickelt [2].

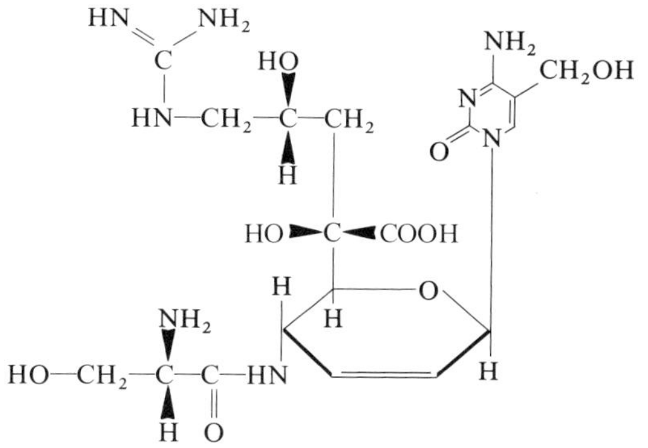

[3]

Da vom Wirkungsspektrum her schwerpunktmäßig die Echten Mehltaupilze erfaßt werden [4], erhielt der Wirkstoff den Namen Mildiomycin und wird seit 1977 in Japan in offiziellen Prüfungen für eine Registrierung zur Bekämpfung dieser Pflanzenkrankheiten getestet.

Im Konzentrationsbereich von 40–80 ppm können u.a. folgende Echte Mehltaupilze wirksam bekämpft werden [2]:

Sphaerotheca spp. an Gurken, Melonen, Rosen, Erdbeeren,
Erysiphe spp. an Tabak, Tomate, Getreide, Erbsen,
Podosphaera leucotricha an Apfel,
Uncinula necator an Wein,
Leveillula taurica an Paprika.

An den geprüften Pflanzenarten zeigten sich auch bei Überdosierungen (an Apfel, Erdbeere und Gurke bis 1280 ppm) keine Pflanzenschäden.

Mildiomycin wird von der Wurzel und von Blättern aufgenommen [2], so daß auch eine kurative Bekämpfung endoparasitischer Mehltaupilze (*Leveillula sp.*) möglich wird. Zum Einsatz kommt der Wirkstoff in einer 80%igen Formulierung für die Spritzbehandlung.

Toxizität [2,5]: Die LD_{50}-Werte bei der Maus betragen: p.o. > 5000 mg/kg, s.c. > 1100 mg/kg, i.v. > 600 mg/kg, i.p. > 1000 mg/kg. Die Fischtoxizität bei Killi-Fischen (LC_{50}: 168 h) liegt bei > 40 ppm. Im 3-Monate-Fütterungsversuch an Ratten wurden 50 mg/kg und Tag ohne Schäden vertragen.

Literatur: 1. DOS 2 513 249, 26. 3. 1975, Takeda Chem. Ind.– 2. Kusaka, T. et al.: Proc. 1979 Brit. Crop Prot. Conf. – Pests and Diseases, *2*, 589 (1979). – 3. Kamiya, K. et al.: Tetrahedron Lett. *44*, 4277 (1978). – 4. Iwasa, T. et al.: J. Antibiotics Ser. A. *31*, 511 (1978). – 5. Harada, S., Kishi, T.: J. Antibiotics Ser. A. *31*, 519 (1978).

Novobiocin
Cathomycin (Meiji Seika Kaisha Ltd.)

Novobiocin wird von dem Stamm *Streptomyces spheroides* und anderen Actinomyceten gebildet [1, 2]. Die Struktur wurde 1958 aufgeklärt [3, 4].

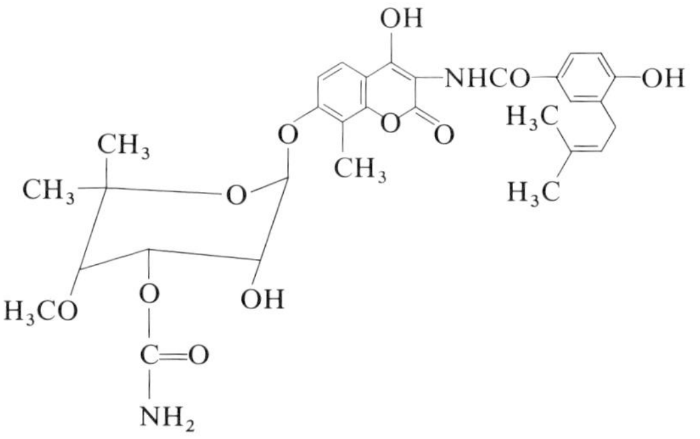

Dieses antibakteriell wirkende Antibiotikum, auch als Cathomycin bekannt, erhielt 1968 die Zulassung in Japan zur Bekämpfung einer bakteriellen Welkekrankheit an Tomaten (*Corynebacterium michiganense*) [5]. Junge Tomatensämlinge werden vor dem Verpflanzen durch eine Tauchbehandlung der Wurzeln in einer 100 ppm Lösung von Novobiocin-Na 83% SL wirksam geschützt [6]. Das Präparat konnte sich aufgrund seines engen Wirkungsspektrums nicht in größerem Umfang durchsetzen.

Literatur: 1. Wallick, H. et al.: Antibiotics Ann. 909 (1955/56). – 2. Smith, C.G. et al.: Antibiotics Chemother. *6*, 135 (1956). – 3. Spencer, C.F. et al.: J. Am. Chem. Soc. *80*, 140 (1958). – 4. Stammer, C.H. et al.: J. Am. Chem. Soc. *80*, 137 (1958). – 5. Huang, K.T., Misato, T.: Rev. Plant Prot. Res. *3*, 12 (1970). – 6. Iwata, Y.: Mededelingen Rijksfaculteit Landbouwwetenschappen Gent *33*, 1113 (1968).

Polyoxine
(Hokko Chemical Industr. Co., Ltd.; Kumiai Chemical Industr. Co., Ltd.; Nihon Nohyaku Co., Ltd.)

Die Polyoxine gehören zur Gruppe der Peptidyl-pyrimidin-nucleosid Antibiotika und werden von dem Stamm *Streptomyces cacaoi* var. asoensis produziert und durch Chromatographie an Kationenaustauschern isoliert [1, 2]. Bisher wurden 13 zur Gruppe der Polyoxine gehörende Verbindungen (A-M) beschrieben [3] und patentiert [4]. Eine Zusammenfassung der Verbindungen ist bei Isono et al. [3] wiedergegeben.

Nach Isono et al. [5] sind alle Komponenten außer C und I wirksam gegen eine Vielzahl pflanzenpathogener Pilze, wobei die verschiedenen Komponenten unterschiedliche Wirkspektren zeigen.

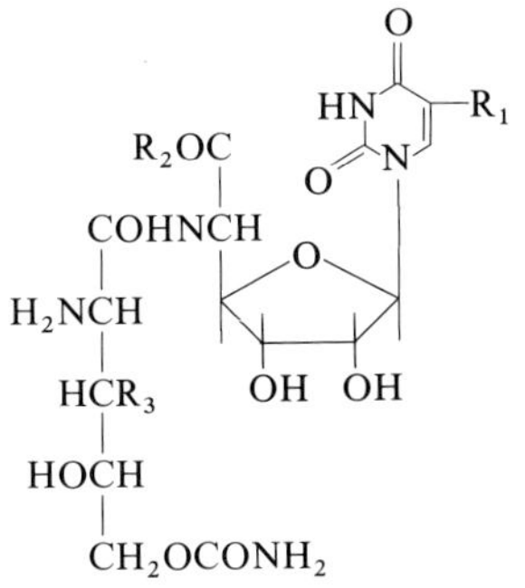

Polyoxin	R_1	R_2	R_3
A	CH_2OH	*	OH
B	CH_2OH	HO	OH
D	COOH	HO	OH
E	COOH	HO	H
F	COOH	*	OH
G	CH_2OH	HO	H
H	CH_3	*	OH
J	CH_3	HO	OH
K	H	*	OH
L	H	HO	OH
M	H	HO	H

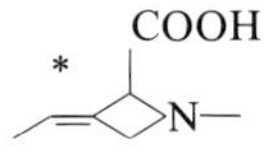

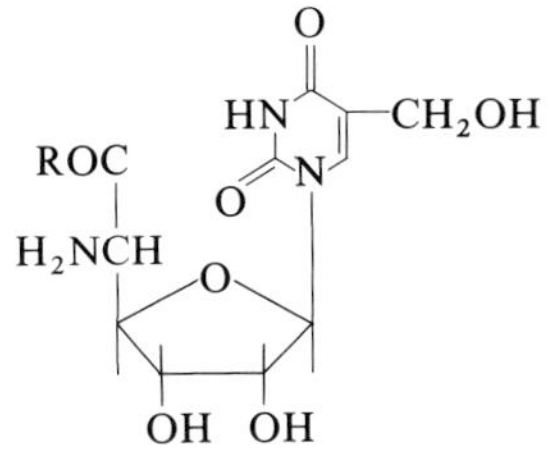

	R
C	HO
I	COOH

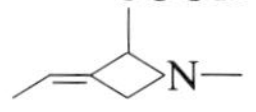

Von den Polyoxin-Komponenten befinden sich nach der Registrierung 1967 in Japan zwei Produkte im Handel. Ein Polyoxin D-reiches Präparat wird gegen *Pellicularia sasakii* an Reis eingesetzt und ist mit 125–500 ppm wirksam. Hingegen wirkt ein Polyoxin B-reiches Präparat besonders gegen *Alternaria* spp. im Obstbau [6].

Bei der Praxisanwendung der systemisch wirkenden Polyoxine [6] treten keine durch Phytotoxizität bedingten Probleme auf [7]. Untersuchungen zum Wirkungsmechanismus von Polyoxin D zeigten, daß der Einbau von N-Acetyl-Glucosamin in Chitin bei der Zellwandsynthese gehemmt wird [8], was sich morphologisch in charakteristischen Anschwellungen der Hyphen äußert [9]. Die Toxizität gegenüber Säugetieren ist aufgrund dieses pilzspezifischen Wirkungsmechanismus niedrig (LD_{50}: Maus p.o. > 15000 mg/kg) [7].

Die Produktion betrug 1977 für den Reisbau incl. Kombinationspräparate ca. 2200 t, fast ausschließlich als Staub formuliert. Für den Obstbau wurden 1977 ca. 650 t Polyoxin und Kombinationen als Sprühpulver formuliert.

1972 erstmals aufgetretene Wirkungsabfälle, bedingt durch das Auftreten von resistenten Stämmen von *Alternaria kikuchiana* in Birnen-Anlagen [10], führten in diesen Gebieten zum Einsatz von Kombinationspräparaten oder anderen Fungiziden.

Literatur: 1. Suzuki, S. et al.: J. Antibiotics Ser. A. 18, 131 (1965). – 2. Isono, K. et al.: Agr. Biol. Chem. *29*, 848 (1965). – 3. Isono, K. et al.: J. Am. Chem. Soc. *91*, 7490 (1969). – 4. JA. 41894, 30.6.1967. – 5. Isono, K. et al.: Agr. Biol. Chem. *31*, 190 (1967). – 6. Dekker, J.: World Rev. Pest. Control *10*, 9 (1971). – 7. Huang, K.T., Misato, T.: Rev. Plant Prot. Res. 3, 12 (1970). – 8. Ohta, N., Kakiki, K., Misato, T.: Agric. Biol. Chem. *34*, 1224 (1970). – 9. Ishizaki, H. et al.: Ann. Phytopath. Soc. Japan *41*, 131 (1975). – 10. Iida, W.: Japan Pesticide Inform. *23*, 13 (1975).

Streptomycin

Agrimycin (Pfizer and Co.); Agri-Strep (Merck and Co., Inc.)

Streptomycin, das zur Gruppe der Aminocyclitole gehört, hat im humanmedizinischen Anwendungsbereich breite Verwendung gefunden und wurde als erstes Antibiotikum im Pflanzenschutz gegen pathogene Bakterien verwendet [1]. Der Wirkstoff wird aus dem Kulturfiltrat verschiedener Streptomyces-Arten (z.B. *Streptomyces griseus*) isoliert und als wasserlösliches Hydrochlorid oder -sulfat eingesetzt. Seine Struktur haben Kuehl et al. 1948 aufgeklärt [2].

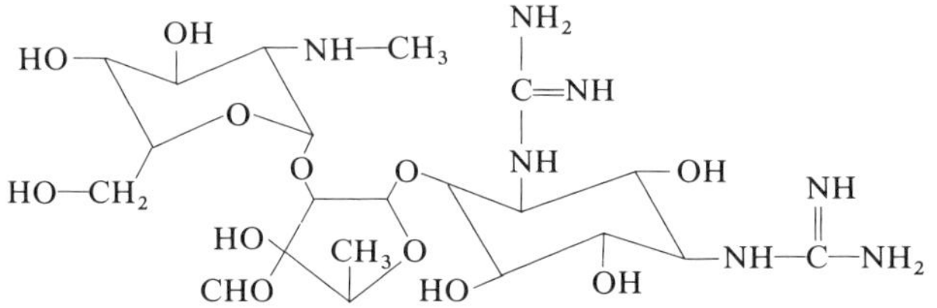

Streptomycin wird seit 1954 in den USA gegen *Erwinia amylovora* (Feuerbrand) an Kernobst sowie gegen Gemüsebakteriosen angewendet. In Japan ist Streptomycin seit 1957 zur Bekämpfung von *Pseudomonas tabaci* an Tabak registriert [3]. Höhere Dosierungen (>200 ppm) können an empfindlichen Pflanzen (z.B. Reis) zu Schäden führen, so daß der breiten Anwendung Grenzen gesetzt sind [4]. Der Wirkstoff wird von der Pflanze aufgenommen und verteilt sich akropetal, allerdings läßt sich diese systemische Eigenschaft bei Gehölzen nicht ausnutzen [5]. Die Wirkung, die durch einen Eingriff in die bakterielle Proteinsynthese hervorgerufen wird [6], wurde in den USA in großem Umfang zur Feuerbrand-Bekämpfung ausgenutzt, obwohl bereits in den 50er Jahren aufgrund der Resistenzgefährdung die Anwendung von Kombinationspräparaten empfohlen wurde [7].

1971 traten in Kernobstanlagen in Kalifornien, wo Streptomycin in dichter Spritzfolge gegen Feuerbrand zum Einsatz kam, die ersten resistenten Stämme von *Erwinia amylovora* auf, so daß inzwischen die Bekämpfung nur noch mit Kombinationspräparaten (mit Oxytetracyclin, zu dem keine Kreuzresistenz besteht) möglich ist [8]. Trotz der Einschränkungen durch Auftreten von resistenten Bakterienstämmen [9, 10, 11] bleibt Streptomycin das am stärksten verbreitete Antibiotikum im Pflanzenschutz. Allerdings hat die Möglichkeit der Resistenzbildung auch bei humanpathogenen Bakterien dazu geführt, daß Streptomycin in einigen Ländern nicht für den Pflanzenschutz zugelassen wurde.

Die akute Toxizität ist mit einer LD_{50} von 9000 mg/kg bei Mäusen p.o. sehr gering.

Literatur: 1. Mitchell, J.W.: Science *115*, 114 (1952). – 2. Kuehl, F.H. et al.: J. Am. Chem. Soc. *70*, 2325 (1948). – 3. Dekker, J.: World Rev. Pest Control *10*, 9 (1971). – 4. Huang, K.T., Misato, T.: Rev. Plant Protec. Res. *3*, 12 (1970). – 5. Brandl, E. in: Die Antibiotica, Bd. 2, 877 (Hrsg. Brunner, R., Machek, F.) Verlag Hans Carl, Nürnberg 1965. – 6. Gottlieb, D., Shaw, P.D.: Ann. Rev. Phytopathol. *8*, 371 (1970). – 7. Beer, S.V., Norelli, J.L.: Plant Disease Rep. *60*, 624 (1976). – 8. Moller, W.J. et al.: Calif. Agriculture, June 1973, 4. – 9. Iida, W.: Jap. Pesticide Inform. *23*,

13 (1975). – 10. Knauss, J.F.: Plant Disease Rep. *56*, 394 (1972). – 11. McMillan, jr., R.T. et al.: J. Agric. Food Chem. *20*, 886 (1972). – 12. Woodcock, D.: Toxicological Considerations, in: Systemic Fungicides (Hrsg. Marsh, R.W.) 2. Aufl., Longman, London, New York 1977.

Tetracyclin

Aureomycin, Terramycin (Ch. F. Pfizer und Co.)

7-Chlor-tetracyclin wurde 1948 im Kulturfiltrat von *Streptomyces aureofaciens* entdeckt, daraus isoliert [1] und 1949 patentiert [2]. Dieses Antibiotikum ist unter dem Handelsnamen Aureomycin bekannt. Seine Struktur wurde von Steffens et al. aufgeklärt [3].

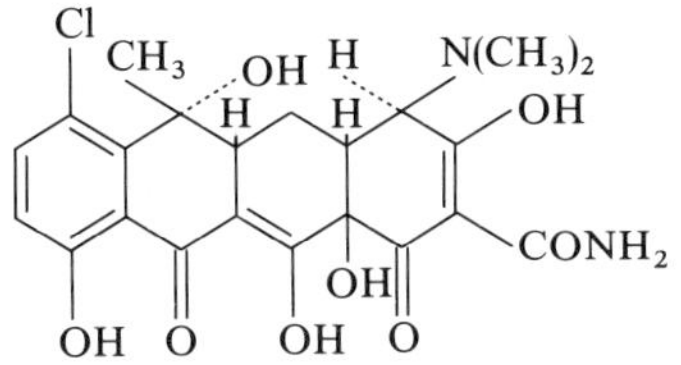

Das unter dem Namen Terramycin bekannte Oxytetracyclin wurde in seiner Struktur von Hochstein et al. [4] aufgeklärt.

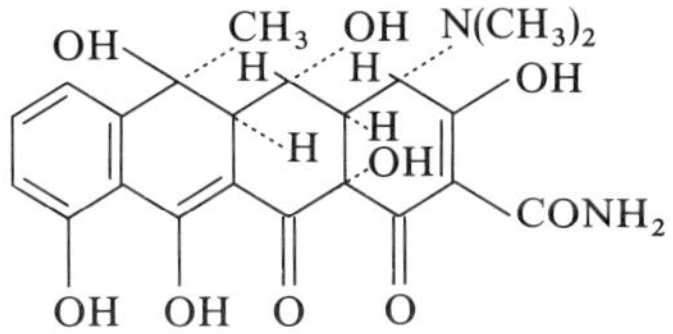

Von den Tetracyclinen wird lediglich das Oxytetracyclin allein (Terramycin: 20% Wirkstoff) oder mit Streptomycin kombiniert (Agrimycin, Streptocycline) in größerem Umfang gegen Pflanzenkrankheiten eingesetzt.

Während die Wirkung gegen pilzliche Krankheitserreger im allgemeinen unbefriedigend ist [5], besitzt Oxytetracyclin ein breites Wirkungsspektrum gegen gram-positive und -negative Bakterien [6]. Im Pflanzenschutz kann diese Wirkung ausgenutzt werden gegen *Xanthomonas pruni*, eine Bakteriose an Pfirsich [7], für dessen Bekämpfung 1968 und 1969 eine Zulassung in den USA erteilt wurde [8]. 400 ppm Terramycin schützen Pecan-Pflanzen gegen *Agrobacterium tumefaciens* (crown gall) [9].

Vergilbungskrankheiten können sowohl durch Viren als auch durch Mycoplasmen hervorgerufen werden. Durch Oxytetracyclin, das gegen Mycoplasmen wirksam ist, wird eine Unterscheidung der von Mycoplasmen hervorgerufenen Pflanzenkrankheiten von den Virosen möglich [10].

Das Birnensterben, das in Kalifornien 1959–1972 zum Verlust von 2 Mill. Birnenbäumen führte, kann durch Stamminjektion ebenso bekämpft werden [11] wie das Lethal Yellowing der Kokospalme, gegen das 1–3 g Wirkstoff/Baum in Langzeitversuchen wirksamen Schutz brachten [12]. In den USA wurde Terramycin für diese Anwendungen 1973 bzw. 1974 registriert [11, 12].

Die Wirksamkeit von Oxytetracyclin beruht auf der Hemmung der Proteinsynthese bedingt durch eine Blockierung der Bindungsstellen für Aminoacyl-t-RNS [6].

Literatur: 1. Onggar, B.M.: N.Y. Acad. Sci. *51*, 175 (1948). – 2. US. 2 482 055, American Cyanamid Co. (1949). – 3. Stephens, C.R. et al.: J. Am. Chem. Soc. *76*, 3568 (1954). – 4. Hochstein, F.H. et al.: J. Am. Chem. Soc. *75*, 5455 (1953). – 5. Müller, H.J.: Arch Pflanzensch. *5*, 83 (1969). – 6. Dekker, J.: World Rev. Pest Control *10*, 9 (1971). – 7. Keil, H.L., Carroll, V.J.: Plant Dis. Rep. *51*, 345 (1967). – 8. Keil, H.L., Weaver, L.O.: ibid *54*, 208 (1970). – 9. Cole, J.R.: ibid *53*, 712 (1969). – 10. Hull, R.: PANS *18*, 154 (1972). – 11. Beutel, J.A. et al.: Calif. Agriculture Aug. 1977, 12. – 12. McCoy, R.E. et al.: Phytopathology *66*, 1148 (1976)

Trichothecin

Tricin (Allunionsinstitut für Antibiotika-Forschung, UdSSR)

Der Antagonismus zwischen *Trichothecium roseum* und pflanzenpathogenen Pilzen wurde zuerst von Whetzel [1], später von Boning [2], Koch [3] sowie Greaney und Machacek [4] beschrieben. Seine hemmende Wirkung auf *Botrytis allii*, die von Brian und Heming [5] gefunden wurde, beruht auf einer Substanz, die von Freeman und Morrison [6, 7] aus dem Kulturfiltrat isoliert und in ihren Eigenschaften beschrieben wurde.

Freeman und Gill [8] fanden, daß Trichothecin der Isocrotonylester des Ketoalkohols Trichothecolon ist. Die genaue Struktur wurde von Fishman et al. [9] aufgeklärt (Physikalisch-chemische Daten s. 4. Herbizide).

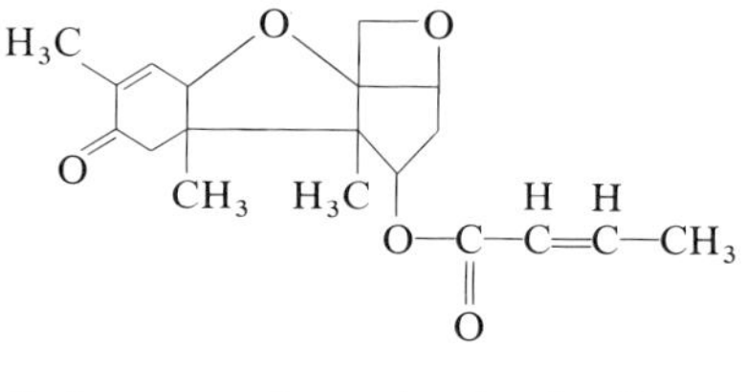

Trichothecin bewies in Laborversuchen eine breite fungizide Wirkung [10], die jedoch im Freiland nur begrenzt ausgenutzt werden kann.

Zwar stellten Ubrizsy und Vörös [11] Wirkungen gegen den Weizensteinbrand (*Tilletia caries*), den Apfelmehltau (*Podosphaera leucotricha*) und eine Anthracnose an Melonen (*Colletotrichum lagenarium*) fest, doch konnte die Wirkung nicht mit der von Handelspräparaten konkurrieren. Zudem zeigten sich bei höheren Anwendungskonzentrationen Schäden und Keimhemmungen bei Pflanze und Saatgut (s. Kap. 4 Herbizide).

Eingesetzt wird das Präparat unter anderem mit der Handelsbezeichnung „Tricin" in Ungarn gegen eine durch *Monilia cinerea* an Kirschen verursachte Krankheit mit einem Wirkungsgrad von 75–90% bei einer Konzentration von 50 ppm. Spaar [12] berichtete von der Anwendung von Trichothecin in der UdSSR als 0,04% ige Suspension gegen Echten Mehltau an Gurken im Gewächshaus.

Literatur: 1. Whetzel, H.H.: Ontario. Nat. Sci. Bull. *5*, 3 (1909). – 2. Boning, K.: Phytopath. Z. *6*, 113 (1933). – 3. Koch, L.W.: Sci. Agric. *15*, 80 (1934). – 4. Greaney, F.J., Machacek, J.E.: Sci. Agric. *15*, 377 (1935). – 5. Brian, P.W., Henning, H.G.: J. Gen. Microbiol. *1*, 158 (1947). – 6. Freeman, G.G., Morrison, R.J.: Nature *162*, 30 (1948). – 7. Freeman, G.G., Morrison, R.J.: Biochem. J. *44*, 1 (1949). – 8. Freeman, G.G., Gill, J.E.: Nature *166*, 698 (1950). – 9. Fishman, J. et al.: J. Chem. Soc., 3948 (1960). – 10. Freeman, G.G., Morrison, R.J.: J. Gen. Microbiol. *3*, 60 (1948). – 11. Ubrizsy, G., Vörös, J.: Acta Phytopath. Hung. *2*, 379 (1967). – 12. Spaar, D.: Nachr.-blatt für Pflanzenschutzdienst DDR *27*, 47 (1973).

Validamycin A

Validacin (Takeda Chemical Industr., Ltd.)

Validamycin A gehört in die Gruppe der Aminocyclitole und wird aus dem Kulturfiltrat des Stammes *Streptomyces hygroscopicus* var. limoneus isoliert [1, 2, 3]. Die Substanz wird durch Chromatographie an sauren Ionenaustauschern gewonnen.

Die chemische Struktur haben Horii et al. [4] aufgeklärt.

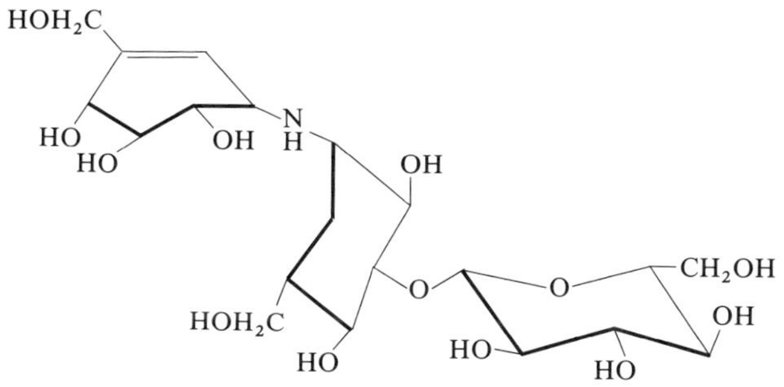

Validamycin A ist vornehmlich wirksam gegen Basidiomyceten, wobei sich die Wirkung primär nicht in einer Wachstumshemmung sondern in einer Veränderung der Morphologie (abnorme Verzweigung der Hyphen) äußert [5]. Das Antibiotikum wirkt gegen *Pellicularia sasakii* an Reis, wobei im frühen Infektionsstadium eine einmalige Spritzung von 30 ppm ausreicht.

Die Substanz ist nicht phytotoxisch und zeichnet sich durch eine äußerst niedrige Toxizität aus (LD_{50}: Ratte p.o. > 20000 mg/kg) [6].

Der Wirkungsmechanismus beruht auf der Hemmung der Biosynthese von Inositol, das für Wachstum und Pathogenität von *Pellicularia sasakii* notwendig ist [7]. Bei der Bekämpfung von *Pellicularia sasakii* stellt Validamycin A einen Marktanteil bis zu 40% und wird überwiegend als Staub (0,3%) mit 30–40 kg/ha eingesetzt. Die Produktion des 1972 registrierten Präparates lag 1977 bei 33000 t Staub.

Literatur: 1. Iwasa, T. et al.: J. Antibiotics *23*, 595 (1970). – 2. Iwasa, T. et al.: ibid *24*, 107 (1971). – 3. Iwasa, T. et al.: ibid *24*, 119 (1971). – 4. Horii, S., Kameda, Y.: J. Chem. Soc. Comm. 747 (1972). – 5. Nioh, T., Mizushima, S.: J. Gen. Appl. Microbiol. *20*, 373 (1974). – 6. Misato, T., in: Environmental Quality and Safety, Global Aspects of Chemistry and Technology as applied to the Environment, Vol. *5*, 48 (Hrsg. Coulston, F., Korte, F.), G. Thieme, Stuttgart, Academic Press New York 1976. – 7. Misato, T., in: Plant Protection in Japan 1976, Agriculture Asia, Spec. Issue *10*, 371 (1976).

2.3. Forschungs- und Entwicklungspräparate

Aabomycin-A [1, 2]

Erzeugerstamm: *Streptomyces* Sp. 325-17 (ATCC 21449) [3]
Isolierung: Extraktion aus dem Kulturfiltrat mit EtOAc. Extraktion aus dem Mycel mit Me_2CO und EtOAc. Chromatographie an Al_2O_3.
Summenformel: $C_{39-40}H_{65-67}NO_{11}$, Mol.-Gew. 724 ~ 738
Elementaranalyse: C:63,67 H:9,17 N:1,95
$[\alpha]_D^{20} = +93,5°$ (c=1, $HCCl_3$) ll: MeOH, EtOH, Me_2CO, EtOAc, $HCCl_3$; nl: Wasser, Hexan.

UV-Spektrum (MeOH) $\lambda_{max} = 280$ nm
IR-Spektrum (Nujol) [2]:

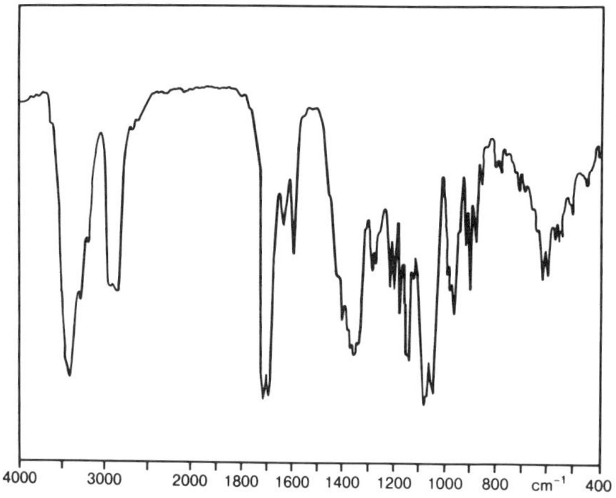

Biologische Wirkung:

in vitro MHK:	*Pyricularia oryzae*	0,001	µg/ml
	Gloeosporium kaki	1,0	µg/ml
	Pellicularia filamentosa	10,0	µg/ml

Die Sporenkeimung bei *Pyricularia oryzae* wird bei Konzentrationen unter 0,1 µg/ml gehemmt.
in vivo: Wirksam gegen *Pyricularia oryzae* an Reis mit 20 ppm [4].

Literatur: 1. DOS 1961 764, Kaken Kayaku KK and Rikagaku Kenkyusho (18.6.1970). – 2. Aizawa, S. et al.: J. Antibiotics Ser. A. *22*, 457 (1969). – 3. Misato, T. et al.: ibid. A. *23*, 204 (1970). – 4. Misato, T. et al.: ibid. A. *22*, 463 (1969).

Ablastmycin [1]

Erzeugerstamm: *Streptomyces aburaviensis* var. ablastmyceticus
Isolierung: Adsorption an Aktivkohle und Amberlite IRA-45 (OH^-)
Summenformel: $C_{18}H_{33}N_5O_{10}$, Mol.-Gew. 479
Elementaranalyse: C:44,7 H:6,4 N:14,6 O:32,9 $[\alpha]_D^{25} = +60°$ (c=0,5, Wasser), ll: Wasser, nl: EtOH, Me_2CO

UV-Spektrum (in 0,1 N HCl): λ max = 274 nm ($E_{cm}^{1\%} = 270$)
(in 0,1 N NaOH): λ max = 235 nm ($E_{cm}^{1\%} = 225$) und
290 nm ($E_{cm}^{1\%} = 180$)

IR-Spektrum [1]:
(Zur Aufnahmetechnik keine Angaben)

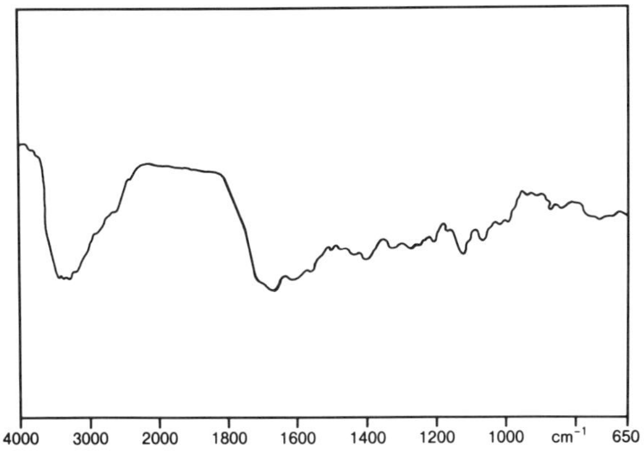

Biologische Wirkung [2]:
in vitro: *Pyricularia oryzae* und *Helminthosporium oryzae* wurden auf einem Reisblattextraktmedium bei saurem und neutralem pH-Wert mit 0,05 µg/ml gehemmt. An der Pflanze konnte diese Wirkung nicht bestätigt werden.

Literatur: 1. Umezawa, U. et al.: J. Antibiotics Ser. A. *21*, 37 (1968). – 2. JA 7207959-R, Nippon Gijutsu Boeki Co., Ltd. 7.3.1972.

Amipurimycin

Gruppe: Nucleosid
Erzeugerstamm: *Streptomyces novoguineensis* sp. nov.
Isolierung: Kombination von Ionenaustausch- und Adsorptionschromatographie. Kristallisation aus EtOH/Wasser (1:1)
Summenformel: $C_{20}H_{27-31}N_7O_8 \cdot H_2O$, Mol.-Gew. 472
Elementaranalyse: C:46,68 H:6,08 N:18,96 O:30,09
Farblose Prismen. $[\alpha]_D = -3,2°$ (c=0,62 in Wasser); ll: Wasser; nl: wenig polare organische Lösungsmittel (1).

UV-Spektrum [1]:

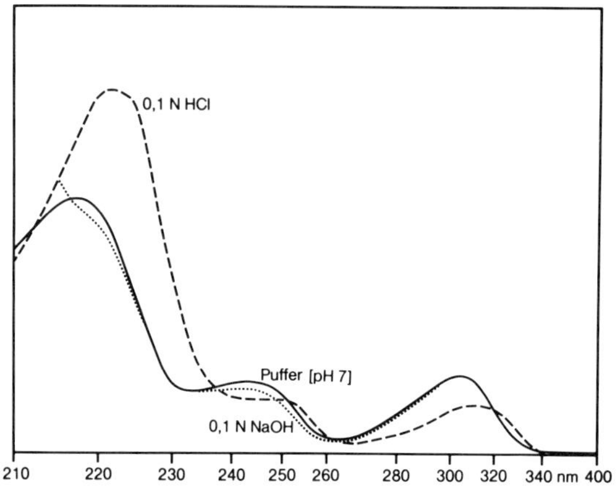

IR-Spektrum (KBr) [1]:

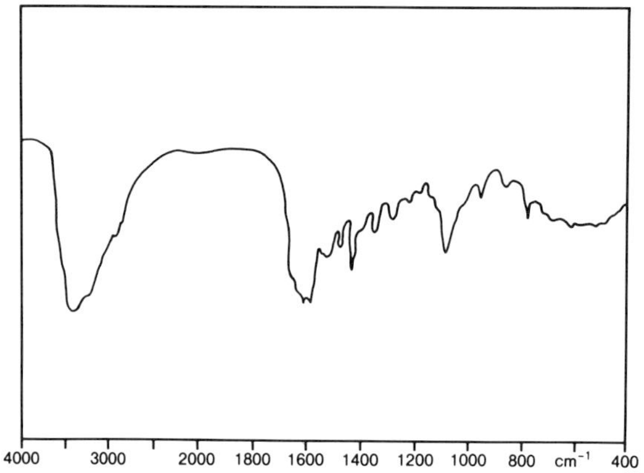

Biologische Wirkung [2]:

in vitro MHK:	*Pyricularia oryzae*	5 µg/ml
	Alternaria kikuchiana	10–20 µg/ml
	Helminthosporium sigmoideum	5–10 µg/ml
	Pythium aphanidermatum	5 µg/ml
	Cochliobolus miyabeanus	10–20 µg/ml
	Sclerotinia sclerotiorum	20 µg/ml

in vivo: Starker kurativer Effekt gegen *Pyricularia oryzae* an Reis im Gewächshaus bei 2,5 ppm. In Feldversuchen wurden hierfür 16 ppm benötigt.
Toxizität: LD_{50}: 1–5 mg/kg i.v., 10–20 mg/kg p.o. bei Mäusen und Ratten.
Fischtoxizität: TLm: (72 h) an Killi-Fischen < 10 ppm.

Literatur: 1. Harada, S., Kishi, T.: J. Antibiotics Ser. A. *30*, 11 (1977). – 2. Iwasa, T. et al.: J. Antibiotics Ser. A. *30*, 1 (1977).

Antifongin

Gruppe: Polyen
Erzeugerstamm: *Streptomyces paucisporogenes* ATCC 12596
Isolierung: Extraktion des Mycels mit MeOH. Chromatographie an Amberlite IRC-50 (Na^+).
Elementaranalyse: C:63,6 H:7,8 N:2,8 Asche:1,3
Gelbes Pulver. ll: Propylen-glycol, Pyridin, DMF; nl: Wasser, Kohlenwasserstoffe, Ketone.

UV-Spektrum (in MeOH:AcOH:Wasser (7:2:1)): λmax = 291 nm ($E_{cm}^{1\%} = 140$)
345 nm ($E_{cm}^{1\%} = 340$)
360 nm ($E_{cm}^{1\%} = 540$)
380 nm ($E_{cm}^{1\%} = 764$)
405 nm ($E_{cm}^{1\%} = 636$)

IR-Spektrum (Nujol) [1]:

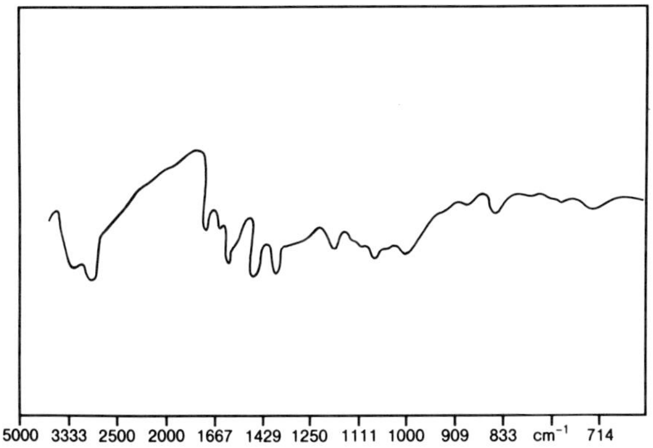

Biologische Wirkung:
in vitro MHK:

Calonectria decora	1,0 μg/ml
Ustilago avenae	1,0 μg/ml
Choanephora cucurbitarum	4,5 μg/ml
Alternaria solani	1,0 μg/ml
Monilia fructigena	10 μg/ml

in vivo: Eine wäßrige Suspension des Rohproduktes (30% des kristallinen Wirkstoffs) hemmt die Infektion von *Venturia inaequalis* (Apfelschorf) an Obstbäumen vollständig bei 250 μg/ml.
Es wurde keine Phytotoxizität beobachtet.

Literatur: 1. US 3 052 605, Roussel-UCLAF, SA., Paris, 4.9.1962.

Antimycin A_1 [1]

Synonyma: Virosin [2], Antipiriculin [2], Antibiotic 720-A [2]
Erzeugerstamm: *Streptomyces kitasawaensis* nov. sp.
Streptomyces griseus, s. sp.
Streptomyces antibioticus NRRL 2838

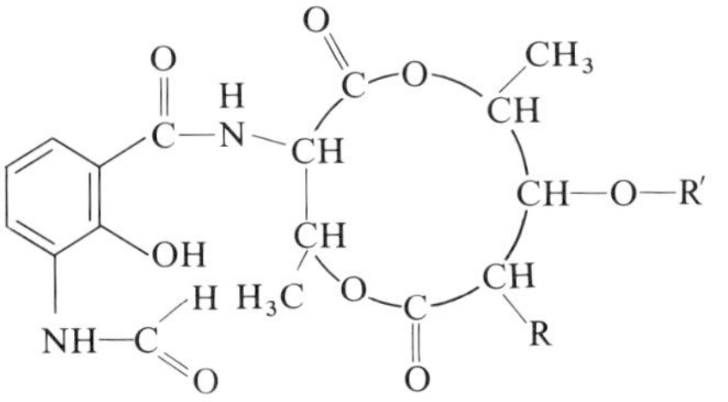

[3]
R = n—C_6H_{13}
R′ = CO—CH_2—$CH(CH_3)_2$

Isolierung: Extraktion aus dem Mycel oder Kulturfiltrat mit EtOAc, Adsorption an Celite bei pH 2,5, Elution mit EtOH, Craig-Verteilung.
Farblose Kristalle, Fp, 144–150 °C, $[\alpha]_D^{24,5}$ +77,2° (c=3,37% in $HCCl_3$), ll: MeOH, Me_2CO, $HCCl_3$.

UV-Spektrum (in 95% EtOH): λmax = 224–228 nm ($E_{1\,cm}^{1\%}$ = 669),
320–323 nm ($E_{1\,cm}^{1\%}$ = 123)
(in 0,1 N NaOH): λmax = 224–228 nm ($E_{1\,cm}^{1\%}$ = 670)
347,5 nm ($E_{1\,cm}^{1\%}$ = 227)
IR-Spektrum (Nujol): 3300, 2950–2880, 1750, 1700, 1667, 1640, 1613, 1540, 1480, 1430, 1360, 1250, 1175, 1150 cm^{-1} [4]
Biologische Wirkung:

in vitro MHK:	*Glomerella cingulata*	0,8 µg/ml
	Stemphylium sarcineforme	0,6 µg/ml
	Verticillium alboatrum	40,0 µg/ml
	Alternaria solani	40,0 µg/ml
	Venturia inaequalis	0,8 µg/ml
	Sclerotinia fructicola	1,6 µg/ml
	Nigrospora sphaerica	0,2 µg/ml
	Chalara quercina	0,8 µg/ml

in vivo: *Alternaria solani* auf Tomaten wurde mit 170 µg/ml zu 90% gehemmt [4]. Bis zu Konzentrationen von 2,9 mg/ml wurden keine phytotoxischen Erscheinungen beobachtet.
Toxizität: LD_{50}: 0,5 mg/kg i.v. bei der Maus.

Die Phyllomycine [5, 6] unterscheiden sich durch Variationen im Alkyl-Rest (R) und möglicherweise auch im Acyl-Rest (R′). Sie sind ebenfalls breit fungizid wirksam und ähnlich toxisch wie die Antimycine; LC_{50}: 10 µg/l bei Goldfischen.

Literatur: 1. Dunshee, B.R. et al.: J. Am. Chem. Soc. *71*, 2436 (1949). – 2. Sagakami, Y. et al.: J. Antibiotics Ser. A *9*, 1 (1956). – 3. Tamelen van, E.E. et al.: J. Am Chem. Soc. *81*, 750 (1959). – 4. Leben, C., Keitt, G.W.: Phytopathology *38*, 899 (1948). – 5. Schmidt-Kastner, G.: Liebigs Ann. Chem. *668*, 122 (1963). – 6. DOS 1 061 964, BAYER AG, 1959.

Aristeromycin [1]

Erzeugerstamm: *Streptomyces citricolor*

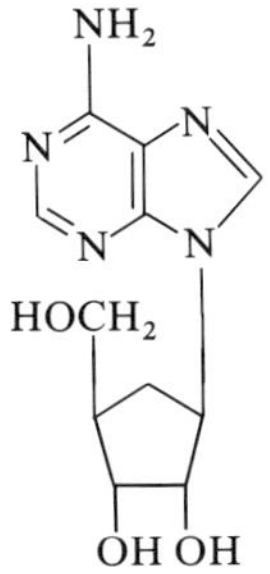

[2, 3]

Isolierung: Aus dem Kulturfiltrat durch Adsorption an Ionenaustauscher, Chromatographie an Al_2O_3 und Kristallisation aus Wasser [1, 2].
Farblose Kristalle, schwach basisch, Fp. 213–215 °C, $[\alpha]_D -52{,}5°$ (DMF).
ll: MeOH, Wasser.

UV-Spektrum (Wasser): $\lambda max = 262$ nm ($E^{1\%}_{1\,cm} = 555$)
IR-Spektrum (KBr) [2]:

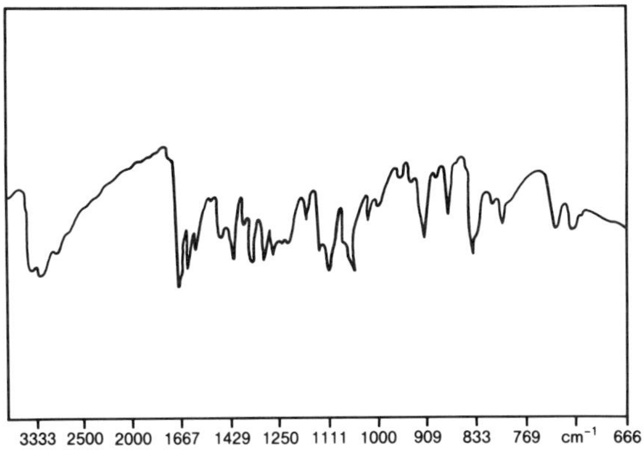

Biologische Wirkung:
in vitro MHK: *Pyricularia oryzae* 5,0 μg/ml
Pellicularia filamentosa 0,2 μg/ml
Cephalosporium gramineum 2,0 μg/ml

in vivo: Kurativer Effekt bei *Pyricularia oryzae* an Reis in Konzentrationen von 100–200 pm.

Die Wirkung beruht auf der Hemmung der Purin-Biosynthese. Dies bedingt eine Phytotoxizität bei Konzentrationen von mehr als 400 ppm an Reis und Tomaten.

Literatur: 1. Kusaka, T. et al.: J. Antibiotics Ser. A. *21*, 255 (1968). – 2. Kusaka, T.: J. Takeda Res. Lab. *31*, 85 (1972). – 3. Kishi, T. et al.: Chem. Communications, 852 (1967)

Ascochytin

Erzeugerstamm: *Ascochyta fabae Spegazzini*

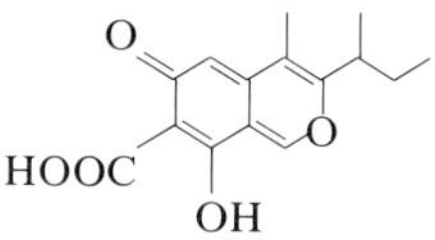

[2]

Isolierung: Extraktion mit $HCCl_3$ bei pH 3–4. Kristallisation aus EtOH.
$[\alpha]_D^{25} = -85{,}99°$ (c=1 in $HCCl_3$), ll: polare organ. Lösungsmittel, nl: Wasser

Absorptions-Spektrum (EtOH) λmax = 286 nm, 415 nm
IR-Spektrum (KBr) [1]:

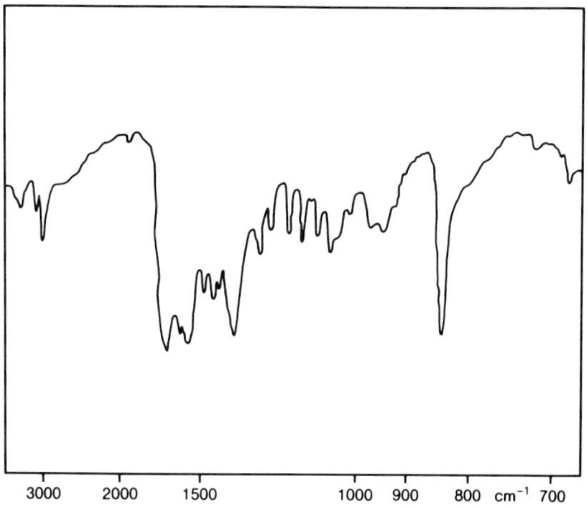

Biologische Wirkung:

in vitro MHK:	*Xanthomonas oryzae*	3	μg/ml
	Pyricularia oryzae	1,5–3,0	μg/ml
	Gloeosporium sp.	12,5	μg/ml

Ascochytin hemmt die Sporenkeimung.

Literatur: 1. Brit. 1 007 800, Sankyo Co., Ltd., 22.10.1965. 2. Galbraith, W.: Chem. Comm. *17*, 620 (1966).

Aureofungin

Gruppe: Polyene (Heptaen)
Erzeugerstamm: *Streptomyces cinnamoneus* var. terricola
Isolierung: Aus dem Mycel durch Extraktion mit MeOH
Elementaranalyse: C:60,4 H:7,9 N:2,2
Hellgelbes Pulver, ll: organ. Lösungsmittel. $[\alpha]_D^{25} = 88{,}7°$ (c=0,3 in DMF), Fp. 170 °C.

UV-Spektrum (in 60% EtOH): λmax = 359 nm, 379 nm, 402 nm

Biologische Wirkung:

in vitro MHK:	*Pyricularia oryzae*	<1	μg/ml [2]
	Alternaria solani	1	μg/ml
	Helminthosporium oryzae	0,1	μg/ml
	Phytophthora palmivora	5	μg/ml

in vivo: Saatgutbehandlung von Reis mit 10–20 ppm Aureofungin schützt gegen den Befall von *Pyricularia oryzae* und *Helminthosporium oryzae*. Konzentrationen von 40 ppm ergeben eine Schutzwirkung gegen *Rhizoctonia solani* bei Tabak. Die Substanz wird über Wurzeln und Blätter aufgenommen [3].

Literatur: 1. Thirumalachar, M.J. et al.: Hindustan Antibiot. Bull. *6*, 108 (1964). – 2. Laxminarayana, P., Reddy, S.M.: ibid. *17*, 29 (1974). – 3. Thirumalachar, M.J. et al.: Bull. Georgia Acad. Sci. Atlanta *29*, 253 (1971).

A-116 SO

Gruppe: Peptid
Erzeugerstamm: *Streptomyces endus* NRRL 2736
Isolierung: Aus dem Mycel durch Extraktion mit MeOH
Summenformel: $C_{85}H_{152}N_{18}O_{26}$, Mol.-Gew. 1840
Elementaranalyse: C:55,4 H:8,33 N:13,8 O:22,4
$[\alpha]_D^{25} = -83{,}4°$ (c=1 in C_2H_5OH)

UV-Spektrum: Endabsorption
IR-Spektrum ($HCCl_3$) [1]:

Biologische Wirkung:
in vivo: 0,2–2 mg/ml gegen *Alternaria solani* auf Tomaten.

Literatur: 1. NL 280040, Eli Lilly and Co., 22.6.1972.

A-150-A

Erzeugerstamm: *Streptomyces hygroscopicus* NRRL 3444
Isolierung: Extraktion aus dem Mycel mit MeOH. Konzentrierter Extrakt mit dem Kulturfiltrat vereinigt und bei pH 3 mit $HCCl_3$ extrahiert; Chromatographie an Kieselgel.
Summenformel: $C_{40}H_{65}NO_{11}$, Mol.-Gew. 735
Elementaranalyse: C:65,06 H:8,87 N:1,89 O:23,39
$[\alpha]_D^{25} = +66,86°$ (c=1 in MeOH).

UV-Spektrum: λmax=274 nm ($E_{cm}^{1\%}=85$) (bei alkalischem pH). Endabsorption bei sauren oder neutralen Bedingungen.
IR-Spektrum ($HCCl_3$) [1]:

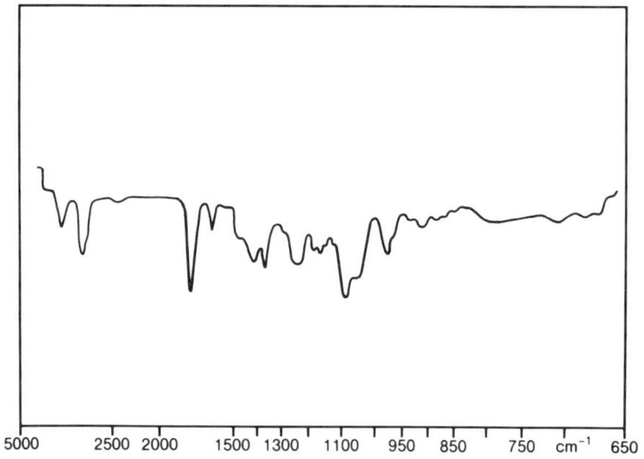

Biologische Wirkung:
in vitro MHK: *Alternaria solani, Botrytis cinerea* 6,25 μg/ml
in vivo: Wirksam gegen Rostkrankheiten an Weizen und Bohne bei 5–50 ppm.

Literatur: 1. US 3711605, Eli Lilly and Co., 16. 1. 1973.

A-201-A

Erzeugerstamm: *Streptomyces capreolus* NRRL 3817
Isolierung: Aus dem Kulturfiltrat durch Extraktions- und Adsorptionsverfahren.
Elementaranalyse: C:55,2 H:6,72 N:10,45 O:28,05, Mol.-Gew.=693
$[\alpha]_D^{25} = -129,6°$ (c=1 in MeOH). ll: EtOH, Me_2CO, nl: Wasser.

UV-Spektrum (in 95% EtOH/Wasser): λmax = 208 nm, 275 nm
IR-Spektrum ($HCCl_3$) [1]:

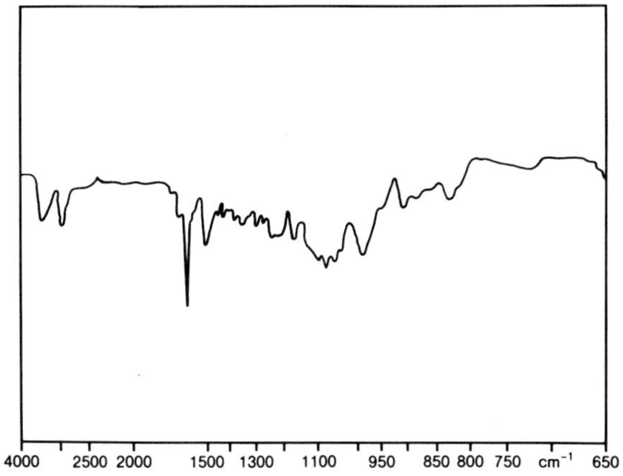

Biologische Wirkung:

in vitro MHK:	*Erwinia amylovora*	12,5 µg/ml
	Xanthomonas phaseoli	50,0 µg/ml
	Botrytis cinerea	100,0 µg/ml

Toxizität: LD_{50}: 400 mg/kg i.p. bei der Maus.

Literatur: 1. NL 7117796, Eli Lilly and Co., 24.12.1971.

A-201-B

Erzeugerstamm: *Streptomyces capreolus* NRRL 3817
Isolierung: wie A-201-A
Elementaranalyse: C:65,4 H:9,03 N:5,76 O:17,78, Mol.-Gew. = 417
ll: EtOH, Me_2CO; nl: Wasser

UV-Spektrum (in 95% EtOH/Wasser): λmax = 222 nm, 242 nm, 327 nm, 340 nm
IR-Spektrum ($HCCl_3$) [1]:

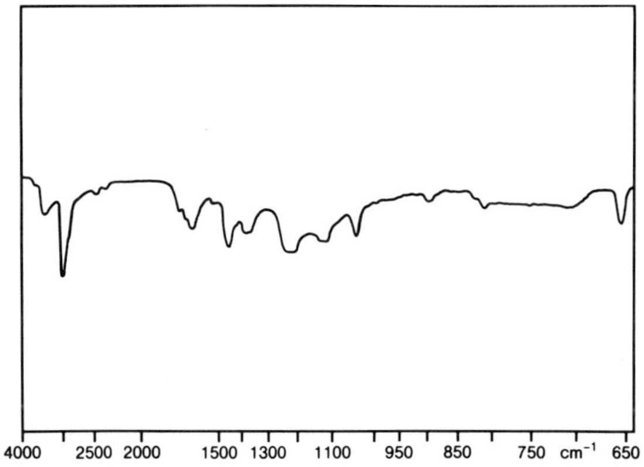

Biologische Wirkung:
in vitro MHK: *Erwinia amylovora* 25,0 µg/ml
Xanthomonas phaseoli 50,0 µg/ml
Botrytis cinerea 0,39 µg/ml

Toxizität: LD_{50}: 400 mg/kg i.p. bei der Maus.

Literatur: 1. NL 7117796, Eli Lilly and Co., 24.12.1971.

A-9145 [1, 2]

Erzeugerstamm: *Streptomyces griseolus* NRRL 3739
Isolierung: Aus dem Kulturfiltrat durch Adsorption an Amberlite IRC-50 (H^+). Elution mit 0,05 N H_2SO_4. Feinreinigung über Sephadex.
Elementaranalyse: C: 46,4 H: 6,8 N: 22,3 O: 26,2. Saure Hydrolyse ergibt Adenin
$[\alpha]_D^{25} = -2{,}61°$ (c = 5 in Wasser), ll: Wasser, nl: MeOH, EtOH

UV-Spektrum: (neutraler pH) λmax = 258 nm ($E_{cm}^{1\%} = 325$)
(saurer pH) λmax = 256 nm ($E_{cm}^{1\%} = 325$)
(basischer pH) λmax = 258 nm ($E_{cm}^{1\%} = 325$)
IR-Spektrum (Nujol) [2]:

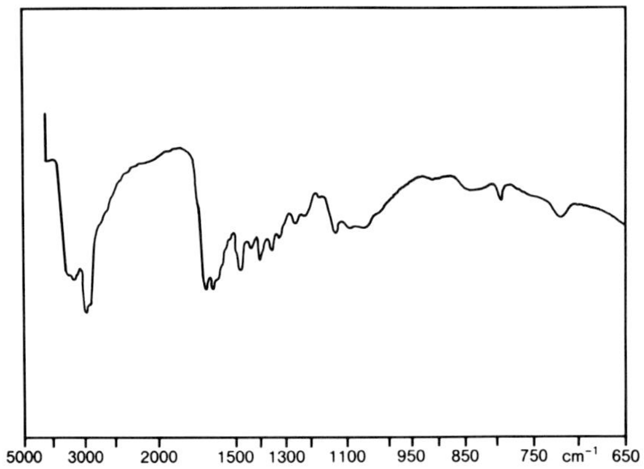

Biologische Wirkung (1):
in vivo: Bei Sprühapplikation mit 50–400 ppm wirksam gegen:
Erysiphe polygoni an Bohnen
Uromyces phaseoli an Bohnen
Colletotrichum lagenarium an Kürbis
Agrobacterium tumefaciens an Tomaten
Xanthomonas phaseoli an Sojabohnen

Toxizität: LD_{50}: 185 mg/kg s.c. bei der Maus.

Literatur: 1. US 3758681, Eli Lilly and Co., 11.9.1973. – 2. Hamill, R.L., Hoehn, M.M.: J. Antibiotics Ser. A *26*, 463 (1973).

A 23187

Erzeugerstamm: *Streptomyces chartreusis* NRRL 3882

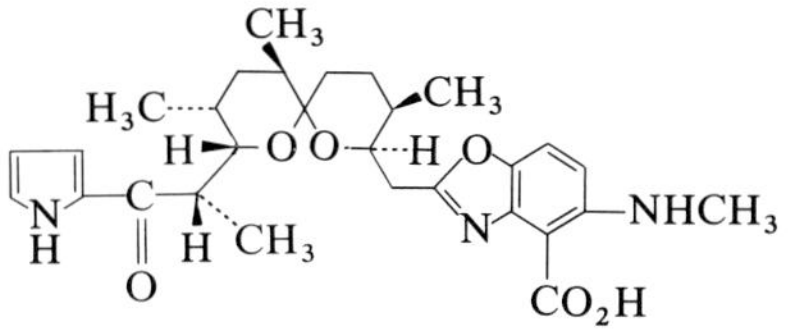

Isolierung: Aus dem Mycel durch Extraktion mit MeOH. Chromatographie auf Kieselgel in Benzol/EtOAc.
$[\alpha]_D^{25} = +362°$ (c = 1 in $HCCl_3$)

IR-Spektrum ($HCCl_3$) [1]:

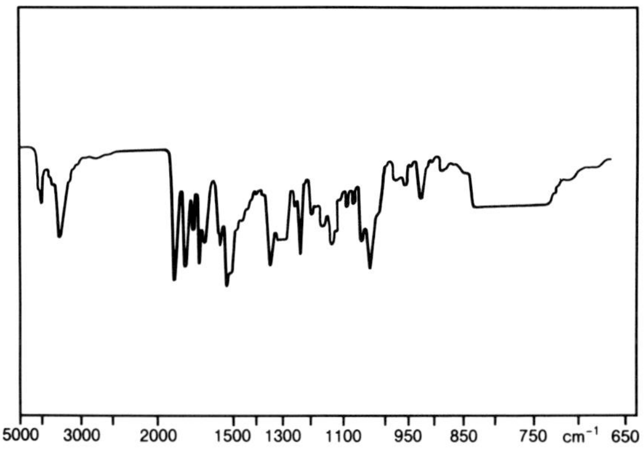

Biologische Wirkung:
in vitro MHK: *Botrytis cinerea* 1,56 µg/ml
Ceratostomella ulmi 6,25 µg/ml
Colletotrichum pisi 6,25 µg/ml
Toxizität: LD_{50}: 10 mg/kg i.p. bei der Maus.

Literatur: 1. US 3960667, Eli Lilly and Co., 1.6.1976.

Antibiotikum A 25822 A [1]

Gruppe: Azasteroide
Erzeugerstamm: *Geotrichum flavobrunneum* NRRL 3862

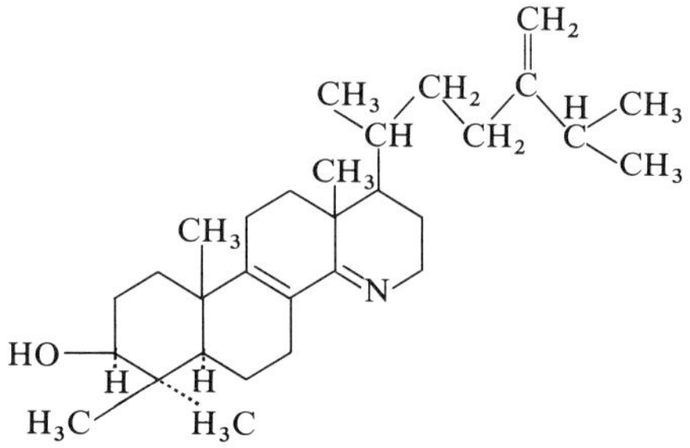

Isolierung: Extraktion aus dem Mycel mit EtOAc. Adsorption an basischem Al_2O_3 in Hexan/EtOAc/Wasser
Farblose Kristalle, Fp. 147 °C, $[\alpha]_{25}^{D} -70°$ (c = 1,15 in MeOH). pKa 8,45 in 66% DMF.
ll: organ. Lösungsmittel, nl: Wasser

UV-Spektrum (EtOH): λmax = 293 nm
(pH 2,0): λmax = 279 nm
IR-Spektrum (KBr) [1]:

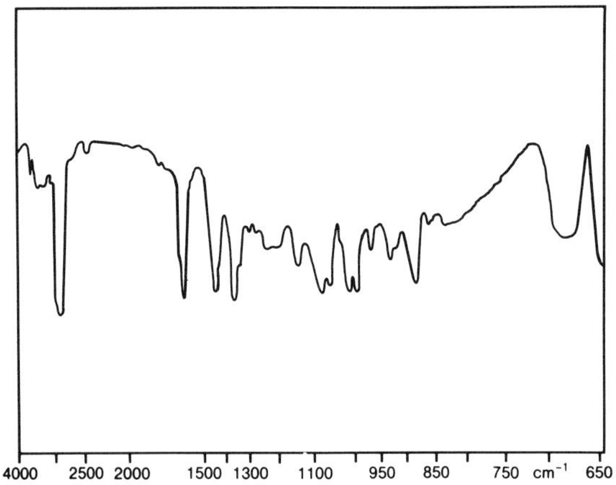

Biologische Wirkung:
in vitro MHK: *Erwinia amylovora* 6,25 µg/ml
Xanthomonas phasaeoli 50,0 µg/ml
Botrytis cinerea 3,12 µg/ml
Fusarium oxysporum 25,0 µg/ml
Toxizität: LD_{50}: 10 mg/kg s.c. bei der Maus.

Neben A 25822 A lassen sich noch die Antibiotika A 25822 B-F aus Mycelien des Stammes *Geotrichum flavobrunneum* isolieren. Sie sind in ihren physikochemischen und biologischen Eigenschaften sehr ähnlich zu A 25822 A.

Literatur: 1. BE 79518 Q, Eli Lilly and Co., 29.2.1972. – 2. DOS 2309108, Eli Lilly and Co., 13.9.1973.

A 30641 [1]

Erzeugerstamm: *Aspergillus tamarii* NRRL 8101

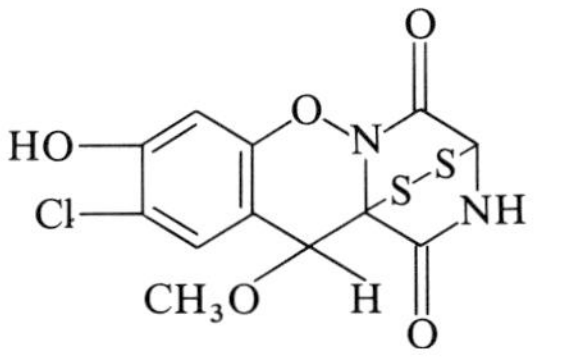

[2]

Isolierung: Adsorption aus dem Kulturfiltrat an Amberlite XAD-2 bei pH 3. Desorption mit MeOH; umfällen aus $HCCl_3$ mit Hexan.
Farbloses amorphes Pulver, Fp. 160–172 °C, $[\alpha]_D^{26} + 73°$ (c = 1 in MeOH), ll: in organ. Lösungsmitteln.

UV-Spektrum (MeOH): λmax = 208 nm (ε = 36800), 300 nm (ε = 6300)
IR-Spektrum ($HCCl_3$) [1]:

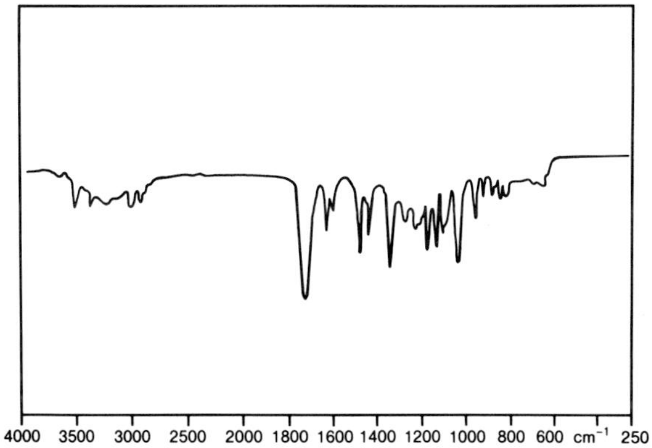

Biologische Wirkung: Die Verbindung ist bakterizid und fungizid wirksam [2].
in vitro MHK: *Erwinia amylovora* 6,25 µg/ml
Ceratocystis ulmi 12,5 µg/ml

Toxizität: LD_{50}: 106 mg/kg i.p. bei der Maus.

Literatur: 1. Berg, D. et al.: J. Antibiotics Ser. A. *29*, 394 (1976). 2. US 3991052, Eli Lilly and Co., 9.11.1976.

Bramycin

Erzeugerstamm: *Streptomyces* No. 658
Isolierung: Extraktion aus dem Mycel mit Aceton, Chromatographie an Al_2O_3 und Kieselgel, Kristallisation aus EtOH.
Summenformel: $C_{32}H_{55}NO_{11}$, Mol.-Gew. 617 (nach Rast)
Elementaranalyse: C:60,98 H:8,62 N:2,13
$[\alpha]_D^{20} = -25°$ (c=0,2 MeOH); ll: organischen Lösungsmittel, nl: Wasser, Hexan.

UV-Spektrum (MeOH): λmax = 244 nm (ε = 6790)
IR-Spektrum (KBr) [1]:

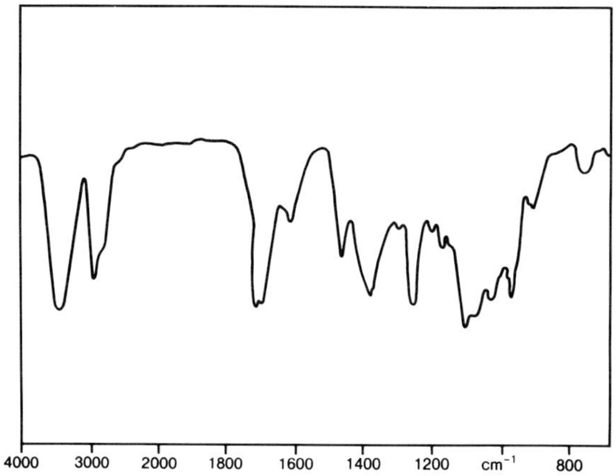

Biologische Wirkung:

in vitro MHK:	*Pyricularia oryzae*	0,1 µg/ml
	Alternaria citri	0,5 µg/ml
	Alternaria kikuchiana	0,1 µg/ml
	Botrytis cinerea	10,0 µg/ml

in vivo: 80%ige Wirkung mit 160 ppm gegen *Pyricularia oryzae* an Reis im Gewächshaus. Kurativer Effekt mit 160 ppm: 62,7% Wirkung.
Toxizität: LD_{50}: 4,8 mg/kg p.o., 0,49 mg/kg i.p. bei der Maus.

Literatur: 1. Sakagami, Y.: J. Antibiotics Ser. A. *19*, 99 (1966).

Bulgerin

Erzeugerstamm: *Streptomyces aburaviensis* var. tuftformis
Isolierung: Säulenchromatographie an Ionenaustauschern, Gelfiltration und präparative Papierchromatographie.
Summenformel: $C_{17}H_{24}N_4O_{10}$, Mol.-Gew. 444
Elementaranalyse: C:45,94 H:5,44 N:12,61
Farbloses amorphes Pulver. $[\alpha]_D^{22} = +24{,}0 \pm 1{,}3°$ (c=0,490, Wasser). ll: Wasser, nl: organ. Lösungsmittel.

UV-Spektrum [1]:

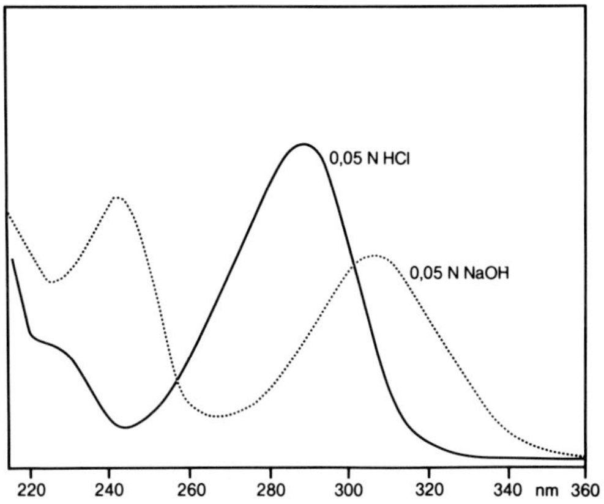

IR-Spektrum (KBr): 3400, 1700, 1650, 1595, 1480, 1420, 1390 cm^{-1}

Biologische Wirkung:

in vitro MHK:	*Pyricularia oryzae*	1,6 µg/ml
	Colletotrichum sp.	6,2 µg/ml
	Pellicularia sasakii	6,2 µg/ml
	Cochliobolus miyabeanus	0,8 µg/ml

Literatur: 1. Shoji, J. et al.: J. Antibiotics Ser. A. *23*, 295 (1970).

Cerulenin [1]

Erzeugerstamm: *Cephalosporium caeruleus* [1, 2]

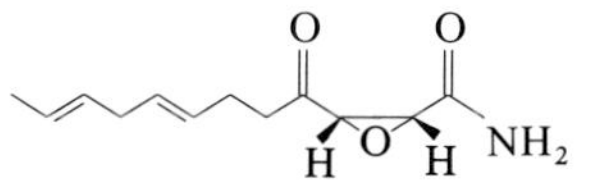

[3, 4]

Isolierung: Extraktion aus der Kulturbrühe mit einem organischen Lösungsmittel wie BuOAc oder $HCCl_3$ bei pH 3, Chromatographie an Al_2O_3 in MeOH/Benzol. Farblose Kristalle, Fp. 90–91 °C, $[\alpha]_D^{16} = -63°$ (c=2 MeOH), ll: $HCCl_3$, Benzol; l: Wasser

UV-Spektrum: Endabsorption
IR-Spektrum (Nujol) [3]:

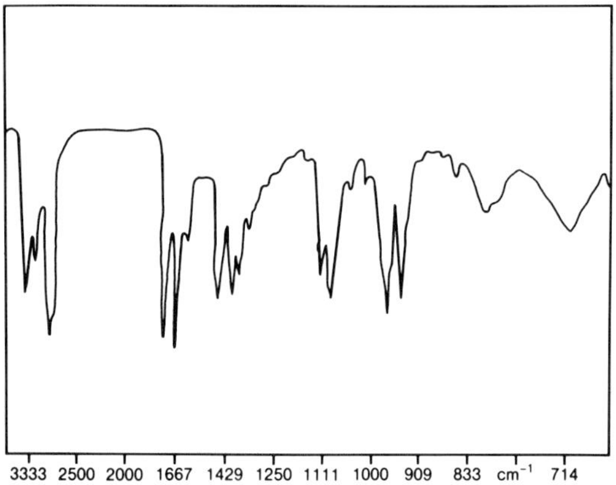

Biologische Wirkung: Die Substanz hemmt die Biosynthese von Fettsäuren, Steroiden und Makroliden [5].
in vitro MHK: *Pyricularia oryzae* 6,25 µg/ml
Cochliobolus miyabeanus 1,65 µg/ml
Gibberella fujikuroi 50 µg/ml

Toxizität: LD_{50}: 211 mg/kg i.p. an der Maus [6].

Literatur: 1. Matsumae, A. et al.: J. Antibiotics Ser. A. *16*, 236 (1963). – 2. Matsumae, A. et al.: ibid. *16*, 239 (1963). – 3. Omura, S. et al.: ibid. *20*, 349 (1967). – 4. Byron, H.A., Omura, S.: ibid. *27*, 28 (1974). – 5. Vance, D. et al.: Biochem. Biophys. Res. Comm. *48*, 649 (1972). – 6. Matsumae, A. et al.: J. Antibiotics Ser. A. *17*, 1 (1964).

Cochliodinol

Erzeugerstamm: *Chaetomium cochliodes* ATCC 10195,
Chaetomium globosum (HLX707 u. HLX819)
Isolierung: Extraktion aus dem Mycel mit Benzol, Chromatographie an Kieselgel. Elution mit Et_2O: Benzol (1:9).
Summenformel: $C_{32}H_{32}N_2O_4$, Mol.-Gew. 500
Elementaranalyse: C:75,6 H:6,4 N:5,5 O:12,6
ll: MeOH, Benzol

UV-Spektrum (MeOH): λmax = 278 nm, 471 nm
IR-Spektrum (KBr) [1]:

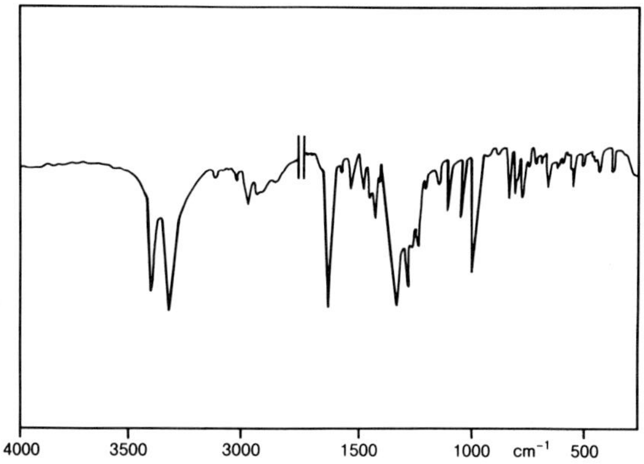

Biologische Wirkung:
in vitro MHK: *Fusarium oxysporum* 10 µg/ml
Botrytis allii 40 µg/ml

Literatur: 1. CN 941323, Canadian Pat. and Development Ltd., 5.2.1974.

Copiamycin [1]

Erzeugerstamm: Streptomyces hygroscopicus var. crystallogenes ATCC 19040
Isolierung: Extraktion mit Alkoholen. Ausfällung mit Me_2CO. Kieselgelchromatographie.
Elementaranalyse: C: 58,30 H: 8,90 N: 3,94
$[\alpha]_D^{25,5} = +14,4$ (c = 4,15% in MeOH).
ll: MeOH, EtOH, nl: Me_2CO, Dioxan.

UV-Spektrum (MeOH): Endabsorption

Biologische Wirkung [2]:
in vitro MHK: Breite Wirkung gegen fadenförmige Pilze bei Konzentrationen < 10 µg/ml.
Toxizität: Bei 1000 mg/kg keine Wirkung an der Maus.

Literatur: 1. Arai, T. et al.: J. Antibiotics Ser. A. *18*, 63 (1965). – 2. DOS 1467757, 30.8.1965, Arai, Tadashi, Nogata-machi, Nakano-ku.

Cryptosporiopsin

Erzeugerstamm: *Cryptosporiopsis*, No. FSC 439

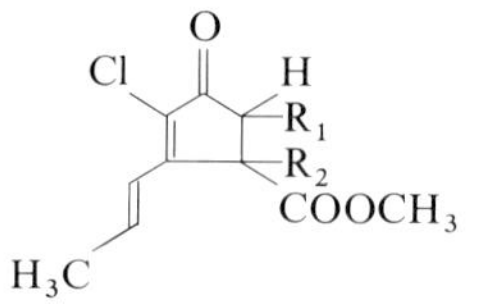

(IA) R_1 = OH und R_2 = Cl
(IB) R_1 = Cl und R_2 = OH

Isolierung: Cryptosporiopsin wird als IA oder IB isoliert. Aus dem Kulturfiltrat durch Extraktion mit organischen Lösungsmitteln; Chromatographie auf Al_2O_3 in $HCCl_3$
$[\alpha]_D^{25} = +129°$ (c = 1,35, in $HCCl_3$).

UV-Spektrum (EtOH): λmax = 292 nm
IR-Spektrum (KBr): 3420, 1755, 1730, 1635, 1570, 965 cm^{-1}

Biologische Wirkung: Hemmt die Keimung der Zoosporen von *Phytophthora infestans* bei einer Konzentration von 0,78 µg/ml vollständig.
Toxizität: Bei 15 mg/kg p.o. bei der Ratte starke Vakuolisierung der Leberzellen.

Literatur: 1. CA 915108, Allied Chemical Corp., 21.11.1972.

Deisovalerylblastmycin

Erzeugerstamm: *Streptomyces* sp. 5140-A_1

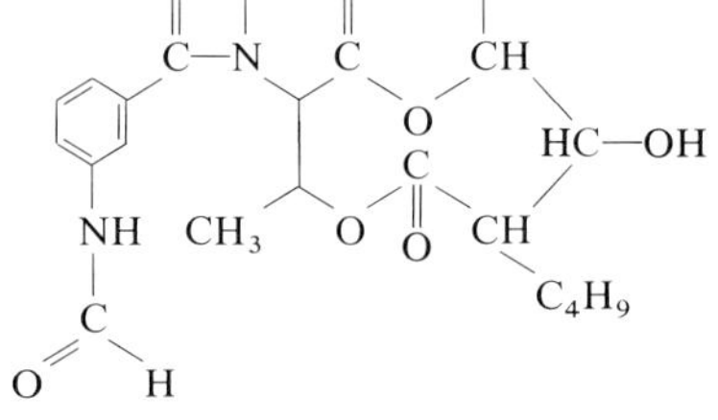

Isolierung: Extraktion aus dem Mycel mit Me_2CO, Chromatographie an Kieselgel mit Benzol/EtOAc [1].
Farblose Kristalle, $[\alpha]_D^{24} = +36°$ (c = 1,0 in MeOH). ll: in organ. Lösungsmitteln, nl: Hexan, Wasser

UV-Spektrum (MeOH): λmax = 226 nm (ε = 30400), 320 nm (ε = 6000)
(N/100 NaOH-MeOH): λmax = 225 nm (ε = 35300), 345 nm (ε = 8500)

IR-Spektrum ($HCCl_3$) [1]:

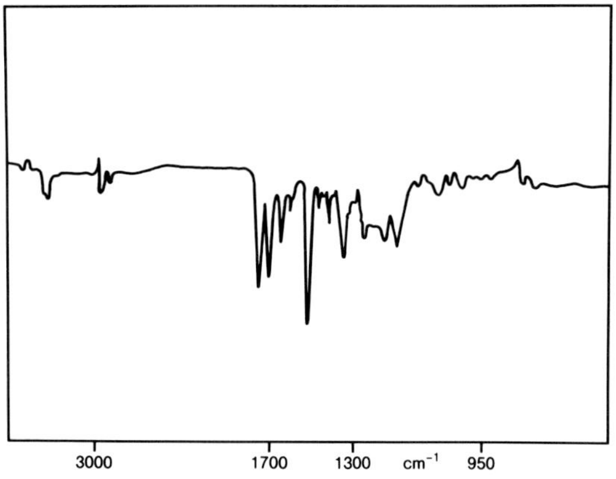

Biologische Wirkung:
in vitro MHK: *Pyricularia oryzae* 0,8 µg/ml
in vivo: 84%ige Wirkung gegen *Pyricularia oryzae* an Reis bei 400 ppm im Gewächshaus. Phytotoxische Wirkungen wurden bei diesen Konzentrationen nicht beobachtet.
Toxizität: LD_{50}: 25 mg/kg i.p. bei der Maus.
Fischtoxizität: 0,5 µg/ml an Killi-Fischen [1].

Literatur: 1. Ishiyama, T. et al.: J. Antibiotics Ser. A. *29*, 804 (1976).

Destomycin [1]

Erzeugerstamm: *Streptomyces rimofaciens*

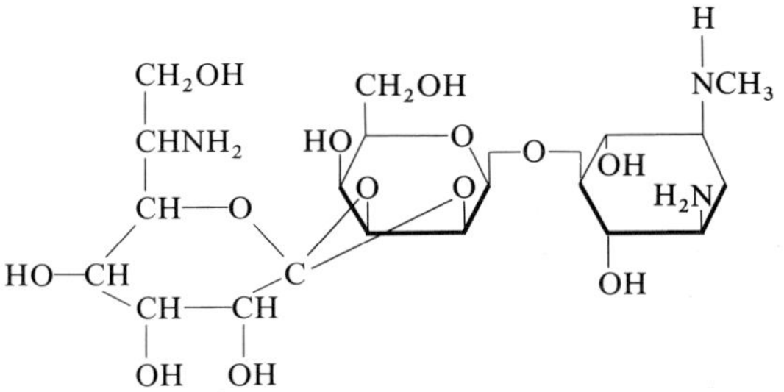

[2]

Isolierung: Adsorption des Kulturfiltrates an Amberlite IRC-50 (Na^+), Aktivkohle, Amberlite IRA-45 (OH^-), Amberlite IRC-50 (NH_4^+) und Dowex 1-X2 (OH^-).
$[\alpha]_D^{22} = 7°$ (c = 2, Wasser), ll: Wasser, niedere Alkohole; nl: wenig polare Lösungsmittel.

UV-Spektrum: Endabsorption
IR-Spektrum (KBr) [1]:

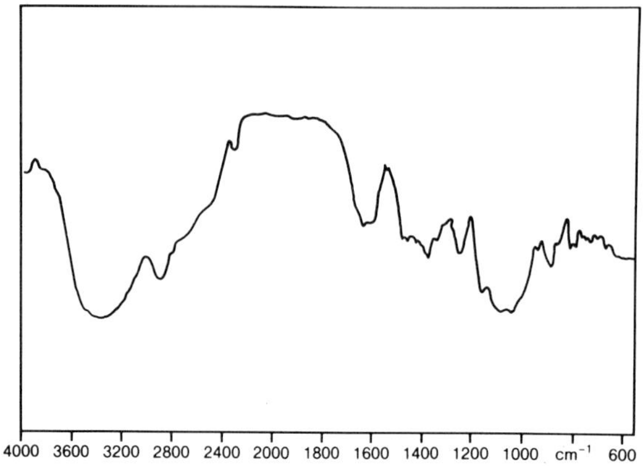

Biologische Wirkung:

in vitro MHK:	*Xanthomonas oryzae*	12,5 μg/ml
	Pyricularia oryzae	12,5 μg/ml
	Pellicularia sasakii	1,56 μg/ml
	Alternaria tenuis	0,39 μg/ml
	Alternaria kikuchiana	3,13 μg/ml
	Fusarium graminearum	1,56 μg/ml
	Cladosporium herbarum	1,56 μg/ml

Toxizität: LD_{50}: 5–10 mg/kg i.v.; 50–100 mg/kg p.o. bei der Maus.

Literatur: 1. Kondo, S. et al.: J. Antibiotics Ser. A, *18*, 38 (1965). – 2. Kondo, S. et al.: ibid. *19*, 137 (1966)

1,6-Dihydroxyphenazin

Erzeugerstamm: *Streptomyces thioluteus*
Isolierung: Extraktion aus dem Kulturfiltrat und Mycel mit Benzol bei pH 6–7.

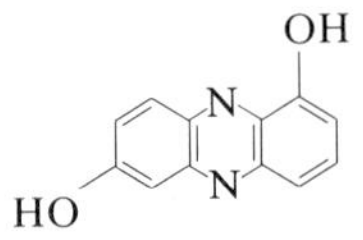

Goldgelbe Kristalle, Indikator: in Säure gelb; in Alkali rot. ll: Alkohole, Me_2CO, EtOAc, Benzol; nl: Wasser, Petroläther.

UV-Spektrum [1]:

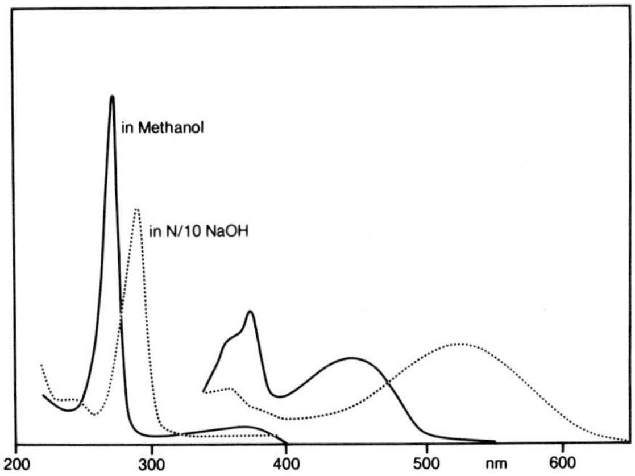

IR-Spektrum (Nujol) [1]:

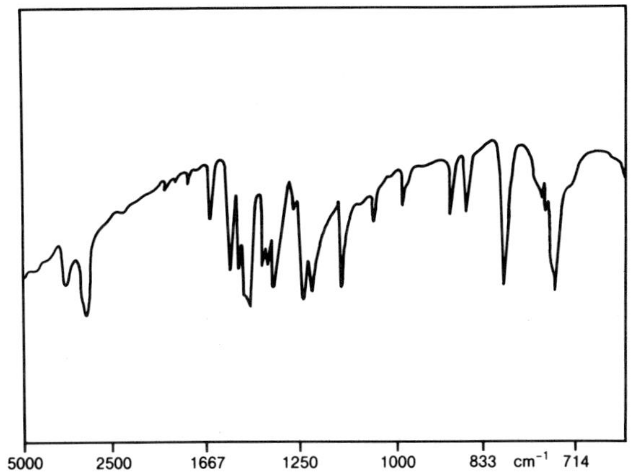

Biologische Wirkung:

in vitro MHK:

	Alternaria kikuchiana	>20 µg/ml
	Fusarium lycopersici	>20 µg/ml
	Gibberella fujikuroi	>20 µg/ml
	Gloeosporium spp.	15 µg/ml
	Cochliobolus miyabeanus	15 µg/ml
	Pyricularia oryzae	5,0 µg/ml

Literatur: 1. Akabori, H., Nakamura, M.: J. Antibiotics Ser. A. *12*, 17 (1959).

Duamycin

Erzeugerstamm: *Streptomyces hygroscopicus* subsp. duamyceticus
Elementaranalyse: C:64,29 H:8,79 O:23,84, Mol.-Gew. = 678
ll: Alkoholen, Estern, $HCCl_3$, nl: Wasser

Biologische Wirkung:
in vitro MHK: *Pyricularia oryzae* 0,08 µg/ml
Pellicularia filamentosa 40,0 µg/ml
in vivo: Gute Wirkung gegen *Pyricularia oryzae* an Reis bei 10 ppm.

Literatur: 1. JA 7123616, Rikagaku Kenkyusho, 6.7.1971.

Filipin

Synonym: 14-Deoxylagosin
Gruppe: Makrolid-Polyen
Erzeugerstamm: *Streptomyces filipinensis* NRRL 2437

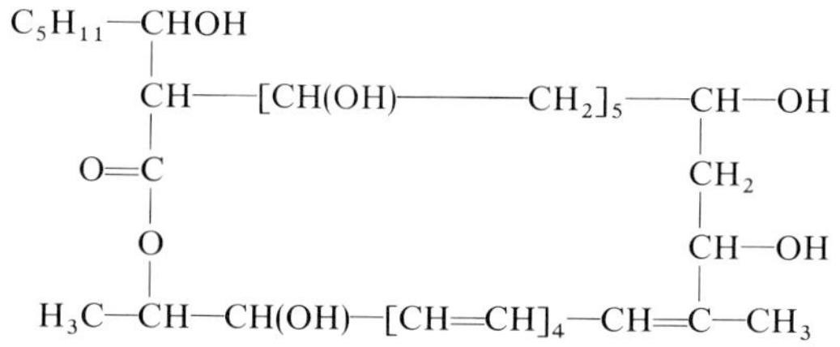

[1]

Isolierung: Extraktion aus dem Kulturfiltrat mit EtOAc, Me_2CO oder BuOH. $[\alpha]_D^{22} = -148{,}3^\circ$ (c = 0,89, MeOH), ll: DMF, Pyridin, EtOH, MeOH, EtOAc; nl: Wasser, $HCCl_3$

UV-Spektrum (EtOH) λmax = 355 nm, 338 nm, 322 nm
IR-Spektrum (Nujol) [2]:

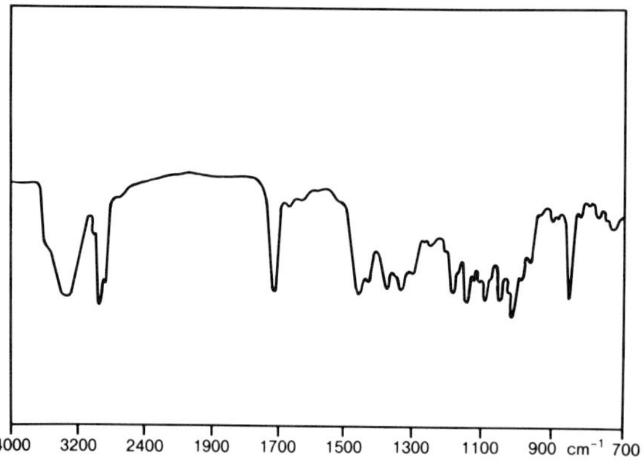

Biologische Wirkung [2]:
in vitro MHK: *Stemphylium solani* 1,0–2,5 μg/ml
in vivo: Wirksam gegen *Stemphylium solani* an Tomaten,
Botrytis allii an Zwiebeln,
Helminthosporium maydis an Mais
Phytophthora cactorum an Tomaten.

Das Antibiotikum bindet an der Zellmembran und hemmt den Transport z.B. von Zuckern, Phosphat und Na^+. Filipin wirkt als Cholesterin-Antagonist [3, 4]. Toxizität: LD_{50}: 17 mg/kg i.p. bei der Maus.

Literatur: 1. Golding, B.T. et al.: Tetrahedron Lett. *37*, 2615 (1964). – 2. US 3188272, Univer. Illinois Foundation, 8.6.1965. – 3. Schröder, F. et al.: Chem.-Biol. Interact. *4*, 239 (1972). – 4. Sweeley, C.C. et al.: ibid. *2*, 247 (1970).

F-1028

Erzeugerstamm: *Streptomyces kagawaensis* nov. sp.

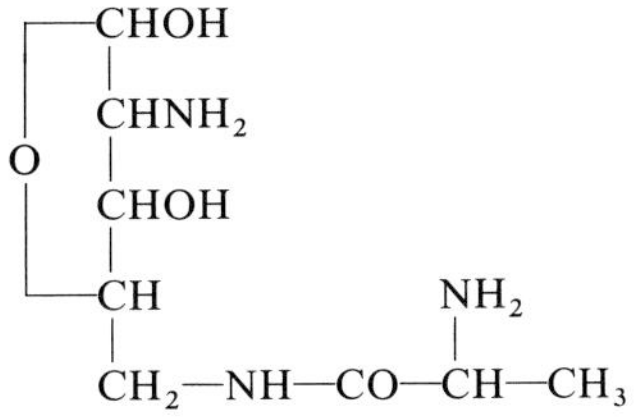

Isolierung: Chromatographie an Amberlite IRC-50 (Na^+)
$[\alpha]_D^{23} = 63{,}2°$ (c = 1; Wasser). ll: Wasser, nl: $HCCl_3$, Et_2O

UV-Spektrum: Endabsorption
IR-Spektrum (KBr) [1]:

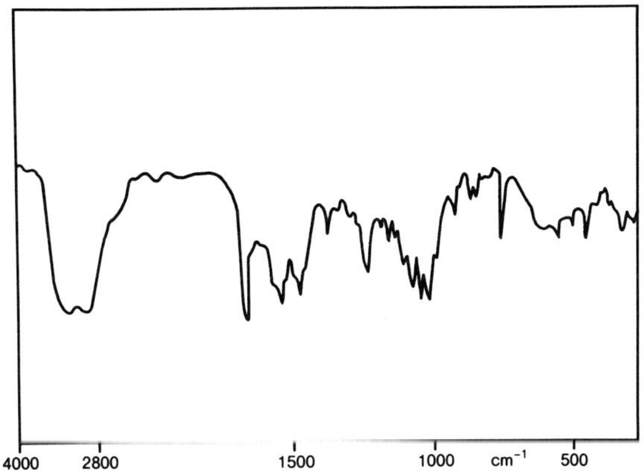

Biologische Wirkung:

in vitro MHK:	*Sclerotinia cinerea*	12,5 μg/ml
	Sclerotinia sclerotiorum	1,56 μg/ml
	Botrytis spp.	6,25 μg/ml

Toxizität: LD_{50}: 160 mg/kg i.v., 750 mg/kg p.o. bei der Maus. Keine Toxizität mit 250 ppm bei jungen Karpfen.

Literatur: 1. DOS 2231979, Kitasato Inst.; Nihon Tokushu Noyaku Seizo, 11.1.1973.

Gangtokmycin

Gruppe: Pentaen
Erzeugerstamm: *Streptomyces gangtokensis* ATCC 15665
Isolierung: Extraktion mit MeOH aus dem Mycel
Elementaranalyse: C:63,70 H:8,39 N:1,34 O:26,57
ll: wäßr. MeOH, wäßr. Me_2CO; nl: Wasser
UV-Spektrum (MeOH): $\lambda max = 317$ nm ($E_{cm}^{1\%} = 670$), 332 nm ($E_{cm}^{1\%} = 960$), 348 nm ($E_{cm}^{1\%} = 840$)
Biologische Wirkung:

in vitro MHK:	*Pyricularia oryzae*	10 μg/ml
	Phytophthora infestans	3 μg/ml
	Ustilago zeae	5 μg/ml
	Rhizoctonia solani	20 μg/ml
	Alternaria kikuchiana	10 μg/ml

Toxizität: LD_{50}: 10–25 mg/kg i.p., 400–600 mg/kg p.o. bei der Maus.

Literatur: 1. DOS 1249456, Takeda Chem. Ind., 7.9.1967.

Grisorixin

Erzeugerstamm: *Streptomyces griseus*

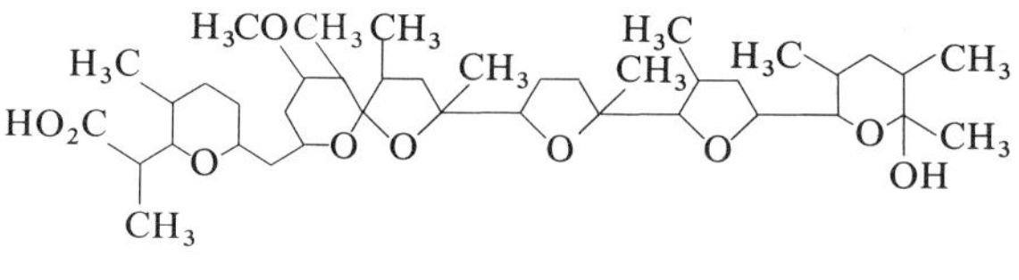

Isolierung: Extraktion aus dem Mycel mit EtOH; Kieselgelchromatographie. $[\alpha]_D^{20} = +16°$ (in Me_2CO), ll: organ. Lösungsmittel.

IR-Spektrum (KBr) [1]:

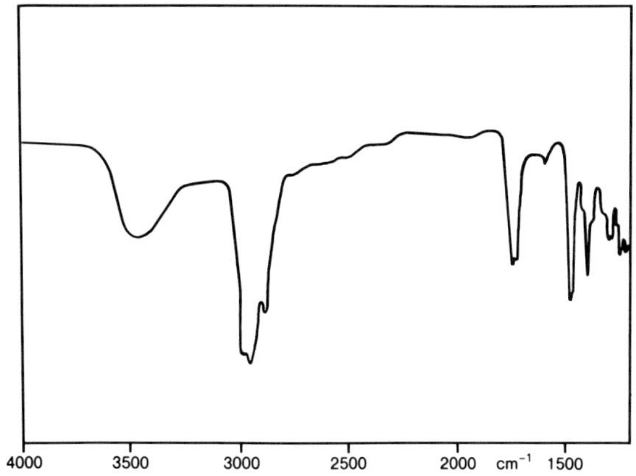

Biologische Wirkung:

in vitro MHK:	*Botrytis cinerea*	0,3 μg/ml
	Rhizoctonia solani	12,0 μg/ml
	Sclerotinia sclerotiorum	0,7 μg/ml
	Monilia laxa	0,7 μg/ml

Toxizität: LD_{50}: 15 mg/kg s.c. bei der Maus.

Literatur: 1. FR. 2091913-Z, Inst. Nat. Agronomique, Inst. Nat. Sante Recherche Med. 21.1.1972.

Helicocerin

Erzeugerstamm: *Helicoceras* spp.
Isolierung: Extraktion mit organ. Lösungsmitteln.
Elementaranalyse: C:64,36 H:7,73 N:6,43, Mol.-Gew. = 229,2
ll: MeOH, Me_2CO; nl: Wasser.

UV-Spektrum: Endabsorption
IR-Spektrum (KBr): 3380, 3210, 3000, 2920, 1720, 1665, 1610, 1450, 1400, 1370, 1340, 1112, 1095, 1053, 968, 940, 857, 810 cm^{-1}

Biologische Wirkung:
in vitro MHK: *Botrytis cinerea* 0,5–1 μg/ml.

Literatur: 1. NL 6801732, Sankyo Co. Ltd., 7.2.1968.

Hondamycin [1]

Synonym: Albimycin [1, 2]
Erzeugerstamm: *Streptomyces griseochromogenes* var. albiens
Isolierung: Extraktion aus dem Mycel mit MeOH, Chromatographie an Kieselgel in Benzol, EtOAc. Craig-Verteilung.
Summenformel: $C_{47}H_{78}O_{13}$, Mol.-Gew. 850
Elementaranalyse: C:66,38 H:9,25 O:24,37
Farblose Kristalle, Fp. 149–150 °C, $[\alpha]_D^{13,5} = -48,1°$ (c = 1,93 in MeOH).

UV-Spektrum (MeOH): λmax = 225 nm (log ε = 4,58), 232,5 (log ε = 4,56)
IR-Spektrum (KBr) [2]:

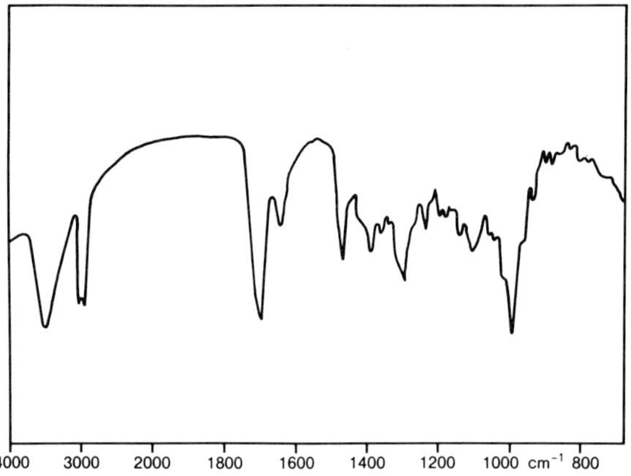

Biologische Wirkung:

in vitro MHK [3]:	*Pyricularia oryzae*	0,1 µg/ml
	Botrytis cinerea	1,0 µg/ml
	Alternaria kikuchiana	0,5 µg/ml
	Cercospora beticola	0,01 µg/ml

in vivo: Wirksam gegen *Rhizoctonia solani* auf Gemüse und gegen *Pyricularia oryzae*, *Ophiobolus miyabeanus* und *Hypochnus sasakii* auf Reis [4].
Toxizität: LD_{50}: 1,2 mg/kg i.v., 500 mg/kg p.o. bei der Maus.

Literatur: 1. Sakagami, Y. et al.: 147th Meet. Japan Antibiotics Res. Ass. Nov. 26, 1965. – 2. Sakagami, Y. et al.: J. Antibiotics Ser. A, *22*, 521 (1969). – 3. Sakagami, Y. et al.: ibid. A. *22*, 528 (1969). – 4. JA 72428 Yoshio Sakanoue, 25.11.1965.

Ikutamycin [1]

Erzeugerstamm: *Streptomyces diastatochromogenes*
Isolierung: Extraktion mit MeOH aus dem Mycel; Kieselgelchromatographie.
Summenformel: $C_{68}H_{114}N_2O_{23}$; Mol.-Gew. 1326
Elementaranalyse: C:61,89 H:8,55 N:2,20 O:27,40
$[\alpha]_D^{15} = -27,8°$ (c = 1,33 in MeOH). ll: MeOH, EtOH; nl: Wasser

IR-Spektrum (KBr) [1]:

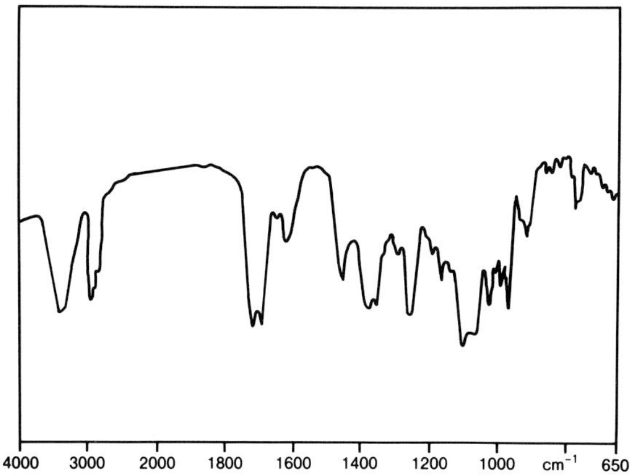

Biologische Wirkung [2]:

in vitro MHK:	*Cercospora beticola*	1,0 µg/ml
	Botrytis cinerea	25,0 µg/ml
	Helminthosporium sigmoideum	0,1 µg/ml
in vivo:	Wirksam gegen *Helminthosporium sigmoideum* an Reis: 25 ppm	

Literatur: 1. Sakagami, Y. et al.: J. Antibiotics Ser. A. *20*, 299 (1967). – 2. JA 7042568-R, Yoshio Sakanoue, 11. 3. 1970.

Ileumycin

Erzeugerstamm: *Streptomyces lavendulae* H 698-SY2
Isolierung: Adsorption des Kulturfiltrates an Amberlite XAD-11, Elution mit wäßrigem MeOH, Chromatographie an SE-Cellulose und Kieselgel. Ileumycin wird als Hydrochlorid erhalten.
Elementaranalyse: C:44,61 H:6,50 N:9,74 Cl:11,45
Amorphes farbloses Pulver.

UV-Spektrum [1]:

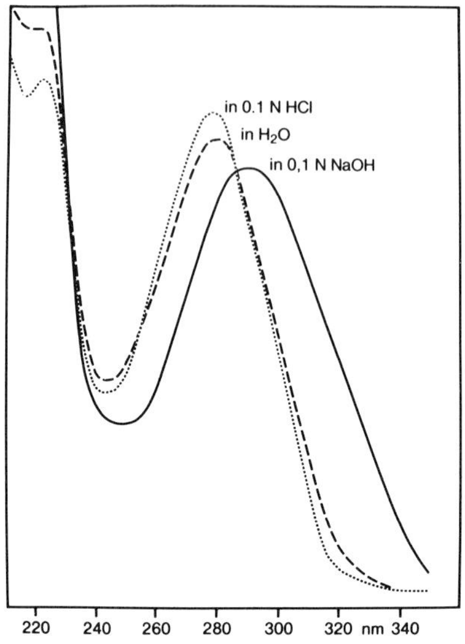

IR-Spektrum (KBr) [1]:

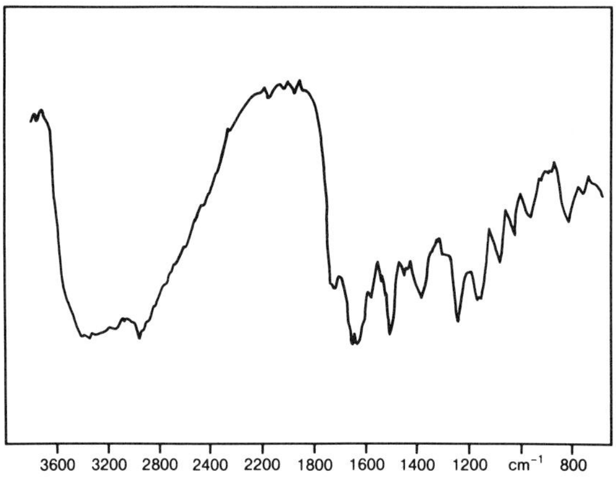

Biologische Wirkung:

in vitro MHK:	*Colletotrichum lagenarium*	0,78 μg/ml
	Glomerella cingulata	0,05 μg/ml

Literatur: 1. Kawakami, Y. et al.: J. Antibiotics Ser. A. *31*, 112 (1978).

IN-183-T

Erzeugerstamm: *Streptomyces lavendulae*
Summenformel: $C_{15}H_{35}N_6O_{13}S$; Mol.-Gew. 539

Elementaranalyse: C:33,35 H:6,50 N:15,59 O:38,58 S:5,94
$[\alpha]_D^{25} = -45°$ (c=1, Wasser); ll: Wasser; nl: EtOH, Me_2CO
Biologische Wirkung:
in vitro MHK: *Xanthomonas oryzae* 5,0 µg/ml
Toxizität: LD_{50}: >200 mg/kg i.v. bei der Maus.

Literatur: 1. JA 7123040, Kaken Chemical Co. Ltd., Kumiai Chemical Ind. Co. Ltd., 1.7.1971.

Lambertellin

Erzeugerstamm: *Lambertella hicoriae*

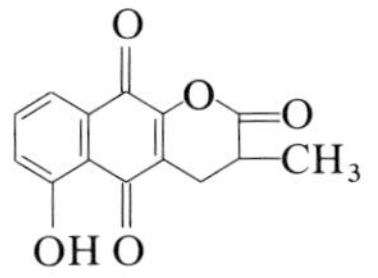

Isolierung: Extraktion aus dem Mycel mit Toluol oder anderen aromatischen Kohlenwasserstoffen; Chromatographie an Magnesiumsilicat in Me_2CO.
Orange Kristalle, Fp. 248–250 °C. ll: Benzol, Hexan, Petroläther

UV-Spektrum (EtOH): λmax = 211 nm, 285 nm, 290 nm, 420 nm.
IR-Spektrum (KBr): 2930, 1740, 1652, 1626, 1375, 1191, 1035, 855, 836 cm^{-1}.

Biologische Wirkung:
in vitro: MHK der Sporenkeimung von

Monilinia fructicola	0,5 µg/ml
Botrytis allii	1,0 µg/ml
Stemphylium sarcineforme	1,0 µg/ml
Glomerella cingulata	1,0 µg/ml

in vivo: Eine 96%ige Wirkung gegen Rost an Bohnen (*Uromyces appendiculatus*) bei 16 ppm im Gewächshaus.

Literatur: 1. US 3438998, Research Corporation, New York, 15.4.1969.

Lysocellin [1]

Erzeugerstamm: *Streptomyces cacoi* var. asonensis

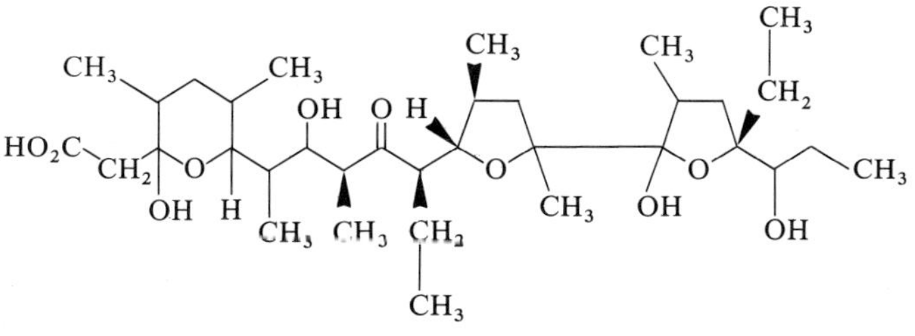

[2]

Isolierung: Extraktion aus dem Kulturfiltrat mit EtOAc, Chromatographie an Kieselgel.
Farblose Kristalle, Fp. 158–160 °C, $[\alpha]_D^{25} = +11{,}5°$ (c=1 in MeOH), ll: organ. Lösungsmittel, nl: Wasser

UV-Spektrum (EtOH): λmax = 292 nm ($E_{1\,cm}^{1\%} = 0{,}68$)
IR-Spektrum (KBr) [1]:

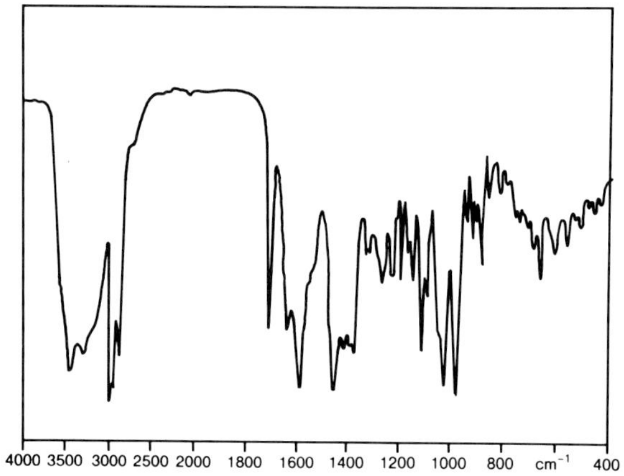

Biologische Wirkung:

in vitro MHK [1, 3]:	*Botrytis cinerea*	50 µg/ml
	Helminthosporum oryzae	50 µg/ml

Toxizität: LD_{50}: 70 mg/kg i.p. bei der Maus.

Literatur: 1. Ebata, H. et al.: J. Antibiotics Ser. A *28*, 118 (1975). – 2. Otake, N.: J. Chem. Soc., Chem. Comm. (*3*), 92 (1975). – 3. US 4033823, Hoffmann-La Roche Inc., 5.7.1977.

Miharamycin [1]

Erzeugerstamm: *Streptomyces miharaensis*
Isolierung: Chromatographie an Amberlite IRC-50 (H^+), Amberlite IRA-45 (OH^-), Aktivkohle und Al_2O_3.
Summenformel: $C_{22}H_{38}N_{10}O_{10} \cdot 2\,HCl$, Mol.-Gew. 674
ll: Wasser, nl: EtOH, Me_2CO
Biologische Wirkung [2]:
in vivo: Wirksam gegen *Pyricularia oryzae* an Reis bei 10–40 ppm. Aufgrund phytotoxischer Wirkung bereits bei 20 ppm wird die Substanz nicht im Feld angewandt.

Literatur: 1. Tsuruoka, T. et al.: Sci. Rep. Meiji Seika Kaisha *9*, 1 (1967). – 2. Noguchi, T. et al.: ibid. *9*, 11 (1967).

Mycobacillin [1]

Erzeugerstamm: *Bacillus subtilis*

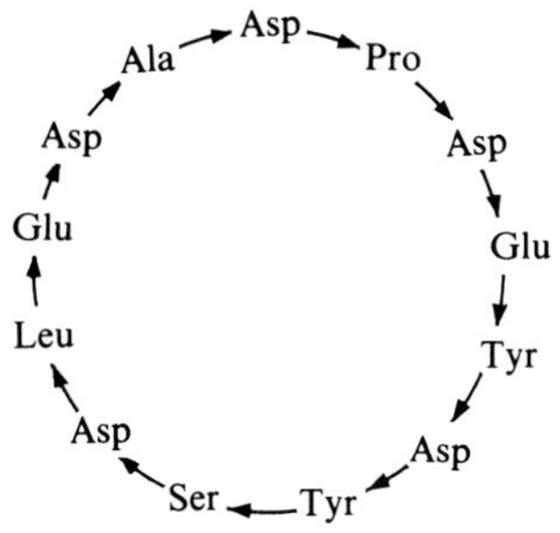

(→, C—N Bindung)

[2,3]

$C_{65}H_{111}N_{13}O_{43}$ Mol.-Gew. 1761

Isolierung: Aus dem Mycel durch Extraktion mit n-BuOH.
Weißes Pulver, ll: MeOH, Me_2CO; l: Wasser

Biologische Wirkung:

in vitro MHK:	*Pyricularia oryzae*	10 µg/ml
	Helminthosporium oryzae	50 µg/ml
	Colletotrichum gloeosporoides	20 µg/ml

Mycobacillin zeigte keine Phytotoxizität bei Reis und Jute bis zu Konzentrationen von 500 ppm [4].

Literatur: 1. Majumder, S.K., Bose, S.K.: Nature *181*, 134 (1958). – 2. Majumder, S.K., Bose, S.K.: Biochem. J. *74*, 596 (1960). – 3. Banerjee, A.B., Bose, S.K.: Nature *200*, 471 (1963). – 4. Nandi, J. et al.: J. Antibiotics Ser. A. *28*, 988 (1975).

Myxin [1]

Erzeugerstamm: Sorangium 3C

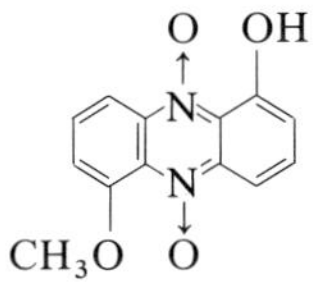

[2, 3]

Isolierung: Aus dem Kulturfiltrat durch Adsorption an Kationenaustauscher und Extraktion des Eluates mit Et_2CO bei pH 1,0.
Dunkelrote Kristalle, Fp. 120–130 °C (Zersetzung). ll: MeOH, Me_2CO, $HCCl_3$, Toluol, l: Wasser, CCl_4

UV-Spektrum (0,1 N NaOH): $\lambda max = 348$ nm, 390 nm und 590 nm
(0,1 N HCl): $\lambda max = 340$ und 505 nm

IR-Spektrum (KBr) [4]:

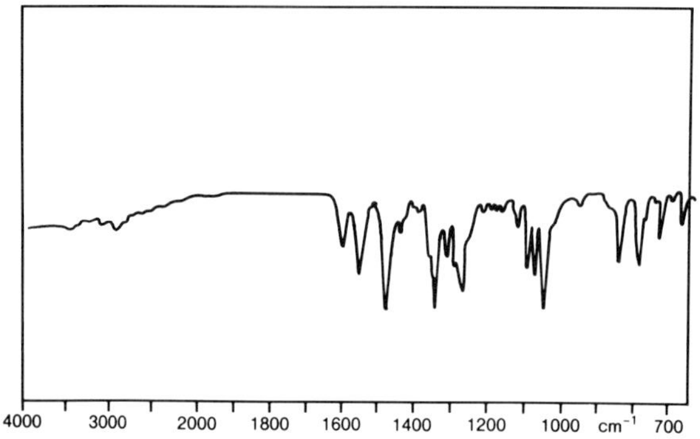

Biologische Wirkung: Im Konzentrationsbereich von 5–10 ppm in vitro breit wirksam gegen Bakterien, z. B. *Erwinia amylovora*, *Xanthomonas campestris*, *Agrobacterium tumefaciens*, und gegen Pilze, z. B. *Botrytis cinerea*, *Fusarium culmorum*, *Helminthosporium sativum*, *Alternaria* sp. [4].

Literatur: 1. Peterson, E.H. et al.: Can. J. Microbiol. *12*, 221 (1966). – 2. Edwards, O.E. et al.: Tetrahedron Lett. 4867 (1966). – 3. Weigele, M. et al.: ibid. 715 (1967). – 4. CA 784213, Atomic Energy of Canada Ltd., 30. 3. 1966.

Nebramycin II

Erzeugerstamm: *Streptomyces tenebrarius* ATCC 17920
Isolierung: Durch Adsorption aus dem Kulturfiltrat an Amberlite IRC-50 H^+, Elution mit Ammoniumhydroxyd.
Summenformel: $C_{21}H_{43}N_5O_{11}$, Mol.-Gew. 541 (1)
Elementaranalyse: C:46,47 H:7,99 N:13,05 O:32,29
Weiße, kristalline Substanz mit basischem Charakter, pKa 5,7; 6,7; 7,7; 8,7. $[\alpha]_D^{25} + 162,5°$ (c = 1 in Wasser). ll: Wasser, MeOH.

UV-Spektrum: Endabsorption
IR-Spektrum (KBr) [1]:

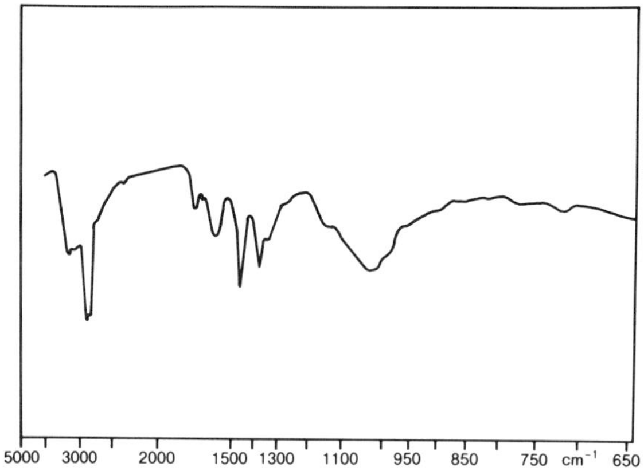

Biologische Wirkung:

in vitro MHK:	*Agrobacterium tumefaciens*	50 µg/ml
	Erwinia amylovora	1,56 µg/ml
	Pseudomonas solanacearum	25 µg/ml
	Xanthomonas phaseoli	12,5 µg/ml

Im Gewächshaus wurde gegen *Erwinia amylovora* an Apfelbäumen in Konzentrationen von 5–50 ppm eine gute protektive Wirkung beobachtet.

Der „Nebramycin-Komplex" besteht aus den Nebramycinen I–XIII [2] (wobei Nebramycin II die Hauptkomponente darstellt), die alle sehr ähnliche physikalisch-chemische Daten und biologische Aktivitäten zeigen.

Toxizität: LD_{50}: 300 mg/kg i.v. bei der Maus.

Literatur: 1. US 3691279, Eli Lilly and Co., 12.9.1972. – 2. Koch, K. F.: J. Org. Chem. **43**, 1430 (1978).

Nikkomycin

Erzeugerstamm: *Streptomyces tendae* Tü 901
ATCC 31160

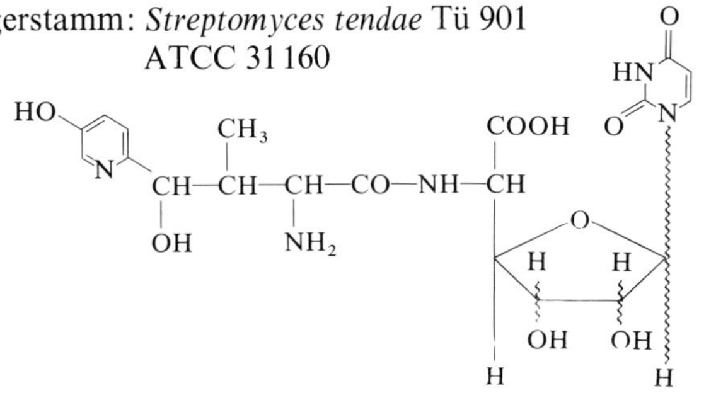

[3]

Nikkomycin Z

Isolierung: Adsorption aus dem Kulturfiltrat an Kationenaustauscher, Gelchromatographie.

Farblose Kristalle. ll: Wasser, Pyridin; nl: organ. Lösungsmittel. Bei pH-Werten unter 3 ist die Substanz weniger stabil [1].

UV-Spektrum [1]:

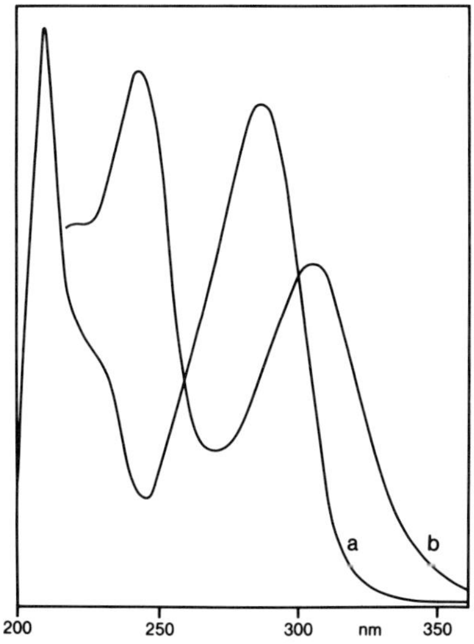

a) in 0,1 N HCl
b) in 0,1 N NaOH

IR-Spektrum (KBr) [1]:

Biologische Wirkung:

in vitro MHK:	*Botrytis cinerea*	5 μg/ml
	Sclerotinia fructicola	10 μg/ml
	Colletotrichum trifolii	5 μg/ml
	Fusarium oxysporum	10 μg/ml
	Rhizoctonia solani	10 μg/ml
	Cercospora beticola	5 μg/ml
	Rhizopus circinans	1 μg/ml

in vivo:	Konz.	Wirkung	
Botrytis cinerea an Bohnen	0,062%	88%	
Puccinia recondita an Weizen	0,001%	100%	
Uromyces phaseoli an Bohnen	0,0031%	69%	[2]

Nikkomycin hemmt die Chitin-Biosynthese, wobei es sehr wahrscheinlich kompetitiv zum UDP-N-Acetylglucosamin die Chitinsynthetase hemmt. Daneben wurde eine starke Hemmung der RNS-Biosynthese beobachtet [1].

1979 wurde ein neues strukturisomeres Nikkomycin beschrieben (Nikkomycin X) [3], das an Stelle des Uracils 4-Formyl-4-imidazolin-2-on enthält. Die biologischen Aktivitäten und Wirkungsspektren der beiden Nikkomycine sind ähnlich.

Literatur: 1. Dähn, U. et al.: Arch. Microbiol. *107*, 143 (1976). – 2. DOS 2537028, Bayer AG, 20.8.1975. – 3. Hagenmaier, H. et al.: Liebigs Ann. Chem. 1494 (1979).

Nonactin [1]

Synonyma: N-329-A [2], FH-3582-A [3], TA 25 M-1 [2]
Erzeugerstamm: *Streptomyces tsushimaensis*
ATCC No. 15141
Streptomyces werraensis

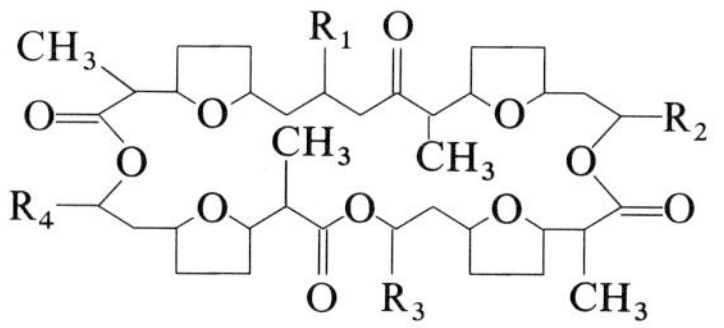

[4]
R_1–R_4 = Methyl; Nonactin
R_1–R_3 = Methyl, R_4 = Äthyl; Monactin
R_1 = Methyl, R_2–R_4 = Äthyl; Trinactin
R_1–R_4 = Äthyl; Tetranactin
R_1, R_2 = Äthyl, R_3, R_4 = Methyl; Dinactin

Isolierung: Durch Extraktion aus dem Mycel mit organ. Lösungsmitteln, Chromatographie an Kieselgel.
Farblose Kristalle, Fp. 148 °C, $[\alpha]_D^{23} = 0°$ (c = 1 in $HCCl_3$). ll: $HCCl_3$, Me_2CO, Benzol

UV-Spektrum: Endabsorption
IR-Spektrum (KBr) [1]:

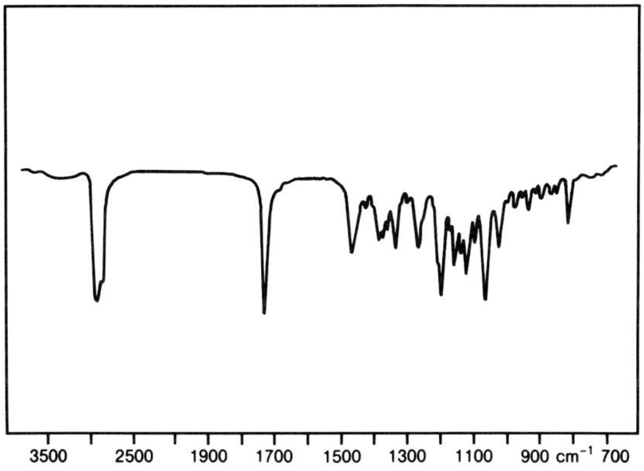

Biologische Wirkung:

in vitro MHK:	*Pyricularia oryzae*	0,1 µg/ml
	Helminthosporium sigmoideum	6,3 µg/ml
	Ophiobolus graminis	6,3 µg/ml

in vivo: Im Gewächshaus wurden gute Wirkungen gegen *Pyricularia oryzae* auf Reis bei Konzentrationen von 50 ppm beobachtet [2, 5].

Das Antibiotikum bindet selektiv NH_4^+ und K^+ Ionen und macht damit Membranen für diese Ionen permeabel [6, 7].

Die Substanzen Monactin, Dinactin, Trinactin und Tetranactin, die sich in den Substituenten R_1–R_4 unterscheiden, zeigen ähnliche biologische Wirkungen. Die insektizide Wirkung des Tetranactins ist im Abschnitt Insektizide ausführlich dargestellt.

Literatur: 1. Corbaz, R. et al.: Helv. Chim. Acta *38*, 1445 (1955). – 2. Nishimura, H. et al.: J. Antibiotics, Ser. A. *27*, 11 (1964). – 3. Wallhäuser, K.H. et al.: Arzneim.-Forsch. *14*, 356 (1964). – 4. Dominguez, J. et al.: Helv. Chim. Acta *45*, 129 (1962). – 5. US 3365363, Shionogi und Co., Ltd., 23.1.1968. – 6. Pioda, L. et al.: Helv. Chim. Acta *45*, 620 (1962). – 7. Pressmann, B.C.: Antimicr. Agents Chemoth. *28*, 1969 (1970).

Oryzacidin A

Erzeugerstamm: *Streptomyces* No. F-35
Isolierung: Aus dem Kulturfiltrat durch Extraktion mit EtOAc oder Benzol bei pH 2,0.
Summenformel: $C_{22}H_{34}N_4O_{12}$, Mol.-Gew. 498
Elementaranalyse: C:58,37 H:7,43 N:12,05
ll: Et_2O, Benzol, EtOH, $HCCl_3$, nl: Hexan, Petroläther

UV-Spektrum (EtOH): λmax = 243 nm ($E^{1\%}_{1\,cm}$ = 122,5)
IR-Spektrum (KBr) [1]:

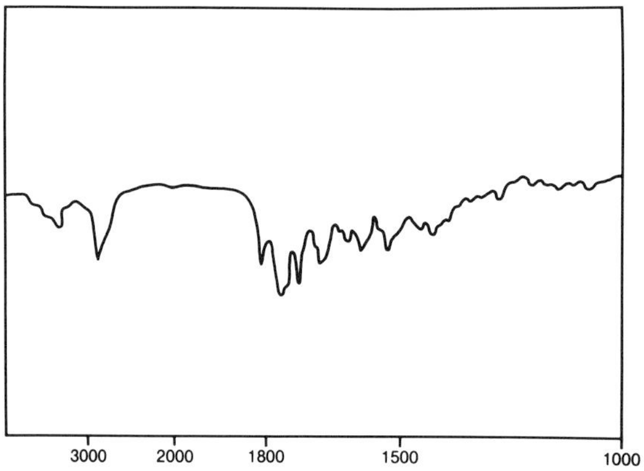

Biologische Wirkung:

in vitro MHK:	*Xanthomonas oryzae*	0,6 µg/ml
	Pyricularia oryzae	10,0 µg/ml
	Physalospora laricina	10,0 µg/ml

Literatur: 1. JA 13198, Chujiro Shimoda, 25.6.1965.

OS-1804

Erzeugerstamm: *Streptomyces michiganensis* OS-1804
Isolierung: Adsorption aus dem Kulturfiltrat an Aktivkohle, Elution mit 60%igem wäßrigem Aceton (pH 2,0). Reinigung über Amberlite IRC-50 (NH_4^+), Amberlite XAD-2, Sephadex G-10 und Avicel.
Summenformel: $C_{18}H_{37}N_7O_2$, Mol.-Gew. 383

Elementaranalyse: C:56,4 H:9,6 N:25,6 O:8,3
Weißes, hygroskopisches Pulver. $[\alpha]_D^{20} = +3°$ (c=1,0, MeOH). ll: Wasser, DMSO, MeOH; nl: EtOAc, Me_2CO, $HCCl_3$

UV-Spektrum: Endabsorption
IR-Spektrum (KBr) [1]:

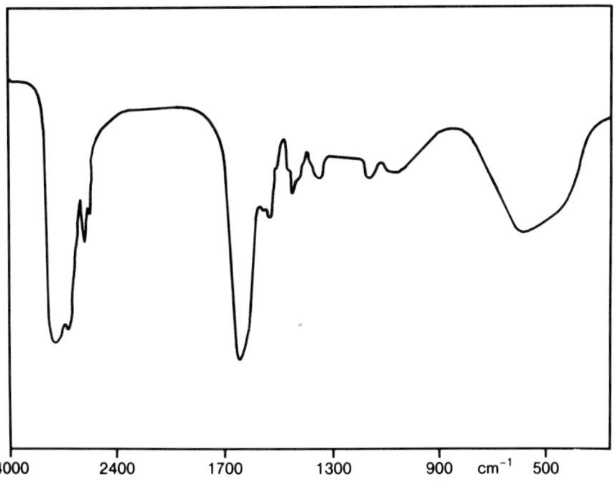

Biologische Wirkung:

in vitro MHK:	*Pyricularia oryzae*	0,78 µg/ml
	Sclerotinia cinerea	<0,09 µg/ml
	Alternaria kikuchiana	<0,09 µg/ml
	Botrytis cinerea	6,25 µg/ml
	Cochliobolus miyabeanus	0,78 µg/ml

Toxizität: LD_{50}: 100 mg/kg i.p. bei der Maus.

Literatur: 1. Shimada, M. et al.: J. Antibiotics Ser. A. *30*, 330 (1977).

Pathocidin (8-Azaguanin)

Gruppe: Purine
Erzeugerstamm [1]: *Streptomyces albus* var. pathocidicus
Isolierung: Extraktion aus dem Mycel mit 80%igem wäßrigem Aceton, Adsorption an Kationen-Austauscherharz (H^+-Form).

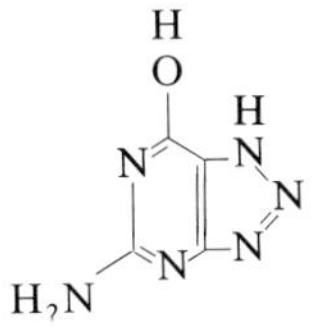

[2]

Farblose Nadeln. nl: organ. Lösungsmittel.

UV-Spektrum [3]:

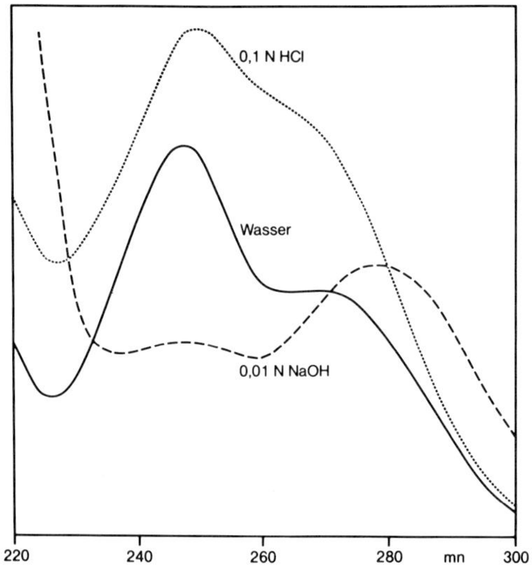

IR-Spektrum (Nujol) [3]:

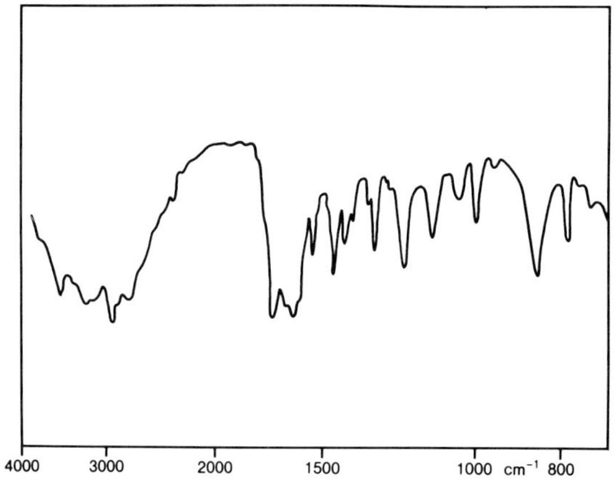

Biologische Wirkung [3]:

in vitro MHK:	*Gibberella fujikuroi*	2,0 μg/ml
	Gloeosporium sp.	2,0 μg/ml
	Alternaria kikuchiana	2,0 μg/ml
	Colletotrichum lagenarium	2,0 μg/ml

Toxizität: Keine Toxizität bei 200 mg/kg i.p. bei der Maus.

Literatur: 1. Nagatsu, J. et al.: J. Antibiotics Ser. A *15*, 103 (1962). – 2. Anzai, K. et al.: ibid. *14*, 253 (1961). – 3. Anzai, K. et al.: ibid. *14*, 340 (1961).

Phytoactin [1]

Gruppe: Polypeptid
Erzeugerstamm: *Streptomyces hygroscopicus* NRRL 2752
Isolierung: Fällung aus der Kulturbrühe bei pH 3, Extraktion mit Methylisobutylketon, Fällung mit Diäthyläther.
Elementaranalyse: C:56,86 H:8,25 N:12,46
Mol.-Gew.: 46000 (Sedimentationsgleichgewicht) [2].
Unscharfer Fp. bei 150 °C, pKa 3,4 in 70% MeOH, $[\alpha]_D^{25} = -86°$ (c = 1 in MeOH); ll: MeOH, $HCCl_3$, 1 N NaOH; nl: Benzol, EtOAc.
Nach Hydrolyse mit 6 N HCl werden die Aminosäuren Valin, α-Alanin, Prolin, Leucin, Arginin, Glycin und Serin gefunden.

UV-Spektrum: Endabsorption
IR Spektrum (KBr):

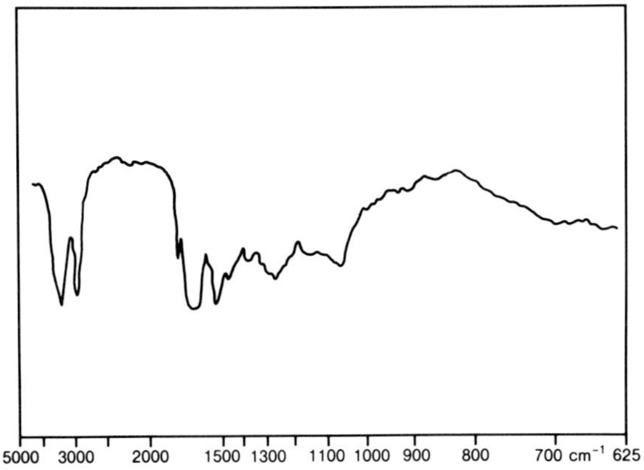

Biologische Wirkung:

in vitro MHK:	*Fusarium roseum*	7,3 μg/ml
	Glomerella cingulata	0,8 μg/ml
	Helminthosporium sativum	0,8 μg/ml
	Alternaria solani	0,8 μg/ml
	Botrytis cinerea	2,4 μg/ml
	Rhizoctonia solani	2,4 μg/ml
	Sclerotinia fructicola	0,8 μg/ml

in vivo: Gute Wirksamkeit im Gewächshaus [2] gegen
Alternaria solani an Tomaten,
Phytophthora infestans an Bohnen,
Uromyces phaseoli an Weizen.

Das Wirkungsoptimum gegen *Podosphaera leucotricha* an Pfirsichen lag bei 100–200 ppm [3].

Literatur: 1. Ziffer, J.S. et al.: Phytopathology *47*, 539 (1957). – 2. DOS 1121276, Pabst Brewing Co., 4.1.1962. – 3. Sprague, R.: Plant Disease Rep. *42*, 1208 (1952).

Pimaricin [1]

Synonym: Natamycin
Erzeugerstamm: *Streptomyces natalensis*

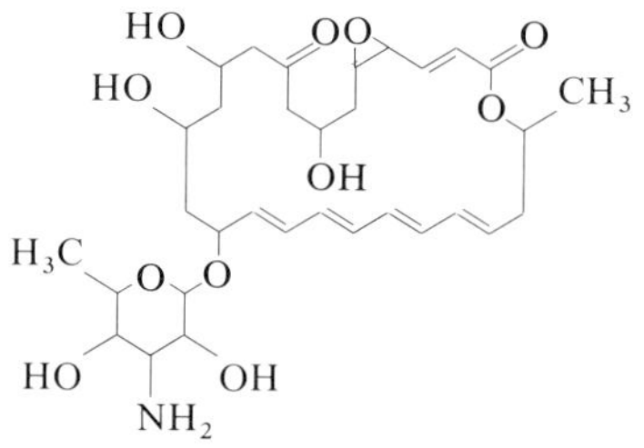

[2, 3]

Isolierung: Aus dem Mycel durch Extraktion mit niederen Alkoholen oder Formamid.
Weiße Kristalle mit amphoterem Charakter, Zers. Punkt 150 °C. $[\alpha]_D^{25} = 250°$ (c = 0,083% in MeOH).

UV-Spektrum (MeOH): λmax = 279 nm, 290 nm, 303 nm, 318 nm
IR-Spektrum (KBr) [4]:

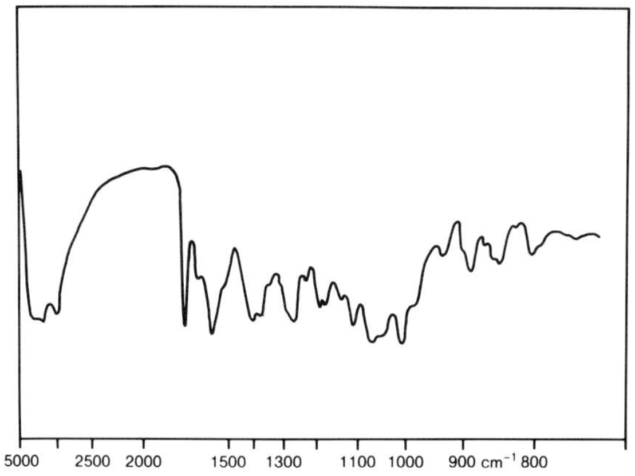

Biologische Aktivität: Bei breiter fungizider Wirkung ergeben sich u.a. folgende MHK-Werte gegen phytopathogene Pilze [4]:

Fusarium spp.	1,2 µg/ml
Verticillium dahliae	1,2 µg/ml
Cladosporium cucumerinum	0,9 µg/ml

Eine Wirkungslücke zeigt sich gegen Pilze, die zu den Oomyceten gehören [5].

Die polyen-spezifischen Eigenschaften wie Lichtinstabilität und Inaktivierung durch Oxidation haben die praktische Anwendbarkeit von Pimaricin auf Saatgutbehandlungen beschränkt [6]. Der Wirkstoff wird über den Samen aufgenommen und bekämpft z.B. *Ascochyta pisi* an Erbsen mit 75–100 ppm, ohne daß sich phytotoxische Nebenwirkungen ergeben.

Als Pflanzenschutzmittel ist Pimaricin in Holland eingesetzt worden zur Bekämpfung von Lagerfäulen an Erdbeeren [7] und zur Zwiebelbehandlung von Narzissen gegen *Fusarium oxysporum* f. sp. narcissi.

Pimaricin greift als Polyen-Macrolid durch Bindungskomplexe mit Sterolen aus Membranen in deren Funktion ein [5].

Toxizität [4]: LD_{50} bei der Maus: p.o. 1500 mg/kg
s.c. >5000 mg/kg
i.p. 250 mg/kg

Literatur: 1. Struyk, A.P.: Antibiotics Annual, 878 (1957–58) (Hsg. Welch, H., Marti-Ibanez, F.) Medical Encyclopadia Inc., New York 1958. – 2. Golding, B.T. et al.: Tetrahedron Lett. 3551 (1966). – 3. Ceder, O. et al.: Chimia *18*, 83 (1964). – 4. US 3892850, Gist-Brocades NV, 3.3.1957. – 5. Dekker, J.: World Rev. Pest Control *10*, 9 (1971). – 6. Dekker, J., in: Fungicides (Hrsg. Torgeson, D.C.) Chemistry and Physiology, Vol. 2, 579, Academic Press, New York, London 1969. – 7. Staden, O.L., Witmondt, M.: Proc. Strawberry Growing 1970's, 74, Baywood Chem. Ltd., Bury St. Edmunds, 1969.

Piomycin

Erzeugerstamm: *Streptomyces piomogenus* ATCC 21137
Isolierung: Adsorption des Kulturfiltrates an Amberlite XE-100, Aktivkohle und Anionenaustauscher (OH^--Form). Das eluierte Konzentrat wird mit EtOH-EtOAc (1:1) versetzt. In der Kälte fällt Piomycin als kristalline gelbe Substanz aus. Chromatographie an Cellulose und erneutes Ausfällen mit EtOH-EtOAc (1:1).
Summenformel: $C_{17}H_{32}N_6O_{15}$, Mol.-Gew. 560
Elementaranalyse: C:36,99 H:5,57 N:13,93 O:43,51
Weißes amorphes Pulver, $[\alpha]_D^{22} = -35°$ (c=1, in Wasser); ll: Wasser; nl: MeOH, Me_2CO, EtOH, Benzol.

UV-Spektrum:
(in 0,05 NHCl): λmax = 260 nm ($E_{cm}^{1\%} = 220$)
(in 0,05 NNaOH): λmax = 262 nm ($E_{cm}^{1\%} = 165$)
IR-Spektrum [1]: (Zur Aufnahmetechnik keine Angaben)

Biologische Wirkung:

in vitro MHK:	*Pyricularia oryzae*	3,13 µg/ml
	Cochliobolus miyabeanus	1,56 µg/ml
	Alternaria spp.	12,5 µg/ml
	Cladosporium fulvum	12,5 µg/ml
	Helminthosporium sigmoideum	12,5 µg/ml

in vivo: In Gewächshausversuchen gute bis volle Wirksamkeit mit 10–40 ppm gegen Blattkrankheiten ohne Pflanzenschäden: *Pyricularia oryzae* und *Cochliobolus miyabeanus* an Reis, *Alternaria kikuchiana* an Birnen.
Toxizität: LD_{50}: 1000 mg/kg p.o., 500 mg/kg s.c., 400 mg/kg i.p. bei der Maus. Bei 100 µg/ml im Wasser keine schädigende Wirkung bei Killi-Fischen.

Literatur: 1. DOS 1695206, Hokko Chemical Industry Co. Ltd., 12.10.1967.

Polyetherin A

Gruppe: Cyclischer Polyäther
Erzeugerstamm: *Streptomyces hygroscopicus* E-749
Isolierung: Extraktion aus dem Kulturfiltrat und Mycel mit organischen Lösungsmitteln, Kieselgelchromatographie.
Summenformel: $C_{42-48}H_{72-74}O_{12}$, Mol.-Gew. 768–782
Farblose Nadeln. $[\alpha]_D^{24} = +36{,}2 \pm 0{,}9°$ (c = 0,842, $HCCl_3$); ll: Alkohole, Me_2CO, EtOAc, $HCCl_3$; nl: Wasser

UV-Spektrum: Endabsorption
IR-Spektrum ($HCCl_3$) [1]:

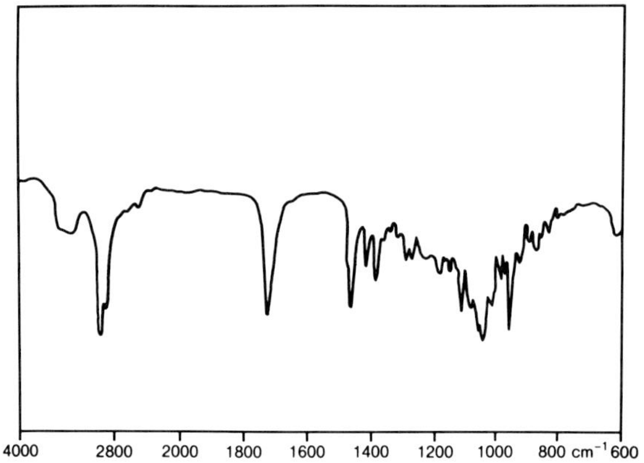

Biologische Wirkung:

in vitro MHK:	*Pyricularia oryzae*	0,8 µg/ml
	Colletotrichum sp.	3,2 µg/ml
	Cochliobolus miyabeanus	1,6 µg/ml

Toxizität: LD_{50}: 10–15 mg/kg i.p. bei der Maus.

Literatur: 1. Shoji, J. et al.: J. Antibiotics Ser. A *21*, 402 (1968).

Prumycin [1, 2]

Gruppe: Glycosid
Erzeugerstamm: *Streptomyces* sp. F-1028

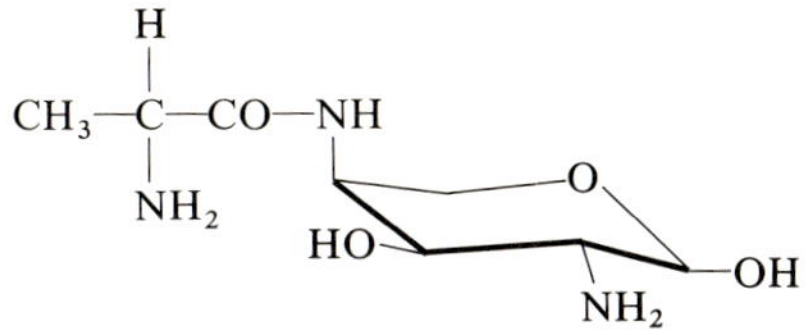

[3]

Isolierung: Chromatographie an Amberlite IRC-50 (Na^+); Aktivkohle, Sephadex G-10.
$[\alpha]_D^{23} = +115°$ (c=0,5, MeOH), ll: Wasser, nl: organ. Lösungsmittel.

UV-Spektrum: Endabsorption
IR-Spektrum (KBr) [2]:

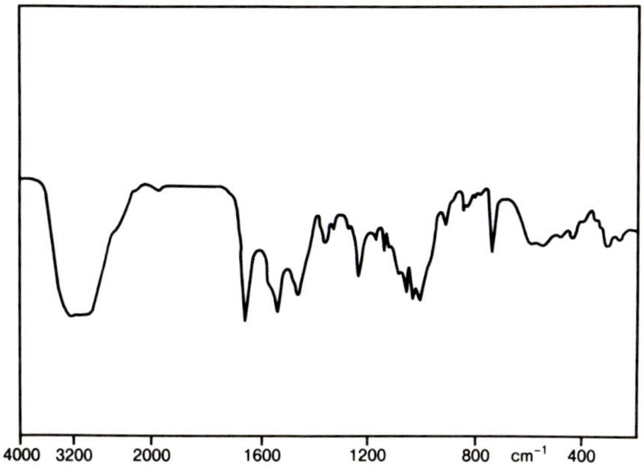

Biologische Wirkung:

in vitro MHK:	*Xanthomonas oryzae*	6,25 µg/ml	
	Sclerotinia cinerea	12,50 µg/ml	
	Botrytis cinerea	6,25 µg/ml	
in vivo:	*Sclerotinia sclerotiorum*	67%	Hemmung bei 50 ppm
	Botrytis cinerea	100%	Hemmung bei 50 ppm an Bohnen.

Toxizität: LD_{50}: 160 mg/kg i.v.; 750 mg/kg p.o. bei der Maus.

Literatur: 1. Hata, T. et al.: J. Antibiotics Ser. A. *24*, 900 (1971). – 2. Omura, S. et al.: Agric. Biol. Chem. *37*, 2805 (1973). – 3. Omura, S. et al.: J. Chem. Soc., Perkin Trans. I *14*, 1627 (1974).

Pyoluteorin [1]

Erzeugerstamm: *Pseudomonas aeruginosa* T 359

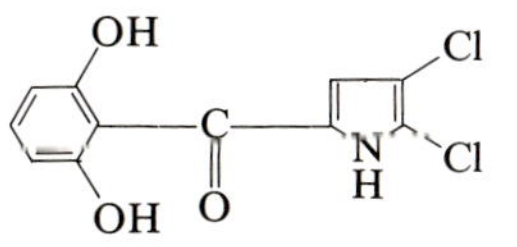

[1]

Isolierung: Durch Extraktion mit Me_2CO aus dem Mycel, Extraktion des konzentrierten Extraktes mit EtOAc und Kristallisation aus Me_2CO/Petroläther. Gelbe Nadeln, Fp. 174 °C, ll: Benzol, Et_2O, $HCCl_3$; nl: Wasser, Petroäther

UV-Spektrum (EtOH): λmax = 255 nm (ε = 4200), 310 nm (ε = 13000)
IR-Spektrum (Nujol): 3311, 1629 cm^{-1}

Im Gewächshaus wurde bei Konzentrationen von 10 ppm eine gute Wirkung gegen *Phytophthora infestans* an Tomaten beobachtet [2].
Toxizität: LD_{50}: 50 mg/kg i.p. bei der Maus.

Literatur: 1. Takeda, R.: J. Ann. Chem. Soc. *80*, 4749 (1958). – 2. US 3298910, Monsanto Chem. Co., 17.1.1967.

Pyrrolnitrin

Pyro green®
Erzeugerstamm: *Pseudomonas pyrrocinia*

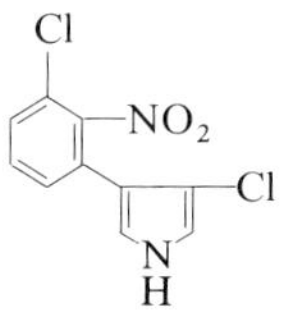

ll: organ. Lösungsmittel, nl: Wasser

UV-Spektrum (EtOH): λmax = 252 nm (ε = 7500)
IR-Spektrum [1]: (Zur Aufnahmetechnik keine Angaben)

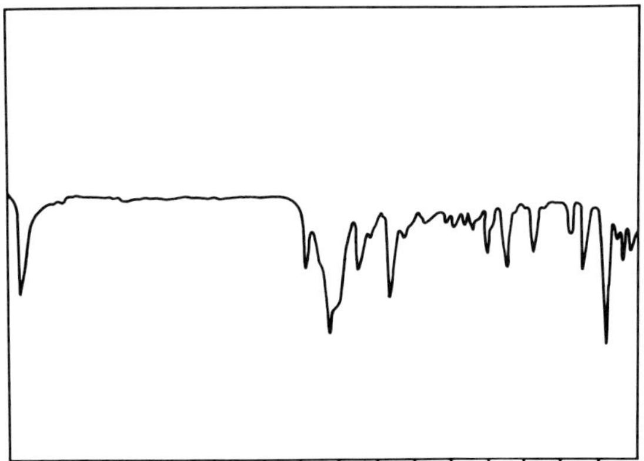

Biologische Wirkung: in vivo: Bekämpfung bodenbürtiger Rasenkrankheiten wie *Rhizoctonia solani*, *Sclerotinia homoeocarpa*, *Helminthosporium* spp. und *Curvularia lunata*. Aufwandmenge = 10–30 kg/ha.
Toxizität: LD_{50}: 90 mg/kg i.p.; >4800 mg/kg p.o. bei der Maus.
Fischtoxizität: TLm: 48 h 13 ppm bei *Oryzias lapitis*.

Literatur: 1. Arima, K. et al.: J. Antibiotics Ser. A. *18*, 201 (1965).

PA 166

Gruppe: Tetraen
Erzeugerstamm: *Streptomyces glaucus* ATCC 12730
Isolierung: Extraktion mit BuOH
Summenformel: $C_{35}H_{53}NO_{14}$, Mol.-Gew. 711
Elementaranalyse: C: 59,59 H: 7,66 N: 2,00 O: 31,50
$[\alpha]_D^{25} = 240–242°$ (c = 1 in 0,1 N HCl in DMF), ll: wäßr. BuOH, nl: $HCCl_3$, Me_2CO

UV-Spektrum (MeOH): λmax = 291 nm ($E_{cm}^{1\%} = 710$), 304 nm ($E_{cm}^{1\%} = 1090$), 319 nm ($E_{cm}^{1\%} = 990$).
IR-Spektrum (KBr): 3333, 2866, 1706, 1563, 1401, 1279 bis 1255, 1085, 1031, 1053, 1005 cm^{-1}.

Biologische Wirkung: in vitro MHK: Breites Spektrum gegen pflanzenpathogene Pilze bei 1–10 µg/ml.
Toxizität: LD_{50}: > 12,5 mg/kg i.p., > 1000 mg/kg p.o. bei der Maus.

Literatur: US 3264185, Ch. F. Pfizer, Co. Inc., 2.8.1966.

PA 1008

Erzeugerstamm: *Streptomyces griseofaciens* ATCC 13179
Isolierung: Extraktion mit organischen Lösungsmitteln
Summenformel: $C_{38}H_{63}NO_{14}$, Mol.-Gew. 787
Elementaranalyse: C: 59,9 H: 8,2 N: 1,99 O: 29,89
$[\alpha]_D^{25} = -36,8°$ (c = 1 in $HCCl_3$)

UV-Spektrum ($HCCl_3$): λmax = 279 nm ($E_{cm}^{1\%} = 289$)
IR-Spektrum ($HCCl_3$) [1]:

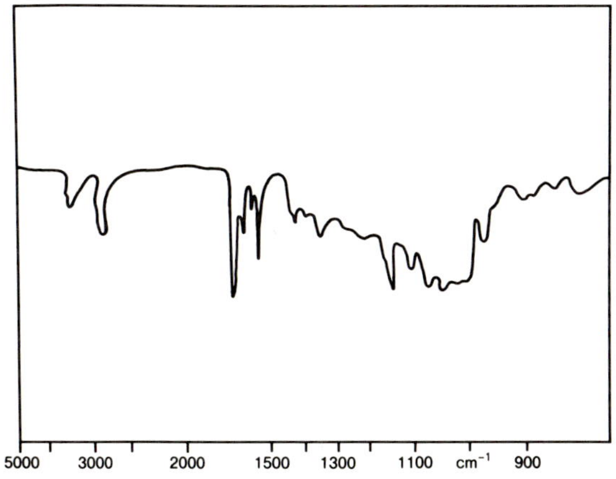

Biologische Wirkung:
in vitro MHK: *Agrobacterium tumefaciens* 0,19 μg/ml
Erwinia amylovora 50,0 μg/ml

Literatur: 1. US 3214339, Ch. F. Pfizer Co. Inc., 26.10.1965.

Quingfengmycin [1]

Erzeugerstamm: *Streptomyces quingfengmyceticus*

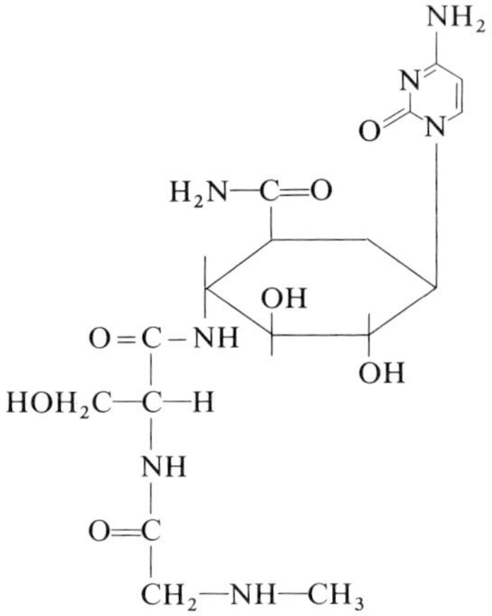

Isolierung: Adsorption aus dem Kulturfiltrat an Dowex IRC-50 (H^+), Chromatographie an Al_2O_3, Kristallisation aus MeOH/Wasser.
Farblose, basische Kristalle, Fp. 200–213 °C, $[\alpha]_D^{23} = +55°$ (c=0,6). ll: Wasser, MeOH.

UV-Spektrum [1]:

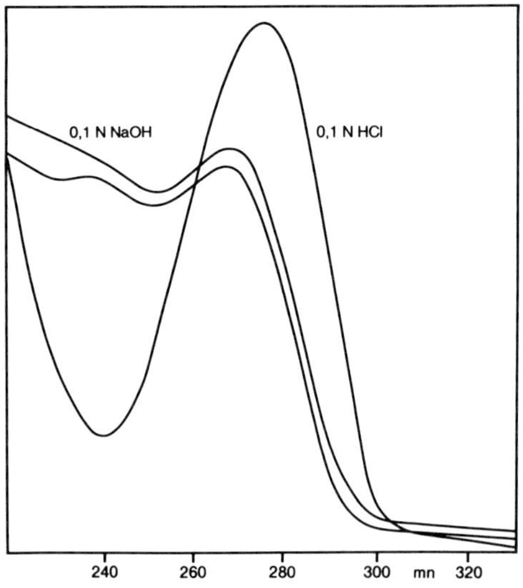

IR-Spektrum (KBr) [1]:

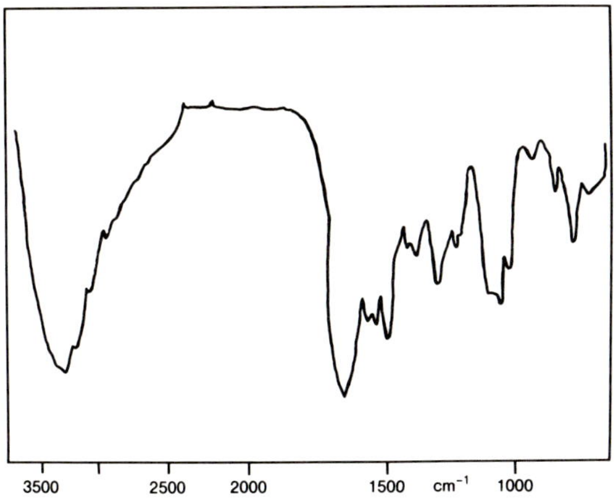

Die Substanz ist gegen Pyricularia oryzae an Reis in den gleichen Konzentrationsbereichen wirksam wie Blasticidin S (s. S. 219).

Literatur: 1. Acta Microbiol. Sinica *15*, 101 (1975).

Ramihyphin A

Gruppe: Peptid
Erzeugerstamm: *Fusarium* Sp. S-435
Isolierung: Extraktion mit EtOAc, Al_2O_3-Chromatographie.
Elementaranalyse: C: 59,7 H: 9,0 N: 12,7, Mol.-Gew. 1192
Saure Hydrolyse ergibt Alanin und Valin.
ll: Alkohole, Me_2CO, nl: Wasser, Petroläther
Biologische Wirkung:

in vitro MHK:	*Stereum hirsutum*	5,0 µg/ml
	Botrytis cinerea	5,0 µg/ml
	Monilia fructigena	10,0 µg/ml

Die Substanz hemmt das Wachstum nach der Sporenkeimung.

Literatur: 1. Barath, Z. et al.: Folia Microbiol. *19*, 507 (1974).

Sclerothricin

Gruppe: Streptothricine
Erzeugerstamm: *Streptomyces sclerogranulatus* [1]
Isolierung: Adsorption des Kulturfiltrates an Aktivkohle, Chromatographie an Amberlite IRC-50 (NH_4^+), Al_2O_3 und Sephadex LH-20

Summenformel: $C_{16}H_{30}N_6O_8 \cdot 2\,HCl$, Mol.-Gew. 434
Elementaranalyse: C:37,87 H:6,31 N:16,57 O:25,25 Cl:14,00
$[\alpha]_D^{23} = -74°$ (c = 1,1, Wasser). ll: Wasser, MeOH.

UV-Spektrum: Endabsorption
IR-Spektrum (KBr) [2]:

Biologische Wirkung [2]:

in vitro MHK:	*Alternaria kikuchiana*	50,0 µg/ml
	Botrytis cinerea	12,5 µg/ml
	Gloeosporium kaki	50,0 µg/ml
	Gibberella fujikuroi	12,5 µg/ml

Toxizität: LD_{50}: 24 mg/kg i.v. bei der Maus.

Literatur: 1. Shimazu, A. et al.: J. Antibiotics Ser. A. *22*, 590 (1969). – 2. Kono, Y. et al.: ibid. A. *22*, 583 (1969).

SF 1293

Gruppe: Peptid
Erzeugerstamm: *Streptomyces hygroscopicus* ATCC 21705

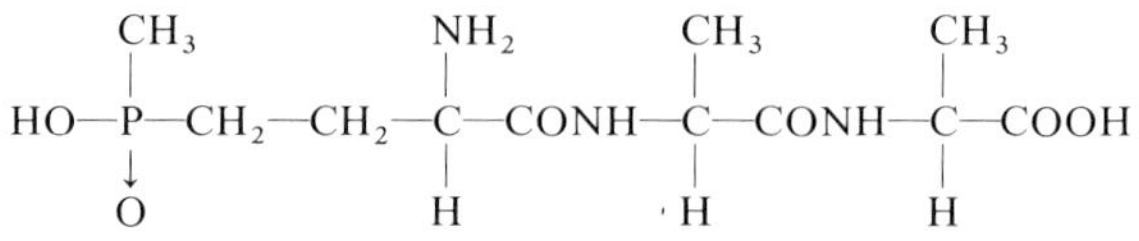

Isolierung: Chromatographie an Kationenaustauschern und Dextran-Gel.
$[\alpha]_D^{25} = -34°$ (c = 1 in Wasser), ll: Wasser, MeOH, nl: EtOH, BuOH, $HCCl_3$.

IR-Spektrum (KBr) [1]:

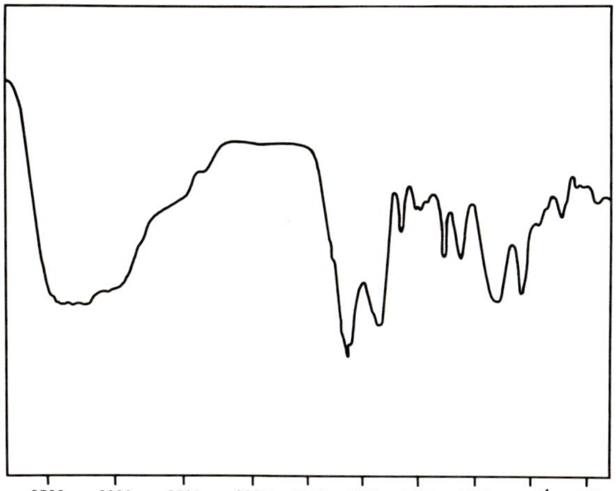

Biologische Wirkung:

in vitro MHK:	*Alternaria kikuchiana*	3,1 µg/ml
	Botrytis cinerea	3,1 µg/ml
	Pellicularia sasakii	0,09 µg/ml
	Sclerotinia sclerotiorum	0,37 µg/ml
	Pyricularia oryzae	25,00 µg/ml

in vivo: Wirksam gegen *Pellicularia sasakii* und *Pyricularia oryzae* an Reis im Konzentrationsbereich von 25–40 ppm im Gewächshaus.

Literatur: 1. NL 7210308, Meiji Seika Kaisha Ltd., 30.1.1973.

SF 1540 [1]

Erzeugerstamm: *Streptomyces hygroscopicus* SF-1540
ATCC 31257

Isolierung: Extraktion aus dem Mycel mit EtOAc, Chromatographie an SiO_2.

Elementaranalyse: C:65,16 H:8,24 N:2,17 O:23,79, Mol.-Gew. 600

Schwachgelbes Pulver, Fp. 131–133 °C, $[\alpha]_D^{25} = +31,6°$ (c=1, in MeOH); ll: MeOH, $HCCl_3$, Benzol; nl: Wasser

UV-Spektrum (MeOH): λmax = 249 nm ($E^{1\%}_{1\,cm}$ = 605), 280 nm ($E^{1\%}_{1\,cm}$ = 270)
IR-Spektrum (KBr) [1]:

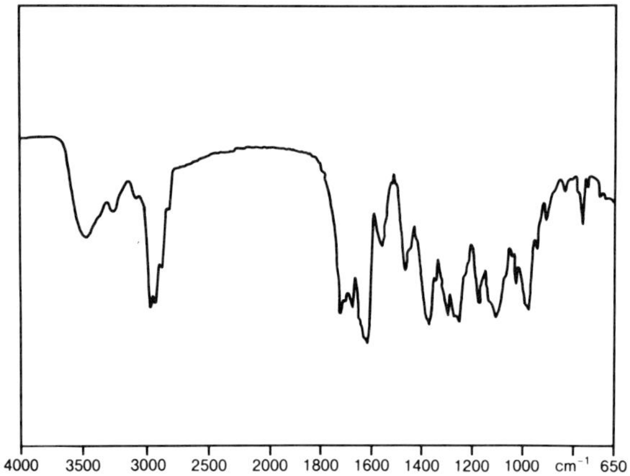

Biologische Wirkung:

in vitro MHK:	*Alternaria kikuchiana*	3,1 µg/ml
	Botrytis cinerea	3,1 µg/ml
	Pellicularia sasakii	0,19 µg/ml
	Pyricularia oryzae	0,78 µg/ml
	Colletotrichum lagenarium	3,12 µg/ml

Toxizität: LD_{50}: 5 mg/kg i.p. bei der Maus.
Der gleiche Stamm produziert noch ein weiteres Antibiotikum, SF 1540 B, das der beschriebenen Substanz in seinen Eigenschaften sehr ähnlich ist [2].

Literatur: 1. DOS 2659180, Meiji Seika Kaisha Ltd., 28.12.1976. – 2. JA 77/118401, Meiji Seika Kaisha Ltd., 31.3.1976.

SF-1836

Erzeugerstamm: *Streptomyces zaomyceticus* SF-1836 [1]

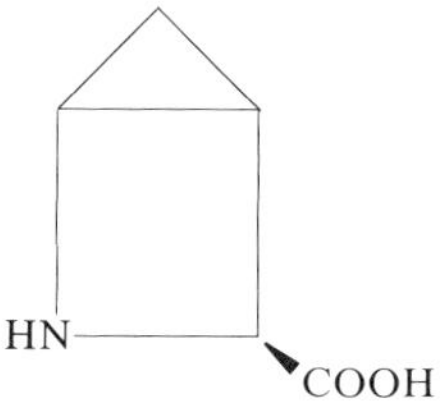

[2]

Isolierung: Aus dem Kulturfiltrat durch Adsorption an Ionenaustauscher
Farblose Kristalle, $[\alpha]_D^{25} = -140°$ (c = 1, Wasser), ll: Wasser, nl: organ. Lösungsmittel

UV-Spektrum: Endabsorption
IR-Spektrum (KBr): 3420, 3050, 2950, 2580, 1630, 1580, 1410, 1360, 1300, 1260, 1080, 1060, 1040, 1010 cm^{-1}

Biologische Wirkung: In Minimalmedien wird die Wirkung gegen Xanthomonas durch L-Prolin aufgehoben. Dieser antagonistische Effekt von L-Prolin wurde in vivo nicht beobachtet.
in vivo: 100%ige Wirkung mit 50 ppm gegen Xanthomonas oryzae an Reis im Gewächshaus.

Literatur: 1. Iwata, M. et al.: Ann. Phytopath. Soc. Japan *45*, 192 (1979). – 2. Shimura, M. et al.: Agric. Biol. Chem. *43*, 2279 (1979).

T-7545

Erzeugerstamm: *Streptomyces hygroscopicus* var. limoneus
Isolierung: Chromatographie an Ionenaustauschern.
Elementaranalyse: C:47,6 H:7,17 N:3,01, Mol.-Gew. 510 ± 25
$[\alpha]_D^{22} = 110 \pm 15°$ (c = 1, Wasser), ll: Wasser, polare Lösungsmittel, nl: EtOAc

UV-Spektrum: Endabsorption
IR-Spektrum [1]: (Zur Aufnahmetechnik keine Angaben)

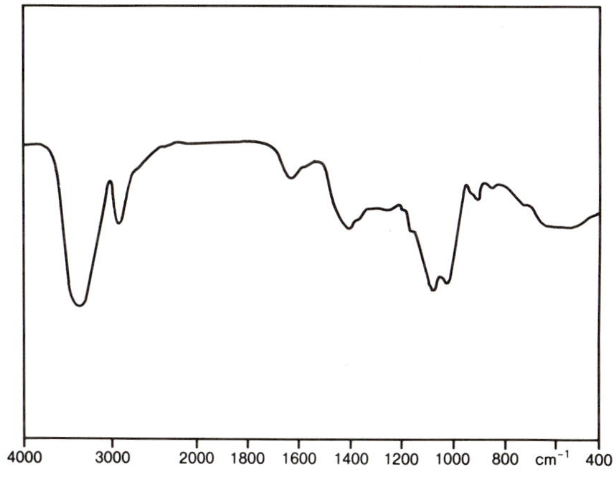

Biologische Wirkung:
in vivo: Volle Wirksamkeit gegen

Pellicularia sasakii	an Reis	mit 20 ppm
Helminthosporium sigmoideum	an Reis	mit 40 ppm
Rhizoctonia solani	an Gurken	mit 10 ppm
Phytophthora infestans	an Tomaten	mit 50 ppm

Literatur: 1. NL 6916326, Takeda Chem. Industr. Ltd., 4.5.1970.

Valacidin

Erzeugerstamm: *Streptomyces lavendulae* NRRL 2675
Isolierung: Extraktion aus dem Mycel mit $HCCl_3$ bei pH 6,0.
Summenformel: $C_{26}H_{24}N_4O_8$, Mol.-Gew. 520
Elementaranalyse: C:60,01 H:4,66 N:10,8 O:24,53
Rotbraunes Pulver, Fp. 261 °C, ll: organ. Lösungsmittel; nl: Wasser und andere polare Lösungsmittel.

UV-Spektrum (MeOH): λmax = 375 nm, 294 nm, 246 nm
IR-Spektrum (KBr) [1]:

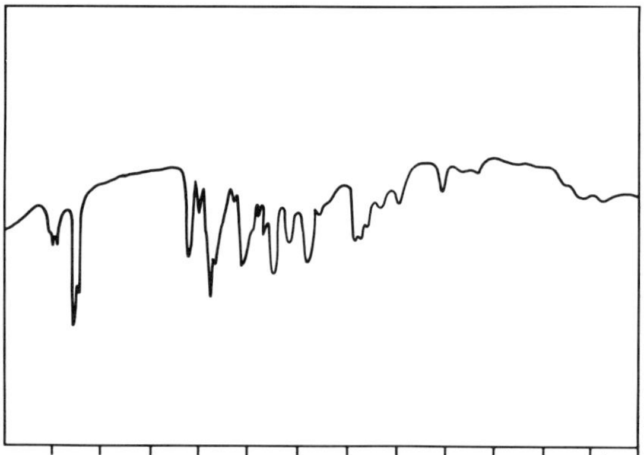

Biologische Wirkung:

in vitro MHK:	*Erwinia amylovora*	1,56 µg/ml
	Pseudomonas solanacearum	6,25 µg/ml
	Pseudomonas syringae	6,25 µg/ml
	Agrobacterium tumefaciens	50,0 µg/ml
	Xanthomonas phaseoli	0,78 µg/ml

Literatur: 1. US 2970943, Eli Lilly and Co., 7.2.1961.

Valinomycin [1,3]

Synonyma: N-329-B [2], Aminomycin
Erzeugerstamm: *Streptomyces tsusimaensis*, *Streptomyces fulvissimus*

```
D-Val→L-Lac→L-Val→D-Hiv→D-Val
  ↑                        ↓
D-Hiv                    L-Lac
  ↑                        ↓
L-Val←L-Lac←D-Val←D-Hiv←L-Val
```

(Val: Valin, Lac: Milchsäure, Hiv: α-Hydroxy-iso-valeriansäure)
Isolierung: Extrakion aus dem Mycel mit organ. Lösungsmitteln, Chromatographie an Kieselgel in Petroläther

Farblose Kristalle, Fp. 192 °C, $[\alpha]_D^{28} = 36{,}2°$ (c = 1,01 in $HCCl_3$). ll: MeOH, Me_2CO, Benzol

UV-Spektrum (Hexan): $\lambda max = 281$ nm
IR-Spektrum (KBr) [4]:

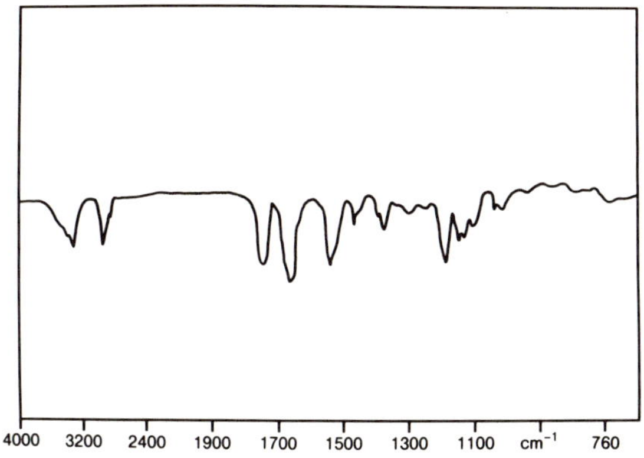

Biologische Wirkung:
in vitro MHK: *Pyricularia oryzae* 0,2 µg/ml
Helminthosporium sigmoideum 6,3 µg/ml
Ophiobolus graminis 6,3 µg/ml
in vivo: Gute Wirkung gegen *Pyricularia oryzae* an Reis im Gewächshaus bei 50 ppm [2, 4].
Entkoppelt die oxydative Phosphorylierung und bindet selektiv Na^+-, K^+- und Cs^+-Ionen [5].
Toxizität: LD_{50}: 0,98 mg/kg i.p. bei der Maus.

Literatur: 1. Brockmann, H., Schmidt-Kastner, G.: Chem. Ber. *88*, 57 (1955). – 2. US 3365363, Shionogi and Co. Ltd., 23.1.1968. – 3. Brockmann, H., Springorum, M.: Naturwiss. *50*, 689 (1963). – 4. Nishimura, H. et al.: J. Antibiotics Ser. A. *17*, 11 (1964). – 5. Grell, E., Funck, T.: Europ. J. Biochem. *34*, 415 (1973).

Venturicidin A [1]

Erzeugerstamm: Streptomyces aureofaciens

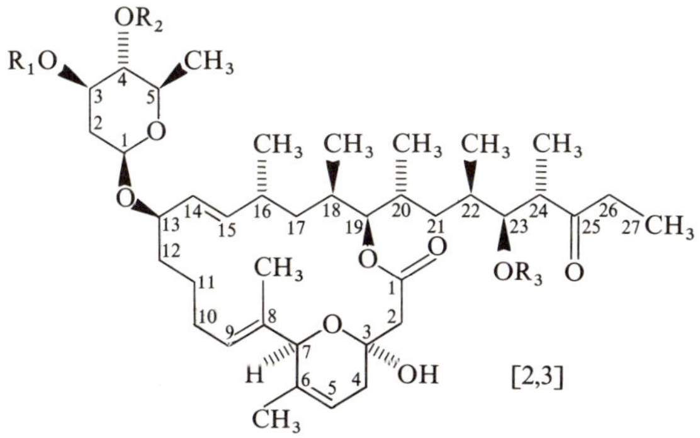

[2,3]

A: $R_1 = NH_2CO$
$R_2 = R_3 = H$
B: $R_1 = R_2 = R_3 = H$

Isolierung: Extraktion aus dem Mycel mit Me_2CO, Extraktion des Konzentrates mit EtOAc, Kristallisation aus MeOH.
Farblose Kristalle, Fp. 142–143 °C, $[\alpha]_D^{23} = +114°$ (c=1% in $HCCl_3$). ll: organ. Lösungsmittel; nl: Hexan, Wasser

IR-Spektrum (KBr) [4]:

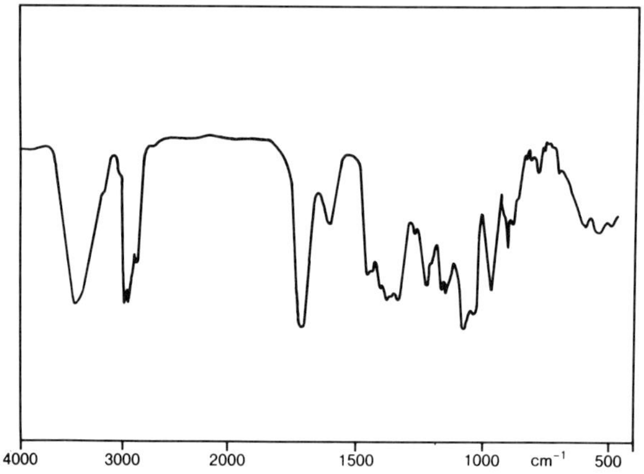

Biologische Wirkung [1]:

in vitro MHK:	*Alternaria tenuis*	5 µg/ml
	Botrytis cinerea	2,5 µg/ml
	Fusarium nivale	2,5 µg/ml
	Glomerella cingulata	10,0 µg/ml
	Verticillium alboatrum	2,5 µg/ml
	Helminthosporium avenae	2,5 µg/ml
	Phytophthora infestans	9 µg/ml
	Venturia inaequalis	1 µg/ml

in vivo:
Gewächshaus: Mit 20 ppm wurden Wirkungen gegen *Venturia inaequalis* und *Podosphaera leucotricha* an Apfel, *Erysiphe graminis* an Gerste und *Botrytis cinerea* an Tomate und Wein erreicht.
Freiland: In Feldversuchen konnte die Wirkung gegen Apfelschorf (*Venturia inaequalis*) mit 100–200 ppm bestätigt werden. Sie war vergleichbar mit der Wirkung einer Captan-Spritzung (0,1% Wirkstoff).
Phytotoxische Wirkungen wurden nicht beobachtet [4].
Toxizität: LD_{50}: 400 mg/kg i.p. bei der Maus.

Das chemisch sehr ähnliche Venturicidin B (s. Formel) unterscheidet sich in den biologischen Wirkungen nur unwesentlich von A.

Literatur: 1. Rhodes, H. et al.: Nature *192*, 952 (1961). – 2. Brufani, M. et al.: Helv. Chim. Acta *51*, 1293 (1968). – 3. Brufani, M. et al.: ibid. *55*, 228 (1972). – 4. DOS 1184458, Glaxo Group Ltd., 22.9.1961.

Versicolin [1, 2]

Erzeugerstamm: *Aspergillus versicolor*

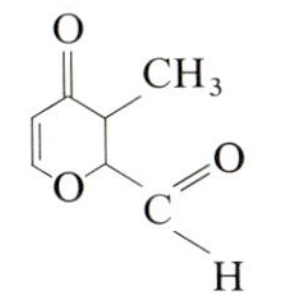

[3]

Isolierung: Extraktion aus der Kulturbrühe mit BuOH oder EtOAc. Chromatographie an Kieselgel in EtOAc [2].
Farblose Kristalle, Fp. 125 °C [2], ll: Wasser, MeOH, $HCCl_3$; nl: Benzol
Absorptionsspektrum (EtOH): λmax = 222 nm (log ε = 4,4), 256 nm (log ε = 3,9), 390 nm (log ε = 2,6), 520 nm (log ε = 2,2)

IR-Spektrum (Nujol): 3345, 1640, 1150 cm^{-1}

Biologische Wirkung [4, 5]:

in vitro MHK:	*Helminthosporium oryzae*	50 µg/ml
	Pyricularia oryzae	40 µg/ml
	Colletotrichum gloeosporoides	2,5 µg/ml

Phytotoxische Wirkungen wurden bei Konzentrationen von 500 ppm beobachtet.
Toxizität: LD_{50}: 150 mg/kg i.v. bei der Maus.

Literatur: 1. Dhar, A.K., Bose, S.K.: J. Antibiotics Ser. A. *20*, 156 (1968). – 2. Dhar, A.K., Bose, S.K.: Appl. Microbiol. *16*, 749 (1968). – 3. Dhar, A.K., Bose, S.K.: Tetrahedron Letters *55*, 4871 (1969). – 4. Chattopadhyay, J.P. et al.: J. Antibiotics Ser. A. *30*, 234 (1977). – 5. Nandi, J. et al.: J. Antibiotics Ser. A. *28*, 988 (1975).

X-4357 B

Bereits ab 5 ppm zeigt X-4157B fungizide Wirkung gegen *Pyricularia oryzae*, *Puccinia* und *Uromyces* spp., *Botrytis cinerea*, *Rhizoctonia solani* an Kulturpflanzen [1].
Toxizität: LD_{50}: 21 mg/kg p.o., 2,45 mg/kg i.p. bei der Maus.
Weiteres: s. Abschnitt Insektizide

Literatur: 1. US 4076802, Hoffmann-La Roche AG., 30. 3. 1977.

156 [1, 2]

Erzeugerstamm: *Streptomyces* 156, NRRL 5319
Isolierung: Extraktion aus der Kulturbrühe mit n-BuOH, Chromatographie an Kieselgel.
Summenformel: $C_{41}H_{70}N_4O_{16-17}$, Mol.-Gew. 874, 890.
Elementaranalyse: C:55,07 H:7,82 N:6,33 O:29,92
Weißes Pulver, Fp. 210 °C, $[\alpha]_D^{18} = +4,3°$ (c=0,2 in n-BuOH). ll: MeOH, BuOH; nl: EtOAc, Benzol

UV-Spektrum (MeOH): λmax = 261 nm
IR-Spektrum (KBr) [1]:

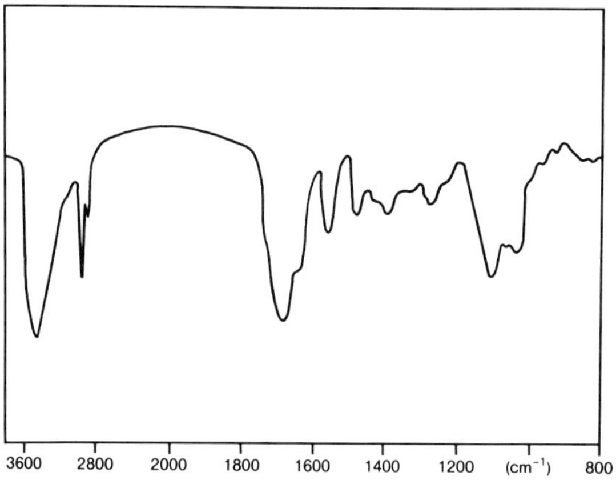

Biologische Wirkung:
in vitro MHK: *Pyricularia oryzae* 1,0 µg/ml
Pellicularia sasakii 20,0 µg/ml

Literatur: 1. US 3825660, Asahi Kasei Kogyo Kabushiki Kaisha, Osaka, 5.11.1970. – 2. JA 7329159, ibid., 7.9.1973.

1308 [1, 2]

Erzeugerstamm: *Streptomyces* 1308, NRRL 5318
Isolierung: Aus dem Kulturfiltrat und dem Mycel durch Extraktion mit Me_2CO, Chromatographie an Kieselgel in Petroläther.
Summenformel: $C_{44}H_{76}O_{11}$, Mol.-Gew. 780
Elementaranalyse: C:67,62 H:9,67 N:9,76 O:22,5
Weißes Pulver, Fp. 128–132 °C, $[\alpha]_D^{18} = -50,5°$ (c=2 in MeOH), ll: MeOH, EtOAc; nl: Hexan, Wasser

UV-Spektrum (MeOH): $\lambda max = 225$ nm, 232 nm, 240 nm
IR-Spektrum (KBr) [1]:

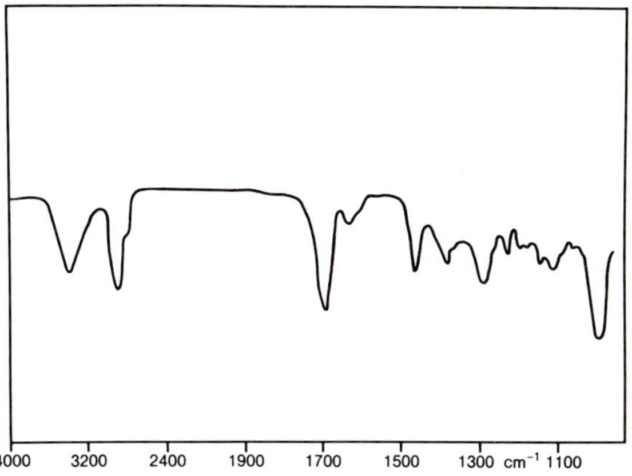

Biologische Wirkung:

in vitro MHK:	*Pyricularia oryzae*	1 μg/ml
	Botrytis cinerea	1 μg/ml
	Alternaria kikuchiana	1 μg/ml
	Pseudomonas solanacearum	1 μg/ml

in vivo: Im Feldversuch wurde gegen *Pyricularia oryzae* an Reis mit 80 ppm eine 80%ige Wirkung erzielt.

Literatur: 1. US 3780172, Asahi Kasei Kogyo Kabushiki Kaisha, Osaka, 18.12.1973. – 2. JA 7330395, ibid. 17.11.1970.

3. Insektizide

3.1. Einleitung

Mikrobielle Insektizide lassen sich aus Mikroorganismen gewinnen, z.B. aus Bakterien, Pilzen, Protozoen und Viren [1, 2]. Eine der wichtigsten Voraussetzungen für eine Kommerzialisierung entspr. Produkte ist wegen des starken Preisdruckes durch die chemisch-synthetischen Präparate gegeben, wenn eine aufwendige biochemische Aufarbeitung und Reindarstellung des betreffenden Wirkstoffs entfallen kann. Bevorzugt wird mit Gattungen von Organismen gearbeitet, die für Menschen und andere höhere Lebewesen sowie Pflanzen unschädlich sind, die aber insektenpathogen wirken. Wenn möglich, werden im Feld Rohpräparate eingesetzt.

Bereits im Jahre 1879 wurde von Miecznikow die Verwendung des Pilzes *Metarrhizium anisopliae* als Insektizid beschrieben. Damals waren die Fermentationstechniken noch nicht ausgereift genug, um Präparate eines standardisierten Wirkstoffgehalts herstellen zu können. Daher konnten mikrobiologische Insektizide nicht in größerem Stil geprüft werden, obwohl ihre Produktion bereits 1884 aufgenommen wurde (Krassilszczik, 1886) [3]. Im Jahre 1891 waren in Europa und in Amerika Präparate erhältlich, die Sporen des Pilzes *Beauveria* spp. enthielten. Die Strukturaufklärung des Wirkstoffs Beauvericin gelang erst 1969 Hamill [4].

In den 30er Jahren gelangte in Frankreich das Präparat „Sporeine“ auf den Markt, das Sporen von *Bacillus thuringiensis* Berliner enthielt. In den 40er Jahren fanden dann die Arbeiten von Dutky über Engerlinge des japanischen Maikäfers (*Popilia japonica* Newm.) besonderes Interesse. Dutky beschrieb im Jahr 1940 die beiden Bakterien-Gattungen *Bacillus popiliae* und *Bacillus lentimorbus* [5], die die sogenannte Milchkrankheit der Engerlinge verursachen.
Nachdem es gelungen war, Sporen dieser Bakterien zu gewinnen, stellten verschiedene Firmen (z. B. Fairfax Biological Laboratories) Präparate her, die diese Mikroorganismen enthielten („Doon“, „Japademex“).

Die Erfolge, die Dutky erzielt hatte, regten zu weiteren Arbeiten an. Als deren Ergebnis wurde 1958 ein mikrobiologisches Präparat von *Bacillus thuringiensis* Berliner unter der Bezeichnung „Thuricide“ von der Fa. Bioferm Corporation produziert. Einige Monate später brachten Merck & Co., Nutrilite Products und andere US-Firmen Insektizide auf den Markt, die unter anderem *B. thuringiensis* enthielten. Bis heute sind solche Präparate Gegenstand intensiver Forschungsarbeiten.

Drei Präparate sollen herausgestellt werden, da sie bereits als Handelsprodukte erhältlich sind: *B. thuringiensis* (Endotoxin mit Exotoxinen als Begleitkomponenten) und Tetranactin. Die übrigen Präparate werden dann in alphabetischer Reihenfolge abgehandelt.

Synonyma

696 s. Phytobakteriomycin
3510 s. Grisein
14-Deoxylagosin s. Filipin
Mycolutein s. Aureothin

Literatur

1. Steinhaus, E.A.: Insect Microbiology, Comstock. Publ. Co., Ithaca, 1946
2. Steinhaus, E.A.: Principles of Insect Pathology, McGraw-Hill Book Co., New York 1949
3. Lipa, J.J.: Postepy Nauk Rolniczych *3*, 63 (1960).
4. Hamill, R.I.: Tetrahedron Lett. *49*, 4255 (1969)
5. Dutky, S.R.: J. Agric. Res. *61*, 694 (1940)

3.2. Handelsprodukte

Bacillus thuringiensis-Endotoxin

Agritol (Merck & Co., Inc.); Bactur (Thompson-Hayward); Bactospeine, Bug Time, Bakthane, Biotrol, BTV, Bio-Guard (Nutrilite Products); Dipel (Abbott); Dendrobazillin, Entobakterin (UdSSR, staatlich); Larvatrol, Lepticide, Leptox, Thuricide (Sandoz); Tribactur (Vondelingenplaat).

B. thuringiensis wurde 1911 von Berliner aus erkrankten oder verendeten Mehlmottenraupen *Ephestia kühniella* Zeller isoliert [1]. Das Bakterium ist verwandt mit der Gattung *Bacillus cereus*. Toumanoff und Corroller [2] bezeichneten *B. thuringiensis* sogar als eine Abart von *B. cereus*. Der systematische Name des heute benutzten Stammes ist *Bacillus thuringiensis* var. thuringiensis.

Schon bei seinen ersten Untersuchungen hatte Berliner beobachtet, daß vegetative Zellen nur sehr geringe Insektentoxizität aufweisen. Dagegen zeigen Zellen von Bakterien, die Sporen und charakteristische Kristall-Einschlüsse bilden, eine außerordentlich hohe Toxizität. Der Kristall befindet sich als parasporaler Einschluß in der Zelle. Form und Größe des Kristalls variieren von Stamm zu Stamm, doch sind die Kristalle im allgemeinen rhomboedrisch.

Die toxischen Eigenschaften der Präparate „Thuricid" [3], „Larvatrol" [1], „Enterobakterin Z" [4] und „Agritol" [5] sind auf die Wirkung der *B. thuringiensis*-Kristalle zurückzuführen. In der Bundesrepublik Deutschland sind Präparate, die das Endotoxin enthalten, unter dem Namen „Bactospeine 6000", „Dipel", „Tarsol" und „Thuricide" zugelassen.

Inzwischen gelang es, den Kristall rein zu isolieren. Er besteht aus einem Glykoprotein mit 3,8% Glucose und 1,8% Mannose, dessen isoelektrischer Punkt bei pH 7,5 liegt. Nach röntgenspektrographischen Untersuchungen haben die Einheiten, aus denen sich der Kristall zusammensetzt, ein Molekulargewicht von 250000 Dalton. Er enthält etwa 30% saure und 70% basische Aminosäuren.

Die toxische Wirkung der Kristalle richtet sich spezifisch gegen Insekten. Fungizide oder herbizide Wirkungen sowie Warmblütertoxizität wurden beim reinen Endotoxin nicht beobachtet. Das bei der Fermentation mitgebildete β-Exotoxin hingegen zeigt ausgeprägte Toxizitäten (s. dort), so daß die Standardisierung eines *B. thuringiensis*-Präparates erhebliche verfahrenstechnische Probleme aufwirft.

Unter den Insekten reagieren die Raupen vieler Schmetterlinge am empfindlichsten. Zahlreiche Untersuchungen beschäftigen sich mit der physiologischen Wirkung von *B. thuringiensis*. Angus und Heimpel [6] und andere [4] gehen davon aus, daß der von einem Insekt aufgenommene Kristall unverändert den oberen Darmabschnitt passiert und sich erst im mittleren Darmabschnitt zersetzt. Dabei ist die Toxizität der erforderlichen Kristallmenge der Körpermasse der Raupe proportional.

Als erstes Anzeichen einer Vergiftung hört die Nahrungsaufnahme infolge einer Darmlähmung auf, die bereits wenige Minuten nach Aufnahme einer entsprechenden Menge Bakterienzellen, die Kristalle enthalten, bzw. von freien Kristallen, eintritt. Als weitere Vergiftungserscheinung ist eine allgemeine Lähmung nach etwa 1 bis 2 h zu beobachten, und schließlich verenden die Insekten.

Nach Heimpel und Angus [6] gibt es drei Gruppen von Insekten, die jeweils unterschiedlich auf die Applikation des *B. thuringiensis*-Endotoxins reagieren.

Zur ersten gehört unter anderem die Maulbeer-Seidenraupe, bei der eine Lähmung des mittleren Darmabschnitts 5 bis 20 min nach der Gabe der Kristalle beobachtet wird. Eine nach 1 bis 7 h eintretende Lähmung führt zum Tod des Insekts.

Die größte Anzahl von Insekten gehört zur zweiten Gruppe, so etwa Spinner der Gattung *Malacosoma*, bei der zwar auch eine Darmlähmung, nicht aber eine generelle Lähmung beobachtet wird. Daher gehen die Insekten dieser Gruppe erst nach 2 bis 4 Tagen ein.

Zur dritten Gruppe gehört nur die Gattung *Ephestia kühniella*, deren besonderem Verhalten zu verdanken ist, daß die Bakterien überhaupt isoliert werden konnten. Bei *Ephestia kühniella* üben nämlich die Kristalle keine direkte toxische Wirkung aus, vielmehr tritt der Tod des Insekts infolge einer Bakterieninfektion ein. Die mit der Nahrung aufgenommenen Sporen keimen im Darm und die sich rasch vermehrenden Bakterien führen zum Tod.

Über den Wirkungsmechanismus von *B. thuringiensis* bei der ersten Insektengruppe ist durch Arbeiten von Angus und Heimpel [7] Näheres bekannt. Bereits 20 bis 25 min nach der Aufnahme hören die Insekten auf zu fressen. Gleichzeitig steigt der pH-Wert ihres Blutes um 1 bis 1,5 Einheiten. Die Darmlähmung tritt etwa 60–70 min nach dem Ausbringen des Insektizids ein. Histochemische Arbeiten [7] zeigten, daß in Gegenwart des Protein-Kristalls das Darmepithel lysiert. Durch die entstehenden Kanäle dringt der alkalische Inhalt des Verdauungstraktes in die Körperhöhle, so daß der pH-Wert des Blutes steigt.

Bei der Herstellung des Präparates werden flüssige Nährmedien verwendet, wie z. B. Fleischboullion, die zudem Zucker und organischen Stickstoff enthalten. Nach erfolgter pH-konstanter Fermentation werden Sporen und Kristalle isoliert. Das Produkt wird eingeengt und anschließend mit Talkum (o. ä.) formuliert. Formulierung und Wahl des Vehikels hängen davon ab, ob das Präparat zum Besprühen oder zum Bestäuben von Pflanzen vorgesehen ist. Dabei enthalten Spritzpräparate je Gramm etwa $70 \cdot 10^9$ Sporen und Kristalle. In Bestäubungspräparaten liegt die Kristallkonzentration zwischen 100 Millionen und $2{,}5 \cdot 10^9$ je Gramm.

Insekten, die besprühte oder bestäubte Blätter fressen, nehmen sehr rasch etwa 40000 bis 80000 Kristalle und Sporen auf. Vergiftung und Tod treten dann in wenigen Stunden ein. Schon wenige Minuten nach Aufnahme hört das Insekt auf zu fressen und fällt auf die Erde, wo dann die Lähmung folgt. Insekten, die nicht gelähmt werden, gehen nach 2 bis 4 Tagen infolge einer allgemeinen Infektion zugrunde.

Die Aufwendungen beim Spritzen betragen zwischen 0,5 und 3 kg je Hektar. Beim Bestäuben müssen etwa 5 bis 30 kg je Hektar ausgebracht werden.

Im Gemüsebau z. B. beträgt bei Spritzmengen von 500–1000 l/ha die Konzentration der Spritzlösung 0,2–0,4 % beim Einsatz von Präparaten mit ca. 10^{11} Sporen je Gramm Präparat [8]. Mehr als 30 landwirtschaftlich wichtige Lepidopteren-Arten wurden bisher erfolgreich bekämpft [9]. Dies führte dazu, daß die Weltjahresproduktion 1978 mehr als 1000 t betrug [10].

Folgende Schädlinge können beispielsweise mit *B. thuringiensis*-Endotoxin bekämpft werden [11, 12, 13, 14]:

Tribolium confusum	*Pieris rapae*	*Lymantria dispar*
Galleria melonella	*Pieris brassicae*	*Aporia crataegi*
Ephestia kühniella	*Heliothis obsoleta*	*Malacosoma neustria*
Pyrausta nubilalis	*Trichoplusia Ni*	*Europoctis chrysorrhoea*
Plutella maculipennis	*Hyphantria cunea*	*Hyponomeuta*

Einer der wichtigsten Vorteile dieses mikrobiologischen Insektizids ist seine völlig fehlende Säugetiertoxizität [13]. Die US-FDA hat dieses Präparat daher sogar von der Pflanzen-Toleranz-Verordnung befreit. Das bedeutet, daß Präparate, die das *B. thuringiensis*-Endotoxin enthalten, sogar noch kurz vor der Ernte ausgebracht werden dürfen.

Zunächst wurde angenommen, daß auch bei längerer Anwendung keine Resistenzbildung auftreten würde [15]. Kürzlich wurde aber dann doch über *B. thuringiensis*-resistente Hausfliegen berichtet [16].

Obgleich Sporen und Kristalle mindestens 5 Jahre lang haltbar sind, haben *B. thuringiensis*-Produkte ebenso wie manche synthetische Insektizide keine lange Wirkungsdauer. Sie sind reine Fraßgifte ohne systemische Wirkung.

Der *B. thuringiensis*-Serotyp H 14, auch als *B. thuringiensis israeliensis* bekannt, zeigt ein erweitertes Wirkungsspektrum. Er ist besonders zur Bekämpfung einiger Dipteren-Larven geeignet, einschließlich Mosquitos der Gattung *Aedes* und *Culex*, die die Überträger von Malaria bzw. Filariasis sind. Eine etwas geringere Wirkung wurde auch gegen *Anopheles* beschrieben [17]. Zur Zeit prüft die WHO einen möglichen Einsatz bei der Bekämpfung der Fliege *Simulium damnosum*, dem Überträger der Flußblindheit (Onchocerciasis). So hat die Franco-Belgische Firma Bellon-Biochemicals die Entwicklung des Präparates eingeleitet. Dabei ist wegen der hohen spezifischen Dichte der Sporen eine der Hauptschwierigkeiten, eine Formulierung zu finden, die ein Absinken der Sporen im Wasser verhindert.

Einer Meldung der FAZ vom 30.10.1979 zufolge sollen mit dieser Variante auch Freilandversuche zur Bekämpfung der Stechmücken im oberen Rheintal durchgeführt werden.

Literatur: 1. Berliner, E.: Z. Angew. Entomol. *2*, 29 (1915). – 2. Toumanoff, C., Le Corroller, Y.: Ann. Inst. Pasteur *96*, 680 (1950). – 3. Anonym: Thuricide, Bioferm. Corporation, Wasco, 1958, 3. – 4. Szwiecowa, O.I.: Zaszczita Rastienij ot Wreditieliej i Boliezniej *5*, 38 (1959). – 5. Anonym: Merck & Co., Inc., Rahway, 1959, 8. – 6. Heimpel, A.M., Angus, T.A.: J. Insect. Pathol. *1*, 152 (1959). – 7. Angus, T.A., Heimpel, A.M.: Canad. Entomol. *91*, 138 (1956). – 8. Franz, J.M., Krieg, A.: Biologische Schädlingsbekämpfung, 2. Aufl., S. 118, Verlag Paul Parey, Berlin/Hamburg, 1976. – 9. Krieg, A.: Nachr.-blatt dtsch. Pflanzenschutzdienst Braunschweig *22*, 97 (1970). – 10. Krieg, A.: Forum mikrobiol. *2*, 60 (1979). – 11. Heimpel, A.M., Angus, T.A.: Proc. 10. Intern. Congr. Entomol. Montreal *4*, 711 (1956). – 12. Steinhaus, E.A.: Ann. Rev. Microbiol. *11*, 165 (1957). – 13. Steinhaus, E.A.: J. Econ. Entomol. *50*, 715 (1957). – 14. Steinhaus, E.A.: Lab. Insect Pathol., Berkeley *4*, 1 (1957). – 15. Lipa, J.J.: Postepy Nauk Rolniczych *3*, 63 (1960). – 16. Agarwal, A.: Nature *279*, 280 (1979). – 17. Collins, P.: Nature *282*, 349 (1979). –

Bacillus thuringiensis-Exotoxine

Neben dem ausführlich beschriebenen Endotoxin finden sich in den Präparationen [1] drei Exotoxine (α, β, γ), von denen die α- und γ-Toxine Phospholipasen sind. Das β-Exotoxin hingegen ist hitzestabil und zeigt insektizide Wirkung gegen Mosquito-Larven, nicht aber gegen Mosquito-Imagines [2].

Shieh et al. beschrieben die mikrobiologische Herstellung und machten auch einen Strukturvorschlag [3], der von Sebesta et al. bestätigt wurde [4]. 1973 gelang Kolvoda et al. [5] die Totalsynthese des α-glycosidischen Adenosin-Derivates:

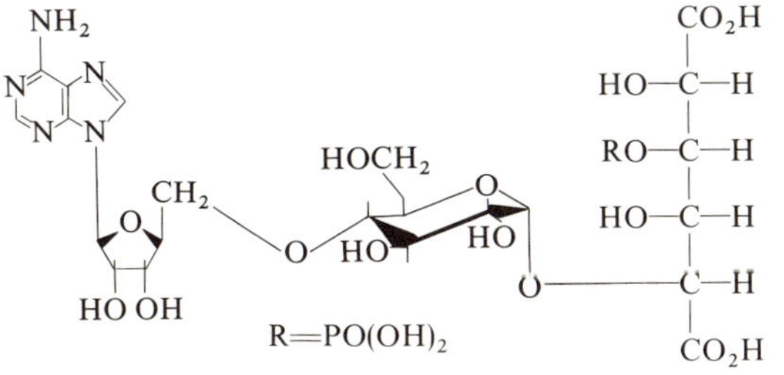

β-Exotoxin

Das *β*-Exotoxin zeigt neben seiner insektiziden Wirkung als Nebenwirkungen Hemmungen von Pflanzen-Mitose [6], teratogene Effekte [7] sowie eine unerwünschte Bienentoxizität [8]. So führt das gleichzeitige Auftreten von *B. thuringiensis*-Endotoxin mit *β*-Exotoxin zu erheblichen Schwierigkeiten bei der Standardisierung und großtechnischen Bereitstellung von Endotoxin-Präparaten.

Literatur: 1. US 127947, John Labatt Ltd., 6.2.1967. – 2. Gurgenidze, T.V.: Tr. Vses. Nauchlo-Issled. inst. Vet. Sanit. *37*, 50 (1970). – 3. US 3758383, Int. Minerals and Chemical Corp. 30.10.1970. – 4. Sebesta, K. et al.: Collect. Czechoslov. Chem. Comm. *34*, 891, (1969). – 5. Kolvoda, L. et al.: Tetrahedron Lett. *47*, 4671 (1973). – 6. Panda, B.B.: Mutation Res. *67*, 161 (1979). – 7. Burges, H.D.: J. Invertebr. Pathol. *26*, 419 (1975). – 8. Krieg, A.: Z. Pflanzenkrankh. *8*, 483 (1973).

Tetranactin

(Chugai Saiyaku Kabushiki Kaisha)

Mitecidin B EC (12% Polynactin-Komplex + 30% BPMC), Mitecidin EC (4% Polynactin in 50% Maschinenöl), Mitecidin C

Das Akarizid Tetranactin läßt sich in Form von rhombischen Prismen aus dem Mycel von *Streptomyces aureus* S-3466 gewinnen [1]. Das Mycel wird mit Aceton extrahiert, anschließend wird zur Trockne eingedampft und das so gewonnene Rohprodukt wird durch Chromatographie an Kieselgel im Fließmittelsystem n-Hexan/EtOAc = 9/1 (v/v) weiter angereichert. Die Feinreinigung [2] des Akarizids gelingt durch eine erneute Chromatographie an Kieselgel im Fließmittelsystem $HCCl_3$/EtOAc = 1/1 (v/v).

Das Tetranactin gehört zu einer homologen Reihe mit Monactin, Nonactin, Dinactin und Trinactin (s. Fungizide-Nonactin) [3, 4, 5].

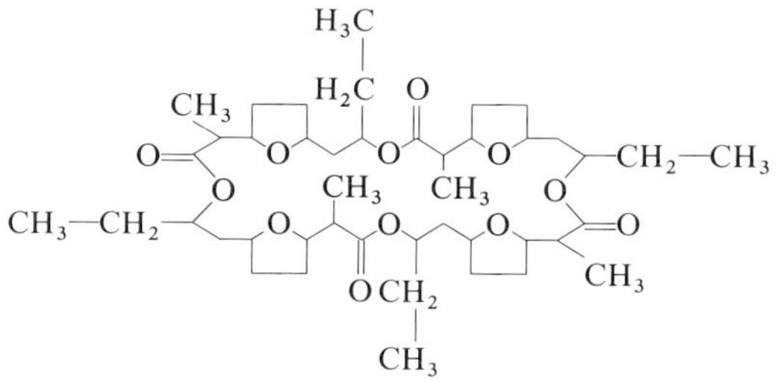

Tetranactin wirkt selektiv praktisch nur gegen adulte *Tetranychus cinnabarius*, *T. urticae*, *T. kanzawa*, *T. viennensis* und *Paranychus ulmi* [6], wobei im Sprühverfahren eine LD_{50} bereits bei 4,8 μg/ml erreicht wird. Eine geringe Empfindlichkeit gegen Tetranactin zeigen noch Bohnenkäfer und Mosquito-Larven, andere Insekten wie Fliegen und Schaben sind hingegen völlig unempfindlich. Als weiterer Nachteil des Präparates muß die fehlende Wirkung am Milbenei gelten.

Tetranactin zeigt bis 1000 ppm keine Phytotoxizität und nur geringe Säugetiertoxizität. Mäuse tolerieren 300 mg/kg (i.p.) und 15 g/kg (p.o.). Laut Herstellerangaben ist Tetranactin toxisch gegen Fische und Muscheln (keine quantitativen Angaben). Außerdem wird vor möglicher Augenreizung gewarnt.

Feldversuche zur akariziden Wirkung von Tetranactin werden in Japan seit 1968 durchgeführt. Als Handelspräparat wurde es dort 1974 registriert.

Literatur: 1. Sagawa, T. et al.: J. Econ. Entomol. *65*, 272 (1972). – 2. NL 7013102, Chugai Saiyaku K.K., 4.9.1970. – 3. Gerlach, H. et al.: Helv. Chim. Acta *50*, 1782 (1967). – 4. Suzuki, K. et al.: J. Antibiotics Ser. A. *24*, 675 (1971). – 5. Haneda, M. et al.: ibid. *27*, 555 (1974). – 6. DOS 2044023, Chugai, Seiyaku, K.K., 25.3.1971.

3.3. Forschungs- und Entwicklungspräparate

Da alle anderen insektizid wirksamen Präparate mikrobieller Provenienz noch keine technische Bedeutung haben, sollen sie nur tabellarisch in alphabetischer Reihenfolge abgehandelt werden.

Antibiotikum 38986 [1]

Gruppe: polycyclisches Äther-Antibiotikum
Erzeugerstamm: *Streptomyces flaveolus* ATCC 31100
Isolierung: Extraktion des Kulturfiltrats mit Methylisobutylketon, Chromatographie an Kieselgel ($HCCl_3$/EtOAc = 1/1), Entfernen von Begleitstoffen an Aktivkohle, Kristallisation aus Me_2CO.
Elementaranalyse: C:67% H:9,5%
$[\alpha]_D^{25} = -2{,}1°$ (c = 1, Me_2CO), l: MeOH, EtOH, Me_2CO, $HCCl_3$, Et_2O; nl: Wasser

IR-Spektrum (KBr) [1]:

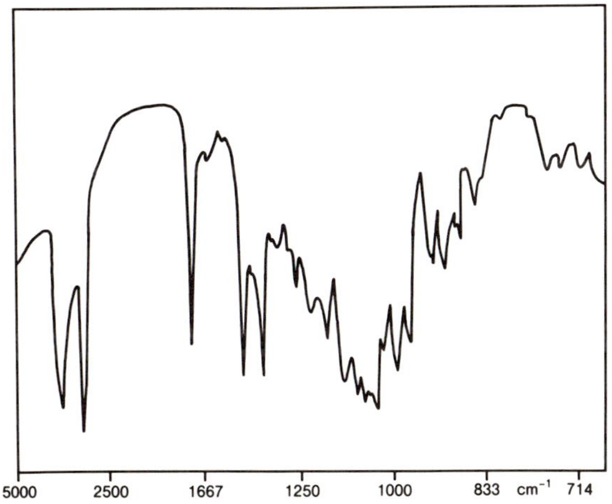

Insektizide Wirkung, geprüft gegen *Musca domestica* (0,25 µg/Insekt) und gegen adulte *Lucilia sericata* (Dosis 25 ppm).
38986 ist sowohl als freie Säure als auch als Salz insektizid wirksam.

Literatur: 1. US 4048304, Pfizer Inc., 13.9.1977.

A-204 I [1, 2, 3]

Gruppe: Polyäther
Erzeugerstamm: *Streptomyces albus* NRRL 3384

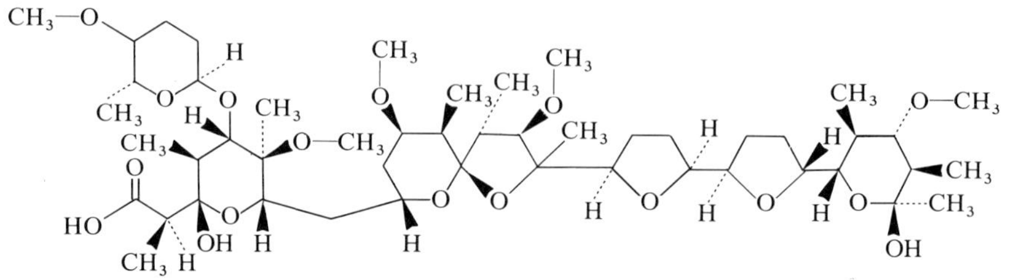

Isolierung: Aus dem Kulturfiltrat und aus dem Mycel durch Extraktion mit 50%igem MeOH. Kulturfiltrat und Mycelextrakt werden vereinigt, auf pH 8,5 eingestellt und mit EtOAc extrahiert. Chromatographie an Aktivkohle mit $HCCl_3$, anschließend Kristallisation aus wäßr. Me_2CO.
pKa = 6,1, $[\alpha]_D^{25} = +68°$ (c = 1, MeOH); ll: DMF, DMSO, MeOAc, EtOAc, Me_2CO, MeEtCO, $HCCl_3$, Benzol; l: MeOH, EtOH, t-BuOH; nl: Wasser

IR-Spektrum (KBr) [2]:

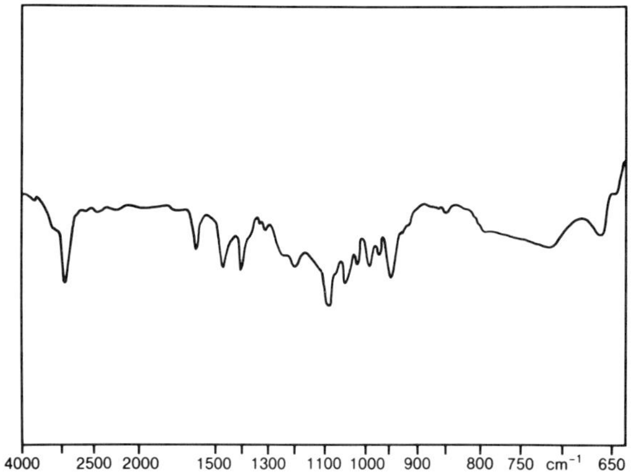

Insektizide Wirkung wurde geprüft gegen [3]:

Epilachna varivestis	1000 ppm	*Tetranychus urticae*	25 ppm
Prodenia eridania	100 ppm	*Oncopeltus fasciatus*	250 ppm
Aphis gossypii	>1000 ppm		

Auch die Kationen-Salze sind insektizid und akarizid wirksam.

Literatur: 1. Jones, N.N. et al.: J. Am. Chem. Soc. *95*, 3399 (1973). – 2. US 3953474, Eli Lilly and Co., 27.4.1976. – 3. US 3705238, Eli Lilly and Co., 5.12.1972.

Aflatoxin B_1 [1]

Erzeugerstamm: *Aspergillus flavus*, *Aspergillus parasiticus*

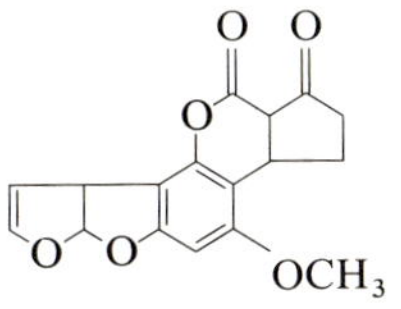

Isolierung: Extraktion des Kulturfiltrats mit $HCCl_3$, Chromatographie an Kieselgel oder Craig-Verteilung.

Absorptions-Spektrum (EtOH): $\lambda max = 362$ nm ($\varepsilon = 21800$),
$\lambda max = 265$ nm ($\varepsilon = 13400$),
$\lambda max = 224$ nm ($\varepsilon = 25600$).

IR-Spektrum ($HCCl_3$) [1]: 3100, 1760, 1684, 1632, 1558 cm^{-1}

Insektizide Wirkung als Larvizid bis zu 1 ppm gegen *Drosophila melanogaster* [2]. Erzeugt „Turkey X-disease", hoch carcinogen [3–6], daher nicht anwendbar.

Literatur: 1. Sargeant, K. et al.: Chem. Ind. *50*, 53 (1963). – 2. Reiss, J.: Chem.-Biol. Interact. *10*, 339 (1975). – 3. van der Merwe, K.J.: Chem. Ind. *50*, 1660 (1963). – 4. Büchi, C. et al.: J. Am. Chem. Soc. *87*, 882 (1965). – 5. Roberts, J.C. et al.: J. Chem. Soc. C., 22 (1968). – 6. Roberts, J.C. et al.: J. Am. Chem. Soc. *89*, 6745 (1967).

Anthelmycin [1]

Erzeugerstamm: *Streptomyces longissimus* ATCC 14562
Isolierung: Aus dem Kulturfiltrat durch Ionenaustausch-Chromatographie an IRC-50, Elution mit 14% NH_4OH.
Summenformel: $C_{25}H_{44}N_5O_{16}$, Mol.-Gew. 670
Elementaranalyse: C:44,39 H:6,53 N:10,33 O:38,69
pK-Werte bei 3,3, 6,8, 7,9.

UV-Spektrum: (Wasser): $\lambda max = 267$ nm, 234 nm
(0,1 N NaOH): $\lambda max = 267$ nm, 234 nm
(0,1 N HCl): $\lambda max = 274$ nm

IR-Spektrum (Nujol) [1]: 3345, 1656, 1605 cm^{-1}

Biologische Wirkung [2]: Quantitativ nicht ausgiebig beschriebene Wirkung an *Apis mellifera* L., gegen *Nosema apis* Zander
Toxizität: LD_{50}: 5 mg/kg Maus i.v.

Literatur: 1. Hamill, R.L. et al.: J. Antibiotics Ser. A. *17*, 100 (1964). – 2. Moffett, J.O. et al.: J. Econ. Entomol. *62*, 886 (1969).

Antimycin

Siehe Fungizide. Antimycin A_1 wurde vergleichend zu Piericidin A und B geprüft, siehe dort.

Aspochracin [1]

Gruppe: Peptid
Erzeugerstamm: *Aspergillus ochraceus* Wilhelm

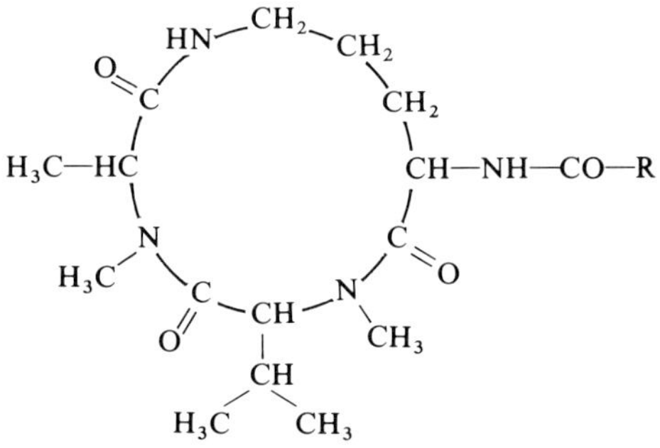

$R = CH_3—CH{=}CH—CH{=}CH—CH{=}CH—$

Isolierung: Extraktion des Kulturfiltrats mit EtOAc, Chromatographie an Kieselgel, 1. mit Benzol/EtOAc/$Me_2CO = 45/40/15$, 2. mit $HCCl_3/Me_2CO = 98/2$.
$[\alpha]_D^{23} = -76{,}0°$ (c = 1, MeOH) [1–4]

UV-Spektrum (MeOH): λmax = 297 nm (ε = 29 500)
IR-Spektrum (Nujol) [2]:

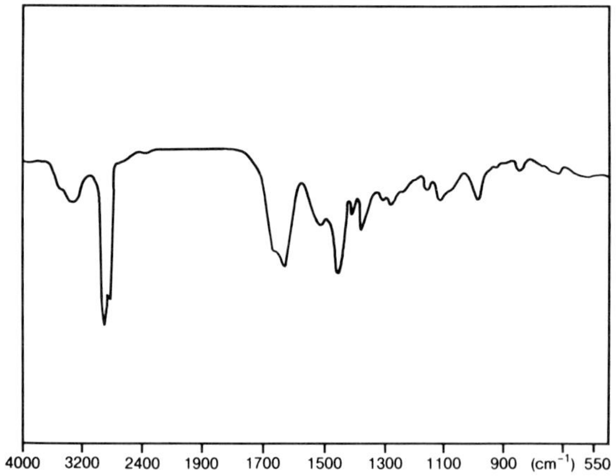

Die insektizide Wirkung wurde gegen Seidenspinnerlarven geprüft [1]. Das letzte Larvenstadium reagiert noch mit 170 µg/g Larve bei Injektion, das erste Larvenstadium beim Tauchen mit $\geqq 200$ ppm und die Eier beim Tauchen mit $\geqq 100$ ppm.

Literatur: 1. Myokei, R. et al.: Agr. Biol. Chem. *33*, 1491 (1969). – 2. Chang, C.F. et al.: Agr. Biol. Chem. *33*, 1501 (1969). – 3. Kodaira, Y. et al.: Tetrahedron Lett. *9*, 695 (1969). – 4. JA 7121789, Sankyo Co. Ltd., 18.10.1968.

Aureothin [1]

Synonym: Mycolutein
Erzeugerstamm: *Streptomyces thioluteus*

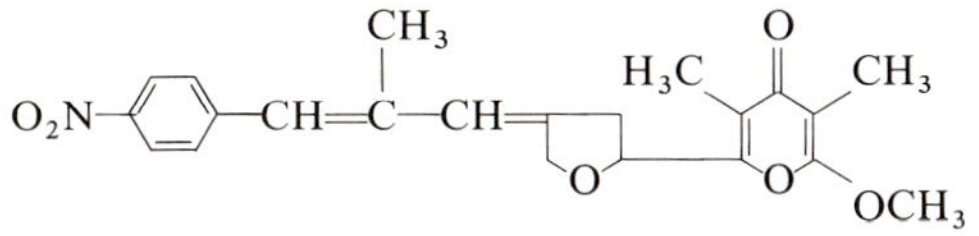

Isolierung: Mycel und Kulturfiltrat werden mit BuOAc extrahiert.
Hellgelbe Kristalle, l: MeOH, $HCCl_3$, THF, Me_2CO. $[\alpha]_D^{18} = +51°$ (c=1, $HCCl_3$) [2, 3, 4, 5].

IR-Spektrum (KBr) [4]:

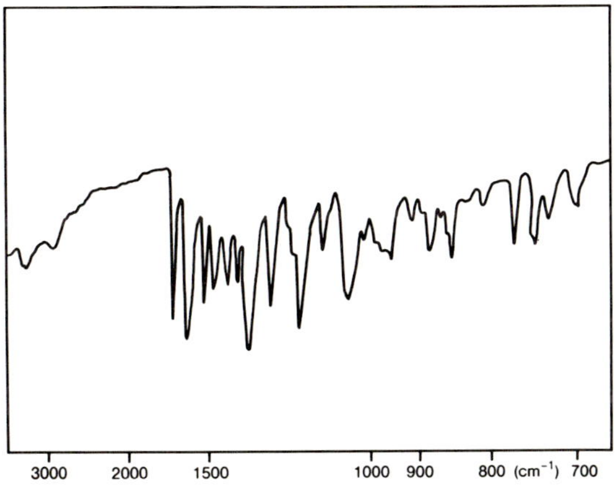

Biologische Wirkung: Cytostatische und insektizide Wirkung, die aber wegen der sehr hohen Säugetiertoxizität nicht weiter verfolgt wird.
Toxizität: LD_{50}: 0,1 bis 1,0 mg/kg Maus, i.v.

Literatur: 1. Maeda, K.: J. Antibiotics Ser. A. *6*, 137 (1953). – 2. Hirata, Y. et al.: Tetrahedron Lett. *14*, 252 (1961). – 3. Oishi, H. et al.: Agr. Biol. Chem. *33*, 1790 (1969). – 4. Umezawa, H. et al.: J. Antibiotics Ser. A. 7, 60 (1954). – 5. Schwartz, U. et al.: ibid. *29*, 237 (1976).

B-41 A, A_1, A_2, A_3, A_4, B_1, B_2

Gruppe: Makrolide
Erzeugerstamm: *Streptomyces* B-41-146

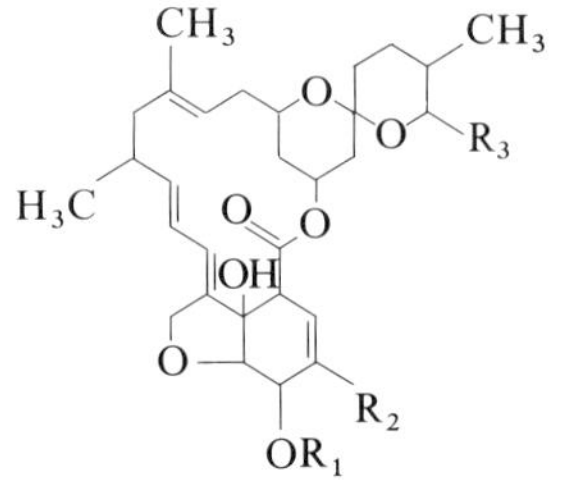

$R_1 = H$ oder CH_3
$R_2 = CH_3$ oder 2-Pyrroyloxymethyl
$R_3 = CH_3$ oder C_2H_5

Isolierung: Extraktion des Mycels mit Me_2CO, Einengen zur Trockne, Extraktion mit n-Hexan, Abdampfen des Lösungsmittels, Chromatographie an Al_2O_3 in $HCCl_3$ Sephadex LH-20-Chromatographie in MeOH.
Die physikalisch-chemischen Eigenschaften aller 7 Verbindungen sind bekannt.
Biologische Wirkung: Akarizide Wirkung besteht bis 0,04 ppm [1].

Literatur: 1. DOS 2329486, Sankyo Co. Ltd., 20.12.1973.

Bassianolid [1]

Gruppe: Cyclopeptid
Erzeugerstamm: *Beauveria bassiana*
Verticillium lecani

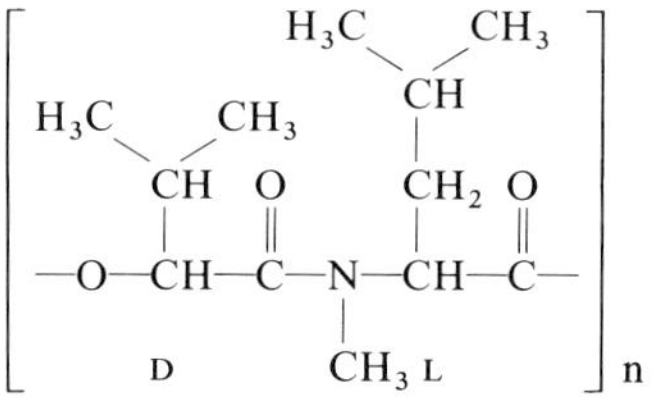

Cyclisches (D-α-Hydroxyisovaleryl-L-N-methylleucyl)$_4$ [2]

n = 4

Die Synthese von Bassianolid haben Kanaoka et al. [5] beschrieben.
Isolierung: Extraktion des Mycels mit MeOH, 1. Chromatographie an Kieselgel, 2. Chromatographie an Al_2O_3 (Abtrennung von Beauvericin), Feinreinigung an Sephadex LH-20 in MeOH.
$[\alpha]_D^{22} = -73°$ (c = 3,3, $HCCl_3$).
IR-Spektrum (Nujol) [2]: max = 1745 (Ester-CO), 1660 (Amid-CO) cm^{-1}
Biologische Wirkung [3, 4]: Die insektizide Wirkung wurde gegen *Bombyx mori* geprüft. Grenzkonzentrationen: 10 ppm im Futter, bei Injektion 5 μg/Larve.

Literatur: 1. Kanaoka, M. et al.: Agric. Biol. Chem. *42*, 629 (1978). – 2. Suzuki, A. et al.: Tetrahedron Lett. *25*, 2167 (1977). – 3. Kanaoka, M. et al.: ibid. *46*, 4049 (1977). – 4. Plattner, Pl.A.: Helv. Chim. Acta *46*, 927 (1963). – 5. Kanaoka, M. et al.: Agric. Biol. Chem. *43*, 1079 (1979).

Beauvericin

Gruppe: Depsipeptid
Erzeugerstamm: *Beauveria bassiana* NRRL 3352

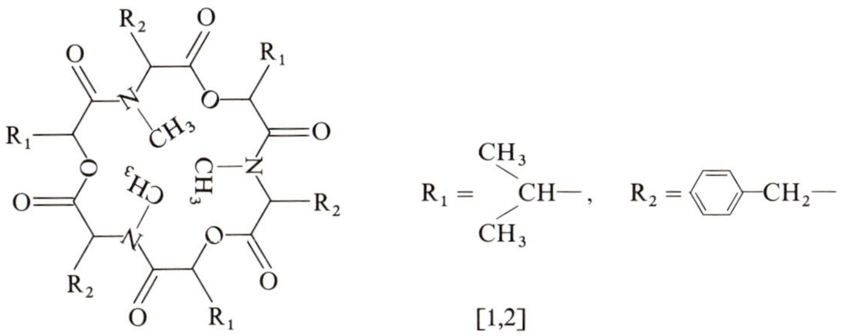

[1,2]

Isolierung: Aus dem Mycel durch Extraktion mit MeOH, Einengen zur Trockne, Verteilung EtOAc-Wasser, Chromatographie an Al_2O_3 (Benzol/EtOAc = 19/1).
$[\alpha]_D^{25} = +65{,}8°$ (c = 1, MeOH)

IR-Spektrum (KBr) [1]:

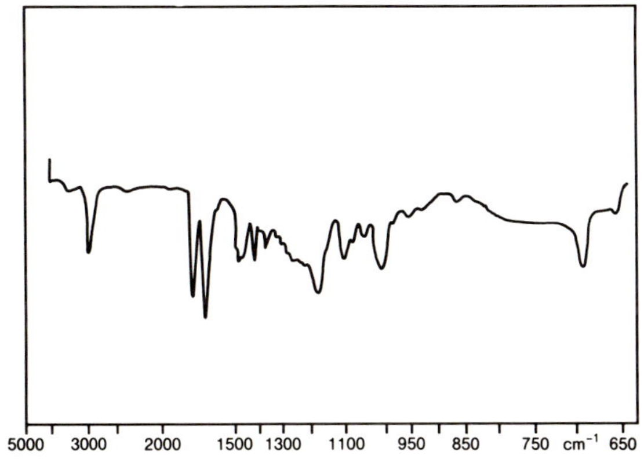

Biologische Wirkung [3, 4]: Insektizid wirksam gegen Mosquito-Larven (keine quantitativen Angaben). Das Präparat zeigt ausgeprägte Garnelen-Toxizität.

Literatur: 1. Hamill, R.L.: Tetrahedron Lett. *49*, 4255 (1969). – 2. Hamill, R.L.: Biochem. Biophys. Res. Commun. *64*, 151 (1975). – 3. SU 507645; 15.8.1974. – 4. SU 313-531, 25.6.1970.

Destruxin A [1]

Gruppe: Cyclisches Peptidlacton
Erzeugerstamm: *Oospora destructor*, ident. Metarrhizium anisopliae

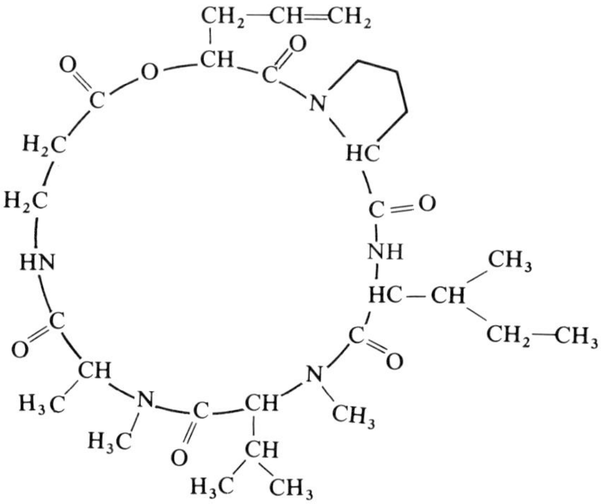

Isolierung: Aus dem Kulturfiltrat durch Adsorption an Aktivkohle, Desorption mit MeOH, anschließend Chromatographie an Kieselgel.
$[\alpha]_D^{15} = -225°$ (c=2,25 MeOH) [2, 3]

IR-Spektrum (KBr) [2]:

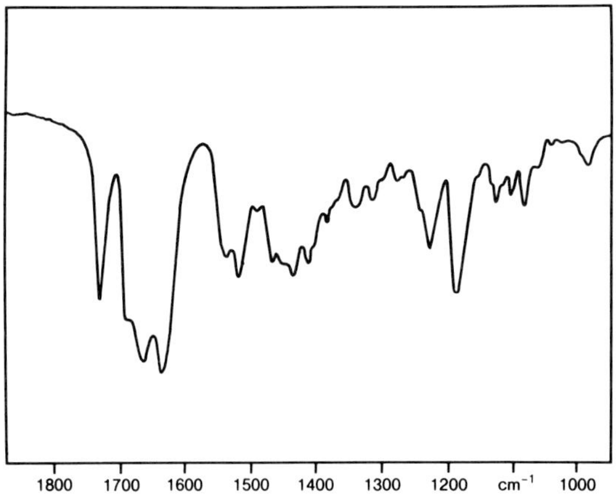

Biologische Wirkung [4]: Insektizid gegen *Anomala costata* und *Scotinophara lurida.*

Literatur: 1. Kodaira, Y.: Agric. Biol. Chem. *25*, 261 (1961). – 2. Suzuki, K.: ibid. *30*, 517 (1966). – 3. Suzuki, K.: ibid. *34*, 813 (1970). – 4. Aizawa, K.: Agric. Asia, Spec. Issue *10*, 362; Tokyo 1976.

Diacetoxy-Scirpenol [1]

Erzeugerstamm: *Fusarium scirpi. var. acuminatum*

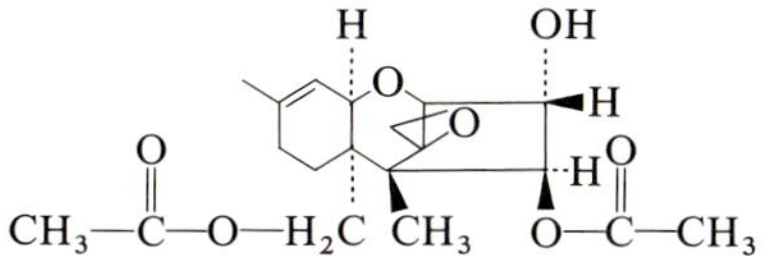

Isolierung: Extraktion des Kulturfiltrats mit EtOAc, Adsorption an Aktivkohle, Kristallisation.
Biologische Wirkung [2]: Insektizide Wirkung gegen *Drosophila melanogaster* bei 1 ppm.

Literatur: 1. Sigg, H.P. et al.: Helv. Chim. Acta *48*, 962 (1965). – 2. Reiss, J.: Chem.-Biol. Interact. *10*, 339 (1975).

Filipin [1]

Siehe Fungizide
Biologische Wirkung: Insektizid geprüft gegen *Musca domestica.*
Toxizität: LD_{50}: 17 mg/kg Maus i.p.

Literatur: 1. Schröder, F. et al.: Chem.-Biol. Interact. *4*, 239 (1972).

Grisein [1, 2]

Synonyma: 3510, Ähnlichkeit mit A-1787, Albomycin
Gruppe: Sideromycine
Erzeugerstamm: *Streptomyces griseus*
Isolierung: Aus dem Kulturfiltrat durch Adsorption an Aktivkohle bei pH 7,5 bis 8,0, Elution mit wäßrigem Pyridin, wäßrigem α-Picolin oder EtOH.
Summenformel: $C_{40}H_{61}N_{10}O_{20}SFe$, Mol.-Gew. 1089
Elementaranalyse: C:44,08 H:5,64 N:12,85 O:29,99 S:2,94 Fe:4,50
Absorptions-Spektrum (EtOH): λmax = 420 nm, 265 nm. Saure Hydrolyse liefert Glutaminsäure, eine nicht charakterisierte Aminosäure und 3-Methyluracil.
Es wurden keine toxischen Nebenwirkungen festgestellt [1, 2]. Insektizide Wirkung gegen Larven von *Culex pipiens* [3] mit 2,5 mg/l.

Literatur: 1. Zähner, H. et al.: Arch. Microbiol. *36*, 325 (1960). – 2. Bickel, V.H. et al.: Experientia *16*, 129 (1960). – 3. Gol'berg, A.M. et al.: Med. Parazitol. Parazit. Bolezn. *44*, 195 (1975).

Hygromycin B [1]

Erzeugerstamm: *Streptomyces hygroscopicus*

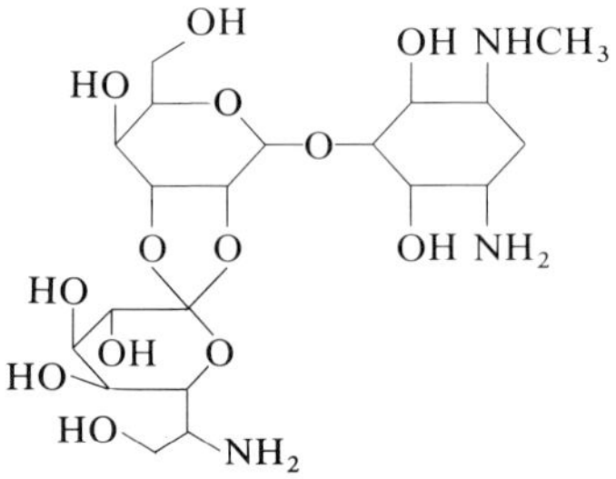

Isolierung: Aus dem Kulturfiltrat durch Adsorption an Amberlite IRC 50 (NH_4^+). pKa: 7,1, 8,8; $[\alpha]_D^{26} = +20°$ (c = 1,0, Wasser). ll: Wasser

IR-Spektrum (Nujol) [1]:

Biologische Wirkung: Insektizid wirksam an *Apis mellifera* L. gegen *Nosema apis* Zander [2].

Literatur: 1. Wiley, P.E. et al.: J. Org. Chem. *27*, 2793 (1962). – 2. Moffett, J.O. et al.: J. Econ. Entomol. *62*, 886 (1969).

Milbemycin β_1, β_2, β_3 [1]

Gruppe: Lacton
Erzeugerstamm: *Streptomyces* B-41-146

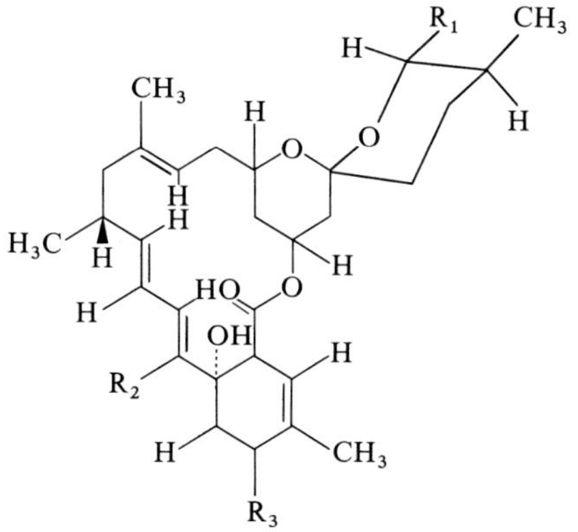

β_1: $R_1 = CH_3$, $R_2 = CH_2OH$, $R_3 = OCH_3$
β_2: $R_1 = C_2H_5$, $R_2 = CH_2OH$, $R_3 = OCH_3$
β_3: $R_1 = CH_3$, $R_2 = CH_3$, $R_3 = OH$

Strukturaufklärung der Akarizide [1].

Isolierung: Extraktion des Kulturfiltrats mit n-Hexan, Kieselgel/Al_2O_3-Chromatographie, Feinreinigung mittels präparativer Dünnschichtchromatographie.
Milbemycin β_1: $[\alpha]_D^{20} = +160°$ (c=0,25, Me_2CO)
Absorptions-Spektrum (EtOH): λmax = 245 nm (ε = 26500)
Milbemycin β_2: siehe β_1
Milbemycin β_3: Fp. 185–7 °C
UV-Spektrum (in EtOH): λmax = 247 nm (ε = 29200)
Biologische Wirkung: Akarizide, geprüft an *Tetranychus urticae.*

Literatur: 1. Mishima, H. et al.: Tetrahedron Lett. *10*, 711 (1975).

MM 4086 [1]

Erzeugerstamm: *Aspergillus versicolor* IMI 129488
IMI 129489
IMI 94159
IMI 96228

Isolierung: Extraktion des Kulturfiltrats mit Benzol, Chromatographie an Kieselgel.
Summenformel: $C_9H_{11}NO_4$, Mol.-Gew. 197
$n_D^{21} = 1,4895$, Kp 120 °C/3 mm Hg; $[\alpha]_D^{20} = +22,3°$ (c=2, $HCCl_3$). ll: Kohlenwasserstoffe.
Absorptions-Spektrum (EtOH): λmax = 204,5 nm (ε = 10260)
IR-Spektrum (Film): νmax = 1780, 1720, 1645, 1440, 1400, 1370, 1300, 1245, 1205, 1190, 1160 cm^{-1}

Biologische Wirkung: 100%ige Abtötung bei 2–20 µg/cm² *a.i.*: *Drosophila melanogaster*, *Musca domestica*, *Aedes aegypti*, *Blatella germanica*, *Plodia interpunctella*, *Ephestia elutella*, *Tribolium confusum*, *Stegobium paniceum*, *Ptinus tectus*. 400 µg/cm² a.i., 14 d: *Tineola bisselliella*.

Literatur: 1. NL 68,14629, Beecham Group Ltd., 14.10.1967.

Nigragillin [1]

Erzeugerstamm: *Aspergillus niger*

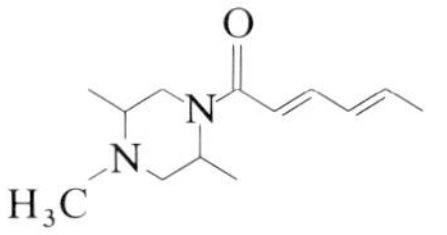

Isolierung: Extraktion des Kulturfiltrats mit Et_2O bei pH 10, Chromatographie an Al_2O_3 (Benzol/EtOAc = 6/1) und an Kieselgel (EtOAc/MeOH = 2/1)
$[\alpha]_D^{26} = +101°$ (c = 1, $HCCl_3$)
Absorptions-Spektrum (MeOH): $\lambda_{max} = 262$ nm ($\varepsilon = 26200$)

IR-Spektrum (Film) [1]:

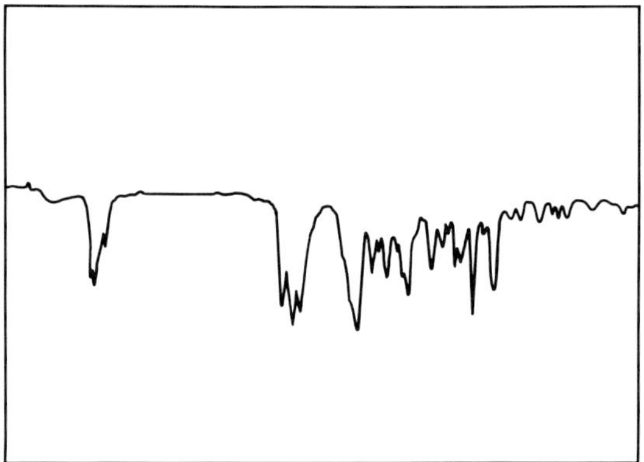

Biologische Wirkung: 100% Abtötung von Seidenspinner-Larven bis 40 ppm, oral.

Literatur: 1. Isogai, A. et al.: Agric. Biol. Chem. *39*, 739 (1975).

Nikkomycin (AMS 0896)

Siehe Fungizide
Biologische Wirkung [1]: AMS 0896 erwies sich als besonders wirksam gegen Spinnmilben und wurde gegen *Tetranychus urticae* (resistent und normal sensibel), im Freiland gegen *Panonychus ulmi* geprüft.

Um die Wirkungsweise aufzuklären, erfolgte eine Untersuchung der stadienspezifischen Wirkung an *Tetranychus urticae.* Selbst bei hohen Dosierungen zeigte AMS 0896 keine akute Wirkung auf Adulte und Eier. Die Entwicklung beweglicher Stadien wurde bei nur 0,004% Wirkstoff stark verzögert und brach im jeweils folgenden Chrysalis-Stadium abrupt ab. AMS 0896 verhindert das Schlüpfen von Nymphen aus allen Chrysalis-Stadien.

Freilandversuche gegen *Panonychus ulmi* zeigten, daß man mit 2 Spritzungen im Abstand von 14 Tagen mit 0,025 bis 0,05% Wirkstoff oder mit 3 bis 5 Spritzungen im Abstand von 14 Tagen mit 0,005 bis 0,01% Wirkstoff Abtötungsgrade von 99% erreicht.

Literatur: 1. Zöbelein, G.: Mitt. Biolog. Bundesanstalt Land- u. Forstwirtschaft *191*, 282 (1979).

Phyllomycin

Siehe Fungizide: Antimycine

Phytobacteriomycin [1]

Synonym: Antibiotikum 696
Gruppe: Streptothricine
Erzeugerstamm: *Streptomyces lavendulae*
Isolierung: Aus dem Kulturfiltrat durch Kationenaustauschchromatographie.
Elementaranalyse: C:39,96 H:5,74 N:6,08
Basisch; ll: Wasser
Biologische Wirkung: Insektizide Wirkung mit 2,5 mg/l gegen *Culex-pipiens* Mosquito-Larven [2].

Literatur: 1. Umezawa, H.: Index of Antibiotics from Actinomycetes, 828, Univers. Tokyo Press, Tokyo 1967. – 2. Gol'berg, A.M. et al.: Med. Parazitol. Parazit. Bolezn. *44*, 195 (1975).

Piericidin A, B [1–4]

Gruppe: Polyen
Erzeugerstamm: *Streptomyces mobaraensis* 1622

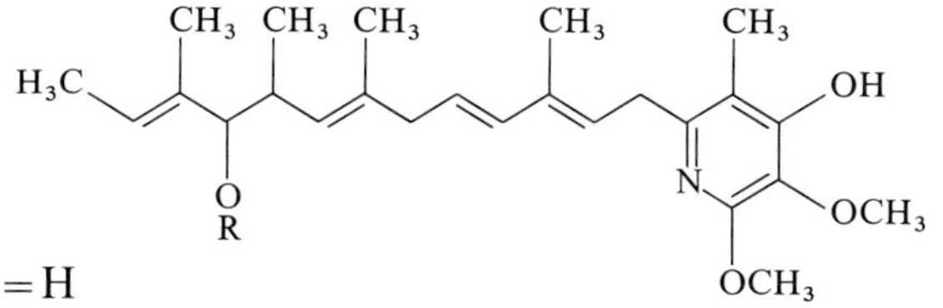

Piericidin A: R = H
Piericidin B: R = CH_3
Isolierung: Aus dem Aceton-Extrakt des Mycels mittels chromatographischer Methoden.

UV-Spektrum (MeOH):
Piericidin A: λmax = 232 nm, 238 nm, 270 nm
Piericidin B: λmax = 230 nm, 267 nm
IR-Spektrum (Film) [1]:

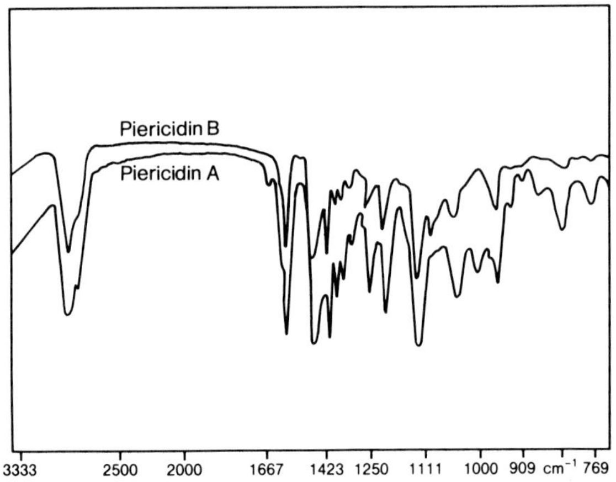

Detaillierte Strukturaufklärung der Piericidine A und B in [5].

Biologische Wirkung [5]	Piericidin A LD_{50}	Piericidin B LD_{50}	Antimycin A LD_{50}
Musca domestica (topikale Appl.)	~ 1 µg/Fliege	~ 1 µg/Fliege	>0,25<0,5 µg/Fliege
Blatella germanica (topikale Appl.)	~ 4 µg/Ins.	> 8 µg/Ins.	8 µg/Ins.
Chilo simplex (topikale Appl.)	>12<24 µg/Larve	>12<24 µg/Larve	—
Bombyx mori (topikale Appl.)	> 1,2<2,4 µg/Larve	> 1,2<2,4 µg/Larve	—
Pieris rapae (topikale Appl.)	> 9,6<24 µg/Larve	~24 µg/Larve	—
Myzus persicae (Sprüh-Appl.)	<10 ppm	<10 ppm	>3,3<75,3 ppm
Tetranychus terrarius (Tauchen)	< 1,25 ppm	~10 ppm	<1,25 ppm

Toxizität: LD_{50}: 8 mg/kg, Maus i.p. Hemmer der Atmungskette in Mitochondrien [6], hemmt die Hill-Reaktion [7].

Literatur: 1. Takahashi, N. et al.: J. Am. Chem. Soc. *87*, 2066 (1965). – 2. Takahashi, N. et al.: Agric. Biol. Chem. *30*, 1, 13 (1966). – 3. Takahashi, N. et al.: Tetrahedron Lett. 1961 (1967). – 4. Tanabe, M. et al.: J. Org. Chem. *35*, 2087 (1970). – 5. Takahashi, N. et al.: Agric. Biol. Chem. *32*, 1115 (1968). – 6. Yoshida, S. et al.: Tetrahedron Lett. *22*, 1863 (1975). – 7. Katoh, S.: Plant Cell Physiol. *13*, 273 (1972).

Polymycin [1]

Gruppe: Streptothricine
Erzeugerstamm: *Actinomyces* sp. 1787-9
Isolierung: Aus dem Kulturfiltrat durch Kationenaustausch-Chromatographie.
Elementaranalyse: C:33,35 H:6,7 N:14,8
Biologische Wirkung: Insektizide Wirkung gegen Larven von *Culex-pipiens* mit 2,5 mg/l [3].
Toxizität: LD_{50}: 50–60 mg/kg Maus i.v. [1, 2].

Literatur: 1. Soloveva, A. et al.: Antibiotiki 5, 5 (1960). – 2. Hermanova, V. et al.: Antibiotiki 6, 293 (1961). – 3. Gol'berg, A.M. et al.: Med. Parazitol. Parazit. Bolezn. *44*, 195 (1975).

Prasinon A, B [1]

Erzeugerstamm: *Streptomyces prasinus* NCIB 10719
Isolierung: Aus dem Kulturfiltrat durch Extraktion mit BuOAc, Chromatographie an Kieselgel (Benzol/MeOH), Sephadex LH-20 in MeOH und präparative Dünnschichtchromatographie ($HCCl_3$/MeOH).
Elementaranalyse:
Prasinon A: C:64,42 H:8,21 N:2,00 O:26,34
Prasinon B: C:65,47 H:8,21 N:2,16 O:24,16
Löslich in: MeOH, EtOH, n-BuOH, Benzol, Toluol, $HCCl_3$, EtOAc, BuOAc, Me_2CO; unlöslich: Wasser

IR-Spektrum von Prasinon A (KBr) [1]:

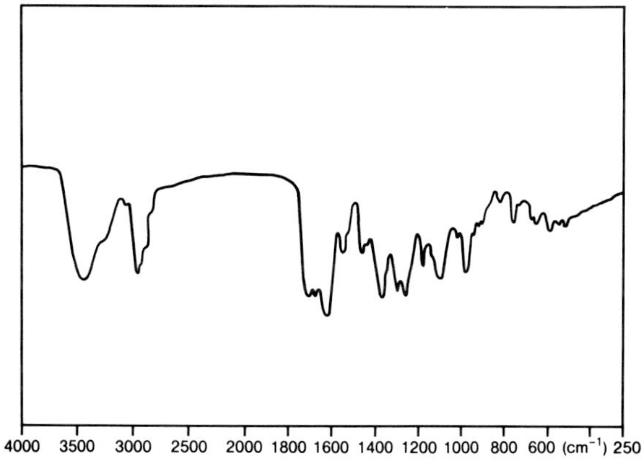

Prasinon A: $[\alpha]_D^{20} = +16{,}3°$ (3,19 g/l, $HCCl_3$)
UV-Spektrum: λmax = 248 nm, $E_{1\,cm}^{1\%} = 570$
Prasinon B: $[\alpha]_D^{20} = +29{,}9$ (2,88 g/l $HCCl_3$)
UV-Spektrum: λmax = 249 nm, $E_{1\,cm}^{1\%} = 540$

Biologische Wirkung: Mit 0,5–1,0 µg/ml wirksam gegen *Lucilia serata*, *L. cuprina*, auch gegen Carbamat- oder Organohalogen-resistente Stämme, unwirksam gegen adulte Stadien, unwirksam gegen *Pieris brassicae*, *Sitophilus granarius*, *Aphis fabae*, *Tetranychus urticae* und *Boophilus decoloratus*-Larven.

Toxizität: Prasinon A: LD_{50}: 500 mg/kg Maus p.o.
Prasinon B: LD_{50}: 90 mg/kg Maus p.o.

Literatur: 1. Box, S.J. et al.: Appl. Microbiol. *26*, 699 (1973).

Rubratoxin B [1–3]

Erzeugerstamm: *Penicillium rubrum* Stoll

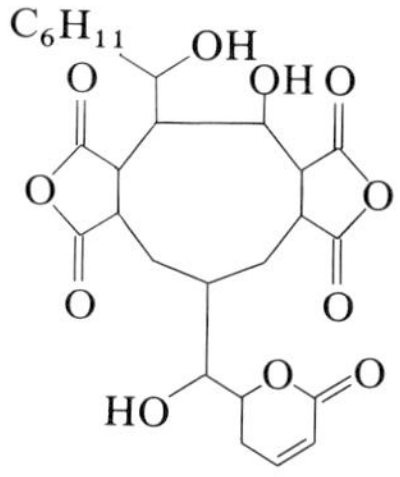

$[\alpha]_D^{20} = +68{,}3°$ (c = 1, Me_2CO)

UV-Spektrum (MeOH): λmax = 249 nm, 207 nm
IR-Spektrum (KBr): λmax = 1842, 1822, 1768, 1701, 1690 cm^{-1}

Biologische Wirkung: Larvizide Wirkung gegen *Drosophila melanogaster* mit 1 ppm [4], sowie gegen *Aedes aegypti*, *Musca domestica* und einige Termitenarten [5].
Toxizität: LD_{50}: 3 mg/kg Maus i.p.

Literatur: 1. Heyes, A.W.: Appl. Microbiol. *16*, 1163 (1968). – 2. Moss, M.O.: Nature *220*, 767 (1968). – 3. Heyes, A.W.: Appl. Microbiol. *20*, 469 (1970). – 4. Reiss, J.: Chem.-Biol. Interact. *10*, 339 (1975). – 5. Matsumura, F. et al.: J. Econ. Entomol. *60*, 871 (1967).

S-3466 A, B, C

Erzeugerstamm: *Streptomyces aureus*
Isolierung: Extraktion des Mycels mit Me_2CO, Einengen zur Trockne, Aufnehmen in n-Hexan, Chromatographie an Kieselgel in n-Hexan/EtOAc (9/1), Feinreinigung mittels stufenweiser Kieselgelchromatographie in $HCCl_3$/EtOAc, 1) 2/1, 2) 1/1, 3) reines EtOAc.

Physikalisch-chemische Eigenschaften [1]:

	Antibiotikumkomplex S-3466 A	Antibiotikum S-3466 B	Antibiotikum S-3466 C
Aussehen und Eigenschaften:	Farblose oder leicht gelbliche prismatische Kristalle; neutral und stabil gegenüber Wärme und Licht	Entspricht S-3466 A	Entspricht S-3466 A
Schmelzpunkt:	73–74 °C (aus n-Hexan umkristallisiert)	79–80 °C (n-Hexan)	105–106 °C (n-Hexan)
Löslichkeit:	Löslich in den meisten organischen Lösungsmitteln, wie Alkoholen, Me_2CO, Benzol, n-Hexan, $HCCl_3$; Petroläther, Diäthyläther und Äthylacetat	wie S-3466 A	wie S-3466 A
Elementaranalyse:	C, H, O; kein Stickstoff, Schwefel oder Halogen C: 65,58 H: 8,85	wie S-3466 A C: 66,48 H: 8,97	wie S-3466 A C: 65,41 H: 9,13
Optische Drehung:	+0,9° in $HCCl_3$	+2,0° in $HCCl_3$	0,0° in $HCCl_3$
Summenformel:	$C_{42}H_{68}O_{12}$	$C_{43}H_{70}O_{12}$	$C_{44}H_{72}O_{12}$

IR-Spektrum von S-3466 A (KBr) [1]:

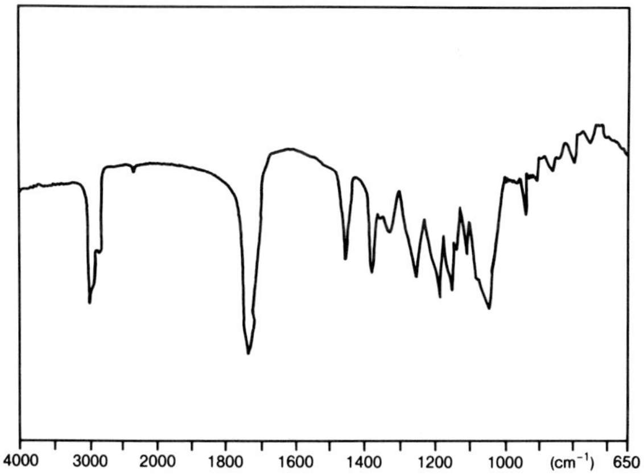

Die drei Präparate zeigen im UV-Spektrum keine charakterischen Absorptionen.

Biologische Wirkung:

a) Wirkung gegen Spinnmilben (A: Prozentsatz der getöteten ausgewachsenen Milben, B: Prozentsatz der getöteten Eier)

Konz. des aktiven Wirkstoffes	S-3466 A Emulsion		S-3466 B Emulsion		S-3466 C Emulsion	
	A	B	A	B	A	B
100 ppm	100	100	100	100	100	100
50 ppm	100	100	98,8	80,5	100	96,4
25 ppm	100	96,0	95,2	63,3	98,2	82,5
12,5 ppm	98,0	90,3	91,2	61,2	92,4	60,4
6,3 ppm	88,3	91,5	75,5	48,9	74,2	50,6
3,1 ppm	79,1	70,3	64,3	36,3	60,5	42,1
1,6 ppm	68,3	69,1	50,1	20,1	51,4	30,5

b) Wirkung gegen Pfirsichblattläuse:

Konzentration des aktiven Agens	S-3466 Emulsion % Sterblichkeit
200 ppm	100
100 ppm	90,5
50 ppm	81,2
25 ppm	62,4

Literatur: 1. DOS 2108739, Chugai Seiyaku Kabushiki Kaisha/Tokio, 23.9.1971. – 2. NL 7102358, Chugai Seiyaku K.K., 24.2.1970. – 3. JA 015358, Chugai Seiyaku K.K., 4.2.1970.

Thiolutin [1, 2]

Gruppe: Pyrrothine
Erzeugerstamm: *Streptomyces albus*, *Streptomyces celluloflavus*

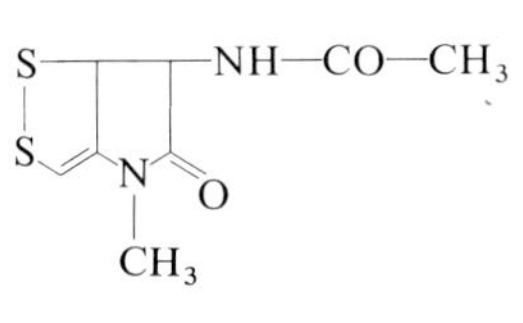

Isolierung: Aus dem Kulturfiltrat durch Extraktion mit BuOH oder $HCCl_3$; ll: BuOH und $HCCl_3$; nl: Wasser

UV-Spektrum: λmax = 245 nm, 315 nm, 365 nm
IR-Spektrum (KBr) [1]:

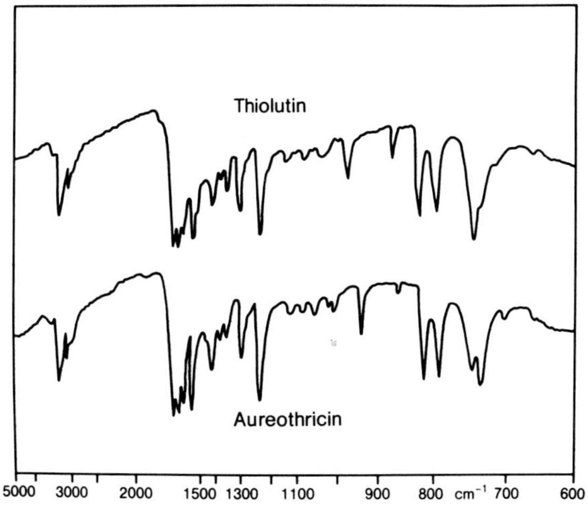

Biologische Wirkung: Als Insektizid gegen *Lucilia sericata* und *Drosophila melanogaster* [3].
Toxizität: LD_{50}: 25 mg/kg Maus s.c. und p.o.

Literatur: 1. Celmer, W.D., Salomons, I.A.: J. Am. Chem. Soc. *77*, 2861 (1955). – 2. Schmidt, U.: Chem. Ber. *97*, 1511 (1964). – 3. Cole, M. et al.: Appl. Microbiol. *24*, 660 (1972).

Trichothecin [1, 2]

Siehe Herbizide, Fungizide – Handelsprodukte. Biologische Wirkung: Insektizide Wirkung gegen *Lucilia sericata* und *Drosophila melanogaster* [3].

Literatur: 1. Bamberg, A. et al.: Microbiol. Toxins *7*, 207 (1971). – 2. Sorenson, L. et al.: Appl. Microbiol. *29*, 653 (1975). – 3. Cole, M. et al.: ibid. *24*, 660 (1971).

X-4356 [1]

Siehe Fungizide. Larvizid: Kartoffelkäfer, Bohnenkäfer, Mosquitos und *Musca domestica.*

Literatur: 1. US 4076802, Hoffmann-La Roche Inc., 30.3.1977.

4. Herbizide und Wachstumsregulatoren

4.1. Einleitung

Eine Besprechung herbizid wirksamer synthetischer Präparate sowie ihre Einteilung nach Wirkungsweise und Anwendung erschien bereits im Band 5 dieses Werkes.

Neben synthetischen Wirkstoffen, die das Pflanzenwachstum beeinflussen, gewinnen Naturstoffe zunehmend an Bedeutung, wobei als Produzenten aktiver Substanzen sowohl höhere Pflanzen als auch Mikroorganismen herangezogen werden. Schwerpunkte früherer Arbeiten waren Wachstumsregulatoren aus pflanzlichem Material wie Abscisinsäure und Gibberelline. Ferner wurde das Gebiet der Cytokinine intensiv bearbeitet. Ein weiterer natürlicher Wachstumsregulator ist die als biologisches Abbauprodukt des Tryptophans vorkommende β-Indolylessigsäure. Die chemische Synthese und schrittweise Weiterentwicklung analoger Verbindungen führte zu einer Reihe von Handelsprodukten, z.B. zu Derivaten der Phenoxyessigsäure (Band 5). Diese Erfolge und die moderne Entwicklung der Biotechnologie waren der Anlaß, bei Mikroorganismen verstärkt nach herbizid wirksamen Sekundärmetaboliten zu suchen. Die Natur liefert eine große Auswahl von molekularen Möglichkeiten. Herbizid und wachstumsregulierend wirkende Naturstoffe gehören den verschiedensten Stoffklassen an; die Herbicidine und Toyokamycin z.B. sind interessante Vertreter der Nucleosid-Gruppe. Eine Einteilung aller Herbizide nach einem chemischen System ist aber noch nicht möglich. Aus diesem Grunde wurden sie in dem folgenden Beitrag alphabetisch abgehandelt. Naturstoffe aus Mikroorganismen, die nur als Phytotoxine, insbesondere Welktoxine bekannt geworden sind, wurden nicht aufgenommen.

Synonyma

Antibioticum E-212 s. Toyokamycin
Antibiotic 720-A s. Antimycin A_1
Antibioticum 1037 s. Toyokamycin
Antipiriculin s. Antimycin A_1
Brefeldin A s. Ascotoxin
Cyanein s. Ascotoxin
Decumbin s. Ascotoxin
Flagecidin s. Anisomycin
Naritheracin s. Toyokamycin
PA 106 s. Anisomycin
Siromycin s. Toyokamycin
Unamycin B s. Toyokamycin
Vengicid s. Toyokamycin
Virosin s. Antimycin A_1

4.2. Forschungs- und Entwicklungspräparate

L-trans-2-Amino-4-(2-aminoäthoxy)-3-butensäure [1]

Erzeugerstamm: *Streptomyces* sp. X-11085, NRRL 5331

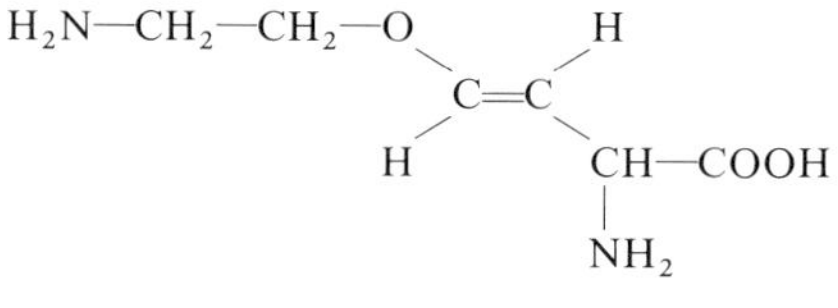

Isolierung: Adsorption aus dem Kulturfiltrat an Kationenaustauscher, Elution mit verdünntem wäßrigem Ammoniak, Säulenchromatographie an Kieselgel.
Farblose Kristalle aus MeOH/Wasser; Fp. 185–86 °C; $[\alpha]_D^{25} = +85{,}8°$ (c=1, Wasser).
Die Verbindung ist mit 1 kg/Hektar im post-emergence-Test wirksam gegen: *Xanthium, Cyperus rotundus, Sorghum halepense, Jimson-weed, Tee weed, Cichorium intybus.*

Literatur: 1. US 3869277, Hoffmann-La Roche Inc. Nutley, N.J., 16.5.1976.

Anisomycin [1]

Synonyma: Flagecidin, PA 106
Erzeugerstamm: *Streptomyces griseolus,*
Streptomyces roseochromogenes,
Streptomyces spez. 638.

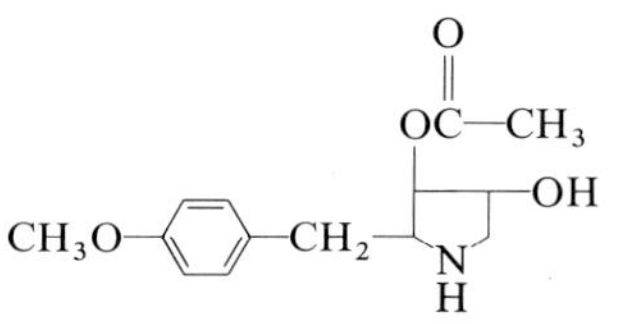

(2)

Isolierung: Extraktion des Kulturfiltrates mit Methylisobutylketon bei pH 9,0.
Farblose Nadeln aus EtOAc, wasserlöslich, basisch; Fp. 140–41 °C; $[\alpha]_D^{20} = -30°$ (c=0,15, MeOH).
IR-Spektrum ($HCCl_3$): 2890, 2800, 1725, 1610, 1582 cm^{-1}
UV-Spektrum (EtOH): λmax. 224 nm, 277 nm, 282 nm
Biologische Wirkung: Hemmt selektiv mit ca. 12,5 ppm das Wurzelwachstum bei Reis, Hirsearten zu 40–60%, bei Luzerne zu 60–80%. Die Desacetyl-Verbindung hemmt mit 25 ppm Trieb- und Wurzelwachstum bei Reis und *Echinochloa* spec., wirkt aber nicht selektiv. Das N-Acetyl-Derivat zeigt bei 100 ppm keine Wuchshemmungen [3].
Toxizität: LD_{50}: 146 mg/kg (Maus, i.v.).
Die Totalsynthese ist gelungen [4]. Anisomycin kommt in *Streptomyces* spez. 638 zusammen mit Toyokamycin vor. Es hemmt kompetitiv Acetylcholinesterase bei Warmblütern [5].

Literatur: 1. Tanner, jr., F.W. et al.: Antibiotics Ann. *1954/55*, 809. – 2. Schaeffer, J.P., Wheatley, P.J.: J. Org. Chem. *33*, 166 (1968). – 3. Yamada, O. et al.: Agric. Biol. Chem. *36*, Nr. 11, 2013 (1972). – 4. Felner, I., Schenker, K.: Helv. Chim. Acta *53*, 754 (1970). – 5. Zech, R., Domagk, G.F.: Life sciences *19*, 1577 (1976).

Ascotoxin [1]

Synonyma: Brefeldin A [2], Cyanein [3], Decumbin
Erzeugerstamm: *Penicillium brefeldianum*,
Ascochyta imperfecta Peck,
Penicillium decumbens

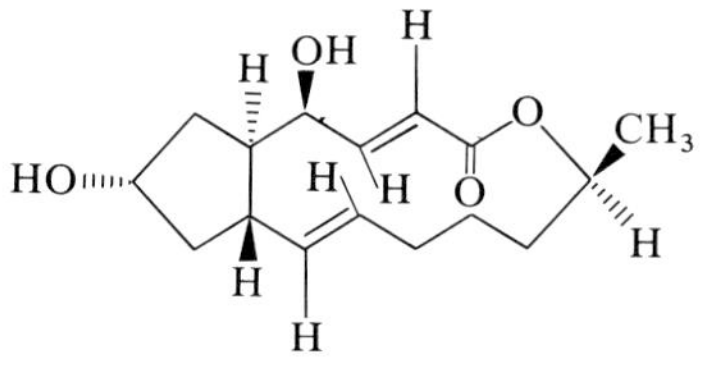

Isolierung: Mycel mit Me_2CO extrahieren, in EtOAc überführen und mit Petroläther fällen. Oder: Das Kulturfiltrat mit Et_2O und EtOAc extrahieren und Chromatographie an Kieselgel.
Neutral, Fp. 204–05 °C; $[\alpha]_D^{25}$ +91,7° (c = 1,92, MeOH).
UV-Spektrum (EtOH): λmax. 215 nm
IR-Spektrum (KBr) [2]: 3400, 2900, 1711, 1646, 1450, 1260 cm^{-1}
Biologische Wirkung: Hemmt mit 20 ppm Wachstum von Luzerne- und Rapssämlingen. Phytotoxisch und fungitoxisch [4].
Biosynthetische Untersuchungen mit P. *brefeldianum* [5].

Literatur: 1. Suzuki, Y. et al.: Agric. Biol. Chem. *34*, Nr. 3, 395 (1970). – 2. Härri, E. et al.: Helv. Chim. Acta *46* (4), 1235 (1963). – 3. Betina, V. et al.: J. Antibiotics Ser. A. *19*, 115 (1966). – 4. NL 7306951 (1972). – 5. Handschin, U. et al.: Helv. Chim. Acta *51*, 1943 (1968).

Citrinin [1]

Erzeugerstamm: die Penicillium-Stämme: *citrinum, expansum, implicatum, citreosulfuratum, lividum, phaeojanthinellum,*
die Aspergillus-Stämme: *terrus candidus, niveus*

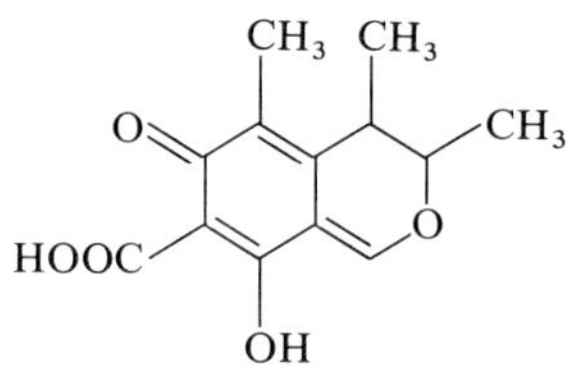

Kristallisiert nach dem Ansäuern des Kulturfiltrates auf pH 2 aus.
Lange gelbe Prismen, Fp. 179 °C (Zers.); $[\alpha]_D^{20} = -34,5°$ (c = 0,6, EtOH); starke Säure; löslich in verdünntem Alkali, Alkoholen, Dioxan, unlöslich in Wasser
UV-Spektrum (EtOH): λmax. 250, 331 nm; Konformation s. [2].

Biologische Wirkung [3]:

Pflanze	Wirkstoffkonzentration in ppm	Wachstumshemmung in %
Erbse	150	100
Flachs	90	100
Weizen	95	30
Weizen	160	100
Centaurea	90	100
Poa	25	30
Sedum	60	100
Sinapsis arvensis	15	30
Spergula	20	100

Toxizität: LD_{50}: 35 mg/kg (Maus subkutan). Reizt Schleimhäute in Konzentrationen von 0,1%.

Literatur: 1. Hetherington, A.C., Raistrick, H.: Trans. Roy. Soc. (London) B *220*, 269 (1931). – 2. Mathieson, D.W., Whalley, W.B.: J. Chem. Soc. (London) *1964*, 4640. – 3. Skorobogätovä, R.A., Mirchink, T.G.: Sel'skokhozyaistvennaya Biologiya 9, 865 (1976); ref. Weed Abstr. *25*, 3 (1976), Ref. 861.

Cycloheximid

Synonyma: Actidion, Acti-aid

Formel und Beschreibung im Kapitel „Fungizide und Bakterizide".
Biologische Wirkung: Hemmt mit 100 μg/ml Luzerne [1]. Dient als Fruchtablöser bei Orangen und Oliven [2].

Literatur: 1. Kirk, J.T.O.: Proc. Biochem. Soc. *1966*, 3. – 2. Ford J.H. et al.: Plant. Diseases Reptr. *42*, 620 (1958).

3,4 Dihydro-3,4,8-trihydroxy-1(2H)-naphtalenon [1, 2]

Erzeugerstamm: *Pyricularia oryzae Cavara*

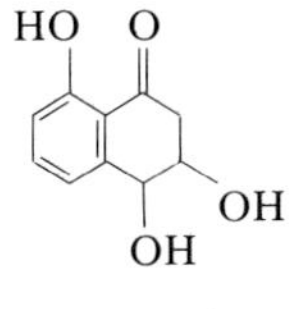

Isolierung: Extraktion aus der Kulturbrühe mit EtOAc, Chromatographie an Kieselgel.
Farblose Kristalle aus Me_2CO; Fp. 191–92 °C; $[\alpha]_D - 36{,}0°$ (MeOH).
UV-Spektrum (EtOH): λmax, 254 nm, 260,8 nm, 335,0 nm
IR-Spektrum (KBr): 3600–2400, 1640, 1580, 1080, 965 cm^{-1}.
NMR-Spektrum in $Me_2CO + D_2O$ [1]
Biologische Wirkung: Wachstumshemmung bei Reissämlingen. Kommt zusammen mit Pyriculol und mit Tenuazonsäure vor.

Literatur: 1. Iwasaki, S. et al.: Tetrahedron Lett. Heft 1, 13 (1972). – 2. Iwasaki, S. et al.: ibid. 3537 (1973).

Discadenin [1]

Erzeugerstamm: *Dictyostelium discoideum*

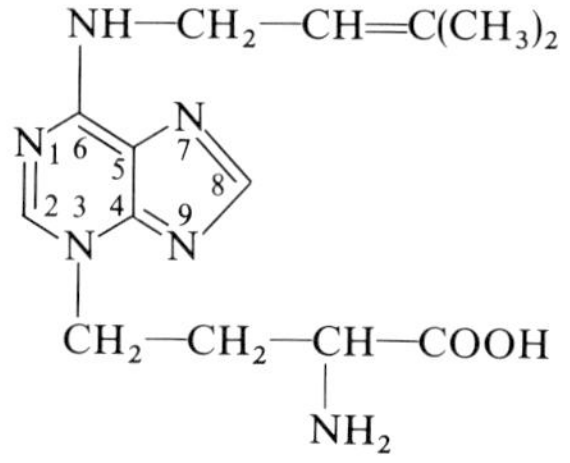

Isolierung: Extraktion der Sporen mit 80% EtOH, nachfolgende Dialyse und Chromatographie des Dialysates an Amberlite XAD 2.
Farblose Nadeln, Fp. 205–07 °C.
UV-Spektrum (MeOH): λmax. 292,7 nm, 248,9 nm
^{13}C-NMR-Spektrum und Massenspektrum siehe [1, 2].
Biologische Wirkung: Signifikante Cytokinin-Aktivität.

Literatur: 1. Abe, H. et al.: Agr. Biol. Chem. *37*, 3807 (1973). – 2. Abe, H. et al.: Tetrahedron Lett. *42*, 3807 (1976).

Fasciculol A [1]

Erzeugerstamm: *Neamatoloma fasciculare*

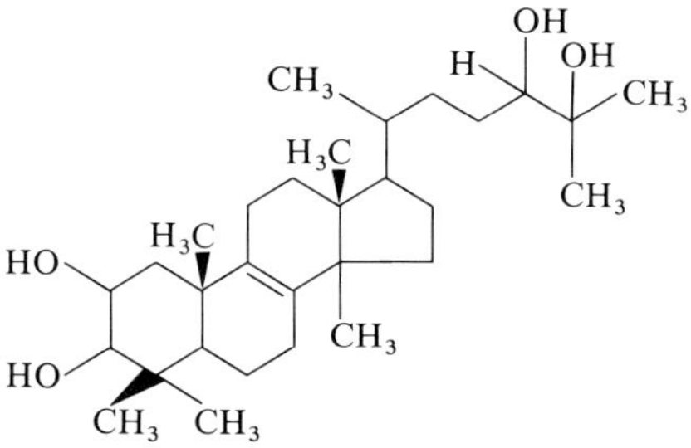

Isolierung: Extraktion der Fruchtkörper mit EtOAc, Kieselgel-Chromatographie der neutralen Fraktion.
Fp. 209 °C, $[\alpha]_D^{21} = -50{,}7°$ (c = 1, MeOH).
Positive Liebermann-Burchard-Reaktion, positiver Tetranitromethan-Test.
UV-Spektrum (MeOH): Endabsorption
Massenspektrum: 476 (M^+), 331, 289, 215, 187, 173, 145
Biologische Wirkung: Hemmt mit 100 ppm das Wurzelwachstum von Chinakohlsämlingen.

Literatur: 1. Ikeda, M. et al.: Agric. Biol. Chem. *41*, 1539 (1977).

Herbicidine A [1]

Gruppe: Adeninnucleosid
Erzeugerstamm: *Streptomyces saganomensis*
Summenformel: $C_{23}H_{29}N_5O_{11} \cdot \frac{1}{2} H_2O$, Mol.-Gew. 551
Isolierung: Adsorption aus dem Kulturfiltrat an Duolite S-30, Elution mit 50% wäßrigem Me_2CO, Gegenstromverteilung, $HCCl_3$ + MeOH + Wasser, Chromatographie an Kieselgel.
Basisch, amorph, l: Wasser, niedere Alkohole, Me_2CO. Saure Hydrolyse ergibt Adenin, milde alkalische Hydrolyse Herbicidin B und 2-Hydroxy-methyl-1,2-butensäure. $[\alpha]_D^{20} = +57{,}7°$ (c = 1, MeOH).

UV-Spektrum [2]:

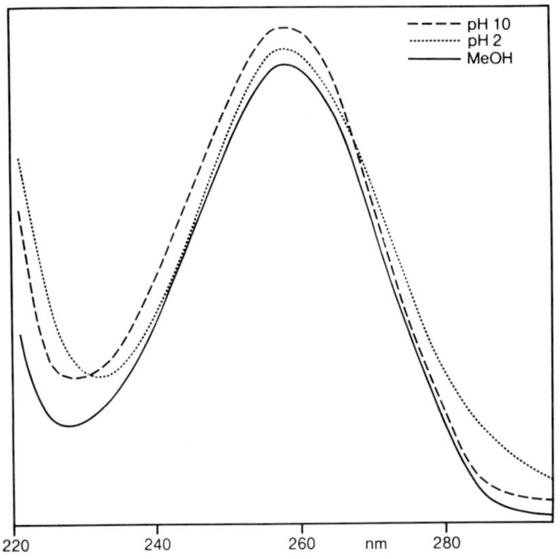

Herbicidin A und B IR-Spektrum (KBr) [2]:

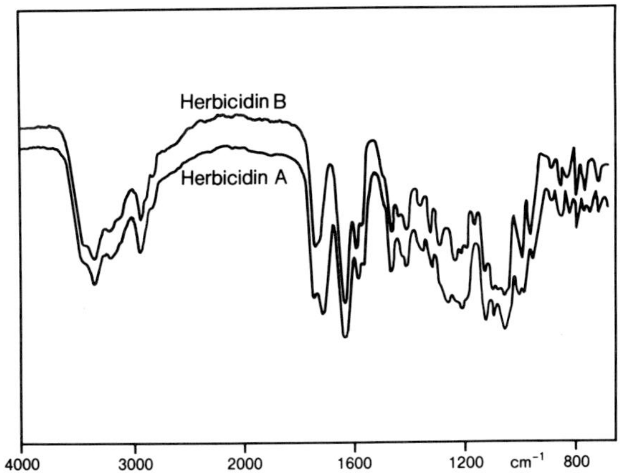

Biologische Wirkung: Ausgeprägte Hemmung bei dicotylen, schwächer bei monocotylen Pflanzen.

	Keimhemmung in µg/ml	
	Herbicidin A	Herbicidin B
Chinakohl	0,78–1,56	1,56–3,12
Reis	6,25	12,5

Herbizide Aktivität

		Herbicidin A (ppm)				Herbicidin B (ppm)			
		300	150	75	37,5	300	150	75	37,5
Monocotyle	Reis	0	0	0	0	3	1	0	0
	Echinochloa	3	3	2	1	4	1	1	0
		4	3	2	1	4	2	2	1
	Eleusine	5	5	4	3	5	4	2	3
	Glyceria	4	4	3	2	5	4	2	2
		3	3	2	0	3	3	2	1
	Panicum	4	2	1	1	3	3	2	1
Dicotyle	Portulaca	5	4	3	3	5	3	1	1
	Achyranthes	5	5	5	4	5	5	4	4
	Chenopodium	5	5	5	4	5	5	4	3
	Polygonum	5	5	5	5	5	4	4	4
	Amaranthus	5	4	4	3	5	4	3	3
	Commelina	5	5	4	3	5	4	3	3
	Tomate	5	5	5	2	4	3	1	2
	Rettich	5	5	4	3	5	4	2	3

0 = keine Wirkung 5 = 100% Wirkung

Toxizität: 800 mg/kg (Maus i.v.) werden toleriert.

Literatur: 1. Haeishi, T. et al.: J. Antibiotics Ser. A. *29*, 870 (1976). – 2. Arai, M. et al.: ibid. *29*, 863 (1976)

Herbicidin B [1]

Gruppe: Adeninnucleosid
Erzeugerstamm: *Streptomyces saganomensis*
Summenformel: $C_{18}H_{23}N_5O_9 \cdot \frac{1}{2} H_2O$, Mol.-Gew. 453
Isolierung: s. Herbicidin A.
Basisch, farblose Kristalle. Löslichkeit ähnlich Herbicidin A. $[\alpha]_D^{20} = +63°$ (c = 1, MeOH)

UV-Spektrum [2]:

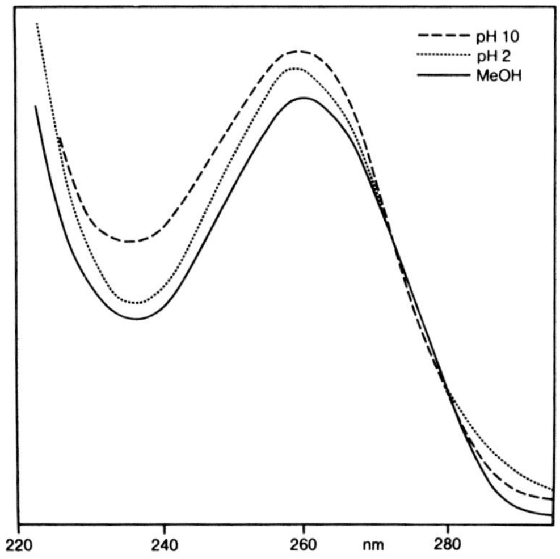

Biologische Wirkung: Siehe Herbicidin A [2].

Literatur: 1. Haeishi, T. et al.: J. Antibiotics Ser. A. *29*, 870 (1976). – 2. Arai, M. et al.: ibid. *29*, 863 (1976).

Herbimycin [1]

Gruppe: Ansamycine
Erzeugerstamm: *Streptomyces hygroscopicus*
Isolierung: Extraktion des Kulturfiltrats mit EtOAc, Einengen zur Trockne, in MeOH-Lsg. auf Aktivkohle, Desorption mit EtOAc, Kristallisation.
Summenformel: $C_{30}H_{42}N_2O_9$, Mol.-Gew. 574
Herbimycin ist ein Benzochinon-Derivat
$[\alpha]_D^{20} = +137°$ (c = 1, $HCCl_3$), ll: Me_2CO, $HCCl_3$, EtOAc, DMSO, DMF; l: MeOH, EtOH, Benzol; nl: Wasser Et_2O, n-Hexan.

UV-Spektrum (MeOH) [1]:

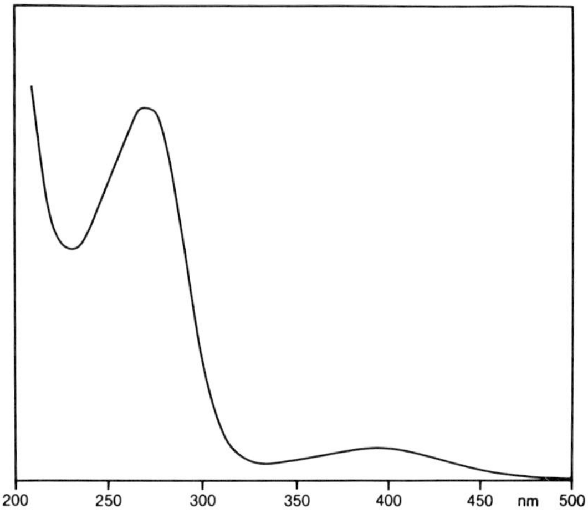

IR-Spektrum ($HCCl_3$) [1]:

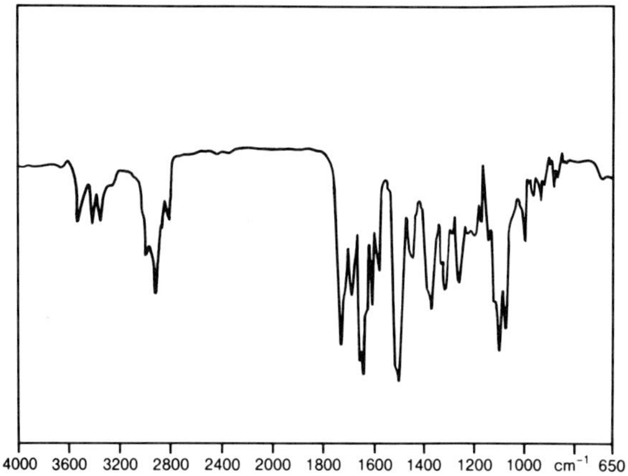

Biologische Wirkung: Breite herbizide Wirkung, besonders im Vorauflaufverfahren, mit Ausnahme von *Oryza sativa.*

Pre-emergence-Test

		Herbimycin (g/ar)			
		100	50	25	12,5
Monocotyle	Oryza sativa	2	1	1	0
	Echinochloa crus-galli	5	5	5	4
	Digitaria adscendens	5	5	5	4
	Cyperus microiria STEUD	5	5	5	5
Dicotyle	Chenopodium ficifolium	5	5	5	4
	Postulaca oleracea	5	5	5	4
	Galinsoga ciliata	5	5	5	4
	Rorippa atrovirens	5	5	4	4

Post-emergence-Test

		Herbimycin (g/ar)		
		100	50	25
Monocotyle	Oryza sativa	0	0	0
	Digitaria adscendens	5	4	4
	Cyperus microiria STEUD	5	5	5
Dicotyle	Chenopodium ficifolium	5	5	5
	Postulaca oleracea	4	4	3
	Rorippa atrovirens	5	4	3
	Galinsoga ciliata	4	3	2

0 = keine Aktivität 3 = 40–70% Wirkung
1 = 20% Wirkung 4 = 70–90% Wirkung
2 = 20–40% Wirkung 5 = 100% Wirkung

Toxizität: LD_{50}: 19 mg/kg, Maus i.p.

Literatur: 1. Omura, S. et al.: J. Antibiotics, Ser. A. *32*, 255 (1979).

Malformin A [1]

Erzeugerstamm: *Aspergillus niger*

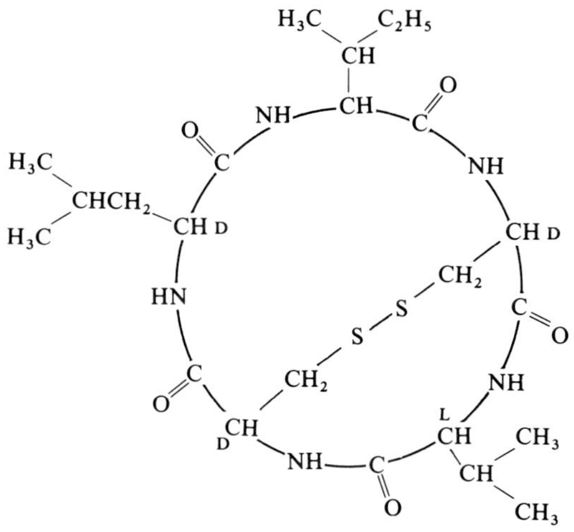

[2]

Isolierung: Adsorption aus dem Kulturfiltrat an Kohle. Elution mit 50% wäßr. Me_2CO. Einengen. Extraktion aus gesättigter $NaHCO_3$-Lösung mit Et_2O. Chromatographie an Cellulose [3].
Biologische Wirkung: Zeigt im Konzentrationsbereich von 10^{-5} bis 10^{-7} M ausgeprägte wachstumsregulierende Eigenschaften [4].

Literatur: 1. Takahashi, N., Curtis, R.W.: Plant. Physiol. *36*, 30 (1961). – 2. Turner, W.B.: Fungal metabolites, p. 338, Academic Press, London 1971. – 3. Yukioka, M., Winneck, T.: Biochem. Biophys. Acta *119*, 614 (1966). – 4. John, W.W., Curtis, R.W.: Experientia *30*, 1392 (1974).

Neosolaniol-monoacetat [1]

Erzeugerstamm: *Fusarium tricinctum* NRRL A 23377

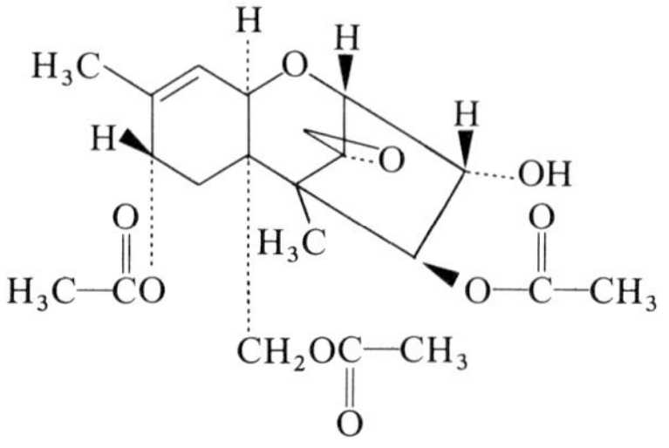

Isolierung: Extraktion von Kulturbrühe und Mycel mit $HCCl_3$, Chromatographie an Kieselgel.
Farblose Nadeln aus Benzol/Heptan, Fp. 190–190,5 °C.
UV-Spektrum: Endabsorption.
IR-Spektrum (KBr): 3370, 3040, 3010, 1730, 1460, 1370, 1265, 1230 cm^{-1}
Massenspektrum: M^+ 424.1735, 422, 406, 393, 365, 306, 247 ^{13}C-NMR s. [1]
Biologische Wirkung: Starke Hemmung mit 10^{-6} M an Weizen-Coleoptilen.
Neosolaniol-monoacetat gehört zur Gruppe der Trichothecin-Mycotoxine (Sesquiterpene), von denen Trichothecin-T-2-toxin ähnliche biologische Wirkungen zeigt [2].

Literatur: 1. Lansden, J.A. et al.: J. Agric. Food Chem. *26*, 246 (1978). – 2. Ellison, R.A., Kotsonis, N.F.: J. Org. Chem. *41*, 576 (1976).

o-Nitrophenylglucosid [1]

Erzeugerstamm: *Rhizoctonia solani* ATCC 10157

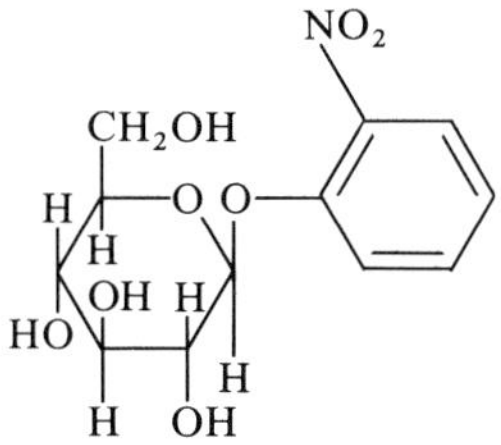

Isolierung: Kulturfiltrat und wäßriger Mycelextrakt werden bei pH 9,6 mit Me_2CO versetzt. Überstand wird nach Einengen mit Et_2O extrahiert und chromatographiert.
Gelbliche Flüssigkeit. Farbreaktion mit diazotierter Sulfanilsäure. Ergibt bei der Hydrolyse Glucose und o-Nitrophenol.
Biologische Wirkung: Hemmt die Keimung und das Wurzelwachstum [2].

Literatur: 1. US 3179653, US Dept. of Agriculture, 20.4.1965. – 2. Sherwood, R.T., Lindberg, C.G.: Phytopathologie *52*, 586 (1962).

Rhizobitoxin [1]

Erzeugerstamm: *Rhizobium japonicum*

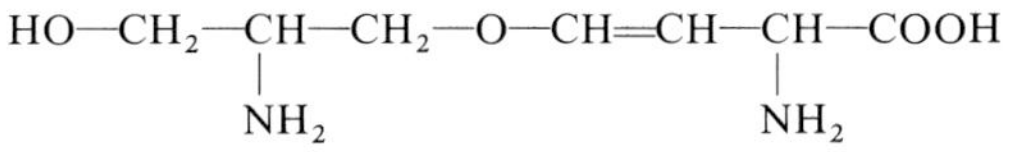

Isolierung: Kulturbrühe von *R. japonicum*, Stamm 94 erst erhitzen, dann klären und einengen; Säulenchromatographie an Dowex 50 W-X4 (NH_4^+ Form); Elution mit 0,02 N Ammoniak [2].
Biologische Wirkung: Im Vor- und Nachauflaufverfahren zeigt Rhizobitoxin eine vergleichbare Wirkung wie Amitrol. Besonders gut wirkt es gegen *Sorghum bicolor* [3].

Literatur: 1. Owens, L.D., Wright, D.A.: Plant Physiol. *40*, 931 (1965). – 2. Giovanelli, J. et al.: Biochim. Biophys. Acta *227*, 671 (1971). – 3. Owens, L.D.: Weed Sci. *21*, 63 (1973).

SF 1293

Formel und Beschreibung im Kapitel „Fungizide und Bakterizide".
Biologische Wirkung: Wirkt als Kontaktherbizid und als systemisches Herbizid gegen eine Reihe von Unkräutern [1].

g/1000 m²	Gehemmte Unkräuter
25– 600	*Artemisia princeps, Rumex japonicus, Cyperus rotundus* und andere mehrjährige unter 1 m Höhe
75– 750	*Sasa nipponica, Miscanthus sinensis* und andere mehrjährige über 1 m Höhe sowie kleine Büsche (z. B. *Rhododendron haempferi*)
150–1500	*Castanea crenata, Quercus serrata* und andere große Gebüsche.

Literatur: 1. DOS 2.848.224, Meiji Seika Kaisha Ltd. Tokio, 7.11.78.

Streptimidon [1]

Erzeugerstamm: *Streptomyces rimosus forma Paromomycinus*

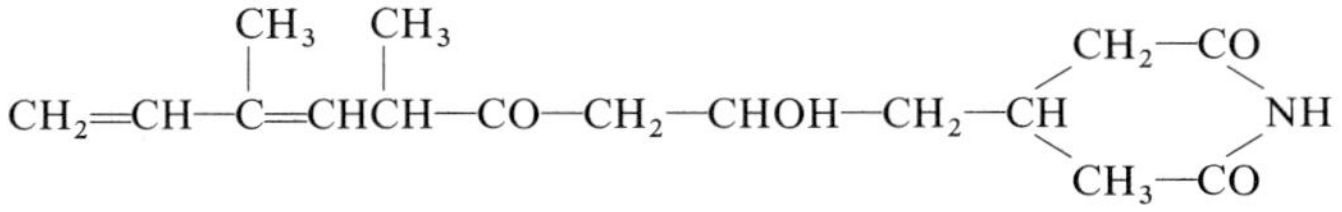
[2]

Isolierung: Extraktion aus dem Kulturfiltrat mit EtOAc, pH 5,0; Fällung mit Petroläther; Chromatographie an Aktivkohle.
Kristallin: Fp. 72–73 °C [3]; löslich in organ. Lösungsmitteln und in wäßrigem Dioxan; $[\alpha]_D^{28}$ +238° (c=0,5, Wasser).

UV-Spektrum (MeOH): λmax. 232 nm, 291 nm
Biologische Wirkung: Nach [1] hohe Wirksamkeit im Vorauflaufverfahren.

Literatur: 1. DOS 1257084, Parke Davis, 11.5.1959. – 2. v. Tamelen, E.E., Haarstad, V.: J. Am. Chem. Soc. *82*, 2974 (1960). – 3. Frohardt, R.P.: J. Am. Chem. Soc. *81*, 5500 (1959).

Tenuazonsäure [1]

Erzeugerstamm: *Alternaria tenuis* Stamm 430 [1]
Pyricularia oryzae [2]

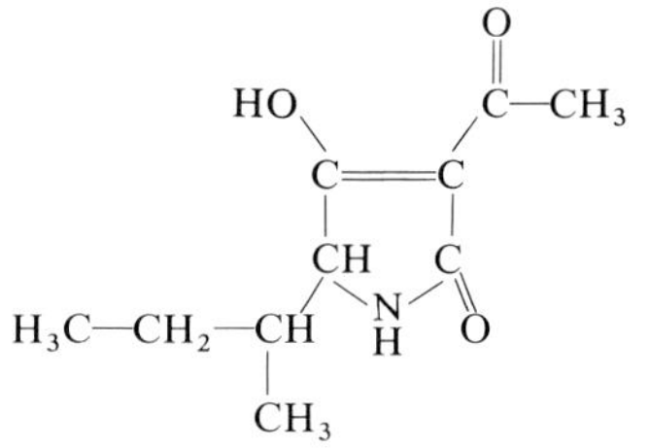

[3]

Isolierung: Extraktion der Kulturbrühe mit EtOAc, Chromatographie an Kieselgel.
Gelbliches, viscoses Öl. Kp 0,035 torr. = 117 °C; l: in organ. Lösungsmitteln inclusive Petroläther. Schwer löslich in Wasser; $[\alpha]_{5461}^{19} = -136 \pm 5°$ (c = 0,2, $HCCl_3$).
Kupfersalz: Grüne Nadeln, Fp. ~ 175 °C.
UV-Spektrum (Wasser): λmax. 239 nm, 279 nm
Biologische Wirkung: Deutliche Keimungshemmung bei Reis, Bohnen, Rettich, Rübe; Wachstumshemmung bei Sojabohne und Reis [4].

Literatur: 1. Rosett, T. et al.: Biochem. J. *67*, 390 (1957). – 2. Iwasaki, S. et al.: Tetrahedron Lett. *1*, 13 (1972). – 3. Stickings, C.E.: Biochem. J. *72*, 332 (1959). – 4. Umetsu, N. et al.: Agric. Biol. Chem. *38*, 791 (1974).

Toyokamycin [1]

Synonyma: Siromycin, Naritheracin, Antibioticum 1037, Unamycin B, Vengicid, Antibioticum E-212 [2]
Erzeugerstamm: *Streptomyces toyocaensis*

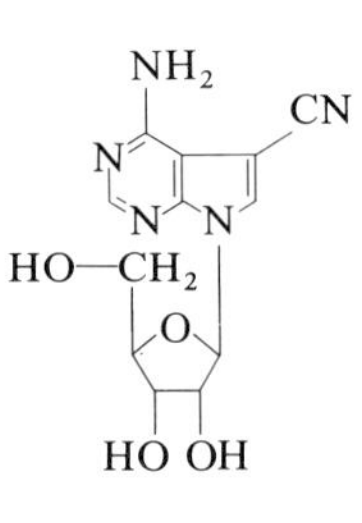

Isolierung: Klärung des Kulturfiltrates mit schwach basischen Ionenaustauschern, Einengung des Eluates und Kristallisation.
Mäßig löslich in Wasser, MeOH, Me_2CO, $[\alpha]_D^{16} = -45{,}7°$ (c = 1,05, 0,1 NHCl).

UV-Spektrum:
λmax: in Wasser: 230 nm, 277 nm
in 0,1 N HCl: 233 nm, 273 nm
in 0,1 N NaOH: 233 nm, 280 nm

Biologische Wirkung: Hemmt bei 25 ppm Trieb- und Wurzelwachstum aller getesteten Pflanzen zu 40–100%, z.B. bei Reis, Bohnen und Rettich. Luzerne wird bei 5 ppm gehemmt [3].
Toxizität: LD_{50}: 10 mg/kg (Maus i.v.) [1].
Die Totalsynthese des Toyokamycins gelang [4]. Behandlung mit Dowex-1 (OH^- Form) ergibt Sangivamycin, ein natürlich vorkommendes Antibiotikum mit einer Carbonamid-Gruppe anstatt einer Cyano-Gruppe [5].

Literatur: 1. Ohkuma, K.: J. Antibiotics Ser. A. *14*, 343 (1961). – 2. Suhadolnik, R.J.: Nucleoside Antibiotics, p. 300, J. Wiley and Sons Inc., 1970. – 3. Yamada, O. et al.: Agric. Biol. Chem. *36*, 2013 (1972). – 4. Tolman, R.L. et al.: J. Am. Chem. Soc. *90*, 524 (1968), *91*, 2102 (1969). – 5. Uematsu, T., Suhadolnik, R.J.: Biochemistry 9, 1260 (1970).

Trichothecin [1]

Erzeugerstamm: *Trichothecium roseum*

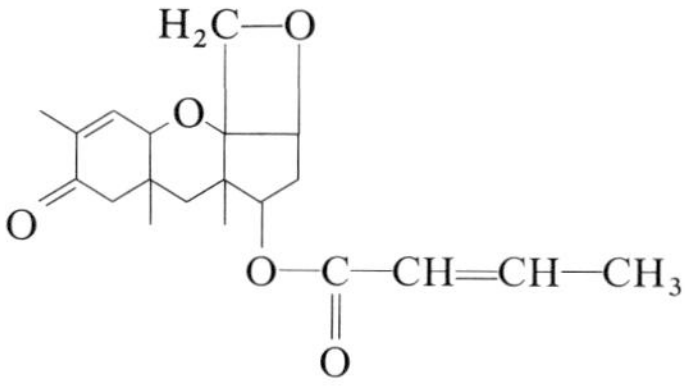

Isolierung: Extraktion Kulturfiltrat mit Chloroform und Chromatographie an Al_2O_3.
Nadeln aus Petroläther. Fp. 118 °C; l: in organ. Lösungsmitteln. Löslichkeit in Wasser: 400 mg/l bei 25 °C; stabil bei pH 1–10; $[\alpha]_D^{18} = +44°$ (c = 1, $HCCl_3$).
Dinitrophenylhydrazon: Orange Nadeln aus EtOH, Fp. 203 °C.
UV-Spektrum (Hexan): λmax. 217 nm
IR-Spektrum (Nujol): 1650, 1670, 1710, 1777, 2900 cm^{-1}
Massenspektrum: $M^+ = 332$
Biologische Wirkung: Mit 50 ppm volle Keimhemmung bei 24 getesteten Unkrautarten. Herbizide Wirkung besonders ausgeprägt bei *Linum austriacum*,

Urtica dioica, Lepidium ruderale [2]. Bei eukaryotischen Zellen wird die Proteinsynthese gehemmt [3].

Literatur: 1. Freeman, G.G., Morrisson, R.I.: Biochem. J. *44*, 1 (1949). – 2. Katar'yan, B.T., Akulyan, L.G.: Dokl. Akad. Nauk. Arm. SSR *53*, 283 (1971). – 3. Sorensen, W.G. et al.: Appl. Microbiol. *29*, 653 (1975).

Diese Arbeit wurde zu einem Teil von dem Bundesministerium für Forschung und Technologie getragen. Für die Durchsicht der Manuskripte danken wir den Herren Dres. Robert Schmidt und Gerhard Zoebelein von der BAYER AG.

Insektizide Phosphorsäureester

Ch. Fest

Bayer AG, Pflanzenschutz – Chemische Forschung
D-5600 Wuppertal-Elberfeld

Inhalt

1. Einleitung

Diese Zusammenfassung basiert auf der im ersten Band dieser Reihe erschienenen Arbeit zum gleichen Thema [1].

Mehr als 40 Jahre sind seit dem Beginn der systematischen Bearbeitung insektizid brauchbarer Phosphorsäureester vergangen. In diesem Zeitraum haben sie ständig steigende Bedeutung erlangt. Ohne chemische Schädlingsbekämpfungsmittel ist die Sicherung der Welternährung, die Bekämpfung der Überträger von Infektionskrankheiten und damit die Verbesserung der Lebensbedingungen nicht mehr möglich. Das gilt besonders für die Entwicklungsländer. Phosphorsäureester spielen hierbei eine wesentliche Rolle. Sie können meist zu erschwinglichen Preisen hergestellt werden und erfordern je Flächeneinheit nur verhältnismäßig geringe Aufwandmengen. Wegen ihrer Ester-Natur können sie leicht hydrolytisch, enzymatisch und biologisch abgebaut werden, sie weisen deswegen nur geringe Rückstandsprobleme auf. Die verschiedenen Liganden am zentralen Phosphor-Atom können mannigfach variiert werden; entsprechend sind die biologischen, toxischen, chemischen und physiologischen Eigenschaften zu beeinflussen. Das wiederum ergibt neue Entwicklungsmöglichkeiten für die Zukunft, was sich bereits in der Auffindung neuer Akarizide, Bodeninsektizide und mindertoxischer Verbindungen usw. gezeigt hat. Da die Entwicklung auf dem Pflanzenschutzgebiet schnell fortschreitet und ständig veränderten Bedingungen angepaßt werden muß, wird es erforderlich, über neue Verbindungen zu berichten. Dabei soll an das Einteilungsschema unserer vorhergehenden Arbeit angeknüpft werden, dem wir die Schrader'sche Acyl-Formel zugrunde gelegt haben. Die meisten neuen Verbindungen fügen sich vollkommen in dieses Schema ein.

2. Insektizide vom Typ der (Thio)Phosphorsäure-arylester

Das Studium der neuentwickelten phosphor-organischen Insektizide lehrt, daß es überwiegend die „klassischen" Phenole der Phosphorsäureester-Chemie sind, die mit abgeänderten Säurechloriden neue Produkte mit interessanter Wirkungsweise ergeben. Belege hierfür sind u. a. die beiden als Insektizide bekannt gewordenen Thionothiolphosphorsäureester Bolstar [O-Ethyl-S-n-propyl-O-(4-methylmercaptophenyl)-dithiophosphorsäureester] (I) und Tokuthion [O-Ethyl-S-n-propyl-O-(2,4-dichlorphenyl)-dithiophosphorsäureester] [2] (II).

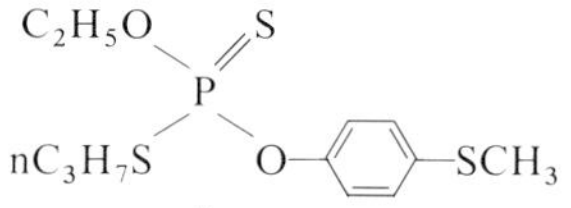

I

®Bolstar, *sulprofos*
LD_{50}: 100–300 mg/kg Ratte per os
(Nitokuno, 1977)

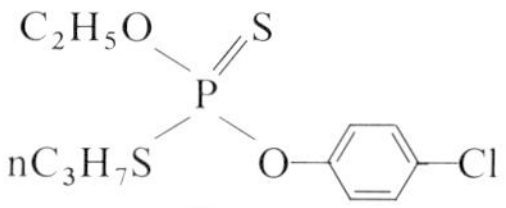

II

®Tokuthion, *prothiofos*
LD_{50}: 1000–2000 mg/kg Ratte per os
(Nitokuno, 1975)

Bolstar zeichnet sich durch besonders gute Wirkung im Baumwollbau gegen Heliothis-Arten aus. Tokuthion ist ein mindertoxisches Insektizid, das hauptsächlich im Gemüsebau eingesetzt wird. Bei den S-Alkyl-O-alkyl-aryldithiophosphorsäureestern in der Art von Bolstar und Tokuthion sind Toxizität und Wirkung stark vom S-ständigen Alkylrest abhängig. Das Optimum liegt in der Regel bei n-Propyl. Zum Beispiel nimmt bei den Bolstar-Homologen die insektizide und akarizide Wirkung in der Reihenfolge S-Methyl, S-Ethyl, S-n-Butyl, S-n-Propyl zu, gegenüber Warmblütern jedoch ist die S-n-Propyl-Verbindung deutlich weniger toxisch als die niedrigeren S-Alkylhomologen [3].

Die Herstellungsverfahren sind für beide Verbindungen analog [2]:

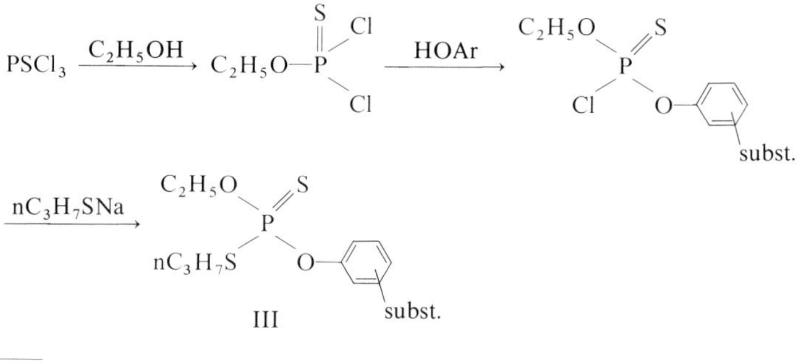

III

® eingetragenes Warenzeichen

oder

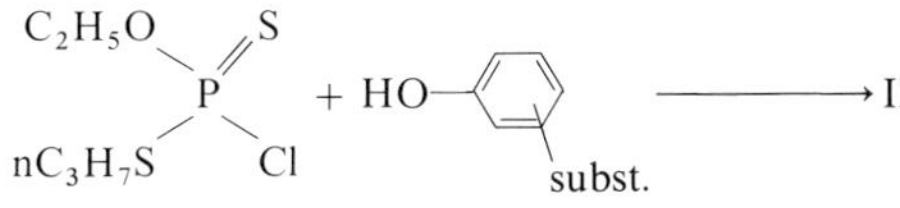

III

IV

Das Säurechlorid (IV) kann man aus O-n-Propylthionophosphorsäureester-dichlorid erhalten, das thermisch zu -thiolodichlorid umgelagert wird [4]:

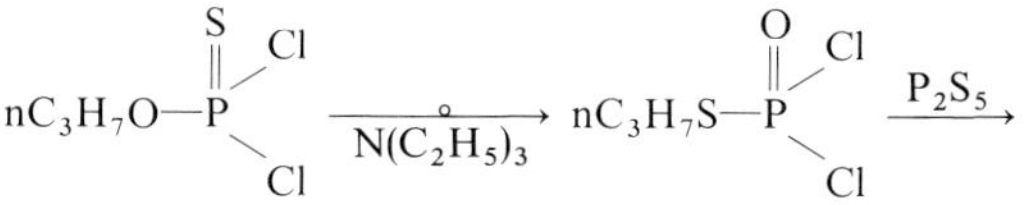

Dieses wird mit Phosphorpentasulfid geschwefelt und anschließend mit einem Äquivalent Alkohol umgesetzt [5]:

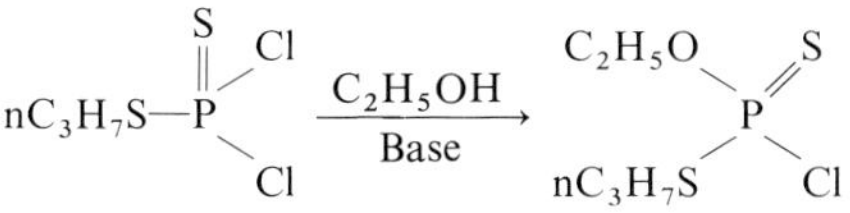

IV

Ein anderer Weg, zum Dithio-phosphorsäureesterdichlorid zu kommen, ist die direkte Reaktion von Phosphorsulfochlorid mit Mercaptan [6, 7]:

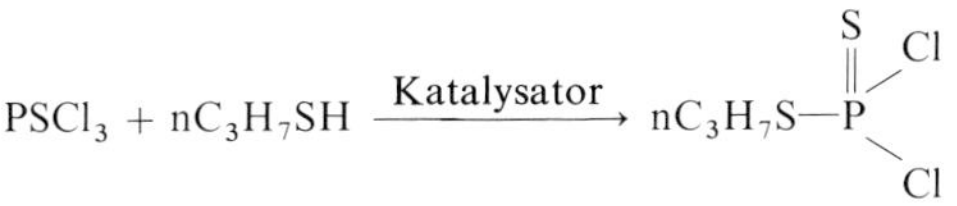

In diesem Zusammenhang ist auch die Synthese des Sauerstoff-Analogen interessant. Triethylphosphit wird mit Phosphortrichlorid behandelt [8] und mit n-Propylsulfensäurechlorid umgesetzt [9, 10]:

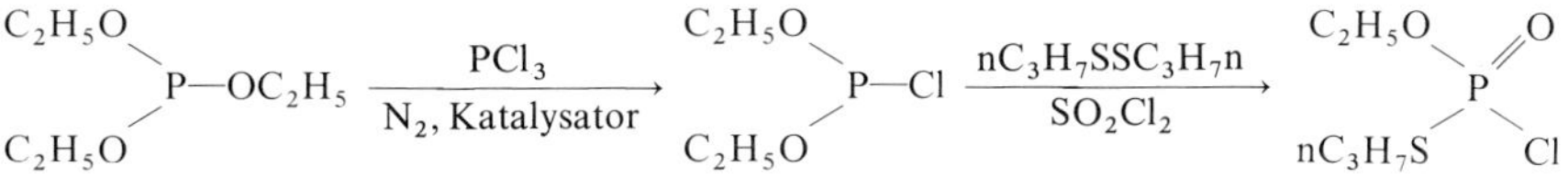

V

Auf diesem Wege erhält man ein Triester-freies Produkt.

Eine andere Synthesemöglichkeit bietet die Reaktion von n-Propylsulfensäurechlorid mit Phosphortrichlorid in Gegenwart einer Carbonsäure [11] und Umsetzung mit einem Äquivalent Alkohol:

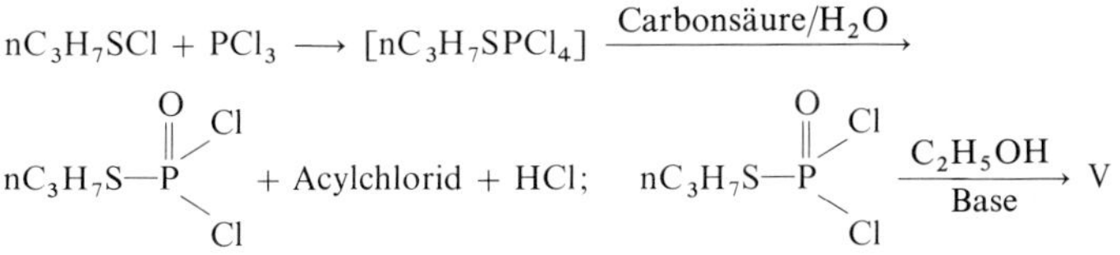

Das Säurechlorid (V) ist die Phosphoryl-Komponente von Curacron [O-Ethyl-S-n-propyl-O-(4-brom-2-chlorphenyl)-thiolphosphorsäureester] [12] (VI), das dem Tokuthion nahe verwandt ist.

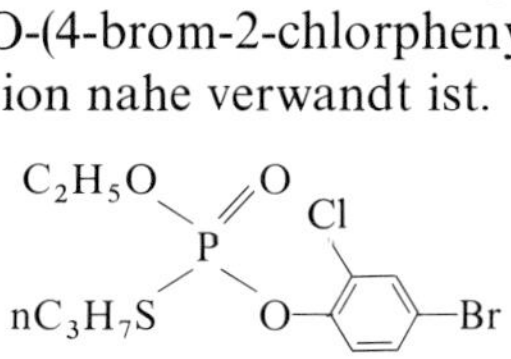

VI

®Curacron, *profenofos*
LD_{50}: 358 mg/kg Ratte per os
(Ciba-Geigy, Ltd.)

Es ist ein Fraß- und Kontaktinsektizid mit breitem Wirkungsspektrum, das im Baumwoll- und Gemüsebau gegen saugende und beißende Insekten, sowie gegen Milben angewandt wird.

Die gleiche Phosphoryl-Komponente enthält das von russischer Seite entwickelte Heterophos [O-Ethyl-S-n-propyl-O-phenyl-thiolphosphorsäureester] [13] (VII).

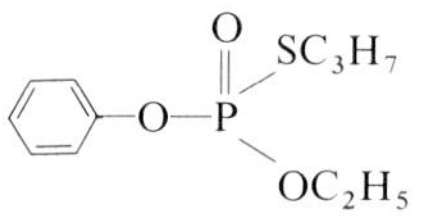

VII

Heterophos
LD_{50}: 295 mg/kg Ratte per os
(UdSSR, JUPAC-Kongreß Zürich 1978)

Es wird als wirksam gegen Wurzelgallennematoden herausgestellt. 3–10 kg/ha (Wirkstoff) ergeben eine Wirkungsdauer von 20–40 Tagen.

Ein weiterer nematizid wirkender Phosphorsäureester mit gleicher Phosphoryl-Komponente ist Mocap [O-Ethyl-S,S-di-n-propyl-dithiophosphorsäureester] (VIII).

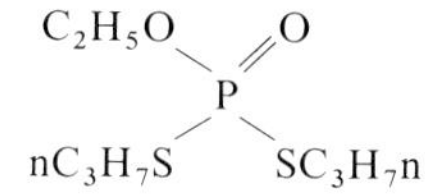

VIII

®Mocap, *ethoprop*
LD_{50}: 61 mg/kg Ratte per os
(Mobil Chemical Co., 1967)

Die Verbindung kann aus Phosphortrichlorid und n-Propylmercaptan, weiterer Umsetzung mit Alkohol zu dem gemischten Phosphit und anschließender Oxidation erhalten werden [14].

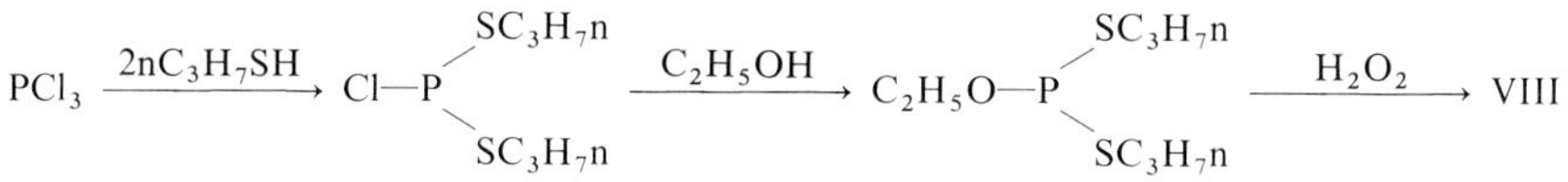

Eine andere Synthesemöglichkeit ist die Umsetzung von O-Ethyl-phosphorsäureesterdihalogenid mit n-Propylmercaptan [15]:

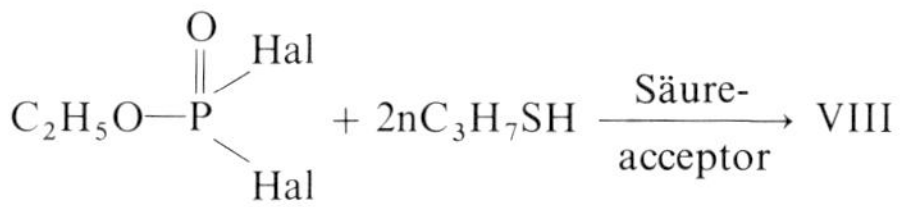

Als rodentizid wirkender Phosphorsäureester ist Muritan [O,O-Diethyl-O-(2-diethylaminomethyl-4-methylsulfinyl-phenyl)-thionophosphorsäureester] (IX) entwickelt worden. Das Präparat ist allerdings nicht mehr im Handel.

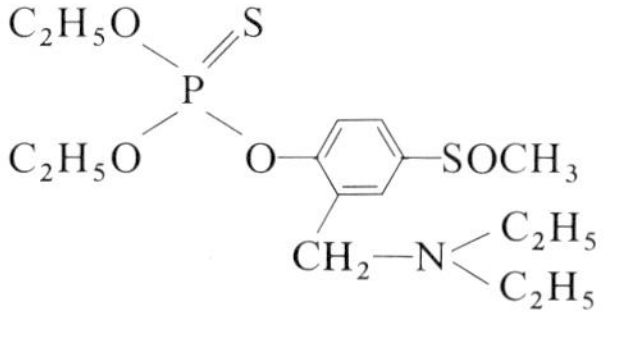

IX

®Muritan, *ethanphenphion*
LD_{50}: 1–2 mg/kg Ratte per os
(Bayer AG)

Zur Synthese wird p-Methylsulfinylphenol mit Diäthylamin und Paraformaldehyd umgesetzt und phosphoryliert [16].

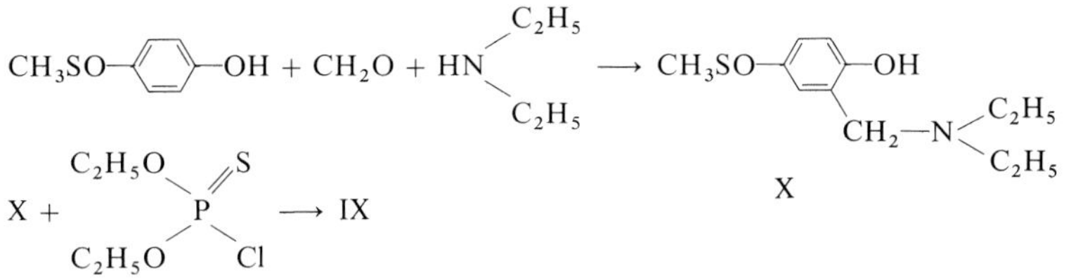

Ein Phosphorsäureesteramid mit insektiziden Eigenschaften ist das Oftanol [O-Ethyl-O-(2-isopropoxycarbonylphenyl)-N-isopropylamino-thionophosphorsäureester] (XI), das hauptsächlich zur Bekämpfung von Bodeninsekten im Handel ist.

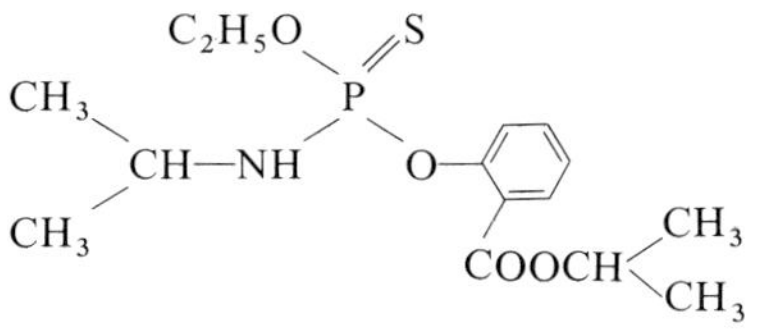

XI

® Oftanol, *isofenphos*
LD_{50}: 30–40 mg/kg Ratte per os
(Bayer AG, 1974)

Die Synthese gelingt durch stufenweises Umsetzen von O-Ethylthionophosphorsäureesterdichlorid mit 2-Isopropoxycarbonylphenol und Isopropylamin [17].

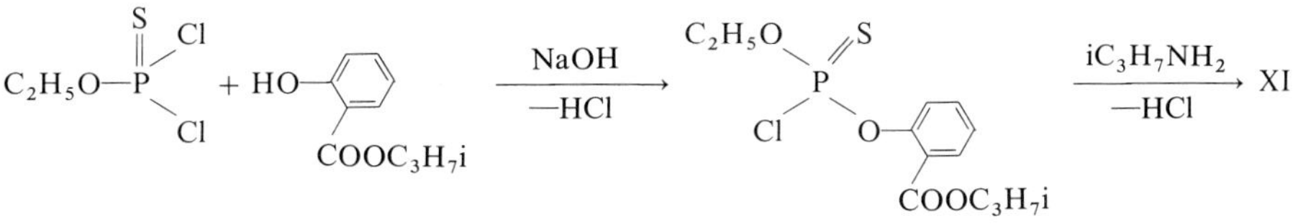

3. Insektizide vom Typ der (Thio)Phosphorsäure-heteroarylester

Heterocyclische Hydroxy-Verbindungen erwiesen sich stets als geeignete „Acyl"-Komponenten, legt man die Schrader'sche Acyl-Regel zugrunde. Auch neu entwickelte Insektizide enthalten immer wieder heterocyclische Hydroxy-Verbindungen, ungeachtet der im Vergleich zu den Phenol-Derivaten oft umständlicheren und teureren Synthesen. Aus der Reihe der Triazol-Derivate ist das Miral [O,O-Diethyl-O-(5-chlor-1-isopropyl-1,2,4-triazolyl-(3)-thionophosphorsäureester] (I) zu nennen.

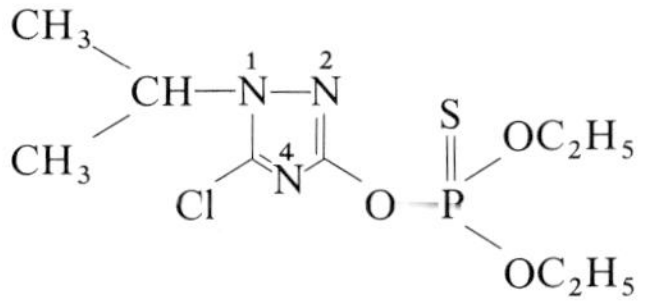

I

® Miral, *isazophos, protriazophos*
LD_{50}. ~60 mg/kg Ratte per os
(Ciba-Geigy, Ltd.)

5-Chlor-1-isopropyl-3-hydroxy-1,2,4-triazol erhält man entweder in einem Reaktionsschritt durch Addition von Isopropylhydrazin an Chlorcarbonyl-isocyaniddichlorid [18]

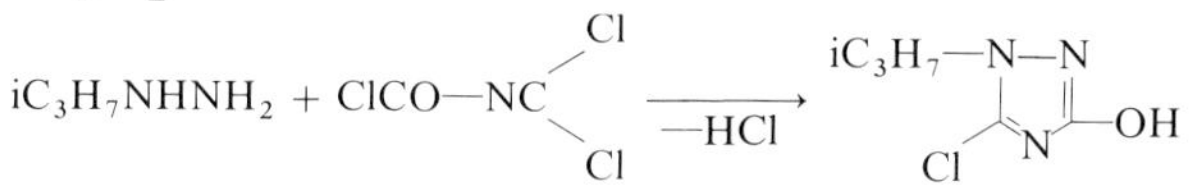

oder durch Kondensation von Isopropylsemicarbazid und Orthoameisensäureethylester [19, 20] und anschließende Chlorierung [21].

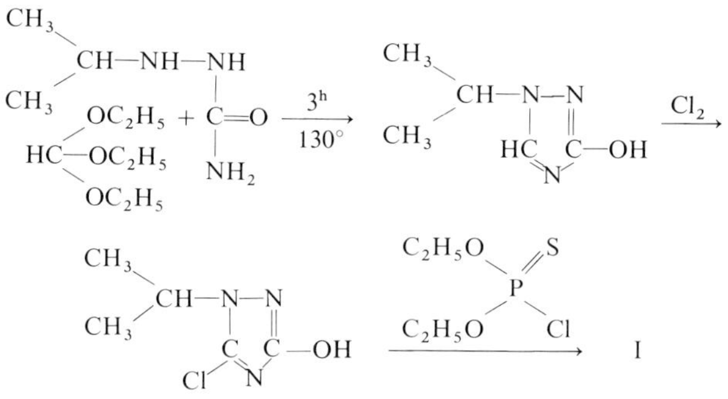

Miral ist ein Bodeninsektizid [22] und Nematizid mit breitem Wirkungsspektrum. Es zeigt wurzelsystemische Wirkung. Die Stabilität im Boden beträgt 4–6 Wochen. Es wirkt gegen Dipterenlarven, gegen *Diabrotica* sp. und Drahtwürmer und gegen freilebende und Wurzelgallennematoden an Gemüse, Mais und Soja. Wegen der systemischen Wirkung kann es auch gegen die Rübenfliege und Blattläuse eingesetzt werden. Es wirkt als Kontakt- und Fraßgift.

Hostathion [O,O-Diethyl-O-(1-phenyl-1,2,4-triazolyl-3)-thionophosphorsäureester] (II) gehört ebenfalls in die Reihe der Triazol-Verbindungen und kann durch Ringschluß von Phenylsemicarbazid mit Orthoameisensäureethylester und anschließender Phosphorylierung [19] hergestellt werden [23].

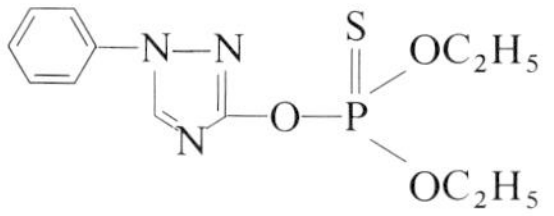

II

®Hostathion, *triazophos*
LD_{50}: 83 mg/kg Ratte per os
(Hoechst AG, 1970)

Es wirkt nicht systemisch, hat aber eine gewisse Tiefenwirkung und eine gute Wirkungsdauer, sowie eine gute Spinnmilbenwirkung. Die Haupteinsatzgebiete liegen im Baumwoll- und Gemüsebau [24].

Ein heterocyclischer Sechsring ist in der Verbindung Ofunack [O,O-Diethyl-O-(3-oxo-2-phenyl-2H-pyridazin-6-yl)-thionophosphorsäureester] (III) enthalten.

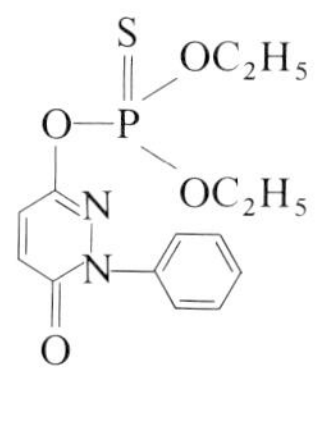

III

®Ofunack, Ofnak, *pyridaphenthion*
LD_{50}: 769 mg/kg Ratte per os
(American Cyanamid Co., 1974)

Das erforderliche Hydroxypyridazinon (IV) wird aus Maleinsäureanhydrid und Phenylhydrazin erhalten und in bekannter Weise

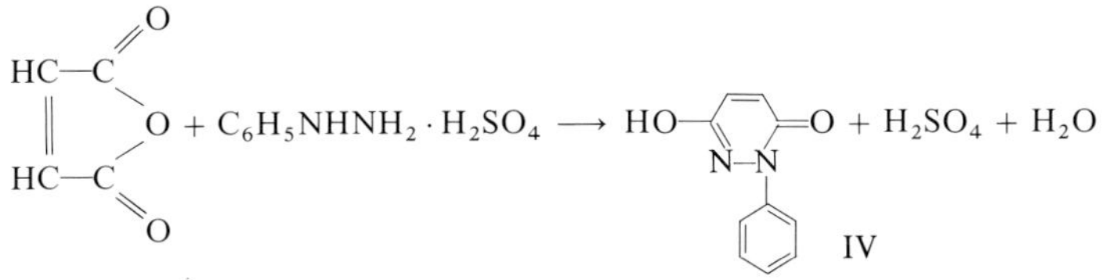

IV

phosphoryliert [25]. Ofunack ist ein aussichtsreiches mindertoxisches Insektizid zur Bekämpfung von *Heliothis* sp. Die Verbindung ist außerdem wirksam gegen Reisstengelbohrer [26].

Ein Kurzzeitinsektizid jüngeren Datums ist Ekamet [O,O-Dimethyl-O-(6-ethoxy-2-ethyl-pyrimidin-4-yl)-thionophosphorsäureester (V) [27].

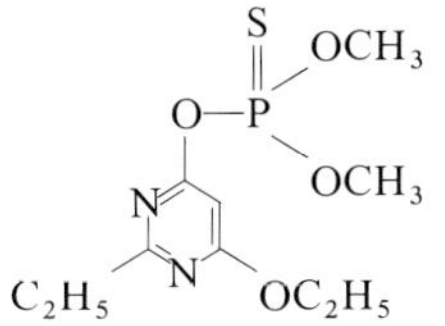

V

®Ekamet, *etrimfos*
LD_{50}: 1800 mg/kg Ratte per os
(Sandoz Ltd., 1972)

2-Ethyl-6-ethoxy-4-hydroxypyrimidin [28–30] ist in mehreren Stufen zunächst durch alkalische Kondensation von Propionamidin mit Malonsäurediethylester, Chlorierung der Dihydroxy-Verbindung, partielle Hydrolyse mit etwa 4%iger wäßriger Salzsäure und schließlich durch Chlor-Substitution mit Na-ethylat erhältlich:

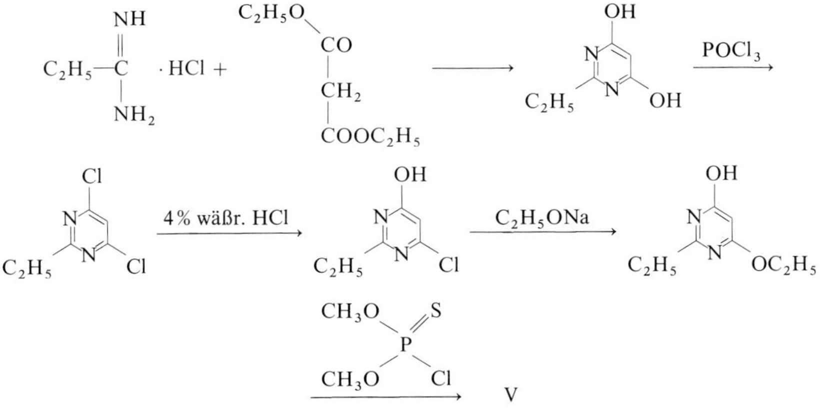

Die Phosphorylierung führt dann zur gewünschten Verbindung. Ekamet hat Kontakt- und Fraßgiftwirkung [31] und ist nicht systemisch, verfügt aber über ein hohes Eindringungsvermögen. Es wirkt gegen beißende und saugende Schädlinge im Obst-, Wein- [32] und Gemüsebau, sowie in tropischen Kulturen. Ekamet ist ein mindertoxisches Insektizid mit kurzer bis mittlerer Wirkungsdauer. Obwohl es kein spezifisches Aphizid ist, zeigt es gute Wirkung gegen einige Blattlausarten. Es ist auch gut pflanzenverträglich. Bemerkenswert ist die geringe Fischtoxizität.

Ein weiteres Pyrimidin-Derivat ist Actellic [O,O-Dimethyl-O-(2-diethylamino-4-methyl-pyrimidin-6-yl)-thionophosphorsäureester (VI) [33, 34].

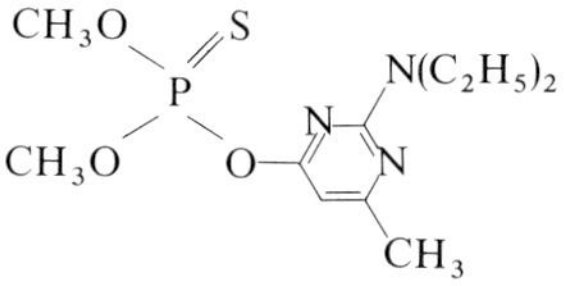

VI

® Actellic, *pirimiphos-methyl*
LD_{50}: 2050 mg/kg Ratte per os (weiblich)
(ICI Ltd., 1970)

Neben dem Methylester ist auch der Ethylester (VII) unter dem Namen Primicid im Handel.

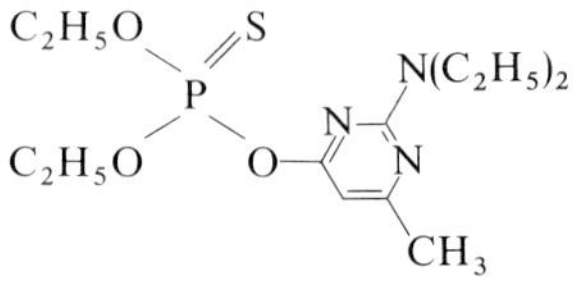

VII

® Primicid, *pirimiphos-ethyl*
LD_{50}: 125 mg/kg Ratte per os
(ICI Ltd., 1971)

Actellic ist ein schnell wirkendes Insektizid und Akarizid mit Kontakt- und Atemgiftwirkung. Außerdem kann es in das Blattgewebe eindringen. Ein großes Anwendungsgebiet ist der Getreidebau. Ebenso ist es für den Vorratsschutz vorgesehen.

Ein bicyclischer Heterocyclus ist Ni 15 [O,O-Diethyl-O-(5-ethyl-6-bromthiazolo[3,2-b][1,2,4]-triazol-2-yl)-thionophosphorsäureester] (VIII) [35, 36].

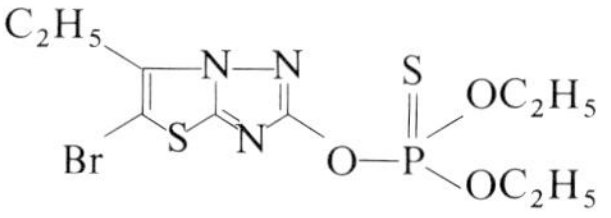

VIII

Ni 15, *thiatriphos*
LD_{50}: 220–230 mg/kg Ratte per os
(Nippon Soda, Versuchsprodukt)

Die einfache Synthese, die zu der heterocyclischen Hydroxy-Verbindung führt, geht von 1-(Methoxycarbonyl)-thiosemicarbazid aus, das alkalisch cyclisiert, mit α-Halogenketon alkyliert, anschließend sauer cyclisiert [37–39] und bromiert wird.

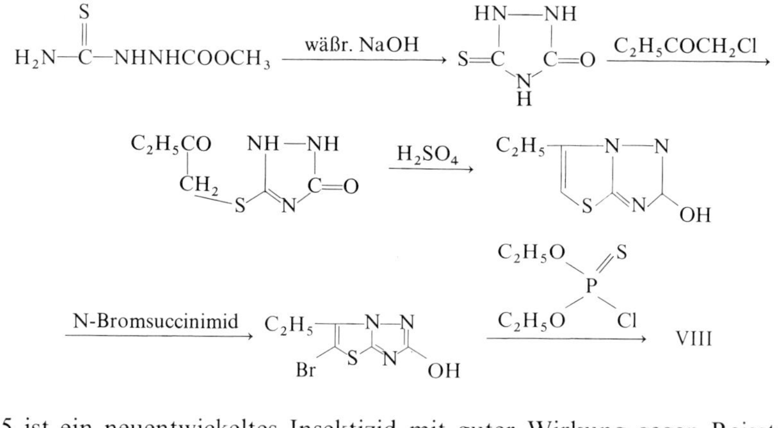

Ni 15 ist ein neuentwickeltes Insektizid mit guter Wirkung gegen Reisstengelbohrer *(Chilo supprenalis)* und die Gemeine Stechmücke. Außerdem kann es auch als Anthelmintikum [40] eingesetzt werden.

Ein anderer bicyclischer Heterocyclus ist *chlorprazophos* [O,O-Diethyl-O-(3-chlor-7-methyl-pyrazol-[1,5a]-pyrimidin-2-yl)-thionophosphorsäureester] (IX) [41].

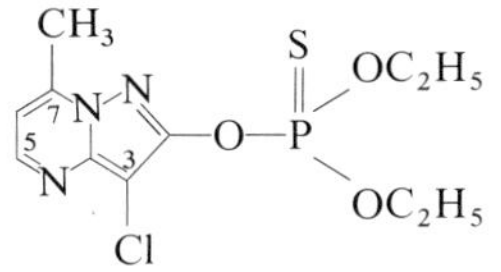

IX

HOX 2709, *chlorprazophos*
LD_{50}: 50–100 mg/kg Ratte per os
(Bayer AG, Versuchsprodukt)

Für die Synthese der Pyrazolopyrimidine [42] wird zunächst der Pyrazolring durch Cyclisieren von Cyanacethydrazid erhalten und mit einer β-Dicarbonyl-Verbindung der zweite Ring ankondensiert.

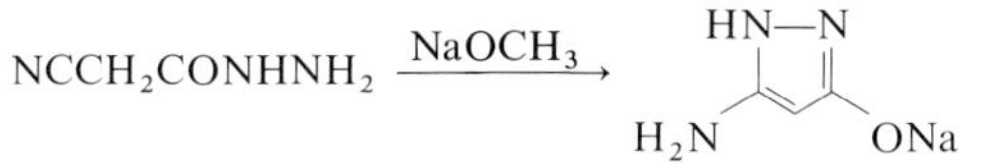

X

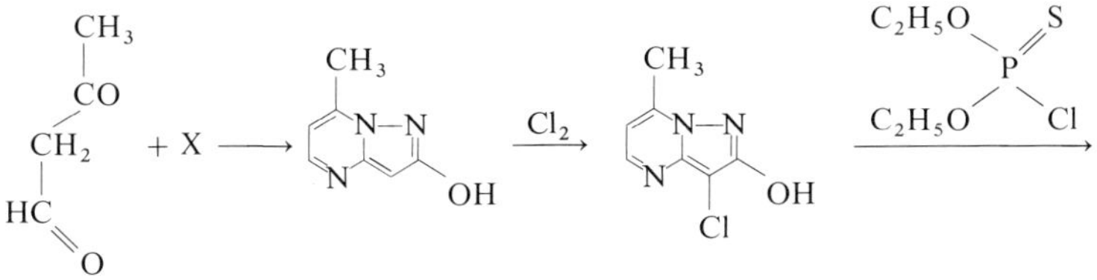

IX

Zur Verbesserung der Wirksamkeit wird noch chloriert. An diesen Verbindungen konnte gut der Substituenteneinfluß auf die Wirkung studiert werden. Die Einführung einer Methylgruppe in 5- oder 7-Stellung vermindert die Toxizität. Die Einführung von Halogen in 3-Stellung senkt die Toxizität und steigert die insektizide Aktivität. Die wirksamste Verbindung aus dieser Reihe ist *chlorprazophos.* Sie ist ein Breitspektrum-Insektizid und ein Akarizid. Es ist nicht systemisch und hat eine gute Pflanzenverträglichkeit.

4. Insektizide vom Typ der (Thio)Phosphorsäure-thioalkylester

Der einfachste Thiolphosphorsäureester, der drei verschiedene Reste an das zentrale Phosphor-Atom gebunden enthält, ist das Tamaron [O,S-Dimethyl-thiolophosphorsäureesteramid] (I) [43, 44].

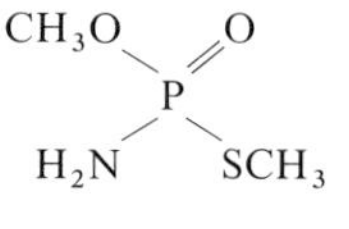

I

®Tamaron, ®Monitor, *methamidophos*
LD_{50}: 30 mg/kg Ratte per os
(Bayer AG, Chevron Chemical Co., 1969)

Für die Lösung der sich daraus ergebenden Syntheseprobleme gibt es mehrere Möglichkeiten. Bei der Methode der Rückalkylierung beginnt man mit O,O-Dimethyl-thionophosphorsäureesteramid. Dieses wird alkalisch verseift, wobei sich ein ambidentes Anion bildet. Dieses wird rückalkyliert [43, 44] und wegen der höheren Nucleophilität erfolgt die Alkylierung am Schwefelatom.

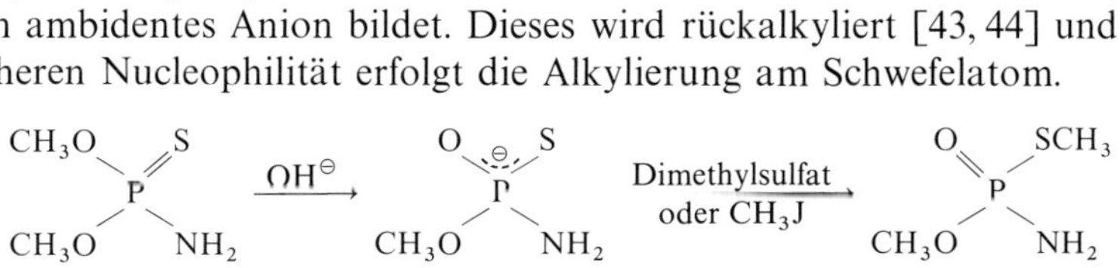

Eine weitere Synthesemöglichkeit ist die Anwendung der Pistschimuka-Reaktion, eine one-step-Isomerisierung [45], die die intermediäre Bildung eines Phosphoniumsalzes enthält.

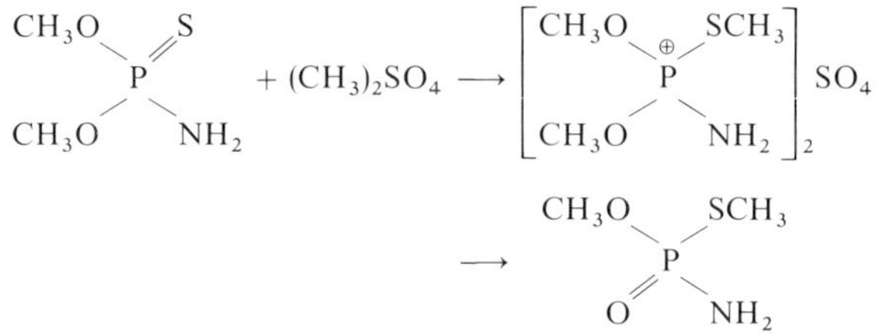

Ferner sei die sog. „Pyrophosphat-Reaktion" genannt [46]. Das Ausgangsmaterial, O,O,O-Trimethylthionophosphat, wird alkalisch hydrolysiert. Das dabei entstehende ambidente Anion lagert sich um. Bei 100 °C reagiert es mit Phosgen zu einem instabilen gemischten Anhydrid, das unter CO_2-Verlust den Pyroester bildet. Dieser wird zu Tamaron aminolysiert. Das mitanfallende Nebenprodukt kann erneut der Phosgenreaktion unterzogen werden.

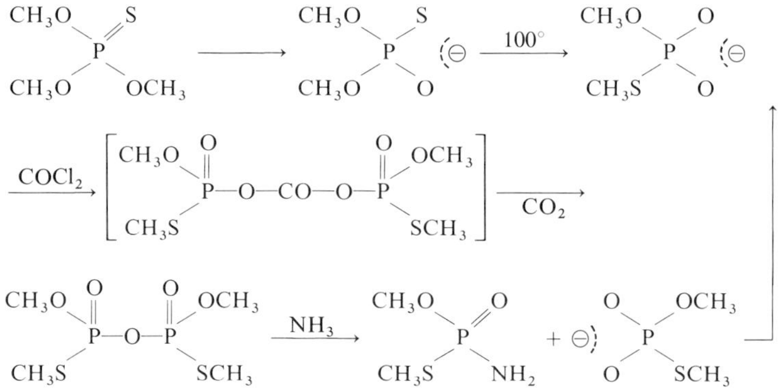

Schließlich sei noch ein Verfahren erwähnt, bei dem die Thiono-Thiol-Umlagerung, ein Reaktionsschritt, der in allen Tamaron-Synthesen enthalten ist, beim Thionoesterdichlorid erfolgt:

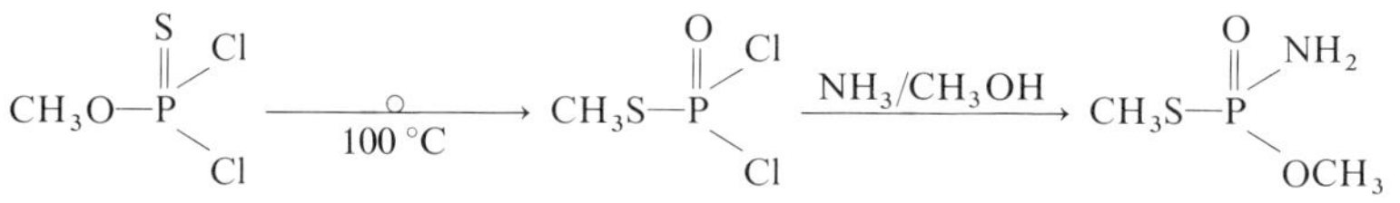

Thiolesterdichlorid wird unter speziellen Bedingungen in Methanol unter Einleiten von Ammoniak in Tamaron überführt [47].

Wegen der 4 verschiedenen Liganden ist Tamaron chiral. Die Trennung in die optischen Antipoden wurde am Beispiel des Ethylmethamidophos studiert [48] und ergab im biologischen Test eine deutliche Überlegenheit der (−)-Form.

Dotan [O,O-Diethyl-S-chlormethyl-dithiophosphorsäureester] (II) ist ein Kontaktinsektizid, das besonders als Bodeninsektizid zur Bekämpfung von

Drahtwürmern, Tausendfüßlern und Engerlingen dient und meist im Mais- und Zuckerrübenbau eingesetzt wird [49].

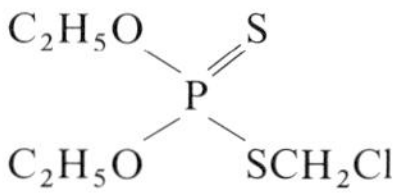

II

®Dotan, *chlormephos*
LD_{50}: 7 mg/kg Ratte per os
(Murphy Chemical Ltd., 1968)
(Rhône-Poulenc in Lizenz)

Wegen der Empfindlichkeit gegenüber Säuren ist es nicht möglich, O,O-Dialkyl-thiono- und -thiolphosphorsäuren zu chlor-methylieren. Vielmehr werden die Chlormethyl-Verbindungen aus den Salzen der Säuren durch Umsetzen mit Chlorbrommethan erhalten [50, 51].

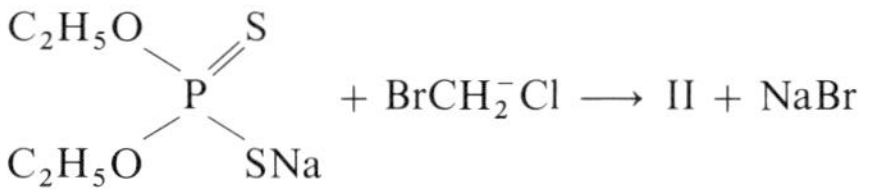

Das systemische Insektizid Phosdon [O,O-Dimethyl-S-2-isopropyl-thioethyl-dithiophosphorsäureester] (III) [52] ist auch in der Dampfphase aktiv und nach Saatgutbehandlung oder Blattspritzung besonders gegen Aphiden wirksam.

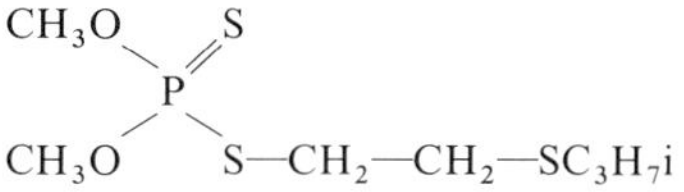

III

®Hosdon, *isothioate*, Phosdon
LD_{50}: 150–170 mg/kg Ratte per os
(Nihon Nohyaku Co. Ltd., 1972)

Ein heterocyclisch substituierter Phosphorsäure-thiolalkylester ist Alfacron [O,O-Dimethyl-S-[6-chlor-oxazolo[4,5-b]pyridin-2(3H)-onyl-(3)]-methyl-thiolo-phosphorsäureester] (IV).

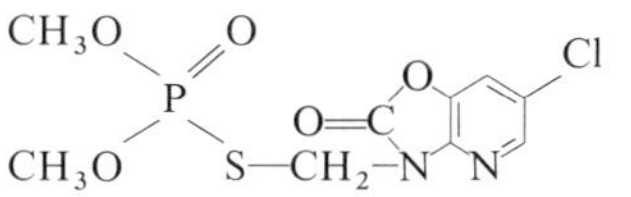

IV

®Alfacron, *azamethiphos*
LD_{50}: 1180 mg/kg Ratte per os
(Ciba-Geigy, Ltd., JUPAC-Kongreß Zürich 1978)

Es ist als das Aza-Analoge des Phosalon aufzufassen. Die Synthese [53–67] dieses Heterocyclus enthält als wichtigsten Reaktionsschritt den von Clauson-Kaas gefundenen oxidativen Übergang eines Furanderivates in ein Pyridinderivat. Das dazu erforderliche Stickstoffatom kann entweder exocyclisch in der Furanverbindung vorliegen oder mit Hilfe von Sulfaminsäure, beispielsweise, eingeführt werden. Auf diese Weise ist das Ausgangsmaterial 3-Hydroxy-2-aminopyridin erhältlich. Es kann aber auch durch nucleophilen Substituentenaustausch aus Halogenpyridin gewonnen werden. Es wird mit Phosgen zu Oxazolo [4,5-b]-pyridin-2(3H)on umgesetzt und chloriert. Die weitere Reaktion mit wäßriger Formaldehyd-Lösung ergibt die Methylolverbindung, die in die Chlormethyl-Verbindung verwandelt und anschließend mit dem Ammoniumsalz der O,O-Dimethyl-thiophosphorsäure phosphoryliert wird. Alfacron wird insbesondere für die Bekämpfung von saugenden und beißenden Insekten und Schädlingen aus der Klasse der Acarina eingesetzt. Es wirkt als Kontakt- und

Fraßgift, hat aber auch systemische Wirkung. Ebenfalls kann es im Vorratsschutz eingesetzt werden. Es wirkt auch gegen Schaben und die Tsetse-Fliege [68].

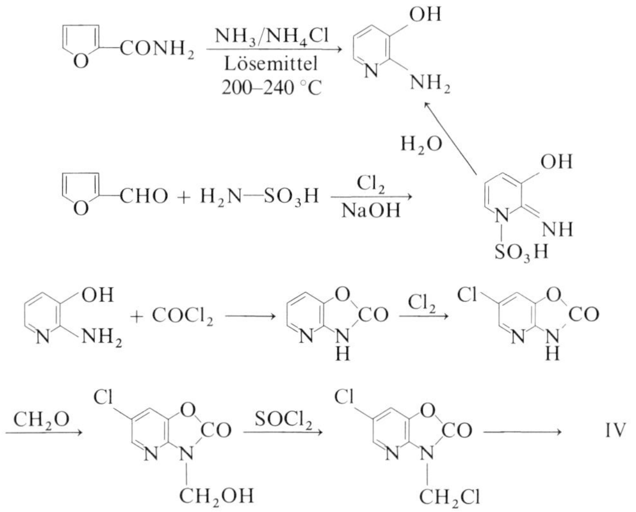

5. Insektizide vom Typ der Enol(thio)phosphorsäureester

Ein neuentwickeltes Enolphosphat ist Hostaquick [6-(O,O-Dimethyl-O-(7-chlorbicyclo[3,2,0]-hepta-2,6-dien-6-yl)-phosphorsäureester] (I) [69].

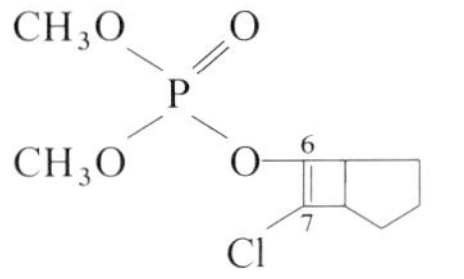

I

®Hostaquick, *heptenophos*
LD_{50}: 98–117 mg/kg Ratte per os
(Hoechst AG, 1970)

Für die Darstellung wird Trimethylphosphit mit einem Dichlorbicycloketon im Sinne einer Perkow-Reaktion umgesetzt:

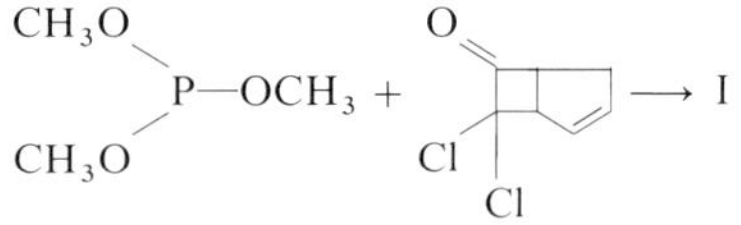

Hostaquick ist ein systemisches Insektizid und Akarizid mit kurzer Wirkungsdauer. Es kann daher als Kurzzeitinsektizid mit kurzer Wartezeit angewendet werden.

Eine weitere Neuentwicklung ist das Insektizid Damfin [O,O-Dimethyl-O-(2-methoxycarbonylprop-1-enyl)-thionophosphorsäureester] (II) [70].

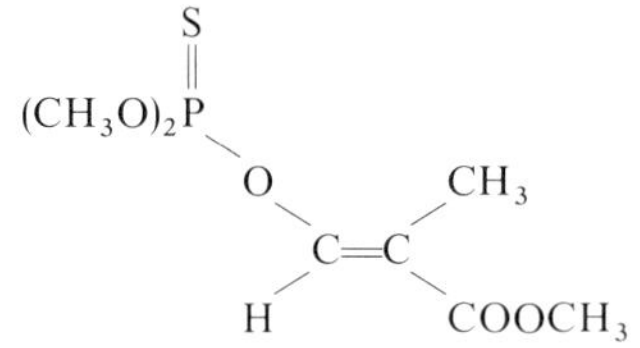

II
(E-Isomeres)

®Damfin, *methacrifos*
LD_{50}: 678 mg/kg Ratte per os
(Ciba-Geigy, Ltd., 1977)

Die Verbindung kann durch Umsetzen von Oxymethylen-propionsäuremethylester-Na-salz mit O,O-Dimethyl-thionophosphorsäureesterchlorid synthetisiert werden. Damfin wirkt als Kontakt-, Atem- und Fraßgift gegen alle wichtigen Arthropoden in Vorratsgütern. Es ist ebenfalls hochwirksam gegen Malathion- und Lindan-resistente Insekten [71].

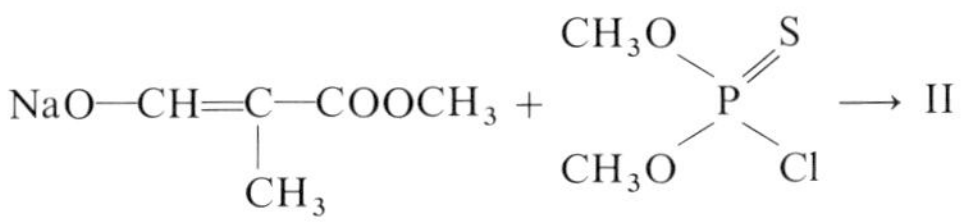

Ein Phosdrin-Derivat ist das Erfolg versprechende Insektizid Safrotin [O-Methyl-N-ethyl-O-(1-carboisopropoxy-1-propen-2-yl)thionophosphorsäureesteramid] (III) [72].

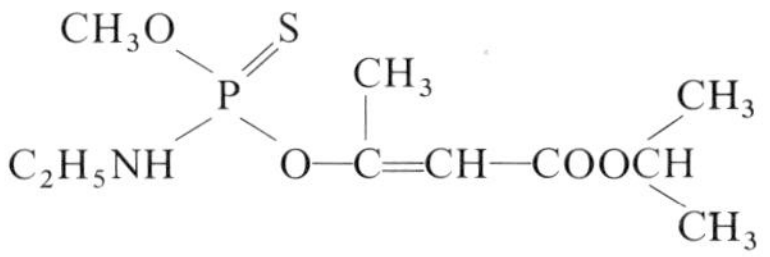

III

®Safrotin, *propetamphos*
LD_{50}: 82 mg/kg Ratte per os
(Sandoz, Ltd., 1977)

Für die Synthese steht folgendes Verfahren zur Verfügung: das Dichlorid ist aus Phosphorsulfochlorid und dem Na-Salz des Acetessigsäureisopropylesters erhältlich. Das besondere Anwendungsgebiet ist die Bekämpfung des Baumwollkapselkäfers *(Anthonomus grandis)* und des Baumwollkapselwurms *(Heliothis zea)*. Außerdem soll es besonders als Haushalt- und Hygieneinsektizid eingesetzt werden [73]. Es wirkt gegen Fliegen, Mücken, Wanzen und Schaben.

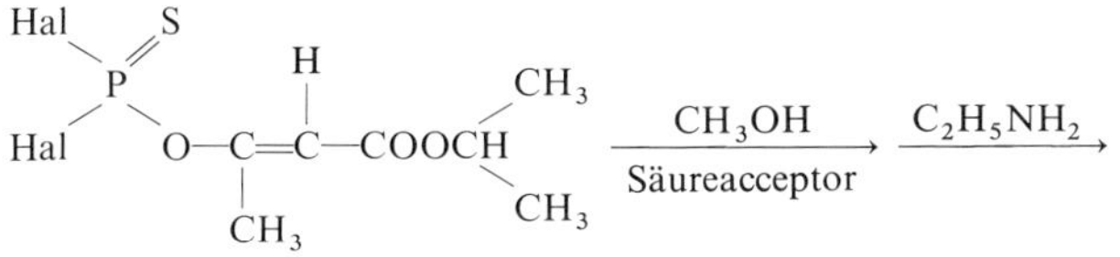

(III)

6. Sonstige

Als „Ausnahmen" von der Schrader'schen Acyl-Regel sollen einige Verbindungen gezeigt werden, die nicht die „klassischen" Acylkomponenten enthalten. Dieses sind z. B. die Phosphorylimine, bei denen der „Acyl"-Rest häufig über Stickstoff gebunden ist. Die Nucleophilität des doppelt gebundenen Stickstoffs ist so reduziert, daß der Imid-Rest als Acyl-Gruppe wirken kann. Beispiele hierfür seien die folgenden Verbindungen:

Das systemische Bodeninsektizid und -nematizid AC 64,475 [2-(Diethoxyphosphinylimino)-1,3-dithietan] (I) [74],

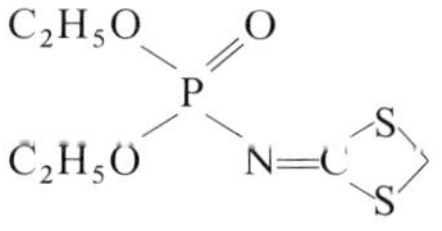

I

®Nem-A-Tak, AC 64,475
LD_{50}: 5 mg/kg Ratte per os
(American Cyanamid Co., 1972)

die Insektizide und Akarizide Cyolane [2-(Diethoxyphosphinylimino)-1,3-dithiolan] (II) und Cytrolane [2-(Diethoxyphosphinylimino)-4-methyl-1,3-dithiolan] (III) [75, 76].

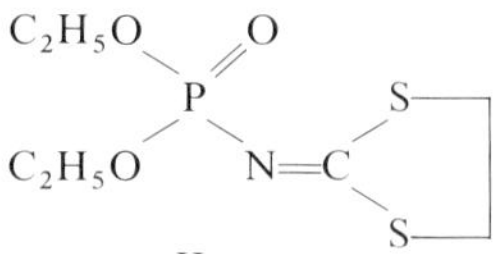

II

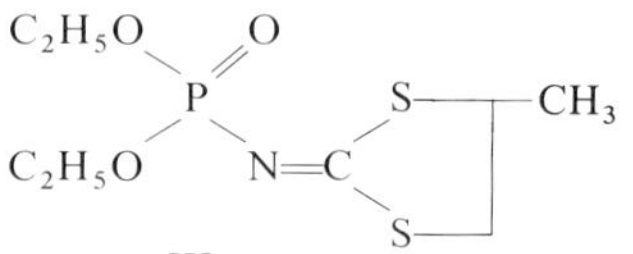

III

®Cyolane, *phosfolan*
LD_{50}: 8,9 mg/kg Ratte per os

®Cytrolane, *mephosfolan*
LD_{50}: 11,3 mg/kg Ratte per os

(American Cyanamid Co., 1963)

Cyolane ist ein systemisches Insektizid. Es kann besonders gut im Baumwollbau eingesetzt werden und wirkt als Kontakt- und Fraßgift. Verbindungen dieses Typs zeigen gute Wirkung gegen beißende und saugende Insekten und gegen Spinnmilben.

Die Darstellung des Heterocyclus kann aus Ethan-1,2-dithiol erfolgen, das in nichtpolaren Lösungsmitteln unter Einleiten von Chlorwasserstoff mit Chlorcyan umgesetzt werden kann

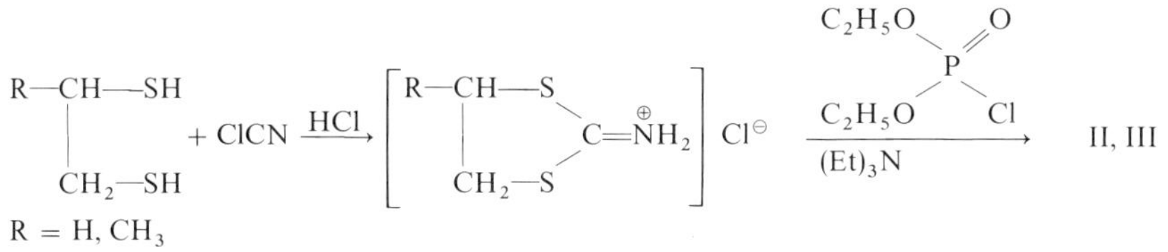

oder durch Addition von Kaliumhydrogensulfid an Diethylphosphorylisothiocyanat [77] und Ringschluß mit Ethylenbromid.

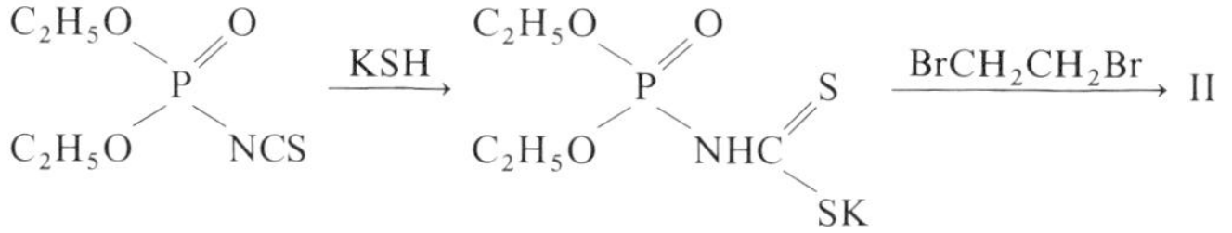

Das Nematizid AC 475 [O-(Diethoxyphosphinyloxy)-2H-1,3-dithiolan-2-oxim] (IV) [78], das sich noch in der Prüfung befindet, ist zwar kein Phosphorylimin, es soll aber, da es sich ebenfalls um ein Dithiolan handelt, an dieser Stelle mit erwähnt werden:

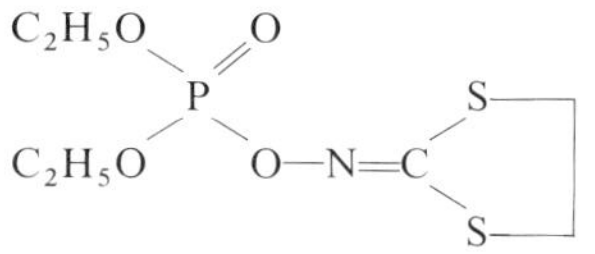

IV

AC 475

Die Synthese (s. o.) beginnt ebenfalls mit Ethan-1,2-dithiol, das zunächst mit Chlorcyan, dann mit Hydroxylaminhydrochlorid umgesetzt wird.

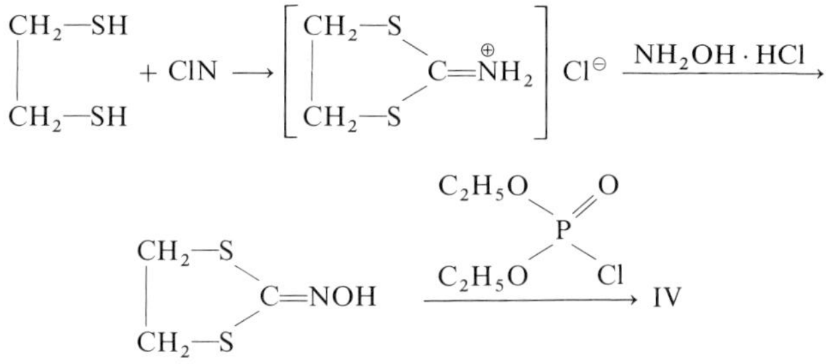

7. Literatur

1. Fest,C., Schmidt,K.-J.: Chemie der Pflanzenschutz- und Schädlingsbekämpfungsmittel (Hrsg. R. Wegler), Springer-Verlag, Berlin-Heidelberg-New York 1970, Bd. I
2. Kishino,Sh., Kudamatsu,A., et al. (Bayer AG): DOS 2111414 (1971/1971)
3. Kudamatsu,A. et al.: Jap. J. Sanit. Zool. Vol. *29*, (4) (1978) S. 317 ff.
4. Hilgetag,G., Teichmann,H.: Angew. Chem. 77, 1001 (1965); Hilgetag,G. et al.: J. prakt. Chem. [4] *12*, 1 (1961)
5. Anonym (Nihon Tokushu Noyaku Sei), JA-OS 49/86347 (1972/1974)
6. Diehr,H.-J., Arold,H. (Bayer AG): DOS 2532396 (1975/1977)
7. Uhing,E.H., Toy,A.D.F. (Stauffer Chemical Co.): USP3879500 (1973/1975)
8. Kleinstück,R., Block,H.D. (Bayer AG): DOS 2636270 (1976/78); Mazour,Z., Brunetti,H. (Ciba-Geigy AG): DOS 2643474 (1976/77); Mazour,Z. (Ciba-Geigy AG): DOS 2643442 (1976/77)
9. Treml,A.G. et al.: (Wsesojusny nautschno-issledowatelskij institut chimitscheskich sredstw saschtschity rastenij, Moskau). DOS 2345069 (1973/1975)
10. Petrov,K.A., Sokolsky,G.H., Polees,B.H.: Zh. Obshch. Khim. *26*, 3381 (1956)
11. Bayer,H.O., Hurt,W.S. (Röhm und Haas Co.): DOS 2635931 (1976/1977)
12. Beriger,E., Drabek,J. (Ciba-Geigy AG): DOS 2249462 (1972/1973)
13. Reisebericht Dr. Fedtge 29.9.1975
14. Wilson,J.H. (Mobil Oil Corp.): USP 3268393 (1965/1966)
15. Mannes,K., Schrader,G., Homeyer,B. (Bayer AG): DAS 1238018 (1965/1967)
16. Schmidt,K.-J., Schrader,G. (Bayer AG): DAS 1211654 (1961/1966)
17. Schrader,G., Hammann,I. et al. (Bayer AG): DOS 1668047 (1967/1971)
18. Hagemann,H. (pers. Mitteilung)
19. Dawes,D., Böhner,B. (Agripat S.A.): DOS 2150098 (1970/1972)
20. Taipale,K.A., Smirnoff,S.A.: Chem. Ber. *56*, 1794 (1923)
21. Dawes,D., Böhner,B. (Ciba-Geigy AG): DOS 2260015 (1972/1973)
22. Weiße Karte: Miral, s. a. Technical Release (Ciba-Geigy)
23. Scherer,O., Mildenberger,H. (Hoechst AG): DAS 1670876 (1967/1972)
24. Finkenbrink,W., Emmel,L.F. (Hoechst AG): DAS 1767970 (1978/1971)
25. Du Breuil,Sh. (American Cyanamid Co.): USP 2759937 (1955/1956)
26. Formigoni,A., Ciampolini,M. Soc. Sipcam-Milano
27. Milzner,K.H., Reisser,F. (Sandoz AG): DOS 2209554 (1972/1972)
28. Henze,H.R., Clegg,W.J., Smart,Ch.W.: J. Org. Chem. *17*, 1320 (1952)
29. Henze,H.R., McPherson,J.L.: J. Org. Chem. *18*, 653 (1953)
30. Van der Plas,H.C. et al.: Rec. Trav. Chim. Pays-Bas *85*, 1101 (1966)
31. Knutti,H.J., Reisser,F.W. (Sandoz Ltd.): Proc 8th Brit. Insecticide and Fungicide Conf. 1975
32. Buess,U., Henauer,A. (Sandoz AG): 28. Internat. Symp. Pflanzenschutz Gent, am 4.5.1976
33. McHattie,G.V. (Imperial Chemical Industries, Ltd.): E.P. 1019227 (1963/1966)
34. Sharpe,St.P., Snell,B.K. (Imperial Chemical Industries, Ltd.): E.P. 1204552 (1968/1970)
35. Kano,S., Nomura,O., Asada,M. et al. (Nippon Soda Co. Ltd.): DOS 2264162 (1972/1973)

36. Asada, M., Hiratsuka, K., Ando, M. et al. (Nippon Soda Co. Ltd.): DOS 2323720 (1973/1974)
37. Kato, K., Morishima, J., Ozawa, K. et al. (Nippon Soda Co. Ltd.): Japan Kokai 77, 111, 575 (16. März 1976/19. Sept. 1977)
38. Kato, K., Morishima, Y., Ozawa, K. et al. (Nippon Soda Co. Ltd.): Japan Kokai 77, 113, 989; C.A. *88*, 550 (1978)
39. Kano, S., Noguchi, T. (Nippon Soda Co. Ltd.): Japan Kokai *71*, 26, 498 (1969/1971); Hoffmann, H., Hammann, I. et al. (Bayer AG): DOS 2361451 (1973/1975)
40. Tanaka, Y., Kano, S., Odawara, K. (Nippon Soda Co. Ltd.): DOS 2416178 (1974/1974)
41. Hoffmann, H., Hammann, I. (Bayer AG): Belg. P. 769702 (1971/1972)
42. Ried, W., Köcher, E.-U.: Liebigs Ann. Chem. *647*, 116 (1961)
43. Schrader, G., Lorenz, W. et al. (Bayer AG): DAS 1210835 (1964/66)
44. Magee, Ph.S. (Chevron Res. Co.): DAS 1246730 (1966/1967)
45. Burn, A.J., Cadogan, J.I.G.: J. Chem. Soc. *1961*, 5532
46. Hofer, W. (Bayer AG): DOS 2156718 (1971/1973)
47. Cölln, R. (Bayer AG): DOS 1668094 (1968/1969)
48. Unveröffentlichte Ingenieurarbeit, J. Wiemer (Bayer AG), 1976
49. Griffiths, D.C. et al.: Ann. appl. Biol. *64*, 21 (1969)
50. Scherer, O., Hahn, H., Stähler, G. (Hoechst AG): USP 3020304 (1957/1962)
51. Pianka, M. (The Murphy Chemical Co.): E.P. 1258922 (1968/1971)
52. Anonym (Nihon Nohyaku Co. Ltd.): Japan-P. 624714
53. Den Hertog, H.J., De Bruyn, J.: Rec. Trav. Chim. Pays-Bas *70*, 182 (1951)
54. Rüfenacht, K., Kristinsson, H.: Helv. Chim. Acta *59*, 1593 (1976)
55. Mattern, G.: Helv. Chim. Acta *60*, 2062 (1977)
56. Greuter, H., Belluš, D.: J. Heterocycl. Chem. *14*, 203 (1977)
57. Clauson-Kaas, N.K.F.W., Elming, N.: USP. 2806852 (1954/1957)
58. Clark, R.L., Pessolano, A.A. (Merck & Co.): USP. 2806853 (1956/1957)
59. Kristinsson, H., Rüfenacht, K. (Ciba-Geigy): USP. 3808218 (1970/1974)
60. Petersen, J.V.B., Clauson-Kaas, N.K.F.W. (Geigy Chem. Co.): USP. 3419570 (1966/1968)
61. Petersen, J.V.B., Clauson-Kaas, N.K.F.W. (Geigy Chem. Co.): USP. 3553224 (1968/1971)
62. Petersen, J.V.B., Clauson-Kaas, N.K.F.W. (Geigy Chem. Co.): USP. 3663555 (1970/1972)
63. Petersen, J.V.B., Clauson-Kaas, N.K.F.W. (Geigy Chem. Co.): USP. 3663556 (1970/1972)
64. Kristinsson, H., Rüfenacht, K. (Ciba-Geigy Co.): USP 3886274 (1974/1975)
65. Mattern, G. (Ciba-Geigy Co.): USP. 4033975 (1976/1977)
66. Mattern, G. (Ciba-Geigy Co.): USP. 4052406 (1976/1977)
67. Greuter, H., Belluš, D. (Ciba-Geigy Co.): USP. 4061644 (1976/1977)
68. Wyniger, R. et al. (Ciba-Geigy Ltd.): Proc. 1977 Brit. Crop Protection Conf. – Pests and Diseases, Vol. 3, 1025
69. Scherer, O., Röchling, H., Mildenberger, H. (Hoechst AG): DOS 1768315 (1968/1971)
70. Beriger, E. (Ciba-Geigy AG): DAS 2147588 (1971/1973)
71. Wyniger, R. et al. (Ciba-Geigy Ltd.): Proc. 1977 Brit. Crop Protection Conf. – Pests and Diseases, Vol. 3, 1033
72. Leber, J.-P., Lutz, K. (Sandoz AG): DOS 2035103 (1970/1971)
73. 27. Ann. Conf. Rep. Cotton Insect Res. and Control, Dallas, Texas, 7.–9. Jan. 1974, S. 46
74. Whitney, W.K., Aston, J.L.: Proc. Eighth Brit. Insecticide and Fungicide Conf. 17.–20.th Nov. 1975 (Brighton) S. 625
75. Addor, R.W.: J. Org. Chem. *29*, 738 (1964)
76. Addor, R.W., Lovell, J.B. (American Cyanamid Co.): F.P. 1327386 (1962/1963)
77. Addor, R.W.,: J. Heterocycl. Chem. *7*, 381 (1970)
78. Addor, R.W. (American Cyanamid Co.): USP. 3281430 (1963/1966)

Bacterial Plant Diseases and Their Control

T. Egli and E. Sturm

CIBA-GEIGY Ltd., Agrochemicals Division
CH-4002 Basel, Switzerland

Contents

1. Introduction

Bacterioses occupy a special position in phytopathology. Being infectious diseases, they naturally have much in common with mycoses and viroses. Nevertheless the microorganisms responsible for them, the phytopathogenic bacteria, differ so widely from fungi and other plant pathogens that the resulting diseases present special features as regards their development and control. Although certain bactericides have already been mentioned in Vol. 4, p. 117ff., bacterioses represent, together with the mycoplasmoses and rickettsioses, a branch of phytopathology that deserves separate treatment.

In warm-blooded animals, including man, bacteria are of greater importance as pathogens than fungi. In plants, exactly the opposite holds true: mycoses are dominant and bacterioses are rarer, even though in some cases they may be of great economic consequence. The reduced suitability of bacteria for a parasitic existence in or on plants is partly due to their morphology. Unlike fungi, which have a vegetative body consisting of threadlike structures known as hyphae, bacteria are normally unicellular microscopic organisms. Growth in bacteria also involves division and separation into individual cells. Single cells of this kind are in a poor position to overcome an obstacle, such as a cell wall, which makes it difficult for them both to penetrate a plant and to spread once inside it. While a fungus generally invades a leaf by way of a germinal hypha developing a flattened tip (appressorium) which becomes attached to the host, a bacterium has no such organ and thus no mechanical means of fixing itself to the cutinized outer layers of the body of a plant. It has only two ways of penetrating to the interior of a plant:
1) via noncutinized parts of the body (root hairs, stigmata, lesions, etc.), or
2) via natural openings (stomata, hydathodes, nectaries, etc.).

To a bacterial pathogen, therefore, a healthy plant offers a smaller area for attack than it does to a fungal agent. Even so, however, an ordinary leaf has per square millimeter fifty to three hundred stomata on its underside, and one maize plant, for example, has anything up to 200,000,000 unprotected entrance points of this kind. Particularly suitable are the nectaries, since they not only meet the conditions required for successful infection with respect to humidity but also supply the hungry parasite with nutrients.

Some species of bacteria rely exclusively on wounds for successful infection (e.g. *Agrobacterium tumefaciens*). For most phytopathogenic bacteria, however, a lesion is merely a welcome opportunity to bypass the specific paths of entry. By this is meant not wide, gaping wounds so much as the innumerable "natural" lesions that are inevitably produced during the growth of the plant. Thus the development of the lenticels on underground organs and bark is always accompanied by fine fissuring; leaf scars, which are innumerable in autumn, also provide natural openings for certain bacteria (e.g. *Pseudomonas mors-prunorum*).

Once a bacterium has successfully penetrated to the inside of a plant, it multiplies in the intercellular spaces or in the dead cells of the xylem vessels. Bacteria, as such, cannot actually penetrate into intact living plant cells; nor is it always possible for them to proliferate continuously in the parenchymal tissue, like certain fungi with their hyphae, because of their unicellular structure. In this respect, however, mycoplasmas and rickettsias form an exception. This special group of bacterial pathogens vegetates in the cytoplasm and, in particular, in that of the phloem cells. For bacteria, which prosper best in a mildly alkaline medium, the interior of a plant, which has a pH of 5–6, is not an ideal environment. Fungi, which prefer a slightly acid pH, thus have the edge over bacteria as plant parasites in this respect as well.

A peculiar feature of phytopathogenic bacteria is that they do not form spores. These organs, so important for fungi both for dissemination and for survival in unfavorable environments, are totally absent in phytobacteriology. As a result, the bacterium, once outside its host plant, must find a moist environment (a watery film or water droplets) in order to survive; despite the protective layer of mucus which surrounds every bacterium, bacterial plant pathogens quickly perish in dry air. This affinity of the bacteria for water is also seen in the fact that many of them are equipped with flagella rendering them motile. This morphology, suited to a humid environment, has the natural consequence that phytopathogenic bacteria are encountered more particularly in regions of heavy rainfall, and it is here that they have their economic impact.

For the transmission of bacterioses over short distances, wind driven water droplets or rain splashes are the most important vectors. In addition, bacterioses can be transmitted by contaminated cutting tools, by insects, and by irrigation water. It is presumed that migratory birds have helped in spreading certain bacterioses over longer distances. For the transport of phytopathogenic bacteria over really great distances, however, the most important agent is surely man, who ships infected seeds, grafts, and entire plants all over the world.

The difficulties encountered by bacteria in their role of plant parasites as regards both dissemination and the infection and colonization of plants led many noted scholars, as recently as the end of the nineteenth century, to rule out as a matter of principle any possibility that bacteria could act as plant pathogens [1]. Today, none doubts that plant bacterioses actually exist. Even though mycoses are more numerous in the vegetable world, it is nevertheless an obvious fact that there is a whole series of important bacterioses capable of causing serious losses in crops, more particularly in moist and hot climatic conditions.

2. The Economic Importance of Bacterioses

The literature only rarely contains properly-founded quantitative data on the economic importance of bacterioses. The writers generally settle for an overall estimate of the potential importance of the individual diseases. One reason for this is that bactericides are still little used today and that the yield losses caused by bacterial infestation can therefore practically never be quantified by comparison with an uninfested field. In addition, it is in the nature of bacterioses that, depending on climatic factors, they often appear unexpectedly and massively only to recede to a milder stage for relatively long periods. For this reason the scanty data in the literature on the economic importance of bacterioses deviate widely from year to year and from region to region (cf. Fig. 1).

Systematic investigations of yield losses caused by bacterial diseases have been undertaken in the United States [2–4]. On the basis of the latest of these studies

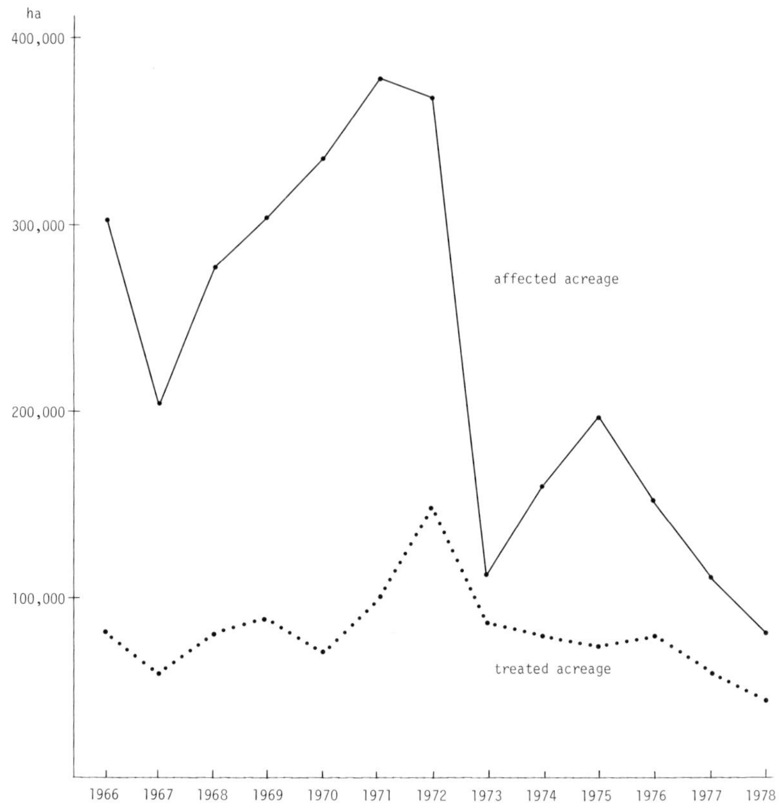

Fig. 1. Rice acreage in Japan affected by and treated against bacterial leaf blight caused by *Xanthomonas oryzae* (source: Pesticide Handbook published by the Japan Plant Protection Association)

the aggregate losses due to bacterioses in the United States reach $ 225 million. Comparison of the figures for the years 1968 and 1976 reveals in addition a marked tendency for these losses to increase (cf. Table 1). This increase in the yield losses may, however, be due partly to improvement in diagnostic methods. It must be stressed that the figures given are based on estimates and thus leave room for some uncertainty.

3. The Causative Agents of Bacterioses

3.1. Classification and Taxonomy

Both research and practice require that bacteria be unambiguously named and properly classified. Only with the help of clear nomenclature and classification will it be possible systematically to assemble and store the vast array of data about microorganisms and their effects.

Table 1. Aggregate estimated losses from major bacterial diseases in USA. Adapted from Alcorn et al. 1971 [3] and Kennedy and Alcorn 1980 [4]

Genus	Species or disease	Loss (1968) ($\$ \times 10^6$)	Loss (1976) ($\$ \times 10^6$)	Main State where losses occur	Main host
Agrobacterium	*tumefaciens*	2.5	23.0	California	Fruits and nuts
Corynebacterium	*insidiosum*	6.9	17.0	Minnesota	Alfalfa
	michiganense	–	0.2	Minnesota	Tomato
	nebraskense	–	3.0	Nebraska	Corn
	sepedonicum	–	1.8	Maine	Potato
Erwinia	*amylovora*	1.7	4.7	California	Pear
	chrysanthemi	–	2.3	N. Carolina	Tobacco
	carotovora	3.8	14.0	Minnesota	Potato
	stewartii	–	0.2	Alabama	Corn
	tracheiphila	–	0.9	Wisconsin	Cucumber
Pseudomonas	*glycinea*	7.2	65.0	Iowa	Soybean
	lachrymans	–	0.5	Wisconsin	Cucumber
	phaseolicola	–	2.0	Idaho	Beans
	solanacearum	–	9.4	S. Carolina	Tobacco Tomato
	syringae	9.6	18.0	Montana	Wheat
Xanthomonas	*campestris*	–	1.0	Florida	Crucifers
	juglandis	–	2.2	California	Walnut
	malvacearum	–	15.0	Louisiana	Cotton
	phaseoli	3.0	5.0	Nebraska	Beans
	pruni	–	2.0	Georgia	Peach
	vesicatoria	1.2	1.5	Florida	Tomato
Mycoplasma	Pear decline	–	1.6	California	Pear
	Peach rosette	–	0.3	S. Carolina	Peach
	Lethal yellowing	–	3.0	Florida	Coconut palm
Spiroplasma	Corn stunt	–	0.06	Mississippi	Corn
	Stubborn disease	–	1.0	Arizona	Citrus
	Aster yellows	–	0.2	N. Dakota	Flax
Rickettsia	Pierce's disease	–	3.0	California	Grape
	Phony peach	–	20.0	Georgia	Peach
	Ratoon stunt	–	10.0	Louisiana	Sugarcane

Because of their small size and similarity of shape it is difficult to describe bacteria exactly and sort them into groups. At first, bacteriologists did in fact attempt to set up a classification based on the structures visible under the microscope; the shape of the bacteria and the number and arrangement of flagella were used as criteria. However, this morphological method quickly found its limits, and especially so in the case of phytopathogenic bacteria, which are nearly all rod-shaped. Physiological characteristics were therefore also brought in to ease description and classification. The tests, which are still often used today, include: shape and color of the bacterial colonies on given culture media; nutrient, oxygen, and temperature requirements; nitrate reduction; production of indole, hydrogen sulfide, and ammonia; acid and gas formation from various carbon sources; etc. Nevertheless, even these tests are not entirely satisfactory, since physiological characteristics vary between certain limits and only imperfectly reflect the natural relationships between bacteria.

A modern taxonomy giving a better picture of these natural relationships will become possible only on the basis of molecular biology. At the present time, studies on the composition of the DNA bases and on DNA hybridization are being carefully evaluated for taxonomic purposes [5]. But the desired goal, a "natural" classification system – in fact a phylogenetic pedigree of the bacteria – is still far off in the future.

The most comprehensive reference work in which bacteria are described and classified is Bergey's Manual of Determinative Bacteriology, of which the eighth edition [6] appeared in 1974. It contains names, descriptions, morphological and physiological characteristics, and bibliographical references, as well as keys for the identification of newly isolated bacteria. This systematics is – like plant and animal systematics – hierarchical in principle. The basic element, the pure culture of an isolated bacterium, is the "strain". Strains are arranged in species, species in genera, genera in families (whose names end in *-aceae*). The families are arranged in orders and classes. While the arrangement of phytopathogenic bacteria in larger groupings (orders, families, and genera) is largely undisputed, considerable difference exists among specialists at the species level. There are at present two fundamentally different concepts about species, with the unfortunate result that different species designations have been attributed to the same organisms. This situation is regrettable, but understandable. Many phytopathogenic bacteria that

Table 2. Classification of genera of bacteria containing plant pathogens according to the scheme of Bergey's Manual (eighth edition)

General grouping	Family and/or order	Genus
Gram-negative aerobic rods and cocci (part 7)	Pseudomonadaceae	*Pseudomonas* *Xanthomonas*
	Rhizobiaceae	*Agrobacterium* (*Rhizobium*)
Gram-negative facultatively anaerobic rods (part 8)	Enterobacteriaceae	*Erwinia*
Actinomycetes and related organisms (part 17)	Coryneform group of bacteria	*Corynebacterium*
	Actinomycetales Streptomycetaceae	*Streptomyces*
Rickettsias (part 18)	Rickettsiales	RLO (rickettsialike organisms)
Mycoplasmas (part 19)	Mycoplasmatales	MLO (mycoplasmalike organisms)

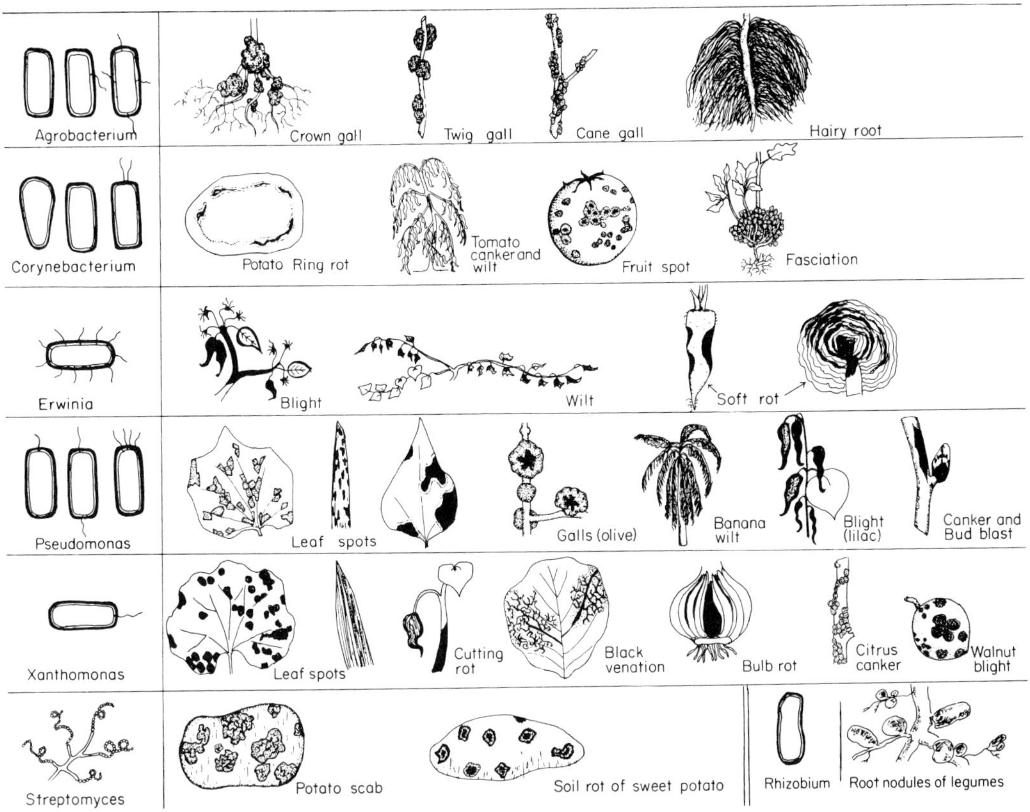

Fig. 2. Genera of bacteria and types of symptoms they cause (from G. N. Agrios: Plant Pathology, 2nd edition, Academic Press 1978)

are clearly distinguishable as regards their host-specific pathogenicity cannot be differentiated in the laboratory with the conventional physiological standard tests. From the point of view of the bacteriologist, therefore, such bacteria belong to one and the same species. The phytopathologist, on the other hand, naturally attaches special importance to pathogenicity and therefore tends to use the range of host plants as the decisive criterion in determining the species. Thus in the eighth edition of the bacteriologically oriented Bergey's Manual of Determinative Bacteriology there are only five *Xanthomonas* and seven *Pseudomonas* species listed as true phytopathogenic taxospecies in these genera, whereas the phytopathological literature speaks of about one hundred *Pseudomonas* species and as many again of *Xanthomonas*. There has been no lack of suggestions for solving this conflict [7, 8], but it can be assumed that the interested layman will be confronted with a confusing situation for some time to come. To avoid misunderstandings, we shall here use the traditional nomen-species designations. Some characteristics of the genera and their arrangement in an overall system have been taken from the last edition of Bergey's Manual (cf. Table 2 and Fig. 2).

3.2. Bacterial Genera to which Causative Agents of Bacterioses are Assigned

Pseudomonas Migula

General characteristics of the genus:

Cells single, straight or curved rods. Dimensions, generally 0.5–1 by 1.5–4 µm. Motile by polar flagella. Do not form spores. Gram-negative. Metabolism respiratory, never fermentation. Strict aerobes. Characteristically but not always a yellowish-green fluorescent pigment (fluorescein) is produced.

The genus as a whole is a diverse one which includes soil bacteria, animal and plant pathogens, food spoilage bacteria, and marine forms. The metabolic diversity found in this genus is remarkable and the activities of its members are important in the mineralization of organic matter in nature. The plant-pathogenic pseudomonads incite a wide variety of symptom types such as leaf spots (e.g. angular leaf spot of cucumber caused by *P. lachrymans*), cankers (e.g. bacterial canker of cherry caused by *P. mors-prunorum*), wilts (e.g. bacterial wilt of several solanaceous crops caused by *P. solanacearum*), galls (e.g. olive knot caused by *P. savastanoi*) and soft rots (e.g. bacterial soft rot of lettuce caused by *P. marginalis*).

Xanthomonas Dowson

General characteristics of the genus:

Cells single, straight rods. Dimensions 0.2–0.8 by 0.6–2.0 µm. Motile by means of one polar flagellum. Do not form spores. Gram-negative. Growth on agar media usually yellow (xanthus = yellow). Metabolism respiratory, never fermentative. Strict aerobes. Temperature optimum 25–27 °C.

The xanthomonads are closely related to the pseudomonads and there are researchers who believe that the two genera should be merged. The generic definition for the genus *Xanthomonas* does indeed not easily differentiate it from the genus *Pseudomonas*. However, the two genera have not been combined because extensive phenotypic comparisons have not yet been made and the yellow pigment, a non-water-soluble carotenoid produced by almost all xanthomonads, characterizes this group of bacteria as standing apart.

All xanthomonads are plant pathogens. Although they can live saprophytically on all sorts of substrates, they are only found in association with plants or plant materials. Reports of xanthomonads from other habitats have not been confirmed. Especially in the warmer parts of the world, the *Xanthomonas* species are quite widespread. They cause tissue necrosis in the form of parenchymal, vascular, or systemic diseases. Galls or tumors (hyperplastic diseases) are not known to be incited by xanthomonads.

Examples of *Xanthomonas* plant diseases are:

- Black rot of crucifers caused by *X. campestris*
- Common bacterial blight of bean caused by *X. phaseoli*
- Bacterial blight (or black arm) of cotton caused by *X. malvacearum*
- Bacterial leaf blight (or kresek) of rice caused by *X. oryzae.*

Agrobacterium Conn

General characteristics of the genus:

Rods 0.8 by 1.5–3.0 µm. Motile by one to four peritrichous flagella (flagella which are uniformly arranged over the surface of the bacterial cell). Non-sporing. Gram-negative. Growth on carbohydrate-containing media usually accompanied by copious extracellular, polysaccharide slime. Colonies non-pigmented. Metabolism respiratory. Aerobic, but able to grow under reduced oxygen tensions in plant tissue.

The genus *Agrobacterium* comprises only 4 species: *A. tumefaciens*, *A. radiobacter*, *A. rhizogenes* and *A. rubi*. They are all successful soil inhabitants. *A. radiobacter* is not pathogenic, but the other three species cause diseases which are characterized by an uncontrolled growth and proliferation of the attacked plant tissue. The resulting disorders are called "crown gall" (caused by *A. tumefaciens* on many plants), galls on raspberries (caused by *A. rubi*) and hairy root (caused by *A. rhizogenes* on nursery stock).

Crown gall tumors have certain similarities to human and animal cancers and therefore *Agrobacterium tumefaciens* is extensively studied not only by phytopathologists but also by medically oriented scientists. The crown gall disease indeed affords a unique chance to investigate at the molecular level how a normal cell is converted into a tumor cell. Recent findings show that for tumor genesis, at least in the case of crown gall, the transfer of genetic material from the bacteria to the plant cells is involved [9].

It is interesting to note that the pathogenic agrobacteria are in many ways similar to the members of the genus *Rhizobium*, whose taxonomic position is quite close. The rhizobia also live in the soil, whence they can attack the roots of certain plants, which then react with hypertrophic growth (root nodules on leguminous plants). However, because the rhizobia can fix nitrogen inside these nodules they are regarded as being beneficial and not pathogenic.

Erwinia Winslow et al.

General characteristics of the genus:

Cells predominantly single, straight rods, 0.5–1.0 by 1.0–3.0 µm. Motile by peritrichous flagella. Gram-negative. Facultatively anaerobic but anaerobic growth by some species is weak.

The genus *Erwinia*, named after the famous American phytobacteriologist Erwin F. Smith (1854–1927), comprises bacteria which cause plant diseases of completely divergent appearance: necrosis, wilts, and soft rots. Many workers feel that the capacity of causing soft rot is something so particular that the soft rot agents should be placed into a special genus. This idea was not adopted by Dye [10], who proposed instead a compromise by retaining one genus but recognizing three clusters of organisms within it. Dye's broad divisions of the genus into the "amylovora", "carotovora", and "herbicola" groups found wide acceptance and were incorporated into the 8th edition of Bergey's Manual.

Within the "amylovora" group we find the dreaded pathogen *E. amylovora* which causes fire blight of apple and pear (see below). Also some other incitants of tree diseases (*E. salicis*, *E. nigrifluens*, *E. quercina*, and *E. rubrifaciens*), as well as *E. tracheiphila*, a species causing vascular wilt of cucurbits, belong to this group.

The "carotovora" group consists of *E. carotovora*, which causes soft rot particularly on potatoes, and some organisms considered to be closely related to this pathogen (among others *E. chrysanthemi*).

The "herbicola" group is named after *E. herbicola*, which is a rather protean species. Its members occur quite commonly on the aerial surfaces of plants where they live as saprophytes. In addition they are frequently isolated from plant disease lesions and this gave rise to the speculation that strains of *E. herbicola* are involved in phytopathological situations as synergists or antagonists of overt bacterial phytopathogens [11]. *E. stewartii*, the incitant of Stewart's wilt of corn, and *E. uredovora*, which attacks uredia of *Puccinia graminis*, are also *Erwinia* species belonging to the "herbicola" group.

Corynebacterium Lehmann und Neumann

General characteristics of the genus:

Cells are rods, pleomorphic (of variable morphology), but often wedge or club-shaped. Motile or nonmotile; motile cells have 1–3 lateral or polar flagella. Do not form spores. Gram-positive. Plant pathogenic species are strictly aerobic. Growth in most media is slow. Growth optimum at about 23 °C.

For diagnostic purposes it is important to remember that the corynebacteria are gram-positive and that, because of their slow growth, colonies do not form on agar plates until after the third or fourth day of incubation. The genus comprises beside the plant pathogenic species also human pathogens (e.g. *Corynebacterium diphtheriae*) and saprophytes. The plant disease symptoms incited by the corynebacteria include vascular wilts, hypertrophy and, in a few instances, slimy heads of grain. Important plant pathogenic representatives of the genus *Corynebacterium* are:

- *C. michiganense*, incitant of bacterial canker of tomatoes.
- *C. sepedonicum*, incitant of ring rot of potatoes.
- *C. insidiosus*, incitant of bacterial wilt of alfalfa.

Streptomyces Waksman and Henric

General characteristics of the genus:

Slender, branched hyphae with few or no cross walls, 0.5–2 μm in diameter. Chains of spores are formed (*Streptomyces* is the only spore-forming genus among the phytopathogenic bacteria). On nutrient agar, colonies are small, at first with smooth surface, later with a weft of aerial mycelium. Many strains produce one or more antibiotics highly active against other microorganisms. All species are aerobic soil inhabitants. Gram-positive.

The *Streptomyces* species do not fit into the general conception of bacteria, because they show filamentous growth similar to the mycelium of fungi. Since, however, they possess nucleoids (non-membrane-bound chromosomal material) rather than true nuclei, these organisms are, in spite of their morphology, more correctly classified as bacteria than fungi. Some species of *Streptomyces* affect subterranean parts of plants, leading to scabby superficial lesions (common scab of potatoes incited by *S. scabies* and soil rot of sweet potato incited by *S. ipomoeae*).

Mycoplasmalike Organisms [12–14]

It was only in 1967 that evidence was put forward about the likely existence of a new bacterial group of plant disease agents appearing "mycoplasma-like". This group of pathogens is associated with diseases characterized mainly by "yellows-type" symptoms, which include chlorosis of leaves, phyllody of flowers, and/or growth abnormalities leading to witches' broom and reduction of leaf size. The causal agents of these diseases, the so-called mycoplasmalike organisms or MLO, grow in their host plants generally in the sap of the phloem sieve tubes or in the cytoplasma of phloem parenchyma cells. They had been overlooked, or rather confused with viruses, for a long time, because they pass through filters that retain bacteria and they cannot be cultivated *in vitro* (an exception to this is *Spiroplasma citri*, the only plant mycoplasma disease agent so far to have been successfully isolated and properly characterized). The erroneously assumed viral etiology was also supported by the fact that the "yellows-type" diseases are transmissible by grafting and in most cases by leafhoppers.

Our information about MLO and the diseases they can cause is still incomplete and based primarily on the following indirect evidence:

1. Plant mycoplasma diseases can be diminished by treatment with antibiotics such as tetracycline (normally, plants infected with viruses do not respond to antibiotics).

2. The bodies found in the phloem of diseased plants are indistinguishable in their fine structure and morphological characteristics from certain well-defined species of animal mycoplasmas. Upon tetracycline treatment these bodies disappear.

The crucial tests, however, concerning identification of MLO in diseased plants require successful isolation and cultivation followed by comparative studies with known mycoplasmas. Until improved isolation techniques and adequate media for cultivation are found, an unequivocal determination and classification of the plant mycoplasmas is not possible and the term MLO rather than mycoplasmas should be used further.

A clear definition of the MLO cannot be given, but the main characteristics of the mycoplasmas in general are the following:

Members of the order Mycoplasmatales are procaryotic microorganisms. They lack a rigid cell wall and their cytoplasm is bounded only by a delicate membrane. This leads to a highly pleomorphic morphology: mycoplasmas range from small spherical particles of about 100 nm diameter to large irregular bodies of 1 μm or more. Also filamentous or branched myceloid structures are formed. Their mode of reproduction is still uncertain. They are gram-negative, insensitive to penicillin and usually nonmotile. The species which can be grown on agar media form minute colonies with typical "fried egg" appearance. They resemble L-form bacteria (mutants of bacteria which have lost their cell wall), but they are probably phylogenetically and taxonomically unrelated to these bacteria.

There is a great number of plant diseases associated with MLO. Many of them, such as aster yellows, corn stunt, citrus stubborn, pear decline and coconut lethal yellowing have serious economic consequences.

Rickettsialike Organisms [15, 16]

The first report in 1967 about the presence of mycoplasmalike bodies in the sieve elements of "yellows-type" diseased plants triggered many investigations on this plant disease category formerly thought to be of viral origin. These studies led to the discovery of yet another new group of plant disease agents: the rickettsialike organisms (RLO).

Together with the mycoplasmas, the rickettsias belong to the smallest microorganisms known. They are also pleomorphic, but they possess, in contrast to the mycoplasmas, a definite cell wall. As with most MLO the RLO cannot be isolated and cultured in vitro.

The phytopathological situation with the diseases caused by RLO is in many respects comparable to the MLO plant disorders. The absolute proof that the RLO are the causal agents of certain plant diseases, i.e. the fulfillment of Koch's postulates, has not yet been accomplished. However, the electron microscopic evidence of RLO in diseased plants and the possibility of curing such plants with certain antibiotics implies strongly that rickettsias can be plant pathogens. Also with respect to symptomatology the RLO resemble the MLO. They also cause chlorosis of the foliage and/or growth abnormalities on the attacked plants. It has been known for a long time that rickettsias live as symbionts or parasites in insects, and it is not surprising therefore that insects, mainly leafhoppers, serve as vectors of RLO.

While the classic phytopathogenic bacteria (belonging to the genera *Pseudomonas*, *Xanthomonas*, *Erwinia*, *Agrobacterium*, and *Corynebacterium*) are incapable of penetrating into living host cells but are found only in the intercellular spaces or the xylem vessels, the RLO reside only in the interior of the cells of their hosts. They are specialized either to the phloem or the xylem cells. Other diseases have non-tissue-restricted RLO associated with them.

The RLO are not yet clearly defined, but the main characteristics of rickettsias can be summarized as follows:

> The majority of Rickettsiales are rod-shaped, coccoid and often pleomorphic microorganisms which have a typical bacterial cell wall, no flagella, and are gram-negative. Their size lies in the order of 200–600 nm by 1000–3200 nm. They multiply only inside host cells and may be cultivated in living tissue such as embryonated chicken eggs or vertebrate cell cultures. All are regarded as parasitic or mutualistic. They are often associated with various organs of arthropods, which may act as vectors or primary hosts. They can cause disease in man (e.g. typhus or related ills) or in other vertebrate and invertebrate hosts.

So far about 30 diseases have been described as being caused by RLO. Perhaps because of its economic impact, Pierce's disease of grapevine is the most widely studied. Among others, phony disease of peach and citrus greening are also RLO diseases of considerable importance.

4. Examples of Bacterial Plant Diseases

4.1. Bacterial Leaf Blight of Rice [17, 18]

History, Distribution and Importance

Bacterial leaf blight (BLB) of rice caused by *Xanthomonas oryzae* was first observed in the latter part of the last century in various localities of Southern Japan. During the decades to come the distribution of BLB increased markedly and by 1960 the disease was known to occur in practically all provinces of Japan. In 1950 a bacterial rice disease called "kresek" was described in Indonesia. Later it turned out that "kresek" is a severe form of BLB and that it occurs in various parts of the tropics. Only in 1959 was BLB reported from India, but it is likely that the disease had been present there before. Today BLB is regarded as one of the most serious and widespread diseases of rice in India and also in other parts of Southeast Asia.

Kauffman et al. (1973) [19] expressed the opinion that BLB had become one of the major rice disease problems in Asia because of the introduction and widespread growing of new, high-yielding rice varieties such as TN1 and IR8. Although the disease had been present in most of the tropical Asian countries earlier, the genetic susceptibility of these new rice varieties and the intensive method of cultivation acted together to aggravate the disease.

In a general discussion on the possible danger of plant diseases presently limited to a few countries but having potential international importance, BLB is listed among the "highly threatening plant diseases" [20]. The speculation that BLB might not remain confined to Asia has already come true: recently the disease has been observed in the Caribbean regions as well as in several countries of South America [21].

A quantitative assessment of the damage done by BLB is difficult. In Japan 100,000 to 400,000 hectares are affected annually by the disease (see Fig. 1) and yield losses range from 20 to 30% in severely infected fields. Under tropical climatic conditions, in the Philippines and Indonesia, the disease can be even more destructive. From India, yield losses of up to 60% have been reported [22]. Recently the potential yield losses of rice due to BLB during the monsoon season were determined in the field for several rice varieties [23, 24]. In comparison to the healthy plots where the plants were protected by sprays of aminothiadiazole bactericides, the highly susceptible varieties "Bala" and "Karuna" suffered losses from BLB attack equivalent to 74% and 56% respectively.

Symptoms (cf. Fig. 3A)

In the temperate regions BLB usually becomes noticeable at the heading stage, but under severe disease conditions the symptoms may also show earlier. BLB is a typical vascular bacterial disease (tracheobacteriosis) and as the bacteria grow and multiply in the xylem the visible appearance of the disease consists of lesions on the leaves along the veins. These lesions begin at the margin of the blades, a few centimeters from the tip, as water-soaked stripes, then enlarge both in length and

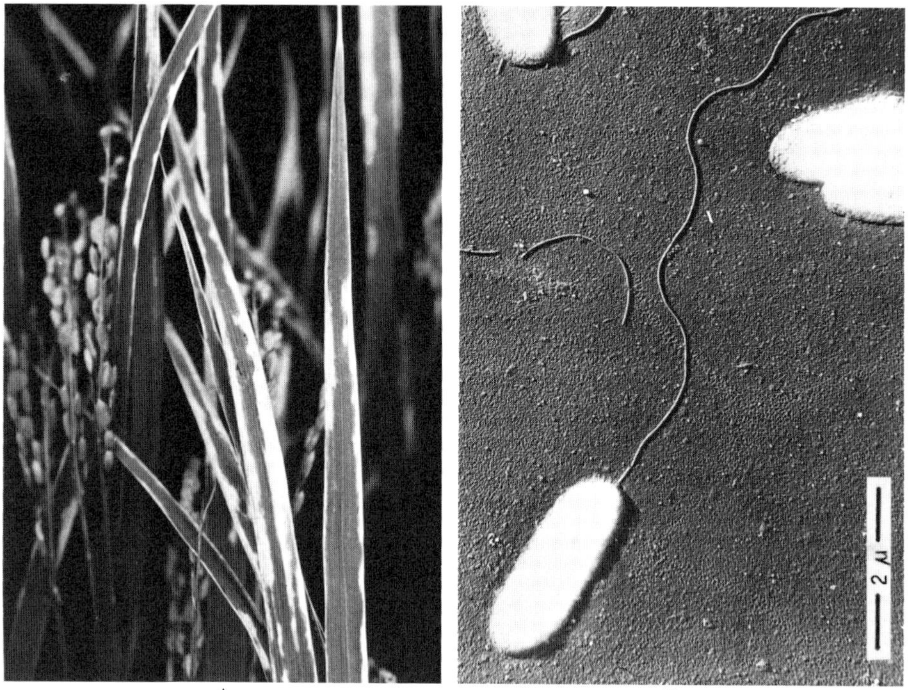

A B

Fig. 3A and B. Bacterial leaf blight of rice. **A** Infected rice plants showing wavy lesions along the edges of the leaf blades. **B** Single cell of the pathogen *Xanthomonas oryzae* equipped with one polar flagellum. (Photos courtesy S. Wakimoto)

width and turn yellow with wavy edges. They may later expand to cover much of the leaf, which turns whitish or greyish and dies. Also the leaf sheaths of the more susceptible varieties may be affected. On such susceptible varieties, infected blades can wilt, roll, and dry up while still remaining green. On the surface of young lesions, milky drops of bacterial masses may be observed in the early morning. In the tropics, a very severe symptom type of BLB, called kresek, can occur. It is the consequence of a strong systemic infection resulting in withering and death of leaves or whole young plants. In older plants, the leaves become pale yellow.

Sometimes the BLB symptoms are not very specific and can be confused with various other leaf diseases or physiological disorders. A simple test may help to identify the disease. If a small piece of a BLB-affected blade is cut out and examined under the microscope in a drop of water, bacteria can be observed streaming from the vascular bundles. In the field one may collect a few diseased plants, cut them, and place them into a test tube filled with water. In case of BLB, bacterial ooze will soon flow from the cut ends into the water, which becomes cloudy.

The Pathogen and Its Transmission (cf. Fig. 3B)

The causal organism of BLB is *X. oryzae*, a typical xanthomonad, as briefly described in Sect. 3.2. Its physiological characteristics can be found in the literature [25]. The species is split up into several races, which differ considerably in virulence. In general, isolates from the tropics are more virulent than those from Japan and Australia. Race variation differing in virulence is an important aspect of BLB, especially with regard to the breeding of resistant rice varieties or screening for bactericides.

X. oryzae can live outside its host in the soil for up to three months; however, soil is not considered to be an important source of infection, as the short survival period means that the pathogen dies before the next year's crop is planted. Also infected seed as a possible source of infection is not believed to be an important factor because the bacteria on seed rapidly decrease, especially at high temperatures. The most important source of primary inoculum is probably the rhizosphere of weed hosts, particularly *Leersia sayanuka* and rice stubble. Here the bacteria survive for a long time and from here they are spread by irrigation water to the rice fields.

Points where the pathogen invades the vascular system of the rice plant are hydathodes (water pores) on the leaves, wounds at leaf blades, and growth cracks caused by new roots at the base of the leaf sheath. Penetration through stomata results in the buildup of bacteria in the intercellular spaces, but it is not until they have been exuded onto the leaf surface and readmitted through the hydathodes and hence into the vascular system that symptoms of the disease appear.

In Southeast Asia, the frequent occurrence of typhoons and rainstorms are major factors in the dissemination of *X. oryzae*. The storms not only effectively spread the pathogen, but also cause wounds on rice plants through which the bacteria can penetrate.

Control

Apart from the use of resistant varieties, control measures include elimination of volunteer plants and weed hosts between crops, careful attention to nursery conditions, especially avoidance of flooding to reduce chances of infection before planting out, and selection of seed from disease-free plants. Chemical control measures are used to some extent, especially in Japan, but the chemicals available at present are not satisfactory. The products most widely applied are Phenadine and Sankel; Oryzemate, a *Piricularia* fungicide with secondary bactericidal properties, and others are also used.

4.2. Fire Blight of Pear and Apple [26–28]

History, Distribution and Importance

Fire blight, caused by *Erwinia amylovora* (Burrill) Winslow et al., is regarded as one of the most destructive diseases of pear and to a lesser extent of apple. Quite a number of ornamental shrubs, all belonging to the rosaceous family, are also susceptible (*Cotoneaster*, *Pyracantha*, *Crataegus*, *Sorbus*, etc.). The disease is much feared because it not only can reduce drastically the yields of a current year's crop, but also has the potential to destroy whole trees and orchards.

Fire blight is believed to be an indigenous plant disease of North America, where it must have occurred for a long time on native plants such as crab apple and hawthorn. Upon the introduction of apple and pear by the European settlers the pathogen spread readily from the wild hosts to the cultivated plants.

For two reasons fire blight represents a landmark in the annals of phytobacteriology:

1) it was the first plant disease in which a bacterium was shown to be the inciting agent (the crucial discovery was made in 1878 by T. J. Burrill), and

2) the impact of the disease was at many places, where it first appeared, of exceptional dimensions.

The fire blight epidemic, which has still not come to an end, started on the east coast of the USA, spread westward over the Allegheny Mountains into the Mississippi River Valley, and about 1840 reached the Midwestern states of Ohio, Indiana and Illinois, where many pear orchards were completely ruined. For many years the vast expanse of the Great Plains, as well as the Rocky Mountains, provided an effective barrier for the movement of the bacterium, preventing its spread to the Pacific Coast. But suddenly, near the end of the 19th century, fire blight made its appearance in the pear orchards of California, causing havoc such as had seldom been seen before. Between 1901 and 1909 about 700,000 pear trees, or roughly one-third of all the trees grown at that time in California, were destroyed. In certain counties the pear industry was practically wiped out and the growers were hit so badly that a major pear industry was never re-established there. In 1957 fire blight was first noticed in England and later it was disseminated (probably by birds) to the Netherlands, Denmark, and Northern Germany. Because of the climatic conditions in these Northern European areas, *E. amylovora* has not caused much damage to Continental pear growers so far; however, this might change in the near future. Recently fire blight has reached a fruit-growing area south of Bordeaux [29] and it is likely that the disease will spread further, wherever the pathogen can thrive. Under these auspices the considerable pear industry in the Rhône Valley of France, in Northern Italy, and in Spain must be regarded as in great danger. Fire blight has been identified also in other countries such as Canada, New Zealand, and certain Central and South American states. Reports from other parts of the world exist, too, but they have not been confirmed and probably involve confusion with *P. syringae.*

Symptoms (cf. Fig. 4)

The first sign of fire blight on pear is the killing of blossoms and/or the distal ends of young shoots. The latter usually bend in a characteristic "shepherd's crook" pattern. As the bacteria move downwards in the bark, dieback extends on twigs and water sprouts, killing the tissue rapidly and turning it a brown color. The dead leaves of such twigs persist, giving the appearance of having been scorched. When the disease progresses to limbs or the trunk of the tree, the green bark turns brown or black and canker-like wounds with cracks are formed. The cankers may girdle twigs and branches, and, as a result, those parts of the trec above the girdle usually die. Especially under warm and humid weather conditions, amber-colored droplets of bacterial exudate can be observed at the margins of cankers and on recently

A B C

Fig. 4A–C. Fire blight in pear and apple. **A** Severe oozing of infected succulent terminal shoot (Bartlett pear). **B** Young blight canker on branch of Golden Delicious apple tree; infection apparently started in young shoot and spread into main limb. **C** Twig blight in apple tree showing necrotic leaves attached to branches. (Photos courtesy T. van der Zwet)

invaded twigs. Diseased fruits can also be observed. They are at first water-soaked, then turn brown, shrivel and mummify, and finally turn black. Like the affected leaves, dead fruits may cling to the tree for several months.

The pathogen and its life cycle (cf. Fig. 5A)

The causal organism of fire blight, *Erwinia amylovora*, is a rod-shaped bacterium. It is facultatively anaerobic and motile, having flagella over its entire surface (peritrichous). Some characteristics of the genus *Erwinia* are described in Sect. 3.2; information on bacteriological aspects of the species can be found in the literature [10, 30].

The pathogen overwinters primarily at the outer edges of cankers formed the previous season. In spring, the bacteria in these "holdover" cankers begin to multiply and exude through lenticels and cracks to the surface of the bark. This slimy, amber-colored exudate, called "bacterial ooze", usually appears first at about the time when the pear blossoms are opening. Various insects, such as wasps, honeybees, and flies, are attracted to the sweet, sticky exudate and pick it up. When these insects visit flowers afterwards, they leave some of the bacteria-containing exudate in the nectar of the flower cups. There the bacteria find ideal conditions for their propagation. They multiply and invade the plant tissue through the nectaries, pistils, and sepals of the flowers.

For the dispersal of *E. amylovora*, especially over short distances, wind-driven rain splashes are also important. Beside flower infections, this type of dissemi-

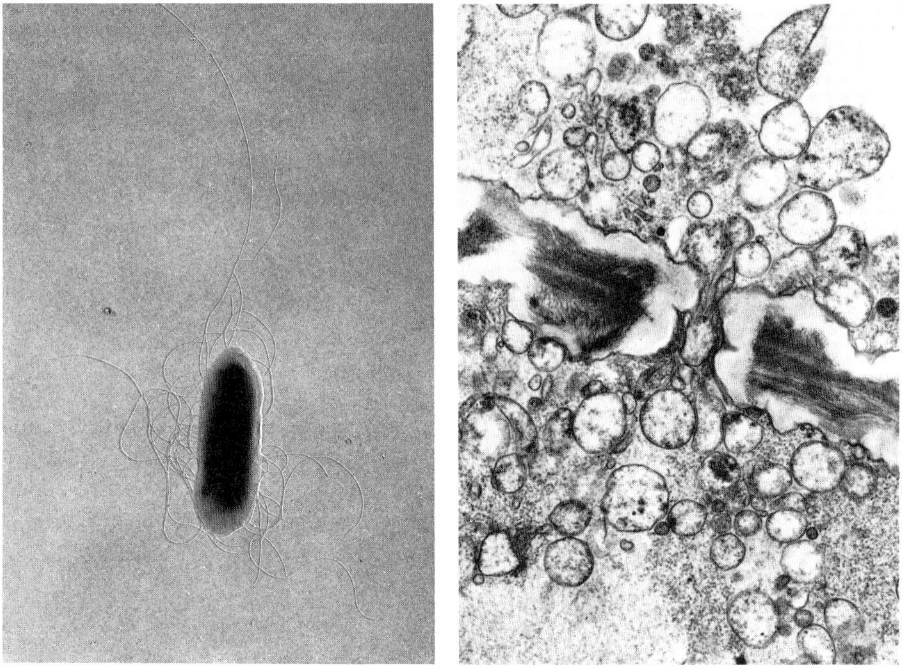

A B

Fig. 5. A Virulent cell of *Erwinia amylovora* with abundant peritrichous flagella (photo courtesy P. Y. Huang). **B** Mycoplasmalike organisms in sieve element of a coconut palm affected by lethal yellowing. The highly plastic bodies of the pathogen are apparently passing through a sieve-plate pore lined with callose. (Photo courtesy M. V. Parthasarathy)

nation leads mainly to shoot infections, as the bacteria can also penetrate through the stomata and wounds on leaves and succulent twigs. Once inside the flower or shoot tissue the bacteria multiply and move quickly, primarily through the intercellular spaces but also through the dissolved middle lamella. In the twigs they soon cause collapse and breakdown of cortical cells, forming large cavities. Invasion of large twigs and branches is restricted primarily to the cortex, where girdling cankers can develop.

In pear the bacteria can travel from the top of a tree to the base of the trunk within a few weeks, killing it to the ground in one season. Apples as a whole are more resistant; usually only the annual shoots become infected. The susceptibility of other host plants (*Cotoneaster*, *Crataegus*, *Pyracantha*, etc.) varies considerably according to the species and/or variety.

Control

Present-day control measures are still based to a large extent on good cultural practices. Since fire blight development is greatly favored by the presence of young, succulent tissue, orchards should be planted on well-drained soils and caution must be exercised with nitrogenous fertilizers to avoid excessively vigorous tree growth. Careful sanitation is of special importance: blighted twigs and overwintering cankers must be removed and destroyed before the blossoms open. The excision of infected plant parts should be done with cuts at least 20–30 cm below the point of visible infection. If this pruning is accomplished during dormancy, there is no need for surface sterilization of tools; during the growth season, however, this should be done by dipping the cutting tools in a disinfectant between cuts. To diminish the source of primary inoculum, alternate host plants, e.g. hawthorn hedges, should also be removed from the vicinity of a pear orchard. In areas where fire blight is common, commercial growers still place heavy reliance on spring blossom sprays with chemicals. The choice of materials is minimal: only copper and certain antibiotics such as streptomycin and oxytetracycline have been used. Both types of chemical, however, have distinct shortcomings. Copper sprays damage the lenticels of developing pome fruit and lead to russeting, especially with valuable pear varieties. Streptomycin can provoke the buildup of resistant *E. amylovora* strains, making this antibiotic completely ineffective in the particular orchard. In addition, for medical considerations, antibiotics are banned in Europe as plant protectants and in the United States streptomycin is permitted for use on apples only up to fifty days and on pears up to thirty days before harvest.

Studies on the influence of climatic factors on fire blight development have led to promising techniques for blight forecasting, enabling pear growers to cut down the number of sprays to the times when they are really needed [31]. The role of saprophytic bacteria in reducing populations of *E. amylovora* has still not been sufficiently elucidated to give farmers the chance of using biological control measures.

4.3. Lethal Yellowing of Coconut Palms

History, Distribution, and Importance

Lethal yellowing is a severe disease of coconut palms caused by a mycoplasmalike organism. The disease has been known for a long time in the Caribbean, but its devastating occurrence has also been reported from West Africa (Togo, Ghana, Nigeria) and from Tanzania. In the United States, lethal yellowing was first noticed in 1955 in Key West, Florida, where in the course of the following five years about three-fourths of the coconut palms were killed. On the mainland of Florida the disease appeared for the first time in the autumn of 1971, and by October 1973 it had killed an estimated 15,000 trees. By 1976 Florida had lost some 300,000 of these majestic trees. The disappearance of the majority of the tall coconut palms in the South of Florida changed the characteristic scenery of this landscape considerably. In the tropics, the sudden loss of coconut palms can be catastrophic, especially to small farmers, who are dependent on coco production and who under normal circumstances expect to crop their valuable trees for several decades.

Symptoms (cf. Fig. 6)

Symptoms of lethal yellowing normally begin with premature drop of coconuts of any size, followed by necrosis of newly developing inflorescences. A yellow "flag" leaf may also appear in the middle of the crown at this time. Then the lower fronds

Fig. 6. Lethal yellowing of coconut palms showing trees in late stage of decline as a result of infection with mycoplasmalike organisms (photo courtesy K. Heinze)

turn yellow and the yellowing progresses upward from the older to the younger leaves until all fronds are affected. While the younger leaves become yellowed, the older ones die prematurely, turn brown, but still cling to the tree. Quite soon all the leaves die, as does the vegetative bud. Eventually the remains of the crown of the palm fall away, leaving nothing but the tall trunk of the tree which by now looks like a telegraph pole. In this way affected palms are usually killed within three to six months of the initial appearance of symptoms.

The Pathogen (cf. Fig. 5B) [32]

The causal agent of lethal yellowing is a mycoplasmalike organism morphologically similar to other such organisms observed in plants. It is highly polymorphic, varies from 0.4 μm to 2 μm in size, and is bounded by a typical triple-layered membrane about 80 Å thick. In diseased palm trees the pathogen is confined to mature sieve elements and probably moves through these with the assimilate stream to the "sink" regions such as young inflorescences and young leaves. Many questions concerning the agent of lethal yellowing are still open, as this microorganism cannot be cultured and studied in cell-free media and its vector is not yet known.

Control [33, 34]

Control of plant mycoplasma diseases relies mainly on indirect methods, that is to say the avoidance of infection. Thus protection of coconut palms against lethal yellowing is attempted primarily by sanitation measures, i.e. removal and burning of diseased trees as soon as symptoms appear in order to reduce the source of inoculum. Certain Malayan dwarf varieties of coconut palms appear to be resistant or immune to lethal yellowing and thousands of such trees are now planted as a replacement for susceptible coconut palms, particularly the Jamaican Tall variety, which is threatened or killed by lethal yellowing.

Initially the application of certain antibiotics was attempted to provide evidence that mycoplasmalike agents are the cause of yellows-type diseases formerly thought to be incited by viruses. In the case of lethal yellowing, however, the results obtained by treating infected palm trees with solutions of oxytetracycline were so encouraging that the use of this chemical was suggested for field control. Extensive trials in Florida did indeed prove oxytetracycline therapy to be gratifyingly successful. The basic method is to drill holes, angled downwards, into the trunk and to inject, under pressure or by gravity flow, an aqueous solution of the antibiotic into the trees (at the rate of 1–6 g of oxytetracycline per tree). The palms respond to this treatment by remission of lethal yellowing symptoms, but the improvement is only temporary; the trees are not cured, and the treatments have to be repeated at four-month intervals. Otherwise the symptoms reappear and the palms will succumb to the disease.

The level of lethal yellowing control thus achieved is considered to be economically viable for Florida, and oxytetracycline (Terramycin) is registered for this use in the United States. However, in Florida the coconut palm is grown primarily as an ornamental tree with a high amenity value; whether chemotherapy of lethal yellowing is feasible on a plantation crop must be carefully evaluated.

5. Control of Bacterial Plant Diseases

5.1. Biological Aspects

Plant disease control measures should take into account the special features of the host-parasite relationship in each case. Not all bacterial plant diseases follow the same biological pattern of development, but some characteristics can be generalized and used successfully for control purposes.

As has been mentioned earlier, the bacterial disease agents form no special resting structures and are therefore short-lived in the absence of their host plants. On the other hand the phytopathogenic bacteria are well-adapted for survival in close association with their hosts: they perennate in or on tubers, bulbs, cuttings, seed, and the rhizosphere of certain plants. In addition they have a tremendous capacity for multiplying under certain climatic conditions inside their hosts. Thus a vast reserve of fresh inoculum can be quickly available for secondary dispersal within a few days. Given the presence of a primary inoculum in a crop, bacterial diseases can very soon reach epidemic proportions when prolonged wet weather prevails.

Because of the difficulty in stopping secondary spread, control measures against bacterial diseases of annual crops tend to focus on the exclusion of primary infection sources. This concept of control has been remarkably successful, especially in those cases where the bacteria perennate in vegetative structures which are used for the propagation of the respective crop. Thus it has been possible to reduce the occurrence of potato ring rot caused by *C. sepedonicum* by field certification schemes and seed production in disease-free areas. The paramount importance of bacteria-free planting material is also well known to the growers of ornamental plants (*Pelargonium*, *Begonia*, *Chrysanthemum*, etc.), which are vegetatively propagated and often affected by bacterial and mycoplasma diseases [35]. The selection of healthy stock for the production of cuttings is not always easy, as latent infections without visible symptoms are quite common. Special diagnostic microscope techniques and also heat treatments or meristem cultures, if uninfected mother plants are not available, have to be used. With seedborne bacterial diseases the exclusion of primary infection sources can be accomplished by concentrating seed production in arid areas, which are adverse for bacterial disease development. In combination with field inspections and seed indexing this has led to substantial reduction of halo blight (*P. phaseolicola*) and other bacterial diseases of beans [36]. The decrease in rice acreage affected by bacterial leaf blight in Japan in recent years (see Fig. 1) is also said to be the consequence of improved cultural techniques (proper water management and elimination of the alternate host *Leersia sayanuka*), resulting in reduced primary inoculum in the paddy fields.

The logical end point of suppression of primary infection sources is eradication, i.e. the elimination of a pathogen from a region by systematic destruction of all diseased plants. One of the best-known examples of successful eradication of a plant disease is citrus canker (*X. citri*). This very serious bacterial disease of citrus was probably introduced from Japan to the United States in about 1910. The eradication campaigns involved the destruction by burning of about twenty million diseased citrus trees which resulted in apparently complete elimination of the pathogen in the southern states of the United States by 1952 [37]. The eradication of a plant disease is a costly and lengthy procedure and can be envisaged only for a limited geographical area. In addition it is likely to be successful as a means for disease control only if the pathogen involved has a narrow host range and its reintroduction can be prevented. Neither proviso was fulfilled when fire blight reached the European continent. The causal organism of this disease, *E. amylovora*, has a particularly broad host range, including hawthorn and other widespread shrubs which is impossible to check and destroy to the very last infected individual plant. And England, where the disease became endemic after its introduction in 1957, has always remained a source of new infection for the neighbouring countries across the Channel. It is not astonishing, therefore, that extensive eradication programs triggered by the first appearance of fire blight in the Netherlands, Denmark, and Germany were only temporarily successful and eventually failed. Only Poland claims that it has been possible to eradicate the disease from its territory, but on the whole fire blight is certainly firmly established in Europe and will never become extinct.

Resistant cultivars may be the best long-term solution for some bacterial diseases and, in the case, for example, of bacterial wilt of forage grasses (*X. graminis*), which is rapidly and efficiently spread by mowing machines, this is probably the only feasible control method [38]. Breeding programs for resistance against major bacterial diseases such as tobacco wilt (*P. solanacearum*), cotton blight (*X. malvacearum*), gummosis of sugar cane (*X. vasculorum*), and bacterial leaf blight of rice (*X. oryzae*) have been carried out, and partly resistant varieties are already being grown with some success. However, the positive results of this breeding work are quite often jeopardized by the great genetic diversity of the bacterial plant pathogens, which enables them to overcome resistance relatively easily [39, 40].

It has been known for a long time that in a natural habitat phytopathogenic bacteria are antagonized by other microorganisms or even parasitized and killed by special viruses called phages. Understandably, many researchers have been fascinated by this situation, which they have tried to exploit in developing biological control techniques. And, indeed, it has been possible to show a clear control effect in quite a few trials with phages; disease suppression has however been achieved only when a mixture of the phage and the bacterium is applied or when the plants are sprayed with a phage suspension prior to infection with the bacteria. Thus disease therapy by means of phages has turned out to be of no practical value and, since in addition phages may transduce genes in an undesirable way, they cannot be recommended at present as control agents [41].

A recent report on biological elicitation of systemic protection (immunization) of cucumber plants against a bacterial disease agent is highly interesting [42], but whether it will contribute to practical disease control is uncertain. Also the role of saprophytic bacteria in reducing populations of bacterial pathogens is still in a speculative stage and considerably more research effort will be needed before the biological control approach can be successfully applied. There is, however, one outstanding exception to this general statement: biological control of crown gall (*Agrobacterium tumefaciens*) by strain K 84 of *Agrobacterium radiobacter* [43, 44]. The exciting research work on this subject was started about ten years ago by an Australian group of scientists and has already led to an effective crown gall control technique widely used by commercial growers. The method involves basically the control of a pathogen, *A. tumefaciens*, by a closely related nonpathogen, *A. radiobacter*. The control mechanism has been largely elucidated, opening up interesting academic insights and new practical perspectives. The antagonist (strain K 84 of *A. radiobacter*) operates through the production of a bacteriocin that is highly toxic to the crown gall organism. Bacteriocins are defined as bactericidal protein-containing substances which are produced by certain strains of bacteria and are active against some other strains of the same or closely related species. The bacteriocin of *A. radiobacter*, agrocin 84, has been purified and characterized as a 6-N-phosphoramidate of an adenine nucleotide analogue [45].

In practice the antagonism between *A. tumefaciens* and *A. radiobacter* is exploited by dipping the roots of plants before transplanting into a bacterial suspension of live cells of strain K 84. Whether agrocin 84 or another bacteriocin can ever be used directly as a plant protectant is an open question. As biogenic and highly specific products, they seem to possess most of the attributes desired by environmentalists for an ideal pesticide. However, the already observed emergence of bacteriocin-resistant mutants [46], their extremely narrow range of activity, and their complicated chemical structure certainly place restrictions on the future of bacteriocins in agriculture.

5.2. Chemical Control – General Considerations

In spite of all biological preventive measures, outbreaks of bacterial diseases reaching catastrophic proportions occur quite often. Changes in farming practices and the development of resistance against certain bactericides are likely to make these bacterial disease attacks of even greater concern in the near future. For the control of sudden outbreaks of bacterial diseases there is therefore a clear need on the part of growers for effective chemicals, but this need is badly met today [47, 48]. Only a few agrobactericides are available to farmers and these are afflicted with distinct shortcomings. They are mostly general but not very active biocides, fungicides with bactericidal side effects or antibiotics used primarily as human and veterinary medicines.

In contrast to the control of fungal diseases, which has been revolutionized in recent years by the introduction of highly active and systemic products, no comparable development has taken place in the chemical control of bacterial plant diseases. This is astonishing, especially in the light of the success in medical bactericide research which has brought forth a multitude of very potent anti-bacterial agents. When looking for the reasons for the deplorable situation in chemical bacterial disease control, two aspects of phytobacteriology should be considered:

1) the relatively small research efforts made by industry in the field of bactericides as a consequence of the limited market for such products, and

2) the special requirements which have to be met by a chemical if it is to be an effective plant bactericide.

An inquiry made in 1970 revealed that out of 120 agricultural chemical companies in the United States and Canada, only eight had screening programs in progress which included plant pathogenic bacteria. Most companies only screened compounds against bacteria if such chemicals were known to have fungicidal activity [3]. In Europe the chemical companies which have *in vivo* bactericide screening tests in operation may be counted on one hand. Considerable efforts to find new agrobactericides are certainly being made in Japan. It is noteworthy that the main targets there are citrus canker and bacterial leaf blight of rice and that the attitude towards antibiotics is more optimistic than in the Western hemisphere. A declining interest in bactericide research in recent years is reflected by the number of published patents. While in 1970 about 100 new bactericide patents were published worldwide (with a Japanese share of about 80%), this number dropped to five in 1974 and has recovered only slightly in the past few years.

In order to increase the success rate in agrobactericide research, closer links between phytobacteriologists and medical bacteriologists have been advocated

[49]. It was felt that antibacterials that are not developed for pharmaceutical purposes might be successfully used in agriculture because plant pathogenic bacteria and some human bacterial pathogens are closely related. This approach, however, must be questioned as many pharmaceutical chemotherapeutic agents (e.g. sulfonamides, nitrofurans, and many antibiotics) have proved to be inactive or only weakly active *in vivo* against bacterial plant diseases.

The real reason for the limited plant protection effect of most clinically used bactericides is no doubt to be found in the very structure of plants. The lack of a circulatory system (analogous to the blood and lymph vessels) and the structure of the plant surface, through which an effective plant protection agent has to pass, clearly demand that the substance not only be bactericidally effective but also possess certain extra physicochemical properties which can be summed up in the expression "systemic activity". The capacity for penetration and translocation, combined with good plant tolerance, is of particular importance for the efficacy of a plant protection bactericide since the phytopathogenic bacteria vegetate inside the plant and must be attacked there. The complexity of these requirements makes it easier to understand that it cannot be a simple task to develop a good agent against bacterioses. At the same time it must be emphasized that the search for ideal bactericides is by no means a hopeless one. The experimental products TF 128 and TF 130 (aminothiadiazoles, see below) show impressively that simple chemical compounds which almost ideally meet the field performance requirements adumbrated above do in fact exist and that substances of this kind can be found by means of a simple screening system.

Present-day knowledge about chemical compounds with relevance to the control of bacterioses is summarized in the following pages.

6. Chemistry of Bactericidal Compounds

The antibacterial compounds that follow are listed according to chemical structure. The sequence does not depend on whether the individual product is commercial or experimental. Some experimental or formerly commercial bactericides with interesting chemical or biochemical aspects may be described in greater detail than more widely used but otherwise obsolete ones. The purpose of this approach is to indicate possible areas of investigation and study in the search for better and safer plant bactericides.

6.1. Inorganic, Nonmetallic Compounds

Chlorine dioxide (ClO_2) in the form of a stabilized aqueous solution is recommended for the sterilization of seeds and the disinfection of agricultural equipment, tools, and containers in order to prevent spreading of bacterial plant diseases. Development product "DI-17" by Bio-Cide Chemical Co. Inc., USA [50].

Sodium hypochlorite (NaOCl), as a 10% aqueous solution (Clorox), may be used for the disinfection of pruning shears to prevent fresh infection of fruit trees with fire blight. This method is claimed to be safe and well suited to replace the dangerous use of mercury disinfectants [51].

6.2. Inorganic and Organic Metal Compounds

6.2.1. *Mercury Compounds*

Mercury compounds have served as fungicidal seed dressings since 1910, when mercury chloride was introduced as the first remedy against snow mold (*Fusarium nivale*) in cereals [52]. Subsequently a great number of organomercury compounds were discovered and sold worldwide as broadspectrum fungicides. Nowadays great efforts are being made to replace these highly toxic and persistent mercurials. Mercury compounds are general biocides and it was thus not surprising that some bacterial plant diseases could be controlled to a certain extent by mercury salts and organomercurials. Owing to their high toxicity and phytotoxicity, most of these substances are not suitable for foliar sprayings. They can however be used effectively as seed dressing agents against various seedborne bacterial diseases, the most important of which is bacterial leaf blight of cotton (*Xanthomonas malvacearum*). This occurs in most cotton-growing countries, where an estimated 6 million hectares of production area are affected, causing yield losses of some 10% [53].

In agricultural practice, mercuric chloride, corrosive sublimate ($HgCl_2$) [54] is still used in solution for seed dips. Typical commercial products for dry seed treatments are Calo-Clor, a mixture of 30% mercuric chloride and 60% mercurous chloride, calomel (Hg_2Cl_2) [54] and Calo-Gran, a 2.7% granulate containing 1.8% Hg_2Cl_2 and 0.9% $HgCl_2$ (both Mallinckrodt) and Abavit (Schering), a mixed formulation of mercuric chloride and mercuric iodide. From the wide range of organomercury derivatives (for a survey of chemistry and use see Vol. 2, pp. 137–147 and Vol. 4, pp. 125–128), mixtures of phenylmercury acetate ($C_6H_5HgOCOCH_3$) and ethylmercury chloride (C_2H_5HgCl) under the trade name Agrosan (Chevron, ICI) are used against cotton blight in the Sudan [53], whereas in India two products under the Ceresan (Bayer) label are preferred, namely 2-methoxyethylmercury chloride ($H_3COCH_2CH_2HgCl$) and 2-methoxyethylmercury silicate ($H_3COCH_2CH_2HgOSiO_2H$). In many other African and Asian countries cotton seed treatments with mercurials are still widely used as well. In Japan, organomercurials have been tested in combination with antibiotics as foliar sprays against bacterial leaf blight of rice *(Xanthomonas oryzae)*. A typical product was Shirahagen (Sankyo K.K.), a mixture of phenylmercury acetate with chloramphenicol which for toxicological reasons is no longer in use. In many countries strict registration laws now ban mercury compounds as foliar sprays or restrict their use to seed dressings. For toxicological and residue aspects of mercury compounds see literature [55, 56].

6.2.2. *Copper Compounds*

In 1898, shortly after the discovery of the fungicidal properties of the classical Bordeaux mixture and its introduction for agricultural use by Millardet [57],

Pierce attempted to control walnut bacterial blight *(Xanthomonas juglandis)* with this mixture in California [58]. The efficacy of copper compounds against bacterioses was subsequently confirmed by a number of experiments, and as early as 1909 copper treatments were first used for the control of bacterial leaf blight of rice in Japan [17]. Although practically all bacterial plant diseases are influenced in some way by copper compounds the success rate is rarely satisfactory. The effect is merely protective and in no way curative, making frequent sprayings necessary. This results in high costs, and plants are likely to be damaged by the phytotoxic effects of copper, e.g. russeting, discoloration and cracks, especially on fruits.

Because of the lack of better products and due to their relative cheapness, coppers remain the remedy of choice in some crops (*X. vesicatoria* in peppers [59], *X. phaseoli* in beans [60]). In Japan about 80% of the citrus orchards afflicted with *X. citri* (some 15% of the total acreage) are treated with copper sprayings. It is not possible to estimate the total amount of copper used worldwide as a bactericide as its fungicidal use is prevalent. The chemistry of the various copper compounds generally used for plant protection is fully treated in Vol. 2, pp. 46–49, and Vol. 4, pp. 120–121, and in the literature [61, 62]. The copper products most frequently used as bactericides are briefly reviewed below:

The traditional Bordeaux mixture has the approximate composition

$$[Cu(OH)_2]_x \cdot CaSO_4$$

and has to be prepared immediately before use by mixing copper sulfate and lime solutions. It has good adhesive properties but because of inconvenience in handling it has been partly replaced by "fixed coppers". These are ready-to-use formulations of basic copper salts with appropriate emulsifiers and stickers; they are of extremely fine particle size and give stable spray suspensions.

Tribasic copper sulfate

$$CuSO_4 \cdot 3\,Cu(OH)_2 \cdot n\,H_2O$$

is prepared by the action of aqueous bases (NaOH, Na_2CO_3, etc.) on copper sulfate solutions. More recent procedures prefer the precipitation of the basic sulfate by ammonia [63]. Typical commercial products in the United States are Top-Cop (Stoller Chem. Co.) and Tri-Basic Copper (Cities Service Co.). Similar formulations are Kocide 101 (Kocide Chem. Co.), a wettable powder, and Kocide 404, a flowable formulation of cupric hydroxide ($Cu(OH)_2$). The latter has a particularly fine particle size and is widely recommended against bacterial and fungal diseases [64].

In Europe, most commercial copper formulations for agricultural use contain copper oxychloride, $CuCl_2 \cdot 3Cu(OH)_2 \cdot nH_2O$. This is made by air oxidation of cuprous salts and scrap copper in sodium chloride solution [62]:

$$6\,CuCl + 3\,H_2O + \tfrac{3}{2}\,O_2 \longrightarrow [CuCl_2 \cdot 3\,Cu(OH)_2] + 2\,CuCl_2$$
$$2\,CuCl_2 + 2\,Cu \longrightarrow 4\,CuCl\,.$$

Some of the more commonly known labels are Cupravit (Bayer), Grünkupfer (BASF), Vitigran (Hoechst), Cuprasol (Spiess-Urania), and Wacker 73.

Besides the inorganic copper salts, some salts of organic acids are also in use as bactericides. A typical product in the United States is Citcop 4 E (Cities Service Co.), a mixture of the copper salts of fatty and rosin acids (e.g. copper abietate, linoleate, oleate). It serves as an emulsifiable, noncorrosive foliar protectant with better plant tolerance than inorganic coppers [65].

The exact mode of action of copper compounds against plant pathogenic bacteria is not known. The activity of these products against fungi and their toxic action on fungus spores are discussed in the literature [66]. It would seem that copper merely hardens the plant surface, making it less penetrable for bacteria.

6.3. Dithiocarbonates and Dithiocarbamates

The chemistry and fungicidal activity of derivatives of carbon disulfide, the dithiocarbonates and dithiocarbamates, as well as the more important commercial products, are described in Vol. 2, pp. 59–69 and in Vol. 4, pp. 134–144. General biochemical and toxicological information on this class of compounds can be found in the literature [67–69].

The reaction of dimethylamine with carbon disulfide in the presence of sodium hydroxide and subsequent precipitation with nickel sulfate yields nickel dimethyldithiocarbamate in the form of a green insoluble powder

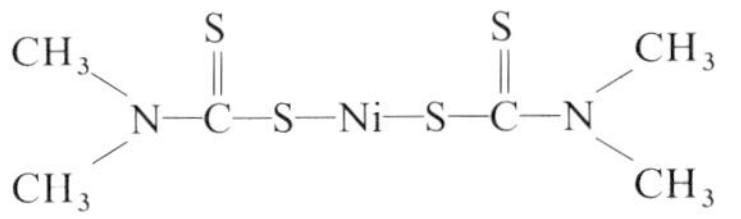

Sankel
acute LD_{50}: 5200 mg/kg (rat)
nontoxic to fish

This compound was developed in 1965 and is sold under the name Sankel by the Japanese companies Mikasa Chemical Industries, Sankyo, and Hokko. It is widely used in Japan against bacterial leaf blight in rice. Although its control of the disease is only 40–60%, nickel dimethyldithiocarbamate is still one of the main rice bactericides, with annual sales of almost 200 t of active ingredient. The standard formulations are a 65% wettable powder and an 8% dust. Early application to nursery beds is reasonably effective in preventing infection. Acrylic polymers have been tried as carriers for the bactericide in an attempt to improve its activity [70].

An analogous compound with similar properties was developed by Bayer AG and Nihon Tokushu Noyaka Seizo K.K. [71].

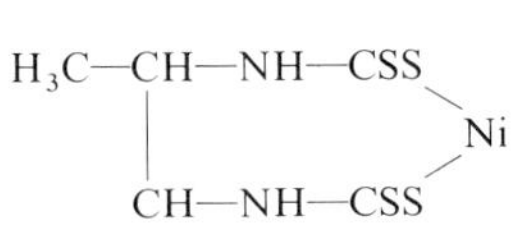

nickel propylene-1,2-bisdithiocarbamate, propikel, Baykel

The specific action of sodium dimethyldithiocarbamate, as a water-soluble model compound, on the *Xanthomonas oryzae* bacterium has been thoroughly investigated by radioactive tracer techniques. The authors have found that the product inhibits the incorporation of nucleosides into nucleic acids and to an even greater extent the process of lipid biosynthesis [72].

Potassium N-hydroxymethyl-N-methyl-dithiocarbamate [73, 74]

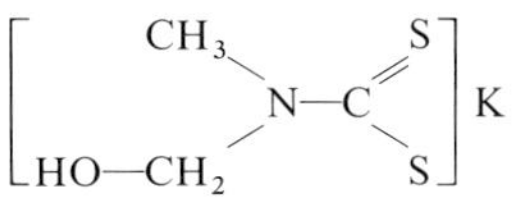

Bunema (Buckman Laboratories)
acute LD_{50}: 590–1032 mg/kg (rat)

This product, with the trade name Bunema, is a water-soluble liquid concentrate containing about 40% active substance. It is recommended for foliar treatments of crops such as tomatoes, celery, peppers, and potatoes against bacteria and also fungi. The chemical is formed, together with some side products, by the reaction of potassium methyldithiocarbamate with formaldehyde and methylamine [75].

For the sake of completeness, cadmium O-n-propyl-dithiocarbonate or cadmium n-propyl-xanthate should be mentioned as a further compound active against *X. oryzae*.

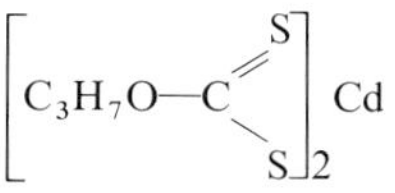

Cadenax
(Tao Agr. Chem. Co., Japan)
acute LD_{50}: 600 mg/kg (rat)

It is no longer on sale owing to poor efficacy and possible toxicological concern linked with the presence of cadmium.

6.4. Alcohols and Other Aliphatic Compounds

2-Bromo-2-nitropropane-1,3-diol [76]

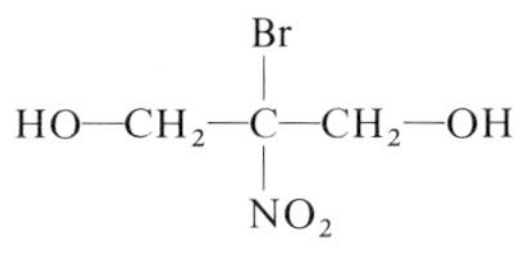

Bronopol
(Boots Co. Ltd.)

This compound, made by bromination of the sodium salt of 2-nitropropane-1,3-diol, is a colorless solid (m.p. 130 °C) with good water solubility [77]. Bronopol is a bacteriostatic substance active against a wide range of plant-pathogenic and other bacteria, with pronounced activity against *X. malvacearum*, the causative agent of bacterial blight of cotton. It is generally recommended as a dry seed dressing (12.5% a.i.) in a concentration of 1 kg per 150 kg of cotton seed. It may be considered a safe substitute (acute LD_{50}: 180–400 mg/kg, rat) for the common mercury treatments of cotton; 10–12% yield increases have been reported from Nigeria [78], where *X. malvacearum* is the only major pathogen of cotton seedlings.

In 1977 a new experimental bactericide was announced by Merck & Co. under the code number ME 135 [79]. Some data from field trials against fire blight (*E. amylovora*) [80] and fungicide tests [81] are reported in the literature.

The compound has the structure

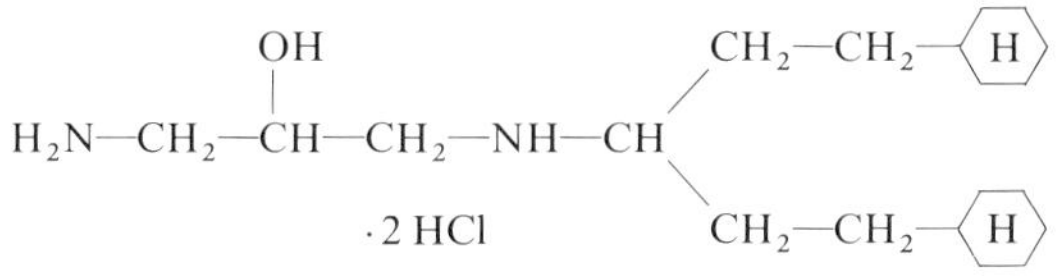

"ME 135"
(Merck)

[1-amino-3-(1,5-dicyclohexyl-3-pentylamino)-2-propanol dihydrochloride] and is prepared as follows [82]:

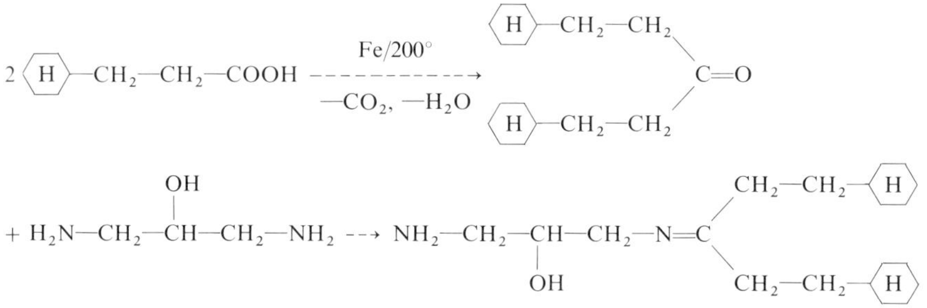

The Schiff base is then reduced catalytically to give the final product. The technical grade compound is water-soluble and contains 80% of active ingredient.

n-Dodecyl-guanidine acetate (dodine, Cyprex, Melprex)

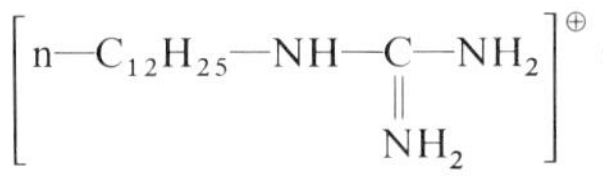

(see Vol. 4, p. 145).

Although this well-known compound is considered primarily a fungicide and is still widely applied against apple scab, it is reported to be active against bacterial leaf spot of stone fruits (*X. pruni*), which can be a serious problem in peach orchards [83]. By using low concentrations and admixing captan the phytotoxic side effects of dodine can be reduced.

6.5. Aromatic Compounds

Phenols are among the oldest disinfectants and antiseptics used in hygiene and surgery. A review of the biological activity and chemistry of phenol disinfectants is to be found in the literature [84, 85]. In plant protection, phenols have only minor importance as most of them are not well tolerated by the plants and the specific bactericidal effect is insufficient.

Bis-(2-hydroxy-3,5,6-trichlorophenyl)-methane, common name hexachlorophene [86],

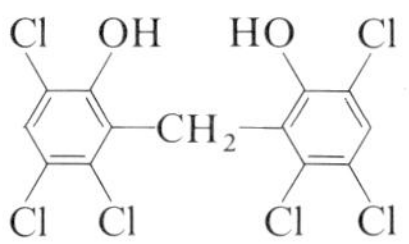

hexachlorophene
acute LD_{50}: 320 mg/kg (rat)

prepared by condensation of trichlorophenol with formaldehyde, is best known as a disinfectant in cosmetics and household products (e.g. G 11, Givaudan). The FDA of the United States has now restricted this use of hexachlorophene because of its toxic properties [172]. In agriculture, hexachlorophene is recommended for the control of some soilborne fungal diseases (e.g. *Rhizoctonia*), as well as bacterial seedling and leaf spot diseases of vegetables and cotton. Agricultural formulations of the product are marketed in the United States by Kalo Laboratories Inc. and Helena Chem. Co. under the trade names Nabac 25 EC or Hexide and Isobac 20 (monosodium salt of hexachlorophene) [87]. The practical value of these products, however, seems to be somewhat limited.

A mixture of 2,4-xylenol and m-cresol

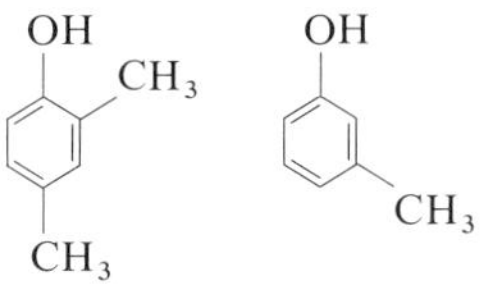

is used for treating bacterial tumors (crown gall, caused by *Agrobacterium tumefaciens*, olive knot, etc.) on woody plants (e.g. almond, apricot, cherry, olive, peach, pear, plum, walnut). The 0.94% paste of this phenol mixture is painted on cut surfaces and galls. The commercial product (now discontinued) is Bacticin (Upjohn) [88].

A new bactericide, N-(2′,3′-dichlorophenyl)-3,4,5,6-tetrachlorophthalamic acid (techloftalam), has recently been introduced by Sankyo Co. Ltd. [89].

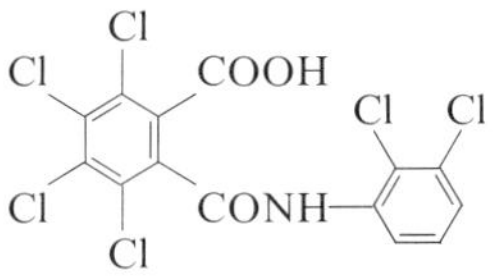

techloftalam,
Codes F 370, SF 7402
(Sankyo)

The compound is active against vascular bacterial diseases, especially leaf blight of rice (*X. oryzae*), but inactive against parenchymatous diseases such as canker of citrus and leaf spots or soft rot of vegetables. Techloftalam is claimed to be a systemic bactericide suitable for soil and foliar treatments. It is not active *in vitro* and only poorly active when the rice plants are artificially infected through wounds. When taken up by the roots, the product becomes slowly active, reaching optimum efficacy about ten days after application. It is said, however, to be highly active in the field under natural disease conditions [90]. In spite of this limited spectrum of activity the product has a considerable market potential for Japan, South Korea, and other rice-growing Asian countries since Sankel and Phenadine are said to be inferior and are ready to be superseded by a new and better product. The production of techloftalam is very simple. It is prepared by the reaction of tetrachlorophthalic anhydride with 2,3-dichloroaniline. The synthesis and the biological use of techloftalam and similar products are covered in patents of Sankyo Co. [91].

6.6. Heterocyclic Compounds

6.6.1. Heterocyclics Containing Nitrogen

Phenazine 5-oxide [92] is the second important rice blight bactericide (after

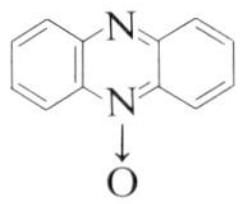

Phenazin
Phenadine
(Meiji Seika K.K. and Nihon Nohyaku Co., Japan)
acute LD_{50}: 12,150 mg/kg (rat)

Sankel) used in Japan, South Korea and other Asian countries. Annual sales in Japan lie between 10 and 20t active ingredient. Japanese exports, mainly to South Korea, exceed 80t annually (figures for 1977, source: Jap. Pesticide Handbook Noyaku Yohran 1978), and Meiji Seika is reported to have exported 1500t of Phenazin to the People's Republic of China in 1978 [93]. The product is used as a 10% wettable powder or as a 1.5% dust formulation. Under severe infection conditions application begins when the rice is in the nursery bed and is followed up with weekly sprays.

The heterocyclic system of the phenazines was discovered almost one hundred years ago and phenazine 5-oxide itself was synthesized as early as 1903 by the reaction of sodium hydroxide with nitrobenzene and aniline at 120 °C [94]:

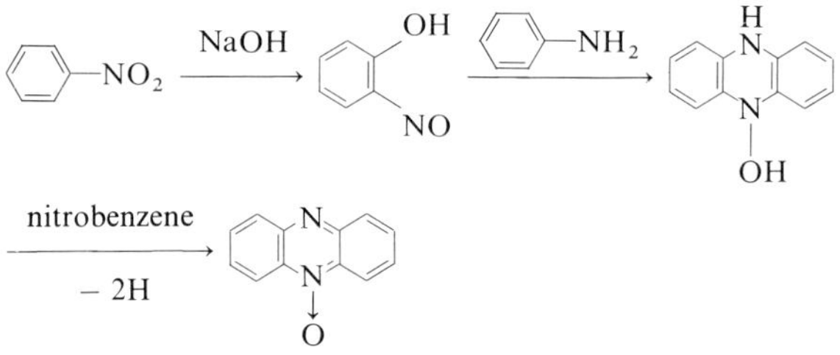

Phenazine 5-oxide forms yellowish-red needles, m.p. 226 °C. The chemistry of the phenazine derivatives is reviewed extensively in a monograph [95]. They generally show biological activity against a wide range of microorganisms and parasites [96, 97] and a group of naturally occurring antibiotics are themselves phenazine derivatives [98]. The practical use of phenazine 5-oxide, in particular, is however confined to the control of *X. oryzae*. Studies on the biochemical mode of action (interference with biological redox systems) have been published [99].

Antibacterial activity is also found in other heterocyclic N-oxides. An intensely investigated and patented class of compounds is represented by the quinoxaline 1,4-dioxides:

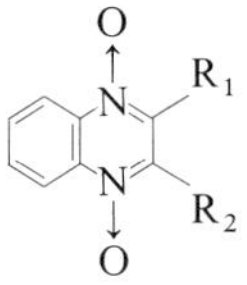

where R_1 and R_2 stand for a variety of functional groups.
Illustrative patents [100, 101].

Although no plant bactericides are known in this particular structural class, some derivatives have gained importance as feed additives against infectious diseases in domestic animals. In other more recent patents similar N-oxides of heterocyclic systems have been claimed to have useful plant bactericidal properties:

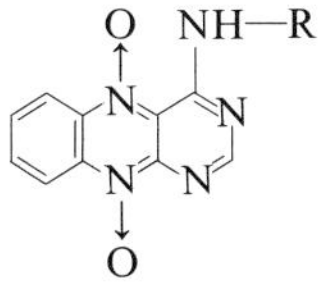

4-alkylamino-pyrimido[4,5-b] quinoxaline-5,10-dioxides or -benzo[g]pteridine 5,10-dioxides [102]

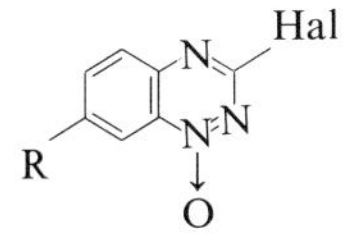

3-halobenzotriazine 1-oxides [103].

No development products have been published to date.

Quinolone-3-carboxylic acids and similar heterocyclics have been known as antibacterial agents for pharmaceutical use since 1962, when Sterling-Winthrop introduced nalidixic acid as a urinary tract disinfectant [104].

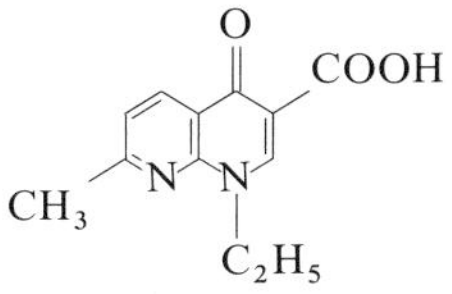

nalidixic acid, Negram (Sterling-Winthrop)

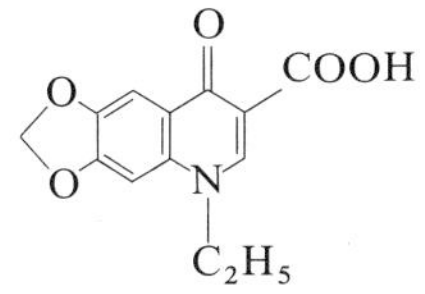

oxolinic acid, Utibid (Warner-Lambert)

Another compound with similar properties, oxolinic acid, had been developed by Warner-Lambert Co. [105]. In the past ten years the pharmaceutical industry has made great efforts to find new analogues in this class of compounds as documented in recent reviews [106, 107]. These materials have recently attracted attention as agricultural bactericides. Sumitomo has claimed the effectiveness of oxolinic acid and analogous compounds against *Erwinia* species on China cabbage and other vegetables [108]. One compound of this group, 5-methoxy-8-oxo-1,3-dioxolo[4,5-g]quinoline-7-carboxylic and (1-methoxy-4-oxo-6,7-methylenedioxy-quinoline-3-carboxylic acid), seems to be of particular interest:

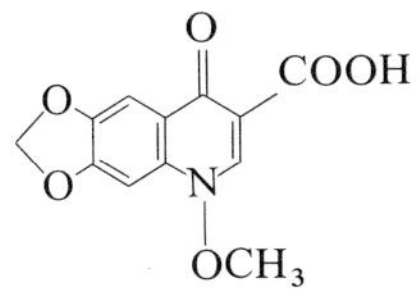

AB-206 (Sumitomo) [109, 110]

Two other types of similar structures have been claimed in new patents to be active against fire blight *(E. amylovora)*:

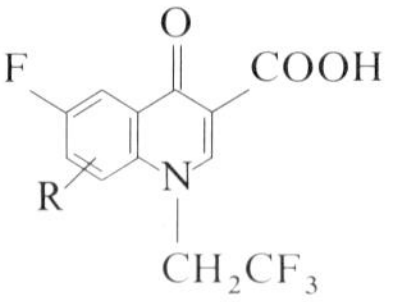

and

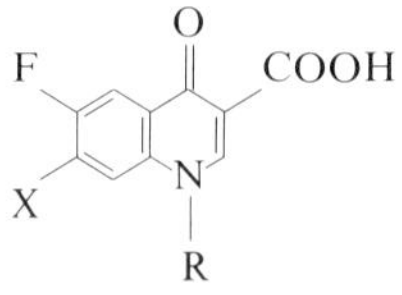

R = CH_3 or F [111]

X = H, Cl, F [112]
R = lower alkyl

In 1973 and 1974 one experimental compound, MBR 10995 (Riker Labs. Inc., division of 3 M), was reported to be unusually active in fire blight field tests in the United States, being more effective than streptomycin [113–115]. This material, the chemical structure of which has not been disclosed, is likely to be 6,7-dihydro-9-fluoro-5-methyl-1-oxo-1H,5H-benzo [i, j] quinolizine-2-carboxylic acid according to a US patent by 3 M [116] in which this very compound is exclusively claimed to be highly effective against *E. amylovora.*

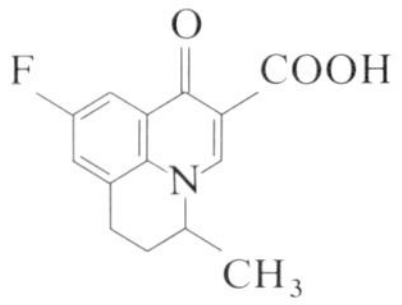

flumequine (Riker Labs.)

An earlier patent [117] describes the preparation and the antimicrobial use, including that against fire blight, of these quinolizine carboxylic acids. The same compound has also been recently introduced by Riker as a chemotherapeutic agent for medical use (common name flumequine) [107, 118]. MBR 10995 is no longer mentioned in the annual "Fungicide and Nematicide Tests" in the United States and its fate as a plant bactericide is uncertain.

6.6.2. *Heterocyclics Containing Nitrogen and Sulfur*

2-Amino-1,3,4-thiadiazoles

In 1970 a research group of Takeda Chemical Industries Ltd. reported that in their screening of compounds against plant diseases 2-amino-1,3,4-thiadiazole (ATDA) had been found highly effective against bacterial leaf blight of rice, citrus canker, and, at fairly high dosages, bacterial wilt of tomatoes *(P. solanacearum)* [119–121].

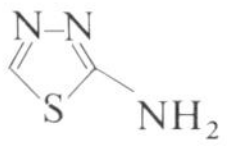

ATDA, TF-128 (Takeda)

ATDA is effective when used for foliar treatment and also as a root drench. As a foliar spray it shows an immediate therapeutic effect which does not however last very long. When applied in paddy water the buildup of long-lasting residual activity is slower. TF-128 and its derivative TF-130 (vide infra) are in many ways unusual compounds. They are fully systemic and highly active *in vivo*, but only against a narrow range of phytopathogenic bacteria (primarily xanthomonads). *In vitro*, however, this activity can be demonstrated only in a special synthetic medium. ATDA is thought to act as a nicotinamide antagonist in sensitive bacteria, where the compound is integrated to nicotinamide adenine dinucleotide (NAD) in the same manner as nicotinamide. The mechanism of its antibacterial selectivity has been demonstrated by the different degree of incorporation of ^{14}C-nicotinamide and ^{14}C-nicotinic acid into NAD by sensitive and resistant strains of *X. oryzae* and by insensitive bacteria (*Erwinia* and *Pseudomonas* spp.) [122]. Changes in the ultrastructure of *X. oryzae* under the influence of ATDA have also been studied [123].

ATDA has long been known [124] and is easily synthesized from thiosemicarbazide and formic acid [125]:

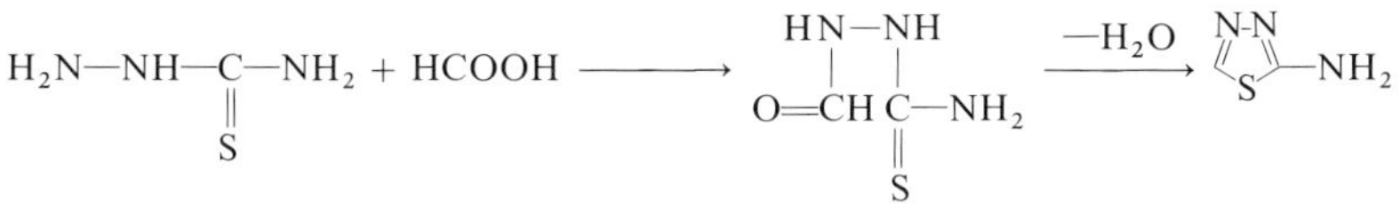

An even more potent derivative of ATDA is N, N′-methylene-bis-2-amino-1,3,4-thiadiazole, readily obtained by acid-catalyzed condensation of 2 moles of ATDA with paraformaldehyde [126].

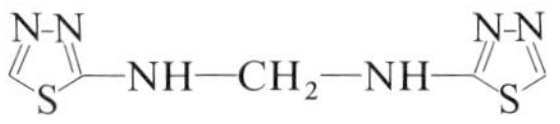

At the time of their discovery a few years ago, the biological activities of TF-128 and TF-130, an officially undisclosed analogue, [127] caused some euphoric reactions. ("Development of these two chemicals is an epochal event in the history of bacterial leaf blight control because they display superlative control effects...") [128]. Neither compound however reached the commercial stage, apparently for two reasons: rapidly developing resistance of the pathogen [129, 130] and marked teratogenic effects on mammals, caused by ATDA and N,N′-methylene-bis-2-amino-1,3,4-thiadiazole [131–133].

Other experimental bactericides with related structures are known from the literature, for example 4-amino (1,3,4-thiadiazol-2-yl)-benzenesulfonamide [24]

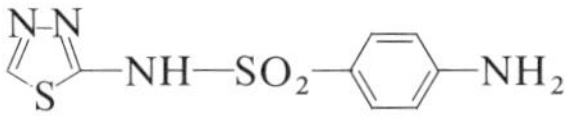

and 2-amino-5-(1-methyl-5-nitro-2-imidazolyl)-1,3,4-thiadiazole [134]

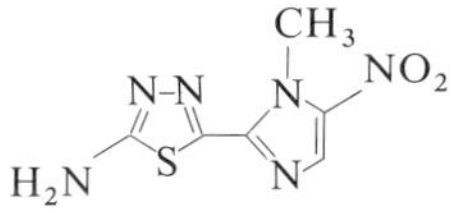

AC-64 855
(American Cyanamid)

Quite a number of patents have been published covering 5-substituted-2-amino-1,3,4-thiadiazoles with an optionally functionalized amino group as illustrated by two examples:

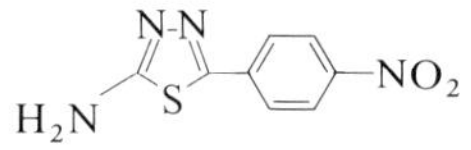
[135]

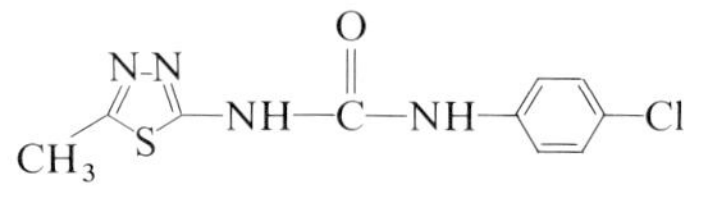
[136]

It is somewhat surprising that even the acyclic precursors of the 2-amino-1,3,4-thiadiazoles, the thiosemicarbazides, have bactericidal activity against *X. oryzae*. In particular, 1-methylthiosemicarbazide

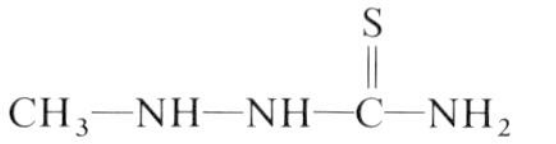
"NK-558"

is highly active when applied to the roots of rice plants or in paddy water. The Japanese authors found that the thiosemicarbazide is partly metabolized in the soil and in the plant to ATDA, which evidently is the active principle [137, 138].

Derivatives of saccharin or 1,2-benzisothiazoline-3-one 1,1-dioxide

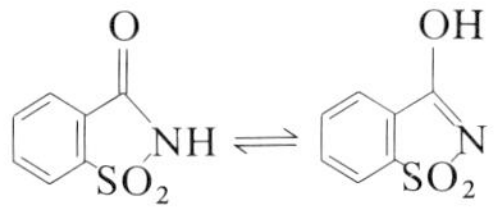

representing another type of heterocyclic compounds, have some activity against xanthomonads. The best known example is 3-propenyloxy-1,2-benzisothiazole 1,1-dioxide [139, 140], a fungicide active against rice blast *(Piricularia oryzae)* with a pronounced antibacterial side effect against leaf blight of rice.

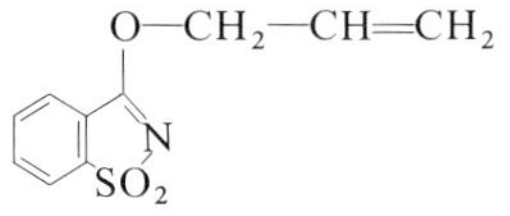
Oryzemate (Meiji Seika)

2-Formyl, 2-acetyl, 2-carbomethoxy and other derivatives of saccharin have also been claimed to act as plant bactericides [141–143]. Even the parent compound seems to have some beneficial effect on the symptoms caused by *X. vesicatoria* [144].

2-(Thiocyanomethylthio)-benzimidazole, TCMTB [145], is a general pesticide active against insects, soilborne and seedborne fungi and bacteria [146].

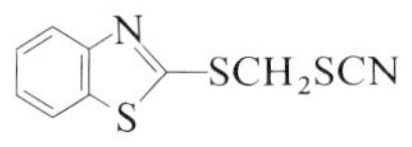
Busan 30, Busan 72 (Buckman Laboratories)
Protector-3L (Agway Inc.)
acute LD_{50}: 1500 mg/kg (rat)

The bactericidal activity of Busan is apparently limited to the genus *Xanthomonas*, which causes bacterial spots on peppers, canker of citrus, and bacterial blight of cotton. It has been recommended as a seed dressing against the last-mentioned disease *(X. malvacearum)* [147].

3-Benzylideneamino-4-phenyl-1,3-thiazoline-2-thione, common name fentiazon, which may be viewed as a cyclic dithiocarbamate, was introduced in 1967 for field sprayings against *X. oryzae* in Japan [148, 149]. Although the compound

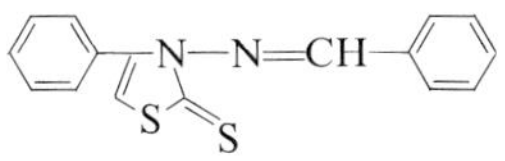

fentiazon, Selzion
Celdion (Takeda, Japan)
acute LD_{50}: 10,000 mg/kg (rat)

shows good protective activity [150] and is not phytotoxic to rice plants, it was dropped from practical use in 1974 probably owing to poor field performance.

6.7. Antibiotics

The outstandingly successful use of antibiotics against bacterial diseases and infections of human beings and domestic animals has led to a large scale screening of this class of natural or synthetic compounds for plant protection. The results obtained with antibiotics or antibiotic-containing culture broths from soil organisms did not, however, fulfill the high expectations expressed. Many of them were too unstable under field conditions or showed toxic side effects on plants. Most antibiotics are, not surprisingly, rather expensive even when used as crude products in agriculture. For fungicidal purposes, specifically for the control of rice blast and sheath blight, both serious fungal rice diseases, antibiotics like blasticidin S, kasugamycin, validamycin and the polyoxins (Vol. 4, pp. 224–231) have proven very effective and today play an important role in Japanese agriculture. Other antibiotics proved useful for the control of bacterial plant diseases and were intensively tested. Yet only one of them, streptomycin, is today sold in ton quantities for agricultural use. Reviews of the use of antibiotics in pesticide practice are found in the literature [151, 152].

Acetylene dicarboximide, cellocidin

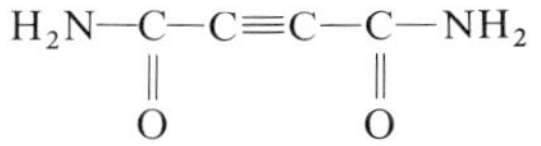

Cellomate
acute LD_{50}: 89–125 mg/kg (rat)

was discovered in the culture medium of *Streptomyces chibaensis* [153]. Its simple chemical structure permits synthesis of the technical grade material either from fumaric acid or from 2-butyne-1,4-diol. The product exhibits a good preventive action against rice bacterial leaf blight when sprayed in concentrations of 100–200 ppm. Higher doses cause phytotoxic effects on rice plants. The antibacterial action of cellocidin is antagonized by cysteine and glutathion, which indicates interaction with SH groups. A study of several metabolic systems in *X. oryzae* disclosed that cellocidin selectively inhibits the biochemical pathway from α-ketoglutaric acid to succinic acid even at a concentration of about 1 ppm. Cellomate was introduced to the market in 1964 by various Japanese manufacturers and dropped in 1970 probably because the safety margin between bactericidal and phytotoxic doses is very narrow [152].

Chloramphenicol, common name for D(–)threo-2-dichloroacetamido-1-(4-nitrophenyl)-1,3-propanediol,

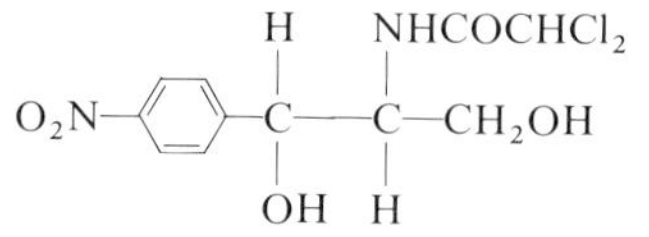

Chloromycetin (Parke-Davis)
Leukomycin (Bayer)
and other labels for
pharmaceutic use

was the first of the so-called broadspectrum medicinal antibiotics to be detected. It was originally isolated from cultures of an actinomycete, *Streptomyces venezuelae*, by Ehrlich and Carter working independently in 1947. It is a neutral crystalline substance with moderate solubility in water and good solubility in organic solvents. Its relatively simple chemical structure permits commercial manufacture by various synthetic routes, the starting materials being, for example, benzaldehyde, p-nitrobenzaldehyde, p-nitroacetophenone or cinnamic alcohol [154, 155].

Chloramphenicol is antimicrobially active against a wide range of gram-positive and gram-negative bacteria and rickettsias. The first agricultural use was against bacterial leaf blight of rice in Japan in 1964. Later, other plant bacterioses were treated with the antibiotic, frequently applied in mixtures with copper, mercury, and arsenic compounds. In 1968, 21 tons of a formulated product was put to agricultural use (trade name Shirahagen) in Japan, but it has since been withdrawn from the market. A combined product containing zineb and chloramphenicol (Bercema-Antispor 6459) has been proposed as a potato seed dressing against fungal and bacterial diseases, being especially active against soft rot caused by *E. carotovora* [156].

Oxytetracycline, 5-hydroxytetracycline

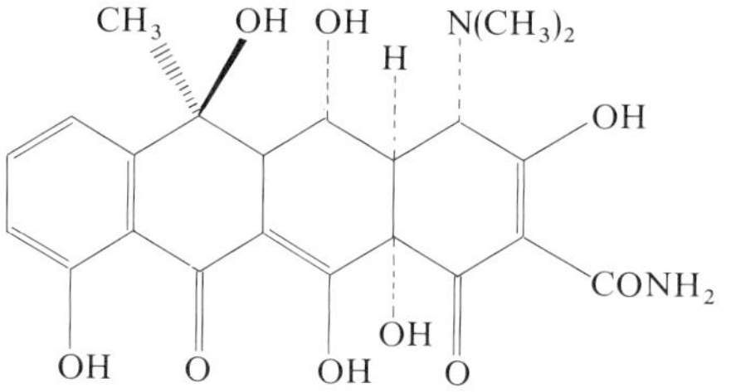

Terramycin,
Mycoshield (Pfizer)

The tetracycline group of antibiotics [157, 158] was discovered between 1948 and 1953 and a number of representatives are used in medicine and as food additives for meat-producing animals. The yellow tetracycline antibiotics are generated by certain soil microorganisms, the actinomycetes. Chlortetracycline, the first derivative discovered, is produced by *Streptomyces aureofaciens* and was put on the market by Lederle Laboratories.

Another derivative, oxytetracycline, produced by *Streptomyces rimosus* on a suitable medium, was introduced in 1950 by Charles Pfizer & Co. (Terramycin). It has been extensively tested and is used to some extent in plant protection. Its activity covers a broad spectrum of plant-pathogenic bacteria. Successful control of black rot in cabbage *(X. campestris)* [159], fire blight on pears [115], and bacterial spot of peach *(X. pruni)* [160] has been reported. Oxytetracycline injections have been used to treat coconut palms in Florida afflicted with lethal yellowing disease, which is caused by a mycoplasmalike organism [161]. Similarly, trunk injections with Terramycin effectively control bacterial spot disease in peach trees [162]. Some of the early agricultural streptomycin formulations (e.g. Agrimycin 100 by Pfizer) contained 1.5% Terramycin to delay the emergence of resistance, but this component was later deleted [28]. The mode of action of the tetracyclines lies in interference with protein synthesis by inhibiting formation of aminoacyl-t-RNA.

Streptomycin [163], the most important substance of the group of aminoglycoside antibiotics, was first isolated by Waksman in 1944 [164] and has since become a valuable, potent drug against diseases caused by aerobic gram-negative bacteria including *Mycobacterium tuberculosis.*

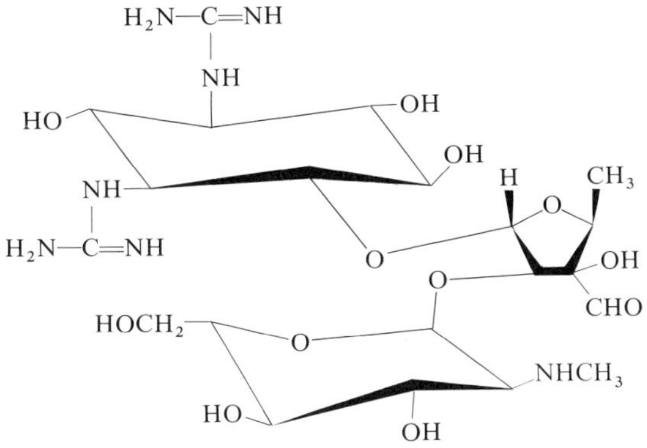

Streptomycin

Streptomycin is recovered from the culture broth of *Streptomyces griseus*, *S. bikiensis*, and *S. mashuensis* and is usually marketed as the tris-hydrochloride or tris-sulfate. The salts form white amorphous powders readily soluble in water. The acute oral toxicity (LD_{50}: 9000 mg/kg in the mouse) indicates low hazard to man. Its mode of action is the inhibition of protein synthesis in the mitochondria [165]. The chemical, biochemical, technological, and pharmacological aspects of streptomycin and related aminoglycosides are reviewed in the literature [166]. Streptomycin was the first antibiotic to be used against plant diseases (1952) [167] and has become perhaps the most effective and versatile plant bactericide. In 1953 the product proved outstandingly successful in the control of fire blight infections of pear and apple blossoms [168] in the United States. Streptomycin sprayings at concentrations of 50–200 ppm are therefore widely used today and have replaced copper compounds in fire blight control. A new problem arose in 1971 when streptomycin-resistant strains of *E. amylovora* were found in California [169]. They spread north to Oregon and Washington, and today streptomycin is virtually ineffective in about 18,000 acres of pear orchards [170]. In Japan the antibiotic is used freely against tobacco wildfire *(P. tabaci)* and citrus canker *(X. citri)* as well as bacterial diseases in vegetables. Bacterial leaf blight of rice cannot be satisfactorily controlled with streptomycin. Annual sales of formulated material are estimated at 200 t in Japan. In Europe the antibiotic is not registered for plant protection (except for ornamentals in some countries) and the market share is thus very small. Streptomycin is produced by a number of companies, the agricultural formulations being generally 1–2% dusts and 8.5–62.5% wettable powders. Among the most common labels are Agri-Strep by Merck, Agrimycin by Pfizer (both containing 21% streptomycin sulfate, equivalent to 17% of the free base), Spikespray by ICI, Embamycin by May & Baker, Hokko-mycin, and others. General drawbacks of streptomycin are the unfavorable price-efficiency ratio, its pronounced phytotoxicity at higher dosages, and restrictive legislation in many countries.

These factors along with the established development of resistance make widespread application of streptomycin in agriculture unlikely and the invention of new and better bactericides even more imperative [47].

Regarding the use of antibiotics for control of bacterial plant diseases, it must be kept in mind that genetically stable and inheritable resistance against this class of chemicals can develop readily and that drug resistance can be transmitted among diverse groups of bacteria, including human pathogens. As yet, there is no documented case in which a human disease has become more difficult to control because of the build-up of antibiotic-resistant phytopathogenic bacteria in a treated field. It could be, however, that such problems have gone unnoticed and it is of great concern to the authorities that certain pharmaceutical agents (specifically antibiotics which are used as drugs of choice in human infections for which few or no alternative treatments are available) are being proposed or used as spray treatments for control of bacterial diseases of plants. The American National Committee NC-135 therefore strongly dissuades government, industry and academia from using antibiotics currently employed in human and animal disease therapy and recommends instead the development of alternative physical, chemical and biological methods for control of plant bacterioses [171].

7. References

1. Braun, H.: Geschichte der Phytomedizin. In: Handbuch der Pflanzenkrankheiten (Sorauer, P., 7th ed.), Vol. 1, Berlin-Hamburg, Verlag Paul Parey 1965
2. Estimates of crop losses and disease-control costs in California 1963. University of California Agricultural Experiment Station and Agricultural Extension Service, 1965
3. Alcorn, S.M., Hoitink, H.A.J., Fulkerson, J.F.: Phytopathology News *5*, 2 (1971)
4. Kennedy, B.W., Alcorn, S.M.: Estimates of US crop losses due to procaryote plant pathogens. Plant Disease (cont. of Plant Dis. Rep.) *64*, 674 (1980)
5. De Ley, J.: Proc. Int. Conf. Plant. Pathog. Bact. (4th.), Angers, *1978*, 347
6. Buchanan, R.E., Gibbons, N.E. (ed.): Bergey's manual of determinative bacteriology, 8th ed., Baltimore, Williams and Wilkins 1974
7. Dye, D.W.: Rev. Plant Pathol. *53*, 953 (1974)
8. Young, J.M. et al.: N. Z. J. Agric. Res. *21*, 153 (1978)
9. Broekaert, D., van Parijs, R.: Crown gall tumorigenesis: 25 years research on the identification of the tumor-inducing principle. Med. Fac. Landbouww. Rijksuniv. Gent *42*, 1955 (1977)
10. Dye, D.W.: N. Z. J. Sci. *11*, 590 (1968); *12*, 81 (1968); *12*, 223 (1968)
11. Gibbins, L.N.: Proc. Int. Conf. Plant Pathog. Bact. (4th), Angers, *1978*, 403
12. Hull, R.: Rev. Plant Pathol. *50*, 121 (1971)

13. Grunewaldt-Stöcker, G., Nienhaus, F.: Mycoplasmaähnliche Organismen als Krankheitserreger in Pflanzen. Acta Phytomedica (supplement to Phytopathol. Z.) *5* (1977)
14. Maramorosch, K.: Plant mycoplasma diseases. In: Physiological plant pathology (Heitefuss, R., Williams, P.H., ed.), Berlin-Heidelberg-New York, Springer-Verlag 1976
15. Maramorosch, K.: Ann. Rev. Microbiol. *28*, 301 (1974)
16. Hopkins, D.L.: Ann. Rev. Phytopathol. *15*, 277 (1977)
17. Ou, S.H.: Rice Diseases. Kew, Surrey (UK), Commonwealth Mycological Inst. 1972, pp. 51–84
18. Wakimoto, S.: *Xanthomonas oryzae*. In: Krankheiten, Schädlinge und Unkräuter im tropischen Pflanzenbau (Kranz, J., Schmutterer, H., Koch, W., ed.), pp. 68–70, Berlin-Hamburg, Verlag Paul Parey 1979
19. Kauffman, H.E. et al.: Plant Dis. Rep. *57*, 537 (1973)
20. Thurston, H.D.: Ann. Rev. Phytopathol. *11*, 27 (1973)
21. Lozano, J.C.: Plant Dis. Rep. *61*, 644 (1977)
22. Singh, R.N.: PANS *18*, 396 (1972)
23. Singh, R.A., Ahmed, K.M.: Indian Phytopathol. *28*, 143 (1975)
24. Rao, P.S., Kauffman, H.E.: Phytopathol. Z. *90*, 281 (1977)
25. Bradbury, J.F.: *Xanthomonas oryzae* (Uyeda et Ishiyama), Dowson, C.M.I., descriptions of pathogenic fungi and bacteria, No. 239, Commonwealth Agricultural Bureaux, Kew, Surrey (UK) 1970
26. van der Zwet, T., Keil, H.L.: Fire blight, a bacterial disease of rosaceous plants, USDA Agriculture Handbook No. 510, 1979
27. Zeller, W.: Der Feuerbrand des Kernobstes hervorgerufen von *Erwinia amylovora* (Burrill) Winslow et al., Mitt. Biol. Bundesanst. Land Forstwirtsch., Berlin-Dahlem No. 158, 1974
28. Schroth, M.N. et al.: Ann. Rev. Phytopathol. *12*, 389 (1974)
29. Fieret, M., Large, M.: Phytoma *1979*, 33
30. Hayward, A.C., Waterston, J.M.: *Erwinia amylovora* (Burrill), C.M.I. descriptions of pathogenic fungi and bacteria, No. 44. Commonwealth Agricultural Bureaux, Kew, Surrey (UK) 1965
31. Thomson, S.V.: 3rd Internat. Congr. Plant Pathology, Munich FRG, 1978. Abstracts p. 66
32. Parthasarathy, M.V.: Phytopathology *64*, 667 (1974)
33. McCoy, R.E. et al.: Phytopathology *66*, 1148 (1976)
34. Markham, P.G.: Proc. 1977 Brit. Crop Protec. Conf. *1977*, 815
35. Lemattre, M.: L'horticulture française *28*, 1 (1973)
36. Butcher, C.L., Dean, L.L., Guthrie, J.W.: Plant Dis. Rep. *53*, 894 (1969)
37. Dopson, R.N.: Plant Dis. Rep. *48*, 30 (1964)
38. Egli, T., Goto, M., Schmidt, D.: Phytopathol. Z. *82*, 111 (1975)
39. Brinkerhoff, L.A.: Ann. Rev. Phytopathol. *8*, 85 (1970)
40. Ezuka, A., Sakaguchi, S.: Rev. Plant Protec. Res. *11*, 93 (1978)
41. Vidaver, A.K.: Ann. Rev. Phytopathol. *14*, 451 (1976)
42. Caruso, F.L., Kuć, J.: Physiol. Plant. Pathol. *14*, 191 (1979)
43. Kerr, A., Htay, K.: Physiol. Plant. Pathol. *4*, 37 (1974)
44. Kerr, A.: Plant Disease (cont. of Plant Dis. Rep.) *64*, 25 (1980)
45. Roberts, W.P., Tate, M., Kerr, A.: Nature (London) *265*, 379 (1977)
46. Süle, S.: Proc. Int. Conf. Plant Pathog. Bact., 4th Angers, *1978*, 255
47. Beer, S.V., Sherf, A.F.: Fungicide and NematicideTests *31*, 1 (1976), Amer. Phytopathol. Soc.
48. Crosse, J.E.: Proc. Brit. Insectic. Fungic. Conf. (6th) *1971*, 694
49. Fox, R.T.V.: Proc. Brit. Insectic. Fungic. Conf. (6th) *1971*, 187
50. Averre, C.W. (ed.): Fungicide and Nematicide Tests *32*, 228 (1977), Amer. Phytopathol. Soc.
51. Keil, H.L., van der Zwet, T.: Plant Dis. Rep. *51*, 753 (1967)
52. Hiltner, L.: Prakt. Blätter f. Pflanzenbau u. Pflanzenschutz *18*, 65 (1915)
53. El Nur, E.: Bacterial blight of cotton. In: Cotton growing in the Gezira environment (Sidding, M.A., Hughes, L.C., ed.), pp. 179–188, Cambridge, Heffer & Sons Ltd. 1970
54. Jönck, P.: Quecksilberverbindungen. In: Ullmanns Encyklopädie der Technischen Chemie (3rd ed.), Vol. 14, pp. 549–553, München-Berlin, Urban & Schwarzenberg 1963
55. Martin, H. (translation by Schmidt, K.J.): Die wissenschaftlichen Grundlagen des Pflanzenschutzes, 5th ed., pp. 469–478, Weinheim, Verlag Chemie 1967
56. Fishbein, L.: Toxicological aspects of fungicides. In: Antifungal compounds, Vol. 2 (Siegel, M.R., Sisler, H.D., ed.), pp. 572–578, New York-Basel, M. Dekker Inc. 1977

57. Millardet, A., Gayon, W.: J. Agric. Prat. *49*, 513, 707 (1885)
58. Pierce, N.B.: Pacific Rural Press. *55*, 404 (1898)
59. Stall, R.E.: Plant Dis. Rep. *43*, 725 (1959)
60. Oshima, N., Dickens, L.E.: Plant Dis. Rep. *55*, 609 (1971)
61. Krämer, W.: Wirkstoffe gegen Pflanzenkrankheiten. In: Pflanzenschutz und Schädlingsbekämpfung (Büchel, K.H., ed.), p. 121, Stuttgart, Georg Thieme 1977
62. Teworte, W., Rabben, H.J.: Kupferverbindungen. In: Ullmanns Encyklopädie der Technischen Chemie (4th ed.) Vol. 15, pp. 559–587, Weinheim, Verlag Chemie 1978
63. FR 1,302,445 (26 May 1961/23 Jul 1962) Compagnie de Saint-Gobain
64. Averre, C.W. (ed.): Fungicide and Nematicide Tests *34*, 229 (1979), Amer. Phytopathol. Soc.
65. Thomson, W.T.: Agricultural Chemicals, IV-Fungicides, 1976–77 Revision, p. 23, Fresno, CA, Thomson Publications
66. Martin, H. (translation by Schmidt, K.J.): Die wissenschaftlichen Grundlagen des Pflanzenschutzes (5th ed.), pp. 218–223, Weinheim, Verlag Chemie 1967
67. Kaars Sijpesteijn, A., Dekhuijzen, H.M., Vonk, J.W.: Biological conversion of fungicides in plants and microorganisms. In: Antifungal compounds, Vol. 2, pp. 95–105 (Siegel, M.R., Sisler, H.D., ed.), New York-Basel, M. Dekker Inc. 1977
68. Fishbein, L.: Toxicological aspects of fungicides. In: Antifungal compounds, Vol. 2, pp. 539–550 (Siegel, M.R., Sisler, H.D., ed.), New York-Basel, M. Dekker Inc. 1977
69. Metcalf, R.L.: Chemistry and biology of pesticides. In: Pesticides in the environment, Vol. 1, Part I, pp. 10–15 (White-Stevens, R., ed.), New York, M. Dekker Inc. 1971
70. JA 7,310,535 (22 Apr 1969/4 Apr 1973) Mikasa Chem. Ind.; C.A. *80*, 141786 z (1974)
71. DT 1,157,027 (1 Mar 1962/7 Nov 1963) Bayer AG, Lehmann, H., Grewe, F., Lautenschlager, W.
72. Yoneyama, K., Misato, T.: Ann. Phytopathol. Soc. Japan *37*, 291 (1971)
73. Averre, C.W. (ed.): Fungicide and Nematicide Tests *34*, 226 (1979), Amer. Phytopathol. Soc.
74. Farm Chemicals Handbook 1979, Willoughby, OH, Meister Publishing Co., p. D-45
75. DT 2,308,807 (22 Feb 1973/8 Nov 1973, US prior. 24 Apr 1972) Buckman Labs. Inc., Buckman, J.D., Pera, J.D.
76. Martin, H., Worthing, C.R. (ed.): Pesticide Manual (5th ed.) p. 56, Brit. Crop Protection Council 1977
77. GB 1,057,131 (19 May 1964/1 Feb 1967), GB 1,193,954 (2 Aug 1967/3 Jun 1970), Boots Drug Co., Clark, N.G., Croshaw, B., Spooner, D.F.
78. Spooner, D.F., Wakerly, S.B.: Proc. Brit. Insectic. Fungic. Conf. (6th) *1971*, 201
79. Averre, C.W. (ed.): Fungicide and Nematicide Tests *32*, 235 (1977). Amer. Phytopathol. Soc.
80. Kooistra, T., Taconis, P.J.: Acta Hortic. *86*, 77 (1978)
81. Cohen, E., Schiffmann-Nadel, M.: Plant Dis. Rep. *62*, 388 (1978)
82. US 4,060,639 (14 May 1976/29 Nov 1977), Merck & Co. Inc., Jacobus, D.P., Dulaney, E.L., Grier, N.
83. Daines, R.H.: Plant Dis. Rep. *44*, 826 (1960); *46*, 251 (1962)
84. Loewe, H.: Desinfektionsmittel und Antiseptica. In: Arzneimittel, Vol. 4, pp. 25–32 (Ehrhardt, G., Ruschig, H., ed.), Weinheim, Verlag Chemie 1972
85. Baker, J.W. et al.: Antiseptics and disinfectants. In: Medicinal Chemistry, Part I, pp. 630–636 (Burger, A., ed.), New York-London-Sydney-Toronto, Wiley-Interscience 1970
86. Kirk-Othmer: Encyclopedia of Chemical Technology, Vol. 2 (2nd ed.) p. 630, New York-London-Sydney, Interscience Publisher 1963
87. Farm Chemicals Handbook 1979, p. D-193, p. D-155, Willoughby, OH, Meister Publishing Co.
88. Thomson, W.T.: Agricultural Chemicals, IV-Fungicides. 1976–77 Revision, p. 123, Fresno, CA, Thomson Publications
89. Nakagami, K., Honda, T.: 3rd Internat. Congr. Plant Pathology, Munich FRG, 1978. Abstr. p. 360
90. Nakagami, K., Tanaka, H.: Ann. Phytopathol. Soc. Japan *44*, 400 (1978)
91. JA 7,326,213 (11 Sep 1970/7 Aug 1973),
JA appl. 0,040,730 (2 Aug 1973/14 Apr 1975),
JA appl. 0,116,640 (20 Feb 1974/12 Sep 1975), all Sankyo Co.
92. Thomson, W.T.: Agricultural Chemicals, IV-Fungicides. 1976–77 Revision, p. 128, Fresno, CA, Thomson Publications
93. Japan Chemical Week, July 6, *1978*

94. Wohl, A.: Ber. Dtsch. Chem. Ges. *36*, 4139 (1903)
95. Swan, G.A., Felton, D.G.I.: Phenazines. In: The chemistry of heterocyclic compounds, Vol. 11 (Weissberger, A., ed.), New York-London, Interscience Publishers Inc. 1957
96. DT 1,094,036 (6 Feb 1959/1 Dec 1960), N.V. Philips Gloeilampenfabrieken, Overeem, J.C., Bijloo, J.D.
97. FR 1,451,606 (1 Sep 1965/25 Jul 1966) Shell International Research Co.
98. Berger, J.: Antibiotics (Phenazines). In: Kirk-Othmer, Encyclopedia of chemical technology (3rd ed.) Vol. 3, pp. 1–21, New York-Chichester-Brisbane-Toronto, John Wiley & Sons 1978
99. Sekizawa, Y., Watanabe, T.: Ann. Phytopathol. Soc. Japan *35*, 208 (1969)
100. DT 1,670,935 (4 Oct 1967/25 Feb 1971) Bayer AG, Ley, K. et al.
101. DT 2,725,023 (31 May 1977/22 Dec 1977) Pfizer Inc., Czuba, L.J.
102. DT 2,745,007 (6 Oct 1977/12 Apr 1979) Bayer AG, Seng, F., Scheinpflug, H., Kraus, P.
103. DT 2,740,887 (10 Sep 1977/22 Mar 1979) Bayer AG, Sasse, K., Kraus, P., Scheinpflug, H.
104. Lesher, G.Y. et al.: J. Med. Pharm. Chem. *5*, 1063 (1962)
105. Kaminsky, D., Meltzer, R.J.: J. Med. Chem. *11*, 160 (1968)
106. Lesher, G.Y.: Antibacterial agents (synthetic). In: Kirk-Othmer, Encyclopedia of chemical technology (3rd ed.), Vol. 2, pp. 782–789, New York-Chichester-Brisbane-Toronto, John Wiley & Sons 1978
107. Albrecht, R.: Prog. Drug Res. *21*, 9 (1977)
108. DT 2,737,298 (18 Aug 1977/23 Feb 1978/Jap. prior. 19 Aug 1976) Sumitomo Co., Hisada, Y. et al.
109. Agui, H. et al.: J. Med. Chem. *20*, 791 (1977)
110. DT 2,134,451 (9 Jul 1971/20 Apr 1972/Jap. prior. 9 Jul 1970) Sumitomo Co., Nakagome, T. et al.
111. US 4,146,625 (13 Feb 1978/27 Mar 1979) DuPont, Lee, K.T.
112. EP 0,000,203 (29 Jun 1978/10 Jan 1979) Ciba-Geigy AG, Sturm, E.
113. Averre, C.W. (ed.): Fungicide and Nematicide Tests *29*, 6, 30 (1974); *30*, 10, 19, 37 (1975), Amer. Phytopathol. Soc.
114. Reil, W.O. et al.: Calif. Agric. *4*, 4 (1974)
115. Thomson, S.V. et al.: Phytopathology *66*, 1457 (1976)
116. US 3,924,042 (26 Mar 1973/2 Dec 1975), Minnesota Mining and Manufacturing, Gerster, J.F.
117. US 3,896,131 (2 Nov 1972/22 Jul 1975), Riker Labs. Inc., Gerster, J.F.
118. Rohlfing, S.R., Gerster, J.F., Kvam, D.C.: Antimicrob. Agents Chemother. *10*, 20 (1976)
119. Mitsukami, T.: Japan Pestic. Inform. *4*, 12 (1970)
120. Yakushiji, K., Okada, Y.: J. Takeda Res. Lab. *29*, 179 (1970)
121. Yakushiji, K., Wakae, O.: Ann. Phytopathol. Soc. Japan *37*, 100 (1971)
122. Yoneyama, K., Sekido, S., Misato, T.: J. Pestic. Sci. *1978*, 379
123. Horino, O.: Ann. Phytopathol. Soc. Japan *39*, 14 (1973)
124. Freund, M., Meinecke, C.: Ber. Dtsch. Chem. Ges. *29*, 2511 (1896)
125. Lauer, R.F., Zenchoff, G.: J. Het. Chem. *13*, 292 (1976)
126. US 3,891,762 (8 Jan 1974/24 Jun 1975/Jap. prior. 13 May 1968), Takeda Chem. Co., Wakae, O., Yakushiji, K., Okada, Y.
127. Yamaguchi, T.: Japan Pestic. Inform. *7*, 17 (1973)
128. Yoshimura, S.: Control of bacterial leaf blight of rice in South-East Asia, p. 349. In: Rice in Asia. Univers. Tokyo Press 1975
129. Plant Protection in Japan (1976) p. 368, Tokyo, Association of Agricultural Relations in Asia 1976
130. Annual Rep. 1971, p. 109. Internat. Rice Research Inst., Los Baños, Philippines 1972
131. Beaudoin, A.R.: Teratology *9*, 179 (1974); ibid. **7**, 65 (1973)
132. Mizutani, M., Ihara, T., Sugitani, T.: Teratology *10*, 90 (1974)
133. Sugino, Y. et al.: Cancer Res. *35*, 2631 (1975)
134. Hoitink, H.A.J., Alcorn, S.M.: Phytopathology *62*, 764 (1972)
135. DT 2,329,134 (8 Jun 1972/20 Dec 1973) Imperial Chemical Inds. Ltd., Balasubramanyan, S., Fox, R.T.V.
136. JA 4,008,253 (1 Oct 1970/25 Feb 1974) Kumiai Chemical Ind.
137. Ohmori, K. et al.: Effect of thiosemicarbazide derivatives against bacterial plant diseases, 1974 Ann. Meet. Phytopathol. Soc. Japan
138. Ohmori, K. et al.: J. Pestic. Sci. (Noyaku Kagaku) *1976*, 95
139. JA 7,014,301 (7 Sep 1967/21 May 1970), Meiji Seika Kaisha Ltd.

140. Uchiyama, M. et al.: Agric. Biol. Chem. *37*, 737 (1973)
141. JA 7,308,500 (1 Dec 1970/15 Mar 1973), Kumiai Chem. Ind.
142. JA 7,305,907 (29 Dec 1970/21 Feb 1973), Kumiai Chem. Ind.
143. JA 7,335,457 (12 Oct 1970/27 Oct 1973), Kumiai Chem. Ind.
144. Kim, S.H., Morton, D.J., Fieldhouse, P.J.: Plant Dis. Rep. *51*, 497 (1967)
145. Krämer, W.: Wirkstoffe gegen Pflanzenkrankheiten. In: Pflanzenschutz und Schädlingsbekämpfung, p. 139 (Büchel, K.H., ed.) Stuttgart, Georg Thieme 1977
146. Pulido, M.L., Silversides, W.H., Marantz, B.B.: Proc. Brit. Insectic. Fungic. Conf. (7th) *1973*, 825
147. Verma, J.P., Singh, R.P.: Cott. Gr. Rev. *48*, 60 (1971)
148. Thomson, W.T.: Agricultural Chemicals, IV-Fungicides, 1976–77 Revision, p. 87, Fresno, CA, Thomson Publications
149. Usui, Y., Yamano, T.: Yakugaku Zasshi (Japan) *89*, 699 (1969); C.A. *71*, 69601 e (1969)
150. Yakushiji, K., Kakuna, I., Usui, Y.: Ann. Phytopathol. Soc. Japan *33*, 117 (1967)
151. Misato, T.: The development of agricultural antibiotics. In: Plant protection in Japan 1976, pp. 371–381, Tokyo, Assoc. Agricult. Relations in Asia 1976
152. Dekker, J.: World Rev. Pest Control *10*, 9 (1971)
153. Suzuki, S. et al.: J. Antibiot. (Tokyo), ser. A *11*, 83 (1958)
154. Ehrlich, J.: Chloramphenicol and its analogues. In: Kirk-Othmer, Encyclopedia of chemical technology (3rd ed.) Vol. 2, pp. 920–929, New York-Chichester-Brisbane-Toronto, John Wiley & Sons 1978
155. Schacht, U.: Chloramphenicol. In: Arzneimittel, Vol. 4, pp. 390–402 (Ehrhardt, G., Ruschig, H., ed.), Weinheim, Verlag Chemie 1972
156. Motte, G. et al.: Nachrichtenbl. Pflanzenschutz DDR *28*, 85 (1974)
157. Boothe, J.H., Hlavka, J.J.: Tetracyclines. In: Kirk-Othmer, Encyclopedia of chemical technology (3rd ed.), Vol. 3, pp. 64–78, New York-Chichester-Brisbane-Toronto, John Wiley & Sons 1978
158. Ullmanns Encyklopädie der Technischen Chemie (4th ed.), Vol. 7, pp. 662–667, "Antibiotica". Weinheim, Verlag Chemie 1974
159. Knösel, D.: Phytopathol. Z. *54*, 31 (1965)
160. Keil, H.L., Weaver, L.O.: Plant Dis. Rep. *54*, 208 (1970)
161. McCoy, R.E., Gwin, G.H.: Plant Dis. Rep. *61*, 154 (1977)
162. Keil, H.L., Civerolo, E.L.: Plant Dis. Rep. *63*, 1 (1979)
163. Martin, H., Worthing, C.R. (ed.): Pesticide Manual, p. 478 (5th ed.), British Crop Protection Council 1977
164. Waksman, S.A.: Science *118*, 259 (1953)
165. Siegel, M.R.: Effect of fungicides on protein synthesis. In: Antifungal compounds, Vol. 2, pp. 426–427 (Siegel, M.R., Sisler, H.D., ed.), New York-Basel, M. Dekker Inc. 1977
166. Daniels, P.J.L.: Aminoglycosides. In: Kirk-Othmer, Encyclopedia of chemical technology (3rd ed.), Vol. 2, pp. 819–852, New York-Chichester-Brisbane-Toronto, John Wiley & Sons 1978
167. Goodman, R.N.: Am. Fruit Grow. *73*, 463 (1953)
168. Winter, H.F., Young, H.C.: Plant Dis. Rep. *37*, 463 (1953)
169. Moller, W.J. et al.: Phytopathology *62*, 779 (1972)
170. Coyier, D.L., Covey, R.P.: Plant Dis. Rep. *59*, 849 (1975)
171. Anonymous: Antibiotic use for control of bacterial diseases of plants. Phytopathology News *12*, 54 (1978)
172. Sax, N.I. (ed.): Dangerous properties of industrial materials (5th ed.), New York, Van Nostrand Reinhold Company 1979, p. 819

Recent Advances in the Chemical Control of Oomycetes

T. H. Staub and A. Hubele

CIBA-GEIGY Ltd., Agricultural Division
CH-4002 Basle, Switzerland

Contents

1. Introduction

Oomycetes are a class of fungi which include some of the most destructive plant-pathogenic species known to man. Most of them belong to the order Peronosporales, and a very few to the order Saprolegniales; the remaining orders Leptomitales and Lagenidiales are of little importance for our purposes, though they contain species which are parasitic on various aquatic life forms [1, 2]. The characteristic features of the oomycetes are: oospores as sexual spore type, the flagellation of the zoospores, the mycelial cell wall composition which includes cellulose but not chitin [3] and the lack of the ability to synthesize sterols [4]. The more advanced oomycetes do not form zoospores but rather conidia which germinate directly.

The economic impact of plant diseases caused by oomycetes is well documented [5, 6]. Efforts to understand and control them contributed significantly to the birth and the advance of both plant pathology as a science and the use of chemicals for the control of plant diseases. The description by De Bary in 1861 of the relationship between the oomycete *Phytophthora infestans* and the potato plant was one of the first reports that demonstrated the causal role of fungi in the development of plant diseases. Similarly, the discovery of the protective activity of Proust's mixture (later called Bordeaux mixture) against grape downy mildew by Millardet in 1882 not only saved the French wine industry but also helped advance the concept of chemical plant disease control. For further details on the history of fungicides see F. Grewe (Vol. 2, p. 17, Vol. 4, p. 68) and [6–8].

2. Biology of Plant-Pathogenic Oomycetes

2.1. Foliar Plant Diseases Caused by Oomycetes

Most oomycetes causing foliar diseases are obligate parasites belonging to the downy mildew fungi (Peronosporaceae, Table 1). Typically they form self-limiting local lesions on various plant parts from which masses of spores are produced (Fig. 1A) and released into the air. Some downy mildew fungi can become systemic in plants under certain conditions (e.g. blue mold of tobacco) and still others are predominantly systemic (e.g. *Peronosclerospora* spp. on graminaceous hosts). Systemic downy mildews and the white rusts (Albuginaceae), which also infect plants systematically, often produce distorted growth of the infected plant parts (Fig. 1C).

A typical life cycle of a downy mildew fungus is that of *Plasmopara viticola* on grapevines (Fig. 2). The fungus overwinters as thick-walled oospores, which represent the sexually-produced stage of this fungus and serve as resting structures to survive adverse climatic conditions or simply periods when no suitable host plant material is available. This fungus also can overwinter as dormant mycelium in infected twigs. In spring, sporangia produced from either overwintering stage can reach young leaves and release motile zoospores. They encyst on the stomata, and infection takes place through the stomatal openings. After an incubation

Table 1. Systematic position of some economically important plant-pathogenic Oomycetes [2] with the diseases they incite

Fungus species	Plant diseases
Order Peronosporales	
Family Peronosporaceae (obligate parasites of plants; downy mildews)	
Peronospora parasitica	Downy mildew of Cruciferae
Peronospora tabacina	Blue mold of tobacco
Plasmopara viticola	Downy mildew of grapevine
Plasmopara halstedii	Downy mildew of sunflower and other Compositae
Pseudoperonospora humuli	Downy mildew of hops
Pseudoperonospora cubensis	Downy mildew of cucurbits
Bremia lactucae	Downy mildew of lettuce
Sclerospora graminicola	Downy mildews of corn, sorghum and grasses
Family Pythiaceae (nonobligate parasites of plants)	
Pythium ultimum	Ubiquitous causes of damping-off, root and seed rot diseases
P. irregulare	
(and many others of a total of 92 Pythium spp.)	
Phytophthora infestans	Late blight of potato and tomato
P. palmivora	Cocoa black pod, leaf fall of rubber
P. citrophthora	Brown rot of citrus fruit
P. parasitica	Root and stem rot of many plants
P. cinnamomi	Root rot of avocado, eucalyptus, and many other trees and shrubs
P. cactorum	Collar rot of rosaceous plants and similar diseases on other plants
P. syringe	Canker and fruit rot of apple
(Total of 48 mostly soilborne Phytophthora spp.)	
Family Albuginaceae (obligate parasites of plants; white rusts)	
Albugo candida	White rust of crucifers
Order Saprolegniales	
Family Saprolegniaceae	
Aphanomyces euteiches	Root rot of peas
A. cochlioides	Damping-off and tip rot of sugar beets

period of one week or more, during which the fungus grows between the host cells, innumerable sporangia form on sporangiophores which grow out of the stomata on the lower leaf surface (Fig. 1A) or on infected young grape bunches (Fig. 1B). These masses of sporangia enable the disease to spread rapidly during favorable moist and warm climatic conditions in summer.

Phytophthora infestans, the causal agent of late blight of potato and tomato, has a similar life cycle (Vol. 4, p. 96, [6, 8]). In addition to causing foliar damage, this fungus can also destroy tubers in the field or in storage.

Fig. 1A–F. Disease symptoms caused by plant-pathogenic Oomycetes. **A and B** *Plasmopara viticola* on grapevine. **A** Symptoms of grape downy mildew on leaf and **B** on young grape bunch just before flowering. **C and D** Systemic downy mildews: **C** *Pseudoperonospora humuli* on hops causing shoot stunting and deformation; **D** *Peronosclerospora maydis* on corn with systemically infected growing tip. **E and F** Soilborne Oomycetes: **E** *Phytophthora parasitica* var. *nicotianae* with early attack on tabacco stem; **F** Pre- and post-emergence damping-off caused by *Pythium ultimum* on sugar beets. Left: Inoculated trays treated with Ridomil (top) and Dexon (bottom); right: Noninoculated (top) and inoculated control trays.

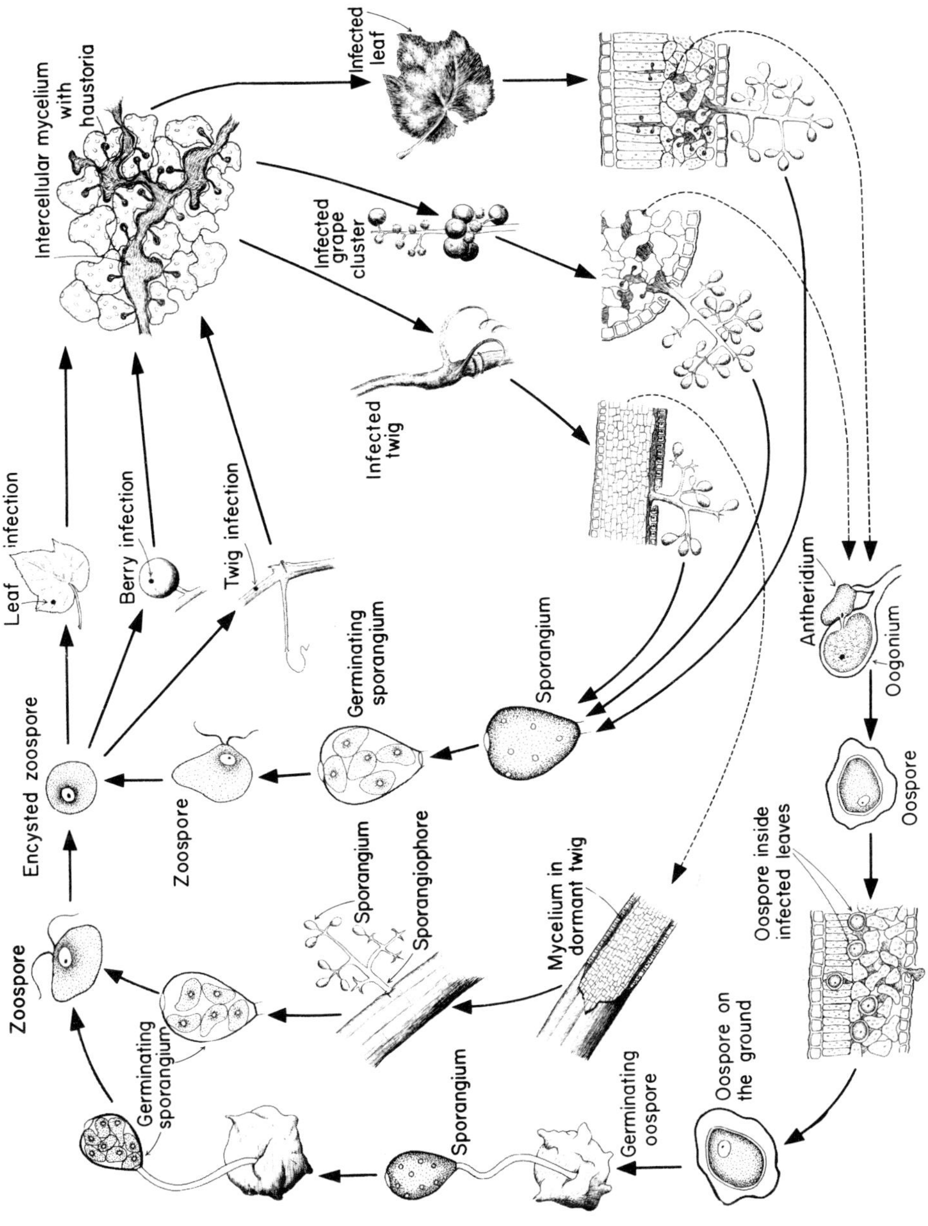

Fig. 2. Disease cycle of downy mildew of grapevine caused by *Plasmopara viticola* (by courtesy of G. Agrios and Academic Press [8])

For some predominantly systemic downy mildews, aerial spread is less important since it usually results in relatively harmless local lesions, whereas the much more damaging systemic form of these diseases is the result of seedborne or soilborne infections [9–11]. It is obvious that disease control strategies must take into account the differences between the various disease cycles.

2.2. Soilborne Seedling and Root Diseases Caused by Oomycetes

The seedborne and soilborne oomycetes causing root and seedling diseases belong to the genera *Aphanomyces*, *Pythium*, and *Phytophthora* (see Table 1). Some of the *Pythium* and *Phytophthora* spp. have an extremely wide host range while others, including *Aphanomyces* spp., are more specific. They all are facultative parasites and can live on dead plant debris. Symptoms caused by soilborne oomycetes include pre- and post-emergence damping-off, root and crown rots, and soft rot of storage organs; occasionally they cause foliar blights (e.g. *Pythium* on pears) or fruit rots (e.g. *Phytophthora syringe* on apple) [6, 12, 13].

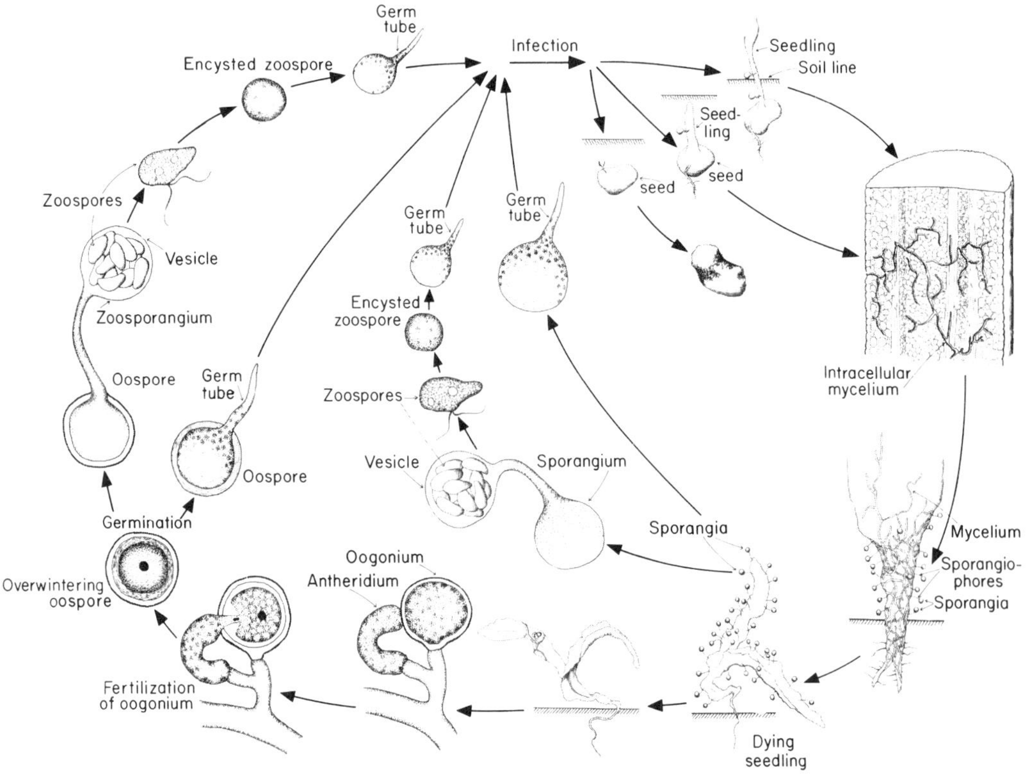

Fig. 3. Disease cycle of damping-off and seed decay caused by *Pythium* spp. (by courtesy of G. Agrios and Academic Press [8])

A typical disease cycle for a soilborne oomycete is given in Fig. 3 with the example of a *Pythium* sp. Infections in this case usually start from oospores which germinate to form sporangia or infect host plants directly. The seedlings can be infected and killed before or after they emerge. Generally, the damage by *Pythium* spp. is most severe when conditions for seedling emergence are suboptimal due to excess soil moisture or unsuitable temperatures. The infected plant parts rot and new oospores are formed again in the debris. These resting structures are very resistant to environmental stress situations such as drought and cold and they remain viable for several years in the soil.

3. Traditional Control of Oomycetes by Nonsystemic Fungicides

The protective fungicides have been covered previously by Schlör (Vol. 2, p. 44) and by Scheinpflug et al. (Vol. 4, p. 117). Table 2 summarizes the major nonsystemic products which are of special importance for the control of oomycetes.

In the past, chemical control of foliar diseases caused by oomycetes has relied on the protection of susceptible plant parts. This was achieved by applying protective fungicides on plant surfaces where ingress of these pathogens could be expected. Once the fungi were inside the host tissue they could no longer be reached by the protective materials, which lacked the ability to penetrate plant surfaces in effective concentrations. This shortcoming of the surface fungicides, as

Table 2. Nonsystemic fungicides used against Oomycetes

Chemical class with typical structure	Common names	Year of introduction[a]
Copper compounds	Bordeaux mixture	1885
	copper oxychloride	early 1900's
Dithiocarbamates	thiram	1931
	zineb	1943
$[-S-C(=S)-NH-CH_2CH_2-NH-C(=S)-S-Mn-]_x$	maneb	1950
	metiram	1958
	mancozeb	1961
	propineb	1962
Polyhalogenalkylthio compounds		
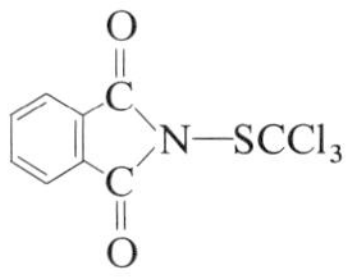	folpet	1949
	captafol	1961
	dichlofluanid	1965
Miscellaneous		
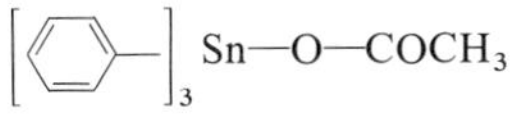	fentin acetate	1954
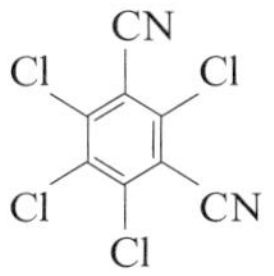	chlorothalonil	1963

[a] Taken from [14]

they might also be called, became even more dramatically apparent in the vain attempts to control systemic downy mildews.

For the control of soilborne oomycetes, nonselective soil fumigation methods as well as seed and soil treatments with selective chemicals have been used [15]. In addition, a wide variety of cultural practices are used to minimize the risk of soilborne plant diseases (e.g. crop rotation) and, more so than against foliar diseases, some success has been achieved with biological control methods [16].

In spite of their above mentioned shortcomings the nonsystemic products still dominate the market of oomycete fungicides. Even inorganic copper still plays a major role in the control of downy mildews and a wide range of diseases in tropical crops [17]. In France, 83 formulations containing copper were available to farmers in 1978; eight were still nothing else than Bordeaux mixture [18]. The expected share of the protective products in the 1980 world-wide fungicide market is another indication of their undiminished importance [19]:

Copper compounds	218 million US $
Dithiocarbamates	570 million US $
Phthalimides	427 million US $
Others	206 million US $

From an environmental standpoint these products are used at relatively high rates. In addition, the major group of the dithiocarbamates, the ethylenebisdithiocarbamates (EBDCs), have come under suspicion owing to the toxicological properties of their common metabolite ethylenethiourea (ETU) (Vol. 2, p. 68). Under certain test conditions ETU was found to be a tumorogenic agent in mice and rats [20–23]. A report also exists on teratogenic properties [24].

EBDCs are unstable in the presence of moisture and oxygen. They are also broken down in biological systems. Several degradation products result from the splitting-off of carbon disulfide or hydrogen sulfide or by oxidative degradation. These include ETU [25] (Vol. 4, p. 142). In a similar way 1,2-propylenethiourea (PTU) may occur in residues of 1,2-propylenebisdithiocarbamate (propineb) [26].

ETU is not the final stage in the degradation of EBDCs and is broken down further in soil, water, plant and animal tissues, and by photo-oxidation.

The degree to which ETU is formed depends on the stability of the parent compounds, the substrates, and the conditions under which degradation takes place. In neutral or alkaline conditions, a rise in temperature favors the formation of ETU. ETU can also be present as an impurity in EBDCs after their preparation or storage [27, 28], or can be formed as a metabolite in crops [29–34]. If EBDCs are still present in food crops they can readily give rise to ETU during the cooking process [35–38].

All nonsystemic fungicides mentioned above are apparently biochemically nonselective cell toxicants, which interfere with the energy metabolism of target organisms [39]. While the derivatives of dithiocarbamic acid form toxic copper complexes, the ethylenebisdithiocarbamates appear to react with sulfhydryl of enzymes and other cell constituants [40]. The latter mechanism has also been proposed for the captan type fungicides [41, 42] and for chlorothalonil [43, 44]. The selectivity of these products seems to be due to their selective penetration into target fungi and to their exclusion from nontarget fungi and plant tissues.

4. Chemistry and Biological Properties of Systemic and Specific Fungicides Against Oomycetes

4.1. Earlier Fungicides with Specific Activity Against Oomycetes

These products have already been reviewed in general terms by Schlör (Vol. 2, p. 44) and Scheinpflug et al. (Vol. 4, p. 117). Here they are of special interest because they already possess some of the attributes of the new era of oomycete fungicides, such as narrow spectrum of activity and systemicity.

4.1.1. Fenaminosulf

Sodium 4-dimethylaminobenzenediazosulfonate

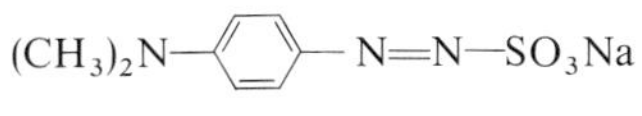

Common name: fenaminosulf
(Vol. 4, p. 167)

At Bayer AG, attempts were made to improve the fungicidal activity of Ceredon® (1,4-benzoquinone 1-benzoyl-hydrazone 4-oxime); acyl derivatives of 4-aminophenylhydrazine were prepared [45] which also exhibited a good activity against oomycetes. A similar fungicidal activity was expected from monodiazotized 4-phenylenediamine, and this led to fenaminosulf [46].

This compound was first synthesized in 1912 [47], and its fungicidal properties were discovered by Urbschat and Frohberger in 1956 [48]. Fenaminosulf is particularly active against *Aphanomyces*, *Pythium*, and *Phytophthora* spp. [49, 50], but has little or no effect against pathogens from other taxonomic groups, with the remarkable exception of *Tilletia caries* [51].

Fenaminosulf is suited only for soil and seed application, due to its instability to light [52]. Incorporated into the soil, it is rather stable and biological activity has been found 1 year after treatments [53, 54]. The systemicity of fenaminosulf is sufficient to protect emerging seedlings from attack [49]; lack of foliage protection after root uptake appears to be due to photolysis rather than lack of systemicity [52, 55], since root uptake is sufficient to protect tomato against *Phytophthora infestans* under artificial light [50].

The selectivity of fenaminosulf appears to be due to penetration barriers rather than biochemical site specificity, since it has been shown to be a multisite inhibitor of respiration in cells [56]. *Pseudomonas fragi* can metabolize fenaminosulf to the equally fungitoxic 4-dimethylaminoaniline [57].

Fenaminosulf is prepared by diazotization of 4-dimethylaminoaniline and subsequent reaction with sodium sulfite ([46, 47], Vol. 2, p. 100).

4.1.2. Etridiazole

5-Ethoxy-3-trichloromethyl-1,2,4-thiadiazole

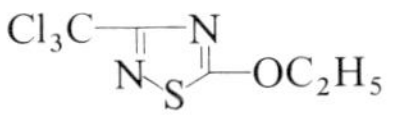

Common name: etridiazole
(Vol. 4, p. 219)

Etridiazole itself [58] and its use as a fungicide [59] were patented by Olin Mathieson in the early 1960s, whereby the good activity against soilborne fungi was disclosed. Its fungicidal properties were first described by Thomas [60], and it was introduced in 1969 by Olin Mathieson under the code number OM 2424 and the trade name Terrazole® as a soil fungicide and a component in seed-dressing formulations.

Like fenaminosulf, etridiazole is used primarily against soilborne oomycetes especially in cotton, turf and ornamentals. Its fungitoxicity in vitro is higher, and its systemicity in plants weaker than that of fenaminosulf [50, 61]. It is thus used most advantageously as a preventive contact fungicide [62]. Unlike fenaminosulf, etridiazole is also active against *Zygomycetes* [50]. There have been occasional reports of protection against other fungi, e.g. against *Rhizoctonia* in cotton [61]. However, the use of mixtures prevails, if higher fungi are also to be controlled.

Etridiazole causes lysis of the mitochondria in *Mucor mucedo* near the ED_{50} concentration for growth inhibition possibly via activation of phospholipases. The basis for its specificity towards oomycetes and zygomycetes is not yet understood [56].

Etridiazole can be prepared by cyclization of trichloroacetamidine hydrochloride and trichloromethanesulfenyl chloride in aqueous sodium hydroxide solution by the method described by Goerdeler [63] to give 5-chloro-3-trichloromethyl-1,2,4-thiadiazole, which is then reacted with sodium ethoxide [58, 64]. In a recent variation, sodium ethoxide is replaced by aqueous sodium hydroxide in ethanol [65]. Another recent process involves the reaction of trichloroacetonitrile in ethanol with excess anhydrous ammonia to give trichloroacetamidine. This is then treated without isolation with trichloromethanesulfenyl chloride, and finally with sodium hydroxide in ethanol, to form etridiazole [66].

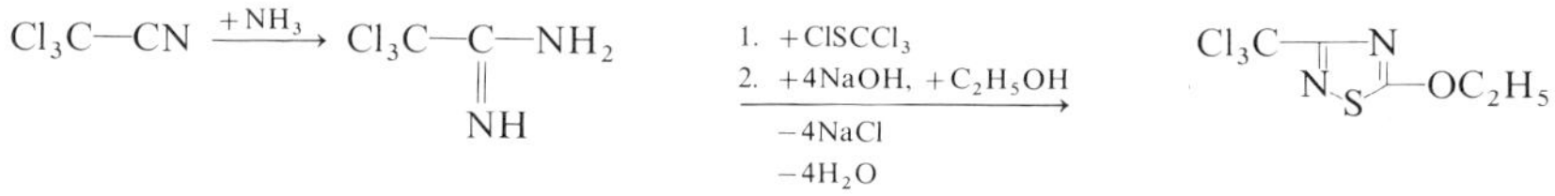

In a further process, trichloroacetamidine is reacted with ethyl chlorothiocarbonate and the resulting intermediate is then treated with gaseous chlorine to give etridiazole [67].

4.1.3. Chloroneb

1,4-Dichloro-2,5-dimethoxybenzene

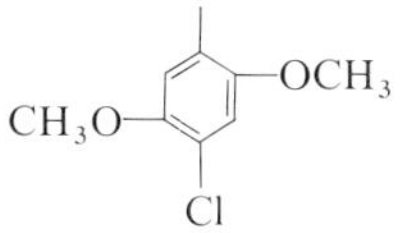

Common name: chloroneb
(Vol. 4, p. 160)

The compound itself has been known for about a century [68]. Its use as a fungicide is covered by a patent filed in 1963 by Du Pont [69]. The use of chloroneb against soilborne cotton diseases was first described by Ryker [70]. It is still primarily used on cotton as a seed dressing and soil fungicide against *Rhizoctonia*, but also on turf against grey snow mold (*Typhula*) and *Pythium* blight [71]. The fungitoxic spectrum of chloroneb is unusual in that it does not follow taxonomic criteria among the lower fungi; e.g. it shows good activity against *Phytophthora cactorum*, *Pythium aphaniderатum*, and *Mucor mucedo*, but not against *P. debaryanum* and *P. ultimum* [50, 72]. In addition, it is active against selected higher fungi, such as *Rhizoctonia solani* and *Colletotrichum lindemuthianum* [50].

Interference of chloroneb with DNA synthesis of *R. solani* [73], as well as lysis of cell organelles and wall thickening around hyphal tips of *M. mucedo* [56, 74], have been described. In *Ustilago maydis*, inhibition of cell division occurred before any effects on DNA could be observed [75].

After both seed and soil treatments of cotton and bean plants with chloroneb, the fungicide is absorbed and accumulated in the roots and lower stem portions, and only a small portion reaches the foliage [76]. In both plants, chloroneb is converted to 2,5-dichloro-4-methoxyphenol, which has a fungitoxicity similar to that of the parent compound.

The half-life in soil of chloroneb is between 3 and 6 months [76]. Soil microorganisms demethylate chloroneb to give the main metabolite 2,5-dichloro-4-methoxyphenol. Some 2,5-dichlorohydroquinone is also formed. On the other hand, some soil microorganisms methylate a part of the 2,5-dichloro-4-methoxyphenol back to chloroneb. This interconversion may explain the relatively high persistence of chloroneb in the soil [77, 78].

Preparation of chloroneb by chlorination of p-dimethoxybenzene according to the usual methods leads to phytotoxic contaminants. Du Pont have, therefore, developed a process in which the chlorination takes place in carbon tetrachloride as solvent; methanol and anhydrous aluminum chloride are used as catalysts. This method reduces the phytotoxic contaminant to less than 1% [79].

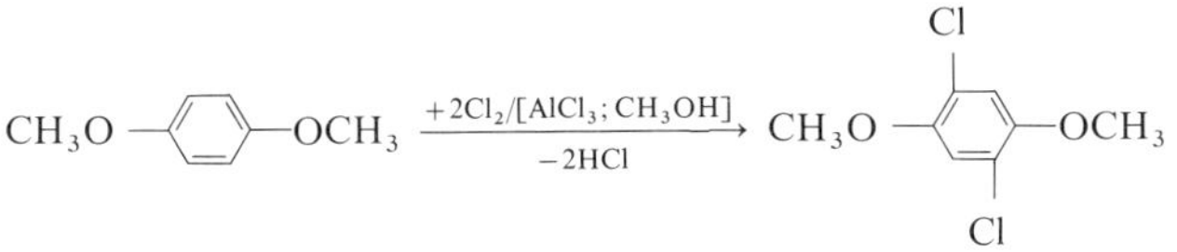

4.1.4. Hymexazol and Related Compounds

3-Hydroxy-5-methylisoxazole

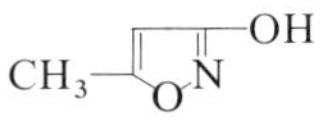

Common name: hymexazol
Other name: Tachigaren®
(Vol. 4, p. 215)

Hymexazol forms colorless, odorless crystals. Its solubility in water at 25 °C is 85 g/l [80]. It is readily soluble in most organic solvents and stable in acids and alkalis. It remains stable in sunlight for up to 20 days but is readily photolyzed by ultraviolet light. The main photolysis product is 5-methyl-4-oxazolin-2-one [81].

The acute oral LD_{50} for male rats is 4678 mg/kg. The acute dermal LD_{50} for rabbits is greater than 2000 mg/kg, and the TLM (48 h) for young carps is 165 ppm [80, 82].

Iwai et al. had been working with acetylenic compounds in the laboratories of the Sankyo Company since 1956 [80] when they developed, in 1964, a new convenient method for preparing 5-substituted 3-hydroxyisoxazoles involving the reaction of α-acetylenic acid esters with hydroxylamine [83]. Of the compounds thus obtained, the 5-methyl-substituted derivative hymexazol proved to be the most effective nonphytotoxic soil fungicide. Replacement of the 5-methyl group by hydrogen, other alkyl groups, or phenyl led to a reduction in fungicidal activity [80]. Patent applications were filed in 1964 and 1965 covering hymexazol itself [84], soilfungicidal compositions containing it [85], and a process for its preparation [86].

The fungicidal activity of hymexazol was described by Iwai et al. and it was introduced by Sankyo in 1970. It can be used as soil or seed treatment for the control of root and damping-off diseases caused by *Aphanomyces*, *Pythium*, *Fusarium*, and *Corticium* spp. [80]. The unique spectrum of activity of hymexazol, which apparently does not include *Phytophthora* spp., has led to its extensive use in media selective for the isolation of *Phytophthora* spp. [87]. Aluminum and iron ions appear to be required for its toxicity to certain fungi. Independently of its fungicidal action, hymexazol exerts a growth-stimulating action on many plants, especially in the seedling stage. Stimulation of rooth growth, including root hairs, is more pronounced than that of shoot growth. In addition, hymexazol has been shown to increase cold hardiness of rice plants [88]. This plant growth stimulation by hymexazol obviously contributes indirectly to its success as a fungicide. It increases the vigor of seedlings and thus makes them less susceptible to damping-off pathogens, even to those which are not sensitive to the direct fungicidal action of hymexazol in vitro.

Hymexazol is taken up rapidly by plant roots and transported acropetally to above-ground parts [89]. Two major metabolites are formed in plants, the structures of which have been confirmed by synthesis:

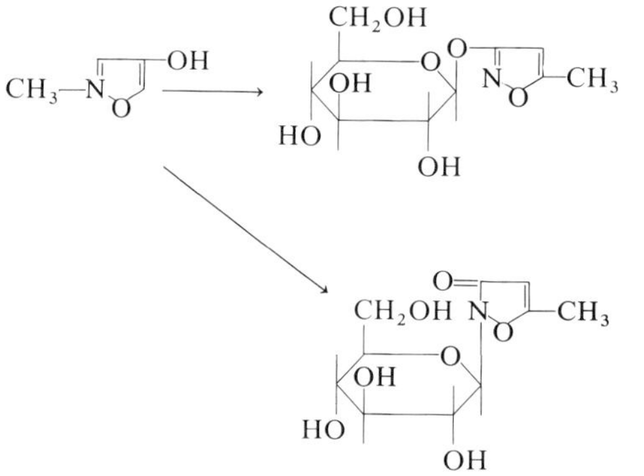

O-β-glucoside

3-(β-D-glucopyranosyloxy)-5-methylisoazole

N-β-glucoside

2-(β-D-glucopyranosyl)-5-methyl-4-isoxazolin-3-one

The O-β-glucoside has similar activity to hymexazol, whereas the N-β-glucoside is not fungitoxic but more active as a growth promoter [89].

In the soil, the effectiveness of hymexazol decreases with time, persisting only a few weeks. The major metabolites are acetoacetamide and 5-methyl 4-oxazolin-2-one, which are in turn broken down to acetone, ammonia, and carbon dioxide [80].

At first, hymexazol was prepared by reacting ethyl tetrolate with hydroxylamine in excess aqueous sodium hydroxide [83, 86]. Replacement of sodium hydroxide by calcium hydroxide produces hymexazol of increased purity [90]. Hymexazol can also be prepared by the reaction of an alkyl 2,3-dihalogenbutyrate or an alkyl 2-halogencrotonate or alkyl 3-halogencrotonate, which are regarded as potential alkyl tetrolates, with hydroxylamine in excess alkali [91]. A very simple, recently developed process starts from ethyl acetoacetate which is converted in the usual manner with phosphorous pentachloride to ethyl 3,3-dichlorobutyrate. This intermediate is then treated with aqueous/methanolic sodium hydroxide containing hydroxylamine [92].

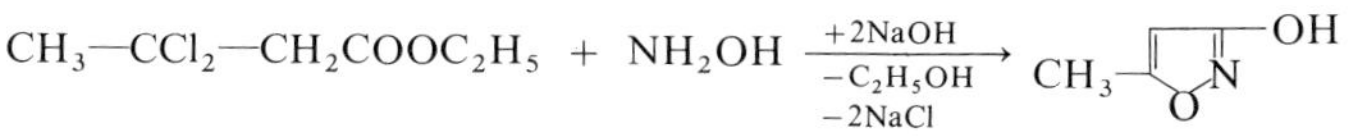

A ring structure related to hymexazol is contained in

drazoxolon 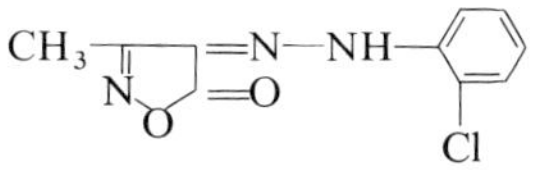 and

(Vol. 4, p. 216)

metazoxolon 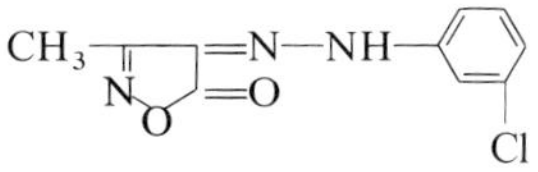

The biological properties and fungicidal uses of drazoxolon were described in 1967 [93], and those of metazoxolon in 1973 [94]. The latter appears to be fairly selective against oomycetes and superior as a seed dressing agent and soil fungicide [95, 96].

4.2. Pyroxychlor and Other Pyridines

Although pyroxychlor is apparently not commercialised, it possesses a number of unique properties as a fungicide which make this chemical group of interest for this review.

The Dow Chemical Company has been working for several years with chlorinated pyridines, discovering such products as the nitrification inhibitor N-serve® [97], the herbicides Tordon® [98] and Lontrel® [99], and the insecticide Dursban® [100]. The three most interesting fungicides to result from this work were Dowco 248, Dowco 263, and especially pyroxychlor (Dowco 269).

2-Chloro-6-methoxy-4-trichloromethyl-pyridine

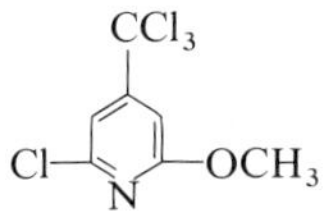

Common name: pyroxychlor
Other names: Dowco 269, Nurelle®, Lorvek®

Pyroxychlor is a light yellow liquid, freely miscible with organic solvents, and has a solubility in water of 11 ppm [101].

The acute oral and dermal LD_{50} of the technical product in rats is 1500 mg/kg. Undiluted material applied to the eyes of rabbits results in slight irritation [101].

Closely related to Dowco 269 is Dowco 248

6-Amino-2-chloro-4-trichloromethyl-pyridine

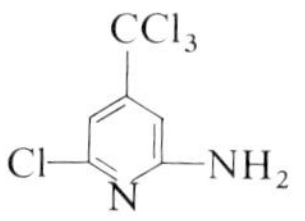

Other name: Dowco 248

Dowco 248 forms pale yellow crystals only sparingly soluble in water. Its solubility in organic solvents is good [102].

Pyroxychlor itself is covered by a Dow Chemical Company patent which claims ethers of trichloromethyl-pyridines and discloses various pesticidal uses for them [103].

A later Dow patent covers the use of certain selected ethers as systemic plant fungicides with downward transport, stressing the dependence of this exceptional activity on the highly specific 2,4,6-substitution with an alkoxy group in the 6-position [104]. Dowco 248 is covered by a patent claiming various substituted aminohalopyridine compounds [102].

Sbragia described how the unique biological properties of pyroxychlor were discovered [105, 106]. In the preliminary structure-activity studies against *Pythium ultimum* the 4-trichloromethyl-group was shown to be essential for fungicidal activity. Dowco 248 was apparently the most active pyridine with this group at the 4-position while pyroxychlor showed only moderate activity. Beside the amino and the methoxy group, chlorine and hydroxy groups in the 6-position also resulted in fungicidal activity. The uniqueness of pyroxychlor became apparent only in a subsequent test against black shank of tobacco (*Phytophthora parasitica* var. *nicotinae*), in which, after drench application, it was more active and had a longer residual effect than all other analogues. Later it was found that the 2-bromo and the 6-ethoxy analogues were equally active and that pyroxychlor controlled tobacco black shank as a foliar spray. This was the first report of a fungicide with symplastic transport and proved that, in principle, soilburne fungi can be controlled by foliar spray applications with suitable chemicals [101, 107, 108].

In further translocation studies it was found that pyroxychlor is transported much less from roots to leaves than vice versa [109, 110], thus placing this fungicide in the small group of xenobiotic chemicals with preferentially symplastic transport. Comparisons with Dowco 248, which showed better apoplastic but weaker symplastic transport than pyroxychlor, led to the suggestion that the substituent at the 6-position determines the symplast-apoplast balance of these pyridines [111]. Basic studies on phloem transport of these chemicals were described by Small et al. [109]. The translocation pattern of pyroxychlor in plants seems to be largely responsible for the lack of activity *in vivo* against *Phytophthora infestans* and other foliar oomycete pathogens and for its outstanding activity against root pathogens through accumulation of the fungicide in the root zone of plants [103]. The in vitro spectrum of activity of pyroxychlor includes *P. infestans*, and it has been found to be active against all *Phytophthora* spp. tested so far [112, 113], whereas *Pythium* spp. appear to vary remarkably in their sensitivity to pyroxychlor both in vitro and in vivo [112]. In addition, activity against *Aphanomyces* pea root rot has been reported [114].

Pyroxychlor and Dowco 248 share the same intermediate, 2,6-dichloro-4-trichloromethylpyridine, which is produced by melting 4-methylpyridine hydrochloride and chlorinating it at 100 °C under irradiation [102]. This common intermediate is reacted with sodium methoxide in boiling methanol to give pyroxychlor [103] or at 100 °C and under pressure with anhydrous ammonia to give Dowco 248 [102].

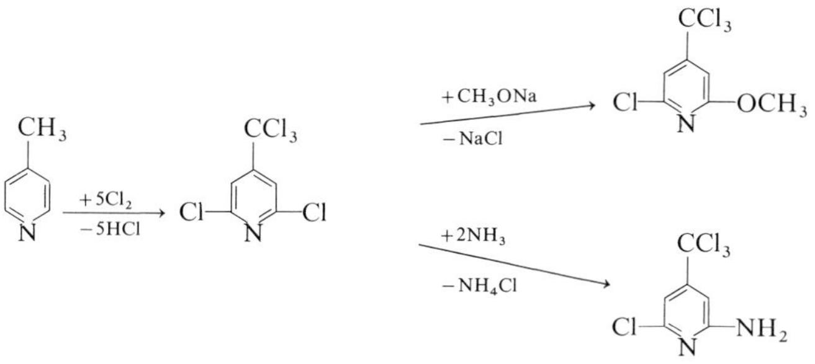

A further pyridine to be described as a fungicide is Dowco 263.

3,4,5-Trichloro-2,6-dicyano-pyridine

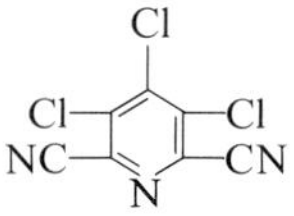

Other name: Dowco 263
(Vol. 4, p. 204)

The compound was first covered by a Diamond Alkali Co. patent [115], which also mentioned both fungicidal and herbicidal activity. A later Dow Chemical Co. patent claimed the use of Dowco 263 as a seed-dressing agent for enhancing plant growth [116]. As a fungicide, it is active against soilborne seedling diseases [117, 118], and also against smut and bunt diseases of cereals [119].

The 2,6-dicyanopyridine required as starting material for Dowco 263 can be prepared by an ammoxidation process involving the reaction of 2,6-lutidine with ammonia and oxygen at elevated temperatures and in the presence of a catalyst. Subsequent chlorination is carried out at a temperature above 300 °C using an activated carbon catalyst impregnated with barium chloride [115].

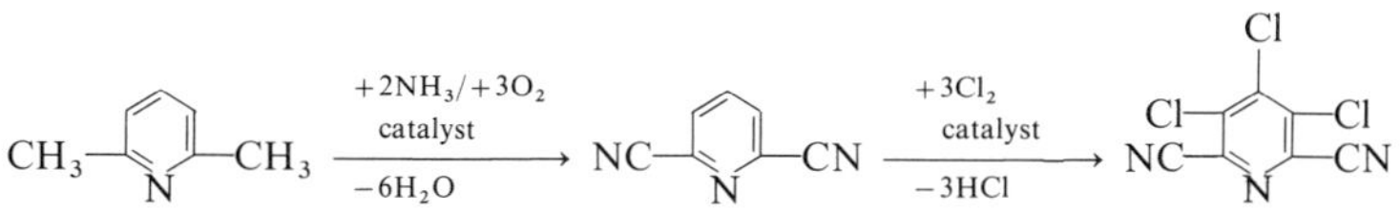

4.3. Prothiocarb and Propamocarb

S-Ethyl-N-(3-dimethylaminopropyl)-thiocarbamate hydrochloride

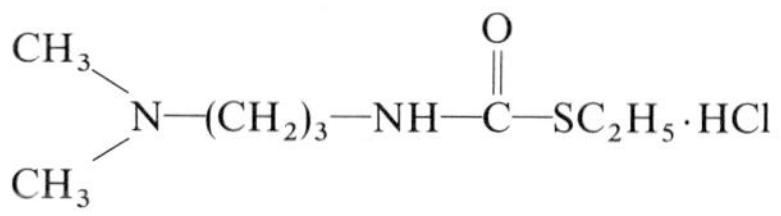

Common name: prothiocarb HCl
Other names: SN 41 703, Previcur®, Dynone®
(Vol. 4, p. 137)

Prothiocarb HCl in the pure state forms colorless, odorless crystals which are strongly hygroscopic. The technical grade product has a faint odor similar to ethyl mercaptan. Its solubility in water at 25 °C is 890 g/l and in benzene and hexane less than 150 mg/l. The acute oral LD_{50} in rats is 1300 mg/kg and the acute dermal LD_{50} of a 70% formulation for rabbits is greater than 1400 mg/kg [14, 120].

Propyl-N-(3-dimethylaminopropyl)-carbamate hydrochloride

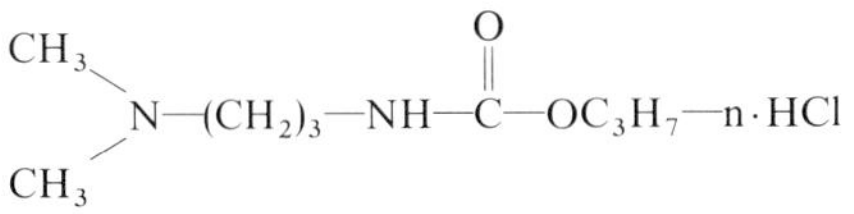

Common name: Propamocarb HCl
Other names: SN 66 752, Previcur® N

The compound forms colorless, odorless crystals which are also strongly hygroscopic. Its solubility in water at 25 °C is greater than 700 g/l. The acute oral LD_{50} for rats is 8600 mg/kg and the acute dermal LD_{50} of a 70% formulation on rats and rabbits is greater than 5000 mg/kg. Propamocarb HCl ist practically nontoxic for fish. The LC_{50} (96 h) of the formulation is 616 ppm for rainbow trout and 234 ppm for carp [121].

In late 1966 Schering AG filed a patent application covering prothiocarb and propamocarb as free bases in fungicidal and fungistatic agents [122]. During 1967 a further application was filed covering inter alia the hydrochlorides of prothiocarb and propamocarb [123].

Prothiocarb HCl was presented in 1974 by Bastiaansen et al. [120] and introduced under the trade name Previcur®. Pieroh et al. reported on propamocarb HCl in 1978 [121]. It bears the trade name Previcur®N.

Prothiocarb HCl ist recommended against soilborne oomycetes on ornamentals and certain horticultural crops [120, 124]. It is used primarily as a soil treatment, but it can also be used as flower bulb treatment against *Pythium* soft rot and root diseases, and it has shown activity against black shank of tobacco [125, 126]. Although there are reports of control of foliar diseases by prothiocarb HCl it seems less effective here than against soilborne pathogens [127–129].

Prothiocarb hydrochloride is transported apoplastically in plants [126, 129]; no evidence has been found for symplastic transport [130] nor for degradation in bean plants. Though acropetal transport is established, the roots seem to represent a substantial barrier for the uptake of prothiocarb, since uptake from a solution into cut stems of tomato plants is far superior to that of intact roots [50]. Prothiocarb HCl, like its somewhat weaker O-analogue propamocarb HCl, is active specifically against oomycetes [131, 132]. The activity in vitro is evident only at a neutral pH and, against some *Phytophthora* spp., only at low temperatures [132]. For most fungi the toxic entity of both prothiocarb HCl and propamocarb HCl appears to be the whole molecule, the action of which can be reversed by sterols [131, 133]. However, certain *Saprolegniales* spp. seem sensitive only to the voltaile degradation product ethyl mercaptan of prothiocarb HCl, the action of which can ge reversed by L-methionine; consequently these fungi are not sensitive to propamocarb HCl [132]. Both in the plants and in soil, prothiocarb HCl ist degraded mainly via hydrolysis to N,N-dimethylpropane-1,3-diamine [134]. In soil an alternative route appears to be demethylation to N-demethylprothiocarb and cyclization of the latter to 1-methyl-tetrahydro-1,3-diazin-2-one. Degradation is slower in more acid soils which contain less organic material than in soils with high organic material content [134].

Prothiocarb HCl ist prepared by reaction of N,N-dimethylpropane-1,3-diamine with S-ethyl chlorothioformate in an inert organic solvent. Tertiary organic amines or inorganic bases can be used as proton acceptors. In a second stage, the free base is converted with hydrogen chloride to prothiocarb HCl [123].

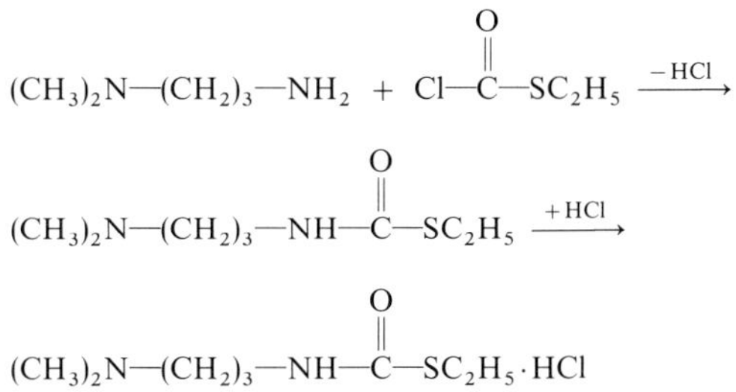

The preparation of propamocarb HCl follows analogously using n-propyl chloroformate [122, 123].

4.4. Curzate

2-Cyano-N-(ethylcarbamoyl)-2-(methoximino)-acetamide

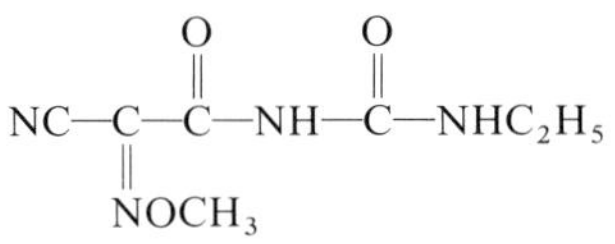

Common name: cymoxanil
Other names: DPX 3217, Curzate®
Mixtures: Remiltine®, Fulvax® } with mancozeb
(Vol. 4, p. 146)

DPX 3217 is a white, crystalline solid. Its solubility in water is approximately 0,1%. It is stable under normal storage conditions and in neutral or slightly acid media, but unstable in alkaline media [135, 136]. It is compatible with most commonly used insecticides and fungicides, but not with compounds such as lime-sulfur or maneb, which increase alkalinity [136].

DPX 3217 has a favorable toxicity pattern with an oral LD_{50} for male rats of 1425 mg/kg for an 80% active formulation. The acute dermal LD_{50} of the active substance alone for male rabbits is 3000 mg/kg. It has been shown to be neither a skin irritant nor a skin sensitizer when tested on guinea pigs as a 50% aqueous dispersion. Very slight eye irritation in rabbits has been observed. DPX 3217 is slightly toxic to bluegill sunfish and rainbow trout, the LC_{50} values (96 h) being 13,5 ppm and 18,7 ppm respectively [135, 136].

Carbamates of 2-cyano-2-hydroximino acetamides and esters had previously been described as fungicidally active [137], when in the early 1970s Du Pont recognized the good systemic and curative activity of related compounds against oomycetes. Other companies also showed interest in such compounds [138–144]. DPX 3217 and its analogues were patented as compounds per se [145] and for use as fungicides, whereby mixtures of DPX 3217 with other fungicides were also protected [146]. Compositions containing DPX 3217 can be stabilized by the addition of acidifying agents [147, 148]. This class of compounds was further developed by Du Pont, for example by making changes in the urea portion of the molecule [149–152].

DPX 3217 is recommended in a number of combinations with protective products against grape downy mildew and potato late blight [135, 136]. DPX

3217 alone loses its protective activity within a few days, especially under hot weather conditions. However, in combinations, low rates of DPX 3217 substantially improve the performance of protective products such as mancozeb, folpet, and copper. This apparently synergistic effect appears to result from the addition of the good curative effect of DPX 3217 up to 3 days after infection to the protective effects of the standard fungicides [135].

Though only local systemicity is claimed for DPX 3217 [135, 153], it seems to be taken up easily by roots of tomato plants and transported acropetally to the leaves in concentrations effective against *P. infestans* [50]. A comparison of systemic uptake via intact roots and via cut stems indicates further that tomato roots are not a major barrier to the uptake of DPX 3217.

The spectrum of activity of DPX 3217 appears to be limited to the order Peronosporales [135, 153]. The fungicide exhibits a remarkable selectivity among *Phytophthora* spp., where *P. infestans* is over 100 times more sensitive in vitro than other species [154].

DPX 3217 is rapidly broken down in plant tissue, and loses its activity within 3–6 days after application [136]. Treated crops contain little or no detectable residues. Stability in soil is also low, more than half the substance being broken down within 2 weeks [136]. Fermentation and the quality of wine from treated grapes are unaffected [135].

DPX 3217 is a derivative of 2-cyano-2-hydroxymino acetic acid. This class of compounds was known from the literature [155–158], as was the conversion of 2-cyano-N-carbamoyl acetamide to 2-cyano-2-hydroximino-N-carbamoyl acetamide [158, 159].

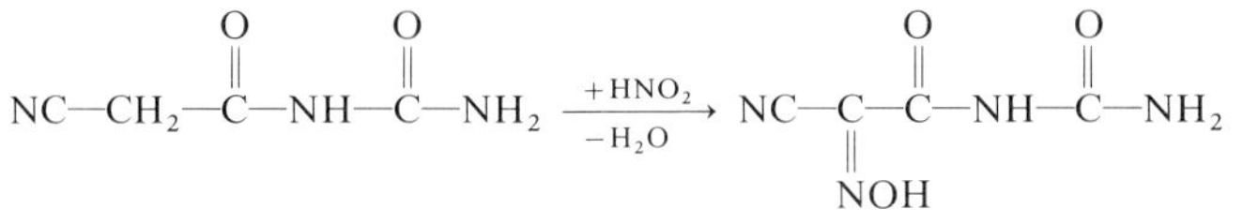

The following is a description of one of the methods by which DPX 3217 can be prepared. Cyanoacetic acid is condensed in the presence of acetic anhydride with ethylurea, and the resulting 2-cyano-N-ethylcarbamoyl acetamide converted by the described process [158] to the hydroximino derivative. The latter is then methylated with dimethyl sulfate [160].

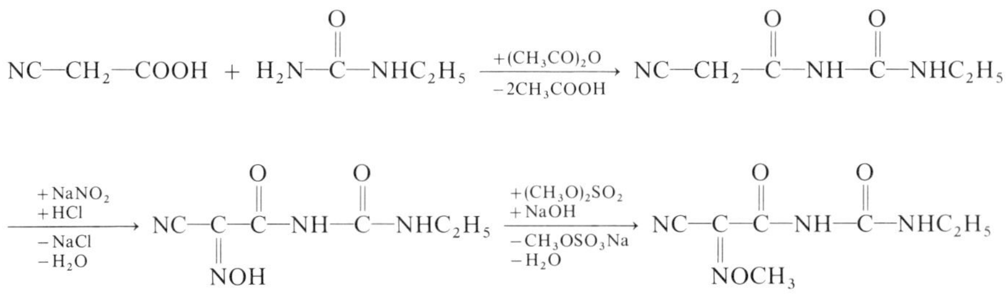

In a variation of the conversion to the hydroximino derivative, less than the stoichiometric amount of acid is used, thus suppressing unwanted side reactions.

The intermediate thus obtained in high yield can then, without isolation, be directly methylated to DPX 3217 after addition of aqueous sodium hydroxide [149].

4.5. Ethyl Phosphites

Aluminum tris(O-ethylphosphite)

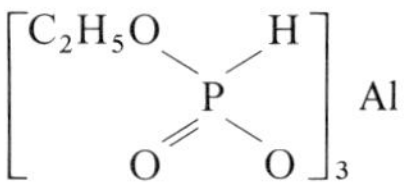

Common name: phosetyl-aluminum
Other names: LS 74 783, "Aliette", phosetal
Mixtures: Mikal® (formulation EXP 1710 = 50% phosetyl Al + 25% Folpet) [161, 162]

Photesyl-Al takes the form of white odorless crystals. Its solubility in water at 20 °C is 120 g/l; in organic solvents it is low. It is stable both as technical grade and as formulated product under normal storage conditions. A 0,1% aqueous solution kept at 20 °C for 100 days still contains more than 50% of unchanged phosetyl-Al [161, 163].

The acute oral LD_{50} of technical grade phosetyl-Al for rats is 5800 mg/kg. Toxicity on contact with skin or mucous membranes is also low, the acute dermal LD_{50} for rats being greater than 3200 mg/kg. The low toxicity to aquatic fauna is shown by LC_{50} values (96 h) for rainbow trout of 428 mg/l and for harlequin fish of 161 mg/l [161, 163].

A related compound is LS 73 1038, which has not, however, been commercially developed:

Sodium O-ethylphosphite

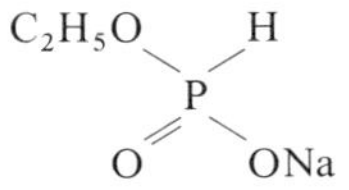

Common name: none
Other names: LS 73 1038, "Nina"

Its acute oral LD_{50} for rats is 6400 mg/kg and the acute dermal LD_{50}, also for rats is greater than 5000 mg/kg [163].

Although both the above compounds had been known for some time [164, 165], there had been only few indications that O-phosphites and related compounds possess fungicidal activity. Monsanto, for example, had patented derivatives of ammonium O-alkylphosphites as defoliants, mentioning fungicidal and germicidal properties in passing [166]. In a further example, a range of compounds including certain secondary phosphites was claimed to be active against *Piricularia oryzae* [167]. The importance of this class of compounds as fungicides was recognized in 1967 by Rhône-Poulenc, who filed a patent application covering their use as fungicides [168]. This application contains field trial reports showing that a compound with the empirical formula $C_2H_8NO_3P$ was more active against *Plasmopara viticola* than zineb. The exact consitution of the compound, prepared by reaction of 0,0-dimethyl phosphite and monoethanolamine, was uncertain.

The following three structures were suggested [168, 169].

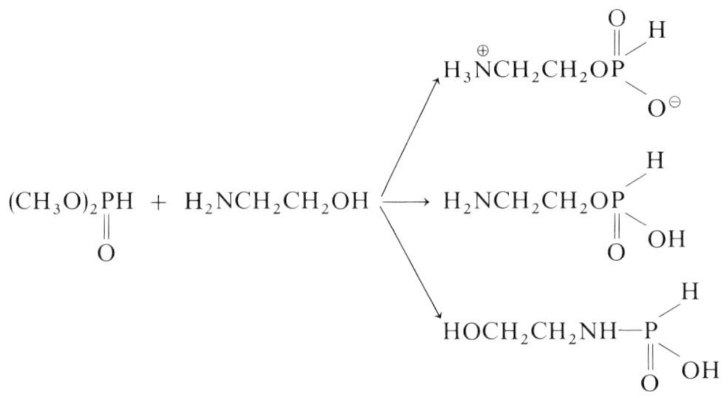

This process and constitution had already been covered by a previous patent, which described the use of this type of compound as flame retardants in aldehyde condensation resins [169].

According to Bertrand et al. [163], the most remarkable property of this chemical class, its pronounced phloem transport was discovered in 1971 by keen observation during a vineyard trial. In one plot, grapevine shoots that had developed after the last spray remained free of downy mildew into the fall, whereas in all other plots these new shoots had lost their leaves due to heavy downy mildew attack. Subsequently, it was found that the sodium and particularly the aluminum salts of O-ethyl phosphite were the most promising representatives of this relatively simple group of compounds for the control of various plant diseases caused by oomycetes. Later, the aluminum salt gained priority owing to its better crop tolerance [163]. This led in 1973 and 1974 to the filing of further patent applications covering the use of such phosphorous acid esters and their salts, e.g. phosetyl-Al and LS 73 1038, as fungicides [170]. Reference was made to their protective, curative, and systemic activity against diseases caused by oomycetes such as downy mildew on grapevine. Phosphorous acid derivatives and related compounds were intensively researched by Rhône-Poulenc, particularly following the filing of the patent application covering phosetyl-Al [171–179]. The major activity of these compounds remained that against oomycetes.

Phosetyl-Al itself and LS 731038 were presented in 1977 by Bertrand et al. [163] and by Williams et al. [161]. The spectrum of dieseases which are controlled by O-ethyl phosphites is remarkable in that it does not appear to follow clear taxonomic lines. Phosetyl-Al is active against most downy mildews as well as most *Pythium* and *Phytophthora* species; however, *P. infestans* does not seem to be sensitive at all, and only weak activity has been reported against blue mold of tobacco. And though *Mucor* and *Rhizopus* are not controlled by phosetyl-Al [180], its spectrum of activity does not seem to be restricted to oomycetes on certain crops: e.g. good activity has been reported against *Phomopsis viticola* (excoriosis) and *Guignardia bidwellii* (black rot) on grapevine, *Cercospora* leaf spot on peanuts, and *Alternaria tenuis* var. *mali* on apples [163, 181]. It appears, therefore, that the spectrum of activity of the O-ethyl phosphites is, at least in part, determined by the host plant.

The activity of the O-ethyl phosphites in vitro is weak and somewhat pH-dependent [182, 183] with ED_{50} values for inhibition of mycelial growth of about 3000 ppm for various *Phytophthora* and *Pythium* spp. This has led some workers to suggest that a more fungitoxic degradation product might be formed in treated plants [184, 185]. However, no such transformation product has been demonstrated so far. The relatively high concentration of O-ethyl phosphites necessary for disease control might in fact be sufficient in many cases to inhibit the target fungi directly, particularly if the relatively higher activities on sporangia formation [183] or at pH value of the host plant tissues are considered. In view of the host plant effect on the spectrum of activity of phosetyl-Al, the possibility of an indirect effect on plant pathogens via host plant metabolism also remains open. First indications of stimulation of plant defense reactions are available [182].

The O-ethyl phosphites, like pyroxychlor, are transported from treated leaves to roots and to new shoots which develop after foliar treatments [163]. On several plants it could be shown that downward transport to roots after foliar treatments was sufficient to protect them against soilborne diseases caused by oomycetes [161, 163, 186]. Often, however, foliar treatments seem to be less effective and more phytotoxic than drench treatments [163, 183]. On grapevine, transport into new shoots formed after foliar treatments also seems sufficient for downy mildew control. The transfer of these fungicides from old to new leaves, however, shortens the protection of the former. This led to combination products (with protective fungicides) such as Mikal® [187, 188], which appears to be superior to both combination partners and which enables the spray intervals for the control of grape downy mildew to be extended [181, 189, 190]. Other protective mixing partners such as copper and mancozeb have shown similar apparently synergistic effects. Such mixtures have also been made the subject of a patent application [188].

Transport from the roots to the foliage of plants does not appear to be as efficient as downward transport, since a rate of 200 kg/ha of phosetyl-Al is needed to achieve good downy mildew control in cauliflowers [191]. The ready penetration of plant parts by the O-ethyl phosphites and the persistence of their activity in the root zone make them suitable for dip treatments of seedlings and vegetative cuttings against soilborne pathogens [192, 193].

Since it is a systemic fungicide, phosetyl-Al reaches the plant tissue by way of the roots or leaves. The main metabolite after leaf and soil application is phosphorous acid. The original compound is only present in barely detectable amounts [161].

The preparation of phosetyl-Al [165] and LS 731038 [164] is simple, and the necessary intermediates are readily available. LS 731038 is prepared according to a process known since 1936 [164] by reacting an aqueous/ethanolic solution of 0,0-diethyl phosphite with an equivalent amount of sodium hydroxide solution [170].

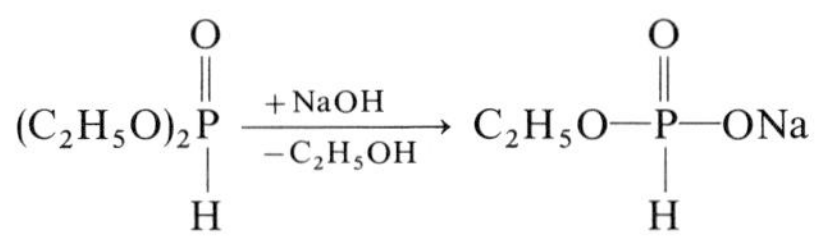

For the preparation of phosetyl-Al, an aqueous solution of sodium O-ethyl phosphite is reacted with aluminum nitrate, whereupon phosetyl-Al is precipitated [170]. Alternatively aluminum chloride is heated to approx. 100 °C with excess 0,0-diethyl phosphite [165].

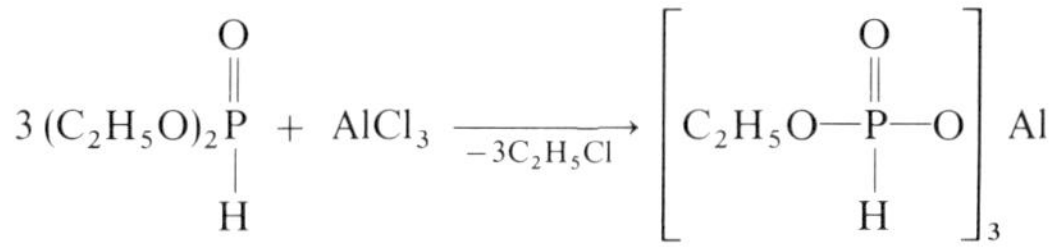

A highly efficient production process has been developed for phosetyl-Al [194]. A mixture of phosphorous acid in excess 0,0-diethyl phosphite is briefly heated to 150 °C, forming a ternary mixture containing, in addition, O-ethyl phosphite.

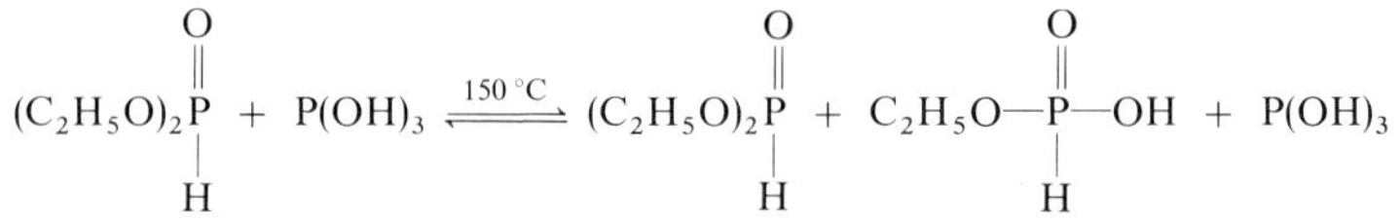

Hydrargillite is then added at 25 °C and the temperature slowly raised to 60 °C. The water formed is distilled off under reduced pressure. Following the addition of ethanol and cooling, the precipitated phosetyl-Al is removed. After distilling off the alcohol, the mother liquor can by recycled into the process.

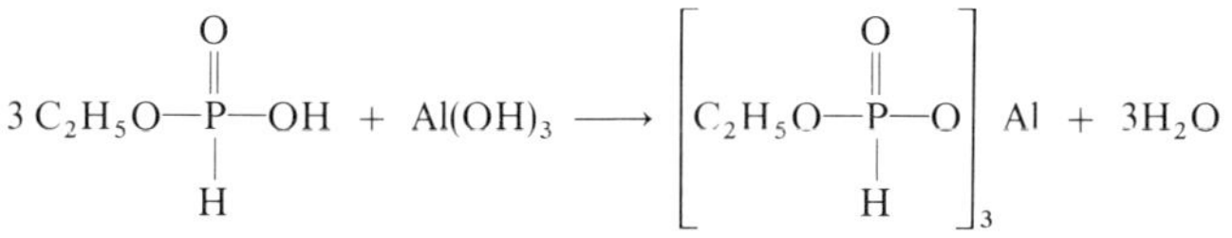

4.6. Acylalanines and Related Compounds

The acylalanines constitute a new group of fungicides with excellent protective, curative, and systemic activity against most plant diseases caused by oomycetes. The most important representatives are metalaxyl and furalaxyl.

Methyl DL-N-(2,6-dimethylphenyl)-N-(methoxyacetyl)-alaninate

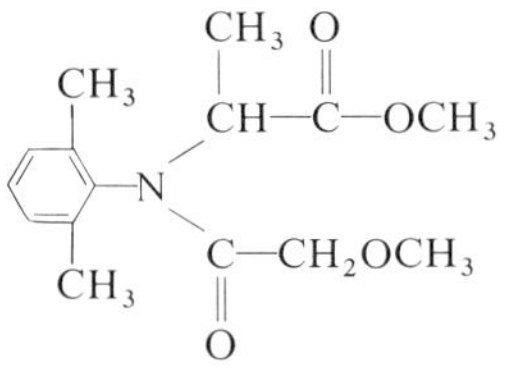

Common name:	metalaxyl
Other names:	CGA 48988, Ridomil®, Apron® 35 SD, Acylon®
Mixtures:	Ridomil plus®, Acylon bleu® — with copper
	Fubol® — with mancozeb

Metalaxyl forms white to beige crystals. Its solubility in water at 20 °C is 0,71 %. It is soluble in most organic solvents, and stable in neutral or slightly acid media. The acute oral LD_{50} for rats is 669 mg/kg for the technical grade product. The acute dermal LD_{50} for rats is greater than 3100 mg/kg for both technical grade product and a 50 % wettable powder. In the rabbit, metalaxyl is a mild skin irritant and minimally irritates the eyes. It has no sensitizing effect in guinea pigs, and is not toxic to fish [195, 196].

Methyl DL-(2,6-dimethylphenyl)-N-furoyl-(2)-alaninate

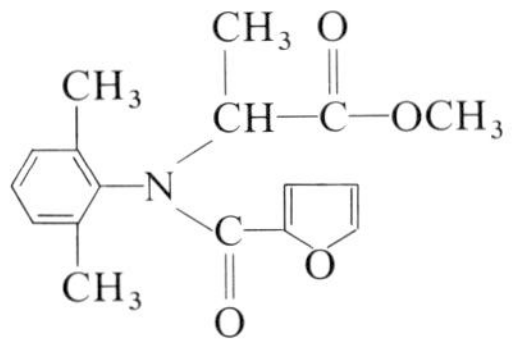

Common name: furalaxyl
Other names: CGA 38140, Fongarid®

Furalaxyl forms white crystals. Its solubility in water at 20 °C is 230 ppm. It is soluble in most organic solvents, and stable in neutral or slightly acid media. The acute oral LD_{50} in rats is approx. 1000 mg/kg for the technical grade product. The acute dermal LD_{50} in rats is greater than 3100 mg/kg for both the technical grade product and a 50% wettable powder. Furalaxyl does not seriously irritate the skin or the eyes of rabbits, and it has no sensitizing effect in guinea pigs. It is slightly toxic to fish, with an LC_{50} of between 10 and 50 ppm in four species [197].

Work on this class of compounds began in 1973 at Ciba-Geigy Ltd. with the discovery of the fungicidal properties of CGA 29212 [methyl DL-N-(2,6-dimethylphenyl)-N-(chloroacetyl)-alaninate], a chloroacetanilide (Vol. 4, p. 166). This compound, which originated from a herbicide project, was notable for its good curvative and systemic activity against oomycetes [198]. The suggestion that the chloroacetyl portion of the molecule was responsible for its fungicidal properties was soon abandoned, and replacement of this group led to the discovery of furalaxyl [199] and metalaxyl [200]. These compounds possess excellent fungicidal activity combined with good crop tolerance. The accepted generic term acylalanines was coined for this class of compounds to avoid confusion with the chloroacetanilides. Further acylalanines and related compounds with marked activity against oomycetes followed and formed the subject of a series of patent applications [201–210]. Metalaxyl and furalaxyl were brought onto the market by Ciba-Geigy Ltd. in 1977 under the trade names Ridomil® and Fongarid®, respectively. Since 1976, Bayer AG has also been working on chloroacetanilides [211] and acylalanines [212] active against oomycetes.

Metalaxyl is a versatile fungicide, and is recommended against a wide range of airborne, soilborne, and seedborne species of oomycetes [195, 196], while furalaxyl is specifically recommended for ornamentals [197, 213, 214], where it has been shown to be the safer of the two chemicals on certain plant species. At low rates of application, metalaxyl acts systemically and affords good curative and lasting protective activity. This has led to spray recommendations with longer intervals than those usually adopted for protective fungicides, which is of particular advantage in situations of high disease pressure [195, 196]. Precipitation, which favors downy mildew diseases and which decreases the efficacy of traditional protective products, has been shown to increase the efficacy of foliar sprays with acylalanines [215], presumably by favoring penetration into the leaves.

Because of its systemic properties, metalaxyl offers new possibilities for the control of systemic diseases such as downy mildew of hops (*Pseudoperonospora humuli*) [196], downy mildews of corn, sorghum, and millet (*Peronosclerospora* spp.) [216], and white rust of crucifers (*Albugo candida*) [217], which previously could not be adequately controlled by fungicides. Good activity by metalaxyl has also been reported against soilborne diseases such as black shank of tobacco [218] and damping-off diseases, as well as against foliar diseases after soil application [219, 220].

Within the class oomycetes, the spectrum of activity of the acylalanines appears to include all the species of the order Peronosporales [195–197]. This specificity is expressed both *in vitro* and *in vivo*. In situations where disease complexes are the rule (e.g. on grapevines), mixtures with fungicides possessing a broad spectrum and a complementary mode of action are used [196]. Such mixtures also improve the protection of senescing foliage, where acylalanines alone are less effective than on young tissue. In addition, the mixtures are thought to decrease the risk of development of resistance to acylalanines [221]. This risk was investigated in preliminary selection experiments with *Phytophthora infestans* in which no strains with in vivo resistance to acylalanines could be produced [222].

Acylalanines are taken up easily by plants through the roots, stems, and leaves, and transported acropetally with the transpiration stream [215, 223]. Phloem transport seems to be less pronounced [215], although control of stem canker and root rot of avocado (*Phytophthora cinnamoni*) has been reported after foliar treatment with metalaxyl [224].

Foliar pathogens are inhibited by metalaxyl only after they have penetrated the leaves, and the fungicide has little effect on spore germination on leaf surfaces [225, 226]. Against soilborne fungi, the formation of chlamydospores and of sporangia represents an additional key step in which metalaxyl can effectively inhibit fungal development [227, 228].

Furalaxyl and metalaxyl can be prepared by reacting 2,6-dimethylaniline at an elevated temperature and in the presence of sodium bicarbonate with an excess of metyhl 2-bromopropionate to give methyl DL-N-(2,6-dimethylphenyl)-alaninate, which is then acylated in the absence of an acid binding agent with furoyl chloride or methoxyacetyl chloride [199, 200].

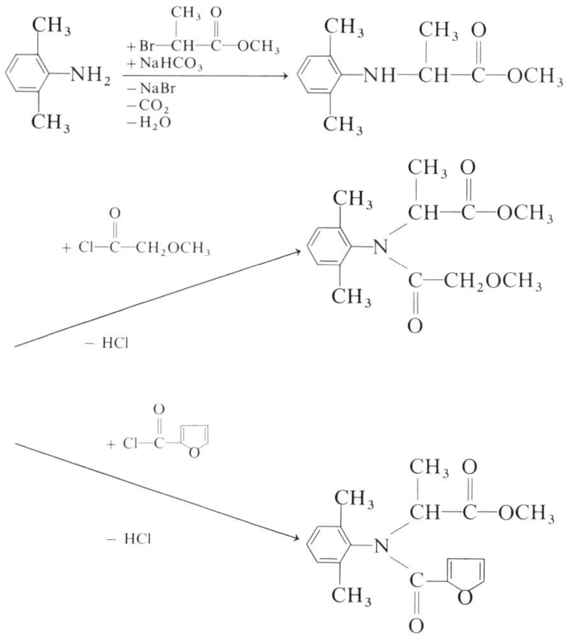

The acylalanines contain a chiralic center, and the technical grade products are obtained as racemates. Preparation of optically pure enantiomers is illustrated for the case of metalaxyl. Initially, DL-N-(2,6-dimethylphenyl)-alanine is resolved by fractional crystallization of its salts with (+) and (−) α-phenylethylamine. Fischer esterification followed by reaction with methoxyacetyl chloride gives the D- and L-form of metalaxyl. The absolute configuration of the intermediate methyl N-(2,6-dimethylphenyl)-alaninate has been ascertained by reacting the p-nitrobenzenesulfonate of methyl L-lactate with 2,6-dimethylaniline to give, by way of a Walden inversion, methyl D-(2,6-dimethylphenyl)-alaninate [229]. It is of special interest to note that the fungicidal activity of racemic metalaxyl stems largely from the D-form.

Acylanilinobutyrolactones
In 1975, Chevron Chemical Co., discovered a related group of compounds, namely the chloroacetylanilinobutyrolactones, with marked activity against oomycetes. One such substance is

DL-3-[N-Chloroacetyl-N-(2,6-dimethylphenyl)-amino]-γ-butyrolactone

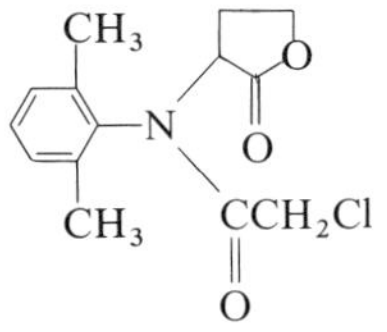

Common name: ofurace
Other name: Ortho 20615

This compound forms colorless crystals. Its acute oral and dermal LD_{50}s in rats are more than 1000 mg/kg and 2000 mg/kg respectively [230].

The compound was patented by Chevron as such [231] and in its use as a fungicide [230, 232]. The corresponding lactams seem to be less active against oomycetes [233]. In 1977, further extensions were made, mainly in the acyl and lactone portion of the molecule [234], and processes for their preparation were developed [235]. Since 1977, other companies have also been working with compounds of this type [236–238].

Ortho 20615 appears to have the same spectrum of activity as the acylalanines mentioned above, i.e. it is predominantly active against oomycetes [239]. Selection experiments on treated grapevines have not produced tolerant strains of *Plasmopara viticola*. Another similarity to the acylalanines is the lower activity of Ortho 20615 on senescing tissue.

To prepare Ortho 20615, 3-N-(2,6-dimethylphenyl)-amino-γ-butyrolactone is acylated in benzene and pyridine under reflux with chloroacetyl chloride [230].

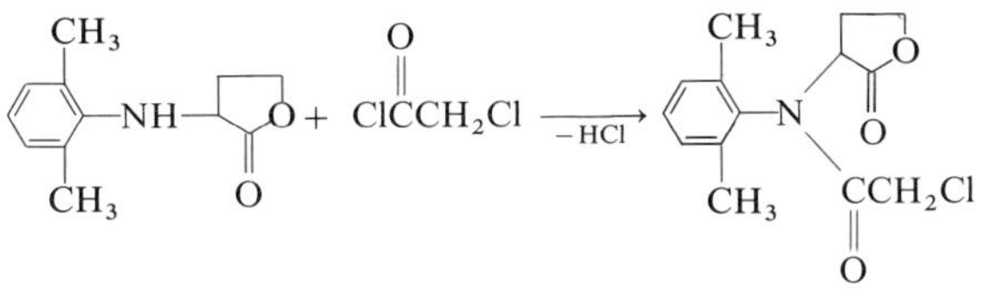

4.7. Miscellaneous New Fungicides Against Oomycetes

Dinitroanilines

The herbicidal activity of the 2,6-dinitroanilines was discovered by Eli Lilly in the early 1960s. This led to the development of the pre-emergence herbicide Treflane® [240, 241].

2,6-Dinitro-N,N-dipropyl-4-trifluoromethyl-aniline

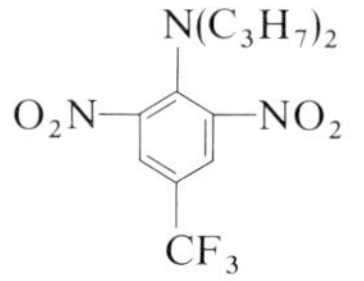

Common name: trifluralin
Other name: Treflan®
(Vol. 5, p. 236)

Field observations of control of root rots in peas caused by *Aphanomyces* and *Pythium* spp. after application of this dinitroaniline herbicide drew the attention of phytopathologists to this class of chemicals [242, 243]. Not only trifluralin but also other related 2,6-dinitroanilines, such as the herbicide oryzalin [3,5-dinitro-4-(dipropylamino)-benzenesulfamide] show this side effect against root rot [242, 244].

Trifluralin itself, and also compounds in which a cyanamino [245], azido [246], cyanoalkylthio [247], or thiocyanato group [248] occupies the 3-position of the trifluralin molecule, are active against *Plasmopara viticola*. This also applies to oryzalin and related structures [249].

Dithiin tetroxides

A more recent compound with activity against oomycetes is

2,3-Dihydro-5-phenyl-1,4-dithiin-1,1,4,4-tetroxide

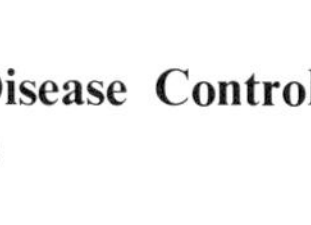

Common name: oxyfenthiin
Other name: P 368

Its use as a fungicide has been protected by Uniroyal Inc. [250]. Reports have been published regarding its activity against *Phytophthora infestans* [251, 252].

5. Discussion of Disease Control Strategies with the New Systemic Fungicides Against Oomycetes

5.1. Novel Disease Control Possibilities with Systemic Fungicides

The systemic properties of the new fungicides against oomycetes make them effective against a number of diseases against which no effective or easy chemical control measures had been known before. Against soilborne fungi seed treatments with nonsystemic products create a fungitoxic zone around the treated seed, but they do not protect roots and shoots which grow out of this zone. Protection with systemic products lasts much longer since they are distributed throughout the susceptible young seedlings. Root and crown rots of perennial crops are another type of disease where some of the new fungicides have shown promise.

Among the foliar diseases the systemic downy mildews have traditionally been particularly hard to control. Especially against the primary infections by these fungi some of the new systemic fungicides offer for the first time good possibilities for easy and effective chemical control.

5.2. Utilization of Curative and Systemic Properties for Integrated Disease Control Strategies

The new systemic fungicides against oomycetes discussed in this review are in many ways tailor-made for integrated control strategies. With their curative properties, they interfere with the disease progress immediately after application, while traditional purely protective fungicides show an effect only after a time-lag of at least one incubation period for the disease in question [253]. This difference is particularly important when the fungicides are used in connection with disease warning systems, which are based on determining infection periods retrospectively. Here protective products cannot prevent the outbreak of the predicted increment of disease progress, while curative products can. Thus, the new fungicides allow for a much more flexible and effective use of predictive systems for spray recommendations.

The systemic properties of the new products offer additional flexibility in spray coverage. It may no longer be necessary to apply a continuous fungicide film over the plant parts to be protected, since the systemic products are distributed in a rather predictable way throughout the plant organs. This internal distribution is especially striking with the phloem-transported fungicides such as pyroxychlor or phosetyl-Al discussed above. The suitability of the new specific fungicides against oomycetes for use in connection with disease warning systems, together with the low use rates and extended spray intervals, offer new possibilities for disease control strategies with lower hazards due to ecological and human exposure and with lower energy requirements.

In this connection it may be well to remember that the use of resistant cultivars and of fungicides should not be thought of as divergent and mutually exclusive methods of disease control but rather as being synergistic [254]. This concept of a combined use of disease resistance and fungicides has recently been further developed [265], and it merits renewed attention in connection with the new systemic fungicides against oomycetes.

5.3. The Risk of Fungicide Resistance

Since the new fungicides discussed are systemic, this review would be incomplete without mentioning the risk of development of resistant strains in the target fungi (M. L. Lacy and J. M. Vargas, Jr., Vol. 4, p. 239). From the preliminary studies done with acylalanines and related compounds [222, 239] it appeared that in vivo resistance does not readily develop in the oomycetes species studied. However, more recent studies [256, 257] show that the development of *in vivo* resistance to these fungicides cannot be excluded. It is advisable, therefore, to avoid use patterns of acylalanines that are favorable for the development of resistance. The question of how generally to cope with a resistance risk was covered by Delp [221]. No information is available concerning the resistance risk connected with the use of the other systemic fungicides included in this review.

Beside the use of mixtures and spray alternations, the effects of the combined use of disease resistance and fungicides on the longevity of both the resistance in the cultivars and the effectiveness of the fungicide are of importance in this connection.

6. References

1. Alexopulos, J.C.: Introductory mycology, 2nd ed. New York: John Wiley & Sons 1962
2. Waterhouse, G.M.: Peronosporales. In: The Fungi, Vol. IVB, pp. 165–185. Ainsworth, G.G. et al. (ed.). New York: Academic Press 1973
3. Cochrane, V.W.: Physiology of fungi. New York: John Wiley & Sons 1958
4. Hendrix, J.W.: Proc. Am. Phytopathol. Soc. *1*, 207 (1974)
5. McCallan, S.A.E.: History of fungicides. In: Fungicides, Vol. 1, pp, 1–37. Torgeson, D.C. (ed.). New York: Academic Press 1967
6. Walker, J.C.: Plant pathology, 3rd ed. New York: McGraw-Hill 1969
7. Horsfall, J.G.: Principles of fungicidal action. Waltham, Mass., USA: Chronica Botanica 1956
8. Agrios, G.N.: Plant pathology. New York: Academic Press 1978
9. Pratt, R.G., Janke, G.D.: Phytopathology *68*, 1600 (1978)
10. Wehtje, G., Zimmer, D.E.: Phytopathology *68*, 1568 (1978)
11. Raghavendra, S., Saffeulla, K.: Seeds & Farms *4* (1978)
12. Harrington, T., Kavenagh, J.A.: Proc. 3rd Int. Congr. Plant Pathology *1978*, 273
13. Podger, F.D.: The role of *Phytophthora cinnamoni* in dieback diseases of Australian Eucalypt forests. In: Biology and control of soilborne plant pathogens, pp. 27–36. Bruehl, G.W. (ed.). St. Paul, Minn., USA: Am. Phytopathol. Soc. 1975,
14. Martin, H., Worthing, C.R.: Pesticide Manual, 5th ed. Brit. Crop Protection Council 1977
15. Rodriguez-Kabana, R., Backman, P.A., Curl, E.A.: Control of seed and soilborne plant diseases. In: Antifungal compounds. I, pp. 117–161. Siegel, M.R., Sisler, M.R. (ed.). New York, Basel: M. Dekker Inc. 1977
16. Baker, K.F., Cook, R.J.: Biological control of plant pathogens. San Francisco: W. H. Freeman & Co. 1974
17. Stover, R.H.: Fungicidal control of plant diseases in the tropics. In: Antifungal compounds. I, pp. 353–396. Siegel, M.R., Sisler, H.D. (ed.). New York, Basel: M. Dekker Inc. 1977
18. Anon.: Index phytosanitaire 1978, Ass. Coord. Technique Agricole, Paris
19. Anon.: Farm Chemicals *140*, 42 (1977)
20. Innes, J.R.M. et al.: J. Nat. Cancer Inst. *42*, 1101 (1969)
21. Ulland, B.M. et al.: J. Nat. Cancer Inst. *49*, 583 (1972)
22. Graham, S.L. et al.: J. Agric. Food Chem. *21*, 324 (1973)
23. Graham, S.L. et al.: Food Cosmet. Toxicol. *13*, 493 (1975)
24. Khera, K.S.: Teratology *7*, 243 (1973)
25. Engst, R. et al.: Pure Appl. Chem. *49*, 675 (1977)
26. Vogeler, K. et al.: Pflanzenschutz-Nachr. *30*, 72 (1977)
27. Czeglédi-Jankó, G.: J. Chromatogr. *31*, 89 (1967)
28. Bontoyan, W.R. et al.: J. Assoc. Off. Anal. Chem. *55*, 923 (1972)

29. Sato, F., Tomizawa, C.: Bull. Nat. Inst. Agric. Sci. Ser. C *12*, 181 (1960)
30. Engst, R., Schnaak, W., Rattba, H.: Nachrichtenbl. Dtsch. Pflanzenschutzdienst *22*, 26 (1968)
31. Vonk, J.W.: Meded. Fac. Landbouwwet. Rijksuniv. Gent *41*, 1383 (1976)
32. Ripley, B.D. et al.: J. Agric. Food Chem. *26*, 134 (1978)
33. Newsome, W.H.: J. Agric. Food Chem. *20*, 967 (1972)
34. Von Stryk, F.G., Jarvis, W.R.: Can. J. Plant Sci. *58*, 623 (1978)
35. Blazquez, C.H.: J. Agric. Food Chem. *21*, 330 (1973)
36. Newsome, W.H.: J. Agric. Food Chem. *24*, 999 (1976)
37. Watts, R.R., Storherr, R.W., Onley, J.H.: Bull. Environm. Contam. Toxicol. *12*, 224 (1974)
38. Casanova, M., Dachaud, R.: Phytiatr. Phytopharm. *26*, 215 (1977)
39. Kaars Sijpesteijn, A.: World Rev. Pest. Control. *8*, 85 (1969)
40. Corbett, J.R.: The biochemical mode of action of pesticides. London, New Yok: Academic Press 1974
41. Lukens, R.J., Sisler, H.D.: Phytopathology *48*, 235 (1958)
42. Siegel, M.R., Sisler, H.D.: Phytopathology *58*, 1123 (1968)
43. Vincent, P.G., Sisler, H.D.: Physiol. Plantarum *21*, 1249 (1968)
44. Tillman, R.W., Siegel, M.R., Long, J.W.: Pestic. Biochem. Physiol. *3*, 160 (1900)
45. DT–PS 1014373 (24 Sep 1954/13 Feb 1958), Bayer AG/E. Urbschat and P.-E. Frohberger
46. Urbschat, E.: Angew. Chem. *72*, 981 (1960)
47. Stollé, R.: Ber. Dtsch. Chem. Ges. *45*, 2682 (1912)
48. DT–PS 1028828 (20 Jan 1956/2 Oct 1958), Bayer AG/E. Urbschat and P.-E. Frohberger
49. Frohberger, P.E.: Pflanzenschutz-Nachr. *22*, 23 (1969)
50. Kluge, E.: Arch. Phytopathol. Pflanzenschutz *14*, 115 (1978)
51. Kniper, J.: Plant Dis. Rep. *58*, 845 (1974)
52. Hills, F.J., Leach, L.D.: Phytopathology *52*, 51 (1961)
53. Alconero, R., Hagedorn, D.J.: Phytopathology *58*, 24 (1968)
54. Mitchell, J.E., Hagedorn, D.J.: Plant Dis. Rep. *50*, 91 (1966)
55. Hills, F.J.: Phytopathology *52*, 389 (1962)
56. Lyr, H.: Effect of fungicides on energy production and intermediary metabolism. In: Antifungal compounds, pp. 301–332. Siegel, M.R., Sisler, H.D. (ed.). New York: Dekker Inc. 1977
57. Karanth, N.G.K. et al.: Appl. Microbiol. *27*, 43 (1974)
58. US 3260725 (15 Mar 1961/12 Jul 1966), Olin Mathieson Chemical Corp./H.A. Schroeder
59. US 3260588 (1 Jul 1963/12 Jul 1966; US prior. 15 Mar 1961), Olin Mathieson Chemical Corp./H.A. Schroeder
60. Thomas, W.D., Eastburg, P.H., Bankuti, M.D.: Phytopathology *52*, 754 (1962)
61. Al-Beldawi, A.S., Sinclair, J.B.: Phytopathology *59*, 68 (1969)
62. Vegh, I., Bourgeois, M.: Phytiatr. Phytopharm. *25*, 335 (1976)
63. Goerdeler, J., Groschopp, H., Sommerlad, U.: Ber. Dtsch. Chem. Ges. *90*, 182 (1957)
64. Schroeder, H.A. et al.: J. Org. Chem. *27*, 2589 (1962)
65. US 3890338 (2 May 1974/17 Jun 1975), Olin Mathieson Chemical Corp./J.A. Wojtowicz and S.K. Bhutani
66. US 3890339 (28 May 1974/17 Jun 1975), Olin Mathieson Chemical Corp./D.F. Gavin
67. DE–OS 1620101 (12 Jul 1966/20 Aug 1970; US prior. 28 Jul 1965), Olin Mathieson Chemical Corp./H.A. Schroeder and R.P. Alexander
68. Habermann, J.: Ber. Dtsch. Chem. Ges. *11*, 1034 (1878)
69. US 3265564 (11 Feb 1964/9 Aug 1966; US prior. 14 Mar 1963), E.I. du Pont de Nemours and Co./R.M. Scribner and E.J. Soboczenski
70. Ryker, T.C.: Proc. Cotton Dis. Council *25*, 137 (1965)
71. Vargas, J.M., Beard, J.B.: Plant Dis. Rep. *54*, 1075 (1970)
72. Erwin, D.C.: Annu. Rev. Phytopathol. *11*, 389 (1973)
73. Hock, W.K., Sisler, H.D.: Phytopathology *59*, 627 (1969)
74. Werner, P., Lyr, H., Gasparson, G.: Arch. Phytopathol. Pflanzenschutz *14*, 301 (1978)
75. Tillmann, R.W., Sisler, H.D.: Phytopathology *63*, 219 (1973)
76. Rhodes, R.C., Pease, H.L., Brantley, R.K.: J. Agric. Food Chem. *19*, 745 (1971)
77. Wiese, W.V., Vargas, J.M.: Phytopathology *62*, 1112 (1972)
78. Wiese, M.V., Vargas, J.M.: Pestic. Biochem. Physiol. *3*, 214 (1973)
79. US 3363005 (30 Jul 1965/9 Jan 1968), E.I. du Pont de Nemours and Co./J.R. Alvarez

80. Tomita, K. et al.: Ann. Rep. Sankyo Res. Lab. *25*, 1 (1973)
81. Nakagawa, M., Nakamura, T., Tomita, K.: Agric. Biol. Chem. *38*, 2205 (1974)
82. Ishitsuka, R.: Jpn. Pestic. Inf. *18*, 27 (1974)
83. Iwai, I., Nakamura, N.: Chem. Pharm. Bull. *14*, 1277 (1966)
84. US 3544584 (8 Sep 1965/1 Dec 1970; JA prior. 14 Sep 1964 and 12 Jun 1965), Sankyo Co. Ltd./I. Iwai, Y. Kishida, T. Hiraoka, and N. Nakamura
85. DE-OS 1567123 (23 Dec 1965/16 Apr 1970; JA prior. 26 Dec 1964 and 14 Dec 1965), Sankyo Co. Ltd./H. Sumi, Y. Takahi, N. Nakamura, and K. Tomita
86. US 3687968 (5 Mar 1970/29 Aug 1972; JA prior. 14 Sep 1964 and 12 Jun 1965), Sankyo Co. Ltd./I. Iwai, Y. Kishida, T. Hiraoka, and N. Nakamura
87. Tsao, P., Guy, S.O.: Phytopathology *67*, 796 (1977)
88. Ota, Y.: Japan Agricultural Research Quarterly *9*, 1 (1975)
89. Kamimura, S. et al.: Phytopathology *64*, 1273 (1974)
90. US 3607880 (9 Feb 1970/21 Sep 1971), Sankyo Co. Ltd./K. Tomita
91. DE-OS 1695762 (18 Aug 1967/29 Apr 1971), Sankyo Co. Ltd./K. Tomita, M. Nagano, and H. Oka
92. DE-OS 2032809 (29 Jun 1970/21 Jan 1971; JA prior. 3 Jul 1969), Sankyo Co. Ltd./K. Ohata, K. Adachi, M. Hashimoto, and A. Kage
93. Geoghegan, M.J.A.: Proc. 4th Brit. Insectic. Fungic. Conf. *1967*, 451
94. Purnell, T.J.: Proc. 7th Brit. Insectic. Fungic. Conf. *1973*, 603
95. Purnell, T.J.: Proc. 2nd Int. Congr. Plant Pathol. *1973*, Abstract 124
96. Logan, I.C., Arst, R.N.: Proc. N.Z. Weed Pest. Cont. Conf. *1974*, 120
97. US 3135594 (1 Sep 1961/2 Jun 1964), Dow Chemical Co./C. A. I. Goring
98. US 3285925 (15 Apr 1964/15 Nov 1966), Dow Chemical Co./H. Johnston and M. S. Tomita
99. US 3317549 (28 Sep 1964/2 May 1967; US prior. 15 Jan 1962), Dow Chemical Co./H. Johnston
100. US 3244586 (30 Apr 1962/5 Apr 1966), Dow Chemical Co./R. H. Rigterink
101. Noveroske, R.L.: Down Earth *30*, 3 (1975)
102. US 3838159 (7 Dec 1972/24 Sep 1974; US prior. 17 Mar. 1969), Dow Chemical Co./H. Johnston
103. US 3244722 (28 Jan 1964/5 Apr 1966), Dow Chemical Co./H. Johnston and M. S. Tomita
104. US 4062962 (9 Feb 1976/13 Dec 1977; US prior. 7 Jul 1972), Dow Chemical Co./R. L. Noveroske
105. Sbragia, R.J.: Annu. Rev. Phytopathol. *13*, 257 (1975)
106. Sbragia, R.J.: Diss. Abstr. Int. B *39*, 470 (1978)
107. Noveroske, R.L.: Phytopathology *65*, 22 (1975)
108. Noveroske, R.L.: Proc. 2nd Int. Congr. Plant Pathol. *1973*, Abstract 123
109. Small, L.W., Martin, R.A., Edington, L.V.: Neth. J. Plant Pathol. *83* (Suppl. 1), 63 (1977)
110. Small, L.W.: Diss. Abstr. Inst. B *37*, 1055 (1976)
111. Small, L.W., Edington, L.V.: Proc. Am. Phytopathol. Soc. *2*, 107 (1975)
112. Richardson, L.T.: Can. J. Plant Sci. *56*, 365 (1976)
113. Small, L.W., Edington, L.V.: Proc. Am. Phytopathol. Soc. *1*, 144 (1974)
114. Pfleger, F.L., Groth, J.V., Reiling, T.P.: Plant Dis. Rep. *60*, 317 (1976)
115. US 3325503 (18 Feb 1965/13 Jun 1967), Diamond Alkali Co./R. M. Bimber
116. US 3578434 (23 Jul 1968/11 May 1971), Dow Chemical Co./R. L. Noveroske and H. K. Tobol
117. Hansing, E.D.: Fungic. Nematic. Rep. *29*, 145 (1975)
118. Jones, J.P.: Fungic. Nematic. Rep. *30*, 135 (1975)
119. Jones, J.P.: Fungic. Nematic. Rep. *30*, 136, 148 (1975)
120. Bastiaansen, M.G., Pieroh, E.A., Aelbers, E.: Meded. Fac. Landbouwwet. Rijksuniv. Gent. *39*, 1019 (1974)
121. Pieroh, E.A., Krass, W., Hemmen, C.: Meded. Fac. Landbouwwet. Rijksuniv. Gent. *43*, 933 (1978)
122. DE-OS 1567169 (17 Dec 1966/6 Aug 1970), Schering AG/G.-A. Hoyer and E. A. Pieroh
123. DE-OS 1643040 (25 Aug 1967/15 Apr 1971), Schering AG/G.-A. Hoyer and E. A. Pieroh
124. Pieroh, E.A., Kohne, S., Ahrens, C.: Mitt. Biol. Bundesanst. Land-Forstwirtschaft *165*, 147 (1975)
125. McIntyre, J.L., Lukens, R.J.: Plant Dis. Rep. *61*, 366 (1977)
126. Prinsloo, G.C.: Phytophylactica *9*, 17 (1977)

127. Beyries, A., Molot, P.M.: Phytiatr. Phytopharm. *25*, 201 (1976)
128. Fletcher, J.T.: Ann. Appl. Biol. *84*, 294 (1976)
129. Ryan, E.W.: Proc. 9th Brit. Insectic. Fungic. Conf. *1977*, 297
130. Iwan, J., Goller, D.: Environ. Qual. Saf. *1975* Suppl., 319
131. Kaars Seijpesteijn, A., Kerkenaar, A., Overeem, J.C.: Meded. Fac. Landbouwwet. Rijksuniv. Gent. *39*, 1027 (1974)
132. Kerkenaar, A., Kaars Seijpesteijn, A.: Neth. J. Plant Pathol. *83* (Suppl. 1) 145 (1977)
133. Papavizas, G.C., O'Neill, N.R., Lewis, J.A.: Phytopathology *68*, 1667 (1978)
134. Iwan, J., Hertel, B., Radzuweit, J.: Neth. J. Plant Pathol. *83* (Suppl. 1), 277 (1977)
135. Serres, J.M., Carraro, G.A.: Meded. Fac. Landbouwwet. Rijksuniv. Gent. *41*, 645 (1976)
136. Douchet, J.P. et al.: Proc. 9th Brit. Insectic. Fungic. Conf. *1977*, 535
137. US 3625987 (7 Nov 1967/7 Dec 1971; CH prior. 10 Nov 1966), Ciba Ltd./A. Hubele
138. DE-OS 2350910 (10 Oct 1973/18 Apr 1974; CH prior. 13 Oct 1972), Ciba-Geigy AG/A. Hubele and M. Kühne
139. DE-OS 2436654 (30 Jul 1974/20 Feb 1975; CH prior. 2 Aug 1973), Ciba-Geigy AG/A. Hubele
140. DE-OS 2436655 (30 Jul 1974/20 Feb 1975; CH prior. 2 Aug 1973), Ciba-Geigy AG/A. Hubele
141. DE-OS 2623847 (28 May 1976/8 Dec 1977), Bayer AG/W. Brandes, W. Daum, and P. Kraus
142. DE-OS 2626828 (15 Jun 1976/22 Dec 1977), Bayer AG/W. Brandes and W. Daum
143. DE-OS 2657145 (16 Dec 1976/22 Jun 1978), Bayer AG/W. Brandes, W. Daum, and P. Kraus
144. DE-OS 2725315 (4 Jun 1977/14 Dec 1978), Bayer AG/W. Brandes and W. Daum
145. US 3957847 (10 Mar 1975/18 May 1976; US prior. 15 Mar 1972), E. I. du Pont de Nemours and Co./S. H. Davidson
146. US 3954992 (21 Mar 1974/4 May 1976; US prior. 15 Mar 1972), E. I. du Pont de Nemours and Co./S. H. Davidson
147. US 4053612 (12 Dec 1975/11 Oct 1977; US prior. 30 Jan 1975), E. I. du Pont de Nemours and Co./F. J. Baude and W. E. Cupery
148. US 4101669 (25 Aug 1977/18 Jul 1978; US prior. 30 Jan 1975), E. I. du Pont de Nemours and Co./F. J. Baude and W. E. Cupery
149. US 3919284 (13 Jun 1974/11 Nov 1975), E. I. du Pont de Nemours and Co./K. Lin
150. US 4051255 (6 Feb 1976/27 Sep 1977; US prior. 15 Mar 1972), E. I. du Pont de Nemours and Co./S. H. Davidson
151. US 3979518 (11 Sep 1975/7 Sep 1976), E. I. du Pont de Nemours and Co./H. L. Klopping
152. US 3984568 (11 Sep 1975/5 Oct 1976), E. I. du Pont de Nemours and Co./H. L. Klopping
153. Roussel, C.: Phytoma *30*, 13 (1978)
154. Richardson, L.T.: Pesticide Res. Rep. *1976*, 385
155. Müller, P.T.: Ann. Chim. Phys. *1*, 507 (1894)
156. Nef, J.U.: Liebigs Ann. Chem. *280*, 331 (1894)
157. Fields, M., Walz, D.E., Rothschild, S.: J. Am. Chem. Soc. *73*, 1001 (1951)
158. Conrad, M., Schulze, A.: Ber. Dtsch. Chem. Ges. *42*, 735 (1909)
159. DRP 227390 (28 Feb 1909/19 Oct 1910), E. Merck
160. Diels, O., Borgwardt, E.: Ber. Dtsch. Chem. Ges. *54*, 1342 (1921)
161. Williams, D.J. et al.: Proc. 9th Brit. Insectic. Fungic. Conf. *1977*, 565
162. Anon.: Phytoma *30*, 28 (1978)
163. Bertrand, A. et al.: Phytiatr. Phytopharm. *26*, 3 (1977)
164. Nylén, P.: Sven. Kem. Tidskr. *48*, 2 (1936); C. A. *30*, 3702[7] (1936)
165. Orlovskii, V.V., Vovsi, B.A., Mishkevich, A.E.: J. Gen. Chem. USSR (Engl. Transl.) *42*, 1924 (1972)
166. US 3415905 (28 Sep 1964/10 Dec 1968), Monsanto Research Corp./A. Y. Garner
167. JA 10639/66 (14 Feb 1964/13 Jun 1966) Ihara Chemical Co. Ltd./M. Yoshido, T. Maeda, and E. Yoshinaga
168. FR 1529738 (20 Apr 1967/21 Jun 1968), Pechiney-Progil/J. Thiolliere
169. FR 1299543 (28 Apr 1961/18 Jun 1962), "OPCIA" and "LA BAKELITE"/D. A. Brown, H. Coates, J. Girard, A. Thiot, and G. Quesnel
170a. FR 2254276 (14 Dec 1973/11 Jul 1975), PEPRO/J. Ducret, G. Lacroix, and J.-M. Gaulliard
170b. FR 2288463, Pat. add to 170a, (23 Oct 1974/21 May 1976) PEPRO/J. Ducret, G. Lacroix, and J.-M. Gaulliard
171a. FR 2213735 (12 Jan 1973/9 Aug 1974) PEPRO/J.-C. Debourge, J.-M. Gaulliard, and J. Thiolliere

171b. FR 2257227, Pat. add. to 171a, (19 Oct 1973/8 Aug 1975), PEPRO/J.-C. Debourge, J.-G. Abblard, G.-B. Lacroix, D. Pillon, J. Ducret, and A. Thizy
172a. FR 2252056 (26 Nov 1973/20 Jun 1975), PEPRO/A. Thizy, D. Pillon, J.-C. Debourge, and G. Lacroix
172b. FR 2285812, Pat. add to 172a, (27 Sep 1974/23 Apr 1976), PEPRO/A. Thizy, D. Pillon, J.-C. Debourge, and G. Lacroix
173. FR 2269862 (8 Mar 1974/5 Dec 1975), PHILAGRO S.A./A. Thizy and J.-C. Debourge
174. FR 2264805 (22 Mar 1974/17 Oct 1975), PHILAGRO S.A./ J. Abblard, G. Marmain, and A. Viricel
175. FR 2267044 (10 Apr 1974/7 Nov 1975), PHILAGRO S.A./G. Lacroix and J.-C. Debourge
176. FR 2287174 (8 Oct 1974/7 May 1976), PHILAGRO S.A./S. Trinh, A. Moraly, and J.-M. Gaulliard
177. FR 2299810 (5 Feb 1975/3 Sep 1976), PHILAGRO S.A./G. Lacroix, J.-C. Debourge, and J. Ducret
178. FR 2303476 (14 Mar 1975/8 Oct 1976), PHILAGRO S.A./P. André, G. Lacroix, and J.-C. Debourge
179. FR 2380286 (14 Feb 1977/8 Sep 1978), PHILAGRO S.A./J. Ducret, J.-M. Gaulliard, and J. Vial
180. Dennis, C., Davis, R.P.: Proc. 9th Brit. Insectic. Fungic. Conf. *1977*, 203
181. Chazalet, M. et al.: Phytiatr. Phytopharm. *26*, 41 (1977)
182. Mai, V.T., Bompeix, G., Ravise, A.: Cont. Rend. Serie D *288*, 1171 (1979)
183. Vegh, I., Baillot, F., Roy, J.: Phytiatr. Phytopharm. *26*, 85 (1977)
184. Clerjean, M., Beyries, A.: Phytiatr. Phytopharm. *26*, 73 (1977)
185. Molot, P.M., Beyries, A.: Phytiatr. Phytopharm. *26*, 63 (1977)
186. Zentmyer, G.A., Ohr, H.D.: Phytopathology News *12*, 269 (1978)
187. Anon.: Farm Chemicals Handbook. Willoughby, Ohio: Meister Publ. Comp. 1979
188. FR 2377155 (14 Jan 1977/15 Sep 1978), PHILAGRO S.A./D. Horriere and M. Chazalet
189. Lafon, R., Bugaret, Y., Bulit, J.: Phytiatr. Phytopharm. *26*, 19 (1977)
190. Marais, P.G., Van der Walt, H.S.: Phytophylactica *10*, 89 (1978)
191. Ryan, E.W.: Proc. 9th Brit. Insectic. Fungic. Conf. *1977*, 297
192. Frossard, P.: Fruits *33*, 183 (1978)
193. Montgomerie, I.G., Kennedy, D.M.: Plant Pathol. *26*, 139 (1977)
194a. FR 2370750 (16 Nov 1976/9 Jun 1978), PHILAGRO S.A./J. Abblard, M. Ghazarian, and R. Viricel
194b. FR 2402663, Pat. add to 194a, (7 Sep 1977/6 Apr 1979) PHILAGRO S.A./J. Abblard, M. Ghazarian, and R. Viricel
195. Schwinn, F., Staub, T. Urech, P.A.: Mitt. Biol. Bundesanst. Land-Forstwirtsch. *178*, 145 (1977)
196. Urech, P.A., Schwinn, F., Staub, T.: Proc. 9th Brit. Insectic. Fungic. Conf. *1977*, 623
197. Schwinn, F.J., Staub, T., Urech, P.A.: Meded. Fac. Landbouwwet. Rijksuniv. Gent *42*, 1181 (1977)
198. DE–OS 2350944 (10 Oct 1973/25 Apr 1974; CH prior. 13 Oct 1972 and 29 Aug 1973), Ciba-Geigy AG/H. Moser
199. DE–OS 2513788 (27 Mar 1975/9 Oct 1975; CH prior. 2 Apr 1974 and 10 Feb 1975), Ciba-Geigy AG/A. Hubele
200. DE–OS 2515091 (7 Apr 1975/23 Oct 1975; CH prior. 9 Apr 1974 and 14 Mar 1975), Ciba-Geigy AG/A. Hubele
201. DE–OS 2513732 (27 Mar 1975/16 Oct 1975; CH prior. 2 Apr 1974 and 10 Feb 1975), Ciba-Geigy AG/A. Hubele
202. DE–OS 2513789 (27 Mar 1975/9 Oct 1975; CH prior. 1 Apr 1974 and 13 Feb 1975), Ciba-Geigy AG/A. Hubele
203. DE–OS 2515113 (7 Apr 1975/23 Oct 1975; CH prior. 9 Apr 1974 and 7 Mar 1975), Ciba-Geigy AG/A. Hubele
204. DE–OS 2643403 (27 Sep 1976/7 Apr 1977; CH prior. 30 Sep 1975), Ciba-Geigy AG/A. Hubele, W. Kunz, and W. Eckhardt
205. DE–OS 2643404 (27 Sep 1976/7 Apr 1977; CH prior. 30 Sep 1975), Ciba-Geigy AG/A. Hubele, W. Kunz, and W. Eckhardt
206. DE–OS 2643445 (27 Sep 1976/7 Apr 1977; CH prior. 30 Sep 1975), Ciba-Geigy AG/A. Hubele, W. Kunz, and W. Eckhardt

207. DE–OS 2643477 (27 Sep 1976/7 Apr 1977; CH prior. 30 Sep 1975), Ciba-Geigy AG/A. Hubele, W. Kunz, and W. Eckhardt
208. DE–OS 2741437 (14 Sep 1977/23 Mar 1978; CH prior. 17 Sep 1976 and 4 Aug 1977), Ciba-Geigy AG/A. Hubele
209. DE–OS 2812957 (23 Mar 1978/12 Oct 1978; CH prior. 28 Mar 1977), Ciba-Geigy AG/W. Eckhardt, W. Kunz, and A. Hubele
210. DE–OS 2813335 (28 Mar 1978/5 Oct 1978; CH prior. 29 Mar 1977), Ciba-Geigy AG/A. Hubele
211. DE–OS 2648074 (23 Oct 1976/27 Apr 1978), Bayer AG/W. Lunkenheimer, W. Brandes, and P. Kraus
212. DE–OS 2702102 (19 Jan 1977/20 Jul 1978), Bayer AG/W. Lunkenheimer, W. Brandes, and V. Paul
213. Wiertsema, W.P., Wissink, G.H.: Meded. Pac. Landbouwwet. Rijksuniv. Gent *42*, 1189 (1977)
214. de Waard, M.D., van Roijen, C.I.: Meded. Fac. Landbouwwet, Rijksuniv. Gent *43*, 951 (1978)
215. Staub, T., Dahmen, H., Schwinn, F.J.: Z. Pflanzenkr. Pflanzenschutz *85*, 162 (1978)
216. Safeeulla, K.M., Venugopal, M.N.: Proc. 3rd Int. Cong. Plant Pathol. *1978*, 363
217. Dueck, J., Stone, J.R.: Can. J. Plant Sci. *59*, 423 (1979)
218. Morton, H.V., Staub, T., Young, T.R.: Proc. 3rd Int. Congr. Plant Pathol. *1978*, 383
219. Urech, P.A., Eberle, J., Ruess, H.: Proc. 3rd Int. Congr. Plant Pathol. *1978*, 359
220. Johnson, G.I., Davis, R.D., O'Brien, R.G.: Plant Dis. Rep. *63*, 212 (1979)
221. Delp, C.J.: Phytopathology *69*, 1026 (1979)
222. Staub, T. et al.: Plant Dis. Rep. *63*, 385 (1979)
223. Cohen, Y., Reuveni, M., Eyal, H.: Phytopathology *69*, 645 (1979)
224. Zentmyer, G.A., Ohr, H.D.: Phytopathology News *12*, 142 (1978)
225. Staub, T., Dahmen, H., Schwinn, F.J.: Z. Pflanzenkr. Pflanzenschutz *87*, 83 (1980)
226. Crute, I.R.: Proc. 3rd Int. Congr. Plant Pathol. *1978*, 384
227. Bensen, D.M.: Phytopathology *69*, 174 (1979)
228. Staub, T., Vandepeute, J.: Plant Physiol. *63*, (Suppl. 5), 136 (1979)
229. Moser, H., Vogel, Ch.: Abstr. 4th Int. Congr. Pestic. Chem. (IUPAC) *1978*, II-310
230. US 4141989 (29 Sep 1977/27 Feb 1979; US prior. 10 Feb 1975), Chevron Research Co./D. C. K. Chan
231. US 3933860 (10 Feb 1975/20 Jan 1976), Chevron Research Co./D. C. K. Chan
232. US 4107323 (12 Oct 1976/15 Aug 1978; US prior. 10 Feb 1975), Chevron Research Co./D. C. K. Chan
233. US 4012519 (12 Nov 1975/15 Mar 1977; US prior. 10 Feb 1975), Chevron Research Co./D. C. K. Chan
234. DE–OS 2845454 (19 Oct 1978/10 May 1979; US prior. 1 Nov 1977), Chevron Research Co./D. C. K. Chan
235. DE–OS 2847075 (28 Oct 1978/10 May 1979; US prior. 1 Nov 1977), Chevron Research Co./R. N. Reynolds, Jr., S. D. Ziman, and D. C. K. Chan
236. DE–OS 2804299 (1 Feb 1978/10 Aug 1978; CH prior. 4 Feb 1977 and 28 Mar 1977), Ciba-Geigy AG/W. Kunz, E. Eckhardt, and A. Hubele
237. DE–OS 2724785 (27 May 1977/14 Dec 1978), Schering AG/U. Bühmann, R. Putter, E.-A. Pieroh, and D. Baumert
238. DE–OS 2274786 (27 May 1977/14 Dec 1978), Schering AG/U. Bühmann, R. Putter, E.-A. Pieroh, and D. Baumert
239. Lukens, R.J., Chan, D.C.K., Etter, G.: Phytopathology News *12*, 142 (1978)
240. US 3111403 (31 Aug 1960/19 Nov 1963), Eli Lilly and Co./Q. F. Soper
241. US 3257190 (10 Dev 1962/21 Jun 1966; US prior. 31 Aug 1960), Eli Lilly and Co./Q. F. Soper
242. Teasdale, J.E., Harvey, R.G., Hagedorn, D.J.: Weed Sci. *26*, 609 (1978)
243. Harvey, R.G., Hagedorn, D.J., D Longherty, R.L.: Crop Sci. *15*, 67 (1975)
244. US 4065559 (19 Mar 1976/27 Dec 1977), Eli Lilly and Co./J. D. Froyd
245. US 3979453 (23 Jun 1975/7 Sep 1976), Eli Lilly and Co./J. A. Yahner
246. US 3948957 (23 Jun 1975/6 Apr 1976), Eli Lilly and Co./J. R. Beck
247. US 4045466 (31 Aug 1976/30 Aug 1977; US prior. 23 Jun 1975), Eli Lilly and Co./J. R. Beck and J. A. Yahner
248. US 3987076 (23 Jun 1975/19 Oct 1976), Eli Lilly and Co./J. R. Beck

249. US 4091096 (31 Aug 1976/23 May 1978; US prior. 23 Jun 1975), Eli Lilly and Co./J. R. Beck and J. A. Yahner
250. US 4004018 (20 Jun 1974/18 Jan 1977), Uniroyal Inc./A. D. Brewer and R. A. Davis
251. Cetas, R.C.: Fungic. Nematic. Rep. *34*, 69 (1979)
252. Rowe, R.C.: Fungic. Nematic. Rep. *34*, 75 (1978)
253. Fry, W.E., Bruck, R.I., Mundt, C.C.: Plant Dis. Rep. *63*, 970 (1979)
254. Van der Plank, J.E.: Epidemiology of fungicidal action. In: Fungicides, Vol. 1, pp. 63–92, Torgeson, D.C. (ed.). New York: Academic Press 1967
255. Fry, W.E.: Fungicides in perspective. In: Antifungal compounds, Vol. I, pp. 19–50. Siegel, M.R., Sisler, M.D. (ed.). New York, Basel: Dekker Inc. 1977
256. Bruin, G.G., Edgington, L.V.: NE meeting APS *1978*, 22
257. Davidse, L.C.: Neth. J. Pl. Path. (in press)

Benzoylphenylurea Insecticides

W. Maas, R. van Hes, A. C. Grosscurt, and D. H. Deul

Duphar B.V.
1243 zg 's-Graveland, The Netherlands

Contents

1. Introduction

The biological properties of substituted phenylurea compounds have been known for a considerable period. Many well-known herbicides belong to this class.

The insecticidal properties of substituted benzoylureas have only been discovered around 1970 at Philips-Duphar. In the framework of a research program around the herbicide dichlobenil, the basic structure of that compound was combined with that of the urea herbicide diuron.

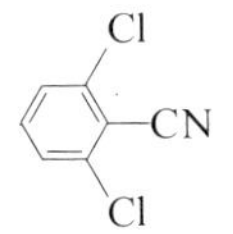

dichlobenil

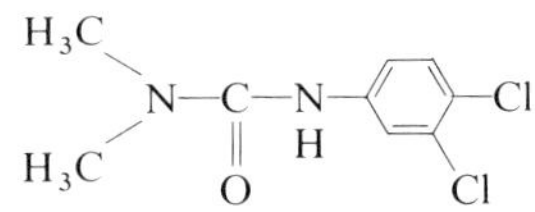

diuron

The resulting compound, 1-(2,6-dichlorobenzoyl)-3-(3,4-dichlorophenyl)-urea, code Du 19.111, appeared to have no herbicidal properties whatsoever. However, the compound showed an interesting insecticidal activity [235].

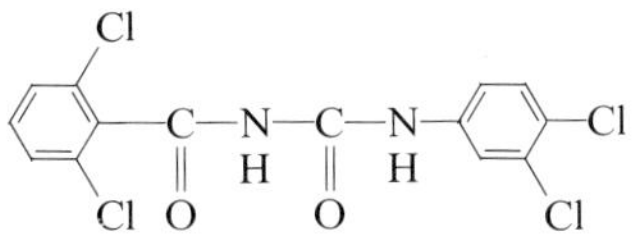

Du 19.111

Apart from the discovery of a new class of insecticides, especially the fact that the phenomena observed in the death of contaminated insects were completely different from those of insects treated with most known insecticides, has created a considerable interest in the benzoylureas. Not only at Philips-Duphar many related compounds have been synthesized [251, 252], but this was and still is the case in various other research laboratories as can be derived from the continuing appearance of new patent applications (see also Sect. 11).

From the Duphar research program, the compound PH 60-40, now known under the common name diflubenzuron, was selected for development [245] and is now being introduced in the market. As diflubenzuron was the first compound from the class of the benzoylureas that was successfully developed, it is obvious that this compound has received by far the most attention. Consequently, the subsequent discussions will deal mostly with diflubenzuron:

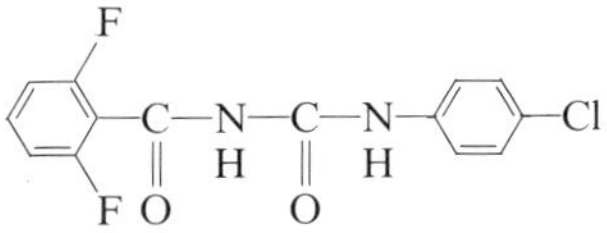

diflubenzuron = PH 60–40

This review covers literature until early 1979. For other summaries regarding benzoylurea insecticides we refer to Verloop and Ferrell [245], Verloop [246], and Grosscurt [97].

2. Synthesis of 1-benzoyl-3-phenylureas

For the preparation of 1-benzoyl-3-phenylureas in general, two routes can be used [251]:

(i) A substituted benzoyl-isocyanate is coupled to a substituted aniline according to the general scheme:

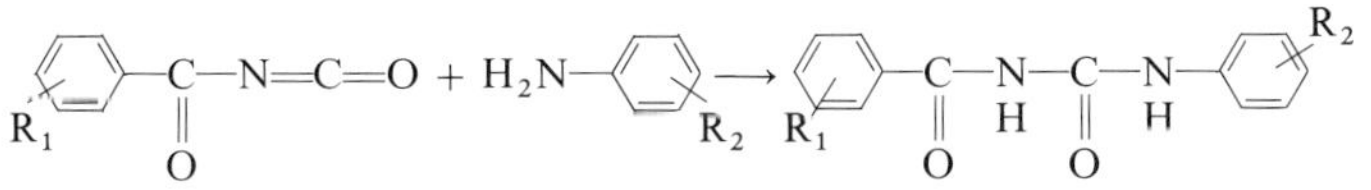

(ii) As an alternative, a substituted benzamide can be reacted with a phenylisocyanate:

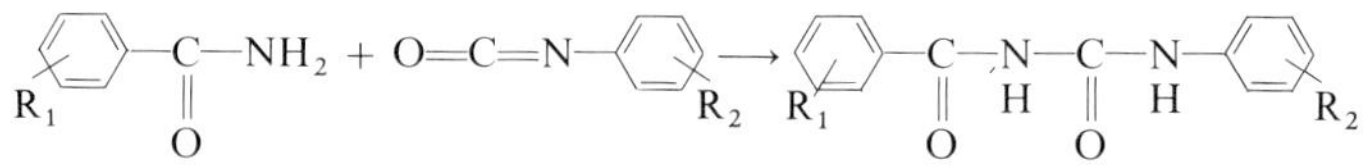

In the case of diflubenzuron, dichlobenil is used as a starting material. Upon reacting with potassiumfluoride, both chlorine atoms are replaced by fluorine to give 2,6-difluorobenzonitril. This compound subsequently is hydrolysed by sulfuric acid. The resulting amide, on treatment with 4-chlorophenylisocyanate in a high boiling organic solvent gives diflubenzuron in an overall yield of 85–90%.

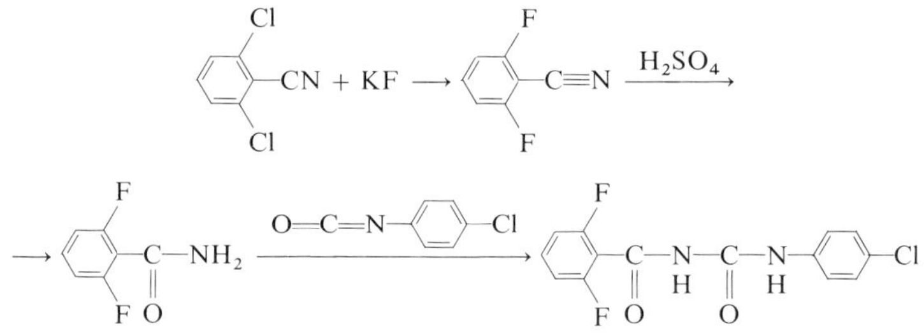

The thio-analogues of 1-benzoyl-3-phenylureas can be synthesized utilizing one of the following reactions [179].

(i)

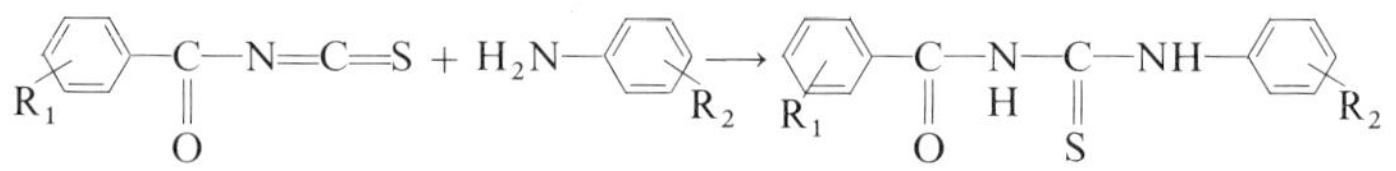

(ii)

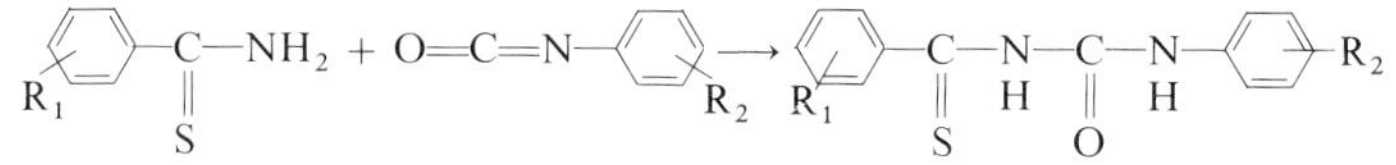

3. Diflubenzuron, Physical/Chemical Properties and Analysis

As already indicated in the introduction, the discussions in this chapter will focuss mainly on the insecticide diflubenzuron:

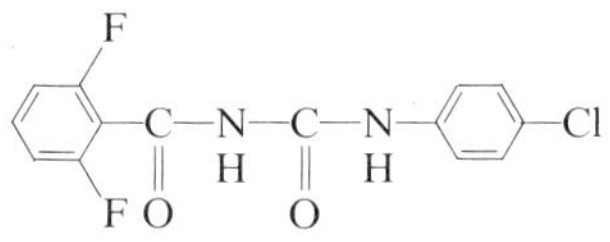

Chemical name (IUPAC): 1-(4-chlorophenyl)-3-(2,6-difluorobenzoyl)-urea.
Chemical Abstracts Index name: N[(4-chlorophenyl)amino carbonyl]-2,6-difluorobenzamide.
Wiswesser line-formula notation: GR DMVMVR BF FF.
Common name: diflubenzuron.
Trade mark: DIMILIN (registered trade mark of Philips-Duphar B.V.).
Code names: PH 60-40, TH 6040, PDD 60-40I, ENT-29 054, OMS 1804, Du 112.307.
Patent applications e.g.: Ref. 59, 172, 173, 221, 223, 225, 226.
Patents: e.g.: Ref. 221, 223, 225, 226.

3.1. Physical and Chemical Properties

Technical diflubenzuron is a white to yellowish-brown crystalline material, m.p. 210–230 °C. The pure compound is white, m.p. 230–232 °C. It is almost insoluble in water (about 0.2 mg/l at 20 °C) and apolar solvents. In polar to very polar solvents, the solubility is moderate to good, e.g., in acetone 6.5 g/l at 20 °C, in dioxane 24 g/l at 25 °C, in DMF 104 g/l at 25 °C. The partition coefficient in dichloromethane-water (pH 5.6) is 50, in 1-octanol-water about 5000. The volatility is very low; 0.5 mg of the compound spread over 50 cm^2 of aluminium foil lost less than 4% of weight during 48 hours at 20 °C. Diflubenzuron is rather stable in acid, but decomposes in an alkaline solution. When stored as an aqueous solution at 20 °C in the dark for three weeks, decomposition amounted to 4% at pH 5, 8% at pH 7 and 26% at pH 9. Both technical and formulated products are stable on storage. Diflubenzuron as such is also rather stable in sunlight. When a solution in either water or methanol is irradiated with UV light, a considerable photolytic breakdown occurs [199].

3.2. Analysis

The methods of analysis for diflubenzuron have recently been summarized by Rabenort et al. [188].

The active ingredient in technical material and in formulations can be determined by high pressure liquid chromatography (HPLC). As an alternative, a UV spectrophotometric method after column chromatography can be used. Due to its greater selectivity, HPLC is preferred.

For the determination of residues of diflubenzuron, two general types of procedures are available, HPLC and gaschromatography. Corley et al. [44] reported the determination of residues in milk by extraction with ethyl acetate, a partition between n-hexane and acetonitrile, followed by detection with HPLC. De Wilde et al. (Philips-Duphar unpublished) described a method for crops, soil, mud and water consisting of extraction with dichloromethane, clean-up of the extract on a Florisil column, and detection by HPLC. An alternate method for crops, soil, mud and water using Celite column chromatography has been described by DiPrima et al. [57]. The determination of diflubenzuron by HPLC in manure has been used by Oehler and Holman [177]. A gaschromatographic method used on the trifluoroacetyl derivative of 4-chloroaniline was described by Woroby and Webster [255, 256] and by DeMilo et al. [55], but this method has not been applied to crop samples. The formation of 4-chloroaniline from diflubenzuron under acidic conditions provides the basis for the recommended GC method. The aniline is derivatized with heptafluorobutyric anhydride. After clean-up by Florisil column chromatography, the derivative is detected by gaschromatography.

4. Formulations

Diflubenzuron is only very slightly soluble in the solvents generally used in pesticide formulations. Consequently, it has to be formulated in the solid state.

Three formulations of diflubenzuron are now commercially available, i.e. a 25% wettable powder, a 45% oil dispersible flowable concentrate and a granule.

One of the factors influencing the effectiveness of a pesticide is the rate at which it is absorbed by the target organism [142]. Since diflubenzuron dissolves poorly and very slowly in water, the particle size of the active material in the formulation has a considerable influence on the biological activity. This is demonstrated in Table 1, from which it can be seen that the effectiveness of diflubenzuron rapidly increases with decreasing particle size [162].

Table 1. Mortality (%) of *Pieris brassicae* larvae after application of diflubenzuron

Particle size range (μm)	mg/l active ingredient in spray liquid					
	3	1	0.3	0.1	0.03	0.01
10–20	50	10	6	0	—	—
4–10	100	50	43	17	—	—
< 4	—	—	100	100	88	3
liquid	—	—	100	100	100	28

In Table 1, the results are included of a liquid formulation, which shows the highest activity. This liquid formulation is a solution of diflubenzuron in a water miscible solvent. After mixing of such a solution with water, a slow crystallization of the active material occurs, initially resulting in very small particles. However, on ageing, the particle size increases. Although with a proper selection of surface active agents some control of the particle size may be achieved, the risk of variable particle size and, consequently, variable biological activity, is unacceptably high. For the same reasons, the addition of a simple acetonic solution to water or soil can easily lead to erroneous results. Hence, the use of solutions of diflubenzuron in solvents like acetone should be avoided.

Not only the biological effectiveness of diflubenzuron is affected by the size of its particles, but also the rate of degradation of the compound in soil depends greatly on the particle size as was demonstrated by Verloop et al. [243, 245], see also Sect. 7.1.

The above mentioned facts led to the development of formulations in which the particles of diflubenzuron are very small, mainly between 2 and 5 μm. The 25% wettable powder is the main formulation. Preliminary tests have shown that this wettable powder is generally compatible with other pesticide products in different formulations. However, it is recommended to preferably use wettable powders in mixture with DIMILIN W-25. In any case, each mixture should first be tested on physical compatibility.

The use of an emulsifiable oil as a tankmix with DIMILIN W-25 may enhance the effectiveness of diflubenzuron, especially in applications in which the ovicidal effects are utilized. Table 2 demonstrates that in the control of the cotton boll weevil the use of oil is essential [143].

Table 2. Percentage female cotton boll weevils with fertile eggs after spraying DIMILIN W-25 at 1.1 kg/ha (1 1b/A)

Treatment	Days after treatment			
	5	7	10	13
Water spray	46	34	86	89
Oil spray	0	0	0	65
Control	60	47	84	80

It was established, that in this application the use of about 10 l of oil per hectare is sufficient.

The 45% w/v oil dispersible flowable concentrate has been developed for ULV and fogging applications. For ULV spraying, the concentrate must be diluted with a low viscous, low volatile and non-phytotoxic oil. For fogging applications, the concentrate must be diluted with a low viscous, but volatile oil.

The granular formulation is used in fly control applications in animal housing. DIMILIN-G may also be used in mosquito control, where the product may have to penetrate through vegetation to reach the water in which the mosquito larvae are present.

5. Toxicology

5.1. Mammalian Toxicity

The vast majority of the results mentioned in this section is the conclusion of studies carried out by or on behalf of Philips-Duphar (unpublished).

5.1.1. Acute Toxicity

Table 3. LD50 in mg/kg bodyweight after 14 days observation

Route	Species	Sex	Technical diflubenzuron	DIMILIN dispersible powder 25%
Oral	Mouse	Male and female	>4640	>40,000
Oral	Rat	Male and female	>4640	>40,000
Intraperitoneal	Mouse	Male and female	>2150	
Percutaneous	Rabbit	Male and female	>2000	>18,650

Acute studies with rats and rabbits, exposed to dust atmospheres of technical diflubenzuron revealed that the product has a very low acute inhalation toxicity. Diflubenzuron technical is only a marginal irritant to the rabbit eye; the 25% dispersible powder formulation is non-irritant. DIMILIN W-25 appears to be non-irritant to rabbit skin.

5.1.2. Sub-acute Toxicity

Sub-acute exposure of rats and rabbits of both sexes to dust atmospheres of DIMILIN W-25 and technical diflubenzuron demonstrated that these products have no appreciable sub-acute inhalation toxicity.

Technical diflubenzuron and DIMILIN W-25 were applied to the intact and abraded skin of rabbits over a 3 weeks' period. Apart from transient erythematous reactions and slightly elevated methemoglobin levels in the highest dosage groups no reactions were found that could be related to the treatment. Both products may be regarded as to be of low sub-acute dermal toxicity.

5.1.3. Semi-chronic Toxicity

In a 90 day study, rats were fed diets containing 3.125, 12.5, 50 and 200 mg/kg of diflubenzuron. The no-effect level with regard to the parameters studied was 50 mg/kg food.

In a similar experiment, diflubenzuron was fed to pure-bred beagle dogs at concentrations of 20, 40, and 160 mg/kg food during 90 days. The only effects observed were marginal increases in methemoglobin and sulfhemoglobin levels in the highest dose level. Consequently, the no-effect level was 40 mg/kg food.

5.1.4. Chronic Toxicity

Diflubenzuron was fed to male and female rats during two years in concentrations of 10, 20, 40, and 160 mg/kg in the food. The only effect observed was a marginal increase in methemoglobin levels among both sexes in the highest dose level group. The no-effect level was 40 mg/kg food.

Diflubenzuron was fed to male and female mice during 80 weeks in concentrations of 4, 8, 16, and 50 mg/kg in the food. There was no evidence of reaction to treatment and in particular the spontaneous tumore profile of the mouse was not influenced. The no-effect level with regard to the parameters studied is 50 mg/kg.

The effect of diflubenzuron on growth and reproductive performance of rats was studied by feeding concentrations of 10, 20, 40, and 160 mg/kg of diflubenzuron in the daily diet through 3 generations. No negative effects were detected, neither on the parent generation nor on the offspring throughout 3 generations with regard to the parameters studied. The no-effect level, therefore, is 160 mg/kg.

5.1.5. Special Studies

In a dominant lethal study, males of albino mice were treated with a single intra-peritoneal injection of diflubenzuron at levels of 1000 and 2000 mg/kg bodyweight. Sequential mating of each male with three females per week was conducted for 6 consecutive weeks. These experiments did not result in dominant lethal mutations. McGregor et al. [146] tested diflubenzuron for possible mutagenic activity, using the micronucleus test in mice, the L5178Y mouse lymphoma forward mutation assay, and the Ames Salmonella reverse mutation test. No mutagenic effects were found.

In the assessment of the effect of diflubenzuron on the pregnancy of the rat, daily dosages of 1, 2, and 4 mg/kg body-weight were administered by intra-gastric intubation during days 1–15 of gestation. Similarly, in the assessment of the effect of diflubenzuron on the pregnancy of the rabbit, daily dosages of 1, 2, and 4 mg/kg body-weight were administered orally during days 6–18 of gestation. In both experiments the treatment did not significantly affect the parameters studied. Therefore the no-effect level for both animals is 4 mg/kg body-weight.

5.2. Toxicity to Birds

The acute oral LD_{50} on red winged blackbirds (*Agelaius phoeniceus*) for technical diflubenzuron is 3762 mg/kg body-weight. In an 8-day dietary LC_{50} study on mallard duck and bobwhite quail using technical diflubenzuron, levels up to 4640 mg/kg in the food showed no observable signs of toxicity. Groups of 10 males and 10 females of one day old chickens of respectively Barred Plymouth Rocks, white Leghorns, Nicholas white turkey, Mallard ducks and Ring-neck pheasants were fed during 91 days following hatching, diets containing respectively 0, 0.25, 1.25, 25, and 250 mg/kg of diflubenzuron. During the feeding period there was no appreciable difference between control and treatment groups with regard to the parameters studied, including body weight, feather, comb and wattle development, skin appearance, food consumption and behaviour.

5.3. Toxicity to Fish

Results of several laboratory studies on the toxicity of diflubenzuron have been reported in the literature. Table 4 summarizes these results, including experiments that lasted from 4 to 30 days.

From the results in Table 4 it can be seen, that even in concentrations far above those to occur in practice, no toxicity to fish is to be expected.

The most severe laboratory test was a 10 month egg to egg study with fathead minnows (*Pimephales promelas*) in a continuous flow system [33]. Even at the highest dose level of 0.1 mg a.i./l, which, in view of the low application rate and the rapid degradation of diflubenzuron, is very unlikely to occur over such a long period, no appreciable difference in survival, growth or behaviour could be observed between control and treatment groups. From field trials, including one in which 5 applications were made at a dosage rate of 56 g a.i./ha (0.05 1b/A) [218], no toxic effects to fish have been reported [127, 148, 218].

Table 4. Toxicity of diflubenzuron and DIMILIN W-25 to fish

Species	Stage	Method	Exposure time	Formulation	mg a.i./liter	Percentage mortality	Ref.
Salmo gairdneri	Juvenile	Static	96 h	W-25	38	0	186
Salmo gairdneri	1.2 g	Static	96 h	W-25	80	50	127
Salmo gairdneri	6 cm	Static	96 h	W-25	75	0	112
Salmo gairdneri	Fingerlings	Flow through	30 days	techn	0.3	0	127
Salmo gairdneri	Fry	Flow through	30 days	techn	0.3	0	127
Oncorhynchus kisuth	Juvenile	Static	96 h	W-25	38	0	148
Gambusia affinis	Fry	Static	96 h	W-25	25	0	157
Gambusia affinis	Fry	Static	240 h ?	techn	1	0	156
Pimephales promelas	0.87 g	Static	96 h	W-25	108	50	127
Ictalurus punctatus	2.2 g	Static	96 h	W-25	93	50	127
Lepomis macrochirus	0.5 g	Static	96 h	W-25	165	50	127
Cyprinus carpio	4 cm	Static	96 h	W-25	75	0	112

6. Mode of Action

Although diflubenzuron chemically has some similarity with both herbicides dichlobenil and monuron, the compound has no herbicidal properties and, even at very high dosages, does not show any phytotoxicity. In this respect it may be mentioned, that Hatzios and Penner [106] found no effect with diflubenzuron on photosynthesis and the ultrastructures of soybean leaves.

Below, both the biological and biochemical aspects of the mode of action of diflubenzuron as an insecticide will be discussed.

6.1. Biological Aspects

6.1.1. Larvicidal Effects

On the majority of the larvae tested, diflubenzuron is mainly active as a stomach poison. The first insecticidal effect observed was that, after oral uptake of diflubenzuron the larvae survive apparently unaffected until their next moult is due, but that at that stage they perish [235]. Histological examination of diflubenzuron-contaminated larvae of *Pieris brassicae* (large white butterfly) by Mulder and Gijswijt [162] revealed a disturbance of the formation of endocuticular tissue. Instead of normal cuticular layers, now globules of apparently coagulated material were present. This malformation of the exoskeleton probably prevents a proper use of the muscles by the new instar to free itself from the exuviae during moult.

With larvae of *Leptinotarsa decemlineata* (Colorado potato beetle), it was demonstrated that after a single sublethal dose of diflubenzuron, these distortions gradually decreased in newly formed tissue. Eventually, even normal layers could be deposited again, indicating that the disturbing effect of diflubenzuron is fully reversible [97]. This may be one of the factors playing a role in the phenomenon that when the period during which diflubenzuron is administered to larvae, is extended, the ultimate mortality is increased [9, 137].

In feeding experiments, the particle size of the active compound is of paramount importance. Smaller particles produce far better effects; an example of this relationship has been given in the section on formulation (Sect. 4). The influence of the particle size on the biological effectiveness is probably due to a better absorption in the insect in the case of smaller particles that have a higher rate of dissolution [142].

A significant contact activity after topical application of a solution of diflubenzuron in acetone has been reported only with larvae of *Musca domestica* (house fly [35]) and *Spodoptera littoralis* (Egyptian cotton leafworm [10]). Slight contact activity was found with *Heliothis virescens* (tobacco budworm [184]) and with *Dacus oleae* (olive fruit fly [70]).

In general, all larval instars of a susceptible species can be controlled by diflubenzuron. However, in many species later instars are less susceptible than earlier ones. This was found with larvae of *Lambdina fiscellaria* (hemlock looper [200]), *Hypera postica* (alfalfa weevil [169]), *Yponomeuta* spp. (small ermine moths [104]) and of the mosquito species *Culex nigripalpus* and *Aedes taeniorhynchus* [190].

No difference in susceptibility of the instars was found in the mosquitoes *Culex pipiens fatigans* and *Anopheles gambiae* [31].

An exception to the general pattern seems to be the spruce budworm (*Choristoneura fumiferana*), for which it has been established [191] that fifth and sixth instar larvae are more susceptible than third and fourth instars, with the fifth instar being the most susceptible one.

As death normally occurs during or shortly after a moult, it is generally advisable to administer the compound as continuously as possible until at least the next moult. A consequence of this delayed death is that feeding damage can still occur after application. From this point of view application against the early instar larvae is preferable. In that case, an additional advantage is that the early instars are generally more susceptible to diflubenzuron than later ones.

6.1.2. Ovicidal Effects

Ovicidal effects of diflubenzuron can be obtained either by topical application to the eggs or by contamination of female insects.

The direct contact activity on eggs is species dependent and is influenced by several factors. In general, younger eggs are more susceptible than older ones. Furthermore, the effect of diflubenzuron increases with smaller particles and higher relative humidity.

The effects on eggs produced after treatment of the female, depend on the concentration of diflubenzuron in the diet, the formulation used, and sometimes on the age of the adult female.

The factors, involved in the ovicidal activity of diflubenzuron, were summarized by Grosscurt [95].

Inhibition of egg hatch was also reported in cases where untreated females were confined with contaminated males [145, 150, 159, 259]. However, the techniques in these experiments did not restrict the possibilities for ovicidal effects to sperm influences. Also, direct uptake of diflubenzuron residues by the female insects and transfer by physical contact between males and females were possible. Moore et al. [160] demonstrated, that in the case of the cotton boll weevil diflubenzuron indeed was transferred from male to female by physical contact and indicated that there was little possibility of sperm effects. To avoid transfer by physical contact, Grosscurt [94] injected males of the housefly and of the Colorado potato beetle with diflubenzuron and allowed them to mate with untreated females. In this case, no effects were found on fertilization and egg hatch, which led to the conclusion that in these two species diflubenzuron had no effect on the sperm quality.

Grosscurt [94, 95, 97] demonstrated that when an ovicidal effect is produced by diflubenzuron, the embryos seemingly develop fully, but are unable to emerge. Electron microscope observations of embryos present in eggs produced by diflubenzuron treated female Colorado potato beetles showed disturbances of the cuticle formation that were similar to those produced in larvae [162]. As a consequence of these disturbances, the exoskeleton of the embryo has insufficient strength and the muscle attachment is inadequate, which prevents the embryo to leave the egg.

Summarizing, it can be concluded that the prevention of egg hatch must be attributed to interference with the formation of endocuticular tissue in the embryo. Although some authors used the word "sterility" for the inhibition of egg hatch after treatment of female insects with diflubenzuron [108, 149, 159, 212], the above mentioned work of Grosscurt shows that this is not justified. Also Moore et al. [160], from their experiments with the cotton boll weevil, arrived at the conclusion that failure of egg hatch was unlikely to be caused by sterility.

6.1.3. Effects on Adult Insects

Diflubenzuron treatment of adult insects generally is not lethal, still the compound can cause effects in adults. Of a number of insects it is known that cuticular tissue is deposited immediately after either moult to the adult stage or emergence from the pupa. This process generally takes place during a restricted period. Hunter and Vincent [111] were the first to show that the rigidity of the cuticle of adult locusts was reduced when diflubenzuron was fed just after moult. Ker [133] found that after a 3-day feeding period of newly moulted adults of the locust, *Schistocerca gregaria*, breaks occurred in the femoro-tibial joints of the hind legs. No such effects were found when the treatment started 4 days after moult.

Grosscurt [96] observed that hardening of elytra of the Colorado potato beetle (*Leptinotarsa decemlineata*) under his experimental conditions took place during about 10 days after adult emergence. At any time during this period hardening of the elytra could be blocked by administration of diflubenzuron.

Microscopic observations revealed stable cuticular layers in elytra of diflubenzuron-treated Colorado potato beetles [96] and in some regions of treated locusts [133]. This is in contrast to observations in larval cuticles of the large white butterfly [162] and the Colorado potato beetle [96], where no distinct cuticular growth took place upon diflubenzuron treatment.

Effects of diflubenzuron were also found on the composition of peritrophic membranes. This membrane is a cylindrical sheath that surrounds the food mass in the midgut and which sometimes also extends into the hindgut. It is composed of chitin fibres in close association with proteins. Clarke et al. [42] examined the peritrophic membrane of the adult locust, *Locusta migratoria.* After feeding the locusts, with diflubenzuron-treated food, an increase in the fibrous appearance and a reduction in weight were observed. This probably also means a weakening of the membrane. These phenomena can be explained by a reduction in chitin contents, also resulting in a reduced capacity to stabilise protein. Consequently, both the amount of chitin and the amount of protein deposited are reduced. In accordance with this, Becker [24] found a reduction of the dry weight and the length of peritrophic membranes of adult *Caliphora erythrocephala* (blowfly) after treatment with diflubenzuron.

6.2. Biochemical Aspects

Insect endocuticles mainly consist of protein and chitin. Consequently, it is obvious that the disturbance of the formation of endocuticular tissue by diflubenzuron and its predecessor Du 19.111 led to biochemical studies on the possible effects on these constituents.

Post et al. [185, 186] and Marks and Sowa [143a, 211] demonstrated that the incorporation of radioactivity from precursors of chitin into insect tissues is inhibited by Du 19.111. Deul et al. [56] found similar results with diflubenzuron in their studies using *Pieris brassicae* larvae. The inhibition process described is a very rapid one. They established, that it is virtually complete 15 min after the injection of diflubenzuron into the larvae. Ker [133] demonstrated inhibition of chitin deposition in locusts within 80 min after injection.

The possible effects of diflubenzuron on the deposition of proteins were first studied by Hunter and Vincent [111], using adult locusts. It was concluded that there were no direct quantitative or qualitative effects of the insecticide. This was confirmed by Ker [133], who found no effects of diflubenzuron on the aminoacid composition of locusts cuticles.

The preliminary conclusion that Du 19.111 and diflubenzuron inhibit the biosynthesis of chitin in insects, was challenged by Ishaaya and Casida [114] who explained the effects on cuticles of larvae of *Musca domestica* by activation of the enzyme chitinase, leading to increased cuticle breakdown. However, Deul et al. [56] found that in *Pieris brassicae* larvae, diflubenzuron did not affect chitinase activity, and furthermore, that chitin deposition was already completely blocked in early stages of fifth instar larvae where chitinase activity was still practically absent.

These and other arguments [245] led to the conclusion that the primary mode of action indeed is a rapid interference with chitin biosynthesis.

In addition to the effects described by Ishaaya and Casida [114], other biochemical influences of diflubenzuron and Du 19.111 have been described in the literature [17, 56, 113, 155, 161, 185, 201, 261, 262]. A common feature of all these effects is that they were found one or more days after treatment with the insecticides. As the effects of benzoyl urea insecticides become visible in the living insects only at the time of the next moult, investigators are prompted to look for defects up to a considerable time after application. However, these studies can at most indicate "secondary" effects [245].

The mode of action of diflubenzuron described above, implies that the insecticide acts directly somewhere in the route leading from glucose to chitin formation and deposition. Several studies have been carried out to identify the exact site of action. Deul (unpublished), in *Pieris brassicae* larvae, Van Eck [236] in housefly body walls, Hajjar and Casida [101, 101 a] in milkweed bug abdomens, and Sowa and Marks [211], in cockroach leg re-generates, found an increase in the level of UDPAG, the ultimate precursor of chitin. Gijswijt et al. [100, 245] and Vardanis [238] found that there is a close analogy between the insecticidal, histological, and biochemical effects of diflubenzuron and of Polyoxin D after injection of these materials into insect larvae. The latter compound is known to be a reversible inhibitor of chitin synthetase in fungi. The above results suggest that diflubenzuron blocks the last step in chitin formation. Whether this implies a direct inhibition of the active site of chitin synthetase, the enzyme involved in this terminal polymerisation reaction, is still an open question, also because of the very complicated nature of this enzyme.

In connection with the specific mode of action of diflubenzuron, it has been suggested that the insecticide might have additional effects on the biosynthesis of biomacromolecules related to chitin, e.g. on hyaluronic acid, which is a copolymer of glucuronic acid and N-acetyl-glucosamine. Crookshank et al. [46] studied the possible inhibition of biosynthesis or deposition of hyaluronic acid by diflubenzuron in chicken, but no effects were found. This was confirmed by Bentley et al. [25] who did not find any effect of diflubenzuron on the biosynthesis of hyaluronic acid and of chondroitin sulfate in mice, rats and chicken, and by Deul (unpublished), who found no effects on hyaluronic acid biosynthesis in chicken. Stoolmiller [215] showed, that diflubenzuron is not an inhibitor of N-acetyl-hexosaminyl-transferases involved in the biosynthesis of glucosaminoglycans in higher animals.

All the above results emphasize the specificity of the biochemical mode of action of diflubenzuron.

7. Distribution and Metabolism

In the course of the research program carried out at Philips-Duphar, the attention concentrated on 1-benzoyl-3-phenyl-ureas in which the benzoyl moiety was disubstituted in the 2 and 6 position, preferably by halogenatoms. This led to the further evaluation of two compounds, i.e. PH 60–38 and PH 60–40 (diflubenzuron).

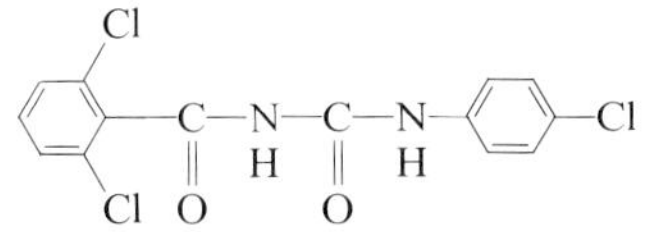

PH 60–38

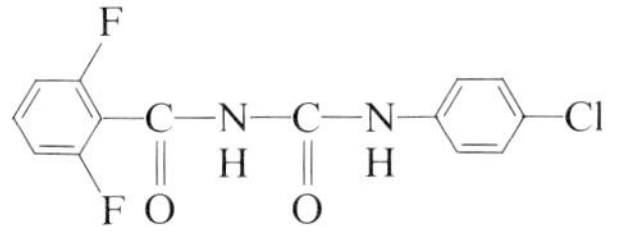

PH 60–40 = diflubenzuron

The metabolic pathway of both compounds was studied in soil and it was established that the compounds degraded along different lines [245].

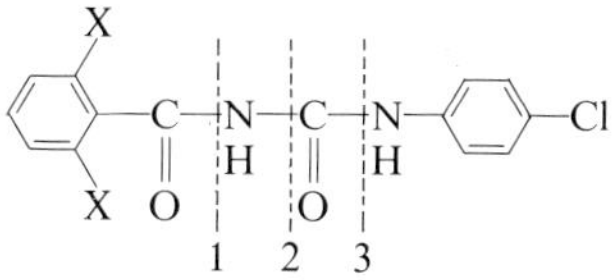

In the case of diflubenzuron, break-down occurred predominantly along line 1, resulting in 2,6-difluorobenzoic acid and 4-chlorophenyl-urea. In the case of PH 60–38, apparently the steric hindrance by the relatively large chlorine atoms is that strong, that degradation takes place along the lines 2 and 3, resulting in 2,6-dichlorobenzamide, carbondioxide and 4-chloro-aniline. Furthermore, the rate of degradation of diflubenzuron was considerably higher than that of PH 60–38, which was one of the factors in the final selection of diflubenzuron.

One of the physical properties of diflubenzuron, that is important in all considerations on its distribution and metabolism, is the very low solubility in water.

In relation to this, in all metabolism studies with diflubenzuron, the particle size of the active material should be carefully controlled.

7.1. Distribution and Degradation of Diflubenzuron in Soil and Water

Verloop et al. [245, 246] reported the results of studies on the fate of diflubenzuron in five agricultural soils and three hydrosoils, including the types recommended by the US EPA (Environmental Protection Agency) and the German BBA (Biologische Bundes-Anstalt). In these studies the compound was mixed through the soil at a concentration of 1 mg/kg. From the results, the following conclusions could be drawn:

(i) The rate of degradation is strongly dependent on the particle size of the diflubenzuron. At an average particle size of approximately 10 μm, the half life in various soils ranged from 8 to 16 weeks, whereas at a mean particle size of about 2 μm, the half life was $\frac{1}{2}$ to 1 week.

(ii) The half life in different types of soil varied by a factor of only 2.

(iii) The main metabolic pathway (over 90%) is hydrolysis leading to 2,6-difluorobenzoic acid (DFBA) and 4-chloro-phenylurea (CPU). A very minor pathway is hydrolysis of the other carbon-nitrogen bond, leading to 4-chloro-aniline (PCA) and 2,6-difluorobenzamide, which is rapidly converted into the corresponding acid, DFBA [245]. DFBA is rapidly degraded further, the half life being approximately 4 weeks. The first step is a decarboxylation, followed by ringdegradation [245]. CPU was found to be converted into bound residues with a half life of 5–10 weeks. In the bound residue, after 2 months, both CPU and PCA are present in roughly equal amounts. As PCA is generated from diflubenzuron in only small amounts, there must be a degradation of CPU into PCA. Free PCA has not been found in these studies, which can be explained by the rapid binding to the soil, found for anilines in general and confirmed for PCA in particular [246]. The soil metabolism of diflubenzuron is summarized in Fig. 1.

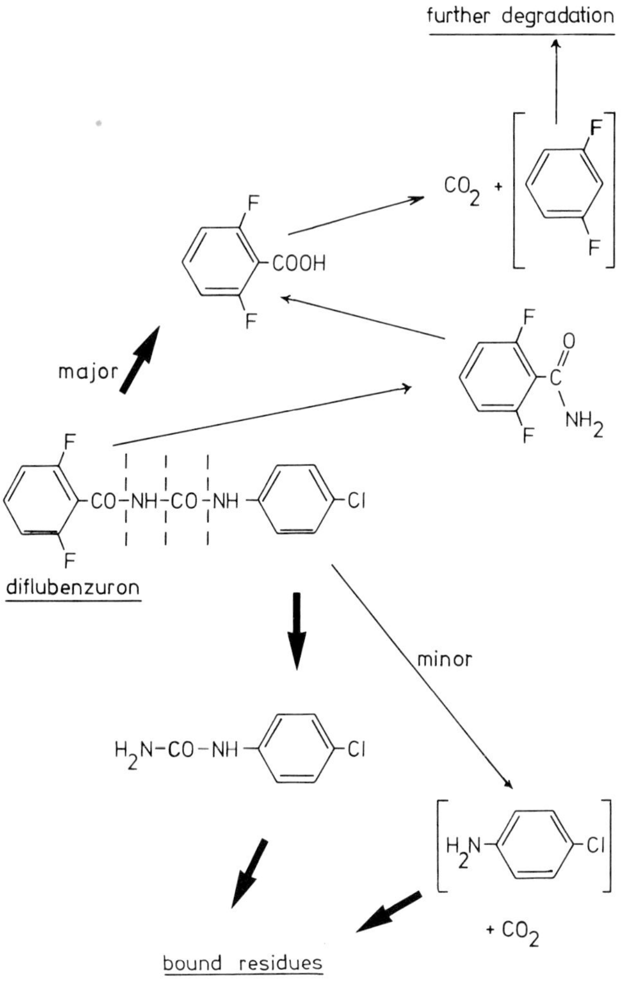

Fig. 1. Metabolism of diflubenzuron in soil

With a sandy loam soil, Verloop et al. [245] demonstrated that the degradation of diflubenzuron is of a microbiological nature: In an experiment with normal soil, after 4 weeks only 2% of the initial amount of diflubenzuron was still present, whereas in steam-sterilized soil 94% of the diflubenzuron was unchanged. Metcalf et al. [151] reported that in Drummer soil no appreciable breakdown of diflubenzuron could be detected after 4 weeks. However, in this study a high concentration of 10 mg/kg was used. Moreover, the diflubenzuron was added to the soil as an acetonic solution, which may have led to the formation of coarse crystals. Schaefer and Dupras [203] studied the persistence of diflubenzuron in field water. It was demonstrated that the diflubenzuron concentration in the water very rapidly decreased to below the detection limit, both by metabolism and absorption onto organic material.

7.2. Plants and Other Organic Matrices

Verloop and Ferrell [245] reported the results of greenhouse studies in which labeled diflubenzuron was applied to the leaves of several plant species as an aqueous suspension of small particles (average size 2 μm).

Table 5 shows the persistence of diflubenzuron on the treated leaves after 2 months.

Table 5. Persistence of diflubenzuron on plants after 2 months

Fraction	Percentage of applied dose			
	Soybean	Apple	Maize	Cabbage
TER	93	86	94	100
Diflubenzuron	96	89	95	100
TBR	4	2	5	5

TER = Total extractable residue
TBR = Total bound residue

It can be seen that more than 95% of the radioactivity recovered was present in the extractable fraction and that it consisted completely of unchanged diflubenzuron.

Bull and Ivie [30] treated cotton foliage under field conditions. In the analysis of the distribution of diflubenzuron, three fractions were discerned: (i) the material removed by rinsing the intact leaves with methanol, (ii) the extractable portion after homogenizing the rinsed leaves, and (iii) the unextractable portion. The results of the analyses are shown in Table 6.

In all methanol rinses and extracts of homogenized leaves, diflubenzuron was the only compound present. These results, as well as other ones described by the same authors indicate that under field conditions no appreciable degradation by light occurs. Verloop [247] studied the persistence of diflubenzuron on leaf litter and established that the degradation of diflubenzuron is closely related to the rate of decay of the leaves. Also in other organic matters, like manure, the degradation of diflubenzuron is rather slow, the half life being close to half a year.

Table 6. Fate of ^{14}C labeled diflubenzuron on cotton leaves in the field

Days post treatment	Percentage of dose in indicated fraction		
	Methanol rinse	Extractable	Unextractable
0	98.2	1.8	0.0
1	89.6	2.8	0.1
3	88.6	2.5	0.1
7	89.2	2.2	0.1
14	87.4	4.8	0.5
21[a]	22.7	6.8	0.4

[a] Between days 14 and 21 a total of 76 mm of rain fell on the treated plants

7.3. Animal Metabolism

In animal metabolism studies in which diflubenzuron is administered orally to the animals, a very important factor is the absorption of the chemical from the intestine. De Lange and Post [53] demonstrated that in rats, the portion absorbed is strongly related to the dosage administered. Whereas at an oral dose of 4 mg/kg, about 42% is absorbed, the absorption is only 3.7% at a dose of 900 mg/kg. Similar results were obtained with mice. In sheep, Ivie [116] found an absorption of about 60% at a dose of 10 mg/kg, whereas at a dosage of 500 mg/kg the percentage absorbed was only approximately 12%. Ivie [116] showed, that diflubenzuron is not metabolized to any significant extent when incubated in vitro with digestive tract fluids of sheep and cattle. On the other hand, the diflubenzuron absorbed was found to be completely metabolized in rats [253] and sheep [116].

Willems et al. [253] established, that in rats the main metabolic pathway is hydroxylation of the parent molecule. The hydroxylation process takes place in both phenyl moieties and yields three different metabolites. The overall hydroxylation process accounts for roughly 80% of the total metabolism. The remaining 20% is metabolized by cleavage of the ureido bridge, yielding 2,6-difluorobenzoic acid and 4-chlorophenylurea as primary degradation products. Of these, 2,6-difluorobenzoic acid could be identified as one of the urinary metabolites. On the other hand, 4-chlorophenylurea as such could not be detected, probably due to further metabolism. Figure 2 summarizes the diflubenzuron metabolism in rats.

In the course of further studies in the rat, it was established that one of the minor metabolites in the urine, is 4-chloro-aniline [52]. This explains the fact, that after ingestion of diflubenzuron, rats show a somewhat elevated level of methemoglobin in the blood, a phenomenon common for aromatic amines. Ivie [115, 116] studied the fate of diflubenzuron in cattle and sheep. Qualitatively, the metabolism in both species was similar to that described for rats. Quantitatively, however, there are considerable differences. The major metabolic transformation in the cow was hydroxylation at the 3-position of the 2,6-difluorobenzoyl moiety. In sheep, the major metabolic pathway is the cleavage of the amide bond at the benzoylcarbon to give 2,6-difluorobenzoic acid, which is excreted either free or conjugated with glycine. 4-Chloro-phenylurea could only be detected in very small amounts. Experiments in which ^{14}C-diflubenzuron was applied dermally to cattle demonstrated that the compound is not absorbed through the skin [116].

7.4. Metabolism in Insects

The distribution and metabolism of diflubenzuron in various insect species has been described in the literature.

Verloop and Ferell [245] indicated that in *Pieris brassicae* no metabolism of diflubenzuron could be detected. Still and Leopold [214] administered diflubenzuron to the cotton boll weevil (*Anthonomus grandis*) by immersion, injection, and feeding. They observed a rapid excretion into the faeces, but found no metabolism. Similar results were reported by Metcalf et al. [151], who used the salt marsh caterpillar (*Estigma acrea*) as a test object. On the other hand, Chang and

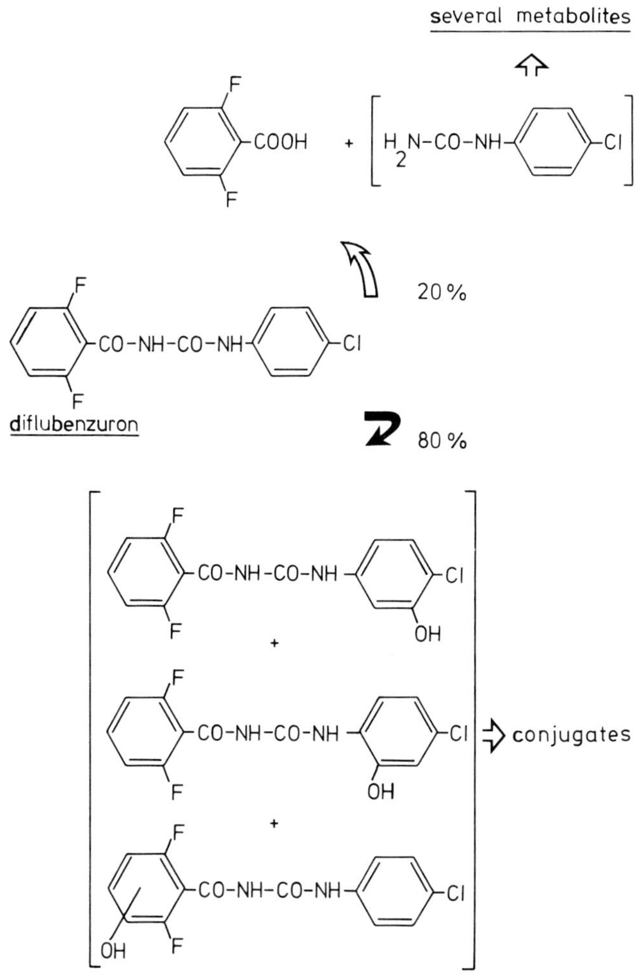

Fig. 2. Metabolism of diflubenzuron in rats

Stokes [40] found that after injection of diflubenzuron in boll weevils, approximately 19% of the injected dose could be recovered from the excreta in the form of conjugates. From the conjugates two compounds could be isolated and these were identified as hydroxylated compounds, the structures of which are shown below

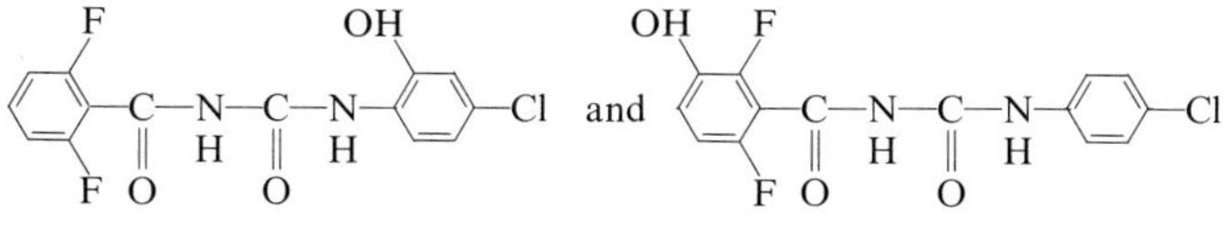

Work with flies has been reported by various authors. Ivie and Wright [117] applied diflubenzuron topically to the stable fly (*Stomoxys calcitrans*) and the housefly (*Musca domestica*, Orlando regular strain, insecticide susceptible). A

fairly rapid excretion was found with both species. Metabolism was found to be less than 3% of the dose applied with the stable fly, whereas the housefly metabolized about 10%. The main metabolic pathway appeared to be the cleavage of the diflubenzuron molecule, giving 2,6-difluorobenzoic acid and 4-chlorophenylurea. Treated stable flies secreted up to 1% of the dose into the eggs as unmetabolized diflubenzuron, but eggs of the housefly contained considerably less. This secretion into eggs likely accounts for the ovicidal effect of diflubenzuron. Chang [38] injected diflubenzuron into houseflies of an unspecified, but susceptible strain. About 20% of the dose appeared to be metabolized. However, in contrast to the results of Ivie and Wright, the metabolites did not result from a cleavage of the diflubenzuron molecule. They appeared to be conjugates, probably of a mixture of "aniline-ring" hydroxylated diflubenzuron. Pimprikar [183] observed both hydrolysis and hydroxylation of the parent molecule by houseflies. As Ivie and Wright on the one hand, and Chang on the other, also reported unknown metabolites, it is probable that both hydrolysis and hydroxylation are common metabolic pathways in the housefly.

7.5. Behaviour in Ecosystems

Metcalf et al. [151] studied the fate of diflubenzuron in a model ecosystem containing soil, water, plants (*Sorghum vulgare*), snails (*Physa sp.*), fish (*Gambusa affinis*) and mosquito larvae (*Culex sp.*). The so-called ecological magnification, defined as the ratio of the concentrations found in the biological object and in the surrounding water, was determined after 33 days. The mosquito larvae, being in the middle of the food chain of this model system, showed magnifications of 600–1,000. However, in the fish, which is at the end of the food chain, the magnification was 14–80, more than an order of magnitude lower than in the mosquito larvae. In the snails, magnifications of 100–200 were found. In the same model system, Metcalf et al. [151a] for DDT found magnifications of 10,000 in fish and 5,000 in snails.

The results of Schaefer et al. [204] indicate that diflubenzuron is hardly accumulated from water into fish tissues. Levels up to 80-fold can be reached when fish is exposed to concentrations of 10 μg/l in the water. However, as diflubenzuron is excreted again, the residue in the tissues is rapidly reduced if the ambient concentration is transient.

The overall conclusion therefore is, that diflubenzuron does not accumulate appreciably in fish.

Booth and Ferrell [26] studied the degradation and accumulation properties of diflubenzuron in various model systems including water, soil, vegetation, algae, bacteria and fish. Following multiple applications, diflubenzuron disappeared from water, sediment and vegetation after several days. Channel catfish did not accumulate diflubenzuron from treated soil in a simulated lake ecosystem. Algae (Plectonema) seemed to degrade diflubenzuron to p-chlorophenylurea and p-chloroaniline. However, the latter compound was identified by thin-layer chromatography with only one solvent system. The experiments described by Booth and Ferrell for algae [26] included a heat sterilization of the aqueous diflubenzuron suspension. Willems (unpublished results, Philips-Duphar) demonstrated that under these conditions diflubenzuron is converted into a compound, coded Du 210.675, the structure of which is shown below.

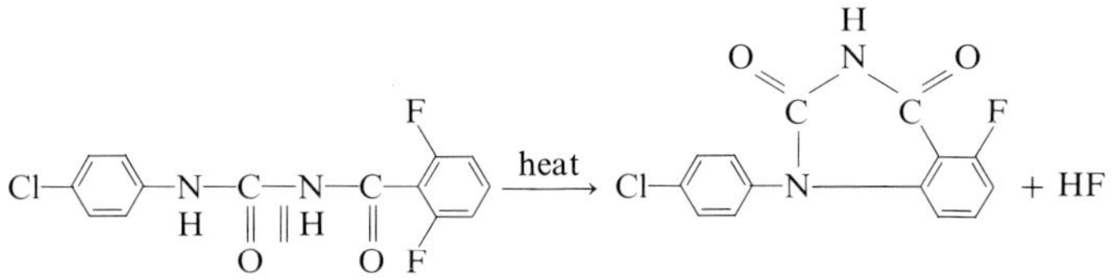

diflubenzuron Du 210.675

The compound Du 210.675 co-chromatographs with p-chloroaniline in some solvent systems, but not in others, as is shown in Table 7.

Table 7. Thin layer chromatography of Du 210.675 and *p*-chloroaniline (PCA)

Solvent system 1: Benzene/dioxane/acetic acid, 90:30:1
Solvent system 2: Chloroform/ethanol/acetic acid, 85:10:5
Solvent system 3: Dichloromethane/methanol, 95:5
Solvent system 4: Dichloromethane

Compound	Rf-values with solvent system			
	1	2	3	4
PCA	0.46	0.60	0.63	0.32
Du 210.675	0.44	0.61	0.51	0.00

As Booth and Ferrell [26] used only solvent system No. 1, they could not distinguish between PCA and Du 210.675. Consequently, the reported metabolism of diflubenzuron to PCA by algae (Plectonema) may be considered as an artefact.

7.6. Residues

In the considerations on residues, a number of different situations can be distinguished, i.e. residues on treated plants, residues in treated animals and as a separate item the residues in mushrooms resulting from the treatment of the growth medium.

7.6.1. Plant Residues

In the section on plant metabolism (7.2), it was described that diflubenzuron after application on plants is not metabolized to any practical extent. Furthermore, the compound does not penetrate into plant tissues. Consequently, it can be predicted that residues will only be present on those parts of the plants that are directly exposed to diflubenzuron during the application. It also is predictable that the only component of the residue will be the parent compound. Examples of edible

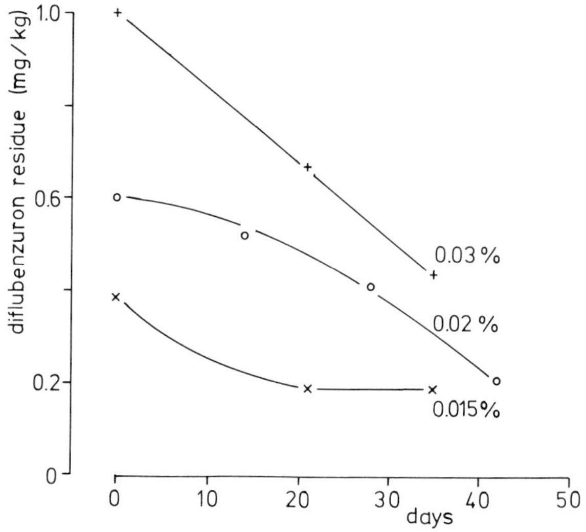

Fig. 3. Decline of diflubenzuron residues on apples after single application at indicated percentage a.i.

parts of crops that are treated directly with diflubenzuron can be found in deciduous fruit and cabbage. In W-Europe, DIMILIN-25 W is used for the control of codling moth and leafminers in apples. The spray scheme includes one to four applications at concentrations of 0.01–0.02% a.i. Figure 3 gives a number of typical degradation curves, representing the decline of the diflubenzuron residues on apple fruits after a single application at the given concentration.

Under practical conditions, following the spray schemes indicated above, the residue level at harvest remains well below 1 mg/kg. Examples of situations in which the edible parts of a crop are not directly exposed to the application of diflubenzuron, can be found in cotton and soybeans. Bull and Ivie [30] described the result of a field study with radio-active diflubenzuron applied to cotton at a rate of 70 g a.i./ha. In one trial, six applications were made. Determination of total radioactivity in the cotton seed showed that no residues could be detected in or on the seed, the limit of detection being 0.01 mg/kg. After 10 applications, a maximum residue of 0.02 mg/kg based on dry weight was found. Numerous residue trials in which non-radioactive material was used have demonstrated, that residue levels in cotton seed remain below the detection limit, which is 0.05 mg/kg. Also in soybeans, no residues can be detected in the beans, the limit of detection being 0.05 mg/kg (Thompson-Hayward Chem. Co., unpublished results).

7.6.2. Residues in Animal Tissues

In various usages of DIMILIN, animals may become contaminated with diflubenzuron or its metabolites. This is most obviously the case in the fly control in manure, where an even distribution of diflubenzuron through the manure may be achieved by feeding the insecticide to the animals. Ivie [116], in his animal metabolism studies on diflubenzuron, determined the level of total radioactive residue in many organs of a lactating cow. The tissues were collected 7 days after a single treatment of 10 mg of diflubenzuron per kg bodyweight. Of the 20 tissues examined, only the liver showed a detectable residue, i.e. 2.9 mg/kg diflubenzuron equivalents. Miller et al. [153] reported the results of experiments in which dairy cows were fed diflubenzuron at rates of 0.25 mg/kg bodyweight/day and 1.0 mg/kg/day for four months; a third cow received rates that were increased from 1 to 16 mg/kg/day, the highest rate being maintained for three months. Of the tissues examined of the cow receiving the 1.0 mg/kg rate, only the omental fat showed a residue at the limit of detection, which was 0.1 mg/kg; in the others no residues were found. The high dose cow showed residues of 0.2 mg/kg in the fat and 0.13 mg/kg in the liver; no residues were found in heart, muscle and kidney. In the milk of this cow, a residue level of 0.02 mg/kg could be established when the rate was 16 mg/kg. No residues could be detected during the period when the feeding rate was 8 mg/kg/day.

Contrary to these low levels of residues found in cattle, in poultry considerably higher residues are present in tissues and also in eggs. Miller et al. [152, 154] described that at concentrations in the feed of chickens that result in adequate fly control in the manure, the residue level in eggs of White Leghorns was in the range of 1 mg/kg. Figure 4 shows, that when White Leghorn hens are kept on a ration containing 10 mg/kg of diflubenzuron the eggs contain a residue that remains at a level of about 0.5–0.6 mg/kg over a prolonged period.

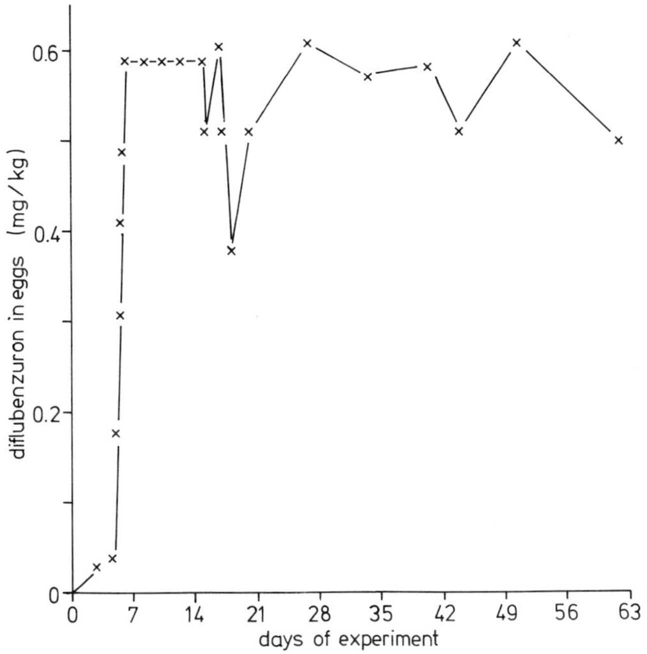

Fig. 4. Residues of diflubenzuron in eggs from White Leghorns fed with a ration containing 10 mg/kg (after [154])

When diflubenzuron is applied to water, part of the material may be taken up by fish. In Sect. 7.5 it was discussed that the bio-accumulation of diflubenzuron is not very high and also that the residues decline in fresh water. A good example of the change of the residue level over a certain period of time can be found in the paper of Apperson et al. [6], which describes the behaviour of diflubenzuron when applied to a lake (surface area 18.6 ha) at a rate equivalent to 5 μg/l in the water. Diflubenzuron residue levels were determined in white crappie, present in the lake. Table 8 shows that relatively shortly after the application the maximum residue level is obtained and that subsequently a rather rapid decline can be observed.

Table 8. Residues of diflubenzuron in white crappie from Lower Blue Lake, treated at 0.005 mg/l [6]

Sampling time	Diflubenzuron (mg/kg)
Pre-treatment	—
1 day post-treatment	0.13
2 days post-treatment	0.31
4 days post-treatment	0.36
7 days post-treatment	0.32
14 days post-treatment	0.20
21 days post-treatment	0.06

7.6.3. Residues in Mushrooms

When mushroom beds are surface-treated with DIMILIN, this may lead to residues in the mushrooms. Such residues can result from direct contact of the mushrooms with the treated compost and also from uptake by the mushrooms from the treated compost. As diflubenzuron degrades in compost in a way similar to that in soil, i.e. giving as primary metabolites 4-chlorophenylurea (CPU) and 2,6-difluorobenzoic acid (DFBA), also these compounds may be part of the residue. Several residue trials have demonstrated that generally diflubenzuron residues are low; after single treatment at 1 g a.i./m^2, they range from the sensitivity limit up to 0.07 mg/kg. CPU residue levels are generally similar to those of diflubenzuron. The major portion of the residue consists of DFBA; single treatments at 1 g a.i./m^2 result in residues from 0.1 to 0.6 mg/kg of DFBA (Philips-Duphar, unpublished).

8. Fields of Application

8.1. Introduction; Selectivity

In the previous sections it has been demonstrated that diflubenzuron has hardly any toxic effect on mammals, birds and fish. In addition to these aspects, diflubenzuron shows considerable selectivity which is due to a number of properties. The compound does not penetrate into leaves, and furthermore is mainly a stomach poison; only on some species a contact activity has been found. As a consequence, sucking insects, e.g. spider mites and aphids, and many hidden feeders, e.g. bollworms, budworms and stem borers cannot be adequately controlled, although the larvae are intrinsically susceptible. Another aspect contributing to the selectivity of diflubenzuron is that the compound acts as a larvicide and an ovicide but not as an adulticide. Even in those cases in which the ovicidal effect is obtained through contamination of the adults, the adults themselves survive the treatment.

Some standardized methods to test effects of pesticides on nontarget insects have been developed by the West-Palearctic Regional Section of the International Organization for Biological Control. In this program diflubenzuron was tested in laboratory experiments on adults of 3 parasitic Hymenoptera: *Trichogramma cacoeciae* [69, 105], *Coccygomimus turionellae* [69] and *Phygadeuon trichops* [168]. As could be expected from the mode of action of diflubenzuron, the compound was classified as harmless in all these tests. Other examples of the selectivity of diflubenzuron will be given in connection with the relevant field of application. A separate section will be devoted to effects of diflubenzuron on non-target organisms which are not strictly connected to a certain field of application, including effects on honey-bees and aquatic organisms (except fish).

8.2. Agricultural Crops

8.2.1. Cotton

The insect problem in cotton is considerably different in various parts of the world. In the USA, as well as in some of the cotton-producing areas of Central and South America, the cotton boll weevil (*Anthonomus grandis*) is an insect pest, that is a major problem in the earlier parts of the growing season.

Another important problem in American cotton is the "boll worm complex", the boll worm (*Heliothis zea*) and the tobacco budworm (*Heliothis virescens*). Bollworm populations may be kept at low levels by various beneficial insects. However, when these beneficials are eliminated in operations that control the boll weevil early in the season, the boll worm populations rapidly build up and bollworm control becomes an absolute necessity.

For diflubenzuron, it was established that when adult female boll weevils ingest the compound, they deposit eggs that do not hatch [159]. This effect is a reversible one, i.e. when the administration of diflubenzuron is discontinued, the hatchability after a certain period, returns to normal levels [149]. In the field, this ovicidal effect results in a drastic suppression of the boll weevil population [218]. Since, much experience has been obtained in the field (e.g: [71, 141]). Ganyard et al. [72] described the use of DIMILIN in a large area field test in which a total area of 262 ha, isolated from other cotton areas was treated from the air at a dosage rate of 0.14 kg a.i./ha. Over a period of two months, 9 applications were made. Results indicated over 90% control of reproduction. At the end of the period, the population of the third generation of boll weevils was no higher than the initial overwintered population, whereas the population increased 12-fold where conventional chemicals were used. Under practical conditions, DIMILIN-25W is now used at a dosage rate of 70 g diflubenzuron per ha (1 oz/A) at a 5–7 day schedule. The product must be applied together with oil (see also Sect. 4).

Very important in the use of DIMILIN for boll weevil control is, that the product has very little effect on beneficial insects. Although some effects were observed in their laboratory experiments, in field trials, Ables et al. [2] found no adverse effects on beneficials after nine aerial applications of diflubenzuron, the highest test rate being 0.13 kg a.i./ha. These beneficials included larvae and adults of lady beetles (Coccinellidae), predacious bugs (Hemiptera), larvae of green lacewings (*Chrysopa* spp.) and spiders. Similar results were reported by Keever et al. [130], who found no significant reduction of the predacious bugs *Nabis* spp. and *Orius insidiosus*, the lady beetles *Hippodamia convergens* and *Coleomegilla maculata*, green lacewings and spiders. Reduction of the population was significant only in the case of the predacious bug *Geocoris punctipes*.

Under field conditions, diflubenzuron has little direct effect on the bollworm complex [218]. This is at least partly due to the fact that the compound cannot reach the larvae because of their hidden life; Wolfenbarger et al. [254] established that intrinsically the tobacco budworm is susceptible to diflubenzuron. Summarizing, diflubenzuron can be an interesting tool in integrated pest management (IPM) programs for control of boll weevil and boll worm complex: The boll weevil population can be effectively suppressed by diflubenzuron, whilst predators may delay the build up of bollworm populations to economically important levels till later in the season.

In the cotton areas in Arizona and California the insect problem is different from that of the rest of the USA, i.e. the main insect pests are the cotton leaf perforator (*Bucculatrix thurberiella*) and the pink bollworm (*Pectinophora gossypiella*). Of the cotton leaf perforator, the young instars mine in the leaves; the fourth and fifth instars feed externally. Flint et al. [66] found excellent control of these 4th and 5th instars in field experiments. In their most recent trials, Flint et al. [67] established, that the rate of diflubenzuron for control of the cotton leaf perforator should be approximately 110 g a.i./ha.

Larvae of the pink bollworm are susceptible to diflubenzuron [65], but as with other boll worms the control under field conditions is insufficient [66].

In the Middle East area, the Egyptian cotton leaf worm (*Spodoptera littoralis*) is an important pest, not only in cotton, but also in other crops. In laboratory experiments, Ascher and Nemny [8] found a high ovicidal contact activity by dipping young eggs in a suspension prepared from DIMILIN-25W, the LC_{100} being equal to 2.5 mg a.i./l. These authors also found a contact activity of an acetonic solution of diflubenzuron on larvae, that was even higher than the activity in the case of oral administration, using the LC_{50}'s as a basis for comparison [9, 10]. In Egypt, DIMILIN is used as a larvicide. Its initial activity is insufficient to prevent unacceptable leaf damage. However, the residual activity of the product is very long-lasting [62]. This has led to the development of combination products that, apart from diflubenzuron, contain a fast but relatively short acting insecticide, like chlorpyrifos or methomyl.

In the rest of Africa the main insect pests are various boll worms, i.e. the American bollworm (*Heliothis armigera*), the red bollworm (*Diparopsis castanea*), the pink bollworm (*Pectinophora gossypiella*), and the spiny bollworm (*Earias insulana*). As with other boll worms, here too, DIMILIN does not provide adequate control in the field [195]. When fed on an artificial diet, larvae of the spiny bollworm were found to be susceptible [11].

In Southern Asia the tobacco caterpillar (*Spodoptera litura*) is a pest on cotton. Sundaramurthy [216], in laboratory experiments established that the larvae are highly susceptible.

8.2.2. Soybeans

Turnipseed et al. [220] reported the results of field trials with diflubenzuron to control insects in soybeans in the U.S. and Brazil. In their trials, diflubenzuron gave excellent control of the velvetbean caterpillar (*Anticarsia gemmatalis*) at a rate of 75 g a.i./ha.

Laboratory experiments by Holst [108] showed that larvae of the Mexican bean beetle (*Epilachna varivestis*) were killed by diflubenzuron after feeding on bean plants sprayed with a suspension of 50 mg a.i./l. In field experiments, the rate was fixed at 35–70 g a.i./ha. The same rates are effective in the control of the green clover worm (*Plathypena scabra*). For control of the soybean looper (*Pseudoplusia includens*), the dosages must be increased to 140 g a.i./ha.

8.2.3. Corn

Büchi [28] found, that when larvae of the European corn borer (*Ostrinia nubilalis*) were fed on an artificial diet containing 2.5 mg/kg of diflubenzuron, a mortality of 96% was obtained. Also when eggs were deposited on corn leaves with a diflubenzuron deposit, a high mortality was obtained. Under practical conditions, however, no satisfactory results were obtained so far.

8.3. Horticultural Crops

8.3.1. Fruit Crops

In apples and pears, diflubenzuron is very active against a number of economically important Lepidoptera. Elings and Dieperink [61] indicated that the codling moth (*Laspeyresia pomonella*) can be effectively controlled.

Audemard et al. in a series of papers [12–14] described the use of DIMILIN in field trials in France. At spray concentrations of 20 g a.i./100 l excellent results were obtained.

Similar results were obtained in the Netherlands by Gruys [98, 99], in England by Cranham [45] and in New Zealand by Wearing and Thomas [249].

Apart from the codling moth, also various leaf miners are well controlled by diflubenzuron, e.g. the apple pygmy moth (*Stigmella malella*) [98]. De Angelis [51] described the control of the leaf miners *Leucoptera scitella* and *Lithocolletis blancardella* in Italian apple orchards. A spray program including two or three applications of diflubenzuron at spray concentrations of 12.5 g a.i./100 l provides adequate control of the insects.

In the family of the leafrollers the susceptibility differs widely. The summer fruit tortrix (*Adoxophyes orana*), the rose tortrix moth (*Archips rosana*) and *Pandemis heparana* show a low susceptibility to diflubenzuron under practical conditions. On the other hand, the eye-spotted bud moth (*Spilonota ocellana*), *Hedya nubiferana* and the fruitlet mining tortrix (*Pammene rhediella*) can be controlled effectively.

Another insect species that can be well controlled with diflubenzuron is the winter moth (*Operophtera brumata*) [98, 187].

Diflubenzuron has become an important component of integrated pest management (IPM) programs in apple orchards. This is due to its effectiveness on a number of Lepidoptera, which have become more important under integrated control than under conventional chemical control. Examples are the codling moth, the fruitlet mining tortrix, and the eye-spotted bud moth. On the other hand, diflubenzuron is highly selective. It has no effect on either spider mites or on two predatory mites, tested so far.

The toxicity of the compound to the predatory mite *Amblyseius potentillae* was investigated by Van de Vrie (personal communication). In laboratory experiments he sprayed apple leaves with 500 mg a.i./l. These were subsequently infested with the predator and the European red spider mite (*Panonychus ulmi*) as a prey. Evaluation 1 week after treatment showed no effect on adults, eggs, or newly hatched larvae of *Ablyseius potentillae*. These results were confirmed in field experiments.

Cranham [45] investigated the effects on the predatory mite *Typhlodromus pyri*, using the European red mite (*Panonychus ulmi*) as a prey. He sprayed potted *Prunus* rootstocks in a greenhouse with concentrations of 250 and 500 mg a.i./l. Assessments up to 5 weeks after application revealed no effect of the compound in either concentration on *Typhlodromus pyri*.

In experiments on pears, in which DIMILIN-25W was applied in a concentration of 0.12% of product, good control of pear suckers (*Psylla piri*) was obtained [237]. In other trials against this insect, however, results were erratic.

The citrus rust mite (*Phyllocoptruta oleivora*) is the only mite presently known to be susceptible to diflubenzuron. Very good control was obtained in field trials at a rate of 350 g a.i./ha and two or three applications per year. Fruit was protected from external blemish for 21 weeks in experiments of McCoy [144]. The compound has neither an ovicidal effect after direct topical application to the eggs nor an effect on eggs laid after adult treatment. In a laboratory experiment [144], however, mortality occurred at the moult from the second nymphal to the adult stage, after diflubenzuron was applied topically to the eggs. Effects after nymphal treatment were not observed.

The West Indian sugar cane rootstalk borer weevil (*Diaprepes abbreviatus*) was first observed in Florida citrus in 1964 and was found in some 1800 ha in 1976. This weevil can be controlled by the ovicidal activity of diflubenzuron via contamination of female adults, similar to the mode of action in the cotton boll weevil. At rates of 700 g a.i./ha the compound has a residual activity of 26 days [206]. Control of larvae does not contribute significantly to the overall effect of the product. Control of larvae is also low when diflubenzuron is mixed through an artificial diet [23].

Schroeder and Sutton reported the results of a field trial in which diflubenzuron was applied from the air. The spray volume was 47 l/ha and the spray liquid was a 50/50 mixture of water and oil. At a dosage rate of 350 g a.i./ha, diflubenzuron gave a reduction in reproductive potential of about 95% for at least 4 weeks [207].

8.3.2. *Mushrooms*

Larvae of some species belonging to the Sciaridae (fungus gnats) and Phoridae (humpbacked flies) live in mushroom compost, mycelium and sporophores. These insects can be controlled by application of diflubenzuron in case mixing and in post-casing drench at a rate of 20 mg a.i./l. Species belonging to the Sciaridae seem to be more susceptible than species belonging to the Phoridae.

In experiments described by Kalberer and Vogel [128], diflubenzuron was applied to the casing soil (thickness 6 cm) and mixed into it. A rate of 1 g a.i./m^2 of casing soil gave good control of the sciarid *Lycoriella auripila.*

After numerous other trials (Philips-Duphar, unpublished), the practical application rate was set at 0.5–1.0 g a.i./m^2.

8.3.3. *Floriculture*

Control of fungus gnats (*Lycoriella solani* and *Bradysia paupera*), of which larvae feed on root hairs of seedlings, cuttings and many potted plants, can be obtained by applying diflubenzuron as a soil drench of 20 mg a.i./l (Van de Vrie, personal communication).

8.3.4. *Cabbage*

In small-scale field experiments by Mulder and Swennen [163], complete control of larvae of *Pieris brassicae* (large white butterfly) was obtained with a spray concentration of 1 mg a.i./l. In field experiments Elings and Dieperink [61] obtained complete control after 14 days at a rate as low as 18 g a.i./ha. Cabbage moths (*Mamestra* spp.) can be controlled at higher dosages.

8.4. Forestry and Woody Ornamentals

The insect pests in forests are considerably different in North America and Europe. Consequently, the DIMILIN possibilities in both areas will be considered separately.

In the USA and Canada one of the most important insects in forests is the spruce budworm (*Choristoneura fumiferana*). Retnakaran and Smith [191] demonstrated that in laboratory and greenhouse experiments larvae of the spruce budworm are susceptible to diflubenzuron. However, field trials have hitherto not shown encouraging results. One of the main problems appears to be the availability of diflubenzuron in the buds, to which feeding is confined.

The Douglas-fir tussock moth (*Orgyia pseudotsugata*) shows periodic extensive outbreaks in the northwestern part of the U.S. and in Canada. In preliminary field tests by Neisses et al. [170], DIMILIN provided excellent population control and foliage protection at a rate of 280 g a.i./ha. There are indications, that this rate can be reduced to about 70 g a.i./ha (Thompson-Hayward Chem. Co., unpublished results).

The effectiveness of diflubenzuron in the control of the forest tent caterpillar (*Malacosoma disstria*) was demonstrated by Retnakaran et al. [192, 193] in greenhouse and field experiments. In field trials, Harper and Abrahamson [103] established that heavy larval mortality and reduction of foliage losses could be obtained at dosages as low as 34 g a.i./ha.

An insect that is found both in N-America and Europe is the gypsy moth (*Lymantria dispar*). From the U.S., Granett and Dunbar [91] reported very promising results from laboratory and field trials with diflubenzuron at application rates as low as 35 g a.i./ha. Laboratory experiments by Granett and Weseloh [93] with *Apanteles melanoscelus*, a gypsy moth larval parasite, revealed a more or less similar susceptibility for the parasite and the host. Diflubenzuron, however, did not affect adults and had no adverse effects on progeny production when fed to adult female parasites. In later field trials, Granett et al. [92] demonstrated, that with a suitable timing of the application, DIMILIN shows an acceptable selectivity and can be used in integrated programs.

In Europe, Skatulla [209, 210] and Donaubauer [58] found complete control of the gypsy moth by aerial application of 75 g a.i./ha. Skatulla also reported that the closely related nun moth (*Lymantria monacha*) could be controlled with the very low rate of 7.5 g a.i./ha. In the experiments of Skatulla diflubenzuron was compared with the organophosphate insecticide trichlorfon. In contrast to the treatment with trichlorfon, the diflubenzuron treatments did not cause significant reductions in various non-target insects. Furthermore, in laboratory tests no effects were found on larvae of the predatory bug *Picromerus bidens* after feeding them diflubenzuron-treated larvae of *L. dispar* [209].

In field trials in Germany by Hartinger [104], last larval instars of *Yponomeuta* spp. were treated with a spray concentration of 7.5 g/100 l. The treatment resulted in normal pupation, but it completely prevented adult eclosion. Surprisingly, emergence of parasitic Hymenoptera and Diptera from the affected pupae was not significantly influenced. However, when younger instars of the host are treated with diflubenzuron, the parasites are unable to complete their development. Again, this is a case in which proper timing of the application results in a good selectivity [104, 209].

An important pest in Southern European forests and also in Northern Africa, is the processionary caterpillar (*Thaumatopoea pityocampa*). Ribrioux and Dolbeau [194] described the results obtained in several field trials. In aerial applications, excellent control of the larvae was obtained at a dosage rate of 150 g a.i./ha. Later trials have shown that this application rate could be decreased to 75 g a.i./ha.

The brown-tail moth (*Euproctis chrysorrhoea*) is very susceptible to diflubenzuron. In some cases the dosage can be as low as 7.5 g a.i./100 l in the spray liquid [61].

The ovicidal effect of diflubenzuron has been demonstrated by Novak and Sehnal [176] for the pine weevil (*Hylobius abietis*) after feeding the compound to the females.

In the case of *Agelastica alni*, ovicidal effects can be found both after feeding diflubenzuron to the females and after direct treatment of the eggs. Weiss [250] showed that the susceptibility of the eggs is higher for younger eggs.

Other insects that have been shown to be susceptible to diflubenzuron include the fall webworm (*Hyphantria cunea* [217]), the western hemlock looper (*Lambdina fiscellaria lugubrosa* [200]), the European pine sawfly (*Neodiprion sertifer* [209]), and the introduced pine sawfly *Diprion similis* [68].

The role of DIMILIN in the protection of forests against insects on the one hand and the forest ecosystem on the other has recently been reviewed by Schwenke [208]. His overall conclusion was that although some of the ecological aspects need further study, the many positive properties of the product render it very suitable for use in forests.

8.5. Medical-hygienic and Veterinary Applications, Including Control of Nuisance Insects

8.5.1. Mosquitoes and Midges

Laboratory experiments with mosquitoes revealed an outstanding activity of diflubenzuron on all larval stages of many species [110, 190, 202]. A summary of the results on sixteen species of mosquitoes was given by Grosscurt [97].

In field trials good control was achieved at dosage rates ranging from around 20 to 50 g a.i./ha at water depths of up to 40 cm [49, 165, 190, 202, 212]. In strongly polluted water, the dosage must be increased.

In mosquito control operations, DIMILIN can not only be applied as the wettable powder, but also in granular formulations. Mulla and Darwazeh [164] compared these formulations of diflubenzuron in the control of the mosquito *Psorophora confinis* in irrigated alfalfa fields and found that the granules were slightly less effective.

To obtain an effective release of diflubenzuron from the granules, the finely ground wettable powder can be coated onto a sand granule [50, 196].

Around artificial lakes in California, nuisance arises from heavy populations of Chironomid midges. When applied as a granular formulation, diflubenzuron at a dosage rate of 110 g a.i./ha of water surface, efficiently controlled larvae of *Chironomus* spp., *Tanytarsus* spp., and *Procladius* spp. during 4–5 weeks in cases that the water depth was less than 1 m and during 3–4 weeks at water depths between 1 and 2 m [3]. *Labrundinia maculata* could not be controlled even at a rate of 280 g a.i./ha of water surface [167].

8.5.2. *Flies*

Control of flies by diflubenzuron can be achieved in several ways.

(i) Surface treatment of the larval breeding medium.

In several countries, DIMILIN has been registered for control of fly larvae in stables. Species that are well controlled include the housefly *(Musca domestica)* and the little housefly *(Fannia canicularis)*. Concentrations used are from 0.5 to 1.0 g a.i./m^2, depending on whether the stables are of the open or closed type. Upon topical treatment of manure in poultry houses with as little as 0.12 g a.i./m^2, an 84% reduction in adult housefly emergence was observed by Ables et al. [1]. When compared with a dimethoate-treated farm, the diflubenzuron-treated one had greater species diversity and abundance of parasites.

Another species that can be controlled effectively by surface treatment of the breeding medium is the stable fly *(Stomoxys calcitrans)* [32, 257].

(ii) Feed-through treatments

Fly control in manure can also be achieved by oral intake by poultry, cattle or swine and subsequent excretion in the faeces. Diflubenzuron, in this technique, can be administered mixed in the feed or via mineral blocks. Miller et al. [152] found that good control of houseflies in chicken manure could be obtained with diflubenzuron at concentrations of 6.2 and 12.5 mg a.i./kg of food. These dosages, however, result in residues in the eggs (see Sect. 7.6.). Hayakawa [107] controlled the stable fly with as little as 1 mg a.i./kg in the food of laying hens. Similar concentrations of diflubenzuron were used as a feed additive to pigs by Kennelly et al. [132]. At 25, 15, and 12 mg a.i./kg of food, the compound gave 100, 98, and 94% inhibition of housefly development.

In a feed-through experiment by Pickens and Miller [182] with cattle, diflubenzuron was mixed with grain to obtain a dosage of 1 mg/kg bodyweight/day. This dosage controlled over 90% of both face fly *(Musca autumnalis)* and housefly larvae in bioassay tests with fecal pats. Non target insects that could be recovered from these pats showed the same species distribution as untreated checks. Only a slight shift in numbers could be observed. Experiments by Cook and Gerhardt [43] revealed that the dosage rate of diflubenzuron could even be decreased to 0.46 mg/kg bodyweight/day. At this dosage, applied via mineral blocks, the numbers of face fly larvae were significantly reduced.

(iii) Ovicidal effects by contamination of adults

In laboratory experiments, ovicidal effects could be obtained on treatment of adult flies by either feeding, topical application, or contact with a treated surface. One or more of these methods of contamination were reported to be successful in achieving ovicidal effects in the housefly [94, 95, 152, 259], the stable fly [258, 259], the horn fly [258], and the screwworm fly (*Cochliomyia hominivorax*) [47].

In field tests the horn fly could be controlled completely by the spraying of range cattle with 2.5 g a.i./l [135, 136, 258]. In laboratory experiments, Kunz et al. [134] applied a dust against the horn fly. A reduction of 82% resulted from 3 mg a.i./animal, while complete control was obtained with 6 mg.

8.5.3. *Blackflies*

Larvae of blood-sucking blackflies (Simuliidae) are difficult to control because they live in running water.

In laboratory experiments in the USA, Lacey and Mulla [137, 138] tested diflubenzuron against larvae of several *Simulium* species and established, that all species tested were susceptible to the compound. At a concentration of 0.02 mg/l and a 1 h exposure time, the ultimate mortalities for the various species ranged from 59 to 88%. In addition to this larvicidal effect, in experiments with *Simulium vittatum* an ovicidal effect was found. At a concentration of 1.0 mg/l and an exposure time of 1 h, almost complete inhibition of egg hatch was observed. The above activity was confirmed in outdoor trials using artificial streamlets containing *Simulium vittatum* [140].

In Canada, McKague et al. [147] used field collected larvae of *S. verecundum* in laboratory tests and found that adult emergence was reduced with 86–100% after a 1 h exposure to a concentration of 0.1–0.2 mg diflubenzuron per liter. Applications to two streams at a concentration of 1 mg a.i./l, maintained over a period of 15 min, resulted in gradual elimination of blackfly larvae [147].

8.5.4. *Tsetse Flies*

Jordan and Trewern [125, 126] performed laboratory experiments with the tsetse fly *Glossina morsitans*. After topical application of diflubenzuron to the females, the egg hatch was normal. However, the larvae died at pupation. The authors found less clear effects after tarsal contact of adult females with deposits. Kaplanis et al. [129] found similar effects upon topical application of females of *Glossina palpalis palpalis*.

8.5.5. *Chewing Lice*

In laboratory experiments by Hopkins and Chamberlain [109], nymphs of the angora-goat-biting-louse (*Bovicola limbatus*) could be controlled by incorporation of diflubenzuron into the diet.

8.6. Effects on Non-targets, not Related to a Specific Field of Application

8.6.1. *Honey Bees*

Diflubenzuron has the advantage that it does not kill adult insects. For honey bees this was confirmed by Stevenson et al. [213] who found an oral and contact LD_{50} which was larger than 30 μg/bee. The toxicity to eggs of honey bees was also low.

Possible effects on the brood were investigated in laboratory experiments by Barker and Taber [15]. They showed that diflubenzuron did reduce the production of sealed brood of the honey bee (*Apis mellifera*), when fed as a mixture with sugar syrup. However, the concentration required for brood reduction was relatively high. Only the test dosage of 59 mg a.i./l in syrup was effective, while the others (5.9 and 0.59 mg a.i./l) had no effect. Laboratory experiments by Barker and Waller [16] confirmed that the compound at high rates could cause harm. They used a diluted sugar syrup containing 60 mg/l of diflubenzuron.

Although the above suggests an intrinsic susceptibility of honey bees to diflubenzuron, the important question is, whether bees in the field can in fact come into contact with such high concentrations of diflubenzuron. In a large-scale field experiment in Canada by Buckner et al. [29], the impact on honey bees of an aerial application of diflubenzuron (87.5 g a.i./ha) was studied during a control program for the spruce budworm (*Choristoneura fumiferana*). No adverse effects were observed. In Germany, honey bees were observed by Egger [60] after treatment of forests with 30 and 100 g a.i./ha against the sawfly *Pristiphora abietina*. During the observation period, effects were monitored 4, 11, 35, and 59 days after the treatment. Only an average of 1.5% of the beelarvae were killed by the treatment. This result is very favourable, especially because that treatment with the insecticide took place in a period when the leaves were covered with honey dew, thus resulting in a more than normal contamination of the bees. The above indicates that diflubenzuron is quite safe for honey bees when used in the field.

8.6.2. Non-target Aquatic Organisms

Diflubenzuron, in some of its applications is directly applied to water, e.g. for the control of mosquitoes, midges and blackflies. In other applications, some of the active material may reach the water as well.

The low toxicity of diflubenzuron to fish was already mentioned in Sect. 5.3.

Below, the effects of diflubenzuron on non-target insects and crustaceans, observed in field experiments, will be discussed.

An extensive study on non-target aquatic insects was done by Farlow et al. [64]. In the trials, two adjoining 20-ha plots were selected in an intermediate marsh habitat. One of these plots received six applications of diflubenzuron at a rate of 28 g a.i./ha over an 18 month period. The other plot remained untreated. During the 18 month period, observations were made at 2 weeks intervals. The treatment resulted in a significant reduction in population numbers of 5 insect species, while for 15 other species the numbers were significantly increased as compared to the untreated plot. The analysis of data on the remaining 27 species indicated no statistically significant differences between the treated and untreated plots. The results showed, that as the population of certain species, that are known as predators, were reduced as a consequence of the treatment, an increase occurred in other species, that are known to be preys.

This change in predator/prey balance was also reported by Steelman et al. [212], who applied diflubenzuron to small rice field plots at a single dose of 30 g a.i./ha. Miura and coworkers, in a series of papers [156–158] described the results of numerous field tests with diflubenzuron. Although it was found that treatment resulted in some reduction of certain non-target insect species, the authors concluded that at rates used for mosquito control, diflubenzuron is relatively safe to organisms associated with the mosquito breeding habitat.

Mulla et al. [166] also observed only slight effects on non-target insects in trials in which diflubenzuron was applied to experimental ponds at rates up to 56 g/ha.

Assessment of reported effects of diflubenzuron on crustaceans reveals that at rates used for mosquito control, water fleas (Cladocera) generally are rather seriously affected. Especially, *Daphnia* spp. proved to be highly susceptible [4, 158, 166]. On the other hand, *Bosmina longirostris* showed a high tolerance to diflubenzuron [4]. Due to the low persistance of diflubenzuron, generally a rapid recovery of the susceptible species can be observed. Pretreatment levels are often reached again within a period of 1–3 weeks [4, 158, 166]. There are indications that water fleas are less severely depressed when a granular formulation is used instead of the wettable powder [166].

Copepods (e.g. *Cyclops* spp. and *Diaptomus* spp.) appear to be far less susceptible to diflubenzuron than water fleas. In some experiments they were slightly depressed, but populations recovered rapidly [156, 166]. In another experiment with only *Cyclops* spp. no effects of the compound were found [4].

On seed shrimps, including *Cyprinotus* spp. and *Cypricercus* spp., little or no effects were found in some experiments [156, 166], whereas from another one a significant reduction of *Cyprinotus* spp. was reported [4].

Cunningham [48] reported that nauplii of the brine shrimp (*Artemia salina*) showed a high mortality in diflubenzuron solutions in which the concentration was higher than 10 µg/l.

Effects of diflubenzuron on the larval development of two estuarine crabs were described by Christiansen et al. [41]. When larvae of *Rhithropanopeus harisii* were exposed to solutions of diflubenzuron, the survival from hatching to the first crab stage was significantly decreased at diflubenzuron concentrations of 1 µg/l and higher. In the case of *Sesarma reticulatum* a significant decrease in survival began at a diflubenzuron concentration of 3 µg/l.

Also Nimmo et al. [171] established that diflubenzuron is acutely toxic to the small estuarine crustacean *Mysidopsis bahia* and that the compound suppressed reproduction at low concentrations.

Alzieu [5] studied the influence of diflubenzuron on several marine organisms, including molluscs, crustaceans and fish. Both on molluscs and fish no adverse effects were found. In the case of crustaceans, population reductions were observed. However, the author arrived at the conclusion that use of DIMILIN in coastal areas will not present a significant risk to the marine fauna.

8.7. Nematicidal Effects

Unlike insects, nematodes are not known to contain chitin in their body wall cuticle. However, chitin is present in the nematode egg shell. Veech [241] found, that when diflubenzuron is incorporated into the dietary medium of various species of free-living nematodes at concentrations of 10 mg/kg and higher, the reproduction of the nematodes is drastically suppressed. At a concentration of 1 mg/kg the effects were negligible.

From the above data it seems very unlikely that the use of diflubenzuron for insect control on crops will present a threat to free-living soil nematodes. To obtain a concentration in the soil of 10 mg/kg throughout a layer of 10 cm, an amount of 10 kg diflubenzuron per hectare would be required. Practical field use of DIMILIN as an insecticide takes place at much lower application rates.

If, however, diflubenzuron is used as either a seed treatment before sowing, or as a root treatment at transplant, the local concentration may be high enough to prevent excessive damage to the root system.

Veech [239, 240] reported, that in the case of cotton, seed treatment with diflubenzuron resulted in a significant reduction in the number of egg masses per root system.

9. Resistance

In considerations on resistance, there are two broad areas, i.e. the cross resistance and the build-up of resistance against a particular compound.

9.1. Cross Resistance

Cross resistance to diflubenzuron, on topical applications to white prepupae of *Musca domestica* (housefly), was first reported by Cerf and Georghiou [35] with strains that were resistant to organophosphorus-, carbamate-, and organochlorine compounds. However, the levels of cross-resistance found (ten-fold to much higher) give an exaggerated view as indicated by Grosscurt [97].

Both the larvicidal and the ovicidal effect, after exposure of female flies, were assessed in resistant strains of *Musca domestica* (housefly) by Keiding et al. [131] and Arevad and Keiding [7]. They collected multiresistant flies from commercial farms in Denmark. In larvicidal tests in which diflubenzuron was mixed with the food, cross resistance factors based on LC_{50} values were found ranging from 1.0 to 2.7, when compared with the susceptible reference strain from the World Health Organization (W.H.O.-S.R.S.). Assessments of ovicidal effects by feeding these flies with milk containing diflubenzuron and also after contact of adults with a diflubenzuron-treated surface, however, indicated the existence of "considerable" cross resistance to the ovicidal effect through the female. This result is in contrast to the very low cross resistance to the larvicidal effect.

In two field-collected strains from The Netherlands, Grosscurt [97a] found that larvicidal cross resistance was either absent or amounted to a factor of 5. Ovicidal cross resistance factors in these strains, however, were 8.4 and 87.5, when based on LC_{50} values. He found that the level of ovicidal cross resistance is probably related to the resistance of adults against some standard insecticides tested.

Low cross resistance factors in larvicidal tests were also found by Rupeš et al. [198] with an organophosphorus-resistant strain of the housefly. Oppenoorth and Van der Pas [180] found somewhat higher factors (ranging from 3–10) in larvicidal tests in which laboratory strains which were resistant to malathion, trichlorphon, diazinon and other organophosphates were compared with the W.H.O.-S.R.S. strain. In the experiments by Oppenoorth and Van der Pas, cross resistance was also apparent on the ovicidal effect after topical application to females.

In larvicidal tests with mosquitoes, cross resistance levels of laboratory and field collected strains are generally low. Rongsriyam and Busvine [197] used DDT-resistant mosquito strains, obtained from several sources and colonised in the laboratory. The following levels based on LC_{50} values, were found: for *Culex pipiens fatigans*, 3.2, for *Aedes aegypti*, 2.1, for *Anopheles gambiae*, 2.9 and for *Anopheles quadrimaculatus*, 2.5.

Georghiou et al. [74] found a cross-resistance level of 1.5 in larvae of an organophosphorus multiresistant field strain of *Culex pipiens quinquefasciatus*. Dame et al. [49] in laboratory experiments found a higher susceptibility to diflubenzuron in larvae of a DDT-resistant strain of *Anopheles quadrimaculatus* (cross resistance level of 0.7) and an unchanged susceptibility in malathion-resistant larvae of *Aedes taeniorhynchus*. Brown et al. [27] observed also no cross resistance to diflubenzuron in a methoprene-resistant strain of *Culex pipiens*.

A malathion-resistant strain of *Tribolium castaneum* (red flour beetle) was used by Carter [34]. He found a higher susceptibility, resulting in a cross resistance level of 0.4.

9.2. Build-up of Resistance Against Diflubenzuron

Keiding et al. [131] attempted selection with diflubenzuron on multiresistant *Musca domestica*, by mixing the compound with the larval medium. After 6 generations, still no resistance developed. Moreover, there was no tendency to a significant decrease of susceptibility to the larvicidal effect of diflubenzuron on farms where treatment of breeding places was the main method of control in 1976–1977 [7].

Oppenoorth and Van der Pas [180] used a mixture of resistant laboratory strains of *Musca domestica* for selection with the compound in the larval medium. After 10 generations, larval resistance, based on LC_{50} values, increased to about 50-fold when compared with the W.H.O.-S.R.S. strain. This factor is only between 5–15 on comparison with the parental strains.

Grosscurt [97a] studied development of both larvicidal and ovicidal resistance in a susceptible and a multiresistant strain of the housefly upon selection with diflubenzuron incorporated in the larval culture medium. A decrease in reproductive potential of adults with progressing selection led to a collapse in the F32 of the multiresistant strain. The final assessment of the development of larval resistance in this strain was made in the F26 where it showed a 13.6-fold increase based upon LC_{50} values. In the F40 of the susceptible strain this factor had developed to an about 40-fold level. The ovicidal activity in the F40 of the susceptible strain had decreased with a similar factor. However, in the F26 of the multiresistant strain even the highest test concentration of 3000 mg a.i./l of diflubenzuron showed an ovicidal activity of only 26% and hence no LC_{50} value could be established.

Brown et al. [27] induced a 7-fold resistance in the F5 generation upon selection with diflubenzuron in a DDT, dieldrin, and malathion susceptible strain of the mosquito *Culex pipiens pipiens*. However, this factor decreased in the following generations, followed by a total reproductive failure of the F11.

In experiments of Georghiou and Lin [75] no resistance to diflubenzuron has evolved in a strain of the mosquito *Culex tarsalis* despite apparently heavy selection for 34 generations and two outcrossings to field populations (in F8 and F29) to further enhance the resistance potential of the strain. In a similar experiment with the same strain, selection with the juvenile hormone mimic methoprene increased resistance to a level of 343-fold in 37 generations.

When applied on larvae of *Tribolium confusum* (confused flour beetle), selection resulted in a two times higher EC_{50} after 8 generations [27].

An interesting phenomenon was found by Radwan [189] with *Spodoptera littoralis* (Egyptian cotton leafworm). During selection in the laboratory of a field-collected strain by topical application at the LD_{15} level, the LD_{50} (also by topical application) decreased from 34 to 7 µg/larva in the F6. This diflubenzuron-selected strain was also more susceptible than the parental strain to all other insecticides tested (Dursban, Gardona, Lannate, Endrin).

10. Structure-Activity Relationship

A large number of benzoylphenylureas were synthesized and tested with respect to larvicidal activity [251, 252], after the discovery of Du 19.111 [235]. From the beginning the Hansch method [102] of quantitative structure-activity relationships (QSAR) was used as a guide line. In these QSAR studies the electronic Hammett σ-constants and the hydrophobic Hansch π-constants, both from linear free-energy relations, were used as substituent constants. Steric effects were described by the STERIMOL constants [242, 244] B_1, B_4, and L. B_1 represents the minimum width, B_4 is an approximation of the maximum width and the L-constant accounts for the length of a substituent.

QSAR for the larvicidal effects of benzoylphenylurea on *Pieris brassicae* and *Aedes aegypti* were studied for the following subseries:

1. derivatives substituted in the aniline ring
2. derivatives substituted in the benzoyl ring
3. derivatives substituted on the "bridge".

As an example, the regression equation for the first subseries mentioned above is given [219]:

$$-\mathrm{Log\,ED}_{50} = +0.90\pi + 2.57\sigma - 0.31\,\mathrm{L\,(para)} - 1.27\,\mathrm{L\,(meta)} + 2.73 \qquad (1)$$

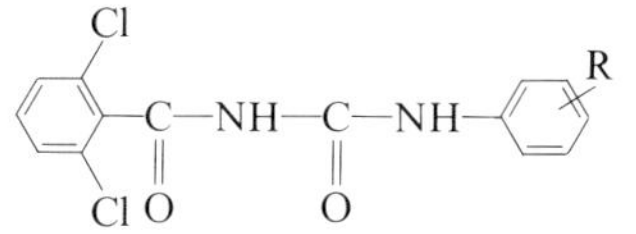

$n = 16 \quad s = 0.44$

$r = 0.871 \quad F = 8.65$

In this analysis, ED_{50} is the concentration required for a 50% reduction of the development of *Pieris brassicae* L., n is the number of compounds in this series, r is the correlation coefficient, s is the standard error of the estimate and the F-value indicates the significance of the correlation found.

In this equation the contribution of the various B-constants was negligible. The equation indicates that lipophilic and electron-withdrawing substituents enhance the larvicidal activity. The longer the substituents, especially at the meta position, the more the activity decreases.

In a subsequent, more elaborate QSAR study, four types of compounds were included that can be characterized as follows

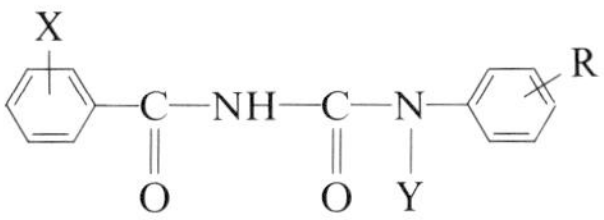

(i)	$X = 2{,}6\,Cl_2$	$Y = H$
(ii)	$X = 2{,}6\,Cl_2$	$Y = CH_3$
(iii)	$X = 2{,}6\,F_2$	$Y = H$
(iv)	$X = 2{,}6\,F_2$	$Y = CH_3$

The regression equation obtained [242], was:

$$-\text{Log}\,ED_{50} = +0.95\pi + 1.99\sigma - 0.34\,L\,(\text{para}) - 0.24\,B_4\,(\text{para}) - 1.30\,L\,(\text{meta}) + 1.40\,D_1 - 0.61\,D_2 + 3.38 \quad (2)$$

$$n = 70 \qquad s = 0.54$$
$$r = 0.892 \qquad F = 34.37\,.$$

In the equation, the dummy parameter D_1 equals zero for the 2,6-dichloro-substitution and $D_1 = 1$ for the 2,6-difluoro-series. The other dummy parameter, D_2, equals zero for $Y = H$ and $D_2 = 1$ for the methyl-substituted compounds. The general form of the Eq. (2) is very similar to that of Eq. (1). From the coefficient of the dummy parameters in Eq. (2) it can be seen that the 2,6-difluoro-compounds are more active on *Pieris brassicae* than the corresponding 2,6-dichloro-compounds. Furthermore, it can be seen that when $Y = CH_3$, the compounds are less active then the unsubstituted ones.

Yu and Kuhr [260] determined the insecticidal activity of 23 substituted 1-benzoyl-3-phenylureas and of 17 substituted 1-benzoyl-3-phenylthioureas by allowing seedcorn maggot larvae, *Hylemya platura*, to feed on lima bean seeds treated with the urea. The most toxic compounds were those containing a 2,6-difluorobenzoyl moiety combined with a 3,4-dichloro, 4-chloro, 4-trifluoromethyl, or a 4-nitrophenyl moiety. Thiourea were less active than the corresponding urea, but similar substitution on the aromatic rings produced parallel effects with both groups. A significant correlation between Hammett's σ-constant (for phenyl substituents) and toxicity suggested that electron-withdrawing groups were important in conferring insecticidal activity.

Hajjar and Casida [101] used 5th instar nymphs of the milkweed bug (*Oncopeltus fasciatus*) as a test organism in their comparison of benzoylurea and established that the activity decreased when the substitution pattern in the benzoyl moiety was changed in the order $2{,}6\,F_2 > 2\text{-}Cl > 2{,}6\,Cl_2$.

Oliver at al. [179] studied the activity of several benzoylureas against four insect species. It appeared, that when in the "bridge" between the two aromatic moieties one oxygen atom was replaced by sulfur, the activity decreased.

De Milo et al. [54] synthesized 26 heterocyclic analogues of diflubenzuron, in which the aniline-moiety was replaced by a heterocyclic ring. When tested as ovicides in the housefly all compounds were inferior to diflubenzuron. When tested as larvicides against the fall armyworm a few compounds reached the level of activity of diflubenzuron.

Walton [248] tested 36 structural analogues of diflubenzuron for insecticidal activity on 4th instar larvae of the mosquito species *Culex quinquefasciatus*. Di-ortho halogen substituents on the benzoyl moiety gave maximum activity. Para-substitution in the anilin moiety results in a better activity than meta- or ortho-substituents. Structure-activity analysis of the 4-atom bridge between the phenylgroups showed the importance of the –C(O)–NH–C(O)–NH group in optimizing the activity.

11. Experimental Products

In addition to Philips Gloeilampenfabrieken many other companies filed patent applications for benzoylurea insecticides: Bayer [76–87, 90], Chugai Pharm. [120, 124], Ciba-Geigy [36, 63], Dow Chem. [174, 228, 229, 232], Eli Lilly [18, 19, 21, 22, 175, 234], GAF [231], Hoechst [88], ICI [73], Ishihara Sangyo Kaisha [89, 119, 263], Monsanto [222], Nihon Tokushu Noyaku Seizo Kabushiki Kaisha [20, 118], Nissan Chem. Ind. [121, 123], Nutrilite Prod. [230], Stauffer Chem. [227], Sumitomo Chem. [122], Upjohn [233], Velsicol [224].

11.1. Experimental Compounds of Bayer AG

In the literature, two experimental compounds of Bayer have been mentioned, BAY SIR 6874 and BAY SIR 8514:

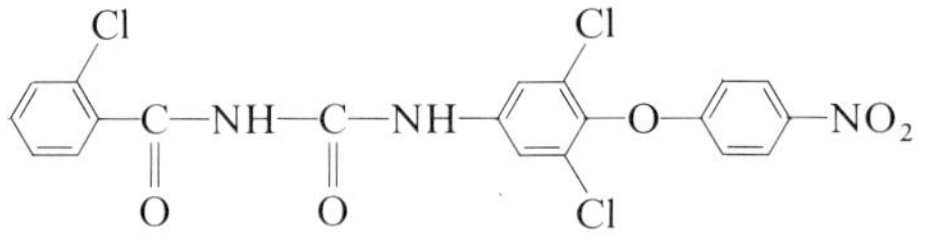

BAY SIR 6874

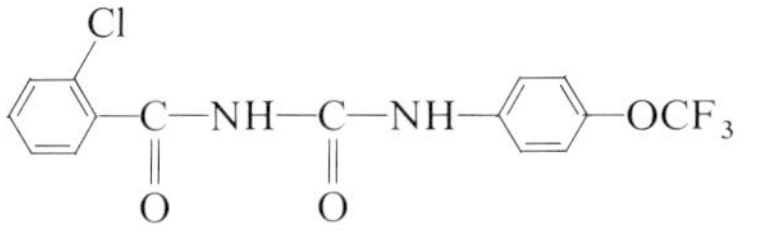

BAY SIR 8514

Activities of these two compounds in larvae of several mosquito species were compared by Schaefer et al. [205]. Larvae of *Aedes nigromaculis* have approximately the same susceptibility to both compounds. Larvae of *Aedes taeniorhynchus* are slightly more susceptible to BAY SIR 6874, while *Culex tarsalis* and *Culex pipiens quinquefasciatus* are much more susceptible to BAY SIR 6874 than to BAY SIR 8514. When compared with diflubenzuron, both experimental products had a shorter residual activity. Lacey and Mulla [139] compared the experimental products with diflubenzuron in their effectiveness against late instars of the blackfly *Simulium vittatum*. At concentrations of 0.02 mg a.i./l and an exposure time of 1 h, the percentage mortality was 70 for diflubenzuron, 86 for BAY SIR 8514 and 100 for BAY SIR 6874. When comparing the ovicidal effect against the same species, a reverse result was found, the percentage mortality at 0.2 mg a.i./l and an exposure time of 24 h, being 100 for diflubenzuron and 50 for BAY SIR 6874. Chang [39] compared diflubenzuron and BAY SIR 8514 in the ovicidal effect in houseflies and found that the experimental compound after injection to the females was about twice as active as diflubenzuron.

11.2. Experimental Compounds of Eli Lilly

Several experimental products of Eli Lilly are known from the literature, including

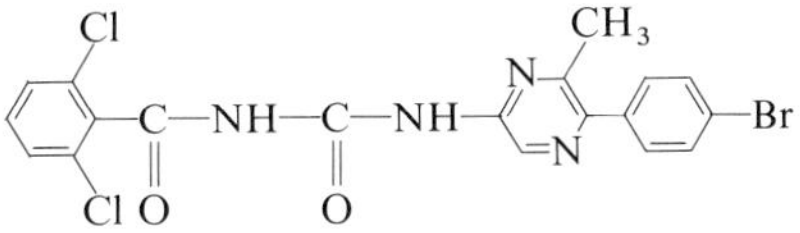

EL 494

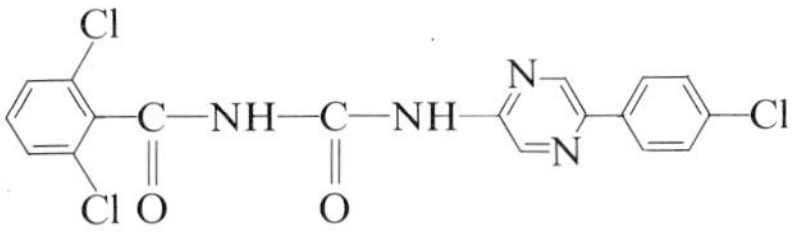

EL 588

Flint et al. [67] tested diflubenzuron, EL 494 and EL 588 in a larval diet for their activity on the pink bollworm, *Pectinophora gossypiella*. All compounds inhibited adult emergence at a concentration of 10 mg/kg in the diet. A further test in field cages using diflubenzuron and EL 494 at a rate of 110 g a.i./ha indicated that diflubenzuron was superior to EL 494 for control of the cotton leaf perforator (*Bucculatrix thurberiella*), but neither compound controlled the pink bollworm.

In April 1979 Chio et al. [37] introduced two other experimental products of Lilly Research Laboratories, LY 127063 and LY 131215.

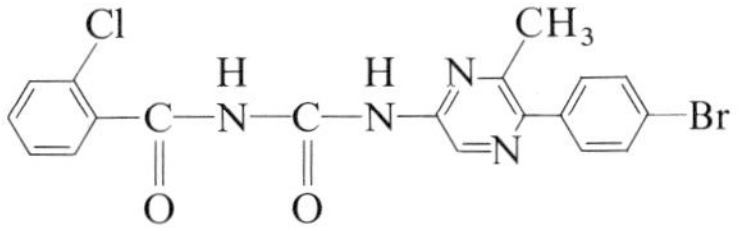

LY 127063

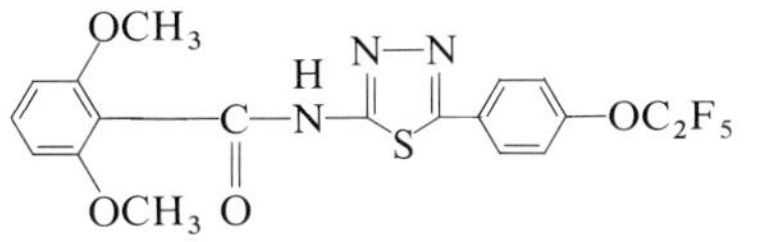

LY 131215

According to the authors, the mode of action of these two compounds may be similar to that of diflubenzuron. The spectrum of activity also seems to be very similar. On larvae of Lepidoptera and Coleoptera application rates are mentioned to range from 280–650 g a.i./ha. For control of mosquito larvae a rate of 28 g a.i./ha is mentioned.

11.3. Experimental Compound of Philips-Duphar

Oliver et al. [178] reported the results of their experiments in which the experimental compound PH 60–44 (= AI3-63223) was compared with diflubenzuron with respect to the ovicidal activity in the cotton boll weevil.

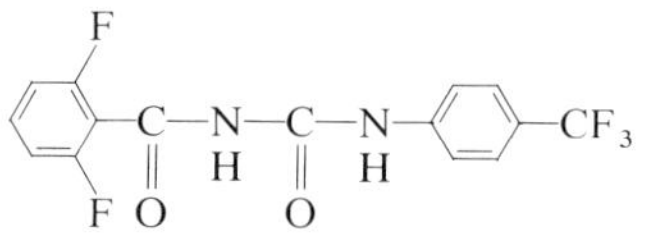

PH 60–44

It was found, that this compound had a longer lasting ovicidal effect than diflubenzuron. Chang [39] compared PH 60–44 with diflubenzuron in the ovicidal effect in the housefly and found that the experimental compound after injection to the females was about twice as active as diflubenzuron. Pickens and DeMilo [181] found that PH 60–44 was not significantly better than diflubenzuron as an ovicide against the face fly.

12. Literature and Patents

1. Ables, J.R., West, R.P., Shepard, M.: J. econ. Entomol. *68*, 622 (1975)
2. Ables, J.R., Jones, S.L., Bee, M.J.: Southwest. Entomol. *2*, 66 (1977)
3. Ali, A., Mulla, M.S.: J. econ. Entomol. *70*, 191 (1977)
4. Ali, A., Mulla, M.S.: Environ. Entomol. *7*, 21 (1978)
5. Alzieu, C.: Rev. Trav. Inst. Pêches marit. *41*, 317 (1977)
6. Apperson, C.S.: J. econ. Entomol. *71*, 521 (1978)
7. Arevad, K., Keiding, J.: Danish Pest Inf. Lab. Ann. Rep. 1977, 46 (1978)
8. Ascher, K.R.S., Nemny, N.E.: Phytoparasitica *2*, 131 (1974)
9. Ascher, K.R.S., Nemny, N.E.: Pestic. Sci. *7*, 1 (1976)
10. Ascher, K.R.S., Nemny, N.E.: Pestic. Sci. *7*, 447 (1976)
11. Ascher, K.R.S., Nemny, N.E., Kehat, M., Gordon, D.: Phytoparasitica *6*, 29 (1978)
12. Audemard, H.: Défense des Végétaux *31*, 303 (1977)
13. Audemard, H.: Mitt. Biol. Bundesanstalt Land-Forstbau *85*, 111 (1978)
14. Audemard, H. et al.: Phytiat.-Phytopharm. *24*, 179 (1975)
15. Barker, R.J., Taber, S.: Environ. Entomol. *6*, 167 (1977)
16. Barker, R.J., Waller, G.D.: Environ. Entomol. *7*, 534 (1978)
17. Bäumler, W., Salama, H.S.: Z. angew. Entomol. *81*, 304 (1976)
18. BE 833.288, 19.9.74/11. 3. 76, Eli Lilly
19. BE 850.712, 9.2. 76/25. 7. 77, Eli Lilly
20. BE 865.757, 6. 4. 77/6. 10. 78, Nihon Tokushu Noyaku Sei
21. BE 868.228, 22. 6. 77/19. 12. 78, Eli Lilly
22. BE 869–674, 15. 8. 77/12. 2. 79, Eli Lilly
23. Beavers, J.B., Schroeder, W.J., Selhime, A.G.: Fla. Ent. *59*, 434 (1976)
24. Becker, B.: J. Insect Physiol. *24*, 699 (1978)
25. Bentley, J.P., Weber, G.H., Gould, D.: Pestic. Biochem. Physiol. *10*, 162 (1979)
26. Booth, G.M., Ferrell, D.: in Pesticides in aquatic environments (Khan. M.A.Q., Ed.) p. 221, Plenum Publ. Cy., New York 1977
27. Brown, T.M., DeVries, D.H., Brown, A.W.A.: J. econ. Entomol. *71*, 223 (1978)
28. Büchi, R.: Z. angew. Entomol. *86*, 67 (1978)
29. Buckner, C.H., McLeod, B.B., Kingsbury, P.D.: Chem. Control Res. Inst., Canada, Rep. CC-X-97 (1975)
30. Bull, D.L., Ivie, G.W.: J. Agric. Food Chem. *26*, 515 (1978)

31. Busvine, J.R., Rongsriyam, Y., Bruno, D.: Pestic. Sci. *7*, 153 (1976)
32. Campbell, J.B., Wright, J.E.: J. econ. Entomol. *69*, 566 (1976)
33. Cannon Res. Lab., Rep. SE-6094, 1976, unpublished
34. Carter, S.W.: J. stored Prod. Res. *11*, 187 (1975)
35. Cerf, D.C., Georghiou, G.P.: J. Agric. Food Chem. *22*, 1145 (1974)
36. CH 34077/15.613.77, 12. 1. 77/13. 7. 78, Ciba-Geigy
37. Chio, E.H., Miesel, J.L., Ward, J.S.: Paper Meet. North Central Branch Entom. Soc. Amer., Indianapolis, April 1979
38. Chang, S.C.: J. econ. Entomol. *71*, 31 (1978)
39. Chang, S.C.: J. econ. Entomol. *72*, 479 (1979)
40. Chang, S.C., Stokes, J.B.: J. econ. Entomol. *72*, 15 (1979)
41. Christiansen, M.E., Costlow, J.D., Monroe, R.J.: Marine Biol. *50*, 29 (1978)
42. Clarke, L., Temple, G.H.R., Vincent, J.F.V.: J. Insect Physiol. *23*, 241 (1977)
43. Cook, C.W., Gerhardt, R.R.: Environ. Entomol. *6*, 589 (1977)
44. Corley, C., Miller, R.W., Hill, K.R.: J. Assoc. Off. Anal. Chem. *57*, 1269 (1974)
45. Cranham, J.E.: Mitt. Biol. Bundesanst. Land-, Forstwirtsch. *180*, 108 (1978)
46. Crookshank, H.R. et al.: Poultry Sci. *57*, 804 (1978)
47. Crystal, M.M.: J. Med. Entomol. *15*, 52 (1978)
48. Cunningham, P.A.: Environm. Entomol. *5*, 701 (1976)
49. Dame, D.A. et al.: Mosquito News *36*, 462 (1976)
50. Darwazeh, A., Mulla, M.S., Dhillon, H.S.: Proc. 45th Ann. Conf. Calif. Mosq. Contr. Ass., 143 (1977)
51. De Angelis, E.: Lotto Antiparasitaria *31*, 51 (1979)
52. De Bree, J., Keet, C.J.M.F., De Lange, N., Overmars, H., Post, L.C.: To be published
53. De Lange, N., Post, L.C.: Incorporated in [253]
54. DeMilo, A.B. et al.: J. Agric. Food Chem. *26*, 164 (1978)
55. DeMilo, A.B., Terry, P.H., Rains, D.M.: J. Assoc. Off. Anal. Chem. *61*, 629 (1978)
56. Deul, D.H., De Jong, B.J., Kortenbach, J.A.M.: Pestic. Biochem. Physiol. *8*, 98 (1978)
57. DiPrima, S.J. et al.: J. Agric. Food Chem. *26*, 968 (1978)
58. Donaubauer, E.: Allgem. Forstzeitg. *87*, 108 (1976)
59. Dos 2123236, 11. 5. 71, N. Prior. 15. 5. 70, Philips Gloeilampenfabrieken
60. Egger, A.: Centralblatt gesamte Forstwesen *94*, 65 (1977)
61. Elings, H., Dieperink, J.G.: Med. Fac. Landbouww. Rijksuniv. Gent *39*, 833 (1974)
62. El-Tantawi, M.A., Belal, M.B., Abdallah, M.D.: Anz. Schädlingsk. Pflschtz. Umweltschtz. *49*, 153 (1976)
63. E.P. 1–203, 28. 7. 77/4. 4. 79, Ciba-Geigy
64. Farlow, J.E. et al.: Environm. Entomol. 7, 199 (1978)
65. Flint, H.M., Smith, R.L.: J. econ. Entomol. *70*, 51 (1977)
66. Flint, H.M. et al.: J. econ. Entomol. *70*, 237 (1977)
67. Flint, H.M. et al.: J. econ. Entomol. *71*, 616 (1978)
68. Fogal, W.H.: Can. Entomol. *109*, 981 (1977)
69. Franz, J.M., Hassan, S.A., Bogenschütz, H.: Nachrichtenbl. Deut. Pflanzenschutzd. *28*, 181 (1976)
70. Fytizas, E.: Z. angew. Ent. *81*, 440 (1976)
71. Ganyard, M.C. et al.: J. econ. Entomol. *70*, 347 (1977)
72. Ganyard, M.C., Bradley, J.R., Brazzell, J.R.: J. econ. Entomol. *71*, 785 (1978)
73. GB 2.954.076, 15. 7. 76/19. 1. 78, ICI
74. Georghiou, G.P. et al.: J. econ. Entomol. *68*, 461 (1975)
75. Georghiou, G.P., Lin, C.S.: Mosquito Control Res., Univ. of Calif. Ann. Rep. 78 (1978)
76. Ger. Offen. 2.438.747, 13. 8. 74/4. 3. 76, Bayer
77. Ger. Offen. 2.504.982, 6. 2. 75/19. 8. 76, Bayer
78. Ger. Offen. 2.504.983, 6. 2. 75/19. 8. 76, Bayer
79. Ger. Offen. 2.504.984, 6. 2. 75/19. 8. 76, Bayer
80. Ger. Offen. 2.528.917, 28. 6. 75/20. 1. 77, Bayer
81. Ger. Offen. 2.531.202, 12. 7. 75/27. 1. 77, Bayer
82. Ger. Offen. 2.531.279, 12. 7. 75/21. 1. 77, Bayer
83. Ger. Offen. 2.531.743, 16. 7. 75/3. 2. 77, Bayer

84. Ger. Offen. 2.537.413, 22. 8. 75/3. 3. 77, Bayer
85. Ger. Offen. 2.601.780, 20. 1. 76/21. 7. 77, Bayer
86. Ger. Offen. 2.637.947, 24. 8. 76/28. 2. 78, Bayer
87. Ger. Offen. 2.638.233, 25. 8. 76/28. 2. 78, Bayer
88. Ger. Offen. 2.726.684, 14. 6. 77/4. 1. 79, Hoechst
89. Ger. Offen. 2.748.636, 29. 10. 76/3. 5. 78, Ishihara Sangyo Kaisha
90. Ger. Offen. 2.801.316, 13. 1. 78/19. 7. 79, Bayer
91. Granett, J., Dunbar, D.M.: J. econ. Entomol. *68*, 99 (1975)
92. Granett, J., Dunbar, D.M., Weseloh, R.M.: J. econ. Entomol. *69*, 403 (1976)
93. Granett, J., Weseloh, R.M.: J. econ. Entomol. *68*, 577 (1975)
94. Grosscurt, A.C.: Med. Fac. Landbouww. Rijksuniv. Gent *41*, 949 (1976)
95. Grosscurt, A.C.: Proc. 9th Br. Insectic. Fungic. Conf. *1*, 141 (1977)
96. Grosscurt, A.C.: J. Insect Physiol. *24*, 827 (1978)
97. Grosscurt, A.C.: Pestic. Sci. *9*, 373 (1978)
97a. Grosscurt, A.C.: Proc. Kon. Ned. Akad. v. Wetensch. C *83* (2), 127 (1980)
98. Gruys, P.: Proc. 8th Br. Insectic. Fungic. Conf. *3*, 823 (1975)
99. Gruys, P.: De Fruitteelt *65*, 176 (1975)
100. Gijswijt, M.J., Deul, D.H., De Jong, B.J.: Pest. Biochem. Physiol. *12*, 87 (1979)
101. Hajjar, N.P., Casida, J.E.: Science *200*, 1499 (1978)
101a. Hajjar, N.P., Casida, J.E.: Pestic. Biochem. Physiol. *11*, 33 (1979)
102. Hansch, C.: In: Drug Design (Ariëns, E.J., Ed.) Vol. 1, 271, Academic Press, New York 1971
103. Harper, J.D., Abrahamson, L.P.: J. econ. Entomol. *72*, 74 (1979)
104. Hartinger, C.: Anz. Schädlingsk., Pflschtz., Umweltschtz. *49*, 156 (1976)
105. Hassan, S.A.: Z. Pflkrankh. Pflschutz *84*, 158 (1977)
106. Hatzios, K.K., Penner, D.: Pestic. Biochem. Physiol. *9*, 65 (1978)
107. Hayakawa, H.: Jap. J. Sanit. Zool. *27*, 261 (1976)
108. Holst, H.: Z. Pflkrankh. Pflschutz *81*, 1 (1974)
109. Hopkins, D.E., Chamberlain, W.F.: J. econ. Entomol. *71*, 25 (1978)
110. Hsieh, M.Y.G., Steelman, C.D.: Mosquito News *33*, 278 (1974)
111. Hunter, E., Vincent, J.F.: Experientia *30*, 1432 (1974)
112. Huntingdon Res. Centre, PDR 242 FT/76702 (1977)
113. Ishaaya, I., Ascher, K.R.S.: Phytoparasitica *5*, 149 (1977)
114. Ishaaya, I., Casida, J.E.: Pestic. Biochem. Physiol. *4*, 484 (1974)
115. Ivie, G.W.: Metabolism of insect growth regulators in animals. In: Fate of pesticides in large animals (Ivie, G.W. and Dorough, H.W., Ed.), p. 111–125, Acad. Press. Inc., New York 1977
116. Ivie, G.W.: J. Agric. Food Chem. *26*, 81 (1978)
117. Ivie, G.W., Wright, J.E.: J. Agric. Food Chem. *26*, 90 (1978)
118. JA 5.3009.743, 16. 7. 76/28. 1. 78, Nihon Tokushu Noyaku Sei
119. JA 5.3103.447, 21. 2. 77/8. 9. 78, Ishihara Sangyo Kaisha
120. JA 5.3105.451, 24. 2. 77/13. 9. 78, Chugai Pharm.
121. JA 5.3107.422, 1. 3. 77/19. 9. 78, Nissan Chem. Ind.
122. JA 5.4012.343, 24. 6. 77/30. 1. 79, Sumitomo Chem.
123. JA 5.4028.824, 5. 8. 77/3. 3. 79, Nissan Chem. Ind.
124. JA Kokai 76.121.515, 14. 4. 75/23. 10. 76, Chugai Pharm.
125. Jordan, A.M., Trewern, M.A.: Nature *272*, 719 (1978)
126. Jordan, A.M., Trewern, M.A.: Bull. ent. Res. *69*, 55 (1979)
127. Julin, A.M., Sanders, H.O.: Mosquito News *38*, 256 (1978)
128. Kalberer, P., Vogel, E.: Z. Pflkrankh. Pflschtz, *85*, 328 (1978)
129. Kaplanis, J.N., Elliott, J., Taber, M.: J. econ. Entomol. *71*, 620 (1978)
130. Keever, D.W., Bradley, J. R., Ganyard, M.C.: Environ. Entomol. *6*, 732 (1977)
131. Keiding, J., Arevad, K., Nielsen, H.: Danish Pest. Inf. Lab. Ann. Rep. 1976, 28 (1977)
132. Kennelly, J.J. et al.: Univ. Alberta, Agric. and Forestry Bull. 57th Ann. Feeders' Day Rep. 48 (1978)
133. Ker, R.F.: J. Insect Physiol. *23*, 39 (1977)
134. Kunz, S.E., Schmidt, C.D., Harris, R.L.: Southwest. Entomol. *1*, 190 (1976)
135. Kunz, S.E., Bay, D.E.: Southwest. Entomol. *2*, 27 (1977)
136. Kunz, S.E. et al.: J. econ. Entomol. *70*, 298 (1977)

137. Lacey, L.A., Mulla, M.S.: J. econ. Entomol. *70*, 369 (1977)
138. Lacey, L.A., Mulla, M.S.: Mosquito News *38*, 264 (1978)
139. Lacey, L.A., Mulla, M.S.: Mosquito News *38*, 377 (1978)
140. Lacey, L.A., Mulla, M.S.: Mosquito News *39*, 86 (1979)
141. Lloyd, E.P., Wood, R.H., Mitchell, E.B.: J. econ. Entomol. *70*, 442 (1977)
142. Maas, W.: Adv. Pestic. Sci., Proc. 4th Int. Congr. Pestic. Chem., Zürich, Part 3, 772 (1978)
143. Maas, W.: ibid. Part 3, 734 (1978)
143a. Marks, E.P., Sowa, B.A.: ACS Symp. Ser. No. 2, 144 (1974)
144. McCoy, C.W.: J. econ. Entomol. *71*, 122 (1978)
145. McGregor, H.E., Kramer, K.J.: J. econ. Entomol. *69*, 479 (1976)
146. McGregor, J.F. et al.: Mutat. Res. *66*, 45 (1979)
147. McKague, A.B., Pridmore, R.B., Wood, P.M.: Can. Ent. *110*, 1103 (1978)
148. McKague, A.B., Pridmore, R.B.: Bull. Environm. Contam. Toxicol. *20*, 167 (1978)
149. McLaughlin, R.E.: J. econ. Entomol. *69*, 317 (1976)
150. McLaughlin, R.E.: Ent. exp. & appl. *23*, 171 (1978)
151. Metcalf, R.L., Po-Yung Lu, Bowlus, S.: J. Agric. Food Chem. *23*, 359 (1975)
151a. Metcalf, R.L., Sangha, G.V., Kapoor, I.P.: Environ. Sci. Technol. *5*, 709 (1971)
152. Miller, R.W., Corley, C., Hill, K.R.: J. econ. Entomol. *68*, 181 (1975)
153. Miller, R.W. et al.: J. Agric. Food Chem. *24*, 687 (1976)
154. Miller, R.W., Corley, C., Shufelt, S.R.: J. econ. Entomol. *69*, 741 (1976)
155. Mitlin, N., Wiygul, G., Haynes, J.W.: Pestic. Biochem. Physiol. *7*, 559 (1977)
156. Miura, T., Takahashi, R.M.: Environ. Entomol. *3*, 631 (1974)
157. Miura, T., Murray, W.D., Takahashi, R.M.: Proc. Calif. Mosq. Control Assn., 43rd Ann. Conf., 79 (1975)
158. Miura, T., Takahashi, R.M.: Mosquito News *35*, 154 (1975)
159. Moore, R.F., Taft, H.M.: J. econ. Entomol. *68*, 96 (1975)
160. Moore, R.F., Leopold, R.A., Taft, H.M.: J. econ. Entomol. *71*, 587 (1978)
161. Moreau, R., Castex, C., Lamy, M.: Ann. Zool.-Ecol. anim. *7*, 161 (1975)
162. Mulder, R., Gijswijt, M.J.: Pestic. Sci. *4*, 737 (1973)
163. Mulder, R., Swennen, A.A.: Proc. 7th Brit. Insectic. Fungic. Conf. *2*, 729 (1973)
164. Mulla, M.S., Darwazeh, A.: J. econ. Entomol. *69*, 309 (1976)
165. Mulla, M.S., Darwazeh, A., Norland, R.L.: J. econ. Entomol. *67*, 329 (1974)
166. Mulla, M.S., Majori, G., Darwazeh, H.: Mosquito News *35*, 211 (1975)
167. Mulla, M.S., Kramer, W.L., Barnard, D.R.: J. econ. Entomol. *69*, 285 (1976)
168. Naton, E.: Anz. Schädlingsk., Pflschtz, Umweltschtz *51*, 136 (1978)
169. Neal, J.W.: J. econ. Entomol. *67*, 300 (1974)
170. Neisses, J., Markin, G.P., Schaefer, R.: J. econ. Entomol. *69*, 783 (1976)
171. Nimmo, D.R. et al.: Bull. Environm. Contam. Toxicol. *22*, 767 (1979)
172. NL 7007040, NL Prior. 15. 5. 70, Philips Gloeilampenfabrieken
173. NL 7105350, 21. 4. 71, Philips Gloeilampenfabrieken
174. NL 7805121, 13. 5. 77/15. 11. 78, Dow Chem.
175. NL 7806678, 22. 6. 77/28. 12. 78, Eli Lilly
176. Novak, V., Sehnal, F.: Acta entomol. Bohemoslovaca *75*, 349 (1978)
177. Oehler, D.D., Holman, G.M.: J. Agric. Food Chem. *23*, 590 (1975)
178. Oliver, J.E. et al.: J. econ. Entomol. *70*, 286 (1977)
179. Oliver, J.E., DeMilo, A.B., Cohen, Ch.R.: J. Agric. Food Chem. *24*, 1065 (1976)
180. Oppenoorth, F.J., Van der Pas, L.J.T.: Ent. exp. & appl. *21*, 217 (1977)
181. Pickens, L.G., DeMilo, A.B.: J. econ. Entomol. *70*, 595 (1977)
182. Pickens, L.G., Miller, R.W.: Environm. Entomol. *4*, 46 (1975)
183. Pimprikar, G.D.: Diss. Univ. Cal., Riverside, december 1977
184. Plapp, F.W.: Ann. Rev. Entomol. *21*, 179 (1976)
185. Post, L.C., Vincent, W.R.: Naturwissenschaften *60*, 431 (1973)
186. Post, L.C., De Jong, B.J., Vincent, W.R.: Pest. Biochem. Physiol. *4*, 473 (1974)
187. Pree, D.J.: Can. Entomologist *108*, 49 (1976)
188. Rabenort, B. et al.: in: Analytical methods for pesticides and plant growth regulators (ed. Günter Zweig), Vol. 10, New and updated methods (ed. G. Zweig and J. Sherma), p. 57 Acad. Press, Inc. 1978

189. Radwan, H.S.A.: Int. Pest Control *20*, 21 (1978)
190. Rathburn, C.B., Boike, A.H.: Mosq. News *35*, 540 (1975)
191. Retnakaran, A., Smith, L.: Can. Ent. *107*, 883 (1975)
192. Retnakaran, A., Smith, L.: Bi-monthly Res. Notes *32* (1), 2 (1976)
193. Retnakaran, A., Smith, L., Tomkins, W.: ibid. *32* (5), 26 (1976)
194. Ribrioux, Y., Dolbeau, C.: Phytiat.-Phytopharm. *24*, 193 (1975)
195. Rizk, G.A.M., Radwan, H.S.A.: Z. angew. Entomol. *79*, 136 (1975)
196. Rogers, A.J. et al.: Mosq. News *36*, 273 (1976)
197. Rongsriyam, Y., Busvine, J.R.: Bull. Ent. Res. *65*, 459 (1975)
198. Rupeš, V., Ždárek, J., Pinterová, J.: Z. angew. Ent. *84*, 328 (1977)
199. Ruzo, L.O., Zabik, M.J., Schuetz, R.D.: J. Agric. Food Chem. *22*, 1106 (1974)
200. Sahota, T.S., Shepherd, R.F.: Bi-monthly Res. Notes *31*, 39 (1975)
201. Salama, H.S., Motagally, Z.A., Skatulla, U.: Z. angew. Ent. *80*, 396 (1976)
202. Schaefer, C.H. et al.: Proc. 13th Ann. Meet. Amer. Mosq. Control Ass. pp. 137–139 (1974)
203. Schaefer, C.H., Dupras, E.: J. Agric. Food Chem. *24*, 733 (1976)
204. Schaefer, C.H. et al.: Bull. Environm. Contam. Toxicol. *21*, 249 (1979)
205. Schaefer, C.H. et al.: J. econ. Entomol. *71*, 427 (1978)
206. Schroeder, W.J. et al.: J. econ. Entomol. *69*, 780 (1976)
207. Schroeder, W.J., Sutton, R.A.: J. econ. Entomol. *71*, 69 (1978)
208. Schwenke, W.: Anz. Schädlingsk., Pflschtz, Umweltschtz *52*, 97 (1979)
209. Skatulla, A.: Anz. Schädlingsk., Pflschtz, Umweltschtz *48*, 145 (1975)
210. Skatulla, A.: Anz. Schädlingsk., Pflschtz, Umweltschtz *48*, 17 (1975)
211. Sowa, B.A., Marks, E.P.: Insect Biochem. *5*, 855 (1975)
212. Steelman, C.D. et al.: Mosquito News *35*, 67 (1975)
213. Stevenson, J.H.: Plant Pathol. *27*, 38 (1978)
214. Still, G.G., Leopold, R.A.: Pestic. Biochem. Physiol. *9*, 304 (1978)
215. Stoolmiller, A.C.: Gen. Pharmac. *9*, 11 (1978)
216. Sundaramurthy, V.T.: Z. Pflkrankh. Pflschutz *84*, 597 (1977)
217. Sźantó, J., Varjas, L.: Növényvédelem, 199 (1973)
218. Taft, H.M., Hopkins, A.R.: J. econ. Entomol. *68*, 551 (1975)
219. Tipker, J., Wellinga, K., Verloop, A.: Symp. Pest. Group S.C.I. London, Febr. 1977
220. Turnipseed, S.G. et al.: J. econ. Entomol. *67*, 760 (1974)
221. US 3748356, pat. 24. 7. 73, NL Prior 15. 5. 70, US Philips Co.
222. US 3793451, 5. 11. 70/19. 2. 74, Monsanto
223. US 3933908, pat. 20. 1. 76, NL Prior. 15. 5. 70, US Philips Co.
224. US 3959478, 17. 3. 75/25. 5. 76, Velsicol
225. US 3989842, pat. 2. 11. 76, NL Prior 15. 5. 70, US Philips Co.
226. US 4013717, pat. 22. 3. 77, NL Prior. 15. 5. 70, US Philips Co.
227. US 4051259, 26. 1. 76/27. 9. 77, Stauffer
228. US 4089975, 13. 5. 77/16. 5. 78, Dow Chem.
229. US 4092421, 13. 5. 77/30. 5. 78, Dow Chem.
230. US 4107294, 27. 6. 77/15. 8. 78, Nutrilite Prod.
231. US 4117009, 9. 6. 77/26. 9. 78, GAF
232. US 4148902, 13. 5. 77/10. 4. 79, Dow Chem.
233. US 4160037, 30. 9. 77/3. 7. 79, Upjohn Co.
234. US 4160834, 19. 12. 77/10. 7. 79, Eli Lilly
235. Van Daalen et al.: Naturwissenschaften *59*, 312 (1972)
236. Van Eck, W.: Insect Biochem. *9*, 295 (1979)
237. Van Frankenhuyzen, A., Meinsma, E.: Gewasbescherming *9*, 53 (1978)
238. Vardanis: Experientia *34*, 228 (1977)
239. Veech, J.: J. Nematol. *9*, 184 (1977)
240. Veech, J.: J. Nematol. *10*, 208 (1978)
241. Veech, J.: Nematologica *24*, 312 (1978)
242. Verloop, A., Hoogenstraaten, W., Tipker, J.: In: Drug Design (Ariëns, E.J., Ed.), Acad. Press New York, Vol. 7, 165 (1976)
243. Verloop, A., Nimmo, W.B., De Wilde, P.C.: VIIIth Internat. Plant Protection Congress, Moscow, August 21–27 (1975)

244. Verloop, A., Hoogenstraaten, W., Tipker, J.: To be published
245. Verloop, A., Ferrell, C.D.: ACS Sympos. Ser. No. *37*, (Plimmer, J.R., Ed.), 237 (1977)
246. Verloop, A.: To be published
247. Verloop, A.: Symp. Insect Growth Regulators, ACS Chem. Congr., Honolulu, 1979
248. Walton, B.T.: Dissertation Abstr. Intern. B *39* (1), 66 (1978)
249. Wearing, C.H., Thomas, W.P.: Proc. N. Z. Weed Pest Control Conf. 31st Meet. 221 (1978)
250. Weiss, M.: Anz. Schädlingsk., Pflschtz, Umweltschtz *50*, 161 (1977)
251. Wellinga, K., Mulder, R., Van Daalen, J.J.: J. Agric. Food Chem. *21*, 348 (1973)
252. Wellinga, K., Mulder, R., Van Daalen, J.J.: ibid. *21*, 993 (1973)
253. Willems, A.G.M., Overmars, H., Scherpenisse, P., De Lange, N., Post, L.C.: Xenobiotica *10*, 103 (1980)
254. Wolfenbarger, D.A., Guerra, A.A., Garcia, R.D.: J. econ. Entomol. *70*, 126 (1977)
255. Woroby, B.L., Webster, G.R.: J. Assoc. Off. Anal. Chem. *60*, 213 (1977)
256. Woroby, B.L., Webster, G.R.: ibid. *60*, 986 (1977)
257. Wright, J.E.: J. econ. Entomol. *67*, 746 (1974)
258. Wright, J.E., Harris, R.L.: J. econ. Entomol. *69*, 728 (1976)
259. Wright, J.E., Spates, G.E.: J. econ. Entomol. *69*, 365 (1976)
260. Yu, Ch.Ch., Kuhr, J.: J. Agric. Food Chem. *24*, 134 (1976)
261. Yu, S.J., Terrière, L.C.: Life Sci. *17*, 619 (1975)
262. Yu, S.J., Terrière, L.C.: Pestic. Biochem. Physiol. 7, 48 (1977)
263. Z.A. 7802–440, 6. 2. 78/15. 3. 79, Ishihara Sangyo Kaisha

Literature Added in Proof

Sect. 3.1: Cruse, W.B.T.: Acta Cryst. B*34*, 2904 (1978)
Ivie, G.W. et al.: J. Agric. Food Chem. *28*, 330 (1980)
Sect. 3.2: Pribyl, J., Herzel, F.: J. Chromatogr. *166*, 272 (1978)
Sect. 5.1: Miller, R.W. et al.: Bull. Environm. Contam. Toxicol. *23*, 482 (1979)
Quarles, J.M. et al.: Bull. Environm. Contam. Toxicol. *25*, 252 (1980)
Sect. 5.3: Madder, D.J., Lockhart, W.L.: Bull. Environm. Contam. Toxicol. *20*, 66 (1978)
Sect. 6.1: O'Neill, M.P. et al.: J. Insect. Physiol. *23*, 1243 (1977)
Ascher, K.R.S. et al.: Pestic. Sci. *11*, 90 (1980)
Grosscurt, A.C., Andersen, S.O.: Proc. Kon. Ned. Akad. v. Wetensch. C*83* (2), 143 (1980)
Moore, R.F.: J. Georgia Entomol. Soc. *15*, 227 (1980)
Sect. 6.2: Ker, R.F.: Pestic. Sci. *9*, 259 (1978)
Bishai, W.R., Stoolmiller, A.C.: Pestic. Biochem. Physiol. *11*, 258 (1979)
Brillinger, G.U.: Arch. Microbiol. *121*, 71 (1979)
Hajjar, N.P.: Mosq. News *39*, 381 (1979)
Cohen, E., Casida, J.E.: Pestic. Biochem. Physiol. *13*, 129 (1980)
Mayer, R.T. et al.: J. Econ. Entomol. *73*, 76 (1980)
Meola, S.M., Mayer, R.T.: Science *207*, 985 (1980)
Sect. 7.1: Mansager, E.R. et al.: Pestic. Biochem. Physiol. *12*, 172 (1979)
Seuferer, S.L. et al.: Pestic. Biochem. Physiol. *10*, 174 (1979)
Madder, D.J., Lockhart, W.L.: Can. Ent. *112*, 173 (1980)
Sect. 7.4: Bull, D.L., Ivie, G.W.: Pestic. Biochem. Physiol. *13*, 41 (1980)
Sect. 7.5: Schaefer, C.H. et al.: Proc. 48th Ann. Conf. Calif. Mosq. & Vector Contr. Ass. 84 (1980)
Sect. 8.2: Henzell, R.F. et al.: N. Z. J. Agr. Res. *22*, 197 (1979)
Murthy, R.L.N.: Pesticides *13*, 33 (1979)
Sundaramurthy, V.T.: Bull. Ent. Res. *70*, 25 (1980)
Sundaramurthy, V.T., Santhanakrisnan, K.: Pestic. Sci. *10*, 147 (1979)
Sundaramurthy, V.T., Balasubramanian, M.: Z. Angew. Entomol. *85*, 317 (1978)
Sect. 8.2.1: Ascher, K.R.S. et al.: Phytoparasitica *6*, 29 (1978)
Abid, M.K. et al.: Z. Angew. Entomol. *85*, 321 (1978)
Fytizas, E.: Med. Fac. Landbouww. Rijksuniv. Gent *43*, 519 (1978)
Radwan, H.S.A. et al.: Z. Angew. Entomol. *86*, 414 (1978)

Zhuravskaya, S.A. et al.: Khim. Sel. Khoz. *16* (5), 41 (1978)
Mitchell, E.B. et al.: J. Georgia Entomol. Soc. *14*, 248 (1979)
Rummel, D.R. et al.: Southw. Entomol. *4*, 315 (1979)
Abo-Elghar, M.R. et al.: Z. Angew. Entomol. *89*, 100 (1980)
Sect. 8.2.2: Lara, F.M. et al.: Ann. Soc. Entomol. Bras. *6*, 276 (1977)
Reed, T., Bass, M.H.: J. Econ. Entomol. *72*, 912 (1979)
Reed, T., Bass, M.H.: J. Econ. Entomol. *73*, 332 (1980)
Sect. 8.3.1: Albajes, R., Santiago-Alvarez, C.: An. INIA/Ser. Prot. Veg. *9* (4), 67 (1979)
Lovestrand, S.A., Beavers, J.B.: Fla. Entomol. *63*, 112 (1980)
Sect. 8.3.4: Mohamad, R.B. et al.: Pertanika *2* (2), 84 (1979)
Sect. 8.4: Vasic, M.: Zastica Bilja *28*, 191 (1977)
Demolin, G.: Ann. Sci. Forest. *35*, 229 (1978)
Gillette, N.L. et al.: J. Econ. Entomol. *71*, 319 (1978)
El-Bahrawi, A. et al.: Med. Fac. Landbouww. Rijksuniv. Gent *44*, 31 (1979)
Fratian, A.: Silvic. Exploatar. Padurilor *94*, 360 (1979)
Lyashenko, P.I., Andpeeva, G.I.: Zashch. Rast. (6), 21 (1979)
Retnakaran, A. et al.: Can. Ent. *111*, 841 (1979)
Sect. 8.5.1: Rettich, F.: Acta Ent. Bohemoslov. *75*, 312 (1978)
Rodrigues, C.S., Wright, R.E.: Can. Ent. *110*, 319 (1978)
Ten Houten, A. et al.: WHO/VBC/80.795 (1980)
Sect. 8.6.2: Ali, A., Mulla, M.S.: Arch. Environm. Contam. Toxicol. *7*, 483 (1978)
Forward, R.B., Costlow, J.D.: Wat. Air Soil Pollut. *9*, 227 (1978)
Gulka, G. et al.: Bull. Environm. Contam. Toxicol. *25*, 477 (1980)
Sect. 8.7: Vrain, T.C. et al.: J. Nematol. *11*, 328 (1979)
Sect. 9.2: Pimprikar, G.D., Georghiou, G.P.: Pestic. Biochem. Physiol. *12*, 10 (1979)
Sect. 10: Grosscurt, A.C., Tipker, J.: Pestic. Biochem. Physiol. *13*, 249 (1980)
Sect. 11.1: Ascher, K.R.S. et al.: Phytoparasitica *7*, 177 (1979)
Cohen, C.F., Marks, E.P.: Southw. Entomol. *4*, 294 (1979)
Schaefer, C.H., Dupras, E.F.: J. Agric. Food Chem. *27*, 1031 (1979)
Zoebelein, G. et al.: Z. Angew. Entomol. *89*, 289 (1980)
Sect. 11.2: Retnakaran, A.: Can. Ent. *111*, 847 (1979)
Abdel-Monem, A.H. et al.: J. Econ. Entomol. *73*, 22 (1980)
Sect. 11.3: Chang, S.C., Borkovec, A.B.: J. Econ. Entomol. *73*, 285 (1980)
Sect. 11: New patent applications:
BASF: EP 8-768
Bayer: EP 8-435, EP 9-762, DT 2848-794, DT 2905-687
Celamerck: EP 13-414
Ciba: EP 4-030, EP 6-184, EP 14-674, EP 14-675, EP 14-676, DT 3003-112, DT 3003-113
Duphar: DT 2926-480
Eli Lilly: US 4173.639, BE 878.048, GB 2028.813
Gulf: US 4164.581, US 4212.870
ICI: EP 5-944
Ishihara: JA 5.4115.3, JA 5.5011.537, JA 5.5038.356, JA 5.5038.357, JA 5.5079.369

Zur Problematik des biologischen Pflanzenschutzes in der Landwirtschaft

H. H. Cramer und Brünhilde Zeller

Ressort Pflanzenschutz-Anwendungstechnik
Stabsabteilung und Biologische Forschung der Bayer AG, D-5090 Leverkusen

Inhalt

1. Problemstellung und Abgrenzung

Über Möglichkeiten, Verfahren, Erfolge, Teilerfolge und Mißerfolge des biologischen Pflanzenschutzes, insbesondere der biologischen Schädlingsbekämpfung, ist in den letzten Jahrzehnten ein umfangreiches Schrifttum erschienen. Es ist nicht die Absicht dieses Beitrages, einen bewertenden Überblick über diese Literatur zu geben, zumal eine sehr gute zusammenfassende Darstellung in deutscher Sprache vorliegt [22]. Nach der Definition dieser Autoren versteht man „unter biologischer Schädlingsbekämpfung ...die Verwendung von Lebewesen, um mit ihnen (durch menschlichen Eingriff) die Populationen bestimmter schädlicher Tiere oder Pflanzen zu begrenzen“. In den Begriff der Schädlingsbekämpfung wird also hier eindeutig der gesamte Pflanzenschutz, einschließlich der Bekämpfung von Unkräutern und pilzlichen Erregern, einbezogen. Wenn dieser Komplex gemeint ist, wird im Zusammenhang dieser Arbeit von „biologischem Pflanzenschutz" gesprochen, während sich „biologische Schädlingsbekämpfung" jeweils auf die Kontrolle von Tieren bezieht, die die Gesundheit oder Wirtschaftsgüter des Menschen bedrohen oder schädigen. Von der biologischen Bekämpfung im hier definierten, engeren Sinne sind die biotechnischen Verfahren des Pflanzenschutzes zu unterscheiden. Sie „nutzen natürliche Reaktionen der Schädlinge auf bestimmte physikalische oder chemische Reize ‚zweckentfremdet' von ihrer natürlichen Bedeutung für den Organismus" [22, S. 143].

In diesen Bereich gehört die Verwendung von Repellentien und Attractants, Pheromonen, Wachstums- und Entwicklungshormonen und dergleichen. Biotechnische Verfahren werden im Rahmen dieses Beitrages nicht besprochen, zumal ihre Verwendung und Verwendbarkeit in anderen Kapiteln dieses Bandes behandelt werden [5, 49].

Die Fragestellung, die hier verfolgt werden soll, geht im wesentlichen von der Situation in der breiten landwirtschaftlichen Praxis aus, die sich angesichts vieler Diskussionen über den chemischen Pflanzenschutz fragt, welche Möglichkeiten des biologischen Pflanzenschutzes im laufenden landwirtschaftlichen Betrieb bestehen oder zu erwarten sind, namentlich im Hinblick auf die wirtschaftlich wichtigen, in großem Stile angebauten Kulturpflanzenarten. Zur Realisierbarkeit gehört hierbei nicht nur die Durchführbarkeit im Betriebsablauf, sondern auch die Wirtschaftlichkeit. Da die Landwirtschaft heute zur Optimierung der Erträge bei möglichst geringem Personalaufwand gezwungen ist, wird sie insbesondere solche Verfahren aufgreifen, die sich in diese Aufgabe einfügen. Für sie ist der Pflanzenschutz – ob chemisch oder biologisch – nicht Selbstzweck sondern Mittel zum Zweck.

2. Problemumfang

Für die Landwirtschaft liegen die wichtigsten Aufgaben des Pflanzenschutzes in der Bekämpfung von Insekten (und Spinnmilben), Schadpilzen und Unkräutern. Welche Bedeutung diese Komponenten in Relation zueinander besitzen, soll in Tabelle 1 wiedergegeben werden. Sie zeigt in Prozentwerten, welche Aufwendungen die Agrarwirtschaft 1977 in einzelnen Regionen für Insektizide/Akarizide, Fungizide, Herbizide und sonstige chemische Pflanzenschutzmittel gemacht hat. Nicht einbezogen sind die Ostblockstaaten. Unter der Rubrik „Sonstige" sind im wesentlichen Präparate gegen tierische Schädlinge, wie Rodentizide, Molluskizide etc. enthalten. Die Aufstellung umfaßt nur den Agrarsektor (einschließlich Forstwirtschaft, jedoch nicht den Bereich der Haushalts- und Hygieneschädlinge). Seit 1977 haben sich die Relationen leicht verschoben, und zwar in Richtung auf einen weiterhin erhöhten Prozentsatz der Herbizide. Es kann eingewendet werden, daß der Markt für chemische Pflanzenschutzmittel zunächst nichts über die potentiellen Anwendungsgebiete biologischer Maßnahmen besagt. Er dürfte aber doch einen Spiegel der Problemverteilung bilden.

Die Tabelle 1 zeigt deutlich, welche dominierende Bedeutung die Unkrautbekämpfung heute besitzt. Sie umfaßt in den landwirtschaftlich hoch entwickelten

Tabelle 1. Relativer Verbrauch von Insektiziden/Akariziden, Fungiziden, Herbiziden und sonstigen chemischen Pflanzenschutzmitteln 1977

Region	Anteiliger Verbrauch (%) von				Summe	Anteil des Gesamtverbrauchs am Weltmarkt (%)
	Insektiziden	Fungiziden	Herbiziden	Sonstigen		
Bundesrepublik Deutschland	13	18	64	5	100	3,4
N-Europa	15	17	66	2	100	5,0
S-Europa	22	31	41	6	100	12,6
Europa	20	26	49	5	100	21,0
Nordamerika	29	6	62	3	100	35,3
Lateinamerika	54	13	32	1	100	14,9
Asien/Ozeanien	45	24	29	2	100	22,6
Afrika	63	10	25	2	100	6,1
Welt	36	15	46	3	100	100,0

Tabelle 2. Anteil der zehn wichtigsten Kulturen am Umsatz chemischer Pflanzenschutzmittel (%)

Mais	13	Gemüse	6
Baumwolle	11	Weinbau	5
Getreide (ohne Reis u. Mais)	11	Kernobst	4
Reis	10	Rüben	3
Soja	9	Kartoffeln	3
	54		21

Ländern nahezu zwei Drittel aller Pflanzenschutzmaßnahmen. Allderdings überlappt sich hier der Einfluß zweier Parameter: Die Bereiche, in denen intensive, ertragsbetonte und arbeitsextensive Landwirtschaft praktiziert wird, liegen überwiegend in den temperierten Regionen, während der Insektenbefall in den warmen Klimaten eine besonders große Rolle spielt. Der unterschiedlich hohe Anteil von Insektiziden, Fungiziden und Herbiziden spiegelt also neben dem Intensitätsgrad der Landwirtschaft auch die ökologischen Unterschiede wieder. So ist z.B. der relativ hohe Insektizidanteil in Nordamerika vorwiegend durch den Anbau subtropischer Kulturpflanzenarten in den Südstaaten der USA bedingt.

Vom gesamten Verbrauch an chemischen Pflanzenschutzmitteln entfallen 56% auf Europa und Nordamerika. Da der Pflanzenschutzmittelmarkt Asiens etwa zur Hälfte in Japan liegt [28], ergibt sich, daß mehr als zwei Drittel des Bekämpfungsaufwandes auf Länder mit hoch entwickelter Landwirtschaft entfallen.

Aufschlußreich ist ferner eine Aufgliederung des Pflanzenschutzmittelmarktes nach Kulturpflanzenarten (Tabelle 2). Hier ergibt sich, daß 54% aller Aufwendungen im Getreidebau (einschließlich Mais und Reis), dem Baumwoll- und Sojabohnenbau liegen; 21% entfallen auf Gemüse-, Wein-, Kernobst-, Rüben- und Kartoffelbau, die restlichen 25% auf sämtliche übrigen Kulturen.

Tabelle 3. Aufteilung der Pflanzenschutzmittelumsätze in den zehn wichtigsten Kulturen nach Indikationen (%) [21]

	Herbizide	Insektizide	Fungizide
Mais	73	25	2
Baumwolle	24	73	3
Getreide (ohne Reis u. Mais)	70	9	21
Reis	36	43	21
Soja	86	11	3
Anteil der fünf Kulturen am jeweiligen Weltumsatz (%)	74	57	27
Gemüse	21	44	35
Weinbau	10	24	66
Kernobst	8	46	46
Zuckerrüben	67	23	10
Kartoffeln	19	40	41
Anteil der fünf Kulturen am jeweiligen Weltumsatz (%)	11	20	46
Gesamtanteil aller zehn Kulturen am jeweiligen Weltumsatz (%)	85	77	73

Von den in Tabelle 2 aufgeführten Kulturen sind nur der Wein- und der Kernobstbau Dauerkulturen, während von der gesamten landwirtschaftlich genutzten Fläche der Welt (außer Wiesen und Weiden) 1977 nur 2,0% Dauerkulturen waren [18]. Eine ins Gewicht fallende Ablösung chemischer Pflanzenschutzmaßnahmen durch biologische Verfahren wird sich mithin nur erzielen lassen, wenn Methoden des biologischen Pflanzenschutzes für die einjährigen Großkulturen gefunden werden und hier wiederum in erster Linie Verfahren zur biologischen Unkrautbekämpfung (Tabelle 3).

3. Praktische Aussichten der biologischen Unkrautbekämpfung

Für die biologische Unkrautbekämpfung gibt es praktische Beispiele und eine große Zahl experimenteller Studien [1, 22, 55, 56]. Bei der überwiegenden Zahl der Fälle handelt es sich hierbei um die Bekämpfung in fremde Florengebiete eingeschleppter Unkräuter oder um die Beseitigung einzelner Unkrautarten, die in bestimmten Biotopen eine dominierende Stellung einnehmen.

Zur biologischen Unkrautbekämpfung werden vornehmlich monophage oder oligophage, pflanzenfressende oder an Pflanzen saugende Insektenarten verwendet, bei eingeschleppten Unkrautarten meist solche, die im heimischen Florengebiet zum natürlichen Vertilgerkreis der Schadpflanzen gehören. In jüngster Zeit werden auch spezifische, phytopathogene Pilze zur Unkrautbekämpfung verwendet [23, 27].

Bekannte Beispiele für das Gelingen biologischer Verfahren zur Bekämpfung eingeschleppter Unkräuter bilden die Eindämmung verschiedener von Zentralamerika nach Australien eingeschleppter Opuntienarten durch nachgeführte Insektenarten, namentlich den Kleinschmetterling *Cactoblastis cactorum* oder die Kontrolle des von Europa nach Australien und Nordamerika eingeschleppten Johanniskrautes *Hypericum perforatum*, besonders durch Blattkäfer *(Coleoptera, Chrysomelidae)*.

Bei nicht eingeschleppten, also im jeweiligen Florengebiet heimischen Unkräutern, sind erfolgreiche Ansätze bislang überwiegend in aquatischen Biotopen gemacht worden, z.B. durch Aussetzen des Grasfisches *Ctenopharyngodon idella*. Dem methodischen Typ nach handelt es sich hier um die Ausnutzung von biologischen Verbreitungsschranken, also um die Kontrolle von Schadpflanzen auf „ökologischen Inseln“.

Auch die Verwendung von Pflanzenpathogenen zur Bekämpfung von aquatischen Unkräutern ist im Gespräch, z.B. die Infektion der Wasserhyazinthe mit *Rhizoctonia solani*. Ein Beispiel für den Versuch der Bekämpfung terrestrischer Unkräuter durch Mikroorganismen bietet die experimentelle Anwendung von *Fusarium oxysporum* f. *orobanche* gegen *Orobanche*-Arten (Sommerwurz) in der UdSSR.

In der landwirtschaftlichen Praxis ergeben sich offensichtlich einige Schwierigkeiten für die Einführung von Verfahren der biologischen Unkrautbekämpfung.

a) Im laufenden landwirtschaftlichen Betrieb besteht die Aufgabe nicht so sehr darin, einzelne Unkrautarten zurückzudrängen, sondern in der Notwendigkeit, die Konkurrenz der gesamten Unkrautflora unter die Schadensschwelle zu drücken. In der praktischen Landwirtschaft handelt es sich also weniger um die Bekämpfung einzelner Unkrautarten, als um die von Unkrautgesellschaften.

b) Zur Bekämpfung einzelner Unkrautarten können im Prinzip biologische Verfahren gefunden werden, weil hier spezifische Antagonisten einsetzbar sind. Die Bekämpfung von Unkrautgesellschaften setzt aber entweder eine Vielzahl jeweils spezifischer oder polyphager Antagonisten voraus. Andernfalls würde der frei werdende Standraum als ökologische Nische alsbald von den verbleibenden Unkrautarten überwuchert werden oder es würde die Gefahr entstehen, daß die nicht an einen Einzelwirt gebundenen Pflanzenvertilger auf Kulturpflanzen übergehen. Das könnten entweder der zu schützende Pflanzenbestand selbst oder die benachbart angebauten Kulturpflanzenarten sein.

c) Die Gefahr, daß für die Bekämpfung eingesetzte biologische Agentien außer Kontrolle geraten, steigt zweifellos, wenn phytopathogene Mikroorganismen in die Betrachtung einbezogen werden. „Spezifische“ Stämme von verbreiteten Pilzarten, z.B. von *Fusarium oxysporum* f. *orobanche*, bleiben nur so lange spezifisch, wie sie genetisch isoliert und stabil sind. Beide Voraussetzungen zu garantieren, ist praktisch fast unmöglich. Das bedeutet aber, daß man potentiell – wie auch im Fall von *Rhizoctonia solani* gegen Wasserhyazinthe – unter Umständen Infektionen bei Kulturpflanzenarten setzt. Eine Sicherheit gegen unkontrollierte Ausbreitung gibt es nicht. Es bleibt darüber hinaus die Frage, in welchem Grade sich die spezifisch pathogenen Eigenschaften bestimmter genetischer Varianten von Ubiquisten auf die Dauer zu halten vermögen. Zusammenfassend läßt sich über die absehbaren Möglichkeiten der biologischen Unkrautbekämpfung für die landwirtschaftliche Praxis sagen, daß für die Bekämpfung von Unkrautgesellschaften im laufenden landwirtschaftlichen Betrieb – etwa im Getreidebau – vorerst keine realisierbaren Aussichten bestehen.

Die erfolgreichen oder im experimentellen Stadium stehenden Verfahren weisen hingegen zwei Charakteristika auf, die sich auch für andere Bereiche des biologischen Pflanzenschutzes als bedeutsam erweisen, nämlich die Anwendbarkeit bei verschleppten Arten und in ökologisch abgegrenzten Räumen.

4. Praktische Aussichten der biologischen Bekämpfung von Schadpilzen

Weltweit werden etwa 15% aller Pflanzenschutzaufwendungen für die Zurückdrängung pflanzenpathogener Pilze getätigt (s. Tabelle 1). Die Möglichkeiten der biologischen Bekämpfung sind sehr beschränkt [46]. Die Überlegungen gehen dahin, den Antagonismus zwischen einzelnen Arten von Pilzen oder die unterschiedliche Virulenz verschiedener Stämme gleicher Arten durch eine Art von Verdrängungswettbewerb auszunutzen. Eine praktische Bedeutung haben diese Konzepte bisher für die Landwirtschaft nicht erlangen können.

5. Biologische Bekämpfung von Insekten und anderen tierischen Schädlingen

Die biologische Schädlingsbekämpfung im engeren Sinne, also die Bekämpfung von Invertebraten und Vertebraten, die die Gesundheit von Menschen oder Nutztieren oder von Wirtschaftsgütern bedrohen, läßt sich hinsichtlich der Objekte in die Kontrolle verschleppter und heimischer Arten einteilen. Hinsichtlich der Methoden kann man unterscheiden
- die Verwendung von Parasiten
- die Verwendung von räuberischen Tieren (Prädatoren)
- die Verwendung von Organismen der eigenen Art (Selbstvernichtungs- oder Autozidverfahren)
- mikrobiologische Bekämpfungsverfahren.

5.1. Verschleppte Schädlinge

Der Vorgang, daß Tierarten aus der Zone ihrer ursprünglichen geographischen Verbreitung in andere Faunengebiete gelangen, ist an sich kein neues Phänomen. Ratten, Schaben, Kornkäfer und andere Vorratsschädlinge sind verschleppt worden, seit es eine Schiffahrt gibt. Die Verschleppung landwirtschaftlicher Schädlinge hat sicher auch eine lange Geschichte und ist wahrscheinlich so alt wie die Verbreitung der wichtigsten Kulturpflanzenarten in neue Anbauzonen. Die Verschleppung kann entweder zusammen mit den Kulturpflanzenarten erfolgen oder – bei hinreichend schnellem Transport bzw. bei Verfrachtung in Ruhestadien der Schädlingsart – unabhängig vom Wirtsorganismus. Die belegten Beispiele häufen sich seit dem letzten Jahrhundert. Beispiele sind die Einschleppung der

Hessenfliege *Mayetiola destructor* von Europa nach Amerika, die Einwanderung des Baumwollkapselkäfers *Anthonomus grandis* von Zentralamerika in die Südstaaten der USA, der Reblaus *Phylloxera vitifoliae*, des Kartoffelkäfers *Leptinotarsa decemlineata*, der San-José-Schildlaus *Quadraspidiotus perniciosus* von Amerika nach Europa und viele andere. Für die Möglichkeit der biologischen Bekämpfung im Einschleppungsland ist es von Bedeutung, ob die Art im Herkunftsland zu massenhafter Vermehrung neigt oder nicht. Ist sie im Herkunftsland wirtschaftlich unbedeutend, tritt aber im neuen Verbreitungsgebiet massenhaft auf, so liegt der Schluß nahe, daß im Heimatgebiet natürliche Feinde – einzelne Arten oder ein Vertilgerkomplex – für eine ständige Dezimierung des Schädlings sorgen. Hier besteht die Möglichkeit, diese natürlichen Feinde im Herkunftsland aufzuspüren, massenhaft zu züchten und im neuen Verbreitungsgebiet einzubürgern. Der Erfolg solcher Maßnahmen ist jedoch an eine Reihe von Voraussetzungen gebunden:

– Die Populationsdynamik sehr vieler Tierarten verläuft in verschiedenen Teilen ihres geographischen Verbreitungsgebietes durchaus unterschiedlich. Einem Optimum-Gebiet, in dem eine hohe Vermehrungsrate mit entsprechenden populationsdynamischen Pendelausschlägen möglich ist, können ökologisch für die Art weniger vorteilhafte Zonen gegenüberstehen, in denen sie also zwar vorkommt, aber nicht oder nur selten in Massenvermehrung tritt (Latenzgebiete). Dies gilt für die Wirts- bzw. Schädlingsarten ebenso wie für deren natürliche Gegenspieler. Nun brauchen die ökologischen Optima für den jeweiligen Schädling und seine natürlichen Feinde durchaus nicht immer deckungsgleich zu sein. Zum Beispiel kann die Latenz einer Art im Herkunftsland darauf beruhen, daß sich die Wirtsart am Rande ihres spezifischen Optimums, der natürliche Feind aber im Zentrum seiner eigenen Optimumzone befindet. Ist ein ganzer Vertilgerkomplex an der Eindämmung der Schädlingsart beteiligt, so potenziert sich die Anzahl der Möglichkeiten. Hat also eine verschleppte Art im neuen Verbreitungsgebiet Fuß gefaßt, so ist es nicht unbedingt sicher, daß die aus dem Herkunftsland nachimportierten natürlichen Feinde sich tatsächlich auch einbürgern lassen oder den gleichen Wirkungsgrad wie im ursprünglichen Verbreitungsgebiet erreichen. Ein Beispiel wird von Franz u. Krieg [22] für die von Nordamerika nach Europa eingeschleppte Blutlaus, *Eriosoma lanigerum* und der zu ihrer Bekämpfung erfolgreich eingebürgerten Zehrwespe *Aphelinus mali* dargestellt: „In wärmeren Lagen und Jahren und vor allem in geschlossenen Obstanbaugebieten verringert die Zehrwespe den Blutlausbefall ganz erheblich. Allerdings sind im Frühling der Wirt und sein Schmarotzer schlecht synchronisiert. Die Wespen erscheinen zu früh, können sich in den wenigen Wirten nach der Überwinterung kaum vermehren und müssen dann im Sommer erst wieder aufholen."

– Die Bewirtschaftung der Kulturpflanzenart muß so erfolgen, daß sich die natürlichen Feinde nach ihrer Einbürgerung auch auf die Dauer zu halten vermögen. Bei der Mehrzahl der einjährigen Kulturen, die noch vor Ende der Vegetationszeit durch Abernten aus dem Biotop entfernt werden, fehlt es mithin an der erforderlichen ökologischen Kontinuität, insbesondere wenn durch Bewirtschaftung in Fruchtfolgen die Wirtsarten in der nächsten Saison nicht zur Verfügung stehen. Dies mag einer der wesentlichen Gründe dafür sein, daß Verfahren der biologischen Bekämpfung in der Mehrzahl der einjährigen Ackerbaukulturen nur sehr begrenzten Eingang gefunden haben. Allerdings spielen hier verschleppte Schädlingsarten auch keine sehr erhebliche Rolle. Für den Getreidebau wird dies von Hagen et al. [24] sehr schlüssig wie folgt beschrieben:

„Cereals have ecological and economic peculiarities which affect the possibilities of utilizing biological control of their pests. Their short cropping period and the rotation practices commonly used in association with their culture create an unstable environment preventing substantial conservation and buildup of natural enemy populations. Moreover, in many Old-World countries the pest fauna of cereal crops is largely indigenous. This limits the prospects of finding and introducing effective exotic natural enemies. In addition, the rather low crop value per acre commonly makes it uneconomical to use expensive, repetitive releases of artificially produced parasites or predators. Perhaps this explains why there have been few attempts and few successes in the biological control of grain pests.“

– Der wirtschaftliche Schutz der Kulturpflanzenbestände verlangt ferner in den meisten Fällen die Bekämpfung nicht nur einer Schädlingsart, sondern eines ganzen Schadkomplexes. Wird innerhalb dieses Schadkomplexes nur *eine* Art durch nachimportierte natürliche Feinde unter Kontrolle gehalten, verursachen aber die anderen Arten schwerwiegende Fraß- oder Saugschäden, ohne daß gegen sie ebenfalls biologische Bekämpfungsmöglichkeiten bestehen, so ist der Landwirt gezwungen, chemische Bekämpfungsmaßnahmen vorzunehmen. Diese können dann das ggf. eingependelte Verhältnis zwischen eingeschlepptem Schädling und eingebürgertem natürlichen Gegenspieler unter Umständen empfindlich stören.

Derartige Komponenten spielen in mehr oder weniger ausgeprägter Form bei einer großen Zahl von Versuchen zur Verwirklichung biologischer Bekämpfungsmaßnahmen bei eingeschleppten Schädlingen eine Rolle. So stößt z.B. die schon erwähnte Bekämpfung der Blutlaus *Eriosoma lanigerum* im Intensivobstbau wegen der gleichzeitig erforderlichen Bekämpfung anderer Schädlinge (Apfelwickler, Fruchtschalenwickler, Blattläuse etc.) auf Schwierigkeiten. In diesen Intensivkulturen ist auch die Schadensschwelle allgemein niedriger anzusetzen als z.B. im Streuobstbau oder im Freizeitgarten.

Ähnlich liegen die Verhältnisse bei dem bisher erfolgreichsten Fall der Einbürgerung eines Parasiten einer nach Europa eingeschleppten Art, nämlich bei der Bekämpfung der San-José-Schildlaus *Quadraspidiotus perniciosus* durch die Schlupfwespenart *Prospaltella perniciosi.* Hier ist es gelungen, durch Massenzucht der Parasiten und ständiges Freilassen im Freiland einen steigenden Parasitierungsgrad zu erzielen, der sich nach 15 Jahren bei 34 bis 43% eingependelt hat. Das reichte aus, um die Populationsdichte der San-José-Schildlaus durchschnittlich auf weniger als 10% der Ausgangsdichte herabzudrücken. Auch hier ergibt sich aber im Intensivobstbau die schon für *Eriosoma lanigerum* dargestellte Problematik der Bekämpfung von Schadkomplexen. So ist es in Gebieten, in denen die durch Mykoplasmen hervorgerufene Apfeltriebsucht vorkommt, zu einer raschen Ausbreitung der Krankheit gekommen, weil sich die Vektoren – vermutlich Zikaden – bei einer verminderten Zahl von Insektizidbehandlungen im integrierten Verfahren stärker vermehren konnten [44]. Charakteristisch ist in beiden Fällen – bei *Eriosoma* und bei der San-José-Schildlaus – daß sich die Erfolge auf Schädlinge an Dauerkulturen beziehen.

War für die Entwicklung der Bekämpfungsvorhaben gegen die eingeschleppte San-José-Schildlaus die Beobachtung maßgebend, daß die Schäden im Herkunftsland weniger ausgeprägt sind als in den neuen Verbreitungsgebieten, so trifft diese Voraussetzung für andere eingeschleppte Arten nicht zu. Bekanntestes Beispiel ist der Kartoffelkäfer *Leptinotarsa decemlineata*. Trotz intensiver Bemühungen ist es bislang noch nicht gelungen, diesen wichtigen, von Amerika nach Europa eingeschleppten Schädling mit biologischen Verfahren wirksam unter Kontrolle zu bringen. Die Gründe hierfür sind einleuchtend: Der „Colorado beetle" ist auch in seiner Herkunftsregion – Nordamerika – ein relativ „neuer" Schädling. Als ursprünglich entomologische Rarität an wildwachsenden Nachtschattengewächsen im Bereich des Staates Colorado (USA) wurde diese Art durch Ausbreitung des Anbaus einer ihrer Wirtspflanzen, der Kartoffel, zur Massenvermehrung angeregt. Der historisch belegbare Vorgang besteht darin, daß beim Bau der transpazifischen Eisenbahnlinie in den USA in der Mitte des vorigen Jahrhunderts entlang der Trasse für die Verpflegung der Arbeiter Kartoffeln angebaut wurden, die dem Käfer quer über den Kontinent eine ideale Ausbreitungsmöglichkeit schufen und ihn schließlich mit Transportschiffen von der Ostküste Amerikas nach Europa gelangen ließen [7]. Bereits um 1860 wurden in den USA erstmals Arsenpräparate verwendet, um den Kartoffelkäfer zu bekämpfen [38]. Wenn, wie in diesem Falle, die Massenvermehrung des Schädlings in seinem Herkunftsland auf dem expandierenden Anbau einer auch für ihn zuvor in diesem Umfang nicht existierenden Wirtspflanze beruht, ist kaum zu erwarten, daß ein natürlicher Vertilgerkreis im Einschleppungsland mehr leistet als im Ursprungsland.

5.2. Autochthone Schädlinge

Theoretisch könnte man annehmen, daß zwischen autochthonen, also im jeweiligen Floren- und Faunengebiet heimischen Arten sich im Laufe der Jahrhunderte oder gar Jahrtausende ein Gleichgewicht eingependelt habe. Entsprechend müßte sich also ein bestimmtes, im langjährigen Mittel konstantes Zahlenverhältnis zwischen der Wirtspflanzenart, deren Schädigern und den natürlichen Feinden dieser Schädiger aufgebaut haben. Dies würde der Theorie des „Biologischen Gleichgewichts" entsprechen [9, 10]. Würde dieses Gleichgewicht tatsächlich in einer streng fixierten Form existieren, so wären Maßnahmen der biologischen Bekämpfung entweder von vornherein zum Scheitern verurteilt, weil das Ökosystem die Zufuhr faunenfremder oder die Anreicherung zur Fauna gehöriger Organismen alsbald eliminieren würde oder weil andernfalls eine nachhaltige Störung des Gleichgewichts unausweichlich wäre.

In der landwirtschaftlichen Praxis ergibt sich jedoch eine deutlich andere Situation: Der Wechsel im Anbau der Kulturpflanzenarten, die Verwendung immer wieder neuer Sorten oder die Entwicklung der Bewirtschaftungssysteme schaffen ständig auch ökologische Veränderungen, die eine Verschiebung der Gleichgewichtssituation bedingen. Damit bestünde, rein gedanklich, auch die Möglichkeit, solche Entwicklungen auch mit Methoden der biologischen Schädlingsbekämpfung nachzuvollziehen. An Versuchen dieser Art hat es nicht gefehlt. Überblickt man jedoch die Liste der Erfolge biologischer Bekämpfungsmaßnahmen gegen autochthone Schädlinge an autochthonen landwirtschaftlichen Kulturen, so sind die Resultate relativ bescheiden.

Neben den dargelegten biozoenotischen Gründen hat diese Tatsache zwei Ursachen.

– Wie das beschriebene Beispiel der San-José-Schildlaus gezeigt hat, bedürfen biologische Verfahren langer Anlaufzeiten. Bedenkt man das enorme Tempo, in dem sich der Ackerbau in den letzten 25 Jahren in Mitteleuropa entwickelt hat, so wird es verständlich, daß sich Verfahren der biologischen Schädlingsbekämpfung nicht mit gleicher Geschwindigkeit haben etablieren können.

– Schädlingsprobleme mit autochthonen Arten an autochthonen Kulturen pflegen – nach anfänglicher Herdbildung – meist großflächig aufzutreten. Entsprechend müßte auch eine biologische Bekämpfung großflächig organisiert werden, es sei denn, das Auftreten beschränke sich auf „ökologische Inseln".

Der letztere Gesichtspunkt bedarf der näheren Erläuterung: Es leuchtet ein, daß die punktuelle Freisetzung parasitischer, mobiler Tierarten in einem großen, nicht abgegrenzten Areal, in dem die Wirtstiere reichlich vorhanden sind, sehr rasch zu einem Verdünnungseffekt führen muß. Die Auffindungswahrscheinlichkeit und damit der Parasitierungsgrad nehmen also einerseits mit zunehmender Entfernung vom Aussetzungsort ab, andererseits bedingt das reichliche Wirtsangebot eine relativ schnelle Dispersion der Parasiten vom Zentrum der Freisetzung her in die Umgebung. Anders liegen die Verhältnisse, wenn ökologische Verbreitungsschranken bestehen. Dies sei an einem praktischen Beispiel aus dem Baumwollbau in Peru erläutert:

An der in den Tälern von Chincha, Pisco und Cañete im Süden von Peru angebauten Baumwolle tritt als Hauptschädling die Baumwolleule *Heliothis virescens* auf. Diese Täler sind nach Südwesten durch den Pazifischen Ozean begrenzt, landeinwärts durch Wüsten und die weitgehend vegetationslose Sierra. In Chincha zwingt der Wassermangel zum mehrjährigen Anbau der Baumwolle ohne Neuaussaat („soca"). Die Beobachtung, daß *Heliothis* in bedeutendem Ausmaß durch *Trichogramma*- und *Encarsia-Arten* parasitiert wird, führte zu dem erfolgreichen Versuch, diese Parasiten durch Zucht auf Weizen- und Maisschädlingen anzureichern. Dadurch konnte der *Heliothis*-Befall unter die wirtschaftliche Schadensschwelle gedrückt werden. Der Versuch, dieses Verfahren in andere Baumwoll-Anbaugebiete zu übertragen, z.B. nach Tolima in Columbien, mißlang jedoch: Das Fehlen geographischer Verbreitungsschranken bewirkte eine so weitgehende Dispersion der Parasiten, daß ausreichende Parasitierungsgrade nicht zu erzielen waren. Darüber hinaus erwies sich die Methode im Norden Perus ebenfalls als unbrauchbar, da hier als wichtigster Baumwollschädling die Wanzenart *Dysdercus ruficollis* auftritt, gegen die es eine biologische Bekämpfungsmethode nicht gibt. Der deshalb erforderliche Insektizideinsatz verbietet es, die biologische Kontrolle des gleichzeitig auftretenden *Heliothis* in ähnlicher Weise wie im Süden des Landes vorzunehmen [6, 42, 51, 53].

In ähnlicher Weise ist die erfolgreiche Bekämpfung der Schraubenwurmfliege, *Calitroga hominivorax*, eines bedeutenden Viehschmarotzers, auf Curaçao und in einigen Gebieten der Südstaaten der USA durch das „Autozidverfahren" dadurch begünstigt worden, daß die Insellage bzw. die ökologisch isolierte Lage einen Zuflug der Schädlinge *von* außen oder eine Abwanderung der sterilisierten Männchen *nach* außen behinderte. Hierauf wird im Zusammenhang mit den „Selbstvernichtungsverfahren" noch eingegangen werden.

5.3. Methoden der biologischen Schädlingsbekämpfung

5.3.1. Die Verwendung von Parasiten

Im Bereich der Entomologie zählt man zu den Parasiten diejenigen Tiere, die für ihre Entwicklung den Stoffwechsel anderer Arten (ihrer „Wirte") ausnutzen. Parasitische Insekten, die von anderen Insekten leben, können hierbei als Ei-, Larven-, Puppen- oder Imaginalparasiten auftreten. Einer der häufigsten Mechanismen der Parasitierung besteht darin, daß die Weibchen der Parasiten mit ihrem Legestachel ihre Eier einzeln oder zu mehreren in das Innere des Wirtes injizieren. Die sich entwickelnden Larven decken dann bis zur Verpuppung ihren Nahrungsbedarf aus dem Wirtsorganismus, der in der Regel abstirbt. Andere Parasitenarten legen ihre Eier an das Integument des Wirtes, in den sich die geschlüpften Larven dann einbohren. Die Formen des Parasitismus sind gerade bei den Insekten außerordentlich vielfältig [17]. Im Zusammenhang mit der biologischen Schädlingsbekämpfung ist von Bedeutung, daß viele Parasiten eng auf einzelne Wirtsarten spezialisiert sind, andere über ein breiteres Wirtsspektrum verfügen. Ferner muß auf die verbreitete Erscheinung des Hyperparasitismus hingewiesen werden, bei dem parasitische Arten von wiederum anderen Parasiten befallen werden (Hyperparasiten 1. Grades), welche ihrerseits von Hyperparasiten 2. Grades dezimiert werden können. Viele Parasiten sind nur in einer Lebensphase an ihren Wirt gebunden und bedürfen in anderen Stadien zusätzlicher Entwicklungsvoraussetzungen, um ihren Generationszyklus schließen zu können. Wie verwickelt die Verhältnisse liegen können, mag die wörtliche Wiedergabe eines Berichtes von Wybou [54] zeigen:

„Von den 22 *Diatraea*-Arten, die Zuckerrohr befallen, ist *Diatraea saccharalis* nicht nur die am weitesten verbreitete sondern auch die wirtschaftlich bedeutsamste Art. Sie ist die einzige Art, die im Cauca-Tal in Kolumbien auftritt.

1969 waren dort 15,1% der Internodien des Zuckerrohrs befallen bzw. geschädigt. Als Resultat biologischer Kontrollmaßnahmen betrug der Befallsgrad 1976 nur noch 3%. Da die Erhaltung der Prädatorenpopulation eine gewisse Befallsschwelle von *Diatraea* voraussetzt, ist mit einem weiteren Zurückdrängen des Bohrers kaum zu rechnen; es hat sich ein neues Gleichgewicht eingestellt.

In den Tropen entwickelt sich *D. saccharalis* das ganze Jahr hindurch, und die Generationen überschneiden sich. In Louisiana jedoch unterbricht der Winter die Entwicklungszyklen des Bohrers, seiner Prädatoren und seiner Parasiten.

1970 wurde eine biologische Bekämpfungskampagne in der 14000 ha großen Zuckerrohr-Plantage Riopaila im Caucatal, Kolumbien, gestartet. Es handelt sich um eine Monokultur in einer isolierten Zone.

Anfangs wurde festgestellt, daß die Eigelege von *D. saccharalis* zu 31,2% durch *Trichogramma perkinsi* und *T. semifumatum* parasitiert waren. Die Aktivität der natürlich auftretenden Tachinide *Paratheresia claripalpis* reichte nicht aus, um die Larvenpopulation von *Diatraea* auf eine wirtschaftlich vertretbare Schwelle zu reduzieren.

Es stellte sich heraus, daß das Larvenstadium der Cauca-Rasse von *Paratheresia claripalpis* viel länger (17–24 Tage) war als das der peruanischen Rasse der gleichen Art (6–9,8 Tage). Die biologische Kontrolle des Bohrers kann nur erfolgreich sein, wenn das Larvenstadium des Parasiten kürzer ist als das des Wirts. Das erklärt die hohe Wirksamkeit der peruanischen *P. claripalpis*-Rasse.

Laboruntersuchungen haben gezeigt, daß bei Kreuzungen von *P.c.* (Peru) × *P.c.* (Kolumbien) Populationen entstehen, deren Larvenstadium 6,6 bis 10,8 Tage beträgt. Die Freilassung von *P.c.* (Peru) hat zur Verbesserung der *P.c.* im Feld und zu einer ausreichenden Parasitierung von *D. saccharalis* geführt.

Darüber hinaus wurden Tachiniden, die in anderen Zuckerrohr-Gebieten mit Erfolg eingesetzt werden, auf ihre Eignung unter Riopaila-Bedingungen untersucht.

Lixophaga diatraea (die Kubanische Fliege) wurde aus dem karibischen Raum eingeführt. Sie paßte sich nicht an die Bedingungen des Cauca-Tales an.

Von *Metagonistylum minense* sind 2 Rassen in Brasilien bekannt:

a) die Amazonen-Rasse

b) die São-Paulo-Rasse.

Da die São-Paulo-Rasse an subtropische Bedingungen angepaßt ist und sie mit Erfolg in Venezuela und Guayana eingeführt worden war, wurde diese für den Einsatz im Cauca-Tal gewählt. Die ersten Freilassungen erfolgten 1971. Bis 1976 wurde nur eine 1,4%ige Parasitierung erreicht.

In der Praxis ist also die ‚verbesserte' *Paratheresia claripalpis* der wichtigste Faktor in der biologischen Kontrolle von *Diatraea saccharalis*. Das Anpflanzen einer Komposite, die im Gebiet nicht vorkam, hat zu einer verstärkten Vermehrung von *P.c.* im Felde beigetragen (Nahrungsquelle der Adulten)."

Es muß in diesem Zusammenhang angemerkt werden, daß das Zuckerrohr in gewissem Sinne als Dauerkultur, zumindest aber als mehrjährige Kultur anzusprechen ist. Es wird jeweils am gleichen Standort und jahrelang ohne nennenswerte Bodenbearbeitung in Kultur gehalten, ohne daß die ökologische Kontinuität unterbrochen wird.

Diese biologisch oft überaus komplizierten Zusammenhänge machen genaue Kenntnisse der Wirts-Parasitbeziehungen und ihrer ökologischen Bedingtheiten zur Voraussetzung für die Verwendung von Parasiten in der biologischen Schädlingsbekämpfung. Andernfalls sind Fehlschläge oder unerwartete Nebeneffekte zu erwarten. Eine völlige Kalkulierbarkeit in der praktischen Anwendung wird ohnehin kaum zu erzielen sein.

Arten mit parasitischer Lebensweise kommen in mehreren Insektenordnungen vor. In der biologischen Schädlingsbekämpfung spielen Hautflügler (*Hymenoptera*) aus der Unterordnung der *Apocrita* die wesentlichste Rolle („Schlupfwespen" im weiteren Sinne, mit z.B. Echten Schlupfwespen, Brackwespen, Erzwespen, Zehrwespen und *Trichogramma*-Arten). Ferner sind Zweiflügler (*Diptera*) aus der Familie der Raupenfliegen (*Tachinidae*) für biologische Bekämpfungsverfahren herangezogen worden.

Laing u. Hamai [33] geben eine ausführliche Liste der bislang erfolgreichen Projekte der biologischen Insektenbekämpfung, in der jeweils der Schädling, seine wichtigsten Wirtspflanzen, die verwendeten natürlichen Feinde und der Erfolgsgrad aufgeführt wird. Hierbei wird unterschieden zwischen „vollem Erfolg" (complete success), „beträchtlichem Erfolg" (substantial success) und Teilerfolg (partial success). Gliedert man diese Liste für die Fälle von vollem und beträchtlichem Erfolg auf, so ergibt sich nach Abzug der nicht landwirtschaftlichen

Tabelle 4. Voll gelungene Projekte der biologischen Schädlingsbekämpfung (errechnet nach Laing u. Hamai [33])

Kulturpflanze	Erfolg auf		Paras.	Räuber	Anzahl Fälle
	Inseln	Festland			
Mais	—	—	—	—	—
Baumwolle	—	—	—	—	—
Getreide (ohne Reis u. Mais)	—	—	—	—	—
Reis	—	—	—	—	—
Soja	—	—	—	—	—
Gemüse	3	2	5	—	5
Weinbau	—	—	—	—	—
Kernobst	1	12	13	—	13
Rüben	—	—	—	—	—
Kartoffeln	—	—	—	—	—
Zwischensumme	4	14	18	—	18
Zuckerrohr	8	1	7	2	9
Zitrus	20	36	31	25	56
Kokos- u. andere Palmen	4	—	1	3	4
Zierpflanzen	2	—	2	—	2
Oliven	—	2	2	—	2
Alle Sonstigen[a]	7	5	8	4	12
	45	58	69	34	103

[a] Mit jeweils einem Fall je Kulturpflanzenart

Kulturen (Forst, Parkbäume usw.), eine Anzahl von 27 voll geglückten Ersteinführungen von Parasiten und Räubern und von 65 Ersteinführungen mit beträchtlichem Erfolg. Von Interesse für die landwirtschaftliche Praxis ist die Frage, in welchem Ausmaß diese Ersterfolge sich haben wiederholen lassen und bei welchen Kulturpflanzenarten dies der Fall war. Hierüber geben die Tabellen 4. u. 5 Auskunft. Sie korrespondieren in ihrem ersten Teil jeweils mit Tabelle 2, in der die zehn Hauptkulturen aufgeführt sind, die derzeit 75% aller chemischen Pflanzenschutzmaßnahmen auf sich vereinigen und knapp zwei Drittel der landwirtschaftlichen Fläche der Welt (ohne Wiesen und Weiden) repräsentieren.

Die beiden Tabellen 4 und 5 geben nun alle Projekte wieder, die zu vollen oder beträchtlichen Erfolgen geführt haben, und zwar für alle Fälle der Einbürgerung, also jeweils auch, wenn die gleiche Insektenart in mehreren verschiedenen Regionen verwendet wurde.

Aus Tabelle 4 geht hervor, daß volle Erfolge der biologischen Bekämpfung in den folgenden Hauptkulturen bislang nicht zu verzeichnen sind: Mais, Reis, alle übrigen Getreidearten, Baumwolle, Sojabohnen, Rüben, Kartoffeln und Weinbau. Diese Kulturen umfassen 60% der landwirtschaftlichen Fläche der Welt und 65% des gegenwärtigen finanziellen Aufwandes der Landwirtschaft für Pflanzenschutzmittel.

Tabelle 5. Beträchtliche (substantial) Erfolge von biologischen Bekämpfungsprojekten (errechnet nach Laing u. Hamai [33])

Kulturpflanze	Erfolg auf		Paras.	Räuber	Anzahl Fälle
	Inseln	Festland			
Mais	—	—	—	—	—
Baumwolle	—	—	—	—	—
Getreide (ohne Mais u. Reis)	—	1	1	—	1
Reis	—	—	—	—	—
Soja	—	—	—	—	—
Gemüse	1	6	7	—	7
Weinbau	—	3	3	—	3
Kernobst	7	14	21	—	21
Rüben	—	—	—	—	—
Kartoffeln	—	—	—	—	—
Zwischensumme	8	24	32	—	32
Zuckerrohr	17	—	15	2	17
Zitrus	5	19	18	6	24
Kokos- u. andere Palmen	11	—	3	8	11
Zierpflanzen	1	2	1	2	3
Luzerne	—	5	4	1	5
Futtergras	—	1	1	—	1
Brotbaum	2	—	—	2	2
Maulbeeren	2	2	4	—	4
Kaffee	2	—	1	1	2
Sonst	4	—	—	4	4
	52	53	79	26	105

Im Gemüse- und Kernobstbau haben sich bislang 18 Projekte verwirklichen lassen und zwar sämtlich mit Hilfe von Parasiten. In vier Fällen ist dies auf Inseln, in 14 Fällen auf dem Festland gelungen.

Das herausragende Beispiel ist hierbei die Kontrolle der Blutlaus *Eriosoma lanigerum*, das schon in Abschnitt 5.1 dargestellt wurde, durch die Zehrwespe *Aphelinus mali*. Die Einbürgerung dieses Parasiten führte im Kernobstbau von 11 verschiedenen Ländern zu vollen Erfolgen. Die beiden anderen Beispiele beziehen sich auf eine Wollausart im Apfelbau Kanadas und auf eine Wanzenart in Hawaii. Im Gemüsebau der Inseln Guam, Mauritius und Hawaii gelang die Bekämpfung je einer Käfer-, Schmetterlings- und Wanzenart durch Parasiten, in Australien bzw. Tasmanien je einer Wanzen- und Blattlausart.

Es muß hier angemerkt werden, daß es sich bei diesen Angaben von Ländern und Regionen nicht darum handelt, daß der Erfolg im gesamten Anbaugebiet erzielt worden wäre. Vielmehr handelt es sich – mit Ausnahme mancher Inseln – meist um Projekte in bestimmten Anbaugebieten der jeweiligen Region.

Betrachtet man die übrigen Kulturpflanzenarten, so ragt der Erfolg im Zitrusbau heraus. Von den 56 voll gelungenen Projekten werden 31 mit Hilfe von

Parasiten erzielt, von diesen wiederum 12 auf Inseln. In allen Fällen sind die Schädlinge Schild- und Wolläuse. Allein in 15 Fällen war die Bekämpfung der schwarzen Mottenschildlaus *Aleurocanthus woglumi* durch Parasiten der Gattungen *Eretmocerus* und *Prospaltella* erfolgreich. *Aleurocanthus woglumi* ist in Südasien beheimatet und wurde von dort in zahlreiche Zitrus-Anbaugebiete der Welt verschleppt.

Aufschlußreich ist auch die Situation beim Zuckerrohr. Von den neun voll geglückten Projekten gelangen acht auf Inseln (Mauritius, Victoria Islands, Mariana Islands, Guadeloupe, Hawaii), eines auf dem Festland (Peru, ökologisch abgegrenztes Gebiet, „ökologische Inseln“). In sieben Fällen, davon sechs auf Inseln, wurde der Erfolg mit Parasiten erzielt. Die Schadorganismen waren eine Maulwurfsgrillenart, drei Schild- bzw. Schmierlausarten, eine Zikaden-, zwei Käferarten sowie – auf Guadeloupe – *Diatraea saccharalis*, die mit einer Raupenfliegenart (*Tachinidae*) unter Kontrolle gehalten wird (vgl. auch [54]). Die übrigen aufgeführten Kulturpflanzenarten sind von geringerer allgemeinwirtschaftlicher Bedeutung. Es überwiegen bei den erfolgreichen Projekten die Dauerkulturen, sowie die positiven Ergebnisse auf Inseln im Vergleich zum Festland und mit Parasiten gegenüber räuberischen Insekten.

Ein sehr ähnliches Bild ergibt sich bei den Projekten mit „beträchtlichem Erfolg“ (vgl. Tabelle 5).

Für die großen Ackerbaukulturen liegt hier nur das Beispiel eines „substantial success“ gegen die Getreidehalmwespe *Cephus pygmaeus* vor, die von Europa nach Kanada eingeschleppt ist. Sie wird durch die Schlupfwespe *Collyria calcitrator* parasitiert. Die größte Anzahl von beträchtlichen Erfolgen findet sich wiederum im Obstbau, wobei die schon beschriebene Kontrolle der Blutlaus *Eriosoma lanigerum* durch *Aphelinus mali* und der San José-Schildlaus durch *Prospaltella* im Vordergrund steht.

5.3.2. Die Verwendung von Prädatoren

Als Räuber oder Prädatoren werden solche Insekten bezeichnet, die andere lebende Tiere aktiv jagen bzw. ihnen auflauern und sie auffressen oder aussaugen. Prädatoren kommen in den meisten Insektengruppen vor. Sie sind überwiegend weniger spezifisch als die Parasiten, d.h. sie leben zum größten Teil oligophag von einem breiten Spektrum von Beutetieren. Zu den polyphagen, sozial lebenden Prädatoren gehören die hügelbauenden Waldameisen, die im biologischen Forstschutz eine gewisse Bedeutung erlangt haben, da sie in hinlänglicher Dichte in den Waldbeständen angesiedelt werden können. In der Landwirtschaft entfällt diese Verwendungsmöglichkeit.

Die meist nicht sehr spezifische Räuber-Beutebeziehung erschwert in vielen Fällen die Verwendung von Prädatoren in der biologischen Schädlingsbekämpfung: Sie veranlaßt die Räuber, namentlich bei sinkender Populationsdichte des Schädlings, auf andere Beuteobjekte auszuweichen. Andererseits kann sich gerade dadurch, nach erfolgter Einbürgerung, stets ein gewisses Reservoir der Prädatoren im Biotop halten, so daß die spezifische Abhängigkeit von der Dichte des Zielorganismus z.T. kompensiert wird.

Das bekannteste und heute „klassische" Beispiel einer gelungenen Nachverfrachtung eines Räubers zur Bekämpfung eines verschleppten Schädlings bildet die Einbürgerung der australischen Marienkäferart *Rodolia cardinalis* gegen die Zitrus-Wollschildlaus *Icerya purchasi* in Kalifornien. Dieser, ebenfalls aus Australien importierte Schädling konnte seither durch die Prädatorenart vollständig unter Kontrolle gehalten werden. Darüber hinaus ließ sich der Erfolg in zahlreichen anderen Zitrusanbaugebieten wiederholen. Die in den Tabellen 4 und 5 angegebenen 25 vollständigen und 6 beträchtlichen Bekämpfungserfolge durch Prädatoren im Zitrusbau wurden sämtlich mit *Rodolia* erzielt.

Insgesamt ist jedoch die Zahl der Erfolge mit räuberischen Arten in der landwirtschaftlichen Praxis geringer gewesen als die Anzahl der mit Hilfe von Parasiten gelungenen Bekämpfungsmaßnahmen. Für die in den Tabellen 4 und 5 aufgeführten 10 Großkulturen liegt kein Erfolgsbericht vor. Außer im Zitrusbau liegen die wesentlichsten positiven Ergebnisse bei den Kokospalmen und anderen Palmen sowie – in begrenztem Umfange – beim Zuckerrohr vor.

5.3.3. Die Verwendung von Mikroorganismen

In der mikrobiologischen Schädlingsbekämpfung werden insektenpathogene Erreger oder deren Stoffwechselprodukte benutzt, um die Schädlingspopulationen zu infizieren oder ihre Infektion zu begünstigen bzw. die Schädlinge durch Stoffwechselgifte der Mikroorganismen zu beeinträchtigen. Als Formen von Erregern kommen hierbei Pilze, Protozoen, Bakterien, Rickettsien und Viren in Betracht.

Die mikrobiologische Schädlingsbekämpfung unterscheidet sich grundsätzlich von allen anderen Verfahren des biologischen Pflanzenschutzes. Während bei der Verwendung von Parasiten und Räubern antagonistische Wechselbeziehungen zwischen einzelnen Arten ausgenutzt werden und es somit gewisser Anlaufzeiten bedarf, bis sich ein Gleichgewicht zwischen den natürlichen Feinden und ihren Wirts- bzw. Beutetieren einpendelt, ist es bei mikrobiologischen Verfahren vielfach möglich, die Erreger bei entstehender oder bereits ausgebrochener Massenvermehrung des Schädlings ähnlich wie chemische Pflanzenschutzmittel zu applizieren. Da jedoch – mit Ausnahme der insektenpathogenen Pilze – alle Krankheitserreger von den Insekten oral aufgenommen werden müssen, verstreicht bis zum Wirkungseintritt jeweils eine gewisse Zeitspanne.

Entomophage Pilze dringen in der Regel durch das Integument der Insekten in die Körperhöhle ein und treiben ihre Hyphen auf umgekehrtem Wege durch die Kutikula nach außen, um dort zu sporulieren. Das bekannteste Beispiel bildet der zu den Phycomyceten gehörende „Fliegenschimmel" (*Entomophthora muscae*). In der biologischen Schädlingsbekämpfung sind indessen vor allem der insektenpathogene Pilz *Beauveria tenella* sowie andere *Beauveria*-Arten versucht worden. Versuchsobjekte waren hierbei neben dem Maikäfer (*Melolontha sp.*) auch der Kartoffelkäfer und andere Arten. Praktische Bedeutung in nennenswertem Umfang hat die biologische entomomykologische Bekämpfungsweise bislang nicht erlangen können. Weder in den USA noch in der Bundesrepublik Deutschland sind bislang insektenpathogene Pilzpräparate amtlich zugelassen. Dagegen können insektenpathogene Pilze als natürliche Begrenzungsfaktoren von Insektenarten eine bedeutende Rolle spielen, z. B. für *Anticarsia sp.* (*Lepidoptera, Noctuidae*) Pilze der Gattungen *Nomuraea, Entomophthora* und *Beauveria.*

Demgegenüber sind aus dem Bereich der mikrobiologischen Schädlingsbekämpfung Bakterienpräparate und Präparate von Kernpolyederviren registriert und von praktischer Bedeutung.

Das weitaus bekannteste und in Diskussionen um die biologische Schädlingsbekämpfung am häufigsten herangezogene Beispiel bietet *Bacillus thuringiensis.* Die Wirkung dieses, weitgehend als Lepidopteren-spezifisch geltenden Mikroorganismus' beruht in erster Linie auf der Toxizität eines Endotoxins, das als parasporales Kristall zusammen mit den Sporen bei Auflösung des Sporangiums frei wird.

Nach strenger Definition wäre die Wirkung des parasporalen Kristalls von *B. thuringiensis* als ein Effekt der chemischen Schädlingsbekämpfung mit biogenen Substanzen einzuordnen. Wegen des biogenen Ursprungs und der parallelen insektiziden Eigenschaften der Sporen wird *B. thuringiensis* jedoch allgemein der biologischen Schädlingsbekämpfung zugerechnet. Aus diesem Bakterium gewonnene Präparate gehören zu den wenigen amtlich registrierten Pflanzenschutzmitteln aus dem Bereich der biologischen Schädlingsbekämpfung. In den USA sind *Bacillus thuringiensis*-Präparate unter mehreren Handelsnamen auf dem Markt. Franz u. Krieg [22] schätzen, daß diese Präparate etwa einen Anteil von 1,5% am Gesamtverbrauch von Insektiziden in den Vereinigten Staaten haben. In der Bundesrepublik Deutschland besteht eine Zulassung von *Bacillus thuringiensis*-Präparaten in der Landwirtschaft zur Bekämpfung des Maiszünslers (*Ostrinia nubilalis*) sowie in verschiedenen anderen Kulturen wie Obst und Gemüse.

Franz u. Krieg führen eine Liste mit 35 Beispielen befriedigender Bekämpfungserfolge mit *Bacillus thuringiensis* gegen Larven von Lepidopteren auf [22]. Dreizehn dieser Erfolge wurden bei Forstinsekten erzielt, zwölf bei Obstschädlingen. Zu den übrigen Arten zählen *Diatraea saccharalis* an Zuckerrohr in den USA, *Ostrinia nubilalis* an Mais, *Pieris brassicae, Pieris rapae* und *Plutella maculipennis* an Kohl sowie je ein Zitrus-, Oliven- und Tabakschädling. Franz u. Krieg [22] kommen zu dem Schluß, „daß in den Sporen-Kristall-Präparaten von *B. thuringiensis* erstmalig in großem Maßstab ein Mittel zur Verfügung steht, das die Selektivität biologischer Bekämpfungsmethoden mit der leichten Anwendbarkeit (wie wir sie aus dem chemischen Pflanzenschutz gewohnt sind) verbindet".

Als zweites Bakterienpräparat ist in den USA *Bacillus popilliae* zur Bekämpfung des Japankäfers *Popillia japonica* zugelassen. Der aus Ostasien eingeschleppte Schädling hat namentlich im Osten der USA eine beträchtliche wirtschaftliche Bedeutung gewonnen. Die Sporen, die sich nur aus erkrankten Wirtstieren gewinnen lassen, werden zur Bekämpfung der Engerlinge in den Boden injiziert und können eine bis 90%ige Mortalität bewirken. In neuerer Zeit ist örtlich ein Wirkungsabfall beobachtet worden, der einesteils mit Resistenzerscheinungen, andererseits mit einer Abnahme der Virulenz durch die zahlreichen Wirtspassagen beim Züchten der Erreger erklärt wird [22].

An weiteren mikrobiologischen Präparaten sind in den USA drei Kernpolyeder-Viren zugelassen, von denen zwei gegen Forstschädlinge verwendet werden, nämlich gegen *Orgyia pseudotsugata* an Nadelholz und gegen den Schwammspinner *Lymantria dispar* an Laubholz. Im landwirtschaftlichen Bereich ist ein spezifisch gegen *Heliothis*-Larven wirksames Kernpolyedervirus amtlich zugelassen.

5.3.4. *Autozidverfahren*

Unter dem Begriff der „Autozidverfahren“ versteht man in der biologischen Schädlingsbekämpfung die Minderung der Populationsdichte von Schadinsekten durch Individuen der eigenen Art. Die hierbei verwendeten Methoden umfassen die unfruchtbare Kreuzung genetisch unverträglicher Herkünfte von Insektenarten, die Verwendung von Chemosterilantien [8] und die Verwendung durch energiereiche Strahlung sterilisierter Geschlechtspartner. Laven [34] hat dargestellt, daß die plasmatische Unverträglichkeit (Inkompatibilität) zwischen verschiedenen Herkünften der Stechmücke *Culex pipiens* die Vereinigung der Spermien mit dem Eikern nach der Copula verhindert. Für die Erklärung ähnlicher Unverträglichkeiten zwischen verschiedenen Rassen von *Rhagoletis cerasi* bieten Boller et al. drei Hypothesen an [4], nämlich eine entweder genetische, zytoplasmatische oder durch Symbionten bedingte Inkompatibilität.

Chemosterilantien haben sich bislang – überwiegend aus toxikologischen Gründen – nicht durchsetzen können.

Hingegen kann die durch energiereiche Strahlung bewirkte Sterilisierung von Insektenmännchen auf ein berühmt gewordenes Beispiel verweisen [30]. Das Konzept dieser Methode besteht darin, durch Freilassen sterilisierter, aber kopulationsfähiger Männchen infertile Kopulationen zu begünstigen, um damit die Schlüpfrate lebensfähiger Larven in der F_1 Generation zu reduzieren.

Dieses Konzept ist in der Tat verwirklicht worden, und zwar bei der amerikanischen Schraubenwurmfliege *Cochliomyia hominivorax*. Diese Art, die ein bedeutsamer Ektoparasit für die Viehhaltung in den Südstaaten der USA, in der Karibik und Zentralamerika ist, war für die Entwicklung der Autozidverfahren ein ungewöhnlich glückliches Objekt und zwar aus folgenden Gründen:

– Es erwies sich als möglich, eine Strahlendosis für die röntgenologische Sterilisierung der Männchen zu finden, die zur Infertilität der Männchen führte, ohne deren sexuelle Aktivität (Virilität) zu mindern.

– Die Weibchen der Schraubenwurmfliege lassen eine Mehrfachbegattung nicht zu, sondern wehren nach einmal vollzogener Copula (ob fruchtbar oder nicht) weitere Kopulationsversuche von Männchen ab.

– Die Weibchen schlüpfen zeitlich in einem gewissen Abstand vor den Männchen (Protogynie).

– Der Befall ist auf ökologisch abgrenzbare Gebiete beschränkt.

Diese Faktorenkombination hat es ermöglicht, ein bislang in dieser Perfektion nicht reproduziertes Verfahren der „sterile male technique“ in die Praxis umzusetzen, denn:

– Es ist nicht der Regelfall, daß die sterilisierende Strahlendosis ohne Einfluß auf das Sexualverhalten bleibt.

– Mehrfachbegattung ist selbst nach fruchtbarer Erstcopula bei den Insekten nicht selten.

– Die Tatsache, daß die Weibchen der Schraubenwurmfliege vor den Männchen erscheinen, ermöglicht es, bei exakter Zeitpunktwahl, sterilisierte Männchen vor dem Erscheinen „natürlicher" fertiler Männchen freizulassen und dadurch Kopulationen zu begünstigen, die eine Ablage fruchtbarer Eier verhindern. Die Schraubenwurmfliege kann sich weder parthenogenetisch vermehren, noch ist die Füllung des receptaculum seminis ausschlaggebend für die Kopulationsbereitschaft.

– Die ökologisch abgegrenzte Verbreitungszone der Schraubenwurmfliege erlaubt es, die Autozidmethode ohne Berücksichtigung eines nennenswerten Zufluges von außen zu praktizieren.

Das Beispiel zeigt, auf welchen differenzierten Verhältnissen der Erfolg der Methode beruht und wie begrenzt die Möglichkeiten der Übertragung der Methode auf andere Arten sind, selbst, wenn man Modifizierungen für jeden gegebenen Fall als selbstverständlich voraussetzt. Interessant ist, daß sich inzwischen auch beim Autozidverfahren für die Schraubenwurmfliegen Resistenzerscheinungen herauszubilden beginnen. Sie beruhen vermutlich auf der Selektion der innerhalb der Populationen ursprünglich sehr wenigen Weibchen, die eine Mehrfachcopula zulassen und entsprechend nach infertiler Erstcopulation mit infertilen Männchen mit fertilen Männchen aus dem Freiland fruchtbar kopulieren können.

Die Methode der „sterile male technique" ist seither bei mehreren Insektenarten versucht worden. Die Erfolge haben sich hierbei meist auf geographisch abgegrenzte Gebiete erstreckt.

6. Nebenwirkungen und Gesichtspunkte der amtlichen Zulassung

Die Mehrzahl der Verfahren des biologischen Pflanzenschutzes zeichnet sich nach dem heutigen Kenntnisstand durch einen hohen Grad von Spezifität aus. Dennoch ist in jedem Einzelfall die Diskussion darüber berechtigt, ob die Verbringung von Organismen in fremde Faunengebiete oder die massenhafte Ausbringung von Mikroorganismen Risiken bergen könnte. Bei sich selbsttätig weitervermehrenden Lebewesen, in deren Populationen sich prinzipiell Mutationen ereignen können, ist in der Tat die Frage zu stellen, mit welcher Sicherheit zukünftige Entwicklungen vorausgesagt werden können, in welchem Maße eine „Standardisierung" möglich ist und welche ökologischen Nebenwirkungen sich ergeben können. Bei Mikroorganismen und ihren Stoffwechselprodukten ist zudem die Frage der Pathogenität für andere Lebenwesen als das jeweilige Bekämpfungsobjekt von wesentlicher Bedeutung. Toxikologische Fragen – einschließlich Kanzerogenität, Mutagenität, teratogener Wirkungen usw. – bedürfen daher einer genauso sorgfältigen Prüfung wie bei chemischen Pflanzenschutzmitteln. Es sei daran erinnert, daß die außerordentlich bedenklichen Wirkungen von Mykotoxinen, die von pflanzenparasitären und auf Vorräten lebenden Pilzen gebildet werden, erst in den letzten zwei Jahrzehnten in ihrem vollen Umfang erkannt worden sind. Diese Gesichtspunkte machen es verständlich, daß die Zulassungsbehörden bei der Registrierung biologischer Agentien große Vorsicht walten lassen. Die Environmental Protection Agency (EPA) der USA bringt in einer vorläufigen Richtlinie zum Ausdruck, daß „there is no reason to assume that a biological agent will not result in adverse effects simply because it is a biological" .... „Viruses, for example, may undergo some alterations which could potentially pose a hazard to humans and other organisms" [20].

In vielen Fällen muß wegen der nicht absehbaren ökologischen Nebenwirkungen von vornherein auf biologische Bekämpfungsverfahren verzichtet werden. Das gilt z.B. für die schon dargestellte biologische Bekämpfung autochthoner Unkrautgesellschaften mit phytophagen Insekten. Bei der Einfuhr von Nützlingen muß ferner durch eine sogenannte Nützlingsquarantäne sichergestellt werden, daß nicht auch Feinde nützlicher Arten mit importiert werden. Daß derartige Gefahren nicht von der Hand zu weisen sind, zeigt das Beispiel des im Zusammenhang mit der biologischen Bekämpfung von Wasserunkräutern schon erwähnten Graskarpfens, *Ctenopharyngodon idella*. Mit diesem Fisch wurden zwei Bandwurmarten aus den Gattungen *Bothriocephalus* und *Khawia*, die noch vor wenigen Jahren in Mitteleuropa unbekannt waren, über die UdSSR, Ungarn und Österreich nach Deutschland verschleppt. Hier blieben sie nicht allein Parasiten der eingeführten Graskarpfen, sondern befielen auch andere Arten, z.B. den heimischen Karpfen, *Cyprinus carpio*. Bereits 1973 gab es Zuchten, die bis zu 100% befallen waren. Namentlich die Karpfenbrut wird geschädigt, da in einem Fisch 20 bis 30, in einem Fall sogar 80 Bandwürmer von bis zu 30 cm Länge im Darm parasitieren [16, 31].

Die insektenpathogenen Mikroorganismen gelten gemeinhin als weitgehend spezifisch, zumindest gruppenspezifisch, wie z.B. *Bacillus thuringiensis* bei Lepidopteren. Doch gibt es offensichtlich auch hier Ausnahmen von der Regel. Hamed [25] hat die Wirkung von *B. thuringiensis* var. *Kurstaki* auf Parasiten und Prädatoren von *Yponomeuta evonymellus* untersucht und stellte keine pathogene Wirkung auf zwei Tachinenarten, eine Schlupfwespenart und auf Prädatoren fest. Dagegen erwiesen sich die Schlupfwespen *Diadegma armillata*, *Pimpla turionellae*, *Ageniaspis fuscicollis* und *Tetrastichus evonymellae* als empfindlich. Bei hohen Konzentrationen des Sporen-Kristallkomplexes und nicht praxisüblichen Langzeitkontakt wies Marchal-Segault [36] toxische Wirkungen auf die beiden Braconiden *Apanteles glomeratus* und *Phanerotoma flavitestacea* nach. Dunbar u. Johnson [15] zeigten, daß die Lebensdauer der Braconide *Cardiochiles nigriceps* nach Aufnahme des Sporen-Kristallkomplexes von *B. thuringiensis* in Zuckerwasser deutlich verkürzt wurde. De Barjac [2] hat ferner bei *B. thuringiensis* var. *israeliensis* eine ausgeprägte Wirkung gegen Stechmückenlarven (*Diptera*) nachgewiesen.

Sporulierte Kulturen von *B. thuringiensis*- und von *B. cereus*-Stämmen enthalten keine insektiziden Exotoxine und sind nach Krieg [32] für erwachsene Honigbienen (*Apis mellifera*) nicht toxisch. Nicht sporulierte Kulturen, die das thermostabile β-Exotoxin enthielten, erwiesen sich in zwei Fraktionen als bienentoxisch.

Die Anwesenheit dieses Exotoxins scheint allgemein die spezifischen Wirkungseigenschaften des Präparates zu durchbrechen. So berichtet Lipa [35] von einem aus dem Exotoxin gewonnenen Präparat, das in Osteuropa zur Bekämpfung des Kartoffelkäfers verwendet wird. Da das Exotoxin eine teratogene Wirkung habe, sei das Präparat schädlich für den Menschen und andere Säugetiere, so daß „Vorsichtsmaßregeln wie bei starken Giften" erforderlich seien.

In diesem Zusammenhang sind ferner die Arbeiten von Sharma u. Sahu [45], Panda et al. [39] von Interesse, nach denen ein Einfluß des β-Exotoxins von *B. thuringiensis* auf mitotische Vorgänge gezeigt wird. Bassand u. Carpy [3] machen jedoch deutlich, daß bei *lege artis* hergestellten Sporen-Kristall-Präparaten dieses Exotoxin nicht vorkommt. Unwidersprochen ist hingegen geblieben, daß das δ-Endotoxin, das in allen Stämmen von *B. thuringiensis* aufträte, sowohl insektizide als auch „antitumor activities" habe. Dies würde zumindest eine zytologische Wirksamkeit andeuten.

Eine gültige Einzelbewertung solcher Befunde ist im Rahmen dieses Beitrages nicht möglich. Die Tatsache, daß die vorliegenden Ergebnisse solche Erscheinungen wissenschaftlich als möglich erscheinen lassen, begründet aber sicher hinlänglich den Anspruch des Gesetzgebers auf entsprechende Zulassungsunterlagen.

Im Falle der entomophagen Pilze *Beauveria bassiana* und *Metarhizium anisopliae* kommt Schaerffenberg [43] zu dem Ergebnis „daß die insektentötenden Pilze vom Muscaridentyp für Mensch und andere Warmblüter weder toxisch noch pathogen sind. Bei empfindlichen Personen beim Umgang mit Sporen dieser Pilze auftretende Reaktionen (Müdigkeit, Kopfschmerzen, Gliederschmerzen, Fieber, Schweißausbrüche) sind nicht genereller, sondern allergischer Natur". Lipa [35] berichtet über das in der UdSSR aus *Beauveria bassiana* hergestellte Boverin, das Präparat habe „allergogene Eigenschaften. Bei Tieren beobachtet man Leukozytose, Eosinophilie, und Monozytose. Bei zweimaliger vierstündiger Inhalation des Präparates in einer Konzentration von 400 mg/m^3 verendeten fünf von zehn Ferkeln am siebenten Versuchstage. Die übrigen Tiere vertrugen dreißig weitere Inhalationen. Bei den mit der Herstellung von Boverin beschäftigten Personen wurde eine Reihe von Erkrankungen der Atemwege vom Bronchitistyp festgestellt" (beglaubigte Übersetzung aus dem Polnischen von Dipl. Slawist Crull [11]). Ganz sicher erheben sich also auch bei derartigen Präparaten Fragen der Zulassungsprüfung.

Hinsichtlich denkbarer ökologischer und toxikologischer Nebenwirkungen insektenpathogener Viren herrscht unter den Fachleuten eine gewisse Unsicherheit, die Tinsley u. Melnik [48] so formulieren: „Das Fehlen von klaren Beweisen für die Gefährlichkeit, auf die Wade [50] hinweist, bedeutet oft nicht, daß gesicherte und grundlegende Untersuchungen vorgenommen und zu negativen Ergebnissen hinsichtlich der Gefährdung geführt hätten, sondern vielmehr, daß gewisse wesentliche Formen der Prüfung ganz einfach nicht vorgenommen worden sind."

Die Bedenken, die sich gegen diese Viren richten, bleiben deshalb vorerst mehr theoretischer Natur und werden von Tinsley u. Melnik [48] in drei Punkten zusammengefaßt:

„1. Das jeweilige im Pflanzenschutz zu verwendende Virus könnte andere Insekten als die Zielorganismen infizieren.

2. Gewisse Insektenviren könnten in der Lage sein, andere Invertebraten und vielleicht sogar Vertebraten zu befallen. Wenn solche Infektionen bislang nicht offenbar geworden sind, dann kann das darauf beruhen, daß derartige Erscheinungen mit den zur Verfügung stehenden Techniken fast nicht aufdeckbar sind. Die Berichte über einen hohen Grad von Artspezifität der Insektenviren sollten mit Vorbehalt betrachtet werden und zwar im Hinblick auf die Tatsache, daß

Antikörper, die mit Insektenviren reagieren, in Haus- und Wildtieren gefunden worden sind, nämlich in Schweinen, Rindern, Schafen, Pferden, Hunden und Hirschen und ebenso in den Seren von Laborkräften, die mit insektenpathogenen Viren umgehen.

3. Bei den meisten der am genauesten untersuchten, bei Vertebraten vorkommenden Viren ist die Wirtsspezifität weder ein grundlegendes noch ein stabiles Charakteristikum. Veränderungen in der Pathogenität und Spezifität insektenpathogener Viren könnten vorkommen und ein breiteres Wirtsspektrum begründen. Dies könnte durch Mutation, Rekombination des genetischen Materials oder durch Akquisition wirtsspezifischer Stoffe während der Passage durch ‚unübliche' Wirte geschehen".

Diese Überlegung ist auch in den WHO-Report 1973 eingegangen [19], der formuliert: „Die Passage eines Virus in irgendeinem Wirt auf unnatürliche Weise oder die Infektion eines nicht naturgegebenen Wirts kann in der Selektion einer Virusmutante resultieren, die veränderte Eigenschaften aufweist."

Wade kommt zu dem Ergebnis [50], daß „wahrscheinlich die wesentlichste theoretische Gefahr, die von insektenpathogenen Viren ausgeht, in der Tatsache besteht, daß, obgleich sie nur Zellen von Insekten anzugreifen scheinen, die molekulare Basis dieser Spezifität völlig unbekannt ist".

Auf dieser Grundlage ist es verständlich, daß seitens der FAO-WHO [19] sehr weitgehende Forderungen an die Prüfung und Zulassung insektenpathogener Viren gestellt werden.

Zu den Fragen nach den Nebenwirkungen biologischer Bekämpfungsverfahren gehören zweifellos auch die nach der Möglichkeit der Resistenzbildung. Bei den mikrobiologischen Methoden ist die Möglichkeit einer Resistenzbildung mit Sicherheit anzunehmen, wenn im Genbestand der Wirtspopulationen Resistenzgene vorhanden sind. Der Aufbau einer solchen Resistenz gegen ein Granulosevirus ist beim Apfelwickler *Laspeyresia* (*Carpocapsa*) *pomonella* auch durch Huber [29] nachgewiesen worden. Hinzu kommt im Wechselspiel zwischen Wirt und Erreger jedoch noch, daß neben Resistenzbildung beim Wirt auch Veränderungen in der Virulenz des Erregers auftreten können, namentlich auch bei permanenter Weiterzucht auf künstlichen Nährmedien. Bei der Möglichkeit eines solchen Virulenzverlustes erhebt sich bei mikrobiologischen Präparaten die Frage nach der „Gewährleistung für gleichbleibende Qualität" durch den Hersteller.

Ist die Möglichkeit der Resistenzbildung gegen insektenpathogene Erreger fachlich eigentlich nie umstritten gewesen, so wurde lange Zeit angezweifelt, ob es eine Resistenzbildung gegen Parasiten geben könne. Dies scheint offenbar der Fall zu sein. In Untersuchungen von Olson u. Pimentel [37] wurden Stubenfliegen *(Musca domestica)* einem starken Parasitierungsdruck durch die Schlupfwespe *Nasonia vitripennis* ausgesetzt. Im Verlaufe von 40 Generationen sank der Parasitierungsgrad von 88% auf 60%, obwohl während der ersten 18 Generationen auf je 10 Wirtstiere nur 1 Parasitenpärchen ausgesetzt worden war, der Parasitenbesatz von dann ab auf 2 Pärchen je 10 Wirtstiere erhöht wurde. Offenbar spielt sich im Verlaufe der Generationen ein Gleichgewicht zwischen Wirt und Parasit ein, das zum Überleben beider Arten auf niedrigerem Niveau der Populationsdichte führt. Als Ursache wird ein genetisches Rückkopplungssystem angenommen.

Ähnliche Erscheinungen sind auch bei der Lärchenblattwespe *Pristiphora erichsonii* und deren Parasiten *Mesoleius tenthredinis* beobachtet worden.

Interessant sind auch zwei Untersuchungen über Resistenzbildung bei zwei Mückenarten, deren Larven von Nematoden parasitiert werden. Woodard u. Fukuda [52] hielten einen Stamm von *Anopheles quadrimaculatus*, um in ihm den parasitischen Nematoden *Diximermis peterseni* zu züchten. Nach vier Jahren hatte sich der Parasitierungsgrad auf die Hälfte dessen der Ausgangspopulation vermindert. Die Autoren zeigen, daß eine Verhaltensresistenz vorliegt, die es den Mückenlarven erlaubt, dem Angriff der Parasiten auszuweichen. An der Art *Culex pipiens quinquefasciatus* zeigte Petersen [41] einen Rückgang der Parasitierung mit dem Nematoden *Romanomermis culicivorax* auf ebenfalls etwa 50% nach 300 Wirtsgenerationen bzw. acht Jahren. Der Resistenzmechanismus konnte in diesem Falle nicht aufgeklärt werden.

7. Diskussion

Wenn man nach der Bedeutung des biologischen Pflanzenschutzes für die landwirtschaftliche Praxis fragt, so stehen ganz offensichtlich folgende Probleme im Vordergrund: Die Möglichkeiten der Anwendung bei den wichtigsten Indikationen im Anbau der bedeutendsten Kulturpflanzenarten, die Praktikabilität im landwirtschaftlichen Betrieb, die Wirtschaftlichkeit und die Nebenwirkungen.

Wie in Abschnitt 5.3.1 und 5.3.2 gezeigt wurde, ist die bisherige Erfolgsquote biologischer Verfahren in den landwirtschaftlichen Großkulturen bislang bescheiden. Die einzige wirkliche Großkultur, in der bislang eine größere Anzahl von Projekten zu Erfolgen führte, ist das Zuckerrohr mit 9 voll gelungenen und 17 weitgehend gelungenen Fällen. Von diesen 26 Projekten wurden 25 auf Inseln realisiert. Die Zuckerrohr-Anbaufläche der Welt beträgt mit 14 Mio ha rund 1% der landwirtschaftlich genutzten Fläche der Welt (ohne Wiesen und Weiden) von 1462 Mio ha. Selbst wenn man ein so großes Anbaugebiet wie Kuba noch zu den Inseln zählt, wird der Zuckerrohranbau auf nicht mehr als maximal 1,7 Mio ha auf Inseln betrieben, das sind höchstens 12% der Zuckerrohr-Anbaufläche der Welt oder 0,1% der landwirtschaftlichen Fläche (ohne Wiesen und Weiden). Die weitaus meisten Beispiele für gelungene Projekte finden sich aus dem Zitrusbau (56 Erfolge und 24 beträchtliche Erfolge). Nimmt man den Kernobstbau mit 13 Erfolgen und 21 beträchtlichen Erfolgen hinzu, so entfallen auf diese beiden Dauerkulturen zusammen mit 69 Erfolgen 67% aller mit Parasiten und Prädatoren erzielten Erfolge und mit 45 beträchtlichen Erfolgen 43% aus dieser Kategorie. Von allen Erfolgen (voll und beträchtlich) entfallen auf die drei Kulturen Zuckerrohr, Zitrus und Kernobst 67,3%. Die Welt-Anbaufläche für Zitrus und Kernobst ist statistisch nicht erfaßt. Da aber alle Dauerkulturen zusammen, also einschließlich Weinbau, Olivenbau, Kaffeebau, Kakaobau usw., nur zu 2% an der landwirtschaftlichen Fläche beteiligt sind, muß die Anbaufläche für diese beiden Kulturen deutlich unter 1% liegen. Nun wird aber nur ein geringer Teil dieser Anbaufläche mit biologischen Verfahren behandelt. Für den Zitrusbau der USA rechnen Starler u. Ridgway [47] mit ca. 3%. Damit käme man auf einen Gesamtanteil der landwirtschaftlichen Fläche der Welt, auf der mit Prädatoren oder Parasiten erfolgreich biologische Bekämpfung betrieben wird oder wurde, der maximal bei 1% liegt.

Mit dieser Zahl ist an sich natürlich nichts über das Potential der Verfahren ausgesagt, sondern nur über den gegenwärtigen Stand. Betrachtet man jedoch die gegenwärtig verfügbaren Methoden und die Bedingungen der erzielten Erfolge, so ist kaum mit einer sehr durchgreifenden Änderung zu rechnen. Als der Idealfall für eine aussichtsreiche biologische Bekämpfung könnte ein in ein isoliertes ökologisches Gebiet eingeschleppter Schädling gelten, der dort eine Dauerkultur befällt, bei der eine relativ hohe Schadensschwelle in Kauf genommen wird und die von keinem einheimischen Schadkomplex befallen wird. Hier wären folgende Voraussetzungen gegeben: Ein natürlicher Feind des Schädlings kann, wenn er erfolgreich eingebürgert worden ist, seine Populationsdichte im Befallsgebiet nicht durch Dispersion verdünnen, Neuzuflug des Schädlings entfällt. Die Kulturpflanze wird nicht durch jährliche Ernte entfernt; es kann sich also unter relativ stabilen ökologischen Bedingungen ein Gleichgewicht zwischen dem Wirt und seinem natürlichen Feind einspielen. Wegen der hohen Schadensschwelle kann dieses Gleichgewicht bei einer relativ hohen Populationsdichte des Schädlings toleriert werden. Das Fehlen eines einheimischen Schadkomplexes macht es nicht erforderlich, anders geartete Bekämpfungsmaßnahmen gegen diesen zu ergreifen, die das sich einpendelnde neue Gleichgewicht zwischen Wirt und Parasit stören könnten.

Die genannten Voraussetzungen treffen nur sehr selten und notwendigerweise nie auf sehr großer Fläche zusammen, doch muß in der Praxis zumindest eine Voraussetzung (z.B. Einschleppung) vorliegen, um einen Teilerfolg zu erzielen, oder es müssen mindestens zwei Voraussetzungen (z.B. Einschleppung und Dauerkultur oder Einschleppung und „ökologische Insel") zusammentreffen, um einen weitergehenden Erfolg möglich zu machen. Bei den landwirtschaftlichen Großkulturen trifft dies nur in Ausnahmefällen zu. Sie sind fast sämtlich einjährige Kulturen, in denen sich während einer Vegetationszeit ein Gleichgewicht zwischen dem Schädling und seinen natürlichen Feinden – wenn überhaupt – dann erst zu einem Zeitpunkt des Jahres einstellt, zu dem die Schäden an der Kulturpflanze bereits gesetzt sind. Die Großkulturen werden ferner überwiegend recht intensiv angebaut, so daß nur eine niedrige Schadensschwelle toleriert wird, ferner bilden eingeschleppte Schädlinge (wie im Fall des Kartoffelkäfers) gegenüber einheimischen Schadkomplexen eher die Ausnahme und isolierte Anbaugebiete sind, schon wegen der Großflächigkeit, selten.

Darüber hinaus muß jedoch auch darauf hingewiesen werden, daß bei den Pflanzenschutzmaßnahmen in den Großkulturen heute die Unkrautbekämpfung eine dominierende und noch weiterhin zunehmende Rolle spielt und daß die Maßnahmen zur Pilzbekämpfung ebenfalls an Bedeutung gewinnen (vgl. Tabelle 3). In beiden Bereichen sind praktikable biologische Bekämpfungsverfahren vorerst nicht zu erwarten. In den Fällen, in denen eine biologische Bekämpfung technisch möglich ist, erhebt sich noch zusätzlich die Frage nach der Durchführbarkeit im landwirtschaftlichen Betrieb. Hier sind zwei Gesichtspunkte von Bedeutung:

1. Die schon dargestellte, auf der meist großen Agilität der Insekten beruhende Schwierigkeit der Dispersion: Ähnlich wie bei einem osmotischen Gefälle tendieren die Schaderreger wie ihre natürlichen Feinde durch Zu- oder Abwanderung zu

einem Ausgleich der Populationsdichten mit der Umgebung. Daher sind entsprechende Maßnahmen nur dann erfolgversprechend, wenn sie entweder in den schon mehrfach angesprochenen ökologisch isolierten (isola = die Insel = island) Regionen oder auf sehr großer Fläche praktiziert werden. Der einzelne Obstbauer, der auf wenigen Hektar Anbaufläche innerhalb eines großen Anbaugebietes Parasiten anreichern oder freilassen würde, hätte kaum eine Chance, wirtschaftlich nennenswerte Wirkungsgrade zu erzielen. Dies wirft rechtliche Fragen auf. Denn entweder muß die großflächige Anwendung per Gesetz oder Anordnung oder per Einigung aller Grundeigentümer, die die jeweilige Kulturpflanze anbauen, herbeigeführt werden, oder der Erfolg ist in Frage gestellt. Dies ist sicherlich einer der Gründe dafür, daß Erfolge des biologischen Pflanzenschutzes in der Forstwirtschaft so stark im Vordergrund stehen. Abgesehen davon, daß es sich hier um Wirtschaftspflanzen handelt, die hinsichtlich der Dauer der Wirtschaftszeiträume (Dauerkulturen) mit Umtriebszeiten (Rotationszeiten) von 80 bis 500 Jahren an der Spitze stehen und nahezu jede Schadensschwelle, die nicht zum Absterben der Bestände führt, heute toleriert wird, wird der Wald überall auf der Welt großflächig bewirtschaftet. Das geschieht entweder durch den Staat als Eigentümer oder als Forstaufsichtsbehörde oder privatwirtschaftlich durch Aktiengesellschaften (durch Cellulose- oder Holzindustrie) wie in Nordamerika oder Skandinavien. Hier sind also großflächige Maßnahmen ohne rechtliche Komplikationen möglich. Im landwirtschaftlichen Bereich, in dem Eigentumsballungen dieser Größenordnung außer bei volkseigenen landwirtschaftlichen Betrieben vom Kolchostyp, sonst nicht der Regelfall sind, bedarf es sonst sehr diffiziler rechtlicher Regelungen. Es darf darauf hingewiesen werden, daß jeder Fachmann, der jemals eine großflächige chemische Pflanzenschutzmaßnahme zu leiten hatte, die enormen Schwierigkeiten kennt, die entstehen, wenn ein einzelner Grundeigentümer von vielleicht 2 ha Fläche sich einer großräumigen Bekämpfungsmaßnahme (an der er kostenmäßig beteiligt werden soll) widersetzt, und sei es auch nur, um die Kosten für seinen Flächenanteil letztlich geschenkt zu bekommen.

2. Maßnahmen der biologischen Schädlingsbekämpfung können überaus kompliziert sein. Ohne hoch spezialisierte Fachleute sind sie in Planung, Durchführung, Überwachung und Erfolgskontrolle nicht praktikabel. Entsprechend müßte für viele Verfahren des biologischen Pflanzenschutzes ein erheblich intensiverer Pflanzenschutzdienst mit hoher Beratungskompetenz und wahrscheinlich zusätzlicher, die Entscheidung des privaten Grundeigentümers weiter einengender Anordnungsbefugnis geschaffen werden. Abgesehen von den einschlägigen sonstigen politischen und rechtlichen Problemen entsteht die Frage, ob die hierfür anfallenden Kosten auf die Grundeigentümer umgelegt oder vom Steuerzahler getragen werden sollen.

Hiermit ist die Frage nach der Wirtschaftlichkeit eng verbunden, die auch dann bedeutungsvoll bleibt, wenn man ökologischen Gesichtspunkten Priorität einräumt. Es gibt zahlreiche Berechnungen und Betrachtungen über die Wirtschaftlichkeit biologischer Pflanzenschutzmaßnahmen. Hierbei differieren die Berechnungsgrundlagen und -ansätze sehr stark. Ein häufiger Ansatz ist der, nach einem erfolgreichen biologischen Bekämpfungsprojekt

a) zu errechnen, was Schadensfeststellung, Ermittlung der natürlichen Feinde, deren Zucht, Transport, Freilassung und Etablierung gekostet haben.

b) zu errechnen, welches Schadensmaß durch dieses Vorgehen verhütet worden ist, und

c) beide Kalkulationen in Beziehung zu einander zu setzen.

Ein solches Vorgehen vermengt jedoch privatwirtschaftliches und volkswirtschaftliches Denken. Unternehmenswirtschaftlich ist es selbstverständlich, daß in die Kosten des Erfolges diejenigen für die Mißerfolge mit eingehen müssen: Ein chemisches Pflanzenschutzmittel muß die Aufwendungen für 10000 bis 12000 erfolglos abgebrochene Entwicklungen mit tragen. Djerassi et al. [12] kritisieren mit Recht, daß dieser Gesichtspunkt bei vielen Erfolgsrechnungen über biologische Pflanzenschutzmaßnahmen in den USA nicht berücksichtigt wird. Darüber hinaus müssen bei derartigen Berechnungen die Gesamtaufwendungen, die erforderlich waren, um ein Projekt zu verwirklichen, mit in Rechnung gestellt werden, nicht nur die Spezialkosten für das Projekt. In einem einfachen Beispiel: Die anteiligen Kosten für alle zentralen Dienste einer Universität müßten ebenso wie die Forschungskosten (einschließlich aller Gehälter) in die Kalkulation eines Instituts eingehen, das ein biologisches Verfahren initiiert und zur Praxisreife entwickelt hat, wie das bei einem privatwirtschaftlichen Unternehmen der Fall wäre (overhead costs). Die volkswirtschaftliche Ersparnis durch Verminderung der Ernteverluste kann zunächst in eine solche Rechnung nicht eingehen, da der Preis von der Nachfrage bestimmt wird. Für diese ist es primär nicht entscheidend, welchen Gebrauch der potentielle Abnehmer vom angebotenen Wirtschaftsgut macht – z.B. einen materiellen oder immateriellen –, sondern lediglich, ob für ihn ein Bedarf besteht.

In der Literatur ist kein Beispiel dafür aufzufinden, das den strengen Berechnungskriterien für die Forschungs- und Entwicklungskosten bei chemischen Pflanzenschutzmitteln, wie sie von Haug [28] dargestellt worden sind, vergleichbar wäre. Da zudem bislang nur Teilbereiche des Pflanzenschutzes durch Forschungsinstitute für biologischen Pflanzenschutz verfolgt werden, muß der Aussage von Franz u. Krieg [22] sicherlich widersprochen werden, daß „gegenwärtig rund 20mal mehr Geld in die Forschung des Pestizidsektors investiert wird als in die zu biologischen Verfahren führenden Vorhaben", und daß „sich dies nur aus dem von der Technik geprägten Denken und Planen der Gegenwart verstehen" ließe, „bei dem lineare Kausalketten mit sofort registrierbaren Resultaten beliebtere Entscheidungshilfen darstellen, als die Berücksichtigung von komplexen und dynamischen Folgerungen".

Aus den vorausgehenden Darlegungen ergibt sich, daß es sich um eine naturwissenschaftliche und wirtschaftliche, nicht um eine weltanschauliche Frage handelt. Zu den bisherigen Wirtschaftlichkeitsberechnungen über biologische Bekämpfungsmethoden sei auf die folgenden Arbeiten hingewiesen: Dulmage [13, 14], Djerassi [12], Starler u. Ridgway [47], Harris [26], Parrella u. Kok [40].

Nur in begrenztem Umfange sind in die vorstehenden Wirtschaflichkeitsberechnungen solche Kosten eingegangen, die durch intensivierte Registrierungsanforderungen für biologische Agentien entstehen werden.

Diese Anforderungen werden sich zweifellos überwiegend auf mögliche Nebenwirkungen beziehen, wie sie unter Abschnitt 6 dargestellt sind. Hierbei werden die Fragestellungen, wenn sie sinnvoll sein sollen, sicher für biologische Verfahren gegenüber chemischen verlagert sein. Die Kosten werden sich in jedem Falle erhöhen, zumal Nebenwirkungsprüfungen hinsichtlich ökologischer, toxikologischer und anwendungssicherer Verfahren erfahrungsgemäß besonders kostspielig sind. Wenn alle in Abschnitt 6 aufgeführten Gesichtspunkte zu Prüfungsvoraussetzungen werden, so müssen sie entweder unerfüllbar oder unbezahlbar werden.

8. Schlußbetrachtung

Die im vorangehenden Abschnitt ausgeführten kritischen Bemerkungen sollen keinesfalls den Eindruck erwecken, als seien biologische Verfahren in der Praxis der Landwirtschaft von untergeordneter Bedeutung oder gar entbehrlich. Ein vernünftiger Pflanzenschutz ist ohne die Ergebnisse der Bemühungen zum biologischen Pflanzenschutz heute kaum mehr denkbar. Der biologische Pflanzenschutz hat zudem die Bedeutung der natürlichen Begrenzungsfaktoren für die Populationsdynamik der Schadorganismen verschärft ins Bewußtsein gerückt und damit die Voraussetzungen für das Konzept des „Integrierten Pflanzenschutzes" geschaffen, in dem alle biologischen, kulturtechnischen und chemischen Möglichkeiten so aufeinander abgestimmt werden sollen, daß der Befall unter die Schadensschwelle gedrückt wird. Daß dies oft im extensiven Anbau von Kulturpflanzen leichter zu verwirklichen sein wird als im Intensivanbau, bedeutet keineswegs, daß nicht auch für diesen integrierte Verfahren gefunden werden können. Die zukünftige Entwicklung wird also auf eine Ergänzung der methodischen Ansätze hinauslaufen, nicht auf eine Polarisierung. Mit dem vorliegenden Beitrag sollte daher vor unfruchtbarem und verkürztem Alternativdenken gewarnt werden. Selbst die auf einzelne Kulturpflanzenarten bezogenen Berechnungen sind nur relativ und als auf den Istzustand bezogen zu sehen. Natürlich kann man die heutigen Anbauflächen nicht nur darauf beziehen, was heute auf ihnen geschieht, sondern muß auch in Rechnung stellen, was bereits auf ihnen geschehen ist. Die unter spezifischen Verhältnissen bewirkte explosionsartige Vermehrung der Opuntien in Australien ist durch biologische Verfahren weitgehend rückgängig gemacht worden und hat damit Standraum für andere Kulturpflanzenarten frei gemacht. Es ist also sicher nicht richtig, biologische Verfahren des Pflanzenschutzes allein alternativ zu chemischen Verfahren zu betrachten. Da mit dem Fortschritt der Naturwissenschaften, zu denen Biologie, Ökologie und Chemie gleichermaßen gehören, zu erwarten ist, daß die Denkweisen verschiedener naturwissenschaftlicher Disziplinen sich immer stärker an einander annähern werden, ist auch zu vermuten, daß sich die Zusammenarbeit dieser Disziplinen im Interesse einer ausreichenden, gesunden und für die Lebensmittelverbraucher in aller Welt vorteilhaften Erntesicherung letztlich durchsetzen wird.

9. Literatur

1. Andres, L.A. et al.: Biological control of weeds (Huffaker, C.B., Messenger, P.S. ed.), Theory and practice of biological control, Academic Press, New York/San Francisco/London, S. 481–499 (1976)
2. Barjac, H. de: Entomophaga *23*, 309 (1978)
3. Bassand, D., Carpy, S.: Experientia *33*, 1545 (1977)

4. Boller, E.F. et al.: Entomologia Experimentalis et Applicata *20*, 237 (1976)
5. Boneß, M.: Die praktische Verwendung von Insektenpheromonen (Wegler, R., Hrsg.), Chemie der Pflanzenschutz- und Schädlingsbekämpfungsmittel, Bd. 6, Springer-Verlag, Berlin/Heidelberg/New York, S. 165 (1980)
6. van den Bosch, R. et al.: Biological control of insect pests of row crops (Huffaker, C.B., Messenger, P.S., ed.): Theory and practice of biological control. Academic Press, New York/San Francisco/London, S. 443–456 (1976)
7. Braun, H.: Arbeitsgem. f. Forsch. d. Landes Nordrh.-Westf., H. 32 (1953)
8. Büchel, K.H.: Chemosterilantien (Wegler, R., Hrsg.), Chemie der Pflanzenschutz- und Schädlingsbekämpfungsmittel, Bd. 1, Springer-Verlag, Berlin/Heidelberg/New York, S. 475–486 (1970)
9. Cramer, H.H.: Habilitationsschr. Freiburg/Brsg. (1962)
10. Cramer, H.H.: Ökonomisch-ökologische Wechselwirkungen. (Wegler, R., Hrsg.), Chemie der Pflanzenschutz- und Schädlingsbekämpfungsmittel, Bd. 3., Springer-Verlag, Berlin/Heidelberg/New York, S. 39–55 (1976)
11. Crull, G. (1974): s. Lipa, J.J. [35]
12. Djerassi, C., Shih-Coleman, C., Diekman, J.: Science *186*, 596 (1974)
13. Dulmage, H.T.: Economics of microbial control (Burges, H.D., Hussey, N.W., ed.), Microbial control of insects and mites. Academic Press, London/New York, S. 581–590 (1971)
14. Dulmage, H.T.: Responsibility for producing and developing insect viruses: individual user, commercial, public. In: Block, D.C.: The effect of patent laws on industrialization of insect viruses. Symp. specificity and development of insect viruses for pest control. Soc. Invertebr. Pathol. Misc. Publ. Entomol. Soc. Am., Tempe, Arizona (1974)
15. Dunbar, J.P., Johnson, A.W.: Environmental Entomology *4*, 352 (1975)
16. Ehlert, W.: Naturwiss. Rundschau *29*, 238 (1976)
17. Eidmann, H., Kühlhorn, F.: Lehrbuch der Entomologie. 2. Aufl., Paul Parey, Hamburg und Berlin (1970)
18. FAO: 1978 FAO production yearbook. Food and Agriculture Organization of the United Nations, Rom Vol. 32 (1979)
19. FAO/WHO: The use of viruses for the control of insect pests and disease vectors. Rep. joint FAO/WHO meet. on insect viruses. World Health Organization Techn. Rep. Ser. No. 91 (1973)
20. Farm Chemicals: Farm Chemicals *142*, 20 (1979)
21. Farm Chemicals: Farm Chemicals *142*, 61 (1979)
22. Franz, J.M., Krieg, A.: Biologische Schädlingsbekämpfung, 2. Aufl., Paul Parey, Berlin und Hamburg (1976)
23. Freeman, T.E., Zettler, F.W., Charudattan, R.: PANS – Pest Articles and News Summaries *20*, 181 (1974)
24. Hagen, K.S. et al.: Range, forage, and grain crops (Huffaker, C.B., Messenger, P.S., ed.), Theory and practice of biological control, Academic Press, New York/San Francisco/London, S. 397–442 (1976)
25. Hamed, A.R.: Z. angew. Entomol. *87*, 294 (1979)
26. Harris, P.: Weed Science *27*, 242 (1979)
27. Hasan, S.: PANS – Pest Articles and News Summaries *20*, 437 (1974)
28. Haug, G.: Pflanzenschutz-Forschung der Industrie (Wegler, R., Hrsg.), Chemie der Pflanzenschutz- und Schädlingsbekämpfungsmittel. Bd. 3, Springer-Verlag, Berlin/Heidelberg/New York, S. 57–84 (1976)
29. Huber, J.: Dissertat. Eidgenöss. Techn. Hochschule Zürich (1974)
30. Knipling, E.F.: J. Economic Entomol. *48*, 459 (1955)
31. Körting, W.: Veterinär-Medizin. Nachr. 152 (1974)
32. Krieg, A.: Z. Pflanzenkrankh. Pflanzenschutz *80*, 483 (1973)
33. Laing, J.E., Hamai, J.: Biological control of insect pests and weeds by imported parasites, predators, and pathogens. (Huffaker, C.B., Messenger, P.S., ed.), Theory and practice of biological control. Academic Press, New York/San Francisco/London, S. 685–743 (1976)
34. Laven, H.: Naturwiss. Rundschau *25*, 391 (1972)
35. Lipa, J.J.: Przeglad bakteryjnych i grzybowych biopreparatów stosowanych w biologicznym zwalczaniu szkodliwych owadów. Biuletyn Institutu Ochrony Roślin *18*, 18 (1974) (beglaubigte Übersetzung ins Deutsche von Crull, G.)
36. Marchal-Segault, D.: Ann. Zoologie – Ecologie animale *6*, 521 (1974)

37. Olson, D., Pimentel, D.: Environmental Entomology *3*, 621 (1974)
38. Ordish, G.: Untaken harvest. Constable et Comp. Ltd., London (1952)
39. Panda, B.B., Sahu, R.K., Sharma, C.B.S.R.: Mutation Res. *67*, 161 (1979)
40. Parrella, M.P., Kok, L.T.: J. Economic Entomol. *72*, 590 (1979)
41. Petersen, J.J.: Environmental Entomol. *7*, 518 (1978)
42. Rabe, W.: Unveröffentl. Reisebericht, Bayer AG, Leverkusen (1959)
43. Schaerffenberg, B.: Entomophaga *13*, 175 (1968)
44. Schuhmann, G.: Wirtschaftsberatungsdienst der Landwirtschaftskammer Weser-Ems *31*, 27 (1979)
45. Sharma, C.B.S.R., Sahu, R.K.: Mutation Res. *46*, 19 (1977)
46. Snyder, W.C., Wallis, G.W., Smith, S.N.: Biological control of plant pathogens (Huffaker, C.B., Messenger, P.S., ed.), Theory and practice of biological control, Academic Press, New York/San Francisco/London, S. 521–539 (1976)
47. Starler, N.H., Ridgway, R.L.: Economic and social considerations for the utilization of augmentation of natural enemies (Ridgway, R.L., Vinson, S.B., ed.), Biological control by augmentation of natural enemies. Plenum Press, New York und London, S. 431–453 (1977)
48. Tinsley, T.W., Melnick, J.L.: Intervirology *2*, 206 (1973/74)
49. Renwick, J.A.A., Vité, J.P.: Biology of Pheromones (Wegler, R., Hrsg.), Chemie der Pflanzenschutz- und Schädlingsbekämpfungsmittel, Bd. 6, Springer-Verlag, Berlin/Heidelberg/New York, S. 1 (1980)
50. Wade, N.: Science *181*, 925 (1973)
51. Wille, J.E., Beingolea, O., Gonzâlez, J.E.: Insectos e insecticidas en la campaña algodonera 1957–58. Estacion Experimental Agrícola de "La Molina", Informe No 106 (1958)
52. Woodard, D.B., Fukuda, T.: Mosquito News *37*, 192 (1977)
53. Wybou, A.P.: Bol. Soc. Entomológica del Perú *5*, 7 (1970)
54. Wybou, A.P.: Biologische Kontrolle des *Diatraea saccharalis* in Zuckerrohr in Riopaila, im Cauca-Tal, Kolumbien. Unveröffentl. Reisebericht, Bayer AG, Leverkusen (1979)
55. Zwölfer, H.: Some aspects of biological weed control in Europe and North America. Proc. 9th Brit. Weed Control Conf. Vol. III, S. 1147–1156 (1968)
56. Zwölfer, H.: Nachrichtenbl. Dtsch. Pflanzenschutzdienstes (Braunschweig) *26*, 98 (1974)

Sachverzeichnis

Chemie der Pflanzenschutz- und Schädlingsbekämpfungsmittel

Herausgeber: R. Wegler

Band 1: **Einführung, Insektizide, Chemosterilantien, Repellents, Lockstoffe, Akarizide, Nematizide, Vogel- bzw. Säugetierabschreckungsmittel, Rodentizide**

1970. 23 Abbildungen, XXIX, 671 Seiten (31 Seiten in Englisch)
ISBN 3-540-04782-4

Aus den Besprechungen:
„...Der Arbeitsumfang, den die 27 Autoren bewältigt haben, war außerordentlich groß, der Benutzer findet etwa 3000 Literaturangaben und außerdem noch eine größere Zahl bibliographischer Angaben. Die konzentrierte Darstellung des Stoffes und eine logische Gliederung gewährleistet in jedem Fall eine Information über die einzelnen Wirkstoffe, wie sie in dieser Form in der Literatur über Pflanzenschutzmittel bisher nicht vorhanden war..."
Biologisches Zentralblatt

Band 2: **Fungizide, Herbizide, Natürliche Pflanzenwuchsstoffe, Rückstandsprobleme**

1970. 24 zum Teil farbige Abbildungen.
XXIX, 550 Seiten
ISBN 3-540-05064-7

Aus den Besprechungen:
„Der Band ist mit seiner Aufgliederung innerhalb der Unterabschnitte so angelegt, daß es dem Studenten wie dem jungen Industriechemiker die Einarbeitung in das vielseitige Gebiet des chemischen Pflanzenschutzes erleichtert. Auch der Fachmann wird ihn gern für spezielle Fragen zu Rate ziehen."
Naturwissenschaftliche Rundschau

Band 3: **Geschichte – Ökologie – Forschung – Tropenkrankheiten. Textilschutz – Insektizid-Resistenz – Materialschutz**

1976. 26 Abbildungen. XIII, 322 Seiten (160 Seiten in Englisch)
ISBN 3-540-07817-7

Aus den Besprechungen:
„Mit dem 3. Band ... legt der Springer-Verlag ein besonders gelungenes Werk vor, das sich nicht nur an den Fachmann wendet, sondern auch von allgemeinem Interesse ist. In neun Beiträgen werden Probleme behandelt, die direkt oder indirekt mit dem chemischen Pflanzenschutz verbunden sind... Die sehr sachliche Darstellung der großen Belastung und Verunsicherung, der sich die industrielle Pflanzenschutzforschung heute ausgesetzt sieht, wird manchen nachdenklich stimmen, der sich Gedanken über die heutige Forschungspolitik macht..."
Forstarchiv

Band 4: **Pflanzenwachstumsregulatoren – Fungizide – Holzschutz**

1977. 17 zum Teil farbige Abbildungen.
XVII, 308 Seiten (18 Seiten in Englisch)
ISBN 3-540-08114-3

Aus den Besprechungen:
„...Der 4. Band ergänzt das umfangreiche Gebiet der chemischen Pflanzenschutzmittel und ist auf Grund seiner exakten und detaillierten Durchsetzung über die einzelnen Wirkstoffe eine außerordentlich wichtige Informationsquelle."
Biologische Rundschau

Band 5: **Herbizide**

1977. 6 Abbildungen. XXI, 752 Seiten
ISBN 3-540-08178-X

Aus den Besprechungen:
„...Für den Forschungschemiker ist dieses Werk eine kaum zu übertreffende Hilfe, um rasch und zu jeder Zeit (unabhängig von der Computerinformation) eine Orientierungsmöglichkeit während seiner wissenschaftlichen Arbeit zu besitzen..."
Qualitas Plantarum

Springer-Verlag
Berlin
Heidelberg
New York